Key Topics in Conservation Biology 2

Key Topics in Conservation Biology 2

Edited by

David W. Macdonald
Wildlife Conservation Research Unit,
Department of Zoology, Recanati-Kaplan Centre, Tubney House,
University of Oxford, UK

Katherine J. Willis
Biodiversity Institute, Oxford Martin School, Department of Zoology,
University of Oxford, UK

WILEY-BLACKWELL
A John Wiley & Sons, Ltd., Publication

This edition first published 2013 © 2013 by John Wiley & Sons, Ltd.

Wiley-Blackwell is an imprint of John Wiley & Sons, formed by the merger of Wiley's global Scientific, Technical and Medical business with Blackwell Publishing.

Registered Office
John Wiley & Sons, Ltd, The Atrium, Southern Gate, Chichester, West Sussex, PO19 8SQ, UK

Editorial Offices
9600 Garsington Road, Oxford, OX4 2DQ, UK
The Atrium, Southern Gate, Chichester, West Sussex, PO19 8SQ, UK
111 River Street, Hoboken, NJ 07030-5774, USA

For details of our global editorial offices, for customer services and for information about how to apply for permission to reuse the copyright material in this book please see our website at www.wiley.com/wiley-blackwell.

Library of Congress Cataloging-in-Publication Data

Key topics in conservation biology 2 / edited by David W. Macdonald & Katherine J. Willis.
pages cm
Includes bibliographical references and index.
ISBN 978-0-470-65876-5 (cloth) – ISBN 978-0-470-65875-8 (pbk.) 1. Conservation biology.
I. Macdonald, David W. (David Whyte) II. Willis, K. J.
QH75.K472 2013
333.95'16–dc23

2012035721

A catalogue record for this book is available from the British Library.

Wiley also publishes its books in a variety of electronic formats. Some content that appears in print may not be available in electronic books.

Cover image: Radio-tagged (see antenna) lion, part of WildCRU's study of the impact of trophy hunting around Hwange National Park, Zimbabwe. Courtesy of D.W. Macdonald.
Cover design by Design Deluxe

Set in 10/12.5pt Meridien by SPi Publisher Services, Pondicherry, India

1 2013

Contents

Contributors

Frank Adriaensen
Department of Biology,
University of Antwerp,
Antwerp, Belgium

Gregory P. Asner
Department of Global Ecology,
Carnegie Institution for Science,
Stanford, CA, USA

Jonathan E.M. Baillie
Zoological Society of London,
London, UK

Christopher B. Barrett
Charles H. Dyson School of Applied Economics and Management
Cornell University,
Ithaca, New York, USA

Yves Basset
Smithsonian Tropical Research Institute,
Apartado 0843-03092, Balboa,
Ancon, Peru

Paul Beier
School of Forestry, Northern Arizona University,
Flagstaff, AZ,
USA

Shonil Bhagwat
School of Geography and the Environment,
University of Oxford, Oxford, UK

Raphaël Billé
Institute for Sustainable Development and International Relations,
27 rue Saint Guillaume,
Paris, France

Luigi Boitani
Department of Biology & Biotechnologies,
Università La Sapienza,
Viale Università 32,
Rome, Italy

Mark Bonta
Center for Community and Economic Development,
Delta State University,
Cleveland, MS, USA

Christina D. Buesching
Wildlife Conservation Research Unit,
Department of Zoology, Recanati-Kaplan Centre,
University of Oxford, Oxford, UK

Erwin H. Bulte
Development Economics Group,
Wageningen University,
The Netherlands

Stuart H.M. Butchart
BirdLife International,
Cambridge, UK

Samuel A. Cushman
USDA Forest Service,
Rocky Mountain Research Station,
Flagstaff, AZ, USA

Amy Dickman
Wildlife Conservation Research Unit,
Department of Zoology, Recanati-Kaplan
Centre, University of Oxford, Oxford, UK

Eric Dinerstein
WWF-US,
1250 24th St. NW,
Washington, D.C. USA

Kingsley W. Dixon
Kings Park and Botanic Garden,
The University of Western Australia,
West Perth, 6005,
Nedlands,
Australia

Andrew Dobson
Keele University,
Keele, UK

C. Josh Donlan
Advanced Conservation Strategies,
Midway, UT, USA
and
Cornell University,
Department of Ecology & Evolutionary
Biology, Ithaca, NY, USA

Adam J. Dutton
Wildlife Conservation Research Unit,
Department of Zoology, Recanati-Kaplan
Centre, University of Oxford, Oxford, UK

Robert M. Ewers
Division of Biology,
Imperial College London,
Ascot, UK

John E. Fa
Durrell Wildlife Conservation Trust, Jersey,
and ICCS, Department of Life Sciences,
Imperial College London,
Ascot, UK

Paul Ferraro
Department of Economics,
Andrew Young School of Policy Studies,
Georgia State University,
Atlanta, GA, USA

Hervé Fritz
Laboratoire Biométrie et Bilogie Evolutive
CNRS UMR 5558,
Université de Lyon,
Villeurbanne Cedex, France

Toby Gardner
Conservation Science Group,
Department of Zoology, University
of Cambridge,
Cambridge, UK

Brendan Godley
Centre for Ecology and Conservation,
University of Exeter,
Exeter, UK

Andrew Gosler
Edward Grey Institute of Field Ornithology
and Institute of Human Sciences,
University of Oxford,
Oxford, UK

Brian Gratwicke
Center for Species Survival,
Smithsonian Conservation Biology Institute,
Washington, D.C., USA

Dennis Hansen
Institute of Evolutionary Biology and Environmental Studies,
University of Zurich,
Winterthurerstrasse 190,
Zurich, Switzerland

David M. Harper
Department of Biology,
University of Leicester,
Leicester, UK

Stephen A. Harris
Department of Plant Sciences,
University of Oxford,
Oxford, UK

Stuart Harrop
Durrell Institute of Conservation and Ecology,
University of Kent,
Canterbury, UK

Peter Henderson
Pisces Conservation,
Lymington, UK

Cameron Hepburn
Smith School of Enterprise and the Environment and James Martin Institute,
Said Business School,
University of Oxford,
Oxford, UK

Emilio A. Herrera
Departamento de Estudios Ambientales,
Universidad Simón Bolívar, Caracas,
Venezuela

Katherine Homewood
Department of Anthropology,
University College London,
Gower Street,
London, UK

Blanca Huertas
Life Sciences Department,
The Natural History Museum,
Cromwell Road,
London, UK.

Joelene Hughes
Wildlife Conservation Research Unit,
Department of Zoology, Recanati-Kaplan Centre, University of Oxford,
Oxford, UK

Susan K. Jacobson
Department of Wildlife Ecology and Conservation,
University of Florida
Gainesville, FL, USA

Julia P.G. Jones
School of Environment, Natural Resources and Geography,
Bangor University,
Bangor, UK

Valerie Kapos
United Nations Environment Programme,
World Conservation Monitoring Centre,
Cambridge, UK

K. Ullas Karanth
Wildlife Conservation Society,
Centre for Wildlife Studies,
India

Aidan M. Keane
Department of Anthropology, University College London and Institute of Zoology,
London, UK

Carolyn King
Department of Biological Sciences,
University of Waikato, Hamilton,
New Zealand

Andrew T. Knight
Division of Ecology and Evolution,
Imperial College London,
Ascot, UK
and

Department of Botany,
Nelson Mandela Metropolitan University,
Port Elizabeth,
South Africa

Dan Laffoley
International Union for Conservation of Nature,
Gland, Switzerland

Tom Le Quesne
WWF-UK,
Godalming, UK

Owen T. Lewis
Department of Zoology,
University of Oxford,
Oxford, UK

Mark Lomolino
Department of Environmental and Forest Biology,
SUNY College of Environmental Science and Forestry,
Syracuse, New York, USA

Margaret D. Lowman
Nature Research Center, North Carolina Museum of Natural Sciences
and
North
Carolina State University, Raleigh,
NC, USA

David W. Macdonald
Wildlife Conservation Research Unit,
Department of Zoology, Recanati-Kaplan Centre, University of Oxford,
Oxford, UK

Yadvinder Malhi
Environmental Change Institute, School of Geography and the Environment,
University of Oxford,
Oxford, UK

Michael Manfredo
Human Dimensions of Natural Resources,
Colorado State University,
Fort Collins, CO, USA

Silvio Marchini
Instituto Pró-Carnívoros,
Av. Horácio Neto,
Atibaia, Brazil

Brad McRae
The Nature Conservancy,
North America Region1917,
Seattle, USA

Thomas Merckx
Wildlife Conservation Research Unit,
Department of Zoology, Recanati-Kaplan Centre, University of Oxford, Oxford, UK

Laurent Mermet
AgroParisTech,
19 avenue du Maine,
Paris, France

Eleanor J. Milner-Gulland
Department of Life Sciences,
Imperial College London,
Ascot, UK

Axel Moehrenschlager
Centre for Conservation Research,
Calgary Zoological Society, Calgary,
Alberta, Canada

Tom P. Moorhouse
Wildlife Conservation Research Unit,
Department of Zoology, Recanati-Kaplan Centre,
University of Oxford, Oxford, UK

Nic Pacini
University of Calabria,
Cosenza, Italy

Nick Polunin
School of Marine Science and Technology,
Newcastle University,
Newcastle, UK

Jules Pretty
Department of Biological Sciences,
University of Essex,
Colchester, UK

Andrew S. Pullin
Centre for Evidence-Based Conservation,
School of Environment, Natural Resources
and Geography,
Bangor University,
Bangor, UK

David Raffaelli
Environment Department,
University of York, York, UK

Tony Rebelo
Threatened Species Research Unit,
South African National Biodiversity Institute,
Claremont, South Africa

Ana S.L. Rodrigues
Centre d'Ecologie Fonctionnelle et Evolutive,
CNRS-CEFE UMR5175,
Montpellier, France

Gary Roemer
Department of Fish, Wildlife & Conservation
Ecology,
New Mexico State University,
Las Cruces, NM, USA

Alex D. Rogers
Department of Zoology, University of Oxford,
Oxford, UK

Chris Sandom
Ecoinformatics & Biodiversity Group,
Department of Bioscience, Aarhus University,
Ny Munkegade 114,
Aarhus, Denmark

Çağan H. Şekercioğlu
Department of Biology, University of Utah,
Salt Lake City, UT, USA

Debra M. Shier, Ph.D.
Applied Animal Ecology Division,
San Diego Zoo Institute for Conservation
Research,
Escondido, CA, USA

Mark Shirley
School of Biology, Newcastle University,
Newcastle upon Tyne, UK

Claudio Sillero-Zubiri
Wildlife Conservation Research Unit,
Department of Zoology, Recanati-Kaplan
Centre, University of Oxford,
Oxford, UK

Jonathan Silvertown
Department of Environment,
Earth and Ecosystems, Faculty of Science,
Open University,
Milton Keynes, UK

Freya A.V. St John
Durrell Institute of Conservation and Ecology,
School of Anthropology and Conservation,
University of Kent,
Canterbury, UK

Mark R. Stanley Price
Wildlife Conservation Research Unit,
Department of Zoology, Recanati-Kaplan
Centre, University of Oxford, Oxford, UK
and
Al Ain Zoo and Aquarium, Abu Dhabi

Niels Strange
Department of Food and Resource Economics,
Centre for Macroecology,
Evolution and Climate,
University of Copenhagen,
Frederiksberg, Denmark

William Sutherland
Conservation Science Group,
Department of Zoology,
University of Cambridge,
Cambridge, UK

Jens-Christian Svenning
Ecoinformatics & Biodiversity Group,
Department of Bioscience,
Aarhus University,
Ny Munkegade 114,
Aarhus, Denmark

Tom Tew
The Environment Bank,
Stamford, UK

Jeremy Thomas
Department of Zoology, University of Oxford,
Oxford, UK

Sonia Tidemann
Batchelor Institute of Indigenous Tertiary
Education, Batchelor, NT, Australia

Derek P. Tittensor
United Nations Environment Programme,
World Conservation Monitoring Centre/
Microsoft Research, Cambridge, UK

Joseph A. Tobias
Edward Grey Institute, Department of Zoology,
University of Oxford, Oxford, UK

F. Hernan Vargas
Peregrine Fund,
Boise, ID, USA

Timothy Walker
University of Oxford Botanic Garden,
Oxford, UK

Peter D. Walsh
Department of Archaeology and
Anthropology, University of Cambridge,
Cambridge, UK

Katherine J. Willis
Biodiversity Institute,
Oxford Martin School, Department of Zoology,
University of Oxford, UK

Kerrie A. Wilson
School of Biological Sciences,
University of Queensland,
St Lucia, QLD, Australia

Richard Wrangham
Department of Human Evolutionary
Biology, Peabody Museum,
Harvard University,
Cambridge, MA, USA

Sven Wunder
Center for International Forestry Research,
Rio de Janeiro, Brazil

Kathy Zeller
Panthera,
8 West 40th Street, 18th Floor,
NY, USA

Preface

It was the best of times, it was the worst of times.

(Charles Dickens, *Tale of Two Cities*)

As we celebrate the completion of this second collection of essays on *Key Topics in Conservation Biology*, a reader might wonder what has changed since the publication of the first volume in 2007. Happily, one thing that has not changed is the relevance of the essays in the first volume. *Key Topics in Conservation Biology 2* does not replace *Key Topics 1*, it complements it. Less happy, however, are the changes in the state of wildlife and nature. These are commemorated in the almost complete failure of the 2010 targets set by the Convention on Biological Diversity to reduce biodiversity loss. Indeed, rare species continue to edge towards the precipice of extinction, more populous ones dwindle in abundance and distribution, environments are degraded and with them go not only beautiful and fascinating organisms but also the ecosystem services on which the human enterprise depends, and so too the quality of life deteriorates for many of the planet's ever more numerous people.

A brief stock-take of the years between *Key Topics 1* and *2* reveals that the global human population has increased by 395 million, the global economy has grown from US$66 trillion to US$77 trillion and CO_2 emissions have increased from 28 billion metric tons to 31.6 billion metric tons. As part of this syndrome of more people using more resources and producing more CO_2, its hardly surprising that 80 million hectares of forests have been converted and, of the 600 marine fish stocks monitored, 52% are fully exploited, 17% over-exploited and 7% depleted. This year, Arctic Ocean sea ice coverage has shrunk to the lowest level since modern records began, smashing the previous record by 293,000 square miles. Arctic summers have been free of sea ice for the first time in thousands of years. Against this dismal backdrop, the need for conservation biology could scarcely be more pressing.

With the 2007 collection of *Key Topics 1* essays safely to hand and still relevant, our response to the escalating threats and associated challenges has been to drive the 2013 collection in *Key Topics 2* to be even more comprehensive, interdisciplinary, frank and focused on solutions. Our aim is that these collections will help a new generation of conservationists to operate at the front line, their feet firmly based on sound theory, their eyes scanning the horizon for innovation and opportunity, and their hands busily at work delivering practical outcomes. This is a vibrant field, and one in which *Key Topics* are bursting on to the world stage with energizing insight and determined to make a difference. These are the subjects of this new crop of essays: ideas, technologies, frameworks for thinking and for action that are beacons of hope amongst the gloom (and peril) of biodiversity loss.

A journey that was already well advanced when *Key Topics 1* was written traced the route through interdisciplinarity, leaving far behind a perception of isolationist conservation that prioritized wildlife over people, in the quest to combine the interests of wildlife and people, reversing biodiversity loss and alleviating poverty. In the 5 years that have passed since its publication, new voices have joined this irresistible mantra of alignment. But facing the facts can be uncomfortable – the interests of biodiversity conservation and development are often at odds, values are often measured in incommensurable currencies, some problems defy easy solution and some alternative priorities are not easily reconciled – so conservation biologists have become not only more ingenious but also more worldly and pragmatic. As conservation biologists take their seats alongside politicians, economists, medics, industrialists, agriculturalists and philosophers to chart a future for the environment and nature, they become expert in trade-offs; they realize that one model does not fit all for biodiversity conservation and that, as we discuss in the concluding essay to this collection (Chapter 25), there are some very hefty 'elephants in the room' whenever the reconciliation of people and wildlife is being discussed.

It is the flushing from cover of metaphorical elephants, along with any bluster, hypocrisy or delusion that conceals them, that explains the philosophy and *modus operandi* behind the structure of this book. As indelicately confided in the Preface to *Key Topics 1*, the working title had been 'Conservation Without Crap', and each of the essays in that volume, and now in this one, cuts to the unvarnished essence of the key topic that it probes. To achieve this, we assembled teams of critical authorities – almost always including people who had not previously collaborated, who had different perspectives and generally came from different parts of the world. This alchemy made things much tougher for the editors (we have herded these scholarly cats relentlessly, even when their claws were unsheathed) and for the authors (whom we thank for tolerating our demands and rising to the challenge). We hope these essays will fascinate, inform and entertain a wide readership from the loftiest authority to the aspirant high school sixth former, from interested layman to policy maker to naturalist. Of this spectrum, our imagined modal reader might well be a Master's student, whose career and world-view may be shaped by these essays – a weighty responsibility with which we repeatedly flogged our authors, and which they have shouldered unstintingly. We thank them, confident that our readers will appreciate the effort.

In the Preface to *Key Topics 1*, we concluded that each essay in the collection was intended to stand alone but that the collection as a whole was more than the sum of its parts, together introducing the nature of the problem, the framework in which it can be understood, some tools that could be used in the quest for solutions and various of the issues that are topical. As such, neither that volume nor this one has pretensions to compendiousness or balance. Thus in 2007, we wrote 'there are many more than 18 key topics in wildlife conservation' – and now, in *Key Topics 2*, we present 25 more of them. This new selection of 25 more key topics was not chosen on a whim. We scanned the landscape, and the researchers navigating it, for emerging ideas and innovative thinkers – in particular, this led to chapters that push the borders of interdisciplinarity, capture the sophistication of finance and culture, and reveal the astonishing potency of new technologies. While the purpose of the essays is unchanged, we have also taken an additional step in grouping them, having in mind the strengthened perspective of viewing a landscape from more than one hilltop. Thus, the first nine chapters explore that landscape, identifying conservation priorities, the levels at which they can fruitfully be approached and paradigms for engagement with society. Next, the essays tackle financial mechanisms to encourage nature conservation, and to prevent unsustainable trade, before exploring theological and cultural dimensions (the latter with respect to conflict between people and carnivores). Still with a societal perspective, two concluding chapters of the framework complete this setting of the scene by tackling the role of citizen science and the capacity of nature to deliver an unsung ecosystem service, namely health and well-being.

Following this framework, next we shuffle the pack in two ways. First, we select four habitats (oceans, fresh water, islands and tropical forests) and then four taxa (butterflies and moths, birds, plants and large mammals – the latter originally intended as two sibling chapters on predators and prey, eventually fused into a bumper chapter on the two together). Having created the framework and then introduced a selection of the actors, we turn to seven chapters that dwell on safeguarding the future, by monitoring, understanding human behaviour, designing solutions, creating biological corridors and the twin aspirations of reintroduction and rewilding. This section concludes with a perspective on intervening to manage wildlife disease. Finally, the book ends with our perspective on a synthesis that encompasses human population, development, sustainability and biodiversity.

As conservation biology matures, at least three different models are now emerging: the first is the original 'protectionist model' where biodiversity conservation takes place in protected areas, sometimes behind a fence and heavily guarded. While the pendulum swung strongly against protectionism throughout the latter years of the 20th century, in recent years it may be swinging back. Nonetheless, protected areas encompass only about 12% of the earth's terrestrial surface and less than 1% of the oceans, and insofar as the protectionist model can apply only in protected areas, it is never going to be enough. Further, within protected areas, how are priorities to be devised amongst species or habitats, or landscapes and corridors? These subjects are amongst those tackled in Chapters 2, 5, 10, 11 and 20.

The second model for biodiversity conservation is community based. The antithesis of the protectionist approach, this model has it that the best people to carry out conservation are the communities that live within biodiverse landscapes. Here, the focus is on participation, not building fences but rather bridges along the road to working with the local communities – once again to find pragmatic solutions. As Chapters 3 and 6 demonstrate, these bridges can involve acknowledging the religious dimension provided by biodiverse landscapes, groves and species, an approach that is based on the premise that those things people value and respect the most, they will conserve. Community conservation highlights the need to understand human behaviour (Chapter 3, 19) associated with, for example, the bush meat trade or addressing conflict with large carnivores (Chapter 7).

The third emphasis, and one that has been rapidly rising up the political agenda in the past decade, is one that values and conserves biodiversity because of the ecosystem services that it provides to human well-being. This is not so much a different way to implement conservation (in the sense that the protected area and community-based approaches are strategically different) but rather an added rationale for conserving biodiversity. This is a complex issue (Chapters 5 and 6) that revolves around the challenging task of valuing the ecosystem services provided by biodiversity (Chapter 4). An acknowledged political reality is that the contribution that biodiversity makes to humankind – be it through direct benefits to health and well-being (Chapter 9), clean water, soil erosion protection, regulation of carbon or maintaining genetic diversity of food stocks – is critical and possibly invaluable. Ideas of this genre, expressed in terms of the agendas of earth security and planetary boundaries, aim at conserving specific ecosystem services that benefit humanity and dominated the agenda at the recent Rio +20 conference. Many biodiversity conservationists are learning the language of economists, and their fluency offers huge potential for delivering conservation. However, not everything can be monetized and much (perhaps most) that is beautiful in nature cannot pay its way but is nonetheless priceless.

Finally, and in a vocabulary where compromise is not a dirty word, another emerging model is that of sustainable landscape and seascape management. This approach, taking some of the working parts from the ethos of each of protectionist and community-based conservation, seeks to fashion a new machinery for nature amidst mankind by managing wildlife sustainably in a matrix landscape.

So where is the future agenda in biodiversity conservation? Is one of these models more promising than the others? Nobody's crystal ball is sufficiently clear to answer definitively. Without societal engagement with wildlife and the wider environment, the battle is lost, which is why citizen science is a key topic (Chapter 8). With the dramatic emergence of technologies from the internet to the smartphone, the gadgetry of engagement is revolutionized, as it is for researchers New ways of managing and restoring nature are also emerging, less focused on static baselines but rather on restoring ecosystem process and the important role of large herbivores as ecosystem engineers (Chapter 23). Whilst the economic model is currently favoured by research councils and governments alike, as we write this chapter the current UK Minister for the Environment has reputedly forbidden his officials from using the phrase 'ecosystem services' on the grounds that it is too jargon laden. Similarly, there is endless carping that the very word 'biodiversity' is too technical and off-putting, yet in news bulletins every day a vast public speedily absorbs much more obscure notions because they appreciate their relevance (it didn't take citizens long to grasp the term 'subprime mortgages' when they realized they might have one!). Leeriness of the, actually rather intuitive, vocabulary of the environment is merely a symptom of not taking the topic seriously. The essays in this book are excitingly diverse but one theme unites them all: everybody should take the environment seriously: it is *the* Key Topic.

David W. Macdonald and Kathy J. Willis
Wildlife Conservation Research Unit
Biodiversity Institute,
Department of Zoology
University of Oxford

With public sentiment, nothing can fail; without it, nothing can succeed.

(Abraham Lincoln)

About the Companion Website

This book is accompanied by a companion website:

www.wiley.com/go/macdonald/conservationbiology

The website includes:

- Powerpoints of all figures from the book for downloading
- PDFs of tables from the book

I

The framework

1

Conservation priorities: identifying need, taking action and evaluating success

Andrew S. Pullin[1], William Sutherland[2], Toby Gardner[3], Valerie Kapos[4] and John E. Fa[5]

[1]Centre for Evidence-Based Conservation, School of Environment, Natural Resources and Geography, Bangor University, Bangor, UK
[2]Conservation Science Group, Department of Zoology, University of Cambridge, Cambridge, UK
[3]Conservation Science Group, Department of Zoology, University of Cambridge, Cambridge, UK
[4]United Nations Environment Programme, World Conservation Monitoring Centre, Cambridge, UK
[5]Durrell Wildlife Conservation Trust, Jersey, and ICCS, Department of Life Sciences, Imperial College London, Ascot, UK

"What I decided I could not continue doing was making decisions about intervening when I had no idea whether I was doing more harm than good"

Archie Cochrane

Introduction

Conserving biodiversity requires identifying and addressing the myriad of problems generated when humans exploit natural resources. This challenge is ongoing and expensive in terms of time, money and access to the necessary expertise. Needs invariably outweigh resources, and actions require prioritization on multiple fronts. Conservation also needs approaches that enable more effective objective setting, as well as critical evaluation of conservation actions and of the extent to which targeted problems are solved.

Although there might seem to be room for some optimism given the increased investment in protected areas, sustainable forest management, and the management of invasive species, the rate of biodiversity loss does not appear to be slowing (Butchart et al. 2010; Secretariat of the Convention on Biological Diversity 2010). In addition, information on the nature and scale of conservation problems is accumulating faster than our ability to process it and respond effectively. Current rates of biodiversity loss

Key Topics in Conservation Biology 2, First Edition. Edited by David W. Macdonald and Katherine J. Willis.

exceed estimates of historical rates by several orders of magnitude (Millennium Ecosystem Assessment 2005). Species extinctions are invariably associated with direct drivers, such as habitat loss and overexploitation, though secondary extinctions can readily be triggered by the initial loss of species that provide key ecosystem functions. Interaction effects between land use and climate change also present increasingly complex challenges for global conservation (Iwamura et al. 2010).

Conservation is part of a continuous cyclical process in which management activities are implemented in spite of uncertainties about their effectiveness. This process typically starts with the detection of the decline or degradation of an aspect of nature that we value. Once this change has been identified, conservation goals can be set, such as an area of habitat to be protected, a wetland area to be restored or species decline to be arrested or reversed. When goals are made clear, interventions can be selected and implemented, and their relative success or failure assessed in order to inform future action. In this cycle of doing and learning, conservation decision making ultimately involves some scientific evaluation of the effectiveness of past efforts to guide future actions (Pullin & Knight 2001; Knight et al. 2006).

Priority setting in conservation research and action will always reflect human-oriented values and be forever changing and contested, not least as baselines of human values shift and other societal priorities change. Nevertheless, science can be a potent guiding force in informing decision making and can help improve the cost-effectiveness of conservation practice. Conservation science is just one component of the overall decision-making process. Economic, social and political considerations also play a role and may determine the outcome. For example, decisions concerning which species and habitats are worth saving are strongly influenced by the necessarily subjective values of individual stakeholders, as well as by the political and socio-economic opportunities and constraints of the region of concern. Science can advise on which are likely to be the most cost-effective solutions for conserving the giant panda, for instance, but this information is only one factor in deciding how much money should be spent on its conservation, or the way in which available funds should be spent.

Table 1.1 Example summary of steps and processes that might be included in a decision-making framework

Steps	Processes
Objective setting: desired trends, targets, time frame	Social process: priority assessment, stakeholder consultation, ethics approval
Solution scanning: identify potential interventions, actions	Expert process: consultation, workshops
Effectiveness assessment: comparison of previous intervention performance	Evidence-based process: evidence synthesis, predictive models
Cost-effectiveness assessment: value from investment	Evidence-based process: economic assessment, planning models
Outcome evaluation: programme evaluation	Mixed methods process: quantitative and qualitative data analysis

In this opening chapter, we first explore ways in which priorities for both conservation action and research emerge and are evaluated. Recognizing that conservation is ultimately a societal process underpinned by values and beliefs, we describe how decisions about resource allocation for conservation actions can be informed by explicit use of scientific evidence in decision-making frameworks. Decision-making frameworks are composed of a set of transparent principles and criteria that can help evaluate the pros and cons of alternative choices, thereby facilitating the identification of cost-effective actions (Table 1.1).We end by outlining future challenges to the development of decision-making frameworks for conservation that encompass policy, management and research.

Identifying need for action

Effective conservation depends on identifying priorities for specific research and/or action. As described in this section, these are typically verified by one of two routes. The first route is more reactive and involves the detection, through surveillance monitoring, of a change in status of a taxon, species group, habitat or ecosystem. The second route is more proactive and works by identifying potential threats that may cause significant negative changes in the future.

Detection of ecological changes

Surveillance monitoring, whether of changes in habitats, species or even life history attributes of particular species, can sometimes detect unexpected and important changes useful for prioritizing conservation activity (whether for action or research). For example, long-term data on the widespread declines of sea turtles (Crouse et al. 1987) have motivated the discovery, development and implementation of innovative solutions such as turtle exclusion devices on shrimp trawlers. In another example, the UK Common Birds Survey (now, with a change in methodology, the Breeding Birds Survey), which was set up in 1962 partly to identify changes in bird populations from direct organophosphate pesticide poisoning, has played an important role in detecting a range of other issues requiring action. These include bird responses to agricultural change and changes in woodland management, as well as to changing conditions in the African wintering grounds (Newson et al. 2009).

Even when ecological changes are detected, the challenge remains of how to interpret and communicate the significance of monitoring data. Biodiversity indices that combine a range of trend data are increasingly used to represent broader changes in the environment, and are often welcomed by policy makers responsible for setting high-level targets. For example, in 2000 the UK government set a target of reversing the decline of farmland birds by 2020. One of the reasons why this target was selected over others was that a single index was available for tracking whether or not the desired changes were taking place. On a global scale, the Living Planet Index (Loh et al. 2005) and other composite indices are being used to track progress towards reducing the current rate of biodiversity loss (Secretariat of the Convention on Biological Diversity 2010). In the last decade, catalysed by the Millennium Ecosystem Assessment (2005) and its political impact, there has been an increase in emphasis on measuring change in ecosystems and the services they provide to human well-being and the global economy. The Economics of Ecosystems and Biodiversity (TEEB) project, for example, has estimated monetary values for many of the headline metrics used to measure environmental change in an effort to help guide conservation policy (Sukhdev et al. 2010). This guidance includes a detailed consideration of subsidies and incentives, environmental liability, national income accounting, cost-benefit analysis, and methods for implementing instruments such as Payments for Ecosystem Services (PES). Adoption of a more ecosystem-based approach to conservation may ultimately encourage a shift in societal values and political priorities far beyond that achieved by traditional species-based conservation approaches.

Identification of the most endangered species has provided a long-standing focus for conservation research and action since the inception of the IUCN Red Lists in the 1960s (IUCN 2011; Mace et al. 2009). Red Lists of species and their conservation status were initially based on subjective expert-based threat assessments for different species groups. The Red Listing process and assessment of extinction risk have now become much more rigorous, and are based on a combination of factors involving population size, rate of decline, size of the distribution range of the species as well as other empirical measures of threat (Mace et al. 2011). More recently, Rodríguez et al. (2011) have argued the need for analogous ecosystem-level threat assessments, suggesting they may be more efficient and less time consuming than species-by-species evaluations, given that ecosystems better represent biological diversity as a whole and require fewer resources to survey. Despite concerted efforts, by 2010 the status of only 47,978 of the world's 1,740,330 known species had been evaluated for potential inclusion on the IUCN Red List (IUCN 2011).

Proactive decisions based on value and threat

Conservation priorities are commonly based on asset value (e.g. total number of species or the number of endemic species in a defined area) and/or potential threat to those assets. Brooks et al. (2006) reviewed nine major approaches for setting global conservation priorities. Most of these approaches prioritize highly irreplaceable regions, with some being reactive (prioritizing high-vulnerability, threatened areas), and others more proactive (prioritizing low-vulnerability wilderness areas). A lack of data means that it is difficult to compare these approaches in terms of their success in generating conservation funding (Halpern et al. 2006), but hot spots alone have mobilized at least $750 million of funding for conservation in these regions (Brooks et al. 2006). More specifically, conservation funding mechanisms have been established for several of the approaches, such as the $100 million, 10-year Global Conservation Fund focused on high-biodiversity wilderness areas and hot spots, and the $137 million Critical Ecosystem Partnership Fund, aimed exclusively at hot spots. The Global Environment Facility, the largest financial mechanism addressing biodiversity conservation, has since 2006 applied a Resource Allocation Framework (RAF) to prioritize its distribution of funds. The RAF allocates resources to countries based on (among other factors) their potential to generate global environmental benefits, which for biodiversity is assessed in relation to the distributions of species and ecosystems and their threat status (GEF 2005).

Given the uneven global distribution of biodiversity, prioritizing conservation efforts makes sense to ensure the 'biggest bang for our buck' (Brooks et al. 2006; Possingham & Wilson, 2005; Wilson et al. 2006). One major challenge is that different measures of conservation value are not always strongly correlated, and as such need to be given joint consideration in any priority setting exercise. For example, Funk & Fa (2010) used global vertebrate distributions in terrestrial ecoregions to evaluate how continuous and categorical ranking schemes target and accumulate endangered taxa within the IUCN Red List, Alliance for Zero Extinction (AZE) and EDGE of Existence programme. By employing total, endemic and threatened species richness as well as an estimator for richness-adjusted endemism, Funk & Fa (2010) showed that all metrics target endangerment more efficiently than by chance. However, each selects unique sets of top-ranking ecoregions, which overlap only partially, and include different sets of threatened species. From these analyses, Funk & Fa (2010) developed an inclusive map for global vertebrate conservation that incorporates important areas for endemism, richness and threat.

Providing information to support prioritization of conservation action has become something of a cottage industry, with many overlapping initiatives collating data on species and habitats, their distribution and status, and the level of protection they are afforded. Some examples are the GEO-Biodiversity Observation Network (www.earthobservations.org/geobon.shtml), the Global Biodiversity Information Facility (www.gbif.org) and the World Database of Protected Areas (www.wdpa.org). While these different databases undoubtedly provide useful information, this plethora of global information providers, well summarized by Brooks et al. (2006), overlap considerably. Such duplication may risk repeating past efforts and wasting valuable resources (Mace et al. 2000). Moreover, they do little to guide decisions on where precisely to allocate resources within large priority areas, and the types of interventions that should be attempted (Wilson et al. 2006).

Systematic conservation planning (SCP) is increasingly widely used to help solve conservation problems at a particular site. At the simplest level, SCP employs analyses of numerical data related to the distribution of biodiversity to aid decision making and optimize allocation of effort (Margules & Sarkar 2007), but it can also involve the application of decision-making frameworks. At the landscape scale, Wilson et al. (2007) have shown that combining information on the spatial distribution of conservation objectives and the cost-effectiveness of actions can achieve more efficient allocation of resources (see section on 'Taking action' below).

However, there is currently a serious mismatch between the development of these methodologies and their use by conservation implementation bodies (global, governmental and non-governmental). Knight et al. (2008) reviewed 88 published conservation plans and found that two-thirds failed to deliver any conservation action. Much of this shortcoming can be attributed to the researchers themselves, as many studies were academic and did not plan for practical and regionally specific implementation (Knight et al. 2008). However, the converse situation is also true in that numerous conservation plans that are implemented are not supported by any systematic or peer-reviewed study. Part of the reason for this is that incentives for conservation bodies to evaluate the success of their investments are often lacking. Achieving the necessary cultural shift in conservation planning will require critical pressure from donors and funders (including the general public) for conservation agencies to adopt and implement more transparent measures of performance (Keene & Pullin 2011).

Scientists are often aware of conservation issues that may be prominent in the future but have attracted little research or policy consideration (Sutherland et al. 2008). Providing mechanisms for the articulation and publication of such issues can become a useful tool. This process, known as 'horizon scanning', is the systematic search for incipient trends, opportunities and constraints that can affect the probability of achieving present and future management goals and objectives. Horizon scanning seeks to inform policy decisions by anticipating issues and accumulating information about them, and is employed by a number of different types of organizations, ranging from the military to, more recently, conservation scientists. As examples, these exercises identified issues such as a step change in pressure on land for agricultural production (Sutherland et al. 2008), high-latitude volcanism (Sutherland et al. 2010) and fracking to remove natural gas (Sutherland et al. 2011a). In each case these have subsequently become high-profile issues (as exemplified by the eruption of Eyjafjallajökull soon after Icelandic volcanoes were discussed), and identifying the issues provides the opportunity to be better prepared (Sutherland & Woodroof 2009).

Since 2009 there has been an annual horizon-scanning exercise to identify global environmental issues (Sutherland et al. 2010). This has involved specialists in horizon scanning, experts in specific areas (e.g. coral reefs, diseases or invasives) as well as representatives from large organizations that have a wide range of conservation interests. The need for this is illustrated by the fact that conservation scientists apparently did not clearly foresee the major shift to biofuel in 2006 by the USA and European Union, with serious consequences for food security, climate change and biodiversity (Fitzherbert et al. 2008; Koh & Wilcove 2008). As a community, we should have seen this coming and been well prepared to contribute to the debate. Issues that are identified as being potentially important but not well recognized are debated and ranked to form a shortlist (Table 1.2). Conservation organizations have taken these issues and identified their responses using a six-point classification, from not planning to track or respond to this issue to committed to

Table 1.2 Examples of horizon-scanning issues given in Sutherland et al. (2010, 2011a)

Example	Issues
Arctic tundra burning	Increased tundra burning associated with climatic conditions, fuel availability and sea ice retreat may impact upon species and human communities, and alter the role Arctic ecosystems play in the global carbon cycle (Hu et al. 2010)
Microplastics	Plastic waste in the sea disintegrates to form tiny fragments to which chemicals may adhere; impact is poorly understood (Barnes et al. 2009)
Hydraulic fracturing (fracking)	Natural gas can be extracted from organic-rich shale basins by pumping in water at high pressure. The impact on hydrology and pollution is poorly understood (Kerr 2010)
Nanosilver	Nanoscale silver is primarily used as an antimicrobial to safeguard human health. Risk to bacteria in ecosystems and aquatic vertebrates is suggested by increased deformities and mortality of exposed zebrafish embryos (Choi & Hu 2009)
Artificial life	These new forms of life could produce vaccines and chemicals, including fuel derived from carbon dioxide. Risks, if the technology becomes widely accessible, include potential interactions with genes and species in natural communities and the potential for malicious use (Lartigue et al. 2009)
Synthetic meat	Muscle stem cells can be taken from live animals, multiplied in a growth medium and stretched to make muscle fibres, potentially shifting meat production from farmland to the factory with considerable impacts for land use (Madrigal 2008)
Assisted colonization	There is considerable debate as to whether this is creating a new wave of invasive species or whether this is an inevitable and sensible conservation measure (Ricciardi & Simberloff 2009; Vitt et al. 2009)
Promotion of biochar	Pyrolysis lessens decomposition, thus may sequester carbon over a long period. However, little is known about the impact on the soil, nor is there a detailed consideration of the source of the wood (Royal Society 2009)

responding now through practice or policy work; in many cases the sensible response is to wait until further developments occur (Sutherland et al. 2012).

Taking action: what to do with limited resources

The concept of prioritizing conservation action is easy to grasp. There are many alternatives for action (interventions) but only limited resources are available to be deployed so difficult choices have to be made. Deciding on how to spend resources depends not only on values but also on what is achievable with current knowledge (Mace et al. 2009).

In approaching this problem, it is useful to list all possible interventions (or candidate solutions) relating to an identified need or problem. One such exercise has been undertaken by Jacquet et al. (2011) who assembled an international team of marine experts to identify potential interventions for protecting the marine environment. The team listed a total of 181 potential interventions, such as 23 cost-effective ways of reducing accidental by-catch of seabirds in fishing nets. Such methods include using streamer lines or spreading shark liver oil in the water to scare birds, deploying acoustic deterrents or setting lines at the side rather than the stern to avoid birds foraging closer to the fishing vessels. Such an exercise can never be fully comprehensive, but it works as a valuable starting point to identify options for action and needs for evidence of their comparative effectiveness in achieving the desired conservation outcome.

Beyond simple listing exercises, a number of decision-making frameworks have been developed to guide the process of moving from conservation goals and potential interventions through the allocation of resources to implementation and conservation monitoring (Wilson et al. 2007; Pullin et al. 2009; Segan et al. 2010). All these frameworks have some common features including:

- a holistic conservation goal that is derived from societal values and concern about undesirable changes and losses to those values. This broad goal (e.g. conservation of tropical forests) may be translated into a more specific conservation target or objective, e.g. to halt loss of tropical forest cover by 2020
- an assumption of a limited budget being available to achieve the stated objectives, and the need to decide which strategies and objectives are most deserving of priority investment, and in which order
- a consideration of all potential interventions that are available to help achieve an objective, assuming that it is invariably necessary to adopt a complementary set of interventions
- explicit use of systematic review and evaluation of effectiveness to inform prioritization of interventions. What do we know about what works and what does not, and how was this learning achieved?
- explicit consideration of the cost-effectiveness of interventions: is the impact of intervention X worth the money compared with intervention Y or no intervention?
- the need to monitor and evaluate resource allocation decisions based on outcomes in relation to objectives (i.e. what impact did our decisions have and why?). See section on 'Evaluation success' below.

One of the most intractable problems is how to allocate funds among alternative conservation actions to address specific threats. To address this, Wilson et al. (2007) proposed an explicit 'ecoaction-specific' framework that focuses on specific objectives, and accounts for the economic costs of interventions (Figure 1.1). The approach goes beyond the decisions to protect areas or species and considers the optimal allocation of resources to specific management interventions in order to address known threats. Wilson et al. demonstrate the utility of this approach by applying it to the management of Mediterranean ecoregions, addressing threats such as invasive species and fire and comparing the likely performance of different interventions based on their cost and the likely biodiversity gain per dollar invested.

A similar approach seeks to determine appropriate interventions for the realization of high-level policy objectives (e.g. halting loss of tropical forest cover or reduction of illegal wildlife trade). Pullin et al. (2009) compared the UK National Service Framework for reducing premature death due to heart disease (the method is well established in the health services) with a potential framework for resolving biodiversity issues, such as the lessening of the impact of alien invasive species. In the health service example, targets for reducing the problem were used to generate strategic actions (such as primary, secondary and tertiary prevention and treatment). Potential interventions contributing to these strategies were then identified based on evidence of their effectiveness (from systematic reviews). For example, thrombolytic therapy has been identified as effective acute treatment, whereas cardiac rehabilitation programmes are effective tertiary prevention for those already suffering from heart disease. The implementation of the National Service Framework enabled the target of reducing premature death from cardiovascular disease by 40% from baseline (1999) to be met 5 years earlier than expected. This equates to thousands fewer premature deaths per year. Such a generic framework that guides decision making from a general policy goal to a set of specific interventions might be useful in conservation. Pullin et al. (2009) concluded that strategic actions (prevention, control and eradication in the case of invasive species) and

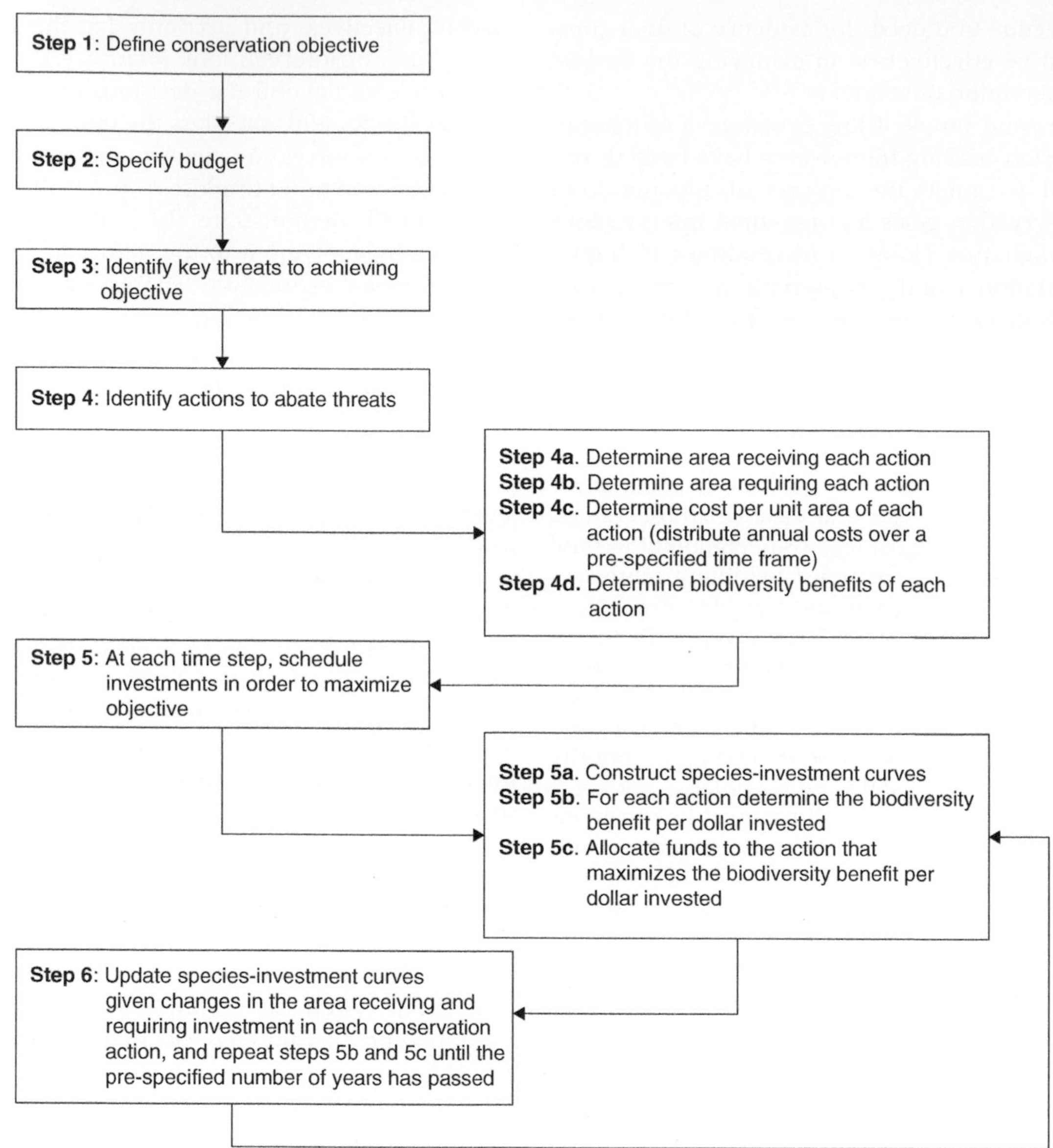

Figure 1.1 Decision steps involved in the conservation investment framework (from Wilson et al. 2007).

potential interventions (e.g. poison baiting for eradication) can readily be identified. Indeed, much of this work has been done by the IUCN Invasive Species Specialist Group. However, evidence for the effectiveness and cost of each intervention is generally lacking. The conservation community has not as yet conducted the necessary research or evidence syntheses (Pullin & Knight 2009). Consequently, it is not possible to make objective, evidence-based decisions among alternative interventions. In a similar way, Segan et al. (2010) considered the methods used by existing government structures, such as the National Institute for Health and Clinical Excellence in England and Wales (NICE), for optimizing resource allocation. In the NICE framework, any intervention is assessed against alternative interventions for achieving a specified goal, in terms of both relative effectiveness of impact and cost. Thus, for example, when protecting fish stocks, the designation of marine protected areas might be assessed against fishing quotas in terms of cost-effectiveness in delivering conservation benefits.

Taken together, the above examples combine: (1) the social process of objective setting, (2) the expert-based process of 'solution scanning', (3) assessing and predicting relative effectiveness of interventions through systematic review and evidence synthesis, and (4) the economic basis of assessing cost effectiveness (see Table 1.1).

A different challenge in planning conservation action is the question of how to measure the comparative value of success among alternative actions, in terms of both the cost of action (cost-effectiveness analysis) and the perceived value of the outcome to society. For example, what might be the comparative value of a successful wetland restoration that requires minimal future management versus the arrested decline of a large mammal that will require considerable continued investment to maintain? In health, there is a standard metric of benefit (the quality-adjusted life-year, QALY) that is recognized as both socially and economically relevant. The QALY is based on the number of years of life added by an action (intervention), and weighted by the quality of life experienced by the patient in each year (i.e. 1.0 for perfect health to 0.0 for death; a debilitating side-effect of treatment might reduce the weight to 0.5). Conservation has no such single simple metric. One useful concept developed by Wilson et al. (2006) is the 'optimal allocation of conservation effort' in which one could use a transferable metric such as the number of species conserved per unit area, but this is both difficult to measure and contestable as a universally valid standard.

Evaluating success

The realities of conservation practice mean that rigorous assessment of the results of projects and programmes is challenging for several reasons.

- *Time frames*: natural systems often take longer than the funding period to undergo detectable change.
- *Scale and context*: conservation actions may have results at different scales from those at which they are implemented and/or from the overall scale of the problem they seek to address.
- *Objective setting*: conservation actions often address multiple objectives, which are sometimes poorly articulated.
- *Attribution*: conservation action (as distinct from research) usually comprises multiple simultaneous interventions, leading to difficulties in attributing outcomes.
- *Resources*: funding agencies and natural resource managers are often reluctant to divert scarce resources from action to monitoring and research, along with uncertainty about what to evaluate and monitor (Gardner 2010).
- *Counterfactual*: rigorously assessing what would have happened without the intervention (a control or counterfactual) can be difficult.

A systematic review of the effectiveness of community forest management programmes in providing global biodiversity and local human welfare benefits found evidence for all of these problems (Bowler et al. 2010), and suggested measures to improve the quality of study designs and provide better evidence of their effectiveness in the future.

In some cases, there is also an insidious disincentive for claiming or demonstrating success in that perceived improvements may reduce the case for public, political and/or financial support for conservation action. Conversely, however, lack of demonstrable success may result in 'donor fatigue'. Furthermore, there are considerable disincentives for assessing and publicizing less successful conservation actions and the problems that explain them (Redford & Taber 2000).

Here we expand on each challenge in turn and identify some possible solutions.

Time frames

Conservation responses usually require time scales much longer than that of the intervention. For example, projects that aim to improve the status of slow-growing trees or large mammals cannot detect population changes during the time frame of the project. Species with long generation times may take decades to respond sufficiently for the effects of an intervention to be measurable above baseline variability.

Scale and context

Scale is important both in geographical terms (where and over how large an area is an intervention expected to be effective, and therefore where should it be monitored and assessed?) and in relation to the scale of the problem. How should practitioners assess the effectiveness of their own actions in relation to the scale of the problem they seek to address? Which is better: a highly effective intervention on a small area or weaker intervention across a wide area?

Objective setting

Despite improvements over recent years in articulating clear objectives for conservation work, it is common that objectives are too poorly formulated to allow rigorous evaluation of success. This is partly due to the reactive 'crisis management' nature of much conservation action. A further contributing factor is that many conservation projects have multiple objectives, some of which may be perceived as less important and/or less attractive to funders and are therefore left unstated or poorly articulated. For example, conservation projects that are supported within development agendas must emphasize their objectives related to human well-being. Similarly, project documents may emphasize the conservation objectives for charismatic species (e.g. large mammals, birds) even when the proposed action is equally or more important for addressing taxa or problems perceived as less attractive to donors (e.g. small brown moths). When there are multiple objectives, the total burden of research and monitoring needed to advance and track progress towards all of them can be very heavy and a major drain on resources. Therefore, evaluation is often limited to a few of the most explicit objectives (or the ones that are easiest to measure). Thus, for example, a project aiming to conserve African birds by improving the livelihoods of local communities and reducing pressures on forest habitats emphasized the livelihoods objectives to obtain its funding from a donor in the development sector. As a result, its monitoring budgets had to be devoted primarily to assessing its livelihoods impacts, and few resources were available for monitoring forest cover or bird populations.

Attribution

Interventions are frequently conducted simultaneously, often by different actors, and it is difficult to attribute changes to particular interventions and actors. Areas and issues that are perceived as of urgent conservation importance, such as the fragmentation of the Atlantic Forest of Brazil or the impacts of ecotourism in the Virunga Volcano region in Central Africa, are the focus of intense conservation effort by many actors and organizations. The Critical Ecosystem Partnership Fund (CEPF) has identified a very large range of funders and organizations working on conservation in the Western Ghats hot spot. These include at least three Indian central government departments, state forest departments, multilateral donors, such as the World Bank and the GEF, and as many as nine bilateral donor agencies (e.g. the UK's DFID), and the CEPF itself. Several international non-governmental organizations (NGOs), including the Ford Foundation, WCS, BirdLife International, and the National Fish and Wildlife Foundation, are also active in the Western Ghats, at least 19 national NGOs have programmes of action and/or research within the hot spot and large numbers of research projects are carried out by academic and technical institutions (CEPF 2007). Of necessity with such a large number of actors, many approaches are employed in parallel, and monitoring is often not unified or strategic across conservation programmes, some of which may differ (financially or physically) by several orders of magnitude. This makes it difficult to link individual changes or outcomes to individual actions or interventions.

Resources

Limited resources and a focus on action mean that practitioners (in conservation and other fields) can be reluctant to divert resources to learning rather than doing (see below), and this is especially true where many objectives are combined and/or many actors involved. In the latter case, collaborative effort may possibly achieve some efficiencies and reduce the total resources needed to assess the achievement of stated objectives.

Counterfactual

However desirable it is to have non-treatment areas for comparison, the realities of conservation practice (limited resources, urgency) mean that most interventions are conducted without an explicit control. Therefore, it is usually impossible to measure or demonstrate success relative to what would happen without (or with a different) intervention. Even where control areas are used, it is often difficult to assess the extent to which the treatment and control samples differ at the beginning of the treatment, potentially confounding the result.

The consequence of failure to cope with these challenges is poor monitoring and evaluation, resulting in future conservation investment decisions that are largely based on belief rather than evidence. In the case of community forest management, this investment amounts to billions of dollars annually (Bowler et al. 2012).

Therefore despite the difficulties, the importance of assessing success has risen higher on the agendas of both practitioners and donors, and researchers have contributed some useful tools. Among the approaches being used to improve the situation are those related to planning and adaptive management, others tied to identifying and assessing outcomes, and new statistical approaches for selecting matched control areas to provide more rigorous assessment of the effectiveness of interventions.

The development of clear frameworks or results chains can help to make objectives explicit, to clarify assumptions about the links between interventions and overarching objectives, and to identify key components of success and their associated monitoring needs. For example, a common intervention in conservation projects is environmental education, frequently targeted at school children, but it is often not clear how these activities relate to project objectives, such as improvement in the status of a population of a particular species. Development of a robust conceptual model of the conservation problem and rigorous application of logical framework or results chain approaches will help to elucidate the conceptual links (or lack of them) between educating school children and the effective conservation of a given target species. These might be that the education programme will build children's awareness and appreciation of the importance of an over-hunted species, and that they will share this with their parents whose choices in the market or the hunt will be affected. By making such logic explicit as a 'theory of change', it will be possible to identify and highlight the constraints and assumptions that apply to each stage in the conceptual chain, such as assumptions about the abilities of the children to understand the main concepts, the degree of influence they have over their parents' actions and the factors influencing whether adults are able to change their behaviour. The use of logical frameworks and related approaches has become standard procedure in other sectors, such as development, and several recent advances have helped to clarify their application in conservation (Salafsky et al. 2001). The Conservation Measures Partnership (www.conservationmeasures.org), in its *Open Standards for the Practice of Conservation*, has provided a consensus among several major conservation organizations (including the Nature Conservancy, African Wildlife Foundation, Wildlife Conservation Society and WWF-US) on the application of these approaches and other aspects of adaptive management in conservation (CMP 2007) and has further provided tools to facilitate their implementation (Miradi software; www.miradi.org).

There has also been considerable recent emphasis on assessing the degree to which the Open Standards and other recognized aspects of good practice are applied, and on using this as an indicator of the likely success of conservation actions. Such 'conservation audits' are described in detail by O'Neil (2007). A similar approach, assessing the degree to which elements of good management are in place, is used in some assessments of management effectiveness in protected areas (Leverington et al. 2008), including the Management Effectiveness Tracking Tool (METT) employed by the WWF and the World Bank to assess management of forest protected areas (Stolton et al. 2007), and the Rapid Assessment and Prioritization of Protected Areas Management – RAPPAM (Ervin 2003).

Another emerging approach builds on results chains to identify *outcomes* (as distinct from *procedures*) that are expected to lead ultimately to conservation success (improved status of target species, ecosystems and sites). A results chain is a tool that clarifies assumptions about how conservation activities contribute to reducing threats and achieving the conservation of biodiversity or thematic targets (O'Connor 2005). It maps out a series of causal statements that link factors in an 'if ... then' fashion – for example, if a threat is reduced, then status of a biodiversity target is enhanced or if an opportunity is taken, then a thematic target might be improved. Results are assessed in relation to those intermediate steps, using rigorous quantitative approaches, expert assessment or self-evaluation.

In developing tools for implementing this approach, the Cambridge Conservation Forum drew on the experience of practitioners from many of its member organizations and on existing categorizations of conservation action (Salafsky et al. 2002, 2008) to define seven broad categories of conservation activity and generic results chain models for each of them (Kapos et al. 2008, 2009, 2010). A scorecard-style questionnaire-based tool was developed to help practitioners assess the results of both past and ongoing conservation projects. Assessing the degree to which intermediate outcomes have been achieved can support adaptive management and provide insights on likely long-term effectiveness of interventions, even where resources may be too limited to support appropriate biological monitoring. Thus, for example, assessing behaviour change among local communities or effective implementation of a regulation adopted as a result of advocacy efforts may provide a clear pointer to eventual improvements in the status of an exploited species.

Recently, several groups have applied novel statistical approaches to provide rigour in assessing the longer term effectiveness of specific conservation interventions (e.g. Andam et al. 2008; Linkie et al. 2008). Particularly promising are the matching techniques used for identifying counterfactual values as a basis for assessing conservation outcomes where no controls have been deliberately established (Ferraro & Pattanayak 2006; Ferraro et al. 2007). These have especially been applied in assessing the effectiveness of protected areas in reducing deforestation (e.g. Nelson & Chomitz 2009, 2011; Joppa & Pfaff 2010, 2011). These analyses have effectively accounted for the fact that many protected areas are remote from drivers of deforestation by matching them with similar controls, and have shown that reductions in deforestation in protected areas are much lower than previously thought. This is important information for planning protected areas systems and targeting other conservation interventions.

While these approaches potentially reduce the need for investing field effort in establishing and monitoring control areas, they are most applicable where the response variables affected by conservation action are readily detectable – they have so far only been applied in assessing effectiveness in reducing deforestation. They require a very careful selection of control areas and often rely upon the availability of, and ability to process very large data sets. Nonetheless, they provide an important opportunity to validate empirically the procedural standards and other proxies that may be used more widely for assessing the effectiveness of conservation actions and programmes.

When and for whom is research a priority?

Beyond efforts to understand patterns of natural variability in the abundance and distribution of species and habitats, there are essentially two areas of conservation research: identifying problems and identifying solutions. The first is concerned largely with measuring human-induced environmental change and understanding the drivers responsible for such changes, while the second is concerned largely with assessing the effectiveness of alternative interventions. To date, a lot more research has been conducted on the former than the latter, and there is still a scarcity of evidence on what types of management or policy intervention have been most successful in tackling key problems (Sutherland et al. 2004). Recent developments in evidence synthesis and the application of systematic review methodology are addressing this gap (Pullin & Knight 2009).

Very occasionally, a problem may be so large, and the method of resolving it so clear and simple, that research is not necessary (e.g. closing down of a sewage outflow that drains directly onto a rare salt marsh). However, conservation problems invariably abound with uncertainty, risk and controversy. There is rarely strong consensus amongst different stakeholders as to the importance of a given problem, and where there is general support for action, there is usually a variety of alternative solutions available, the relative effectiveness of which is unclear (see section on 'Taking action' above). Research can provide much needed clarity as to the most pressing problems or rewarding solutions, but the challenge comes in deciding whether further research is really needed or whether enough information is already available to take informed action. There are two important reminders that should be considered when making this decision.

First, it may be the case that there is enough information available to make adequately informed decisions without further research. Instead, what is lacking is sufficient political will and resources, or the opportunity for making a strategic investment. As discussed earlier in this chapter, a surprising amount of information is already available to help guide conservation decision making, both regarding the distribution of biodiversity at global (e.g. the Key Biodiversity Areas assessment of the IUCN) and regional (e.g. the www.natureserve.org portal which compiles information on rare and endangered species and ecosystems in the Americas) scales, and the relative effectiveness of alternative intervention strategies (see evidence syntheses and databases at the Conservation Evidence project – www.conservationevidence.

com – and the Collaboration for Environmental Evidence – www.environmentalevidence.org). It is common to hear the call that more research is needed before any recommendations can be made. Such advice can be problematic because funding for continued research and conservation action often comes from the same pot, and opportunities for successful interventions may be short-lived. Because access to species data is only one of the factors that determines the overall success of a conservation plan (Knight et al. 2006), there comes a point where the collection of ever more field data is redundant. Grantham et al. (2008) provide a convincing example of this by demonstrating that a database of South African protea (the regionally endemic plants for which South African *fynbos* vegetation is so famous) could be reduced in size by a factor of 25 without any marked impact on the effectiveness of spatial conservation prioritizations.

Second, it is important to consider that academic performance is not always correlated with the generation of practical and effective conservation guidance. As such, investment in research, unless carefully managed, may not produce the desired results. In 2001 Tony Whitten and colleagues challenged the conventional views on the value of conservation research, claiming that much of conservation biology is 'a displacement behaviour for academics' and that many scientific priorities are far removed from what is really needed to safeguard the future of biodiversity (Whitten et al. 2001). While this essay was deliberately provocative, there is often an irrefutable gap between the indicators used to measure success in academia (e.g. numbers of publications and impact factors) and those used to measure success in on-the-ground conservation programmes (e.g. reductions in loss of habitat, increased numbers of an endangered species) (Chapron & Arlettaz 2008), and much research is criticized for failing to provide practical management recommendations (e.g. Meijaard & Sheil 2007). One possible solution is to assess research papers by their contribution to the knowledge that is considered a priority by policy makers and practitioners; this approach has been used to evaluate papers testing interventions to enhance bee populations (Sutherland et al. 2011b).

As Arlettaz et al. (2010) illustrate in a project to restore endangered hoopoes in the Swiss alps, ensuring that research is capable of developing useful recommendations requires the involvement of scientists in the actual implementation process. However, there are opportunity costs to investing time and resources in tackling messy on-the-ground conservation problems in a timely fashion. It is unrealistic to expect that individuals can operate effectively as both researchers and managers, but it is vital that scientists are given professional recognition for their practical as well as academic impact.

Identifying priority research questions and knowledge gaps

Identifying the kind of research that is likely to deliver the most useful and cost-effective results is deceptively difficult. Foremost is the need to be explicit and transparent about the choices available in deciding what to study, and about the ultimate purpose of different types of research, given overall conservation objectives and alternative ways of spending our time and resources. Choices of research questions are commonly decided based not on any objective or systematic framework, but rather on some combination of circumstance, personal interest, past experience and personal appeal. To try and get around this, it can be useful to step back and appraise the different kinds of motivation that underpin choices about what to study.

First, what is the likely value of a new piece of research for delivering measurable outcomes for biodiversity conservation? What is the scale of the problem or threat that is being studied, and can results be generalized to other places and species or are they confined to a specific site or taxon? What is the relative importance of the problem being studied, given other threats? For example, many biologists have focused their research efforts on understanding the role of infectious diseases, climate change and acidification on declining amphibian populations. These are justifiable concerns, yet a much bigger problem facing the survival of amphibians worldwide is habitat loss and degradation, which has received disproportionally little attention (Gardner et al. 2007).

As well as thinking about likely impacts, it is important to consider the probability that results will be adopted by relevant decision makers. If you are charged with advising a small island state in the Pacific on how to mitigate population declines of endemic species, it is arguably more useful to focus efforts on threats that fall under their jurisdiction (e.g. habitat loss, eradication of exotic species) than factors almost completely outside their control (e.g. climate change).

Second, and as mentioned previously, it is important to reflect on what information is already available and where evidence is critically lacking. If enough biological data are already collected to identify (if only crudely) areas of high conservation priority, it may be more worthwhile collecting information on the costs of different conservation activities and availability of land to achieve a particular goal (Wilson et al. 2009).

Finally, it is important to contemplate the feasibility of a conservation project, where feasibility includes access to necessary background data, appropriate experimental treatments (whether natural or directly manipulated) and study site conditions, and technical expertise necessary to tackle adequately the posed research question, and sufficient funds to ensure that the research is conducted with necessary rigour. Sometimes a project may be highly desirable but impossible to implement, as the conditions just do not exist in the area of interest.

Evaluation and learning from experience as forms of research

While it is important to recognize and think about trade-offs between research and management, as well as between competing research objectives, it is also important to avoid extended procrastination. When it comes to conservation research, the most valuable learning can often be achieved only once the first measurements are made, and at least a few different interventions have been tried and responses evaluated. As highlighted at the start of this chapter, research should not be conceived as a linear process with clear start and end points but rather a continuous cycle of observation, evaluation and recommendation that feeds back into future changes in management, which in turn require their own evaluation. In this sense, research and management are two sides of the same coin; neither exercise makes proper sense unless accompanied by the other. This notion has its roots in adaptive management (Holling 1978; Stem et al. 2005), but also highlights the importance of many less formal or indirect ways of 'learning by doing', or experiential learning as it is termed by educationalists, including implicit and tacit knowledge, and expert observation (Fazey et al. 2006). Experienced observers and conservation practitioners may, without any formal training, be able to recognize emergent properties and make good predictions of environmental change without being able to explain precisely how they do it. Fazey et al. (2006) present an example of this by eliciting the expert knowledge of on-the-ground managers – including both local government staff and cattle ranchers – of wetland systems in Australia.

Conclusions and recommendations

Whilst past conservation actions may well have saved many species and habitats, biodiversity continues to decline. To date, conservation has fallen short in providing evidence of measurable benefits and value for money. This is true for both science and action. You cannot manage what you do not measure, and without measurement, the effectiveness of conservation research and action is fraught with uncertainty both within and outside the community. That said, measurements cost time and money and need themselves to be justified. To demonstrate value for money, the conservation community needs to put in place a more transparent system for identifying priorities, deciding on appropriate actions and measuring effectiveness. The concept is simple but the actors and actions are so many and complex that a major improvement in information and knowledge sharing is vital if we are to progress and become more effective. Decision makers need to be able to identify knowledge gaps and communicate those to science funders, not just as a wish list but with evidence that the gaps are real, are of the highest priority, and are inhibiting their ability to make decisions. Conservation scientists need to balance their (often healthy) tendency toward challenging current wisdom and emphasizing uncertainty with a need to provide critical evidence syntheses and evaluations of the success of conservation actions and find scientific consensus in a way that is useful to decision makers. These latter roles are undervalued in academic institutions that depend on traditional publication-based reward systems.

It is most likely that decision-making frameworks will be more widely adopted in conservation if they can be first demonstrated on a real-world conservation problem, such as those articulated in the Convention on Biologi cal Diversity Aichi Targets (Convention on Biological Diversity 2011). Whether policy makers and funders will take the risk of investing in such a programme remains to be seen. Nevertheless, many of the components described above are already being used and are arguably increasing in frequency. To date, we are not aware of any conservation programme in which all these pieces have been combined. There may be many reasons for this, but here are a few.

- *Lack of vision and time*: many conservation organizations work on a largely responsive mode and dedicate much of their time to dealing with short-term problems and fundraising. Framing any conservation programme within a broader strategic framework requires an ability to step back and calmly evaluate the costs and benefits of different choices. Ideally this requires dedicated staff members or a department.
- *Lack of data*: decision-making frameworks are only viable if there is sufficient accessible information to inform the steps in the process. Rigorous evidence syntheses in the form of systematic reviews are few and this inhibits the key step of comparing effectiveness of alternative interventions.
- *Perceived cost*: employing the range of skills and conducting the range of tasks implied in the use of decision-making frameworks is expensive. Data mining, evidence synthesis and programme evaluation are all resource-hungry activities.
- *Fear of objective processes outside organizational control*: for smaller conservation organizations, these processes might seem to threaten their very existence by questioning the programmes for which they raise money, such as buying land for protection or for species-based management and recovery programmes. For government departments, the process might suggest the adoption of policies that would be unpopular with the electorate or powerful pressure groups.
- *Lack of incentive*: there is no incentive for organizations in which financial performance does not rely on the effectiveness of conservation action *per se*, despite it being an implicit objective. Many organizations provide little information on their effectiveness (nor is it demanded by any independent body); instead they appeal to their members and donors in terms of ongoing threats to species and habitats and of the ability of the organization to act on their behalf (i.e. they are action oriented, not performance oriented).

There have been many individuals in conservation who have sought to facilitate the increased use of good science in conservation decision making and to provide the tools to inform priority setting and resource allocation. Some reside in academia, whilst others populate conservation organizations, government departments and the business community. Some come from other disciplines such as economics, public health, information science and programme evaluation. A major social challenge is to marshal this diverse group to the benefit of conservation at large. Informal networks and collaborations, such as the Environmental Evaluators Network and the Collaboration for Environmental Evidence, are developing interdisciplinary communities capable of generating information that will inform conservation decision making in a co-ordinated and transparent way. There are also some signs of intergovernmental demand for assessments of changes in biodiversity, but at present there is often no clear structure in which that community should work (and be provided with the resources). The emergence of an Intergovernmental Science-Policy Platform on Biodiversity and Ecosystem Services (ipbes.net) might provide such a structure but its remit is still to be determined. Whatever the major structures and organizations turn out to be, a new culture of effective, rather than simply well-intentioned, conservation practice is essential if tough conservation challenges are to be met with continuingly scarce resources.

Acknowledgements

WJS is funded by Arcadia and ESRC and thanks Stephanie Prior for useful input. TAG thanks the Natural Environmental Research Council (NE/F01614X/1) and the Instituto Nacional de Ciência e Tecnologia -Biodiversidade e Uso daTerra na Amazônia (CNPq 574008/2008-0) for funding. ASP is funded by the Centre for Integrated Research in the Rural Environment.

The difficulty lies not so much in developing new ideas as in escaping from old ones

John Maynard Keynes

References

Andam, K.S., Ferraro, P.J., Pfaff, A., Sanchez-Azofeifa, G.A. & Robalino, J.A. (2008) Measuring the effectiveness of protected area networks in reducing deforestation. *Proceedings of the National Academy of Sciences USA*, **105**, 16089–16094.

Arlettaz, R., Schaub, M., Fournier, J., *et al.* (2010) From publications to public actions: when conservation biologists bridge the gap between research and implementation. *BioScience*, **60**, 835–842.

Barnes, D.K.A., Galgani, F., Thompson, R.C. & Barlaz, M. (2009) Accumulation and fragmentation of plastic debris in global environments. *Philosophical Transactions of the Royal Society, Series B*, **364**, 1985–1998.

Bowler, D., Buyung-Ali, L., Healey, J.R., Jones, J.P.G., Knight, T. & Pullin, A.S. (2010) The evidence base for community forest management as a mechanism for supplying global environmental benefits and improving local welfare. CEE review 08-011. *Environmental Evidence*: www.environmentalevidence.org/SR48.html

Bowler, D., Buyung-Ali, L., Healey, J.R., Jones, J.P.G., Knight T.M. & Pullin, A.S. (2012) Does community forest management provide global environmental benefit and improve local welfare? *Frontiers in Ecology and the Environment*, **10**, 29–36.

Brooks, T.M., Mittermeier, R.A., da Fonseca, G.A.B., *et al.* (2006) Global biodiversity conservation priorities. *Science*, **313**, 58–61.

Butchart, S.H.M., Walpole, M., Collen, B., *et al.* (2010) Global biodiversity: indicators of recent declines. *Science*, **328**, 1164–1168.

CEPF (2007) *Ecosystem Profile Western Ghats & Sri Lanka Biodiversity Hotspot, Western Ghats Region*. Critical Ecosystem Partnership Fund, Washington, DC.

Chapron, G. & Arlettaz, R. (2008) Conservation: academics should "conserve or perish". *Nature*, **451**, 127.

Choi, O.K. & Hu, Z.Q. (2009) Nitrification inhibition by silver nanoparticles. *Water Science Technology*, **59**, 1699–1702.

CMP (2007) *Open Standards for the Practice of Conservation*, Version 2.0. Conservation Measures Partnership: www.conservationmeasures.org/wp-content/uploads/2010/04/CMP_Open_Standards_Version_2.0.pdf

Convention on Biological Diversity (2011) *Strategic Plan for Biodiversity 2011–2020 and the Aichi*

Biodiversity Targets. Convention on Biological Diversity, Montreal, Canada.

Crouse, D.T., Crowder, L.B. & Caswell, H. (1987) A stage-based population model for loggerhead sea turtles and implications for conservation. *Ecology*, **68**, 1412–1423.

Ervin, J. (2003) *WWF: Rapid Assessment and Prioritization of Protected Area Management (RAPPAM) methodology*. WWF, Gland, Switzerland.

Fazey, I., Fazey, J.A., Salisbury, J.G., Lindenmayer, D.B. & Dovers, S. (2006) The nature and role of experiential knowledge for environmental conservation. *Environmental Conservation*, **33**, 110.

Ferraro, P. & Pattanayak, S.K. (2006) Money for nothing? A call for empirical evaluation of biodiversity conservation investments. *PLoS Biology*, **4**, 105.

Ferraro, P., McIntosh, C. & Ospina, M. (2007) The effectiveness of the US endangered species act: an econometric analysis using matching methods. *Journal of Environmental Economics and Management*, **54**, 245–261.

Fitzherbert, E.B., Struebig, M. J., Morel, A., *et al*. (2008) How will oil palm expansion affect biodiversity? *Trends in Ecology and Evolution*, **23**, 538–545.

Funk, S.M. & Fa, J.E. (2010) Ecoregion prioritization suggests an armoury not a silver bullet for conservation planning. *PLoS ONE*, **5**, e8923.

Gardner, T. (2010) *Monitoring Forest Biodiversity: Improving Conservation through Ecologically Responsible Management*. Earthscan, London.

Gardner, T., Barlow, J. & Peres, C. (2007) Paradox, presumption and pitfalls in conservation biology: the importance of habitat change for amphibians and reptiles. *Biological Conservation*, **138**, 166–179.

GEF (2005) *The GEF Resource Allocation Framework*. Global Environment Facility, Washington, DC.

Grantham, H.S., Moilanen, A., Wilson, K.A., Pressey, R.L., Rebelo, T.G. & Possingham, H.P. (2008) Diminishing return on investment for biodiversity data in conservation planning. *Conservation Letters*, **1**, 190–198.

Halpern, B.S., Pyke, C.R., Fox, H.E., Haney, J.C., Schlaepfer Martin A. & Zaradic, P. (2006) Gaps and mismatches between global conservation priorities and spending. *Conservation Biology*, **20**, 56–64.

Holling, C.S. (1978) *Adaptive Environmental Assessment and Management*. Wiley, Chichester.

Hu F.S., Higuera P.E., Walsh J.E., *et al*. (2010) Tundra burning in Alaska: linkages to climatic change and sea ice retreat. *Journal of Geophysical Research*, **115**, G04002.

IUCN (2011) *IUCN Red List of Threatened Species*, Version 2011.1: www.iucnredlist.org

Iwamura, T., Wilson, K.A., Venter, O. & Possingham H.P. (2010) A climatic stability approach to prioritizing global conservation investments. *PLoS ONE*, **5**, e15103.

Jacquet, J., Boyd, I., Carlton, J.T., *et al*. (2011) Scanning the oceans for solutions. *Solutions*, **2**, 46–55.

Joppa, L.N., & Pfaff, A. (2010) Re-assessing the forest impacts of protection: the challenge of non-random location & a corrective method. *Annual Review of Ecological Economics*, **1185**, 135–149.

Joppa, L.N. & Pfaff, A. (2011) Global protected area impacts. *Proceedings of the Royal Society B Biological Sciences*, **278**, 1633–1638.

Kapos, V., Balmford, A., Aveling, R., *et al*. (2008) Calibrating conservation: new tools for measuring success. *Conservation Letters*, **1**, 155–164.

Kapos, V., Balmford, A., Aveling, R., *et al*. (2009) Outcomes, not implementation, predict conservation success. *Oryx*, **43**, 336–342.

Kapos, V., A. Manica, R., Aveling, P., *et al*. (2010) Defining and measuring success in conservation. In: *Trade-Offs in Conservation: Deciding What to Save*. (eds. N. Leader-Williams, W.M. Adams & R.J. Smith), pp. 73–93. Wiley-Blackwell, Oxford.

Keene, M. & Pullin, A.S. (2011) Realizing an effectiveness revolution in environmental management. *Journal of Environmental Management*, **92**, 2130–2135.

Kerr, R.A. (2010) Natural gas from shale bursts onto the scene. *Science*, **328**, 1624–1626.

Knight, A.T., Cowling, R.M. & Campbell, B.M. (2006) An operational model for implementing conservation action. *Conservation Biology*, **20**, 408–419.

Knight, A.T., Cowling, R.M., Rouget, M., Balmford, A., Lombard, A.T. & Campbell, B.M. (2008) Knowing but not doing: selecting priority conservation areas and the research–implementation gap. *Conservation Biology*, **22**, 610–617.

Koh, L.P. & Wilcove, D.S. (2008) Is oil palm agriculture really destroying tropical biodiversity? *Conservation Letters*, **1**, 60–64.

Lartigue, C., Vashee, S., Algire, M.A., *et al*. (2009) Creating bacterial strains from genomes that have been cloned and engineered in yeast. *Science*, **325**, 1693–1696.

Leverington, F., Hockings, M., Pavese, H., Lemos Costa, K. & Courrau, J. (2008) *Management Effectiveness Evaluation in Protected Areas – A Global*

Study. Supplementary Report No. 1. University of Queensland, Gatton.

Linkie, M., Smith, R.J., Zhu, Y., *et al.* (2008) Evaluating biodiversity conservation around a large Sumatran protected area. *Conservation Biology*, **22**, 683–690.

Loh, J., Green, R.E., Ricketts, T., *et al.* (2005) The Living Planet Index: using species population time series to track trends in biodiversity. *Philosophical Transactions of the Royal Society B Biological Sciences*, **360**, 289–295.

Mace, G.M., Balmford, A., Boitani, L., *et al.* (2000) It's time to work together and stop duplicating conservation efforts. *Nature*, **405**, 393.

Mace, G.M., Possingham, H.P. & Leafer-Williams, N. (2009) Prioritizing choices in conservation. In: *Key Topics in Conservation Biology* (eds. D.W. Macdonald & K. Service), pp. 17–34. Wiley-Blackwell, Oxford.

Mace, G.M., Collar, N.J., Gaston, K.J., *et al.* (2011) Quantification of extinction risk: IUCN's system for classifying threatened species. *Conservation Biology*, **22**, 1424–1442.

Madrigal, A. (2008) Scientists flesh out plans to grow (and sell) test tube meat. www.wired.com/science/discoveries/news/2008/04/invitro_meat

Margules, C. R. & Sarkar, S. (2007) *Systematic Conservation Planning*. Cambridge University Press, Cambridge.

Meijaard, E. & Sheil, D. (2007) Is wildlife research useful for wildlife conservation in the tropics? A review for Borneo with global implications. *Biodiversity and Conservation*, **16**, 3053–3065.

Millennium Ecosystem Assessment (2005) World Resources Institute, Washington, DC.

Nelson, A. & Chomitz, K.M. (2009) *Protected Area Effectiveness in Reducing Tropical Deforestation: A Global Analysis of the Impact of Protection Status*. 42 Independent Evaluation Group. Communications, Learning and Strategy. World Bank, Washington, DC.

Nelson, A. & Chomitz, K.M. (2011) Effectiveness of strict vs. multiple-use protected areas in reducing tropical forest fires: a global analysis using matching methods. *PloS ONE*, **6**, e22722.

Newson, S.E., Ockendon, N., Joys, A., Noble, D.G. & Baillie, S.R. (2009) Comparison of habitat specific trends in the abundance of breeding birds in the UK. *Bird Study*, **56**, 233–243.

O'Connor, S. (2005) *Basic Guidance for Tools: Results Chains. Resources for Implementing the WWF Standards*. WWF, Gland, Switzerland.

O'Neil, E. (2007) *Conservation Audits: Auditing the Conservation Process – Lessons Learned, 2003–2007*. Conservation Measures Partnership, Bethesda, MD.

Possingham, H.P. & Wilson, K.A. (2005) Turning up the heat on hotspots. *Nature*, **436**, 919–920.

Pullin, A.S. & Knight, T.M. (2001) Effectiveness in conservation practice: pointers from medicine and public health. *Conservation Biology*, **15**, 50–54.

Pullin, A.S. & Knight, T.M. (2009) Doing more good than harm – building an evidence-base for conservation and environmental management. *Biological Conservation*, **142**, 931–934.

Pullin, A.S., Knight, T.M. & Watkinson, A.R. (2009) Linking reductionist science and holistic policy using systematic reviews: unpacking environmental policy questions to construct an evidence-based framework. *Journal of Applied Ecology*, **46**, 970–975.

Redford, K.H. & Taber, A. (2000). Writing the wrongs: developing a safe-fail culture in conservation. *Conservation Biology*, **14**, 1567–1568.

Ricciardi, A. & Simberloff, D. (2009) Assisted colonization is not a viable conservation strategy. *Trends in Ecology and Evolution*, **24**, 248–253.

Rodríguez, J. P., Rodríguez-Clark, K. M., Baillie, J. E. M., *et al.* (2011) Establishing IUCN red list criteria for threatened ecosystems. *Conservation Biology*, **25**, 21–29.

Royal Society (2009) *Geoengineering the Climate: Science, Governance and Uncertainty*. The Royal Society, London.

Salafsky, N., Margoluis, R. & Redford, K.H. (2001) *Adaptive Management: A Tool for Conservation Practitioners*. Biodiversity Support Program, Washington, DC.

Salafsky, N., Margoluis, R., Redford, K.H. & Robinson, J.G. (2002) Improving the practice of conservation: a conceptual framework and research agenda for conservation science. *Conservation Biology*, **16**, 1469–1479.

Salafsky, N., Salzer, D., Stattersfield, A.J., *et al.* (2008). A standard lexicon for biodiversity conservation: unified classifications of threats and actions. *Conservation Biology*, **22**, 897–911.

Secretariat of the Convention on Biological Diversity (2010) *Global Biodiversity Outlook 3*. Secretariat of the Convention on Biological Diversity, Montréal.

Segan, D.B., Bottrill, M.C., Baxter, P.W.J. & Possingham, H.P. (2010) Using conservation evidence to guide management. *Conservation Biology*, **25**, 200–202.

Stem, C., Margoluis, R., Salafsky, N. & Brown, M. (2005) Monitoring and evaluation in conservation: a review of trends and approaches. *Conservation Biology*, **19**, 295–309.

Stolton, S., Hockings, M., Dudley, N., MacKinnon, K., Whitten, T. & Leverington, F. (2007) *Reporting Progress in Protected Areas: A Site Level Management Effectiveness Tracking Tool*, 2nd edn. World Bank/WWF Forest Alliance, WWF, Gland, Switzerland.

Sukhdev, P., Wittmer, H., Schroter-Schlaack, C., *et al.* (2010) *The Economics of Ecosystems and Biodiversity. Mainstreaming the Economics of Nature: A Synthesis of the Approach, Conclusions and Recommendations of TEEB*. TEEB, Bonn.

Sutherland, W.J. & Woodroof, H.J. (2009) The need for environmental horizon scanning. *Trends in Ecology and Evolution*, **24**, 523–527.

Sutherland, W.J., Pullin, A.S., Dolman, P.M. & Knight, T.M. (2004) The need for evidence-based conservation. *Trends in Ecology and Evolution*, **19**, 305–308.

Sutherland, W.J., Bailey, M.J., Bainbridge, I.P., *et al.* (2008) Future novel threats and opportunities facing UK biodiversity identified by horizon scanning. *Journal of Applied Ecology*, **45**, 821–833.

Sutherland, W.J., Clout, M., Côté, I.M., *et al.* (2010) A horizon scan of global conservation issues for 2010. *Trends in Ecology and Evolution*, **25**, 1–7.

Sutherland, W.J., Bardsley, S., Bennun, L., *et al.* (2011a) A horizon scan of global conservation issues for 2011. *Trends in Ecology and Evolution*, **26**, 10–16.

Sutherland, W.J., Goulson, D., Potts, S.G. & Dicks, L.V. (2011b) Quantifying the impact and relevance of scientific research. *PLoS ONE*, **6**, e27537.

Sutherland, W.J., Allison, H., Aveling, R., *et al.* (2012) Enhancing the value of horizon scanning through collaborative review. *Oryx*, **46**(3), 368–374.

Vitt, P., Havens, K. & Hoegh-Guldberg, O. (2009) Assisted migration: part of an integrated conservation strategy. *Trends in Ecology and Evolution*, **24**, 473–474.

Whitten, T., Holmes, D. & MacKinnon, K. (2001) Conservation biology: a displacement behavior for academia? *Conservation Biology*, **15**, 1–3.

Wilson, K.A., McBride, M.F., Bode, M. & Possingham, H.P. (2006) Prioritizing global conservation efforts. *Nature*, **440**, 337–340.

Wilson, K.A., Underwood, E.C., Morrison, S.A., *et al.* (2007) Conserving biodiversity efficiently: what to do, where and when. *PLoS Biology*, **5**, e223.

Wilson, K.A., Carwardine, J. & Possingham, H.P. (2009) Setting conservation priorities. *Annals of the New York Academy of Science*, **1162**, 237–264.

2

Levels of approach: on the appropriate scales for conservation interventions and planning

Jonathan E.M. Baillie[1], David Raffaelli[2] and Claudio Sillero-Zubiri[3]

[1]Zoological Society of London, London, UK
[2]Environment Department, University of York, York, UK
[3]Wildlife Conservation Research Unit, Department of Zoology, Recanati-Kaplan Centre, University of Oxford, Oxford, UK

"Let our advance worrying become advance thinking and planning."

— **Winston Churchill**

Introduction

As seen in the previous chapter, there are many ways to identify conservation priority species (Mace and Collar 2002), sites or regions (Stattersfield et al. 1998; Myers et al. 2000; Olson et al. 2001; Mittermeier et al. 2004; Mace et al. 2007). However, once the priority has been identified, decisions must be made about the most appropriate scale for interventions and conservation planning. This chapter defines the most common scales of conservation intervention and introduces a number of approaches for conservation planning. It highlights the need for the conservation community to work together to develop a more co-ordinated and inclusive planning process that integrates conservation interventions at a broad range of scales.

The formation of the IUCN – the International Union for Conservation of Nature (originally called the International Union for the Protection of Nature) – in 1948 was an important moment in the birth of the modern conservation era. At that point, conservation efforts tended to focus on either protected areas or individual iconic species, usually large mammals or birds (Scott et al. 1987), and interventions were undertaken

Key Topics in Conservation Biology 2, First Edition. Edited by David W. Macdonald and Katherine J. Willis.

Table 2.1 Conservation planning as well as spatial priority setting approaches and the scales at which they are most often implemented

Level of intervention	Conservation planning approaches	Spatial priority setting approaches	References
Population	SSC Action Plans		IUCN/SSC 2008; Lacy 1994
Species	SSC Action Plans		www.redlist.org; www.iucn.org
	Alliance for Zero Extinction		www.zeroextinction.org
			Isaac et al. 2007
	EDGE species		www.edgeofexistence.org
		WCS Range-Wide Priority Setting	Sanderson et al. 2002
Protected Areas	Protected Area Management Plans		Thomas & Middleton 2003
Landscape/ Ecosystem	Ecosystem Approach		CDB 2000, Waltner-Toews et al. 2008
	TNC Conservation Action Planning		TNC 2007
	National Biodiversity Strategies and Action Plans		Prip et al. 2010
		Important Bird Areas	Bennun & Fishpool 2000
		Important Plant Areas	Anderson 2002
		Key Biodiversity Areas	Eken et al. 2004
		WWF's Global 200	Olson and Dinerstein 1998
		CI Hotspots	Mittermeier et al. 2004
		EDGE Zones	
		National Character Areas (UK)	Natural England 2009
		GIS/Map-based approaches	Nelson et al. 2008; Naidoo et al. 2008; Eigenbrod et al. 2009

primarily at the population or species level. However, since the formation of the IUCN, the human population has more than doubled, as has water and food consumption (Pollard et al. 2010), resulting in greater anthropogenic changes to the landscape than at any previous time in human history (Millennium Ecosystem Assessment 2005). Now roughly one-fifth of the world's vertebrates and plants are threatened with extinction (Baillie et al. 2010; Hoffmann et al. 2010), and instead of individual species, entire lineages or ecosystems are threatened, such as amphibians (Stuart et al. 2004) or coral reefs (Carpenter et al. 2008). As dominant threats, including habitat destruction, overexploitation, invasive alien species, nitrogen pollution and climate change impacts, continue to increase in scale and intensity (Butchart et al. 2010), so must the response. This has resulted in a trend towards much larger-scale strategies or action plans and interventions.

To explore the utility of interventions ranging from a single population to the landscape scale, it is important first to have a clear definition of each scale of intervention (Table 2.1). In this chapter, we introduce four levels of intervention: populations, species, protected areas, and landscapes or ecosystems. We selected these scales as they are the most commonly used.

Populations

The most common definition of a population is a geographical entity within a species that is distinguished either ecologically or genetically (Hughes et al. 1997). This definition is applicable to evolutionarily significant units (a group of individuals that are considered distinct such as subspecies or a geographic race) (Moritz 1994; Crandall et al. 2000), which are often the focus of conservation attention, but does not necessarily encompass management units. For example, fish stocks are often referred to as populations but may not be isolated ecologically or genetically, and so are not 'populations' in the strictest sense. In practice, conservation biologists commonly use the term 'population' to identify a group of individuals within their zone of intervention; this could range from elephants in a specific protected area to a small group of crickets in a single field. In this chapter we use a relatively broad definition: a group of individuals of the same species, occupying a defined area and usually isolated to some degree from other similar groups (Millennium Ecosystem Assessment 2005). For conservation interventions, this will range from a group of individuals within a specific management unit, to an isolated discrete population, to disjunct populations with some exchange of individuals between them (known collectively as a metapopulation; Wells and Richmond 1995), and to one continuous population where the terms 'species' and 'population' may be interchangeable.

Species

There has been no consensus on how best to define a species since naturalists started classifying organisms (Hey 2001). The most common definition is the biological species concept, which defines a species as a group of interbreeding natural populations whose members are unable to reproduce successfully with members of other such groups (Mayr 1963). Other species definitions include the morphological, genetic or phylogenetic species concepts (Mallet 2006). The approach of classifying species will obviously influence both conservation priorities and the scale of intervention (Mace 2004). However, the important thing for conservationists is that commonly understood units can be defined and interventions can be implemented and monitored. Although the species-based approach would imply that it is being implemented throughout the species distribution, it is often used to describe interventions undertaken at the population level. For example, focusing on grey wolves (*Canis lupus*) in Yellowstone National Park might be referred to as a species-based approach, but clearly it only represents one small part of the global wolf distribution. However, there is a growing trend towards developing conservation action plans and implementing conservation interventions at the scale of the species entire distribution (e.g. range wide priority setting – Sanderson et al. 2002; species conservation strategies – IUCN/SSC 2008). These approaches are more consistent with a landscape-scale approach, but use a particular species or group of species as flagships or indicators.

Protected areas

Protected areas are encompassed by generally arbitrary boundaries that signal a formal intention to reduce human impacts. They have long been the most common conservation intervention for conserving land cover and ecosystem services (Joppa et al. 2008). Roughly 12.2% of the earth's terrestrial land surface is under some form of legal protection, comprising more than 120,000 protected areas (Secretariat of the Convention on Biological Diversity 2010). In contrast, less than 0.5% of the total ocean area is classified as fully protected marine reserves (Wood et al. 2008; Jay Nelson, personal communication). These numbers are likely to increase as the world's governments have now set ambitious targets through the Convention on Biological Diversity (CBD) to ensure that by 2020, at least 17% of terrestrial and inland water and 10% of coastal and marine areas are protected (CBD Decision X/2; Convention on Biological Diversity 2010).

Landscapes/ecosystems

The fact that all species are part of complex, interactive systems and exist as populations that occur over extensive spatial extents has lead to a focus on the ecosystem as a unit for conservation. The term 'ecosystem' can have multiple meanings in the context of approaches to conservation. To many, it simply implies working at a very large spatial scale, perhaps including many species of conservation interest with their habitats, such as in hotspots (Mittermeier et al. 2004). As defined by Conservation International, a hotspot is an area that meets the criteria of supporting 'at least 1,500 species of vascular plants (>0.5 % of the world's total) as endemics, and it has to have lost at least 70 % of its original habitat'.

However, the Ecosystem Approach, as articulated by the CBD (CBD Decision V/6; Convention on Biological Diversity 2000), is radically different from the species or population approaches over large areas. The Ecosystem Approach puts people and natural resource use at the centre of decision making and attempts to address both human and biological aspects of ecosystems with a focus firmly on sustainability, achieved by ensuring the conservation of ecosystem structure and function (not species) in order to maintain ecosystem services. At its core, the Ecosystem Approach has 12 statements, often called the Malawi Principles (Raffaelli & Frid 2009): the first and second are that 'the objectives of management of land, water and living resources are a matter of societal choice' and that 'management should be decentralised to the lowest appropriate level', respectively. These principles may not always sit well with species-based approaches to conservation, unless the latter are enshrined in superior legislation, and there is clear potential for tensions to arise within the Ecosystem Approach. For instance, the 11th principle states that 'an appropriate balance needs to be sought between conservation and the use of biological diversity', leaving stakeholders to decide what that balance may be, in line with the first principle concerning societal choice. The Ecosystem Approach has now been adopted by the CBD as the main framework for action and to achieve its three objectives: conservation, sustainable use, and the fair and equal distribution of goods and services (CBD Decision V/6; Convention on Biological Diversity 2000).

The Ecosystem Approach is more of a conceptual framework of guiding principles than a management tool for conservation and is therefore open to broad interpretation. Implementing the Ecosystem Approach is not easy, but an excellent account of how this might be done for conservation within the context of the Malawi Principles is provided by Rice et al. (2005) for the European marine fisheries environment, by Shepherd (2004) on behalf of the IUCN for a broad range of regions across South East Asia, Africa and South America, and for agricultural landscapes by Charron & Waltner-Toews (2008) and Ecoagricultural Partners (2008). It should be noted that the Ecosystem Approach is applied to the wider ecosystem and for that reason, many of the current applications deal with broader issues of sustainability and the need to manage environments within the context of coupled socio-ecological systems (Waltner-Toews et al. 2008), rather than focusing on species of conservation interest.

The Ecosystem Approach, with its acknowledgement of societal choice, the focus on ecosystem service sustainability and the placing of monetary and non-monetary values on natural elements for decision making, raises tensions in some parts of the conservation community but 'win–win' solutions for regions important for both ecosystem services and biodiversity can be identified, both among ecoregions and at finer scales within them (Naidoo et al. 2008; Nelson et al. 2008; Daily et al. 2009; Eigenbrod et al 2009). Implementing the Ecosystem Approach in practice can be done in a wide variety of ways, but central is the participation of stakeholders and the adoption of a holistic view of the ecosystem, so that the concept of ecosystem health (White et al. 2010; Wiegand et al. 2010) is a recurrent feature of many real-world applications (see case studies in Waltner-Toews et al. 2008). For the purposes of the present review, it is best to distinguish the CBD Ecosystem Approach from the conservation and management of large-scale units, here termed landscape-scale conservation, which often encompass a matrix of land-use types, including protected areas (Gutzwiller & Forman 2002).

IUCN Red Lists and conservation planning

Producing lists of threatened species was one of the first initiatives of the IUCN and remains one of its major contributions to conservation today. The first lists contained a small number of animals that experts believed were endangered (IUCN/UNEP 1987). Today this list, known as the IUCN Red List (www.iucnredlist.org), contains nearly 60,000 (version 2011.1) evaluated species, including entire taxonomic or functional groups, such as mammals, birds, amphibians, cycads and reef-forming corals (Baillie et al. 2010; IUCN 2011) as well as comprehensive coverage of some taxonomic groups in specific regions such as freshwater fish in Africa (Darwall et al. 2005).

The Red List assessment process is now based on more objective quantitative and qualitative criteria (Mace & Lande 1991; IUCN 1994, 2001; Mace et al. 2007), and all species assessed are classified into categories of extinction risk (Figure 2.1). Threatened species are listed as Critically Endangered, Endangered or Vulnerable. The Red List also includes information on species categorized as Extinct or Extinct in the Wild, on taxa that cannot be evaluated because of insufficient information (i.e. are Data Deficient), and on species that are either close to meeting the threatened thresholds or that would be threatened were it not for an ongoing taxon-specific conservation programme (i.e. are Near Threatened). The assessments are generally conducted by a specialist or groups of specialists that pull together available information on rate of decline, population size, distribution and threat processes. Information is also collected on conservation responses to identify interventions that are in place and conservation measures that may be needed in the future. The IUCN categories and criteria (IUCN 2001) are then applied and all species slot into one of the categories (see Figure 2.1). Assessments

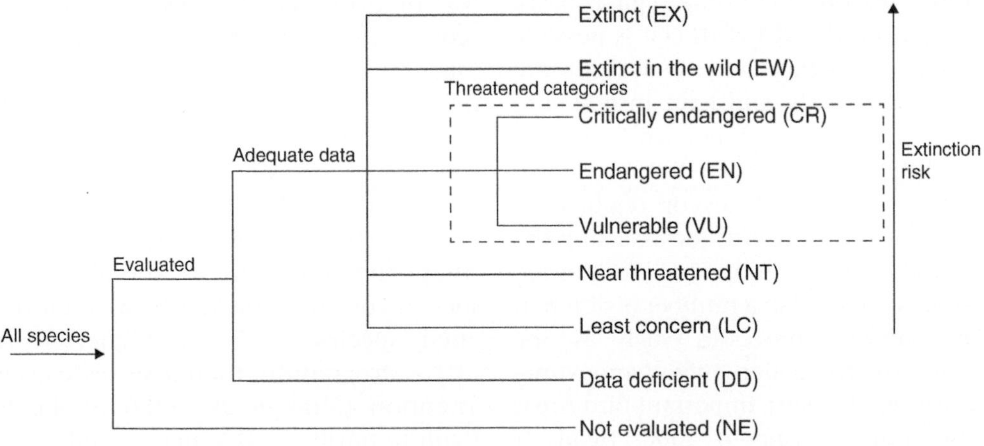

Figure 2.1 The IUCN Red List of Threatened Species™ provides taxonomic, conservation status and distribution information on plants and animals that have been globally evaluated using the IUCN Red List categories and criteria.

should be updated on a regular basis and assessments over 10 years old are considered outdated.

All species assessments now require species distribution maps (there are currently roughly 30,000 species with distribution data), which assist in the identification of sites and the appropriate scale for conservation interventions for species, subspecies and aggregates of species. This level of detail has enabled the production of maps highlighting patterns of threat, endemism, richness and even areas where little is known about the conservation status of species (e.g. Rodrigues et al. 2006). For each species, direct and indirect threat processes are also identified, such as exploitation or the impact of agriculture, enabling threats to be mapped spatially, which provides additional insight into the necessary scale of intervention.

Regular assessments of specific taxonomic groups, such as birds and mammals, provide information on trends in extinction risk (Butchart et al. 2005; Schipper et al. 2008; Hoffmann et al. 2010), known as the IUCN Red List Index (RLI). The bird RLI runs over the longest time period, starting in 1994, and indicates a slow and continual decline. This approach can also be applied to much larger taxonomic groups, such as fishes and butterflies, by selecting a random sample of 1500 species and assessing them on a regular basis. By disaggregating the IUCN RLI, it is possible to assess specific regions where declines are occurring most rapidly and gain insight into the threat process driving the decline (Butchart 2008). While the IUCN RLI is important for identifying the sites or regions for intervention, it is also useful for assessing the extent to which interventions are working.

Red List data are used in a number of different site-based planning processes such as the Alliance for Zero Extinction (www.zeroextinction.org) (Parr et al. 2009), Important Bird Areas (IBAs) (Bennun & Fishpool 2000; Heath & Evans 2000), Important Plant Areas (Anderson 2002) and Key Biodiversity Areas (KBAs) (Eken et al. 2004; Knight et al. 2007). These approaches use Red List data on status and distribution with information on appropriate management units to identify specific areas for conservation interventions. Red List data are also combined with information on evolutionary history to produce a list of the most Evolutionarily Distinct and Globally Endangered (EDGE) species (Isaac et al. 2007). EDGE species are highly threatened and have few or no close relatives. This approach currently focuses on conservation interventions at the species level, although it is now being developed to highlight EDGE zones – sites of disproportionate amounts of evolutionary history (Armour-Marshall et al. in prep).

Red Lists and conservation planning: the EDGE approach case study

To generate an EDGE score, the clade (a group of species sharing a common ancestor) of interest must have a phylogeny, or super-tree, and the species in the group must have been assessed with the IUCN categories and criteria (Mace & Lande 1991) with a status category ascribed (e.g. Endangered). Using the phylogeny, it is possible to generate a score representing the evolutionary distinctiveness (ED) of each species (Isaac et al. 2007). Species with few or no close relatives will have a higher score as they have evolved independently for a longer period of time. These species have unique genetic information (Redding & Mooers 2006) and also appear to be different from the rest of their clade in terms of morphology, behaviour and ecology (Redding et al. 2010). To identify the most urgent priorities, the ED value is combined with the species conservation status, with more threatened species having a higher score. The EDGE Programme then assesses conservation attention (Sitas et al. 2009) of the top 100 Evolutionarily Distinct and Globally Endangered species and focuses on those receiving little or no conservation attention. EDGE Fellows, in-country scientists or conservationists, are then identified to work on the focal species to help develop an initial status report that defines the species distribution, status, threats and conservation needs. Where appropriate, a detailed action planning process is then initiated for the species' entire range, following the IUCN guidelines. The action plan implementation schedule and the status of the target species are then reviewed on a regular basis to monitor success.

The evolution of species action plans

Another major IUCN contribution to conservation was the development of species action plans, designed to assess the conservation status of species and their habitats and specify conservation priorities. More than 60 such plans have been published by specialist groups of the IUCN Species Survival Commission (SSC) since 1987, and their implementation has been at the core of the SSC's work. The majority of the plans have covered mammals – especially the larger species – but there are also action plans for several groups of birds, fishes, dragonflies, orchids and conifers (IUCN/SSC 2008).

The SSC is a science-based network of some 7500 volunteer experts from almost every country in the world, working together towards achieving species conservation. Deployed in more than 100 specialist groups and task forces, these experts address conservation issues related to particular groups of plants or animals, with some tackling cross-taxonomic topical issues, such as invasive species, reintroduction of species into former habitats or wildlife health. SSC members provide scientific advice to conservation organizations, government agencies and other IUCN members and support the implementation of multilateral environmental agreements.

The SSC action plan series is one of the world's most authoritative sources of species conservation information available to natural resources managers, conservationists and government officials around the world (IUCN/SSC Cat Specialist Group 2006a,b). Action plans collate large quantities of useful information on the distribution, status and habitats of species or groups of species and identify priorities and gaps. They also provide scientifically based recommendations for those who can promote and support species conservation, establish priorities in species conservation, serve as a baseline record against which to measure change and aid fundraising.

Most of these plans, however, have had only a limited effect on the ground (IUCN/SSC 2008). For instance, most action plans lacked a conceptual framework to guide the actions required to deliver goals and objectives. The wide scope of most plans made it difficult to encompass within a single document the detailed sets of actions needed to conserve a large number of diverse species within an entire taxonomic group. Furthermore, most plans were expert based, the result of one or a small set of hard-working species specialists, resulting in little or no buy-in from relevant stakeholders (IUCN/SSC 2008).

The SSC action plan guidelines recognized early on that simply publishing information on species was not sufficient to ensure conservation results and therefore recommended that plans should 'make prioritized recommendations specifically designed for key players'. The great complexity of implementing realistic and sustainable conservation programmes has become increasingly clear, and as a result the SSC's species conservation planning activities are evolving to reflect the changing world. In particular, there is a need to draw on new approaches and techniques from a variety of fields and to identify clearly and then engage with a much wider community of stakeholders. It has also become evident that conservation strategies are largely about changing human behaviour and therefore require a much stronger focus on identifying relevant human behaviours and incorporating clear strategies to make sure this is accomplished.

The process of endangered species conservation planning is often quite difficult,

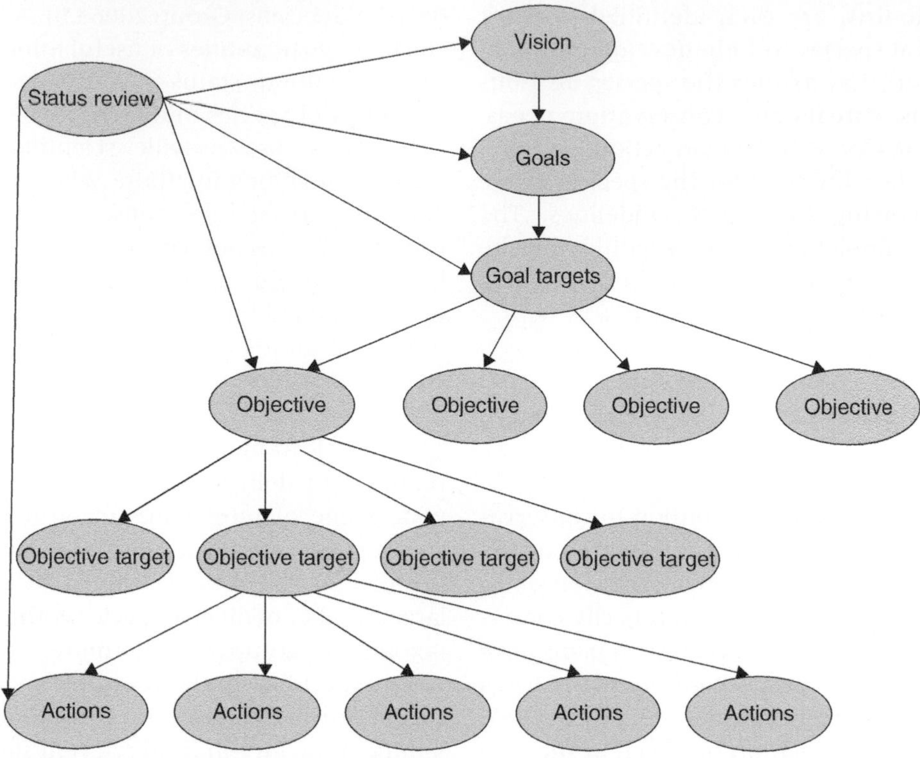

Figure 2.2 Relationship between the various components of a species conservation strategy (adapted from IUCN/SSC 2008).

with highly complex problems leading to species threats, multiple stakeholders with different value systems and incomplete information on the species and habitat of interest. IUCN/SSC (2008) provides a framework for planning species conservation strategies (SCSs) based on a conceptual framework with clear linkages between its component parts (Figure 2.2). Such strategies may be formulated for single species, groups of species or regional scales, but in general strategies with a taxonomic focus allow for the development of the specific actions needed to ensure successful conservation. SCSs must be based on sound conservation biology approaches and prepared through participatory processes that lead to broad ownership, improving prospects for implementation and, ultimately, sustainability of results.

The species conservation strategy approach: a case study

A recent example of an SCS at the species level comes from the endangered Ethiopian wolf (*Canis simensis*), which is endemic to a few Afroalpine montane enclaves of Ethiopia (Marino 2003; Sillero-Zubiri et al. 2004a). The threats facing this rare and threatened species have been previously dealt with by IUCN species actions plans, both at species (Sillero-Zubiri & Macdonald 1997), and multi-species scales (Sillero-Zubiri et al. 2004b). While these plans helped guide field conservation actions, with hindsight it became apparent that being largely expert based, they lacked the necessary stakeholder buy-in.

To resolve this shortcoming, stakeholders, including non-governmental and governmental agencies, community representatives and members of international conservation bodies, met in a 2-day workshop to draft a species conservation strategy and national action plan (IUCN/SSC Canid Specialist Group 2011). The process began with a status overview of the Ethiopian wolf, its distribution, ecological requirements and the threats it faces; however, it would have been preferable for this overview to have been prepared in advance of the meeting to provide an update on the current state of knowledge of the taxon in question.

The next step was creating a threat tree and then consolidating the various threats and their primary causes into key drivers. This threat tree provided a useful and visual framework for considering which components of the network of threats and constraints could be most easily and effectively addressed with logical objectives and implementable actions. A vision statement – a description of the participants' collective wish for the future state of the species – was produced and broken down into its various components. Targeted goals were developed to meet a specific component over the next 10 years, a time scale that was deemed appropriate to implement many actions and evaluate their effectiveness. These goals rephrased the vision in operational terms to capture in greater detail what needed to be achieved and where to save the species. Both the vision and the goals had the same geographical and temporal scale, with the goals ascribing to the SMART philosophy – Specific, Measurable, Achievable, Realistic and Time-bound. Objectives that outline how the vision and goals will be achieved should be clear, understandable and realistic, and should allow actions to follow on from them. Each action was given a priority, a time frame and an implementing body and was associated with the human and financial resources potentially needed to achieve that specific action.

For instance, rabies outbreaks are a direct threat to wolves resulting from the presence, and poor control and husbandry, of domestic dogs in the Ethiopian wolf range. A goal that addresses this threat has the aim that all existing Ethiopian wolf populations are secured, not at risk of immediate extinction, with genetic diversity maintained, and the number of wolf packs increased. A logical objective linked to this goal requires reducing the incidence of rabies in small and medium wolf populations to zero (no outbreaks) within 5–12 years, and ensuring no losses of breeding units within large wolf populations from rabies within 5–12 years. Actions designed to deliver this objective include (1) testing feasibility, safety and efficacy of oral rabies vaccines in wolves; (2) developing a rabies management plan, based on trial results; (3) implementing the plan, with priority in smaller populations; (4) continuing to vaccinate dogs living in wolf habitats in Bale Mountains (until rabies management plan is operational) (IUCN/SSC Canid Specialist Group 2011).

Other action planning processes for species

There are several other competing species-focused conservation planning processes aimed at developing strategic plans for endangered species conservation. These efforts typically have either a strong quantitative scientific focus (e.g. population viability analysis (PVA) or GIS-based habitat assessment) or alternatively they emphasize a more discussion-centric approach to encourage largely qualitative input from a broad range of stakeholders (IUCN/SSC 2008). Each will here be described in turn.

The Population and Habitat Viability Assessment (PHVA) workshop process, designed and primarily implemented by SSC's Conservation Breeding Specialist Group (CBSG), is a clear example of an integrated approach to species conservation planning (Boyce 1992; Lacy 1994). The PHVA workshop process is designed to encourage creative thinking and open communication among all participants, from the local village representative to the academic scientist to the government official, collectively moving a group of people through a process of information analysis and decision making. The PHVA explicitly incorporates methods for PVA modelling to assess the risks of threatened population decline or extinction in the presence of destabilizing human activities. This model can then be used to make basic predictions of future population demographic behaviour in the presence of those threats known to affect the species or thought to impact them in the future. A good example was the PHVA for African wild dogs (*Lycaon pictus*) in southern Africa (Mills et al. 1998), which eventually resulted in a successful metapopulation programme for the species in South Africa (see Akçakaya et al. 2007).

Other species-focused planning approaches may be specific to a country or region. For instance, under the United States Endangered Species Act (ESA) of 1973, US government agencies are charged with developing species recovery plans for listed endangered and threatened species. These recovery plans guide US agencies in restoring listed species and the ecosystems upon which they depend and safeguarding them so that further listing under the ESA is no longer required (National Oceanic and Atmospheric Administration/ National Marine Fisheries Service 2006; Crouse et al. 2002). Recovery plans typically include delineating the aspects of the species' biology, life history and threats pertinent to its endangerment and recovery, identifying goals and criteria by which to measure recovery and outlining a strategy, including site-specific actions, to achieve recovery. Recovery plans serve as communication tools for stakeholders, plans for monitoring success and fundraising documents. They may also carry restricted legal mandates for species conservation. Many other countries around the world have similar legislation, such as India's Wildlife Conservation Act of 1972, the Species at Risk Act (SARA) of Canada and the Biodiversity Action Plans (BAP) of Australia, Sweden and the United Kingdom.

An expert-based, geographically explicit planning methodology developed for widely distributed species, often occurring across international borders, is the range-wide priority setting (RWPS). It was first applied to jaguars (*Panthera onca*; Medellin et al. 2002; Sanderson et al. 2002) and since has been applied to other species including American crocodiles (*Crocodylus acutus*; Thorbjarnson et al. 2006), North American bison (*Bison bison*; Sanderson et al. 2008) and African lion (*Panthera leo*; IUCN/SSC Cat Specialist Group 2006a,b). RWPS draws significantly on past priority-setting efforts for species (i.e. tiger, *Panthera tigris*; Wikramanyake et al. 1998, and several IUCN action plans) and regions (ecoregions, Olson et al. 2001; site portfolios, Nature Conservancy 2007). The process is based on the premise that conserving species requires consideration of the species across its historical range, recognition that populations exist in different ecological settings (capturing not only genetic distinctiveness but also ecological and behavioural distinctions) and identifying those populations for restoration where the potential for long-term conservation is greatest, based on population factors and threats. The RWPS process consists of a geographically based, systematic evaluation of the status and distribution of the species across its historical range, followed by a prioritization of identified populations and/or restoration opportunities based on their ecological importance. This process is well illustrated by the American crocodile RWPS (Thorbjarnson et al. 2006).

Range-wide priority setting: a case study

American crocodiles are widely distributed in wetlands of the northern Neotropics, but their populations shrank dramatically in the 20th century due to commercial hunting for their skins (Thorbjarnson et al. 2006). More recently, due to hunting and trade controls, the species has been recovering in several areas (see Chapter 5). Most effort for recovering crocodilian populations had been devoted to regulating commercial use, but these approaches were limited in their ability to deal with issues such as habitat loss and fragmentation. Since these habitat limitations are expected to be the most critical for crocodile conservation, there was a call for alternative strategies that prioritize the most critical habitats for this wide-ranging species.

Using information from a group of American crocodile experts to identify the most important areas for the conservation of this species, 69 crocodile conservation units (CCU) were classified in eight distinct crocodile bioregions. The relative importance of these CCUs in each bioregion was quantified using an algorithm that weighted the factors that the experts considered to be most important for the long-term viability of crocodile populations (Thorbjarnson et al. 2006). Two bioregions given a high priority for the creation of protected areas were the Dry Pacific South America (northern Peru and southern Ecuador) and the Northwest and Central Pacific Mexico. This effort in turn will hopefully lead to the development of a regional conservation plan for the American crocodile.

Practical prescriptions for area or landscape approaches to conservation planning

Many of the above strategies and approaches recognize the need to embed the protection and conservation of species within their broader ecosystem, but they tend to be bespoke for particular species. A more generic approach which can be used as a toolbox by any group is the Conservation Action Planning (CAP) process promoted by the Nature Conservancy (TNC). This process is designed to help conservation projects develop strategies, take action and measure success over time in an adaptive framework (Nature Conservancy 2007). CAP developed from earlier project-level planning approaches such as Site Conservation Planning, Conservation Area Planning and the 5-S Framework. CAP chooses the conservation targets (i.e. species, communities or ecological systems) that best represent biodiversity in the project area and includes mechanisms for defining the conservation team and project scope. It assesses the viability of the focal conservation targets, identifying threats and developing strategies, including specific objectives, actions and measures of success. TNC has produced decision support tools to guide the CAP process (www.conservationgateway.org/cap), as illustrated by the Condor Bioreserve case study described in the next section.

Area conservation planning may result from directives of the 1973 US Endangered Species Act. Under the ESA, Habitat Conservation Plans (HCP) are prepared to form partnerships between private individuals and the US government to 'minimize or mitigate' reductions in endangered species. More than 200 approved HCPs collectively cover millions of hectares in the US. HCPs must include an assessment of impacts likely to result from actions that might reduce species abundance, suggestions of measures that the landowner could take to monitor, minimize and mitigate impacts, alternatives of actions that could be taken that would not result in any take and additional regulatory measures required by the appropriate federal agency (Vogel & Hicks 2000; Audubon Society 2008).

The Landscape Species Approach is a planning tool developed by the Wildlife Conservation Society (WCS), that can be applied in any country or region, and builds conservation efforts around 'landscape species'. Landscape species are ones that use large, ecologically diverse areas and have a significant impact on the structure and function of natural ecosystems (Sanderson et al. 2002). They are selected using a standardized evaluation system that seeks to represent all the major habitat types and threats within a landscape area

(Coppolillo et al. 2004; Stokes et al. 2010). The landscape potential for each species is mapped in GIS and assessed against maps representing the threats to the species producing a 'conservation landscape' and linked to defined population target levels for each species (Sanderson et al. 2006). Based on this analysis, conservation actions are planned to restore or conserve the species. By conserving the entire suite of landscape species, conservationists aim to conserve not only those species and the species on which they directly depend, but the landscape as a whole.

While the various approaches to species conservation illustrated here would imply some level of competition between agencies for the *right* tool to tackle conservation planning, it is important to emphasize that for many approaches, the end results may be very similar. For example, in a workshop where conservation practitioners from TNC, World Wildlife Fund, WCS, the African Wildlife Foundation and Conservation International were tasked with applying their own landscape approaches to Samburu/Laikipia in northern Kenya and to then present the pros and cons of the different approaches, the outcomes of the first four were broadly congruent despite the various approaches (Bottrill et al. 2006). This may reflect common goals concerning ecosystem integrity and functioning; Conservation International's stated goals are more species-centric, so it is not surprising that their action plan was more restricted in its ability to achieve landscape conservation (Bottrill et al. 2006).

Landscape-orientated approaches have also been adopted in the UK, through the identification of National Character Areas (Natural England 2009). These areas are delimited by those landscape characters that comprise a distinct, recognizable and consistent pattern of natural and cultural elements. Such landscape units provide an obvious social-ecological-environmental grouping to allow identification of the relevant local responses for conservation planning.

The Conservation Action Planning (CAP) approach in practice: a case study

The CAP approach has enjoyed considerable application within the United States, and there is a wealth of case study material in the US and elsewhere which is useful as training material for new actions and users (www.conserveonline.org). The approach has been largely based on consensus amongst a defined group of core project team members, including stakeholders, of the geographical scope of the plan, the conservation targets and their viability, the threats to each target, the conservation strategies needed to meet those targets, how success and failure can be measured, how to implement the actions and how best to analyse, reflect and, most importantly, adapt, as new data become available and objectives are modified (Nature Conservancy 2007).

The CAP process is well illustrated by the Condor Bioreserve in Ecuador (Hamberg & Martin 2007), a multifunctional landscape that embraces 760 bird, 150 mammal and 120 amphibian species and comprises a mixture of protected areas, buffer zones and unprotected, but highly biodiverse, areas. A period of intense planning in 2002 allowed the main partners to be identified and brought on board, followed by conservation target selection that included three species of threatened mammals, 16 species of amphibians and four major habitat types. A threat analysis was achieved through a highly participatory approach involving stakeholders from the Ministry of the Environment, local municipalities, indigenous communities, small- and medium-scale farmers and large landowners as well as NGOs working in the area. This process identified inappropriate activities and planning in agriculture, forestry and infrastructure projects as well as hunting – especially of Andean bear – as the main threats. A monitoring programme was then put in place based on indicators such as rates of deforestation and land-use change and numbers of Andean bears killed. This is a clear example of planning at the landscape scale, and the reflective process highlighted the importance of having a landscape strategy for reaching consensus about targets, threats and indicators for measuring success of implementation measures. It stressed the importance of developing the overall conservation strategy and establishing measurable results, rather than becoming bogged down in perfecting the results of each step in the process (Hamberg & Martin 2007).

Judging from the level of uptake and its long history, the CAP approach works well. However, the approach has been criticized for aspects that others see as its strengths in the absence of data and the need to take action: the empirical basis of many of the initial steps. To address such criticisms, Low et al. (2010) have proposed the inclusion of GIS-based elements to enhance CAPs by including remote sensing data that should add rigour to many of the CAP steps, particularly in assessing departure of landscapes from desired reference conditions and for monitoring the success of management strategies that may have the greatest ecological pay-off for the least cost. An analogous spatial planning approach has been taken by Valenzuela-Galván et al. (2007) for carnivores in North and Central America, allowing more efficient and optimal use of limited resources for conservation planning.

Action plans for entire taxonomic groups or functional groups

Recently action plans have been produced for entire taxonomic groups, such as the Amphibian Conservation Action Plan (Gascon et al. 2007), or functional groups, such as the GLOBE Action Plan for Coral Reefs (Harding et al. 2010). In both cases the groups of species are expected to experience major extinctions over the next few decades. In the case of amphibians, the most recent extinctions appear to be driven by chytrid fungus, although the most dominant threat to the species as a group is habitat loss (Stuart et al. 2004). Coral reefs are increasingly threatened by warming of the oceans, acidification and siltation (Carpenter et al. 2008) (see Chapter 10). The approach of addressing an entire group of species is effective when the threat processes are similar (see Chapter 14). A focused research agenda, usually to better understand the threat processes, and co-ordinated conservation efforts will benefit a broad range of species.

Another more recent approach, still in the experimental phase, is sector-based conservation intervention mapping that does not focus on species or ecosystems but on particular sectors that are threatening biodiversity such as the forestry, palm oil or fisheries sectors. With this approach, the conservation community works together and reviews the sector as a whole from financing to production or harvesting to sales and then maps the various conservation interventions across the sector. Gaps can be identified and addressed and potential synergies realized. With such limited resources, it is essential that the conservation community does not duplicate efforts (Mace et al. 2000). This co-ordinated and strategic approach is one of the few ways in which conservation activities can be leveraged.

Conservation action plans for countries

The Convention on Biological Diversity (CBD) has greatly increased conservation planning at the national level. Member countries of the CBD are required to produce National Biodiversity Strategies and Action Plans (NBSAPs) or equivalent. These are meant to be the primary mechanism for the implementation of the CBD and its strategic plan. The strategy is intended as a road map of how a country can meet the objectives of the convention, and the action plan is meant to identify the steps necessary to meet the goals of the strategy (Prip et al. 2010). The action plans are also intended to assess progress towards the Aichi Biodiversity Targets (2011–2020) agreed at Nagoya, Japan (CBD Decision X/2; Convention on Biological Diversity 2010). These targets include five strategic goals.

- Address the underlying causes of biodiversity loss by mainstreaming biodiversity across government and society.
- Reduce the direct pressures on biodiversity and promote sustainable use.
- Improve the status of biodiversity by safeguarding ecosystems, species and genetic diversity.
- Enhance the benefits to all from biodiversity and ecosystem services.
- Enhance implementation through participatory planning, knowledge management and capacity building.

While the NBSAPs provide an excellent opportunity for effective biodiversity planning at the national level and for monitoring progress through time, they have had minimal impact to date. This is clearly illustrated by the spectacular failure of governments to meet the 2010 CBD Biodiversity Target 'to achieve by 2010 a significant reduction of the current rate of biodiversity loss …'. Under COP 10 Decision X/5, governments call to 'prepare a further and in-depth analysis of the main reasons why the 2010 biodiversity target has not been met despite the activities undertaken by Parties, drawing upon the third edition of the Global Biodiversity Outlook, the fourth national reports and other relevant sources of information.' However, it does not take an in-depth analysis to reveal that the 2010 CBD Biodiversity Target and the NBSAPs are not legally binding and therefore a low priority for governments. The resolution for the 2020 Aichi Biodiversity Targets stated that 'parties are invited to set their own targets within this flexible framework, taking into account national needs and priorities …' (CDB Decision X/2). This level of commitment will ensure that another in-depth analysis will be required in 2020 to assess the main reason for failure.

Many countries also lack the technical and financial resources needed to both develop and implement strategies and actions plans. However, as the NBSAP process evolves, it continues to improve and there is the potential for NBSAPS to become much more effective if countries can be persuaded to report against a standardized template with explicit guidelines and clear reporting requirements. This would facilitate the comparison of National Action Plans and allow for data to be aggregated, providing insight at the regional and global level. Combined with this, greater commitment is required from countries to integrate the NBSAP process into national legislation and across government sectors. For example, reporting on the status and trends of key ecosystems or ecosystem services such as forest, freshwater systems and fisheries could also be integrated into the country's natural capital accounts.

The Canadian National Report: a case study

All countries that are signatories to the CBD are expected to develop National Reports every 4 years, outlining the current status of the Biodiversity Strategies and Action Plans. However, many countries failed to achieve this target. Here we focus on Canada's Fourth National Report to the CBD (Canadian Government 2010), one of the strongest submitted. This report follows the guidelines of the Conference of Parties (Decision VIII/14) and is divided into four main areas:

- Chapter I: Overview of Biodiversity Status, Trends and Threats
- Chapter II: Current Status of National Biodiversity Strategies and Action Plans
- Chapter III: Subnational Planning and Mainstreaming of Biodiversity
- Chapter IV: Conclusions – Progress Towards the 2010 Target.

The first chapter is primarily based on a relatively new initiative within Canada called the Ecosystem Status and Trends Report (ESTR), a federal, provincial and territorial initiative under the Canadian Councils of Resource Ministers. The ESTR aims to provide a baseline assessment of Canada's major ecosystem status and trends for future reporting. Such a process is essential for countries that wish to monitor natural capital and develop green accounting frameworks. A large series of indicators are presented in the CBD's framework of goals, targets and indicators (CBD Decision VIII/15) to help measure Canada's progress toward the overall 2010 CBD Biodiversity Target. Examples of indicators include terrestrial and marine protected area coverage, trends in summer sea ice in the Arctic, trends in grasslands, species status, sustainable forest certification, knowledge of aboriginal languages, expansion of urban land and change in temperature.

Chapter II is based on a Biodiversity Outcomes Framework that was approved by the Canadian federal, provincial and territorial ministers in 2006. Central to the framework is a set of desired national outcomes: healthy and diverse ecosystems, viable populations of species, genetic resources and adaptive potential and sustainable use of biological resources. The social benefits in association with these outcomes are highlighted, such as clean water, air and soil, sustainable food supply, healthy communities and sustainable lifestyles. It is meant to be presented in an adaptive management framework based on the Ecosystem Approach and linked back to the CBD targets. However, under each of the desired outcome headings, the report tends to just list a broad range of conservation interventions and their current status, with minimal links to the indicators in Chapter I or the framework of the CBD.

Chapter III focuses mainly on 'mainstreaming' biodiversity and how various sectors and audiences are being engaged. It is largely a list of actions that have been taken in each sector or cross-sectoral initiatives under way. Examples of the various sectors or constituencies targeted include the provincial and territorial governments, urban areas, aboriginal peoples, academic and scientific institutions, non-governmental organizations, industry and business.

Finally, Chapter IV synthesizes the indicators in Chapter I and assesses progress towards the 2010 target. It also identifies lessons learned and future directions. The report recognizes that although Canada has very good data relative to other countries, long-term data sets are lacking and it is therefore very difficult fully to understand trends in species or ecosystems. Canada's first ESTR was completed in 2010, forming what is now a solid baseline from which to monitor and report on trends. Unfortunately, most signatories of the CBD have not even reached this point and for most, it will be a number of years before even a baseline is set. It is of course difficult to assess whether the current rate of biodiversity loss has been reduced at the national level if there are insufficient data to define the recent rates of biodiversity decline.

The growing grey area between the various levels of approach

We have explored the various scales for conservation interventions and planning. In practice, there are three main units or scales at which people think about interventions. The first is the population or species level, which are often interchangeable; the second is the scale of the protected area; and the third is the landscape scale, which is generally larger than a protected area and encompasses a matrix of land-use types. Not surprisingly, conservation planning also tends to be at these scales, but national- or global-level planning, such as NBSAPs, action plans for entire taxonomic groups or sector-based conservation intervention mapping, are increasingly common. It is clear from the case studies in this chapter that interventions at these various scales are not mutually exclusive. For example, RWPS focuses on a particular species, but often involves interventions from the population to the landscape scale, resulting in conservation benefits for a broad range of species. Thus the division between the various approaches is becoming increasingly grey. When asked which is the most effective approach or scale for conservation, the response depends on the specific objective and resources available. All the approaches discussed in this chapter play an important role in biodiversity conservation; the conservation community should therefore focus less on trying to identify the most appropriate scale for interventions (e.g. species or ecosystems) and more on co-ordinating and developing larger scale strategic conservation plans around specific objectives. This process will help naturally identify the most appropriate interventions at the relevant scale, given the resources available.

"A goal without a plan is just a wish."
— Antoine de Saint-Exupéry

Acknowledgements

Thanks to Karolyn Upham from ZSL for reviewing an earlier draft. Claudio Sillero-Zubiri's work at Oxford is funded by the Born Free Foundation.

References

Akçakaya, H.R., Mills, M.G.L. & Doncaster, C.P. (2007) The role of metapopulations in conservation. In: *Key Topics in Conservation Biology*. (eds D.W. Macdonald & K. Service), pp. 64–84. Blackwell Publishing, Oxford.

Anderson, S. (2002) *Identifying Important Plant Areas*. Plantlife International, London.

Armour-Marshall, K., Baillie, J.E.M., Isaac, N.J.B., *et al.* (in prep) Global patterns of evolutionary distinct and globally endangered amphibians and mammals.

Audubon Society (2008) *Report of the National Audubon Society Task Force on Habitat Conservation Plans*. http://web4.audubon.org/campaign/esa/hcp-report.html

Baillie, J.E.M., Griffiths, J., Turvey, S.T., *et al.* (2010) *Evolution Lost: Status and Trends of the World's Vertebrates*. Zoological Society of London, London.

Bennun, L.A. & Fishpool, L.D.C. (2000) The Important Bird Areas Programme in Africa: an outline. *Ostrich*, **71**, 150–153.

Bottrill M., Didier K., Baumgartner J., *et al.* (2006) *Selecting Conservation Targets for Landscape-Scale Priority Setting: A Comparative Assessment of Selection Processes used by Five Conservation NGOs for a Landscape in Samburu, Kenya*. World Wildlife Fund, Washington, D.C.

Boyce, M.S. (1992) Population viability analysis. *Annual Review of Ecology and Systematics*, **23**, 481–506.

Butchart, S.H.M. (2008) Red List Indices to measure the sustainability of species use and impacts of invasive alien species. *Bird Conservation International*, **18**, S245–S262.

Butchart, S.H.M., Stattersfield, A.J., Baillie, J.E.M., *et al.* (2005) Using Red List Indices to measure progress towards the 2010 target and beyond. *Philosophical Transactions of the Royal Society of London, Series B*, **360**, 255–268.

Butchart, S.H.M., Walpole M., Collen B., *et al.* (2010) Global biodiversity: indicators of recent decline. *Science*, **328**, 1164–1168.

Canadian Government (2010) *Canada's 4th National Report to the United Nations Convention on Biological Diversity*. Canadian Government, Ottawa.

Carpenter, K.E., Abrar, M., Aeby, G., *et al.* (2008) One-third of reef-building corals face elevated extinction risk from climate change and local impacts. *Science*, **321**, 560–563.

Charron, D. & Waltner-Toews, D. (2008). Landscape perspectives on agroecosystem health in the Great Lakes Basin. In: *The Ecosystem Approach* (eds D. Waltner-Toews, J.J. Kay & N-M.E. Lister), pp. 175–189. Columbia University Press, New York.

Convention on Biological Diversity (2000) *Decision V/6. Ecosystem Approach*. Fifth Conference of the Parties, Nairobi.

Convention on Biological Diversity (2010) *Decision X/2. Aichi Biodiversity Targets*. Tenth Conference of the Parties, Aichi.

Coppolillo, P.B., Gomez, H., Maisels, F. & Wallace, R. (2004) Selection criteria for suites of landscape species as a basis for site-based conservation. *Biological Conservation*, **115**, 419–430.

Crandall, K.A., Bininda-Emonds, O.R.P., Mace, G.M. & Wayne, R.K. (2000) Considering evolutionary processes in conservation biology. *Trends in Ecology & Evolution*, **15**, 290–295.

Crouse, D.T., Mehrhoff, L.A., Parkin, M.J., Elam, D.R. & Chen, L.Y. (2002) Endangered species recovery and the SCB study: a US Fish and Wildlife Service perspective. *Ecological Application*, **12**, 719–23.

Daily, G.C., Polasky, S., Goldstein, J., *et al.* (2009). Ecosystem services in decision making: time to deliver. *Frontiers in Ecology and Environment*, **7**, 21–28.

Darwall, W., Smith, K. & Vié, J-C. (eds) (2005) *The Status and Distribution of Freshwater Biodiversity in Eastern Africa*. IUCN/SSC: http://data.iucn.org/dbtw-wpd/edocs/ssc-op-031.pdf

Ecoagriculture Partners (2008) Applying the ecosystem approach to biodiversity conservation in agricultural landscapes. *Ecoagriculture Policy Focus*, **1**, 1–4.

Eigenbrod, F., Anderson, B.J., Armsworth, P.R., *et al.* (2009) Ecosystem service benefits of contrasting conservation strategies in a human-dominated region. *Proceedings of the Royal Society B – Biological Science*, **276**, 2903–2911.

Eken, G., Bozdogan, M., Karatas, A. & Lise, Y. (2004) Key biodiversity areas as site conservation targets. *BioScience*, **54**, 1110–1118.

Gascon, C., Collins, J.P., Moore, R.D., *et al.* (eds) (2007) *Amphibian Conservation Action Plan*. IUCN/SSC Amphibian Specialist Group, Gland, Switzerland.

Gutzwiller, K. & Forman, R.T.T. (2002) *Applying Landscape Ecology in Biological Conservation*, Springer-Verlag, New York.

Hamberg, S.E. & Martin, A.S. (2007) *Conservation Action Planning. Innovations in Conservation Series. Parks in Peril Programme*. The Nature Conservancy, Arlington, VA.

Harding, S., Clark, E., Gardiner-Smith, B., *et al.* (2010) *GLOBE Action Plan for Coral Reefs*. GLOBE International, London.

Heath, M.F. & Evans, M.I. (2000) *Important Bird Areas in Europe: Priority Sites for Conservation*. BirdLife International, Cambridge.

Hey, J. (2001) The mind of the species problem. *Trends in Ecology and Evolution*, **16**, 326–329.

Hoffmann, M., Hilton-Taylor, C., Angula, A., *et al.* (2010) The impact of conservation on the status of the world's vertebrates. *Science*, **330**, 1503–1509.

Hughes, J.B., Daily, G.C. & Ehrlich, P.R. (1997) Population diversity: its extent and extinction. *Science*, **278**, 689–692.

Isaac, N.J.B., Turvey, S.T., Collen, B., *et al.* (2007) Mammals on the EDGE: conservation priorities based on threat and phylogeny. *PLoS One*, **2**(3), e296.

IUCN (1994) *IUCN Red List Categories*. Prepared by the IUCN Species Survival Commission. IUCN, Gland, Switzerland.

IUCN (2001) *IUCN Red List Categories and Criteria: Version 3.1*. IUCN Species Survival Commission, Gland, Switzerland.

IUCN (2011) *IUCN Red List of Threatened Species*. Version 2011.1: www.iucnredlist.org

IUCN/SSC (2008) *Strategic Planning for Species Conservation: A Handbook*. Version 1.0. IUCN Species Survival Commission, Gland, Switzerland.

IUCN/SSC Canid Specialist Group (2011) *Strategic Plan for Ethiopian Wolf Conservation*. IUCN/SSC Canid Specialist Group, Oxford. www.canids.org

IUCN/SSC Cat Specialist Group (2006a) *Regional Conservation Strategy for the Lion* Panthera Leo *in Eastern and Southern Africa*. IUCN/SSC Cat Specialist Group, Gland, Switzerland. www.catsg.org

IUCN/SSC Cat Specialist Group (2006b) *Conservation Strategy for the Lion in West and Central Africa*. IUCN Regional Office of Central Africa (BRAO) and the West and Central African Lion Conservation Network (ROCAL), Yaoundé, Cameroon.

IUCN/UNEP (1987) *The Road to Extinction*. IUCN, Gland, Switzerland.

Joppa, L., Loarie, S. & Pimm, S. (2008) On the protection of 'protected areas'. *Proceedings of the National Academy of Science*, **105**, 6673–6678.

Knight, A. T., Smith, R.J., Cowling, R.M., *et al.* (2007) Improving the key biodiversity areas approach for effective conservation planning. *BioScience*, **57**, 256–261.

Lacy, R.C. (1994) What is population (and habitat) viability analysis? *Primate Conservation*, **14/15**, 27–33.

Low, G., Provencher, L. & Abele, S.L. (2010) Enhancing conservation action planning: assessing landscape condition and predicting benefits of conservation strategies. *Journal of Conservation Planning*, **6**, 36–60.

Mace, G.M. (2004) The role of taxonomy in species conservation. *Proceedings of the Royal Society B – Biological Sciences*, **359**, 711–779.

Mace, G.M. & Collar, N.J. (2002) Priority-setting in species conservation. In: *Conserving Bird Biodiversity: General Principles and their Application*. (eds K. Norris

& D.J. Pain), pp. 61–73. Cambridge University Press, Cambridge.

Mace, G.M. & Lande, R. (1991) Assessing extinction threats – toward a re-evaluation of IUCN Threatened Species categories. *Conservation Biology*, **5**, 148–157.

Mace, G.M., Balmford, A., Boitani, L., *et al.* (2000) It's time to work together and stop duplicating conservation efforts. *Nature*, **405**, 393.

Mace, G.M., Possingham, H.P. & Leader-Williams, N. (2007) Prioritizing choices in conservation. In: *Key Topics in Conservation Biology* (eds D. Macdonald & K. Service), pp. 17–34. Blackwell Publishers, Oxford.

Mallet, J. (2006) Species concepts. In: *Evolutionary Genetics: Concepts and Case Studies* (eds C.W. Fox & J.B. Wolf), pp. 367–73. Oxford University Press, Oxford.

Marino, J. (2003) Threatened Ethiopian wolves persist in small isolated Afroalpine enclaves. *Oryx*, **37**, 62–71.

Mayr, E. (1963) *Animal Species and Evolution*. Belknap Press of Harvard University Press, Cambridge.

Medellin, R.A.., Equihua, C., Chetkiewicz,C.L.B., *et al.* (eds) (2002) *El Jaguar en el Nuevo Milenio*. Ediciones Científicas Universitarias, Mexico.

Millennium Ecosystem Assessment (2005) *Ecosystems and Human Well-Being: Synthesis*. Island Press, Washington, D.C.

Mills, M.G.L., Ellis, S., Woodroffe, R., *et al.* (eds) (1998) *Population and Habitat Viability Assessment: African Wild Dog (Lycaon pictus) in Southern Africa*. Final Workshop Report. IUCN/SSC Conservation Breeding Specialist Group, Apple Valley, MN.

Mittermeier, R.A., Gil, P.R., Hoffmann, M., *et al.* (2004) *Hotspots Revisited: Earth's Biologically Richest and Most Endangered Terrestrial Ecoregions*. Cemex Books on Nature, Washington, D.C.

Moritz, C. (1994) Defining 'evolutionarily significant units' for conservation. *Trends in Ecology and Evolution*, **9**, 373–375.

Myers, N., Mittermeier, R.A., Mittermeier, C.G., *et al.* (2000) Biodiversity hotspots for conservation priorities. *Nature*, **403**(6772), 853–858.

Naidoo, R., Balmford, A., Costanza, R., *et al.* (2008). Global mapping of ecosystem services and conservation priorities. *Proceedings of the National Academy of Sciences*, **105**, 9495–9500.

National Oceanic and Atmospheric Administration/ National Marine Fisheries Service, Highly Migratory Species (2006) *SEDAR 11. Stock Assessment Report: Large Coastal Shark Complex, Blacktip and Sandbar Shark*. National Oceanic and Atmospheric Administration, Silver Spring, MD.

Natural England (2009). *Experiencing Landscapes: Capturing the 'Cultural Services' and 'Experiential Qualities' of Landscape*. Study Report. Natural England, Sheffield.

Nature Conservancy (2007) *Conservation Action Planning: Developing Strategies, Taking Action and Measuring Success at Any Scale. Overview of Basic Practices*. Nature Conservancy, Arlington, VA.

Nelson, E., Polasky, S., Lewis, D.J., *et al.* (2008). Efficiency of incentives to jointly increase carbon sequestration and species conservation on a landscape. *Proceedings of the National Academy of Sciences*, **105**, 9471–9476.

Olson, D.M. & Dinerstein, E. (1998) The global 200: a representation approach to conserving the Earth's most biologically valuable ecoregions. *Conservation Biology*, **12**, 502–515.

Olson, D.M., Dinerstein, E., Wikramanayake, E.D., *et al.* (2001) Terrestrial ecoregions of the world: a new map of life on earth. *Bioscience*, **51**, 933–938.

Parr, M.J., Bennun, L., Boucher, T., *et al.* (2009) Why we should aim for zero extinction. *Trends in Ecology and Evolution*, **24**, 181.

Pollard, D., Almond, R., Duncan, E., *et al.* (2010) *Living Planet Report 2010 Biodiversity, Biocapacity and Development*. WWF International, Gland, Switzerland.

Prip, C., Gross, T., Johnston, S. & Vierros, M. (2010) *Biodiversity Planning: An Assessment of National Biodiversity Strategies and Action Plans*. United Nations University Institute of Advanced Studies, Yokohama, Japan.

Raffaelli, D.G. & Frid, C.L.J. (2010) *Ecosystem Ecology: A New Synthesis*. Cambridge University Press, Cambridge.

Redding, D.W. & Mooers, A.O. (2006) Incorporating evolutionary measures into conservation prioritization. *Conservation Biology*, **20**, 1670–1678.

Redding, D.W., Dewolff, C.V. & Mooers, A.O. (2010) Evolutionary distinctiveness, threat status, and ecological oddity in primates. *Conservation Biology*, **24**, 1052–1058.

Rice, J., Trujillo, V., Jennings, S., *et al.* (2005). *Guidance on the Application of the Ecosystem Approach to Management of Human Activities in the European Marine Environment*. ICES Cooperative Research Report 273. ICES, Copenhagen.

Rodrigues, A.S.L., Pilgrim, J.D., Lamoreux, J.F., *et al.* (2006) The value of the IUCN Red List for conservation. *Trends in Ecology and Evolution*, **21**, 71–76.

Sanderson, W.E., Redford, K.H., Chetkiewicz, C.B., *et al.* (2002) Planning to save a species: the Jaguar as a model. *Conservation Biology*, **16**, 58–72.

Sanderson, E., Forrest, J., Loucks, C., *et al.* (2006) *Setting Priorities for the Conservation and Recovery of Wild Tigers: 2005-2015. The Technical Assessment.* WCS, WWF, Smithsonian and NFWF-STF, New York and Washington, D.C.

Sanderson, E., Redford, K.H., Weber, B., *et al.* (2008) The ecological future of the North American bison: conceiving long-term, large-scale conservation of wildlife. *Conservation Biology*, **22**, 252–266.

Schipper, J., Chanson, J.S., Chiozza, F., *et al.* (2008) The status of the world's land and marine mammals: diversity, threat, and knowledge. *Science*, **322**, 225–230.

Scott, P., Burton J.A. & Fitter, R. (1987) Red Data Books: the historical background. In: *The Road to Extinction* (eds R. Fitter & M. Fitter), pp. 1–5. IUCN, Gland, Switzerland.

Secretariat of the Convention on Biological Diversity (2010) *Global Biodiversity Outlook 3*. Convention on Biological Diversity, Montréal, Canada. www.cbd.int/doc/publications/gbo/gbo3-final-en.pdf

Shepherd, G. (2004) *The Ecosystem Approach: Five Steps to Implementation*. IUCN, Gland, Switzerland.

Sillero-Zubiri, C., & Macdonald, D.W. (eds) (1997) *The Ethiopian Wolf: Status Survey and Conservation Action Plan*. IUCN/World Conservation Union, Gland, Switzerland.

Sillero-Zubiri, C., Marino, J., Gottelli, D., & Macdonald, D.W. (2004a) Afroalpine ecology, solitary foraging and intense sociality amongst Ethiopian wolves. In: *Canid Biology and Conservation* (eds D.W. Macdonald & C. Sillero-Zubiri), pp. 311–322. Oxford University Press, Oxford.

Sillero-Zubiri, C., Hoffmann. M, & Macdonald, D.W. (eds) (2004b) *Canids: Foxes, Wolves, Jackals and Dogs: Status Survey and Conservation Action Plan*, 2nd edn. IUCN Canid Specialist Group, Gland, Switzerland.

Sitas, N., Baillie, J.E.M., & Isaac, N.J.B. (2009) What are we saving? Developing a standardised approach for conservation action. *Animal Conservation*, **12**, 231–237.

Stattersfield, A.J., Crosby M.J., Long A.J. & Wege, D.C. (1998) *Endemic Bird Areas of the World: Priorities for Biodiversity Conservation*. BirdLife International, Cambridge.

Stokes, E.J., Strindberg, S., Bakabana, P.C., *et al.* (2010) Monitoring great ape and elephant abundance at large spatial scales: measuring effectiveness of a conservation landscape. *PLoS One*, **5**(4), e10294.

Stuart, S.N., Chanson, J.S., Cox, N.A., *et al.* (2004) Status and trends of amphibian declines and extinctions worldwide. *Science*, **306**, 1783–1786.

Thomas, L & Middleton, J. (2003) *Guidelines for Management Planning of Protected Areas*. IUCN, Gland, Switzerland.

Thorbjarnarson, J., Mazzotti, F., Sanderson, E., *et al.* (2006) Regional habitat conservation priorities for the American crocodile. *Biological Conservation*, **128**, 25–36.

Vogel, W. & Hicks, L. (2000) Multi-species HCPs: experiments with the Ecosystem Approach. *Endangered Species Bulletin*, **25**, 20–22.

Valenzuela-Galván, D., Arita, H.T. & Macdonald, D.W. (2007) Conservation priorities for carnivores considering protected natural areas and human population density. *Biodiversity and Conservation*, **17**, 539–558.

Waltner-Toews, D., Kay, J.J. and Lister, N-M.E. (2008) *The Ecosystem Approach*. Columbia University Press, New York.

Wells, J.V. & Richmond, M.E. (1995) Populations, metapopulations, and species populations: what are they and who should care? *Wildlife Society Bulletin*, **23**, 458–462.

White, P.C.L., Smart, J.C.R., Renwick, A.R. & Raffaelli, D. (2010) Ecosystem health. In: *Ecosystem Ecology: a New Synthesis* (eds C.L.J. Frid & D. Raffaelli), pp. 65–93. Cambridge University Press, Cambridge.

Wiegand, J., Raffaelli, D., Smart, J.C.R.S. & White, P.C.L. (2010). Assessment of temporal trends in ecosystem health using an holistic indicator. *Journal of Environmental Management*, **91**, 1446–1455.

Wikramanayake, E.D., Dinerstein, E., Robinson, J.G., *et al.* (1998) An ecology-based method for defining priorities for large mammal conservation: the tiger as a case study. *Conservation Biology*, **12**, 865–878.

Wood, J.L., Fish, L., Laughren, J. & Pauly, D. (2008) Assessing progress towards global marine protection targets: shortfalls in information and action. *Oryx*, **42**, 1–12.

3

Five paradigms of collective action underlying the human dimension of conservation

Laurent Mermet[1], Katherine Homewood[2], Andrew Dobson[3] and Raphaël Billé[4]

[1]AgroParisTech, 19 avenue du Maine, Paris, France
[2]Department of Anthropology, University College London, Gower Street, London, UK
[3]Keele University, Keele, UK
[4]Institute for Sustainable Development and International Relations, 27 rue Saint Guillaume, Paris, France

Theory is itself a practice, no less than its object is. It is no more abstract than its object. It is a conceptual practice, and it must be judged in terms of the other practices with which it interacts.

Gilles Deleuze, quoted by François Cusset as the apposite quotation to head his book on *French Theory*, University of Minnesota Press, 2008

Introduction

Conservation of biodiversity rests on changing some human activities, projects, plans and policies, so as to stop or limit negative impacts on valued ecosystem features, and foster positive ones (Mascia et al. 2003). It has to be based on an effective understanding of ecosystems, of the ecological consequences of damaging activities, and of the ways these can be alleviated and positive benefits enhanced through conservation. This first imperative gives biology a pivotal role in conservation research. But achieving changes in human behaviour entails a second imperative: an understanding of what kind of societal actions (economic, legal, political, educative, etc.) can bring about specific desired changes. Conservation biologists work to meet this 'human dimension' imperative in three ways:

- *reflection*: they actively reflect on the accessibility and relevance of their own, essentially biological work, to society (Robinson 2006)
- *involvement in practice*: they collaborate directly with conservation practitioners involved in the field
- *involvement in interdisciplinary collaboration*: they work with social scientists.

Key Topics in Conservation Biology 2, First Edition. Edited by David W. Macdonald and Katherine J. Willis.

Since the beginnings of conservation biology (see for instance Soulé's (1985) founding paper or Ehrenfeld's (1987) editorial for the first issue of *Conservation Biology*), there has been a vivid perception both of the centrality of biology to research on conservation and of the need for biologists to collaborate with disciplines dealing with human choices and activities.

Twenty-five years later, neither the reflection of conservation biologists on how to achieve a higher impact on society nor their collaboration with social scientists has yet reached a level and a relevance that most of them would consider satisfactory. As stated by Meffe (2006), 'we are facing a fundamental problem relative to human behaviour, and the solution ultimately will need to take human behaviour into account. This is the great challenge that confronts us in the next decades'. This challenge is felt by many in the conservation field with a mounting sense of urgency, but not because little has been done: many efforts have been made to reach out to the public and to decision makers, and there has been growing, often fruitful, collaboration between conservation biology and various disciplines of social sciences. Nevertheless, biodiversity loss continues to accelerate and despite increasing conservation knowledge and efforts, the threats are larger than ever. Conservation biology remains a 'crisis discipline' and the passing of time continually challenges its very goal – 'to provide principles and tools for preserving biological diversity' (Soulé 1985). The sense that society is 'knowing but not doing' (Knight et al. 2008) leads conservationists and conservation biologists to feel that they are 'doing their part but not getting there'.

There is a pressing need for new concepts and methodological resources to improve the three approaches by conservation biologists to connect more with society: (1) to feed the reflection of conservation biologists, (2) to guide more effective involvement with practitioners, and (3) to orient and enrich interdisciplinary collaboration.

In addressing this need, a dual question plays a central role: *who* acts for conservation, and *how*, i.e. of what does such action consist? This is essentially a question of agency: who has a capacity to act, and what kind of activities does that action entail? In conservation issues, individual action *per se* is usually not enough and, even though the role of individuals is important, the most effective action is collective – that is, it involves forming groups, networks, organizations or institutions that will exercise some capacity to act. In this chapter, we will propose a clarification centred on the question of collective action in conservation. When 'we' say that 'we' should act for conservation and sustainability, who is 'we'? When stating that 'society' should act, what is our understanding of how society is organized to act? Whom do we see as the definer of goals? Whom do we see as taking action? Whom do we see as accountable to whom?

The question of collective action – who has the right to act and how can they act? – is at the centre of the practice of conservation. When debating on how to act, or choosing a strategy, conservation operators explicitly or implicitly choose one or another model of collective action. But the question is also very important for researchers, in particular conservation biologists, because it conditions how ecological knowledge can translate into conservation action. To give just a single example, if conservation biology research indicates what would be an optimal size and location for protected areas, who is going (or supposed) to take action on whom with that information? In other words, the relevance to society of conservation biology as a whole, or of a given conservation biology project, can be conceived of in very different ways, depending on how one sees collective action operating in the conservation field. Throughout the chapter, we will have both conservation practice and conservation research in mind, as appropriate in different sections.

Although we will focus here on the question of collective action, we are well aware that addressing the human dimension of conservation raises other, very different but also essential questions: ethical (what obligations do we have to conserve?), cultural (what concepts of nature underpin management of ecosystems, and how do they differ between societies?), and social (who benefits and who suffers from conservation policies?). They each deserve in-depth investigation and debate in their own right, but here we will focus on the question of agency, and especially on collective action as a core dimension of the gap between knowledge and action in conservation research.

The first part of the chapter will explain the five fundamental paradigms of collective action that, in our view, underpin both lay and academic discourses on action for conservation. Each one offers a very different answer to the question of collective action (Who defines goals? Who takes action? Who is accountable to whom?). In the course of this chapter, it will become increasingly clear that deep differences in the way these questions are answered underpin both the theoretical and the practical debates about conservation, and that making such differences explicit can contribute greatly to clarifying such debates.

The second part of the chapter will provide an illustration of such clarification. It will introduce current controversies about community-based conservation in Africa – more particularly, in East Africa's Maasailand – and show how the five-paradigms model proposed here can shed light on them.

The chapter will end with a discussion of some possible misunderstandings hindering the effort to work on collective action across conservation biology and social sciences, and offer some suggestions for further learning and research.

Divided we must act: five paradigms of collective action on environmental issues

Robinson (2006) observes that conservation biology should 'derive conclusions and generalizations in a context that is more accessible and more relevant to society'. Conservation biology produces findings that have an intention and a potential to be relevant to human interests. Much of its work bears on indicators to identify the problems and their extent, on the precise goals that could be pursued in given cases, on the consequences of various human activities and thus on who is responsible for various aspects of biodiversity loss and on the means and action plans that could be effective. Problems, goals, responsibilities, action: this is, *prima facie*, knowledge ready to be picked up by 'society'.

But who exactly can speak for society in the discussions the conservation biologist needs to have in order to bridge the 'knowing to doing' gap? Many different answers are possible. Not only are they extremely diverse, but they often contradict one another. In practical debates, as the example of community-based conservation will illustrate, what is seen by some as the solution, others see as being the problem. In theoretical debates, for instance, there is little common ground between those (like environmental economists) who study optimal 'instruments' that governments may use to control biodiversity loss and those (like Latour 2004) who see environmental action as a vast system of negotiations in which even 'non-humans' play an active political role. Rarely are these differences in our concepts of agency ('who is the subject of conservation action?') and collective action ('what is the basic organizational pattern of our action for conservation?') clearly explicit. Rather, they express themselves as puzzlement, irritation or even anger when we make little progress in discussing how "we" might act to conserve biodiversity or evaluate past and current actions.

Therefore, it is important to map out the various fundamental concepts of action underlying such debates. To do so, one has to realize that 'society' is fundamentally divided. The utopian view held implicitly by many, that we 'are all in it together' and thus ought, as it were, to act in unison, is an illusion. There is no unity of aims, no close coincidence of interests, no consensus on responsibility, and there is no such thing as action that would be literally 'collective action' if that were to mean that we all act together. What we do have is a set of partial, contradictory concepts and tools for organized joint action. How can we map them beyond the bewildering variety of scales, disciplinary languages and practical controversies?

If one examines the controversies, the practical and theoretical discourses on how 'society' could or should manage the environment, one can identify five distinct paradigms underpinning what is seen to be the main organizational source of the problem, how the discussion of aims should be organized, and who should lead the action (Mermet 2013). Each paradigm is like a fundamental cultural perspective on collective responsibility and action, with its likes and dislikes, its heroes and its tools, its buzzwords and its particular feeling of what can be both right as well as effective in a practical sense (Table 3.1).

Government paradigm

The *government* paradigm rests on the conviction that to overcome the innumerable, intense divisions and conflicts in a human group, power has to be handed over to a single legitimate actor whom all should obey: a government (national or local). However this delegation is established (democratic or otherwise), it provides the basis for an authority to set goals, identify responsibility and carry out action on behalf of society. Here, the buzzwords are decision makers, official targets, legitimacy, implementation, policy instruments. If one focuses straight away on the instruments of action – on what kind of 'carrots, sticks and sermons' (Bemelmans-Videc et al. 1998) may be used to alter behaviour in favour of conservation – the general assumption is that one knows in advance who is in charge of the action, and this is usually some kind of public authority. A commercial summary of Sterner's book *Policy Instruments for Environmental and Natural Resources Management* (2002) eloquently expresses where the problems lie and who is really looking for solutions in this paradigm: '[the author] is careful to distinguish between the well-designed plans of policy-makers and the resulting behaviour of society' (quote from Amazon). The government paradigm is familiar to conservation biologists, who tend to be convinced that if we can really show what must be done for the common good, then it ought to be converted by 'political will' into appropriate regulations and economic tools. Here, the relevance to society of conservation research depends on providing convincing and solid advice for the use of authorities – and then crossing fingers.

Co-ordination paradigm

In the *co-ordination* paradigm, the problem is not seen as susceptible to solution by a purposeful power, but as a set of differences and misunderstandings to be addressed and discussed directly by the stakeholders themselves. These actors do differ on how to manage the resources, so they end up with severe problems. But potentially they have the capacity to solve such problems by themselves if they can only co-ordinate better. The main obstacle to overcome is a lack of the sort of communication that can allow them to realize their joint interest in co-operating. What is required here are procedures for such communication, for negotiation, for joint action. A reference book written from that perspective would be Elinor Ostrom's *Governing the Commons* (1990). Ostrom provides examples of how this has been repeatedly achieved in managing a range of common resources, explores in depth the rationales and conditions for success, and warns how government intervention often makes the problem worse instead of better.

Bringing everyone to the negotiation table, mediation and co-construction, and replacing management by an administration with

Table 3.1 Five paradigms of collective action for conservation

Paradigms	Who is the main operator of conservation action?	What does collective action essentially consist of?	Typical buzzwords	Conservation research is relevant if ...
Government	A government that has a delegation to act for the collective	Intervention to modify behaviour through various tools and policies	Decision makers, official targets, legitimacy, implementation, policy instruments	... it provides government with reference goals, indicators, objective choice of tools
Governance	A complex set of government and stakeholders	Complex procedures combining policy and stakeholder participation	Participation, participatory planning, stakeholders involvement, public–private co-operation	... it provides information to and participates in complex, multilevel decision-making processes
Co-ordination	Stakeholders themselves	Co-ordination and direct collaboration between stakeholders	Actors around the table, co-construction, mediation, collaboration, community	... it engages all stakeholders in a collaborative way
Minority action	An actor focused on a specific conservation goal and acting to reach it	Strategic action to obtain changes from specific actors whose activities impact biodiversity	Environmental groups, activism, innovators and advocates, legal or political challenging of decisions	... it provides compelling facts and arguments to support environmental advocacy confronting other interests
Revolution	Masses and their leaders in opposition to 'the system'	Mass action for wholesale systemic change addressing a whole range of societal and environmental issues	Globalization, commodification, capitalism, ecological crisis, colonialism, growth as the systemic cause of environmental problems	... it participates in the overall efforts of the vast coalition of those who oppose 'the system' because of the whole set of its negative effects on society and nature

management by a community are important concepts here. In conservation research, this has been a rising paradigm for the last two or three decades. Here, relevance to society may mean, for instance, research that engages all stakeholders in a collaborative way.

Revolution paradigm

The approaches of the *revolution* paradigm, far from expecting conservation action from the powers that be, consider them to be the very cause of the ecological crisis and loss of biodiversity. The key concept here is that we are all entangled in a system (political, economic and/or cultural) which both destroys nature (and many other human concerns) and hides the process behind a constant barrage of ideological rhetoric. The title of Joel Kovel's book *The Enemy of Nature: The End of Capitalism or the End of the World* (2002) summarizes in a nutshell one such revolution-oriented diagnostic. In revolutionary approaches, the main issue for action is to bring the masses to a renewed awareness that would allow them to be conscious of their entanglement and its consequences. This would lead to such a massive shift in values and practice that the 'system' would become untenable and the major obstacles to ecologically sound lifestyles would be overcome.

This is an important but troubling paradigm for conservation biology. Should conservation be mainstreamed in a fundamentally unchanged mode of development? Or is it part of a wider environmental and sociopolitical agenda for deep change? Are the more radical movements of political ecology possible allies or are they a threat? And what could bring about the 'global change in worldview' (Meffe 2006), the massive shift in priorities for which many conservation biologists think the current accumulation of conservation projects can be only a temporary, transitional substitute? Here, relevance to society means research that goes beyond government framings and contributes to a much wider shift in society.

Governance paradigm

The *governance* paradigm is a hybrid between 'government' and 'co-ordination'. Here, government is seen as both overambitious and insufficiently effective. The key to more efficient action is then to be found in reinforced co-operation between government and civil society, i.e. both public and civil society organizations, including NGOs and the private sector. In a governance perspective, government must open its decision-making processes to stakeholders – the discussion on goals, the allocation of responsibility, the choice and implementation of means – and, conversely, the initiatives taken by civil society have to be gradually taken up to become forms of government. Here, relevance to society means conservation research that finds its niches and provides the right types of information, packaged in the right way, at the right moments among the multiple stages and scenes of multiscale, multiactor, semi-open, complex decision-making processes that have proliferated over the last two decades.

Minority action paradigm

The *minority action* paradigm assigns responsibility for biodiversity erosion not to society as a whole, or to 'the system', but to some clearly identifiable human causes, specific powerful actors, activities or sectors. The question then is how actors committed to conservation can act to obtain changes from other actors in behaviours that threaten biodiversity (Mermet 1992; Mermet 2011). Since conservation actors usually start from a minority position, and want to obtain changes in the behaviour of powerful actors, action is fundamentally strategic. *Silent Spring* (Carson 1962) is an emblematic book here, an outstanding example of how powerful one single voice and a civil society movement starting from a minority position can be in the face of overpowering social, political and economic forces. The issue is not seen as revolution but as transformation. The system is not to be toppled over wholesale, but to be transformed from the inside by minority actors. The process is not co-ordinated 'around a table', but is both pluralistic and strategic. The many examples of conservation struggles that have finally become conservation successes show the importance of this model in the practical experience and culture of conservation experts on the ground. Here, relevance to society means research that contributes to the ongoing struggle of the conservation and environmental sector, confronting specific biodiversity damaging industries and policies.

Illustrating the five paradigms of collective action for conservation: community-based conservation in East Africa's Maasailand

The intense debate about community-based conservation in Africa, and in East Africa's Maasailand in particular, illustrates how these different perspectives conflict with and complement one another, and how their recognition may form a first step towards 'change we can believe in' (Lund et al. 2009).

Present-day Maasailand straddles the Tanzania/Kenya border. This ~150,000 km² area of semi-arid rangelands has strong ecological continuities, with dryland, wetland and higher, more mesic montane rangeland and forest habitats repeated either side of the border. Migratory wildlife and livestock species move around the system seasonally. The rural population is predominantly made up of a single, relatively cohesive ethnic group of people all speaking the same language, sharing a common demography distinct from national patterns (Coast 2002) and practising the same age-set customs and land use. However diversified their livelihoods, most rural Maasai remain semi-sedentary, mobile transhumant agropastoralists (Homewood et al. 2009).

In both Kenya and Tanzania, tourism is among the top three contributors to GDP, accounting for nearly 1 billion USD annually in each. In both Kenya and Tanzania, the highest-earning protected areas are situated within, and effectively excised from, Maasailand, as is a high proportion of the two countries' conservation estate overall. Yet poverty is widespread and severe within Maasai communities in both Kenya and Tanzania, as measured against national rural poverty thresholds, let alone international datum lines (Oxfam 2006; Thornton et al. 2006; Tenga et al. 2008). Thus Maasailand is of particular relevance to the three dimensions of conservation, poverty and community-based conservation (CBC) initiatives.

Across Maasailand, CBC initiatives are expanding quickly. Amboseli in southern Kenya Maasailand is considered by many to be the birthplace of community conservation, with early initiatives established as far back as 1975 (Western 1994), leading up to the present proliferation of community conservation and conservancy models across the region. The promotion of these models followed a variety of motivations: pragmatic conservation concerns seeking to enlist the support of reserve adjacent dwellers, in the context of structural adjustment and the loss of public enforcement capacity; social justice, human rights, and poverty alleviation imperatives; green development. The overall intention was to deliver both conservation and development as 'balanced', 'sustainable', 'win/win' outcomes (Hughes & Flintan 2001; Roe et al. 2009). Principles of rational choice pointed to collective action solutions and to working examples of sustainably managed commons (Ostrom 1990). Initial evaluations tended to be uncritical endorsements, often by people heavily involved in the process. More recently, evaluations of CBC initiatives based on detailed and independent datasets have emerged (e.g. Western et al. 2006; Homewood et al. 2009), including systematic reviews of the growing numbers of case studies available worldwide (Bowler et al. 2010; Waylen et al. 2010).

Overall, most authors and actors look back on CBC initiatives with a sense of disillusionment about their actual ecological and economic outcomes. But the grounds for such disillusionment and the proposed remedies are quite different from one author, actor or school of thought to the next. Frustration with the results of CBC has driven some conservationists 'back to the barriers' (Gartlan 1997; Oates 1999); that is, back to less participatory and collaborative practice in protected areas management. Another outcome in East African rangelands has been a shift to private conservancies where land tenure and market revenues favour conservation-friendly choices by landowners (Western et al. 2006). Some political ecologists and social scientists documenting such trends in decentralization and devolution increasingly express disappointment with the way that elite members of a society can capture resources or influence and dispossess others in the name of community conservation (Blaikie 2006; Peet et al. 2010). Others, aware of the complex and long-term nature of the processes involved, expect positive ecological and economic outcomes to be slow to emerge, recognizing that the evolution of institutional structures fostering more sustainable and legitimate collective action mechanisms is an oft-overlooked but essential part of the process (Brechin et al. 2002).

Increasingly, those who support the potential of CBC for positive change adopt a more qualified defence of it, for instance by exploring the qualitative dimensions of meaningful as opposed to token participation (Lund et al. 2009). Overall, there is a vague consensus across most schools of thought that CBC has had disappointing outcomes to date. There remains, however, a 'dialogue of the deaf' between different contradictory interpretations of the current situation regarding conservation in Maasailand and the respective implications of these interpretations for further conservation action.

We ask, can a clarification of the underlying collective action paradigms help move analysis – and maybe action – beyond this stalemate?

The government paradigm discussed above precisely captures official attitudes and practice in northern Tanzania. Tanzania has numerous, high-profile national parks and reserves (conferring total protection across 25% of the national land surface area, and partial protection to ~40%: Homewood et al. 2009). In response to donor pressure to move towards decentralized natural resource management and conservation and to share revenue with local people, the Tanzanian state slowly and reluctantly developed the Wildlife Management Area (WMA) model, with considerable support from international conservation scientists (Leader-Williams et al. 1996). Under this model, groups of registered villages would come together to set aside land, and to negotiate contracts with tourism entrepreneurs who would pay for access to the set-aside. The government established criteria for eligibility and procedures, which were complex and hard for poorly literate village leaders to navigate. Even so, a number of WMAs were established with the aid of NGOs such as the African Wildlife Foundation (AWF). Conflicts arose between villages, but also between game viewing and hunting enterprises. The Tanzanian state, which had always retained central control of hunting licences and revenues, supported hunting enterprises, and in 2007 issued a ministerial decree criminalizing any deals between WMA villages and game-viewing entrepreneurs (TNRF 2007). Such arrangements are now negotiated centrally. This development illustrates how from a government perspective, direct co-ordination between actors, such as the management by Maasai communities without state interference, or direct deals between communities and game-viewing companies, are part of the problem, not of the solution.

Alternative approaches, based on the co-ordination paradigm, advocate and put into practice precisely the opposite. The example of two villages in North Tanzania, documented by Yann Laurans (in Laurans et al. 2011), shows the relative (ecological and social) success of such a direct deal. The case shows clearly the importance of the simplicity of the negotiation and of the agreement, as well as their vulnerability to potentially counterproductive government interference. Indeed, in the government-run scheme, revenues theoretically trickle down via central state, regional, district and WMA governing bodies to individual village governments. In practice, however, with a proportion being legitimately retained at each stage, and more being corruptly diverted, *no* revenues flow to participating villages or households. People have effectively given up about half of the farming and grazing lands on which their central livelihoods depend, for no return, creating hardship and conflicts (Homewood & Thompson 2010).

Another example from Maasailand, the intervention by the Tanzania Natural Resources Forum (TNRF), provides a further illustration of the co-ordination paradigm, this time in the form of conflict resolution, rather than direct payment for ecosystem services. TNRF was constituted with support from the US-based Sand County Foundation as a forum where all stakeholders – government, civil society and entrepreneurs – can meet to discuss and hopefully resolve potential conflicts between conservation and development. The TNRF intervenes, for instance, to defuse the clashes that have escalated in Loliondo, Northern Tanzania, since recent legislation has rendered incompatible two land-use categories that formerly overlapped without serious conflict: Game Controlled Areas (GCA) and Village Land. The interests of hunting companies now align with the former, and those of the 20,000-strong local community of Maasai agropastoralists with the latter.

The TNRF promotes a compromise through designating as Wildlife Management Areas (a more flexible statute than the GCA) the remaining Village Land. The TNRF's approach in this case reflects its wider strategy of pursuing 'a long-term, innovative and adaptive process of advocacy and capacity-building, based on collaboration and collective interests' (www.tnrf.org). Favoured means of action are (1) 'increasing the flow of information' and (2) 'facilitating collective action'. This approach reflects the view that 'members see the need to come together of their own accord to work on key issues that affect the way they are able to use, manage and conserve natural resources. Working groups collaboratively develop rounded and innovative solutions…'.

Although TNRF's approach clearly espouses the vocabulary, philosophy and methods of the co-ordination paradigm, it also touches on the governance paradigm when describing its third means of action: 'Being effective advocates – in compellingly communicating our ideas and solutions to government … Very often they need our support, and we need theirs. Without government support and better governance, we simply won't succeed in realizing our vision'.

In this example, the hybrid or ambiguous nature of the governance approach is evident in the arduous, top-down/bottom-up negotiations between the Tanzanian government and TNRF.

The minority action paradigm, so forcefully articulated by Gartlan (1997) and Oates (1999) against CBC in west/central African forest ecosystems, is less prominent in East African rangelands. Norton-Griffiths (2007; see also Norton-Griffiths & Said 2010) exemplifies the lone champion seeking to achieve conservation through changing the behaviour of a few powerful actors responsible, as he sees it, for the catastrophic decline observed in Kenya's wildlife populations from the mid-1970s to the present. Norton-Griffiths attributes this decline to the fact that economic returns from wildlife per unit area are consistently lower than returns from livestock, and returns from livestock are themselves lower than returns from commercial cultivation. This creates an overwhelming incentive to convert rangeland habitat to commercial cultivation. Norton-Griffiths argues for economic incentives to landowners, potentially through allowing consumptive use of wildlife (including hunting, currently banned in Kenya) and through ensuring better distribution of profits along the tourism commodity chain, which currently leaves landowners bearing the risk and capturing little of the profits (Norton-Griffiths 2007). His argument segues into a case for private conservancies, in practice the preserve of wealthy landowners, often international investors, and increasingly the best remaining strongholds of Kenyan wildlife.

Most conservation organizations try to work out positions that link together co-ordination, minority action and governance models of action, while presenting themselves in a light that potential donors (but also the governments whose approval they require) will find favourable. A good example is the African Wildlife Foundation (AWF). This international conservation organization portrays itself as promoting working from the bottom-up by local actors themselves:

> 'Who better to protect their land and resources than Africans themselves? Living on the land we strive to protect, Africans are in touch with both its potential and its challenges. They have witnessed the draw of tourists to their land. And, they have come face-to-face with the sometimes destructive consequences of sharing land with Africa's wildlife. Empowering Africans to be Africa's stewards is at the core of our strategy'. (www.awf.org/section/about).

The AWF also participates in coalitions opposing mainstream development projects, for example, by opposing the proposal of the Tanzanian government for a road across the Serengeti national park. At the same time, it puts forward as part of its mission 'Working with governments at every level to shape and support conservation policy' (www.awf.org/section/people). Clarifying such variable positions towards government is an essential element of the five-paradigm framework. When conservation action is seen as supporting those parts of government that act for conservation and/or opposing other parts of government that act *de facto* against conservation, the underpinning model is minority action (including minority action within the government, e.g. by an environment ministry on other ministries). When conservation action is seen as best conducted without government intervention, this signals the co-ordination paradigm. When conservation action is presented as best served by intense dialogue and joint action of all stakeholders, governmental or not, from all activity sectors involved, the underpinning model is governance. Many conservation organizations today, whether strategically pragmatic or simply opportunist, switch between models or combine them.

However, observers with a more radical view interpret such positions very differently. In the case of AWF, Sachedina (2008) sees it as having focused on fundraising and rewards to its directors and staff at the expense of any downward accountability to the people and wildlife of the areas where it supposedly supports community initiatives. This revolutionary perspective is strongly articulated by political ecologists analysing events in Tanzania Maasailand. Igoe (2007), Igoe & Brockington (1999), and others point to the systematic marginalization of Maasai, the blanket failure to acknowledge their ecological knowledge as a valid and vital part of understanding and managing the landscapes created by millennia of pastoralist use (Goldman 2003), and an unholy alliance between conservation, international investors and celebrity figures in the progressive alienation of pastoralist rangelands to outside investors. From this perspective it is hard to see the history of conservation in Maasailand as anything other than progressive accumulation by dispossession (Nelson et al. 2009; Peet et al. 2010) as predicted by political economy and political ecology theory (Jones 2006). Where analyses based on other paradigms claim that considerable benefits flow to communities from CBC, livelihoods studies suggest otherwise for the great majority of reserve-adjacent dwellers and CBC participants (Homewood et al. 2009). Garland (2008) presents the African-mediated component of such conservation as operating through the purchase or co-opting of loyalties by what remains a very colonial form of conservation.

Discussion

Deeply held, incompatible perspectives ... that complement each other

The example of Maasailand illustrates how identifying the five paradigms of collective action can help to map the different perspectives that underpin significant disagreements and ambiguities in debates over conservation action. Such disagreements and ambiguities can be found in both the practice of and academic debate about conservation action. In-depth discussion of the theoretical issues involved in this five-paradigm model is beyond the scope of this chapter, but we will flag some possible misapprehensions.

First, these five paradigms should not be taken as superficial tools and perspectives between which we could jump back and forth at will from some all-encompassing position. They reflect deeply held worldviews and convictions that organize both lay and scholarly thinking about conservation action or the relevance to society of conservation research. They can sometimes be combined, or one can sometimes move from one to the other, but only to a limited extent, and this involves important tensions and generates problems of coherence (both in practice and conceptually). As for an all-encompassing position, it does not exist. However difficult this may be to accept, there is no sense of progress between or through the five paradigms: they simply co-exist. Their respective detailed contents, as well as their scientific and political weight, vary over time but none is the silver bullet that would make the others obsolete. Progress in one direction is often seen as damage in another: these are discontinuous and often contradictory perspectives. However, they play a complementary role in developing the democratic processes of acting on public issues.

Since its very birth in ancient Athens, democracy itself can be seen as the tense co-existence of antagonistic political models and forces, none of which has the ability to overcome the others (Ober 1991). In this context, a broad acceptance of co-existing contradictory paradigms, of the fact that the conservation community is moved not only by shared but also by divisive viewpoints, is essential. Here, deep reflection about one's own position and some literacy on the positions of others are essential. The five-paradigm framework helps to reveal deeply rooted differences. They need to be heard and accommodated, but not in the sense of the 'governance' paradigm which seeks to reconcile or, better, overcome what it sees as essentially the misunderstandings of others. Rather, they should be perceived as the basis of a political process which acknowledges that, while differences may not be reconcilable, mutual recognition is a positive step in the pursuit both of action and of academic debates about conservation.

Second, the concepts used in analyses of the human dimension of conservation are dynamic, not stable. Moreover, they cannot be stable, because these analyses and the concepts themselves (and thus the very words they use) are moves in the political games they describe and so are subject to frequent changes as the proponents of each view strategically use evidence and argument. In the example of Maasailand, narratives centering on ecosystem degradation, for instance, underpin justifications for governments or international agencies to take control of contested resources, while stewardship narratives underpin justification for co-ordination and/or governance paradigms, and so on. A more general example is the multiple meanings of 'governance'. Many authors and actors use the concept with a specific meaning, describing a collective action model where state and other actors work together, as in Floranoy's definition:

> 'Environmental governance can be defined ... as multi-level interactions ... among, but not limited to, three main actors, i.e. state, market, and civil society, which interact with one another ... in formulating and implementing policies in response to environment-related demands ...'. (http://ecogov.blogspot.com/2007/04/)

But other authors employ an all-inclusive notion of governance as 'the act of governing – it relates to decisions that define expectations, grant power, or verify performance' (http://en.wikipedia.org/wiki/governance). With that definition, governance encompasses the entire field of collective action for conservation.

Lemos & Agrawal (2006) then propose the term 'hybrid governance' for governance in the more specific sense. The issue is that using 'governance' to designate the whole scene covertly generalizes the perspective of governance in its more specific sense and in so doing, promotes a pervasive sense that multi-institution, multilevel dialogue is at the centre of conservation action. This is why we deliberately use the more specific sense here. We wish to underline the fact that although some supporters of each paradigm think that theirs is able to subsume the others, this is really not the case and it is necessary to acknowledge the co-existence of paradigms that are irreducible to one another.

Third, although the mapping of collective action paradigms that we have suggested here can be useful in its own right, it is no substitute for more specific theories and conceptual frameworks. To give just two examples: decisive minority action may be understood through quite different conceptual languages, analyses and practical perspectives, depending on whether one sees it as a matter of innovation (Callon 1986), of advocacy (Sabatier & Jenkins-Smith 1993), of whistle-blowing (Chateauraynaud & Torny 2000) or of strategic action for change (Mermet 1992). Similarly, the co-ordination paradigm leads to contrasting views according to whether one seeks better co-ordination in private–social partnerships (e.g. direct payments for ecosystem services), in conflict resolution (Susskind 2009), in collaboration (Gray 1989) or in the subsidiary construction of institutional arrangements (Ostrom 1990, 2007).

Fourth, as we stated in the introduction, we have covered here only one of the main issues in the human dimension of conservation: agency or the organization of collective action, but others are essential too. For example, there has been much debate regarding the *why* of conservation. Conservation is by no means a moral imperative for everybody, and those who adopt an ethical perspective on conservation must make good arguments for it. Such arguments tend to fall into one of two camps: either we should conserve nature because it is of some benefit to humans to do so, or because it benefits the animal or plant in question whether there is any human benefit involved or not. In environmental ethics these are called anthropocentric and ecocentric reasons respectively. Each of these very generic terms, however, needs to be further broken down into subcategories. On the anthropocentric side, the most important subcategory is the distinction between 'crude' and 'enlightened' anthropocentrism. The former fails to take the non-human natural world into account when deciding what is in humans' interest, while the latter recognises that human interests can often not be realised without taking the (conservation of) the non-human natural world into account. On the ecocentric side, on the other hand, debates continue between those who argue that the objects of concern should be biotic or abiotic – and whether the focus should be on individuals or on collectives, such as species or ecosystems.

With the exception of crude anthropocentrism, we find virtually all these positions represented in arguments for conservation, whether in a research or a practical context. Thus there will be those who believe that conservation is important because some current human benefit can be derived from it, like income from 'ecotourism', or because it is a form of 'intergenerational injustice' to deprive future generations of humans of the joy of encountering a given species and thereby depriving them of something that can contribute to human flourishing. On the ecocentric side, conservationists might argue that a given species should be conserved because of its intrinsic value, i.e. it has value irrespective of whether it is useful for humans.

Anthropocentric and ecocentric views are often regarded as opposites, in that they appear to pit the interests of the human and the non-human world against one another. This is why, in practice, one version or another of 'enlightened anthropocentrism' tends to hold sway, in that it recognizes an interdependence of human and non-human interests. Important tensions still exist, though, for not all 'enlightened anthropocentric' reasons for conservation point in the same direction, particularly with respect to winners and losers within the human community itself. Thus ecotourism projects involving conservation (safari parks, for example) are often criticized for not taking sufficiently into account the needs of the human population occupying the same space as the animals.

These conflicts of interest are endemic to the project of conservation. There is no determinate way of resolving them; that is, there is no one objective reasoning, no one set of data that might possess sufficient authority from its own logic to establish an indisputable normative reference (for instance, a biological or human values baseline) that would compel adherence from all stakeholders and establish a universally shared set of goals for conservation efforts. Although the setting of goals, the interests at stake and the means of action are distinct aspects of conservation action, they cannot be separated in an absolute way, as if goals could be established in an objective manner in the arena of science and then action could be turned over to the political sphere (Latour 2004).

This very lack of determinacy points, of indisputable baselines, is viewed very differently from each of the various paradigms presented in this chapter. The governance or the co-ordination paradigms, or some combination between them, are increasingly promoted as providing the appropriate context and method for driving collective action, precisely because their focus on co-ordinating a variety of views seems a reasonable answer to the lack of determinacy points. From such perspectives, the government, minority and revolutionary paradigms all seem inappropriate because they apparently require excessive degrees of certainty in both information and in the value basis for decisions. Advocates of procedural approaches (governance, co-ordination) claim that the range of values is so wide and the knowledge available so provisional that only discursive and contingent processes are best suited to conservation-related decision making. But seen from a revolutionary or minority action for change perspective, such process-based models of action are themselves a problem because they usually (and covertly) play in favour of the status quo. By providing only a process for co-ordination, they reinforce *de facto* the existing cast of actors and issues and do not alter power balances that are decisive for conservation. Or if they attempt to tilt existing balances of power in favour of given social groups, for instance, poor farmers, they then exhibit just as high a level of very debatable certainty (in the choice of the social interests they choose to defend), and so make themselves just as vulnerable to political manipulation (by interests vested behind the displays of defending this or that social group), as those others whom they reproach for their choices of specific biodiversity baselines and interests to defend.

This brief discussion on the 'why' of conservation shows that while it opens a very different discussion space than the 'who can act?' and 'how can they act?' questions, there are also deep links between the two, and the five-paradigm framework can help explore those too.

What orientations for human dimension research in conservation?

The differing paradigms of collective action not only underpin the discussion of each of the many issues in conservation. They also bear strongly on the research agenda in conservation science. A currently influential stream of work (Salafsky et al. 2002; Sutherland et al. 2004) recommends an evidence-based clinical approach to the choice of conservation options and tools. The underlying metaphor is the recent development of evidence-based medicine. Treating a conservation problem is seen as the application of tools that have proven, evidence-based (i.e. empirically proven, usually based on statistics) capacity to counter well-identified threats. Amongst the merits of such an approach are the quest for a larger view above the multiplicity of cases, an obstinate effort to identify and classify problems, and an ordered review of tools. This approach is typical of the government paradigm: adopting a wide, ostensibly neutral view, choosing rationally from a comprehensive range of instruments, and guiding from above the choice of solutions to be implemented at lower levels of action.

A discussion of the limits of this approach could start, as a first step, with the limits of the evidence-based clinical approach in general. It is not medicine in all its aspects (including, for example, the close relationship of trust between clinical patient and doctor, or the individual's choice of lifestyle and health) that is taken as the organizing metaphor in evidence-based conservation, but one current model of rationalization of medical expenditure by government. Whatever its strengths, this is only one approach to the complexity and ambiguity of real-world, social problems (for an overview of the challenge and of possible alternative perspectives, see Denzin & Lincoln 2005). A second step for a critique would be consideration of the strategic, minority action perspective. Is conservation only about suppressing impersonal threats, or is it about competing or struggling with other stakeholders and policy sectors? If, as Salwasser claims (in Jacobson 1998), 'the business of fish and wildlife conservation is in competition with all other businesses for access to the ... land and water resources', conservation action is not addressing impersonal threats but is strategic. By definition, it rests on interaction with intelligent actors with other interests, who will actively strive, through counter-strategies, to make conservation action fail (or at least, have effects as limited as possible). Those stakeholders who threaten conservation are not microbes or blind forces; some of them employ consultants who may read such books as *Key Topics in Conservation Biology*, to find out what conservation advocates are up to, and how to keep them in check (Rowell 1996). Such strategic interaction is quite different from the neutral, technical or clinical type of action problem addressed by evidence-based approaches. Each of the paradigms could serve as a basis for enlarging and diversifying the focus in a similar way. Other, more explicitly strategic approaches will have to be considered, alongside evidence-based clinical choice of tools, and they deserve an investment on a comparable scale.

What forums for discussion of the human dimension?

In what arenas are we to discuss tools and approaches to the human dimension of conservation? Redford & Taber (2000) advocate the need for a 'fail-safe' environment of discussion where conservation researchers and experts could openly discuss failures of conservation projects so as to learn from them. They show how the Pollyanna bias of bureaucratic decision processes (for a striking example in another, tragic field, see the classic book on the Vietnam war by Sheehan 1989) stifles much needed open discussion of the raw reality of conservation action. They are right. But other, further impediments are to be considered too. For instance, how does one discuss publicly (that is, in full view of competitors or opponents) the issues of one's own strategies, in a really strategic (that is, competitive, or adversative) context of action? This is an obvious question routinely addressed by parties organizing ways to consider and evaluate strategies in business, in politics, in communication and public relations, etc. It is clearly one that the conservation research community will have to address as well.

A significant part of the current uneasiness in developing work on the human dimension of conservation may be rooted in the difficulty of establishing arenas for cumulative discussion of really strategic issues. It is true that this is difficult to reconcile with the current dominance of the governance and co-ordination paradigms. The two of them exercise considerable pressure towards co-operative views and discussion arenas open to all. The need for, and difficulty of, designing arenas of discussion that will allow enough room for the various paradigms is reflected by the conflicting metaphors regularly found in the literature, such as when Knight et al. (2008) rightly stress that 'to collaborate with people' (including stakeholders) is central for the conservation planner, and a moment later describe 'real-world conservation activities' as 'trenches', a war-like metaphor which is also used by Jacobson (1998). The tensions implied in all these metaphors reflect the difficult co-existence of paradigms that are both contradictory and complementary. If we want conservation biology both to hold its part in a struggle over conservation and also to invest in trust building and collaborative approaches, to guide government policy and also to participate in revolutionary societal shifts, and to switch from one role to another within complex governance processes whilst also remaining a clear and distinctive voice in academic and public debate, we need to enlarge and differentiate the arenas in which conservation issues are discussed, so that they can more clearly include contradictory perspectives and fruitfully accommodate their momentous relationships.

Acknowledgements

The writing of this chapter was made possible by Laurent Mermet's invitation by the Smith School of Enterprise and the Environment (Oxford University) as a visiting research fellow.

References

Bemelmans-Videc, M.L., Rist, R.C. & Vedung, E. (eds) (1998) *Carrots, Sticks and Sermons. Policy Instruments and Their Evaluation*. Transaction Publishers, London.

Blaikie, P. (2006) Is small really beautiful? Community-based natural resource management in Malawi and Botswana. *World Development*, **34**, 1942–1957.

Bowler, D., Buyung-Ali, L., Healey, J. *et al.* (2010) The evidence base for community based forest management as a mechanism for supplying global environmental benefits and improving local welfare. *Systematic Review*, **48**. www.unep.org/stap/Portals/61/pubs/STAP%20CFM%20document%20 2010.pdf

Brechin, S.R, Wilshusen, P.R., Fortwangler C. & West P. (2002) Beyond the square wheel: toward a more comprehensive understanding of biodiversity conservation as social and political process. *Society and Natural Resources*, **15**, 41–64.

Callon, M. (1986). Eléments pour une sociologie de la traduction – la domestication des coquilles saint-jacques et des marins pêcheurs dans la baie de Saint-Brieuc. *L'Année Sociologique*, **36**, 169–201.

Carson, R. (1962) *Silent Spring*. Houghton Mifflin, Boston.

Chateauraynaud, F. & Torny, D. (2000). *Les Sombres Précurseurs: Une Sociologie Pragmatique de L'Alerte et du Risque*. Editions de l'EHESS, Paris.

Coast, E. (2002) Maasai socio-economic conditions: cross-border comparison. *Human Ecology*, **30**(1), 79–105.

Denzin, N.K. and Lincoln Y.S. (eds) (2005) *The Sage Book of Qualitative Research*, 3rd edn. Sage, London.

Ehrenfeld, D. (1987) Editorial. *Conservation Biology*, **1**, 6–7.

Garland E. (2008) The elephant in the room: confronting the colonial character of wildlife conservation in Africa. *African Studies Review*, **51**(3), 51–74.

Gartlan, S. (1997) Every man for himself and God against all: history, social science and the conservation of nature. Presented at the Worldwide Fund for Nature Annual Conference on People and Conservation.

Gray, B. (1989) *Collaborating: Finding Common Ground for Multiparty Problems*. Jossey-Bass, San Francisco.

Goldman, M. (2003) Partitioned nature, privileged knowledge: community based conservation in Tanzania. *Development and Change*, **34**(5), 833–862.

Homewood, K. & Thompson, M. (2010) Social and economic challenges for conservation in east African rangelands. In: *Wild Rangelands* (eds J. du Toit, R. Kock & J. Deutsch). Wiley-Blackwell, Oxford.

Homewood, K., Kristjanson, P. & Trench P.C. (2009) *Staying Maasai? Livelihoods, Conservation and*

Development in East African Rangelands. Springer Verlag, New York.

Hughes, R. & Flintan, F. (2001) *Integrating Conservation and Development Experience: A Review and Bibliography of the ICDP Literature*. International Institute for Environment and Development, London. www.oceandocs.org/bitstream/1834/805/1/HUGHES,%20R1-24.pdf

Igoe, J. (2007) Human rights, conservation and the privatization of sovereignty in Africa – a discussion of recent changes in Tanzania. *Policy Matters*, **15**, 241–254.

Igoe, J. & Brockington, D. (1999) *Pastoral Land Tenure and Community Conservation: A Case Study from North-East Tanzania*. Pastoral Land Tenure Series No 11. IIED, London.

Jacobson, S.K. (1998) Training idiots savants: the lack of human dimensions in conservation biology. *Conservation Biology*, **12**, 263–267.

Jones, S. (2006) A political ecology of wildlife conservation in Africa. *Review of African Political Economy*, **33**(109), 483–495.

Knight, A.T., Cowling, R.M., Rouget, M., Balmford, A., Lombard, A.T. & Campbell, B.M. (2008) Knowing but not doing: selecting priority conservation areas and the research-implementation gap. *Conservation Biology*, **22**, 610–617.

Kovel, J. (2002) *The Enemy of Nature: The End of Capitalism or the End of the World?* Zed Books, London.

Latour, B. (2004) *Politics of Nature: How to Bring the Sciences into Democracy*. Harvard University Press, Harvard, MA.

Laurans, Y., Leménager, T. & Aoubid, S. (2012) *Payments for Ecosystem Services, from Theory to Practice - What prospects for Developping Countries?* Agence Française de Développement, collection A Savoir n°7, avril 2012, Paris.

Leader-Williams, N., Kayera, J.A. & Overton, G.L. (1996) *Community-based Conservation in Tanzania*. Occasional Paper of the IUCN Species Survival Commission No. 15. IUCN, Gland, Switzerland.

Lemos, M.C. & Agrawal, A. (2006) Environmental governance. *Annual Review of Environmental Resources*, **31**, 297–325.

Lund, J.F., Balooni, K. & Casse, T. (2009) Change we can believe in? *Conservation and Society*, **7**, 71–82.

Mascia, M.B., Brosius, J.P., Dobson, T.A., *et al.* (2003) Conservation and the social sciences. *Conservation Biology*, **17**, 649–650.

Meffe, G.K. (2006) *Conservation Biology* at twenty. *Conservation Biology*, **20**, 595–596.

Mermet, L. (1992) *Stratégies pour La Gestion de L'Environnement. La Nature Comme Jeu de Societe?* L'Harmattan, Paris.

Mermet, L. (2011) Strategic Environmental Management Analysis: Addressing the Blind Spots of Collaborative Approaches. IDDRI, collection "Idées pour le débat", n°5-2011, 34p. http://www.iddri.org/Publications/Collections/Idees-pour-le-debat/ID_1105_mermet_SEMA.pdf

Mermet, L. (2013, in press) Les paradigmes contradictoires de l'action organisée en matière de conservation de la biodiversité. In: *Sciences de La Conservation* (eds M. Gauthier-Clerc, F. Mesleard & J. Blondel). De Boeck Université, Bruxelles.

Nelson, F., Gardner, B., Igoe, J. & Williams, A. (2009) Community-based conservation and Maasai livelihoods in Tanzania. In: *Staying Maasai? Livelihoods, Conservation and Development in East African Rangelands* (eds K. Homewood, P. Kristjanson & P.C. Trench). Springer Verlag, New York.

Norton-Griffiths, M. (2007) How many wildebeest do you need? *World Economics*, **8**(2), 41–64.

Norton-Griffiths, M. & Said, M. (2010) The future for wildlife on Kenya's rangelands: an economic perspective. In: *Can Rangelands be Wild Lands?* (eds J. Deutsch, J. du Toit & R. Kock). Zoological Society of London, London.

Oates, J.F. (1999) *Myth and Reality in the Rain Forest: How Conservation Strategies are failing in West Africa*. University of California Press, Berkeley, CA.

Ober, J. (1991) *Mass and Elite in Democratic Athens: Rhetoric, Ideology, and the Power of the People*. Princeton University Press, Princeton.

Ostrom, E. (1990) *Governing the Commons: The Evolution of Institutions for Collective Action*. Cambridge University Press, Cambridge.

Ostrom, E. (2007) A diagnostic approach for going beyond panaceas. *PNAS*, **104**(39), 15181–15187.

Oxfam (2006) *Delivering the Agenda: Addressing Chronic Under-Development in Kenya's Arid Lands*. Oxfam International Briefing Paper. Oxfam, London.

Peet, R., Robbins, P. & Watts, M. (2010) *Global Political Ecology*. Taylor and Francis, Oxford.

Redford, K.H. & Taber, A. (2000) Writing the wrongs: developing a safe-fail culture in conservation. *Conservation Biology*, **14**, 1567–1568.

Robinson, J.G. (2006) Conservation biology and real-world conservation. *Conservation Biology*, **20**, 658–669.

Roe, D., Nelson, F. & Sandbrook, C. (2009) *Community Management of Natural Resources in Africa: Impacts, Experiences and Future Directions*. Natural Resource Issues No. 18. International Institute for Environment and Development, London.

Rowell, A. (1996) *Green Backlash – Global Subversion of the Environmental Movement*. Routledge, London.

Sabatier, P.A. & Jenkins-Smith, H.C. (eds) (1993) *Policy Change and Learning - An Advocacy Coalition Approach*. Westview Press, Boulder, CO.

Sachedina, H.T. (2008) *Wildlife is our oil: conservation, livelihood and NGOs in the Tarangire ecosystem, Tanzania*. DPhil thesis, University of Oxford.

Salafsky, N., Margoulis, R., Redford, K.H. & Robinson, J.G. (2002) Improving the practice of conservation: a conceptual framework and research agenda for conservation science. *Conservation Biology*, **16**, 1469–1479.

Sheehan, N. (1989) *A Bright Shining Lie: John Paul Vann and America in Vietnam*. Vintage Books, New York.

Soulé, M E. (1985) What is conservation biology? *BioScience*, **35**, 727–734.

Sterner, T. (2002) *Policy Instruments for Environmental and Natural Resources Management*. Resources for the Future, Washington, DC.

Susskind, L. (2009) Twenty-five years ago and twenty-five years from now: the future of public dispute resolution. *Negotiation Journal*, **25**(4), 551–557.

Sutherland, W.J., Pullin, A.S., Dolman, P.M. & Knight, T.M. (2004) The need for evidence-based conservation. *Trends in Ecology and Evolution*, **19**, 305–308.

Tenga, R., Mattee, A., Mdoe, N., Mnenwa, R., Mvungi, S. & Walsh, M. (2008) *A Study on the Options for Pastoralists to Secure Their Livelihoods in Tanzania*. TNRF. www.tnrf.org/node/7487

Thornton, P., Burnsilver, S., Boone, R., & Galvin, K. (2006) Modelling the impacts of group ranch subdivision on agro-pastoral households in Kajiado, Kenya. *Agricultural Systems*, **87**, 331–356.

TNRF (2007) New regulations signed for all non-consumptive wildlife use in game reserves and on 1621 village lands. Tanzania Natural Resources Forum. www.tnrf6-org/node/6529

Waylen, K.A., Fischer, A., McGowan, P.J.K., Thirgood, S.J. & Milner-Gulland, E.J. (2010) Effect of local cultural context on the success of community-based conservation interventions. *Conservation Biology*, **24**, 1119–1129.

Western, D. (1994) Ecosystem conservation and rural development. In: *Natural Connections: Perspectives in Community-Based Conservation* (eds D. Western & M. Wright). pp.15–52. Island Press, Washington, DC.

Western, D., Russell, S. & Mutu, K. (2006) *The status of wildlife in Kenya's protected and non-protected areas*. Paper commissioned by Kenya's Wildlife Policy Review Team presented at the First Stakeholders Symposium of the Wildlife Policy and Legislation Review, 27–28th September, African Conservation Centre, Nairobi, Kenya.

4

Economic instruments for nature conservation

Christopher B. Barrett[1], Erwin H. Bulte[2], Paul Ferraro[3] and Sven Wunder[4]

[1]Charles H. Dyson School of Applied Economics and Management, Cornell University, Ithaca, New York, USA
[2]Development Economics Group, Wageningen University, The Netherlands
[3]Department of Economics, Andrew Young School of Policy Studies, Georgia State University, Atlanta, GA, USA
[4]Center for International Forestry Research, Rio de Janeiro, Brazil

'The labour of Nature is paid, not because she does much but because she does little. In proportion, as she becomes niggardly in her gifts, she exacts a greater price for her work. Where she is magnificently beneficent, she always works gratis.'

(Ricardo, 1817)

Introduction

The degradation of natural systems worldwide requires an appropriate policy response. The Millennium Ecosystem Assessment showed that degradation is progressing at a rapid pace, and argued that reduced flows of ecosystem services (especially regulatory and cultural services) are likely to affect adversely the welfare of future generations. The World Bank has also documented ongoing processes of ecosystem destruction, and couches its interpretation in terms of capital depletion. Capital comes in different forms, and economists often distinguish between man-made capital, human capital and natural capital. Ecosystems are a specific form of natural capital assets, and provide a host of services. They maintain a genetic library, preserve and regenerate soils, fix nitrogen and carbon, recycle nutrients, control floods, filter pollutants, pollinate crops, operate the hydrological cycles, etc. Degradation of ecosystems is much like the depreciation of physical capital (e.g. roads, buildings, machinery) but with two big differences: damages are frequently hard to reverse, and ecological processes tend to be non-linear, so that ecosystems can collapse abruptly, without much prior warning. Another important difference is

Key Topics in Conservation Biology 2, First Edition. Edited by David W. Macdonald and Katherine J. Willis.

that property rights over natural capital are often unclear, which substantially complicates the matching of costs and benefits within and across generations.

Empirical work demonstrates that aggregate *capital stocks* may be falling, even if *income* is growing (e.g. Dasgupta 2001). This may happen, for example, when forests are cut and the proceeds are used for consumption (rather than investment in alternative forms of capital, such as infrastructure). During the period 1970–2000 this was true for all Asian countries, except China. Africa was even worse off, with both income and capital stocks falling. Looking at national income as a proxy for welfare is misleading, as income growth financed by capital depletion ('selling the family silver') will be only a temporary reprieve. Capital stocks are a measure of true wealth, and are the basic source from which future income is derived. This is one of the main reasons why economists are interested in the conservation of nature. In addition to the conventional 'efficiency' concerns that are the bread and butter of economists, issues like intergenerational equity also come into play.

Conservation of nature might 'pay' from an economic perspective: it may be rational to conserve natural capital, rather than to convert it into alternative forms of capital or to consume it today. Some forms of natural resource use leave natural habitats more or less intact (think of reduced-impact logging or sustainable fishing) while others convert natural systems into something completely different (plantation cropping, intensive farming, etc.). An overview paper that compares the profitability of various forms of land use (Balmford et al. 2001) identifies many examples where investing in nature pays: sustainable management of ecosystems yields a greater discounted flow of benefits than going for short-term gains (where discounting a flow of benefits implies carrying future benefits forward to the present, so that they are commensurate with current benefits and costs – for this purpose economists employ a discount factor).[1]

To observe that nature conservation pays does not imply that there are no incentives for resource managers to destroy it (or to convert it into other forms of capital). An important reason why nature may be destroyed is that many of the valuable services and benefits nature provides are 'free' in the sense that they are not marketed. This is an obvious case of *market failure*. Even if nature is valuable, as long as it does not command a flow of money, its value tends to be ignored by private parties. Examples include the air-purifying qualities of ecosystems but also non-use values associated with biodiversity conservation. Environmental economists have developed methods to attach monetary values to ecosystem services, but that is not the same as money in the bank. The point is that even if the *social returns* to conservation and sustainable management are high, from the perspective of a *private* individual it may be better to ignore some of these benefits (i.e. those accruing to others) and focus on the small subset of tradable benefits only. Hence the conversion of nature into alternative forms of capital may be privately optimal and socially non-optimal at the same time. Examples include farmers who have incentives to intensively manage their lands to the detriment of the conservation of meadow birds, or whalers who believe that whales are more valuable in the bank than in the water.

Market failure implies that governments and other intermediaries can play a role to improve social welfare – the sum of pay-offs for producers and consumers (citizens). Since market failure occurs at multiple scales, from the local to the global level, the appropriate intermediary varies from case to case. However, history has

[1] Of course, counter-examples also exist; sometimes converting natural capital to financial capital provides higher returns. This is particularly probable when natural resource stocks are plentiful, and when the perspective of private investors or resource owners is adopted (see below).

taught us that there are many cases where public policies aggravate problems (or even create them). Examples of *policy failure* include 'perverse subsidies' – subsidies that are bad both for the economy and nature. The total amount of perverse subsidies, in both developed and developing countries, may be as high as US$2 trillion per year (Myers & Kent 2001). A vivid example of a perverse subsidy relevant to biodiversity conservation would be investment subsidies and tax holidays for fishers, or subsidies to convert tropical forests into rangeland.

Nevertheless, it is understood that government has a role to play in promoting nature conservation. In the past, two approaches to conservation have been extensively tried: traditional 'command-and-control' approaches emphasizing protected areas (bordered by fences or otherwise) and other disincentives to degradation, and an indirect approach called 'integrated conservation and development projects' (ICDP). ICDPs try to reconcile community development and conservation by promoting sustainable resource use or alternative sources of livelihood (which explains why ICDPs are occasionally referred to as 'conservation by distraction'). Both approaches have yielded some positive effects but it is becoming increasingly clear that a third wave of more direct, targeted incentives is needed to make further progress on the conservation and development front (Ferraro & Kiss 2002).

Increasingly, policy makers turn to economists for advice on how to design efficient and effective conservation policies. Cap-and-trade programmes are one solution, combining command and control with market mechanisms. This only works if the regulator is able to set the cap and has the power to force producers (traders) to play according to the rules. People now trade fish quotas the world over, trade the right to emit sulphur in the USA, and trade the right to emit carbon in Europe. Often markets for trading did not exist; they had to be created.

Consider the example of fishing. People have been trading fish for a long time, but trading the right to go out and catch fish is more novel. How does this work? Policy makers have to put an upper limit (or 'cap') on the quantity of fish that may be harvested. These harvest rights are transferred to the fishermen, through any of a variety of allocation mechanisms, who are then allowed to trade amongst themselves, thereby determining where the value of the rights is highest, i.e. who can harvest at lowest cost. Through trade, firms themselves decide who will harvest, this is not done by relatively uninformed regulators (who are only enforcing aggregate harvesting, so that the cap is not exceeded). Trading harvest quotas implies maximum flexibility in choosing the distributional implications of management, and also provides incentives to improve technologies.

Subsidy or tax systems are alternative market-based instruments. One important difference between cap-and-trade systems versus tax-and-subsidy schemes is that the former typically imply redistribution of rents within the sector (if rights are 'grandfathered' or simply given to existing producers), while taxes imply a transfer of rents between sectors, e.g. from producers or consumers to the government. A third market-based instrument is payments for ecosystem services (PES).

In this chapter we describe various well-known conventional and market-based instruments for nature conservation. In the light of the recent attention to PES as a tool to both promote conservation and alleviate poverty, we will pay special attention to this instrument. As will become evident, all instruments have advantages and disadvantages. However, intelligent application of economic instruments is likely to be a precondition for efficient and effective conservation of valuable natural capital.

Economic growth, poverty reduction and conservation

The relationship between economic growth, or poverty reduction more specifically, and environmental protection has long been central to economic debates about nature conservation. As economies grow, do increasing pressures on natural resources necessitate the use of economic instruments to protect the environment? Or will conservation emerge endogenously as a by-product of increased incomes and well-being?

For many years, roughly from the dawn of the modern environmental movement until the 1980s, most scholars and policy makers perceived a fundamental trade-off between environmental protection and economic growth. Conservationists commonly opposed economic development efforts and pushed for measures to address the environmental consequences of economic growth, in part by restricting or taxing resource use. Thus the gazetting of parks and protected areas was perhaps the central policy instrument for at least a century from the establishment of the world's first park, Yellowstone in the United States, in 1872.

By the mid-1980s, however, some evidence and theory had built behind the notion of synergies between environmental protection and economic development. As reflected in the well-known 1987 Report of the World Commission on Environment and Development, the so-called Brundtland Report, the prior belief in an intrinsic tension between growth and development gave way to a more hopeful belief that technological progress could stimulate productivity, growth and thus economic development while simultaneously reducing pressure on natural resources. This view gained support as researchers recognized the strong correlation between areas with high rates of endemism in fauna or flora and high levels of poverty (Fisher & Christopher 2007). The geographic co-location of efforts to reduce poverty through economic development and to conserve biodiversity naturally fed hypotheses that poverty and environmental degradation have common drivers that, if effectively addressed, could yield 'win–win' results on both fronts simultaneously. Indeed, appropriate alignment of environmental protection and poverty reduction goals has long been a concern of many in the conservation community (Macdonald et al. 2006, 2010).

This was the spirit behind the Convention on Biological Diversity (CBD), an international treaty agreed in 1993 that formally declared the conservation of biological diversity 'a common concern of humankind' and an integral part of the development process. The possibility of discovering commercially valuable biochemical substances through 'bioprospecting' is one of the many values the CBD associates with biodiversity conservation. Some early bioprospecting cases had been highly contentious, perhaps most notoriously the pharmaceutical multinational Eli Lilly's use of compounds found in Madagascar's rosy periwinkle to develop blockbuster cancer treatments, without compensation to the original source nation, one of the world's poorest.[2] Then in 1991 Merck reached an agreement that gave Costa Rica's National Institute for Biodiversity US$1 million and undisclosed royalties for any useful products derived from bioprospecting samples. The CBD then enshrined this principle of sharing of benefits between bioprospectors and source countries.

Bioprospecting has largely failed to deliver on its promise. While biodiversity is valuable in aggregate, the economic value of the marginal species and of habitat conservation associated with the marginal species is likely low. Either there are multiple species that can produce the same commercially valuable compound, rendering the marginal value of any one source species low, or the probability of finding a unique source species is very low.[3] In either case, the expected profitability of bioprospecting, and thus the payments one could reasonably expect firms to make in order to

[2] Dwyer (2008) describes this and several other alleged cases of 'biopiracy' in which multinational firms tapped indigenous ethnobotanical knowledge to generate commercially profitable products with little or no compensation to the communities from which insights and plant material were originally extracted.

[3] In addition, many species are native to various tropical countries at the same time, so that these countries would have to act as a cartel to be able to reap the bulk of the benefits.

help conserve biodiversity, are likewise low (Simpson et al. 1996; Barrett & Lybbert 2000; Costello & Ward 2006). It is little surprise to economists, therefore, that bioprospecting has never generated significant funds for conservation activities nor tangible poverty reduction benefits.

So-called integrated conservation and development projects (ICDPs) temporarily flourished in the 1990s for reasons similar to those that fuelled enthusiasm for bioprospecting. The essence of an ICDP is the merging of conservation with poverty reduction and rural development goals, commonly advanced by attempting to promote livelihoods compatible with sustainable resource use in, or more typically around, parks and protected areas or by providing compensatory transfers to rural residents who use resources sustainably (Brandon & Wells 1992). The core idea was that by giving local communities a stake in maintaining biodiversity, the prospects for conservation success would increase relative to approaches that relied on government, or some other external agent, imposing resource use constraints. The problem is that ICDP designs, and community-based conservation designs more broadly, have commonly been based on untested assumptions about both human and ecosystem response (Brandon & Wells 1992; Barrett & Arcese 1995). As a consequence, common designs can actually hasten, rather than avert, ecosystem collapse if the increased demand for natural products induced by local income growth outpaces the disincentive effects built into the quasi-contract between the conservation agency managing the ICDP and the local community, as Barrett & Arcese (1998) demonstrate using the example of the Serengeti Regional Conservation Strategy that introduced an ICDP intended to reduce poaching and thereby to reduce pressure on wild ungulates in northern Tanzania.

In spite of their obvious appeal, efforts to achieve 'win–win' solutions, such as bioprospecting or ICDPs, have rarely fully delivered on their promise. The general problem, well known to economists as the Tinbergen Principle, is that typically each policy objective requires its own policy instrument. Thus trying to achieve both conservation and development goals simultaneously through a single instrument is difficult at best. Economists typically think of 'win–win' objectives as too lofty. Our more modest criterion for improvement, Pareto improvement, only requires making at least one person better off without making any worse off. The Pareto criterion implies that one should consider as a success any effort that achieves biodiversity conservation without imposing suffering on the poor, or sustainable improvement in living standards without compromising ecosystem function (Barrett et al. 2011).

A parallel literature attempts to reconcile disparate observations that suggest economic development and environmental protection are sometimes complementary and sometimes antithetical processes. The Environmental Kuznets Curve (EKC) hypothesis posits an inverted U-shaped relationship between various measures of environmental degradation and per capita income. Under the EKC hypothesis, economic growth in poor societies is initially associated with environmental degradation leading to increased effluent and waste discharge, habitat loss, etc. but as incomes grow further, beyond some turning point, economic growth begins to drive environmental improvement as pressures to produce more natural resource-based commodities decrease and society comes to value and protect nature. The idea is that income growth brings greater capacity and willingness to pay for environmental protection, as well as technological advances that are less wasteful in resource use. After some unspecified threshold level of income, these pro-conservation effects start dominating the environmentally damaging effects of increased resource consumption and waste generation. The powerful implication of the EKC is that while economic growth may temporarily despoil the environment, development is not a long-term threat: it is complementary to, or even necessary for, long-term conservation success.

The empirical evidence in support of the EKC is, however, at best mixed. Most studies that support the EKC focus on measures of pollution (Grossman & Krueger 1995; Barbier 1997), although even those findings are contested (Harbaugh et al. 2002; Deacon & Norman 2006). Other studies focused on deforestation, energy use and other indicators typically find empirical little support for the EKC hypothesis (Koop & Tole 1999; Dasgupta et al. 2002; Mills & Waite 2009).

In the end, no compelling theoretical or empirical case exists for either universal trade-offs or synergies between economic and environmental goals. Complementarities between these desired outcomes appear feasible in many cases, but are too rarely achieved. The key determinants of successful synergy have proved rather elusive to identify. Because few observational or experimental data exist to describe and test hypotheses about interactions in closely coupled human and natural systems in developing countries, we have only a sparse set of convincing empirical studies, which collectively suggest tremendous site specificity of results, as well as dependence on the environmental problem at hand. Few truly generalizable lessons have appeared and associated policy debates often exhibit an ideological tone. The empirical evidence base and the theoretical literature underlying competing claims of trade-offs or synergies among economic and environmental goals yield few unambiguous, generalizable conclusions (Lee & Barrett 2001). Perhaps the best guidance we economists can offer at this juncture is that one should anticipate trade-offs but work for synergies and be content to make advances in one dimension without worsening conditions in the other. We cannot be confident that nature conservation will 'spontaneously' emerge as economies develop.

Regulatory approaches

Environmental economists distinguish between economic instruments for environmental conservation – manipulating incentives by altering relative prices — and so-called 'command-and-control' approaches. Textbook economics tend to view command and control with suspicion. In the domains of pollution abatement and resource extraction, it is well established that command and control is inefficient (i.e. does not achieve its objectives at lowest cost, partly because of information asymmetries between regulator and private parties) and fails to provide appropriate incentives for technological development (Perman et al. 2003). Nevertheless, while economic instruments such as taxes, subsidies and tradable permits are widely employed in pollution control and resource harvesting, command-and-control style regulation is still the dominant strategy in the domain of nature conservation.[4]

Global biodiversity conservation efforts rely heavily on protected areas, such as parks and reserves (Millenium Ecosystem Assessment 2005). In 1965, there were approximately 2,000,000 km^2 of national terrestrial protected areas. Forty years later, there were approximately 14,000,000 km^2, as well as millions more square kilometres of local, regional, indigenous and private protected areas. In total, more than 13% of the terrestrial surface of the earth is formally protected, and almost 4,000,000 km^2 more are managed as marine protected areas (IUCN and UNEP-WCMC 2009).

The theory of protected areas is simple: if governments legally restrict human access to an ecosystem or its components, they decrease anthropogenic pressures, such as conversion and hunting, that otherwise would have taken place. Despite the simple theory, protected areas may fail to generate as much *avoided* pressure as anticipated. First, governments may be unwilling or unable to enforce compliance with protected area restrictions (so-called 'paper parks'). The African 'bushmeat crisis', in which illegal hunting in protected forests rivals habitat loss as a threat to endangered animal species, is a good example of legal protection failing to

[4] Note that tradable permits are a combination of command and control and economic instruments; before trade can start, aggregate harvesting (or pollution) should be 'capped'.

stem human pressures. Second, land users may recognize the correlation between regulations and the presence of intact habitat and wildlife populations, and thus pre-emptively destroy habitat and wildlife to reduce the probability of future regulations (Lueck & Michael 2003). Third, protection may displace pressures as demand for the protected ecosystem products (e.g. timber, crops, fish/meat) induces spatial shifts in production to unprotected locations.

Fourth, protection is often assigned to low-pressure areas, whose ecological trajectory may be little affected by the assignment of protection. Protected areas are often assigned to low-pressure areas because such assignment is politically expedient: influential citizens and firms do not protest the protection of areas that are of little productive use to them. For example, Andam et al. (2008) found that, in the well-known protected area system of Costa Rica, more than 90% of *unprotected* forests are on high- or medium-productivity lands yet only 10% of *protected* forests comprise such lands. This so-called 'rock and ice' or 'high and far' phenomenon has been documented globally (Millennium Ecosystem Assessment 2005; Joppa & Pfaff 2009).

Despite the long history of protected areas and the vast conservation planning literature dedicated to determining where to put them, the empirical evidence base on their effectiveness in reducing pressures is scant and not particularly credible. Much of the empirical evidence falls into two categories: (1) measures of indicator trends (e.g. forest cover) inside protected areas, and (2) cross-sectional comparisons of indicators inside and outside protected areas. When no ecosystems would be destroyed in the absence of protection, when there is displacement or when protection is assigned conditional on baseline ecosystem and community characteristics that also affect human use of the ecosystems, such approaches tend to make protected areas look more effective than they are. For example, Andam et al.'s (2008) estimates of avoided deforestation from Costa Rica's protected area network range from 20% to 50% of the forest protected in an analysis that fails to control for baseline characteristics that affect both deforestation and where protected areas were established. The estimated range falls to 8–12% after controlling for these characteristics.

Fewer than a dozen studies of terrestrial and marine protected areas in developing and developed nations make credible efforts to control for confounding characteristics that affect both the environmental outcome indicators and where protection is assigned. They generally conclude that protection reduces ecosystem disturbance, but at much lower levels than conservation scientists might claim because protection tends to be assigned to ecosystems at below-average risk of disturbance. Only one study considers the potential displacement as a result of protection (as well as potentially positive enforcement spillovers; Andam et al. 2008) and finds no evidence of such displacement. Little empirical research has been conducted on the effects of heterogeneous land-use restrictions across a protected area system (e.g. Sims 2010) or of different protected area manager types, such as comparisons of protected areas run by government, non-profit organizations and indigenous communities (Somanathan et al. 2009). No study has examined the cost-effectiveness of protected areas compared to other conservation strategies.

Likewise, most studies estimating the socio-economic impact of protected areas are simple case study narratives or *ex ante* projections based on extrapolations of historical economic activity, which merely confirm that protected areas are established near poor people and restrict access. Fewer than a half dozen studies observe socio-economic status and control for confounding factors. Most find no adverse impacts, on average, largely because protection is assigned to unproductive areas (for references, see Andam et al. 2010 and Sims 2010).

The other most common regulatory approaches include land-use restrictions and zoning laws, such as Brazil's laws that dictate how much on-farm forest can be deforested, and species-focused protection laws, such as the United States Endangered Species Act (ESA) or the Convention on International Trade in Endangered Species (CITES). As in the case of protected areas, the theory underlying these regulatory approaches is quite simple but lack of compliance, pre-emptive destruction, spatial displacement and poor administrative targeting can reduce their effectiveness. Likewise, the evidence base for these regulatory approaches is weak. In one study that controls for some aspects of the non-random assignment of protection, Ferraro et al. (2007) find no effect of the US ESA on species recovery in the absence of substantial recovery funding. In fact, their results imply that in the absence of such funding, legal protection leads to adverse consequences for species. One possible explanation is that the ESA encourages private landowners to undertake pre-emptive actions to eliminate listed species, and thus regulations, on their land (the so-called 'shoot, shovel and shut up' response – see Lueck and Michael 2003). Species-specific funding overcomes these perverse incentives by creating a sufficient level of perceived monitoring and thus credible enforcement.

Regulatory approaches have increasingly been combined with economic incentives to harness the advantages of each approach, building a bridge between command and control and economic approaches to conservation. For example, developed nations have experimented with Tradable Development Rights (TDR), which combine a regulatory cap on the amount of habitat that can be converted with tradable development permits to encourage land users to meet the regulatory target at least cost. In other nations, protected areas are leased from local communities (e.g. Richtersveld National Park in South Africa; National Park of American Somoa). In Costa Rica, a legal ban on deforestation exists side by side with forest protection incentive payments. With the exception of a few TDR programmes in developed countries that are focused on open space or farmland (rather than habitat) and Costa Rica's payment system (see review in Pattanayak et al. 2010), little evidence exists about the performance of these hybrid regulatory-incentive approaches.

Payments for environmental services

What form might a pure economic incentive approach to conservation take? In recent years policy makers and academics have embraced the payments for environmental services (PES) approach to conservation. For example, Dickman et al. (2011) studied PECS (Payments to Encourage Co-existence with, for example, big predators). A key point is that one needs a blend of mechanisms, and that amongst these payments for conservation deliverables are increasingly looking promising.

Wunder (2008a) defined a payment for an ecosystem service as (1) a *voluntary* transaction where (2) a *well-defined* ecosystem service (ES) is (3) being bought by a (minimum one) ES buyer from (4) a (minimum one) ES supplier, (5) if and only if the provider secures ES provision. The transaction should be voluntary and the payment should be conditional on the service being delivered; this is no handout to the poor and needy but payment for a genuine service. Note, however, that paying for an ecosystem service is not necessarily the same as trading nature on a market. Markets may play a role, as will become clear, but because many of the ecosystem services come in the form of a public good, we cannot rely on markets alone. Governments and intergovernmental organizations are essential to the package.

It is useful to work towards some form of PES classification system. We can distinguish between cases where the ecosystem service benefits a small group of agents versus cases where it benefits a large and presumably more diverse group of agents. If the number of people increases, for example because we are considering regulatory services that affect everybody, the ecosystem services start to resemble a club good or a public good. Another useful distinction is between cases where service 'suppliers' and 'demanders' are geographically located close together, and those cases where they are

Table 4.1 Different PES cases

	Local service linkage	**Cross-border service linkage**
Few beneficiaries	*CASE I*: Water and tourism companies	*CASE III*: International users of watershed services
Many beneficiaries	*CASE II*: Urban drinking water users	*CASE IV*: Biodiversity conservation and climate change mitigation

not. Table 4.1 provides a few examples, highlighting the broad variety of possible cases.

Case I (local, few users) occurs, for example, when a micro-watershed holds a single main corporate user (e.g. a hydroelectric company or brewery), or when an ecotourism company pays a local community for not hunting in a forest used for wildlife viewing, as happened in the Cuyabeno Wildlife Reserve in Ecuador (Wunder 2000). Many such examples exist, especially in watershed management (e.g. Porras et al. 2008; Wunder et al. 2008). Case II (local, many users) is common in macro-watersheds and/or cases with multiple users, e.g. drinking water users in a mega-city combined with irrigating farmers and water-using companies. The challenge of aggregating user interests into payment flows can be met by user associations (the case of 'club goods'), by water utilities (forcing individual consumers to pay) or the public sector (municipalities). Municipal watershed schemes have been mushrooming particularly in Latin America (Southgate & Wunder 2009). Another well-cited example is PES in the Catskills watershed, from which New York City gets most of its drinking water (Ashendorff et al. 1997).

Case III (few users of a cross-border ranging environmental service) is the rarest category. There are some ongoing efforts to generate PES systems, for example for conservation of the Danube river system and its biodiversity, or avoiding the siltation of the Panama Canal, which would hurt the international shipping industry. However, in most cases international payments also link to global-level environmental services (Case IV), such as carbon trading around the climate change mitigation services or biodiversity conservation initiatives. Just like in Case II, PES schemes will only emerge here if the multiple beneficiaries find means and ways to aggregate their interests into a functional service-buying organ. This can be the case of private companies (e.g. the Dutch FACE Foundation buying carbon credits), multilateral organizations (e.g. the Global Environmental Facility paying for carbon- and biodiversity-enhancing actions), non-governmental organizations (e.g. a nature conservation organization running a biodiversity protection PES) or national governments (e.g. Costa Rica's national PES paying for biodiversity and carbon benefits).

As we can see, the universe of existing and emerging PES initiatives is defined by both the range of the underlying externality (local versus global) and the number of beneficiaries (few versus many). Empirically, we see among real-world PES a clear dominance of local-range club services (watersheds, recreation) and global-range, quasi-public services (biodiversity, carbon). Since most PES schemes emerge from the buyer side, aggregating multiple service users into functional service buyers thus becomes a main challenge for PES developers. For instance, nature conservation organizations like the Nature Conservancy (TNC) and Conservation International (CI) have in Latin America invested in starting up water funds and payment schemes designed to jointly conserve watersheds that are strategic for both water and biodiversity purposes (for examples, see Wunder & Wertz-Kanounnikoff 2009).

Why has PES lately received so much attention? Basically, decision makers worldwide have increasingly realized that politically convenient win–win projects, allegedly benefiting both local land users and the environment, are hard to implement in practice. Environmental degradation tends to have a local economic rationale, and counteracting it is hence associated with local losers. PES are a tool for natural resource management that explicitly recognizes these contradictions, and uses compensation to bridge between losers and winners.[5] Secondly, by focusing on a specific utility to beneficiaries, rather than nature in its broader and abstract form, PES are also believed to have the potential of eventually becoming more efficient than more indirect conservation tools. Thirdly, compensating the losers is also an equitable way of doing conservation, so PES should have a greater probability of being adopted than command-and-control approaches that directly inflict costs on environmentally degrading stakeholders. Finally, compared to the bipolar character (legal versus illegal) of most command-and-control measures or the predetermined investments related to ICDPs, PES are a more flexible conservation tool, since the agreed-upon compensations can be renegotiated in response to the rapidly changing benefits and costs in a dynamic world.

Economic incentives as the road forward?

The discussion until now suggests that economic incentives are a valuable instrument in the toolkit of policy makers interested in promoting sustainable resource management and nature conservation. However, this chapter would not be complete without a short discussion of several challenges that should be considered. Some of these may, however, apply to virtually all forms of environmental regulation. For example, often we don't fully understand the 'production function' of ecosystem services. Consider an effort to re- or afforest agricultural land in order to obtain watershed benefits. How does forest conservation contribute to various hydrological benefits? This is a topic clouded by myths. In the absence of sufficient ecological understanding, both command-and-control and incentive-based regulatory efforts may be doomed to fail.

Conservation efforts may also be undermined by slippage or leakage. This captures the idea that 'successful' conservation schemes may lead to their own demise because they trigger behavioural changes. Consider the effect of a policy where African landowners are paid to remove fences (enabling seasonal migration of wildlife). If such transfers enable landowners to purchase or hold more livestock, then competition for food between wildlife and cows or goats intensifies and the conservation gains from new habitat are eroded (Bulte et al. 2008). Similarly, a law prohibiting fuelwood uses from natural forests could increase pressures on plantation forests, which is another example of leakage.

Occasionally leakage may occur via the market or other mechanisms. If farmers are encouraged to convert farmland back into nature, then local food prices may increase if local markets are not integrated into regional or national ones, inducing other farmers to convert new areas of habitat into agricultural fields. So, gains of habitat in one place may be (partially) offset by losses elsewhere (e.g. Wunder 2008b).

In addition to these generic challenges, each type of intervention has its own strengths and weaknesses. Since PES is gaining momentum as the instrument of the future, it is worth paying special attention to some specific challenges relevant for this tool. First, farmers (and plots) are heterogeneous, and information about local conditions or preferences is privately held. That is, farmers have an information advantage over the regulator that they can use to their advantage. They are better informed about the quality of their land, and will strategically try to retire their least valuable lands, or the ones that they would have retired even in the absence of the payment ('zero additionality'). Generally, this asymmetrical information will give rise to so-called information rents, and these will accrue to the better informed party – the

[5] Economists often refer to the Coase Theorem in this context: PES try to facilitate exchanges that internalize externalities where property rights are clearly defined and where transactions costs are relatively low.

farmers (see Ferraro 2008). The cost per unit of nature conserved therefore goes up. Basically, the asymmetry problem occurs as a mirror of the voluntariness of PES. Because farmers themselves have to decide to sign up, there may be an adverse selection problem that reduces additionality. These problems can be overcome by increasing environmental service buyers' information and by careful spatial targeting of PES, often combined with differentiated payment rates per land unit.[6]

There can also be a problem related to the voluntariness of payments on the buyer side. It is not always obvious who will pay for the ecosystem services. This is less of a problem for the cases with 'few demanders' who can agree among each other, but negotiations may fail in the presence of public goods with many users, where transaction costs are high and free-riding incentives prevail. When considering services within a nation's borders, it is thus obvious that governments can play an important co-ordinating role. Using tax dollars to pay for public goods can in those cases be a solution, so that the government purchases the ecosystem service from the supplier on behalf of society at large. An example is agri-environment schemes in many European countries. When we move to the global sphere, this role could be played by international organizations but we currently lack good institutional mechanisms to broker deals between suppliers of ecosystem services and the rest of the world.

Payments for environmental services require that land or resource managers can be identified who, through the provided incentive, will change behaviour, compared to business as usual, and thus provide additional environmental services. In developing countries, however, rules of tenure and access to land and natural resources resource may not be clearly defined, especially in agricultural frontier areas where weak governance and environmentally sensitive land areas overlap. It may thus not be possible to find out whom to pay, in ways that are legitimate and efficient, in which case PES may not be feasible.

Finally, it is important to remember the difficulty of simultaneously alleviating poverty and promoting conservation, per the Tinbergen Principle mentioned earlier. While PES may be best suited to achieving such win–win outcomes, compared to other regulatory approaches, unfortunately the link is not automatic.[7] Some 1.2 billion people are living on less than a dollar a day (Chen & Ravallion 2007), and many of these poor are found in rural areas, especially in marginal areas like steep slopes of the upper watershed that should be tackled from a PES perspective. Indigenous groups and impoverished rural communities sometimes own or manage sizeable shares of tropical forests. The distribution of gains is obviously directly related to the *ex ante* distribution of control over resources generating the services for which compensation is paid.

Regarding PES and poverty, we can distinguish between two perspectives: (1) the micro-perspective of payments and the fate of poor households (participating in the scheme or otherwise), and (2) the macro-perspective of nature conservation as part of an economically viable development trajectory for the economy as a whole. Not much work has been done to address the issue of PES and the national economy (presumably because PES efforts are currently too small to have significant aggregate impacts). As just one example, Norton-Griffiths

[6] Additional issues emerge when, from a nature conservation perspective, we care about securing congruent plots of land. How can we convince neighbouring farmers to retire adjacent plots of land for the PES scheme? Presumably this will also involve additional costs (in the form of an agglomeration bonus). For example, aggregated farmers deliver better moth biodiversity under agri-environment schemes – see Chapter 14.

[7] Note that some of the effects discussed above in the context of leakage are also relevant when analysing the impact on poverty. The poverty effects are generally ambiguous, depending on whether labour demand goes up or down (affecting local wages), and whether or not the price of food crops changes (affecting the purchasing power of households depending on such crops).

& Southey (1995) estimated the opportunity cost of biodiversity conservation to the Kenyan economy at 2.8% of national income. However, we may expect that nature protection produces little in terms of spillover benefits from which other sectors in the economy benefit. Matsuyama (1992) developed a model of economic growth of a two-sector economy: the manufacturing sector, with increasing returns to scale at the sector level, and the agricultural sector, subject to constant returns to scale. If we consider 'production of ecosystem services' on a par with agriculture, PES could have the adverse and unintended effect of 'locking' economies into economically suboptimal development paths.[8]

The micro-perspective gives rise to somewhat more optimism, but here too we should be cautious not to expect too much (e.g. Pagiola et al. 2005; Wunder 2008a). A precondition for beneficial effects is that the PES programme actually reaches the poor: they should (1) be in the 'right place', (2) want to participate (e.g. it should 'fit' into the poor's prevailing livelihoods), and (3) be able to participate (e.g. they should be able to make the necessary investments, have sufficiently secure tenure, etc.). It is quite easy to design programmes that fail on one of these conditions, so that PES does not help in fighting poverty. Next, *if* PES reaches the poor, we may infer that it will make them better off. Participation in a PES scheme is voluntary so participants should gain, at least in expectation. The proper net measure of benefits for the ecoservice seller is payment received minus opportunity cost. When payments are aligned with opportunity cost, the net benefits may be small (although the stabilizing effect of payments on household income is welcome). But additional effects may exist. In the case of imperfect tenure security, PES schemes could potentially induce powerful stakeholders to 'muscle out' poor households; conversely PES could help poor people to consolidate tenure *vis-à-vis* external introducers, and stimulating better community organization. Empirically, the latter, positive impacts have been dominating (Pagiola et al. 2005; Wunder 2008a).

In the light of these challenges it is perhaps no surprise that the empirical evidence on the efficacy of PES is rather ambiguous.

[8] But note that, in other cases, where PES pay for the restoration of environmental services (e.g. on degraded lands), it may also have positive multiplier effects, turning this argument on its head.

Application: REDD

Climate change mitigation is high on the international agenda, and so is the most favoured forestry response – Reduced Emissions from Deforestation and forest Degradation in developing countries (REDD). In the latest multilateral negotiations around the Kyoto Protocol, the concept has been widened to REDD+, where forest regeneration and enhancement of the carbon stocks of standing forests are also being integrated. The Stern Review (Stern 2006) and subsequent work done in particular by the consultancy firm McKinsey (see www.redd-monitor.org/tag/mckinsey/ for a critical assessment) have pointed to REDD+ projects as some of the cheapest mitigation actions that could be taken in the short run: buying out extensive converted uses of forestlands with low profitability is allegedly much cheaper than many of the mitigation options in other carbon-emitting sectors. Hence, with the UN, the World Bank and other agencies as facilitators, many tropical countries are currently getting ready to receive REDD+ assistance on a supposedly large scale.

Conceptually, we can think of REDD+ as a kind of international PES scheme, with conditionality as a key design feature (Angelsen et al. 2009). The idea is that forested countries receive periodic economic compensations for deforesting less than they otherwise would have done. The size of REDD+ payments will depend on the achievements the country has made in stabilizing its forest cover, compared to a (yet to be defined) baseline counterfactual. Some incipient bilateral mechanisms of this type are already in place, such as the agreements Norway has struck with Brazil, Tanzania and Indonesia. Only time will tell how seriously the *quid pro quo* design will be taken by donors and recipients. As with government-co-ordinated PES programmes, the political costs for donors of withholding money are often very high, and therefore some non-compliance is likely to be ignored.

How would tropical REDD+ recipient countries then make sure that deforestation is actually being curbed, so as to qualify for a continuous flow of conditional REDD+ funding? Certainly, PES agreements with landowners on the ground could be one possibility for REDD+ recipient countries to directly pass on carbon credits (and land-use obligations) to the subnational level. However, in those cases where forestland tenure is ill defined, PES may not be the right solution. One option will be facilitating investments that better enhance forest conservation, such as land reforms and improved governance. Yet disincentive measures, such as increased forest law enforcement, or creating new protected areas (and better managing existing ones) are also options to spend REDD+ money in ways that can actually curb forest clearing. Finally, REDD+ resources also make it possible to achieve leverage at the policy level, for instance compensating municipalities for not receiving a new road into the forest or incentivizing the cattle farmer association to co-operate on REDD+ with subsidized credits for pasture intensification. The international debate has so far focused more on benefit-sharing questions, widely bypassing the fact that deforestation will not just stop by itself, making targeted investments necessary to make REDD+ work.

Conclusions

The conversion of natural capital into alternative forms of capital continues, threatening the supply of valuable environmental services. While part of this conversion process is economically rational (alternative forms of capital simply yield greater discounted returns), undoubtedly some destruction is socially wasteful and may be accelerated by market or policy failures. From both intergenerational welfare and ethical perspectives (including non-anthropocentric concerns), policy interventions to slow down or reverse natural degradation are warranted.

But which policy instruments should be used for this purpose? Traditionally, command-and-control style regulation has been dominant. In the domain of biodiversity conservation, regulation has mostly relied on protected areas, banning certain forms of land and resource use altogether. However, mounting pressures, including population growth and economic expansion, have spurred increasingly active searches for alternative, supplementary approaches to conservation. Economic instruments are one such supplement, and in general a promising one, we believe. Standard economic instruments, such as taxes, subsidies and tradable permits, have been applied (with varying success) in the domains of pollution control or the regulation of resource extraction. One specific economic instrument, payments for environmental services, has recently entered the domain of biodiversity conservation.

Payments for environmental services have many advantages, and their voluntary nature makes them more socially palatable. They can be applied at the local level between private parties, for example when local consumers or an electricity company compensate upland farmers for providing valuable watershed services. The same principles can also be applied at the global level, even if this typically involves public parties as well as private ones. For example, via REDD+ schemes PES may come to play a significant role in global efforts to mitigate the adverse consequences of the greenhouse effect.

Obviously, economic instruments are not a panacea. Leakage may partially erode conservation gains, information asymmetries between private parties and regulators may introduce information rents (raising the costs per unit of nature conserved) and institutional weaknesses, especially at the international level, may limit the scope for large-scale and successful application. However, the flexibility of economic instruments and the implied efficiency benefits suggest they will have an increasingly large role to play in the portfolio of conservation activities. Moreover, many of the potential shortcomings of economic instruments can be overcome through specific targeting and more fine-grained regulation, as we learn to improve the design of these instruments. Yet for this purpose, we will also need rigorous impact assessments and modelling to better understand 'what works when, and for whom'.

References

Andam, K,, Ferraro, P., Pfaff, A., Sanchez-Azofeifa, G. & Robalino, J. (2008) Measuring the effectiveness of protected area networks in reducing deforestation. *Proceedings of the National Academy of Sciences*, **105**(42), 16089–16094.

Andam, K, Ferraro, P., Sims, K., Healy, A. & Holland, M. (2010) Protected areas reduced poverty in Costa Rica and Thailand. *Proceedings of the National Academy of Sciences*, **107**(22), 9996–10001.

Angelsen, A., Brockhaus, M., Kanninen, M., Sills, E., Sunderlin, W. & Wertz-Kaunonnikoff, S. (eds) (2009) *Realizing REDD: National Strategy and Policy Options*. Center for International Forestry, Bogor, Indonesia.

Ashendorff, A., Principe, M., Seeley, A., *et al.* (1997) Watershed protection for New York City's supply. *Journal of American Water Works Assessment*, **89**, 75–88.

Balmford, A., Bruner, A., Cooper, P. *et al.* (2001) Economic reasons for conserving wild nature. *Science*, **297**, 950–953.

Barbier, E.B. (1997) Environmental Kuznets Curve special issue. *Environment and Development Economics* **2**(4).

Barrett, C.B. & Arcese, P. (1995) Are integrated conservation–development projects (ICDPs) sustainable? On the conservation of large mammals in Sub-Saharan Africa. *World Development*, **23**(7), 1073–1085.

Barrett, C.B. & Arcese, P. (1998) Wildlife harvest in integrated conservation and development projects: linking harvest to household demand, agricultural production and environmental shocks in the Serengeti. *Land Economics*, **74**(4), 449–465.

Barrett, C.B. & Lybbert, T.J. (2000) Is bioprospecting a viable strategy for conserving tropical ecosystems? *Ecological Economics*, **34**(2), 293–300.

Barrett, C.B., Travis, A.J. & Dasgupta, P. (2011) On biodiversity conservation and poverty traps. *Proceedings of the National Academy of Sciences*, **108**, 13907–13912.

Brandon, K. & Wells, M. (1992) Planning for people and parks: design dilemmas. *World Development*, **20**(4), 557–570.

Bulte, E.H., Boone, R., Stringer, R. & Thornton, P. (2008) Elephants or onions? Paying for nature in Amboseli, Kenya. *Environment and Development Economics*, **13**, 395–414.

Chen, S. & Ravallion, M. (2007) Absolute poverty measures for the developing world, 1991–2004. *Proceedings of the National Academy of Sciences*, **104**, 16757–16762.

Costello, C. & Ward, M. (2006) Search, bioprospecting, and biodiversity conservation. *Journal of Environmental Economics and Management*, **52**(3), 615–636.

Dasgupta, P. (2001) *Human Well-Being and the Natural Environment*. Oxford University Press, Oxford.

Dasgupta, S., Laplante, B., Wang, H. & Wheeler, D. (2002) Confronting the environmental Kuznets Curve. *Journal of Economic Perspectives*, **16**(1), 147–168.

Deacon, R.T. & Norman, C.S. (2006) Does the environmental Kuznets curve describe how individual countries behave? *Land Economics*, **82**(2), 291–315.

Dickman, A.J., Macdonald, E.A. & Macdonald, D.W. (2011) Paying for predators: a review of financial instruments to encourage human–carnivore coexistence. *Proceedings of the National Academy of Sciences.*, **108**(34), 13937–13944.

Dwyer, L. (2008) Biopiracy, trade, and sustainable development. *Colorado Journal of International Environmental Law and Policy*, **19**(2), 219–259.

Ferraro, P. (2008) Asymmetric information and contract design for payments for environmental services. *Ecological Economics*, **65**(4), 810–821.

Ferraro, P. & Kiss, A. (2002) Direct payments to conserve biodiversity. *Science*, **298**, 1718–1719.

Ferraro, P., McIntosh, C. & Ospina, M. (2007) The effectiveness of listing under the U.S. Endangered Species Act: an econometric analysis using matching methods. *Journal of Environmental Economics and Management*, **54**(3), 245–261.

Fisher, B. & Christopher, T. (2007) Poverty and biodiversity: measuring the overlap of human poverty and the biodiversity hotspots. *Ecological Economics*, **62**, 93–101.

Grossman, G. & Krueger, A. (1995) Economic growth and the environment. *Quarterly Journal of Economics*, **110**(2), 353–377.

Harbaugh, W.T., Levinson, A. & Wilson, D. (2002) Reexamining the empirical evidence for an environmental Kuznets curve. *Review of Economics and Statistics*, **84**(3), 541–551.

IUCN & UNEP-WCMC (2009) *The World Database on Protected Areas (WDPA)*. UNEP-WCMC, Cambridge.

Joppa, L. & Pfaff, A. (2009) High and far: biases in the location of protected areas. *PLoS ONE*, **4**(12), e8273.

Koop, G. & Tole, L. (1999) Is there an environmental Kuznets curve for deforestation? *Journal of Development Economics*, **58**(1), 231–244.

Lee, D. & Barrett, C. (eds) (2001) *Tradeoffs or Synergies? Agricultural Intensification, Economic Development and the Environment*. CABI Publishing, Wallingford, UK.

Lueck, D. & Michael, J. (2003) Preemptive habitat destruction under the Endangered Species Act. *Journal of Law and Economics*, **XLVI**, 27–60.

Macdonald, D.W., Collins, N.M. & Wrangham, R. (2006) Principles, practice and priorities: the quest for alignment. In: *Key Topics in Conservation Biology* (eds D.W. Macdonald & K. Service). Blackwell Publishing, Malden, MA.

Macdonald, D.W., Loveridge, A.J. & Rabinowitz, A. (2010) Felid futures: crossing disciplines, broders, and generations. In: *The Biology and Conservation of Wild Felids* (eds D.W. Macdonald & A. Loveridge). Oxford University Press, Oxford.

Matsuyama, K. (1992) Agricultural productivity, comparative advantage and economic growth. *Journal of Economic Theory*, **58**, 317–334.

Millennium Ecosystem Assessment (2005) *Ecosystems and Human Well-Being: Policy Responses*. Island Press, Washington, DC.

Mills, J.H. & Waite, T.A. (2009) Economic prosperity, biodiversity conservation, and the environmental Kuznets curve. *Ecological Economics*, **68**(7), 2087–2095.

Myers, N. & Kent, J. (2001) *Perverse Subsidies: How Misused Tax Dollars Harm the Environment and the Economy*. Island Press, Washington, DC.

Norton-Griffiths, M. & Southey, C. (1995) The opportunity costs of biodiversity conservation in Kenya. *Ecological Economics*, **12**(2), 125–139.

Pagiola, S., Arcenas, A. & Platais, G. (2005) Can payments for environmental services help reduce poverty? An exploration of the issues and the evidence to date from Latin America. *World Development*, **33**, 237–253.

Pattanayak, S., Wunder, S. & Ferraro, P.J. (2010) Show me the money: do payments supply ecosystem services in developing countries? *Review of Environmental Economics and Policy*, **4**(2), 254–274.

Perman, R., Ma, Y., McGilvray, J. & Common, M. (2003) *Natural Resource and Environmental Economics*, 3rd edn. Pearson, Harlow.

Porras, I., Grieg-Gran, M. & Neves, N. (2008) All that glitters: a review of payments for watershed services in developing countries. In: *Natural Resource Issues No. 11*. International Institute for Environment and Development, London.

Ricardo, T. (1817) *On the Principles of Political Economy and Taxation*. John Murray, London.

Simpson, R.D., Sedjo, R.A. & Reid, J.W. (1996) Valuing biodiversity for use in pharmaceutical research. *Journal of Political Economy*, **1041**(1), 163–185.

Sims, K. (2010) Conservation and development: evidence from Thai Protected Areas. *Journal of Environmental Economics and Management*, **60**, 94–114.

Somanathan, E., Prabhakar, R. & Mehta, B. (2009) Decentralization for cost-effective conservation. *Proceedings of the National Academy of Sciences*, **106**, 4143–4147.

Southgate, D., & Wunder, S. (2009) Paying for watershed services in Latin America: a review of current initiatives. *Journal of Sustainable Forestry*, **28**(3-4), 497–524.

Stern, N. (2006) *Stern Review: The Economics of Climate Change*. Cambridge University Press, Cambridge.

Wunder, S. (2000) Ecotourism and economic incentives – an empirical approach. *Ecological Economics*, **32**, 465–479.

Wunder, S. (2008a) Payments for environmental services and the poor: concepts and preliminary evidence. *Environment and Development Economics*, **13**(3), 279–297.

Wunder, S. (2008b) How do we deal with leakage? In: *Moving Ahead with REDD. Issues, Options and Implications* (ed. A. Angelsen). Center for International Forestry, Bogor, Indonesia.

Wunder, S., & Wertz-Kanounnikoff, S. (2009) Payments for ecosystem services: a new way of conserving biodiversity in forests. *Journal of Sustainable Forestry*, **28**(3),576–596.

Wunder, S., Engel, S. & Pagiola, S. (2008) Taking stock: a comparative analysis of payments for environmental services programs in developed and developing countries. *Ecological Economics*, **65**(4), 834–852.

5

Tackling unsustainable wildlife trade

Adam J. Dutton[1], Brian Gratwicke[2], Cameron Hepburn[3], Emilio A. Herrera[4] and David W. Macdonald[5]

[1]Wildlife Conservation Research Unit, Department of Zoology, Recanati-Kaplan Centre, University of Oxford, Oxford, UK
[2]Center for Species Survival, Smithsonian Conservation Biology Institute, Washington, D.C., USA
[3]Smith School of Enterprise and the Environment and James Martin Institute, Said Business School, University of Oxford, Oxford, UK
[4]Departamento de Estudios Ambientales, Universidad Simón Bolívar, Caracas, Venezuela
[5]Wildlife Conservation Research Unit, Department of Zoology, Recanati-Kaplan Centre, Tubney House, University of Oxford, UK

'Once Confucius was walking on the mountains and he came across a woman weeping by a grave. He asked the woman what her sorrow was, and she replied, "We are a family of hunters. My father was eaten by a tiger. My husband was bitten by a tiger and died. And now my only son!". "Why don't you move down and live in the valley? Why do you continue to live up here?" asked Confucius. And the woman replied, "But sir, there are no tax collectors here!". Confucius added to his disciples, "You see, a bad government is more to be feared than tigers".'

(Lin Yutang)

Introduction

Humans have bartered wildlife products for goods and services since before currency even existed. Wildlife is used for traditional medicine, wild meat for human consumption, exotic pets, luxury fashion items, trophy hunting and tourist curios. Patterns of consumption vary geographically with human cultural landscapes and patterns of wildlife distribution. The size and scope of wildlife trade networks vary today from hunting and gathering barter systems to massive commercial industries like the commercial fishing industry that captures 75 million tonnes of fish from wild stocks each year (FAO 2010). Illegal wildlife trafficking networks of products like rhino horn, caviar, tiger parts and wild birds is estimated to be worth between US$5 billion and US$20 billion annually (in

Key Topics in Conservation Biology 2, First Edition. Edited by David W. Macdonald and Katherine J. Willis.

2011), ranking alongside illegal drugs and arms trafficking as one of the most lucrative illicit industries (Wyler & Sheikh 2009).

Wildlife resources are renewable and it is theoretically possible to harvest wildlife sustainably, meeting the needs of the present without compromising the ability of future generations to meet their own needs (WCED 1987). Yet as wild populations collapse, it is clear that humanity is running an unsustainable environmental deficit and we are on course for bankruptcy as wild populations collapse, fundamentally damaging the global economy (Sukhdev et al. 2010). To name a few: fisheries are overfished, with 80% of European Union (EU) fish stocks overexploited or at risk of collapse (Press Association 2011); the bushmeat trade in combination with habitat loss is likely to cause the extinction of sensitive species, including some primates (Wright & Muller-Landau 2006; Fa & Brown 2009); and trade in their bones and skins is the greatest threat to the remaining effective global population size of 1225–2026 tigers (*Panthera tigris*) (Chundawat et al. 2010).

Illegal wildlife trade can have adverse impacts beyond the direct effects of overharvesting on the sustainability of the wildlife resource: for example, wildlife trade can spread disease, and this also needs to be considered alongside sustainability in ecological risk assessments (Karesh et al. 2005; Gratwicke et al. 2009). We also know that illicit trades including alcohol and drugs or prostitution share the same networks, and profits from illicit wildlife trade may also help these organized crime syndicates to grow (Wyler & Sheikh 2009). Given these problems, it is important that we are able to appraise qualitatively the policies available to control these trades.

It is important to understand how human, governmental and technical limitations and abilities associated with a trade policy can best be ameliorated or employed to protect a wild resource. This chapter describes and illustrates the issues surrounding wildlife trade policies. Wildlife trades are complex and heterogeneous, requiring different management approaches although the issues discussed may be common. It should be possible to develop a suite of investigations based on the issues raised in this chapter to estimate the impacts of each policy on wild populations, each being judged relative to the others. One key message which must be conveyed is that total protection of the wild resource may prove impossible (at least in the short term) under any trade regulation system. We need a quantitative appraisal of impacts so that investigators might be better placed to find what we might call 'the least leaky bucket'.

Law and Policy

Trade bans versus regulated trade

The two primary legal instruments used to address unsustainable exploitation are outright bans and conditional allowances permitting limited trade. International wildlife trade is primarily regulated through the Convention on the International Trade in Endangered Species (CITES) under the United Nations. Like most UN conventions, CITES is a legally binding international framework that countries adhere to voluntarily and policies are debated at regular Conference of the Parties (CoP) meetings. Countries must decide how to reconcile their CITES commitments with domestic laws, and are responsible for enforcing any CITES commitments. For example, in the USA, the US Fish and Wildlife Service is responsible for issuing CITES permits and inspecting international wildlife shipments. In the UK there is a blend of enforcement from the UK Borders Agency, administration through the Department for Environment, Food and Rural Affairs (DEFRA) and other support from the Joint Nature Conservation Council (JNCC). However, being in the EU, the UK is also subject to a further set of international wildlife trade rules.

While the individual EU nations are signatories of CITES, the EU also has a separate set of wildlife trade rules. EU wildlife trade laws largely reflect CITES obligations but are stricter than CITES in some cases; the EU also has its own scientific authority with the power to ban trade which it deems unsustainable but which is not yet banned by CITES. For instance, in 2005 (fully implemented in 2007), the EU banned trade in wild birds (CITES has no such blanket ban). The initial ban in 2005 was enacted to prevent the spread of the H5N1 bird flu strain (BBC 2007). However, the permanent ban was driven by a mixture of concern for animal welfare (60% of birds die in transit (Anonymous 2006)) and conservation, as the trade was deemed to be unsustainable (Cooney & Jepson 2006).

If CITES agrees to ban a commercial trade then it lists that species in Appendix I of its lists. In 2011 there were just under 900 species listed in Appendix I of CITES. Appendix II of CITES is a list of 33,000 species for which international commercial trade is legal but monitoring is required to examine its sustainability and quotas may be set.

In the 1970s and 1980s, before the ban, the wild African elephant (*Loxodonta africana*) population fell from approximately 1.3 million to closer to 600,000, with an estimated 70,000 killed a year (Shoumatoff 2011). In response to the alarming declines, it was listed on CITES Appendix I in 1989, alongside Asian elephants which had been listed in 1973 (Stiles 2004). Many southern African countries opposed the ban, and petitioned for downlisting to Appendix II. They cited growing elephant populations and a need for revenues from their increasing stockpiles of confiscated or recovered ivory to put into conservation efforts (Stiles 2004). Before the ban, it is estimated that 80% of ivory traded was illegal, there was a limited legal quota and it was listed Appendix II (Khanna & Harford 1996), underlining how difficult it can be to police a legally controlled trade.

A ban is an unambiguous purchasing guide for consumers, and a legal guide for enforcement authorities. A ban avoids the need, for instance, to differentiate between Asian or African elephant sources or to account for the variable management practices in each source country and the quotas associated with them (Cooney & Jepson 2006). Nonetheless, in 1999 and 2004, CITES allowed an experimental, regulated ivory sale from southern African stockpiles in the face of strong opposition from groups who warned that this would lead to a surge in elephant poaching in other parts of Asia and Africa (Stiles 2004). Analyses of the effectiveness of the ban have produced ambiguous results (Bulte & van Kooten 1996; Stiles 2004). The debate over whether to overturn the ban continues and so does poaching; on average, it is estimated that 100 elephants are killed, and two Chinese nationals are arrested with illegal ivory, every day (Shoumatoff 2011).

The capacity of range states to control poaching varies with the resources available and corruption. This variation means that while some range states may be able to police a controlled trade, others struggle. The variation in management capacity leads many experts to believe that the simplicity of a ban on ivory trade remains the only viable option (Wasser et al. 2010) and CITES has continued to uphold the ban.

Critics of bans or prohibition draw on analogies with the demand for recreational drugs. Cigarette demand, for instance, is relatively inelastic, meaning that it changes little as prices change (Gallet & List 2003). Desire for cigarettes or alcohol is therefore unlikely to fall in the short term following a ban. Investigations into the trade of alcohol during its prohibition in the US indicate that the ban had impacts that were either small and ephemeral (Roumasset & Thaw 2003) or insignificant and potentially counterproductive (Dills et al. 2005). As production and supply chains developed during prohibition, consumption of alcohol recovered to pre-prohibition levels (Thornton 1991).

It has also been demonstrated that just before a prohibition, consumption is likely to increase, as it did marginally before the American prohibition on the sale of Cuban goods. An aide to President J.F. Kennedy, named Pierre Salinger, was allegedly sent to buy the President a supply of his favourite cigars. Upon the aide's return with 1200 Petit Upmann Cuban cigars, Kennedy signed the bill banning their sale in the United States (Salinger 2002). There is also evidence to show that before a CITES restriction becomes active, trade levels increase. Rivalan et al. (2007) found that during the year before a restriction became active, on average, trade would increase by 135%. For instance, peak volumes of 2800 Kleinmann's tortoises (*Testudo kleinmanni*) and 5500 Geoffroy's cats (*Leopardus geoffroyi*) were imported during the transition period between decision and enactment of a ban.

Once a ban is implemented, a lucrative trade is often left in the hands of criminals, and criminalizes those who demand the goods. Not only did the US prohibition of alcohol fail, it also empowered violent criminal gangs by presenting them with a lucrative business (Thornton 1991; Demleitner 1994). The criminal gangs would bribe members of the police and judiciary, who may have had little respect for prohibition, undermining the rule of law (Thornton 1991). This trade ban germinated a powerful criminal profession and damage to the criminal justice system, which lasted long after it was repealed. There are a number of examples where illegal trade in wildlife has been linked to organized crime and terrorism (Grieser-Johns & Thomson 2005; Lin 2005); for example, Al-Shabaab, a youth militia from Somalia linked to al-Qaeda, cross the border into Kenya to poach African elephant ivory (Shoumatoff 2011).

Law enforcement

Regulations must be enforced by CITES range countries, and the effectiveness of enforcement varies widely depending on the capacity of both law makers to write effective regulations and enforcement officials to enforce existing laws, and on the deterrent nature of the penalties for wildlife crime. The duties of detecting and prosecuting wildlife crimes involve military personnel, parks rangers, coastguards, police forces and customs and excise departments. In some ways, legislating against wildlife trafficking takes a wildlife problem and turns it into a law enforcement problem, often without the additional resources needed for the enforcers to do their job properly (Oldfield 2003).

Stories of failed regulation due to a lack of political will or ability to enforce it are legion. They range from high-profile international disputes between developed nations to a lack of capacity to enforce laws, which is especially pronounced in developing countries. Internationally, the USA has in the past threatened to use the Pelly amendment to restrict imports from nations which are damaging wildlife populations. One example was with Norway, Denmark and West Germany due to overharvesting of Atlantic salmon (Charnovitz 1993). These threats can work temporarily but when they are not carried through, as they have not been with the Pelly amendment, these nations revert to unsustainable harvesting (Hawes 1994). Developing country examples include Equatorial Guinea's introduction of a ban on the taking and possession of primates. Initially this ban led to a significant fall in the harvest of primates but over time, it became clear that the law's reach exceeded its grasp and that nobody would be prosecuted for these crimes and the harvest returned (BBPP 2011). Enforcement agencies are often understaffed: the 1256 km^2 of Korup National Park in Cameroon is protected by only 17 guards, and the park continues to provide a source of wildlife for the bushmeat trade (Robinson 2011, personal communication).

Even where political will exists and wild harvesting restrictions are in place, harvesters may take little note. Many enforcement difficulties are common between trade policies; for instance, disrespect for a given law is likely to undermine that law (Sutinen & Kuperan 1999; Tyler 2006) and law makers need to consult more with the stakeholders affected by proposed laws, including enforcement officials (Oldfield 2003).

Enforcement agencies working on wildlife trade have the added difficulty that it crosses national borders. The fact that wildlife habitat and markets are often separated by international boundaries does not appear to interfere with trade as much as it does with enforcement (Oldfield 2003). The movement of bushmeat from Korup National Park in Cameroon to the urban markets of Nigeria disrupts attempts to regulate the trade (Robinson 2011, personal communication).

INTERPOL's Environmental Crime Programme was established in 1992 to assist its 188 member countries in the effective enforcement of national and international environmental laws and treaties. Regional initiatives include the Association of Southeast Asian Nations' Wildlife Enforcement Network (ASEAN-WEN), supported by the US State Department and USAID, which was created in order to build capacity within existing legal frameworks to tackle wildlife crime in member countries. In some cases, law enforcers are assisted by non-governmental organizations such as the Environmental Investigation Agency and the Wildlife Protection Society of India. These NGOs gather information, train officials and convene multiple law enforcement departments to co-operate in stings on highly sophisticated wildlife crime networks or to develop computerized management information system for the collection of data on wildlife crime (Stokes 2010).

Even assuming that wildlife laws are well framed and that there is sufficient capacity to police them, the technical difficulty of investigation is immense, particularly when parallel to a legal trade. If a legal trade exists, once fishermen or hunters have taken their unlicensed harvest they can sometimes slip it into the legal trade. A bleak example of the 'blurring' impact of a controlled trade on enforcement is found in legalized sex trades. Trafficking of women into sex slavery cannot be measured or easily discovered when it is hidden within an illegal trade, and those who support legalizing prostitution believe that women in a regulated trade can be better protected from slavery and violence (Korvinus et al. 2005). Conversely, the original purpose of criminalizing aspects of the sex trade was to prevent sex slavery (Outshoorn 2005). In The Netherlands brothels are legal, regulated, private enterprises. Those who believe legal prostitution in fact masks trafficked women as economic migrants (US Department of State 2007) point out that The Netherlands was listed in 2007 amongst the most common destinations for human trafficking (CATW 2010).

The debate over prostitution shows that legalizing a trade can make enforcement of controls more difficult and also complicate the task of measuring the size of the trade. Illegal, unreported and unregulated (IUU) fishing is a case in point. Fishing itself is often legal and so identifying illegally caught fish becomes more difficult and IUU fishing is estimated to be worth between $10bn and $23.5bn per year (UNEP 2010).

Livelihoods and incentives

Certification and labelling

Voluntary management schemes are sometimes used to encourage sustainable practices instead of legal obligations. Today there are many different certification schemes, e.g. Marine Stewardship Council and Rainforest Alliance, that work with private industries to create and set standards for sustainability and provide independent evaluations of sustainability practices (Searle et al. 2004). These certification schemes help companies to set policies and standards voluntarily that can help a business achieve the twin aims of sustainability and profit (Treves & Jones 2009). Motives for participation can include a genuine interest in developing a long-term, sustainable business model, a desire to limit liability for environmental impacts, or a desire to project an environmentally friendly image to customers. All these motives can be tied up into what is known as corporate social responsibility. It is unclear, however, whether these schemes are successful in terms of protection of long-term sustainable wildlife populations. Certification systems' management requirements are not always rigorously policed and where they are, they still often require the public to choose the certified products over non-certified. As a result, these schemes are usually accompanied by environmental labelling practices identifying the product as sustainably harvested.

Eco-labelling uses a value that a potential consumer can identify with as an incentive to buy an eco-friendly product. The label provides credibility to the sustainability claims and carefully calibrates and simplifies marketing messages to reduce confusion (Treves & Jones 2009). There are three types of labelling:

- supportive labels allow the vendor to charge a premium for the product and donate a portion of the proceeds to conservation efforts, e.g. endangered species chocolate
- persuasive labels indicate that a product was produced in a certain environmentally responsible way, and provide production verification mechanisms such as third-party inspections, e.g. Smithsonian Migratory Bird Center's bird-friendly coffee label
- protective labels claim that purchasing the product will actually help to protect the species or ecosystem that the product was derived from, e.g. Wildlife Conservation Society's tiger-friendly label (Treves & Jones 2009).

On the flip side of the coin, labelling can be a huge problem following a trade ban. For example, many tiger products such as tiger-bone plasters (a widely used poultice used to treat rheumatism in traditional Chinese medicine – TCM) continued to be sold under tiger-bone labels and prosecutions were avoided by claims that, in fact, the products were fake, prompting further legislation that made it illegal even to list tigers as an ingredient (IFAW 2007). This has created an enforcement nightmare, with products like tiger-bone wine being sold in bottles resembling tigers sending a clear message to the consumer, but with lion bone listed as the ingredient which is a separate message for law enforcement officials (IFAW 2007).

Substitutes and preferences

Proponents of bans argue that the stigma attached to an illegal trade will reduce demand (Cameron 2009). There are few data to support this as yet, but reduction in supply of the illegal commodity might lead to a search for alternatives. At the height of the 19th-century ivory trade, 12,000 elephants were killed each year to make British billiard balls alone (Clare 1982). As supply dwindled, prices rose and in 1865 the Phelan and Collender Company of New York City offered a $10,000 prize for a better alternative material. John Wesley Hyatt won with celluloid and pool and snooker balls are no longer made from ivory (Clare 1982). As may be true for bushmeat species, the substitution of a product with no conservation impact was not difficult to effect.

Detailed knowledge of preferences and possible substitutes can help law makers to frame laws which both protect wildlife and acknowledge the needs of local human populations. We often read, for example, that trade in 'bushmeat' is unsustainable. And so it is for some species; observation of markets in West Africa confirms that the number of primates being consumed almost certainly exceeds their reproductive capacity (Fa et al. 2006; Albrechtsen et al. 2007). It has also been observed that demand for bushmeat does not vary appreciably between taxonomic groups; carcass prices are determined primarily by their size (Macdonald et al. 2011). This suggests that different species should be readily 'substitutable', i.e. that incentives to consume selected rapidly breeding species (rodents, for example) rather than primates do not need to be extravagant, provided enforcement is adequate.

Unsurprisingly, though, consumers are not homogeneous. In urban areas, where average wealth is higher, bushmeat is consumed more often by the wealthier, while in rural areas wealth does not affect bushmeat consumption (Fa et al. 2009). Bushmeat seems to be a luxury in urban areas, but it is all that is available for the rural poor. This has clear implications for policies aimed at providing alternative sources of protein. Fish, for example, seems to be considered inferior and is consumed less by the wealthier in both rural and urban areas, at least in Equatorial Guinea, where it is relatively plentiful (Fa et al. 2009). The nuances of individual preferences in different regions can have significant impacts on policy choice.

Consumers' wildlife value systems can be an important factor in the management of wildlife. Sport hunting for trophies in Africa is carefully regulated by licensing. The cost of hunting species is, as might be expected, related to the characteristic body size of the species (Johnson et al. 2010). Large antelope cost more than small ones. But it is also observed that hunters value some groups more highly than others for reasons that are related to 'charisma'. Felids are, on average, more expensive to hunt than ungulates, for example, and their cost increases as a function of body size more steeply than occurs for antelope. There is also a clear effect of real or perceived rarity, as advertised by IUCN status (Prescott et al. 2011). All else being equal, endangered species cost more than less endangered species to hunt. There is a clear lesson here for policy makers considering a change in status, particularly for unlicensed markets. A perverse consequence of listing might be increased demand, and a poorer conservation outcome.

Livelihood incentives to protect wildlife

Given the limited funds available to conservation (Pearce et al. 2007; Parker & Cranford 2010), a strategy that satisfies the twin aims of conservation and profit clearly represents a holy grail. If a non-damaging alternative livelihood can be undertaken by communities which might otherwise be poachers then a sustainable alternative is created and the hunters have an incentive to stop hunting. For example, greater cane rats (*Thryonomys swinderianus*) from North Africa (Adu et al. 1999) and corals in coastal regions (Delbeek 2001) have both been farmed on a small scale by communities in the hope that the income would reduce the need to poach, but with only minor success.

The money from trade can simultaneously fund economic development and encourage habitat and resource protection as well as providing funds for conflict mitigation (Swanson & Barbier 1992). One example of an incentive to protect wildlife is trophy hunting of the Suleiman markhor (*Capra falconeri jerdoni*) and Afghan urial (*Ovis orientalis cycloceros*) that began in Pakistan in 1986 with the stated aim of preventing their population decline (Frisina & Tareen 2009). Promises of jobs guarding the wild populations convinced local people to give up traditional hunting rights. Between 1986 and 2006, the project earned $1.3 million, and between 1994 and 2006 urial numbers rose from 1173 to 3146, whilst markhor rose from 695 to 2541. The project not only provides employment but healthcare and agricultural development funds in an isolated land. Dickman et al. (2011) refer to such incentives as payments to encourage co-existence (PEC) (see Chapter 4).

The costs of co-existence can be significant. Loveridge et al. (2010) present a range of studies of livestock predation illustrating the scale of wildlife conflict facing some farmers. Notably, data from a study in Zimbabwe between 1993 and 1996 found that on Gokwe communal lands, felid predation of livestock cost US$13/household which is 12% of their annual income. Another two studies (1995–1996 and 2002–2003) in areas of India with low levels of natural prey indicate that 12% of livestock are lost each year to snow leopard (*Uncia uncia*). The average cost of livestock losses to snow leopards for each family in these areas amounts to 50% of the average per capita income. In order for a trade ban to work, employment must be found for those who previously were harvesters, and mitigation and recompense must provide for potentially increased real and perceived conflict. The money for conservation projects must come from either non-extractive exploitation of the resource, direct tax receipts or charitable donation (Macdonald et al. 2007).

On the flip side of this coin, incentives have been used to perpetuate unsustainable practices. For example, $27bn in perverse subsidies is injected into fishing each year despite predictions of a collapse in global fisheries (Sumaila et al. 2010). UNEP estimate that 20 million jobs would be lost globally if fishing were reduced to sustainable levels (though more would be lost if stocks collapse). Fishing communities may be isolated, with scant opportunity for alternative employment, and the fishing industry can be highly politically active (Naughton 2009). Reductions in fishing harvest to sustainable levels can therefore seem politically impossible.

Incentives to hunt are not always purely financial, meaning that hunters are not always pure profit maximizers (Muth & Bowe 1998), making them difficult to influence through economic or legal means alone. For instance, even as a trade becomes economically unprofitable, hunters and fishermen may continue to hunt and fish because of feelings of cultural heritage (Forsyth & Marckese 1993) or simple fun – hence the use of the English word 'game'.

Wildlife management

Wild harvest from a common resource

When multiple individuals, acting independently and rationally consulting their own self-interest, share a resource, they will ultimately deplete it, even when it is clear that it is not in anyone's long-term interest for this to happen. This scenario is known as the tragedy of the commons (Hardin 1968). The traditional response to prevent such a tragedy of the commons is for people to surrender some of their own rights to a shared resource and abide by limits based upon the maximum sustainable yields, dividing the rights to capture them in the form of quotas that are often legally enforced.

A properly managed off-take from any wild population must be able to predict likely future numbers as it is these which the harvest will affect. It is usually impossible to count all current individuals and yet detailed knowledge of a population is required to predict future numbers with accuracy. Knowledge of population dynamics must then be combined with potential behavioural impacts of losses such as perturbation effects (Tuyttens et al. 2000), infanticide (Packer & Pusey 1983) or increases in problem animals (Bradshaw et al. 2005). For these reasons, even with a well-regulated trade, sustainable off-takes are not certain. For many years the number of Zimbabwean trophy licences available for hunting lions in Hwange National Park was far in excess of the entire lion population in the region, in part because the size of the population was unknown (Loveridge et al. 2007). In order to avoid complex management requirements, it might be possible to use a rule of thumb, such as taking only animals of a certain sex or over a given age (Whitman et al. 2004). Options to control fishing have included gear regulation, income quotas, tonnage quotas or species quotas. None is perfect and they often result in causing damaging by-catch and significant costs to the state in enforcement (Lewison et al. 2004). The North East Atlantic fishery produces an estimated 1.3 million tonnes of discards each year (Suuronen & Sard 2007). These issues are clearly so difficult that even if it proves impossible to overcome these technical issues entirely, it is important that we recognize that improvements are valuable. Very often, conversations will point to these numbers as failures of policy, but what we must start to do is to look instead at trends and pay less attention to absolute numbers.

In Zimbabwe, about half of the land is under a system of communal ownership where communities have rights to the land governed by tribal leaders. An effective ban on harvesting wildlife that was the property of the state meant that villagers only viewed wildlife as a pest that raided crops or preyed on livestock. The ban was lifted in 1975, resulting in a tragedy of the commons (Frost & Bond 2008). In 1989 a novel scheme known as CAMPFIRE (Communal Areas Management Programme for Indigenous Resources) recognized the need for a more sustainable solution by developing a sense of ownership and custodianship of wildlife living on communal lands. The programme provided trophy hunters with access to wildlife on communal lands for a premium that provided new income for the community, but a portion of the proceeds was used to pay for wildlife management practices to ensure sustainability, such as setting quotas, monitoring off-take and hiring game scouts. Despite management problems, the programme did receive significant support from communities because there was a financial incentive (Child 1995). CAMPFIRE is widely credited with having extended the boundaries for conservation beyond national parks and is a model that offsets the costs to the state of managing wildlife (Oldfield 2003).

It is thus conceivable that people can avoid a tragedy of the commons by setting aside their short-term individual interests to exploit the resource sustainably and preserve their own livelihoods. There will, however, remain an incentive for individuals within the group to quietly cheat. Much as in the famous Prisoner's Dilemma, all are better off if they co-operate. So we would hope that all would co-operate and only take the agreed quota. However, if everybody else co-operates, an individual is better off cheating (taking above their quota). Each individual knows that their neighbours are better off cheating. If their neighbours are cheating then they are much worse off if they co-operate and their best reaction is to cheat also. So we know that if everybody else is co-operating, the best reaction is to cheat and if everybody else is cheating then the best reaction is to cheat. As such, the optimum outcome of everybody co-operating is never reached and instead circumstances tend towards a Nash Equilibrium, with everybody cheating.

Cultural and psychological preferences can overcome the Prisoner's Dilemma as in experiments for these kinds of games, the optimum outcome is often the result. Alternatively, strong laws and enforcement can prevent cheating. However, even where property rights are enforceable, and a single owner governs the exploitation of the resource, sustainable exploitation is not always the most rational economic decision. Blue whale populations grow more slowly than interest rates in the bank so the option of driving the species to extinction and banking the profits is a rational economic approach to maximize the financial benefits to the harvester (Clark 1972).

Wildlife ranching

Some forms of ranching are more like farming, with significant management of the stock, whilst at the other extreme it may appear more like a wild harvest and a number of methods of ranching lie along this gradient. Private land ownership prevents a tragedy of the commons scenario and managing wildlife sustainably is a straightforward choice for the landowner. King William's proclamation that *'Whoever killed a hart or a hind should be blinded'* (Anonymous 1087) was not a ban on hunting but a control for his own benefit. Aristocracies have often maintained hunting grounds for their own pleasure, excluding others. The incentive for the privileged to jealously guard hunting stock of wild deer seems to have led many deer species to fare better than other large mammals in Europe.

In the early 1960s in Zimbabwe, Ray Dasman conducted an experiment where he showed that wildlife ranching produced more meat per unit area than cattle production, arguing that native herbivores were best suited to the environment and therefore could be maintained sustainably (Dasman 1964). This argument was used to lift the ban on selling wildlife products commercially in 1975. When land owners also owned wildlife and could market the products, it spawned a vibrant game-ranching industry that spread elsewhere in Africa. By the mid-1990s game ranching occupied 6% of the land surface area of Zimbabwe, while the game ranchers' association had over 300 members and managed more than a quarter of a million head of game (Gratwicke & Stapelcamp 2006). In 2000, the Zimbabwean government confiscated most land from private ranchers and this effectively destroyed this promising industry but it has continued to thrive in other places, including Namibia and South Africa (Gratwicke & Stapelcamp 2006).

Capybara (*Hydrochoerus hydrochaeris*) harvesting in Venezuela is another example of wildlife ranching (Ojasti 1991; Herrera & Barreto 2012). A pioneering measure taken in the late 1960s, to commission a baseline study of the biology of a species with the explicit aim of designing a sustainable harvest plan, led to a successful programme which provides extra income to cattle ranch owners and workers in the seasonally flooded savannas, or llanos (Ojasti 1973). The characteristics of the species, a large, native, gregarious grazer well adapted to the region, with large and fast-growing litters (Ojasti 1973; Herrera 1992), have certainly helped in the success of the annual harvest. The fact that they coalesce in dwindling water pools during the dry season (Macdonald 1981), while females are at the lowest point in their yearly reproductive cycle, has made the plan effective and ecologically sound (Herrera & Barreto 2012), although illustrating important ethical questions that are general to wildlife trade (Macdonald et al. 2012).

Wildlife farming

In contrast to ranching, wildlife farming may be defined as the intensively managed production of a non-domesticated species for commercial purposes. In recent years, wildlife farming has been considered as a method to displace trade in wild products in the market. Farming is an ancient practice beginning around 11,000 years ago (Chessa et al. 2009). Given its long history, the impact of farming on the protection of wild species has left well-established historical examples of where farming has or has not protected the wild species.

The wild aurochs (*Bos primigenius*), for instance, has not survived. Aurochs were first farmed around 8000–10,000 years ago and were the ancestors of the dominant European, north and west African and Middle Eastern domesticated cattle (*Bos taurus*) (Bruford et al. 2003). The domesticated cattle provided easier access to blood, meat, leather, milk, traction power, central heating and even company. When an animal is bred in farms then no new positive value is placed on the species in the wild, there is only a possible decrease in direct consumption. Lacking any positive economic incentives for protection, the wild aurochs was lost in Europe 400 years ago, when the last population died in Poland under the protection of their kings (IUCN 2010). Farming does not provide any positive monetary incentive to protect wild stocks and the aurochs provides the first example of farming failing to prevent human-driven extinction.

In contrast with the aurochs, farming of crocodile and alligator skins is now acknowledged to have helped the recovery of their populations in the wild (Ross 1998), but this did not happen quickly. Crocodile farming began in Thailand in the 1950s. Managed harvests of crocodilians followed in Florida, Zimbabwe, Papua New Guinea and Venezuela from the mid-1960s to 1970s (IUCN 1989). Recoveries of populations in the United States were well documented and led to the downgrading of alligators to CITES Appendix II, to allow controlled international trade. In the 1980s the illegal trade was still prolific to the extent that caimans (considered to produce low-quality skins) were being exploited in the New World in response to declines in wild alligator numbers (Thorbjarnarson 1999). Over two decades, suppliers from the new sustainable harvests and farms displaced the illegal and unsustainable hunts. The trade in illegal skins was slowly reduced through close collaboration between sustainable harvesters/farms, the leather industry and regulators (Ross 1998). Wild crocodiles continued to make up the majority of the supply of skins for 2–3 decades after the Floridian and Zimbabwean farms

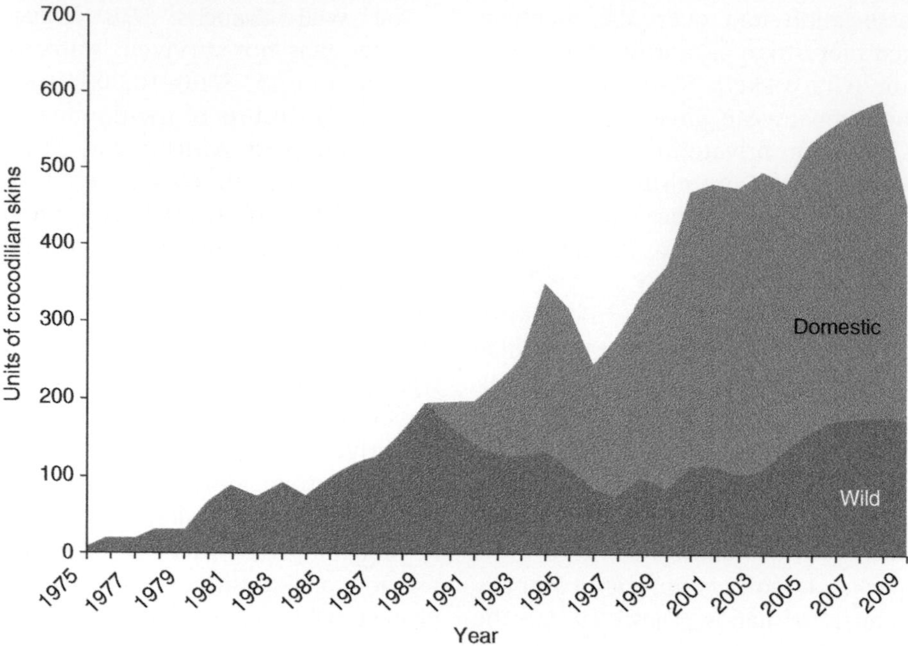

Figure 5.1 From CITES trade data describing total numbers of skins supplied by source to the global market between 1977 and 1999.

started and four decades after Samutprakan crocodile farm began breeding in Thailand (Figure 5.1). The number of skins being traded as recorded by CITES increased over that time. It took more than 20 years for the farmed crocodilian skins to overtake wild skins in quantities traded.

Economists referring to the 'short term' have in mind the time over which the markets and legal authorities adjust to the new product, and the likely length of that period and the potential impact on the wild population ought to be considered before embarking upon a farming policy intended to take pressure off the wild. If the wild crocodilian population had been less able to survive, over the decades it took for the farmed trade to take over, the result of opening this trade may have been disastrous.

Farmed products are often indistinguishable from the illegal wild-harvested products, leading to arguments that legalizing trade from farmed sources will open markets to launder unsustainably harvested illegal products, as in the trade of tiger bones (Gratwicke et al. 2008a). In order to combat this, authorities may attempt to tag or certify legal products but no system is likely to be watertight. It might help if there are clear bottlenecks in the trade process so that regulatory intervention can be focused on these critical points in the supply chain. Crocodile skins need to be tanned by highly skilled professionals. While many tanners exist, many crocodile skins pass through one of three tanning companies, which provides a pinch point in the production chain. Furthermore, crocodilian farms can usually produce a more consistent level of skin quality and quantity. This is particularly useful because regular-sized scales can be produced consistently to order for a high-value market. This enables the manufacturers to plan their business more efficiently and makes farmed skins preferable to producers.

If wildlife farming is to benefit wild populations, the capacity to produce captive stock must be able to meet market demands rapidly. Musk deer (*Moschus* spp.) have been farmed for over half a century but cannot yet fulfil market demand (Yang et al. 2003). Musk deer are threatened throughout their range and the musk the males produce is worth more by weight than gold (Yang et al. 2003). Used regularly in TCM, it is a popular and 'potent' ingredient. Attempts to take advantage of this prized good through farming began in the 1950s. More recently, musk deer breeding success has improved and there are thought to be around 2000 individuals in farms, mostly of the alpine species (*Moschus chrysogaster*). Males between 3 and 5 years old can produce an average of 18 g/year (Xiuxiang et al. 2006). They would therefore produce a maximum of approximately 30 kg/year. Estimates of the total demand for musk range from 500 to 1500 kg/year (Yang et al. 2003; Xiuxiang et al. 2006). Xiuxiang et al. (2006) estimate that musk from the farmed musk deer population can only satisfy 0.5–2% of total demand. Farming, therefore, is unlikely to benefit wild populations.

Bear bile is described in the earliest official pharmacopoeia of TCM in AD 659 (Huang 1994). TCM practitioners use bile for a variety of illnesses including liver disease, epilepsy and eclampsia (Bensky et al. 2004). Bear bile would historically have been a scarce and costly product reserved for the wealthy or for serious illness (Lee 1999). A wide variety of alternatives is available, depending upon the illness, and Huang (1994) lists 27 alternative species whose bile was said to mimic the effect of bear bile on specific conditions. A WSPA report (WSPA 2005) lists 39 species of flora which might replace bear bile. A non-random survey of 50 TCM practitioners found that 8% felt that bear bile was an irreplaceable and vital part of the pharmacopoeia (Mills 1994). Bears have been farmed for their bile in East Asia since the 1980s when Korean scientists developed a method for extracting bile from live bears through a cannula to the bear's bile duct (WSPA 2002). Proponents believe that low production costs of bear bile from these farms offset demand for consumption of bear bile from the wild.

Dutton (2010) investigated bear bile poaching through 78 interviews in Cambodia with 100 members of hunting communities. Respondents stated that hunting equipment can be borrowed from wildlife traders, reducing to zero their fixed costs. The full list of equipment mentioned in interviews was guns, wire, rope, lamps and batteries, dogs, bows, axes, knives and cooking equipment. Much of the equipment might be considered to be 'general tools' kept by farmers, such as axes and knives, and therefore of little further cost. Costs of hunting, where costs were incurred, were estimated between US$4 and US$5 per bear, but hunters could earn US$200 for a bear's gall bladder. Thus, even with a very low hunting success rate, hunting bears will be profitable for those living at the poverty line. In 2004 35% of Cambodians lived below the poverty line (US 34 cents per day in rural areas) (IMF 2006). In addition, Cambodian hunters were generalists and would take bears when they were available, further increasing the potential profit per trip.

Dutton et al. (2011) interviewed 1677 respondents in mainland China. Respondents were presented with hypothetical situations in which they must choose a treatment from a list including wild and, sometimes, farmed bear bile, given the prices of different treatments and an illness. From these interviews, the authors modelled the relationship between price and demand (known as the demand function) for wild bear bile both when competing with farmed bear bile and when alone in the market. They found a willingness to pay considerably more for wild bear bile than for farmed. They found that demand for wild bear bile changes little when its price changes or when the farmed price changes, suggesting that the ability of farmed bear bile to reduce demand for wild bear bile is at best limited.

Trade in tiger parts has been illegal in China since 1992 (Gratwicke et al. 2008a). Products from tiger bone can retail at $1250–3750 per kilogram (Gratwicke et al. 2008b). Tiger farms were developed in the 1980s to take advantage of this lucrative trade but were not yet operational at the time of the ban. These farms survive as tourist attractions, and meanwhile they campaign to have the tiger parts trade ban lifted in order to realize their full profitability. Those opposed to tiger farms believe they will fail to protect and may endanger wild tigers as wild tiger parts are not only preferred to farmed tiger parts by the public, but the farm's costs are much higher than those faced by poachers (Macdonald et al. 2010).

Lapointe et al. (2007) estimate the cost of raising a tiger at around $4000 but this is likely to be a gross underestimate. A report from a farm in Guilin (Anonymous 2007) stated that they have a US$4.9 million annual shortfall after receipts to the business as a tourist attraction. The farm claims that if the ban was lifted if could produce 600 cubs per year. This suggests that in order to make up this shortfall, they would need to make US$8160 per tiger to cover current costs and would need more to finance the debts they have accrued. The poacher's costs are not easily estimated but are likely to be low. The costs of trafficking are unknown but unlikely to be high and the poison required to kill a tiger is estimated at $20 (Gratwicke et al. 2008a). If tiger farmers are unable to undercut the poachers and illegal traffickers, then, in the absence of the ability to produce a more desirable substitute, any potential disincentive for the poacher from the less desirable and more expensive competitor is most likely undermined.

An important figure, which may be very hard to estimate, is the number of individuals that might be removed illegally from the wild in order to bolster farm stocks once trade from farms is legalized. The initial stock for a farm is likely to come from the wild, and provided it remains cheaper to take them from the wild than to breed stock, there will be an incentive for farms to poach to increase stocks. Brooks et al.(2010) examined the farming of the South East Asian porcupine (*Hystrix brachyura*) in Vietnam; the species is severely threatened by trade and populations declined 20% in the 1990s. Through interviews with farm owners, the authors found that 58% of examined farms took wild founder stock and at least 19% continued to buy wild founder stock.

Damania & Bulte (2007) point out that the wild suppliers may not be in 'perfect competition'. Under perfect competition, any attempt by one supplier to reduce supply in order to increase the price they receive would create an incentive for another supplier to increase supply to sell more goods. In a theoretically perfectly competitive market, each supplier produces as much as they can profitably produce at market prices. Without competition, it is likely that suppliers are restricting output to increase profits. Governments encourage competition not because they hope to remove incumbent firms but because neo-classical economic theory suggests that increasing competition lowers costs and increases output. If traffickers are currently restricting output in order to maintain higher prices, then increasing competition may actually encourage an increase in off-take from the wild as their ability to keep prices up is reduced.

Education

Training

Even if staffing exists to tackle wildlife crime, effective enforcement is an area that requires extensive knowledge and skills that can only be obtained through training and experience. For example, customs agents inspect goods at borders in many countries but many of them lack the training to recognize the illegal wildlife products that pass through their ports each day, prompting NGOs to provide training support and develop identification manuals to help improve the effectiveness of existing capacity (Nowell 2000). The US facilitates policy development and shares technical expertise to assist developing countries in building their capacity to combat the illegal wildlife trade through CITES, the Coalition Against Wildlife Trafficking (CAWT) and the ASEAN-WEN Wildlife Law Enforcement Network (McMurray 2008).

Outreach campaigns

The reality of conservation is that it is inextricably linked with human value systems. Conservation scientists often use the scientific method as an objective tool to inform actions that ensure sustainability; however, science itself seldom changes human value systems. Our values tend to be formed on the basis of human emotions. Outreach campaigns are crucial tools to influence people's value systems and to mobilize them, but they have the potential to be very polarizing. An illustration of this point is a campaign mounted by Greenpeace to 'Save the whales' in 1973. One leader of that early campaign is quoted as saying:

> 'The scientific debate about whether whales really are in danger of extinction is not one we want to get reduced to. The general public is not going to understand the science of ecology, so to get them to save the whale you have to get them to believe that whales are good.' (Pearce 1996)

Indeed, the images of a malevolent whaling ship with an explosive harpoon pointed at a fleeing whale and being blocked by a dinghy emblazoned with the words 'Greenpeace' resonated hugely with the public (Pearce 1996). The campaign was not just anti-whaling; its specific goal was a moratorium on whaling, something it eventually achieved at a meeting of the International Whaling Commission in 1982 (Pearce 1996). The international moratorium on whaling has angered many nations with whaling traditions and has produced a vicious cultural battleground between nations that have no whaling traditions and countries like Japan who flagrantly violate the moratorium under the banner of scientific research (Clapham et al. 2007).

We have in this chapter attempted to cover the key issues which drive the success or failure of a given wildlife trade policy. The key messages are that whilst markets may be affected by a common set of issues, there is unlikely to be a universal formula for intervention, and that failure to entirely prevent damaging trade is not the same as total failure. We hope in the future to improve capacity to measure the impacts of these interventions and so choose either the best or perhaps least-worst solution in each instance.

Whilst pursuing this empirical goal, we must also remember that conservation actions are complicated by varying ethical issues regarding the form wildlife management takes. Murders occur too frequently, but nobody suggests legalizing murder so that it might be regulated and controlled. The idea is fundamentally abhorrent, just as many feel that prostitution is a fundamentally barbaric attack on women. Still others feel that hunting whales or bear bile farming are unconscionable acts of cruelty that destroy beauty and degrade living beings to commodities. Conversely, nationalism or potential for profit has driven desires to open trade. That is not to say that we ought not to attempt to calm the furies and reduce obfuscation through categorization of the issues and collation of facts, though Bertrand Russell was referring to an even more divisive debate when he stated that: 'The most savage controversies are those about matters as to which there is no good evidence either way' ('An outline of intellectual rubbish' in *Unpopular Essays*, 1950).

Acknowledgements

Part of the work on which this chapter draws was funded by a grant from the World Society for the Protection of Animals (WSPA) to DWM and AJD, which we gratefully acknowledge. Paul Johnson made helpful comments on an earlier draft and Sandra Baker did likewise on a later draft.

References

Adu, E.K., Alhassan, W.S. & Nelson, F.S. (1999) Smallholder farming of the Greater Cane Rat, *Thryonomys swinderianus*, Temminck, in Southern Ghana: a baseline survey of management practices. *Tropical Animal Health & Production*, **31**(4), 223–232.

Albrechtsen, L., Macdonald, D.W., Johnson, P.J., Castelo, R. & Fa, J.E. (2007) Faunal loss from bushmeat hunting: empirical evidence and policy implications in Bioko Island. *Environmental Science & Policy*, **10**(7–8), 654–667.

Anonymous (1087) *The Rime of King William*. The Peterborough Chronicle, Peterborough.

Anonymous (2006) BVA calls for a permanent ban on wild bird imports. *Veterinary Record*, **159**(5), 130.

Anonymous (2007) *The Status Quo of Captive Bred Tigers and Difficulties*. Guilin Xiong Sen, Bear and Tiger Garden, Guilin.

BBC (2007) EU to ban imports of wild birds. BBC News. London: http://news.bbc.co.uk/1/hi/world/europe/6253543.stm

BBPP (2011) *Commercial Bushmeat Hunting*. Bioko Biodiversity Protection Program: www.bioko.org/conservation/

Bensky, D. & Gamble, A. (eds) (2004) *Materia Medica: Chinese Herbal Medicine*. Eastland Press, Seattle, OR.

Bradshaw, G.A., Schore, A.N., Brown, J.L., Poole, J.H. & Moss, C.J. (2005) Elephant breakdown. *Nature*, **433**(7028), 807.

Brooks, E.G.E., Roberton, S.I., & Bell, D.J. (2010) The conservation impact of commercial wildlife farming of porcupines in Vietnam. *Biological Conservation*, **143**(11), 2808–2814.

Bruford, M.W., Bradley, D.G. & Luikart, G. (2003) DNA markers reveal the complexity of livestock domestication. *Nature Reviews Genetics*, **4**, 900–910.

Bulte, E.H. & van Kooten, G.C. (1996) A note on ivory trade and elephant conservation. *Environment and Development Economics*, **1**(04), 433–443.

Cameron, A. (2009) *Saving the Wild Tiger: Enforcement, Tiger Trade and Free Market Folly*. Environmental Investigation Agency, London.

CATW (2010) *Coalition Against Trafficking in Women: Netherlands, Trafficking*. www.catwinternational.org/

Charnovitz, S. (1993) Environmental trade sanctions and the GATT: an analysis of the Pelly amendment on foreign environmental practices. *American University Journal of International Law and Policy*, **9**, 751–808.

Chessa, B., Pereira, F., Arnaud, F., *et al.* (2009) Revealing the history of sheep domestication using retrovirus integrations. *Science*, **324**(5926), 532–536.

Child, G. (1995) *Wildlife and People: The Zimbabwean Success*. Wisdom Foundation, Harare.

Chundawat, R.S., Habib, B., Karanth, U., *et al.* (2010) *Panthera tigris*. IUCN Red List: www.iucnredlist.org/apps/redlist/details/15955/0

Clapham, P.J., Childerhouse, S., Gales, N.J., Rojas-Bracho, L., Tillman, M.F. & Brownell, R.L. Jnr. (2007) The whaling issue: conservation, confusion, and casuistry. *Marine Policy*, **31**(3), 314–319.

Clare, N. (1982) The Balls. *DAYS OF OLD No. 5*. www.normanclare.co.uk/DOY_No5_Balls.html

Clark, C.W. (1972) Profit maximisation and the extinction of animal species. *Journal of Political Economy*, **81**, 950–961.

Cooney, R. & Jepson, P. (2006) The international wild bird trade: what's wrong with blanket bans? *Oryx*, **40**(1), 18–23.

Damania, R. & Bulte, E.H. (2007) The economics of wildlife farming and endangered species conservation. *Ecological Economics*, **62**: 461–472.

Dasman, R. (1964) *African Game Ranching*. Pergamon Press, New York.

Delbeek, J.C. (2001) Coral farming: past, present and future trends. *Aquarium Sciences and Conservation*, **3**(1), 171–181.

Demleitner, N. V. (1994) Organized crime and prohibition: what difference does legislation make? *Whittier Law Review*, **15**, 613–646.

Dickman, A., Macdonald, E. & Macdonald, D.W. (2011) A review of financial instruments to encourage predator conservation and human–carnivore coexistence. *Proceedings of the National Academy of Sciences*, **108**(34), 13937–13944.

Dills, A K., Jacobson, M. & Miron, J.A. (2005) The effect of alcohol prohibition on alcohol consumption: evidence from drunkenness arrests. *Economics Letters*, **86**(2), 279–284.

Dutton, A.J., (2010) Hunting Bears, out in Cambodia. In: *Examining when and why farming might reduce demand for wildlife products and by extension extractive pressure on wild populations*. DPhil Thesis, University of Oxford.

Dutton, A.J., Hepburn, C. & Macdonald D.W. (2011) A stated preference investigation into the Chinese demand for farmed vs. wild bear bile. *PLoS ONE*, **6**(7), e21243.

Fa, J.E. & Brown, D. (2009) Impacts of hunting on mammals in African tropical moist forests: a review and synthesis. *Mammal Review*, **39**(4), 231–264.

Fa, J.E., Seymour, S., Dupain, J., Amin, R., Albrechtsen, L. & Macdonald, D.W. (2006) Getting to grips with the magnitude of exploitation: bushmeat in the Cross-Sanaga rivers region, Nigeria and Cameroon. *Biological Conservation*, **129**(4), 497–510.

Fa, J.E., Albrechtsen, L., Johnson, P.J. & Macdonald, D.W. (2009) Linkages between household wealth, bushmeat and other animal protein consumption are not invariant: evidence from Rio Muni, Equatorial Guinea. *Animal Conservation*, **12**(6), 599–610.

FAO (2010) *State of World Fisheries and Aquaculture*. Fisheries and Aquaculture Department Food and Agriculture Organization of the United Nations, Rome.

Forsyth, C.J. & Marckese, T.A. (1993) Folk outlaws: vocabularies of motives. *International Review of Modern Sociology*, **23**, 17–31.

Frisina, M.R. & Tareen, S.N.A. (2009) Exploitation prevents extinction: case study of endangered Himalayan sheep and goats. In: *Recreational Hunting, Conservation and Rural Livelihoods: Science and Practice* (eds B. Dickson, J. Hutton & B. Adams). Wiley-Blackwell, Oxford.

Frost, P.G.H. & Bond, I. (2008) The CAMPFIRE programme in Zimbabwe: payments for wildlife services. *Ecological Economics*, **65**(4), 776–787.

Gallet, C.A. & List, J.A. (2003) Cigarette demand: a meta-analysis of elasticities. *Health Economics*, **12**(10), 821–835.

Gratwicke, B. & Stapelcamp, B. (2006) *Wildlife conservation in an outpost of tyrrany. ZimConservation Opinion Report*, **3**, 1–39.

Gratwicke, B., Bennett, E.L., Broad, S., *et al.* (2008a) The world can't have wild tigers and eat them, too. *Conservation Biology*, **22**(1), 222–223.

Gratwicke, B., Mills, J., Dutton, A., *et al.* (2008b) Attitudes toward consumption and conservation of tigers in China. *PLoS ONE*, **3**(7), e2544.

Gratwicke, B., Evans, M.J., Jenkins, P.T., *et al.* (2009) Is the international frog legs trade a potential vector for deadly amphibian pathogens? *Frontiers in Ecology and the Environment*, **8**(8), 438–442.

Grieser-Johns, A. & Thomson, J. (2005) *Going, Going, Gone: The Illegal Trade in Wildlife in East and Southeast Asia*. World Bank, Washington, D.C.

Hardin, G. (1968) Tragedy of the commons. *Science*, **162**(3859), 1243–1248.

Hawes, C.E. (1994) Norwegian whaling and the Pelly Amendment: a misguided attempt at conservation. *Minnesota Journal of Global Trade*, **3**, 97–130.

Herrera, E.A. (1992) Growth and dispersal of capybaras, *Hydrochaeris hydrochaeris*, in the Llanos of Venezuela. *Journal of Zoology*, **228**, 307–316.

Herrera, E.A. & Barreto, G. (2012) Capybara as a source of protein: utilization and management in Venezuela. In: *Capybara: Biology, Use and Conservation of a Valuable Neotropical Resource* (eds J.R. Moreira, K.M.P. Ferraz, E.A. Herrera and D.W. Macdonald). Springer, New York.

Huang, J. (1994) Asian perspectives on the therapeutic value of bear bile and alternatives. Proceedings of the International Symposium on the Trade of Bear Parts for Medicinal Use, Washington D.C.

IFAW (2007) *Made in China – Farming Tigers to Extinction*. International Fund for Animal Welfare, London.

IMF (2006) *Cambodia: Poverty Reduction Strategy Paper*. International Monetary Fund County Report, **6**(266).

IUCN (1989) *Crocodiles: Their Ecology, Management and Conservation*. IUCN, Gland, Switzerland.

IUCN (2010) *Bos primigenius*. www.iucnredlist.org/apps/redlist/details/136721/0

Johnson, P.J., Kansky, R., Loveridge, A.J. & Macdonald, D.W. (2010) Size, rarity and charisma: valuing African wildlife trophies. *PLoS ONE*, **5**(9), e12866.

Karesh, W.B., Cook, R.A., Bennett, E.L. & Newcomb, J. (2005) Wildlife trade and global disease emergence. *Emerging Infectious Disease*, **11**(7), 1000–1002.

Khanna, J. & Harford, J. (1996) The ivory trade ban: is it effective? *Ecological Economics*, **19**(2), 147–155.

Korvinus, A.G., van Dijk, E.M.H., Koster, D.A.C. & Smit, M. (2005) *Trafficking in Human Beings*. Third Report of the Dutch National Rapporteur. NRM Bureau, The Hague.

Lapointe, E., Conrad, K., Mitra, B. & Jenkins, H. (2007) *Tiger Conservation: It's Time to Think Outside the Box*. IWMC, World Conservation Trust, Lausanne, Switzerland.

Lee, Y. (1999) The use of bear bile as medicine versus tonic. Proceedings of the Third International Symposium on the Trade of Bear Parts for Medicinal Use, National Insititute of Environmental Research, Seoul, Republic of Korea.

Lewison, R.L., Crowder, L.B., Read, A.J. & Freeman, S.A. (2004) Understanding impacts of fisheries bycatch on marine megafauna. *Trends in Ecology and Evolution*, **19**(11), 598–604.

Lin, J. (2005) Tackling Southeast Asia's illegal wildlife trade. *Singapore Year Book of International Law and Contributors*, **9**, 191–208.

Loveridge, A.J., Searle, A.W., Murindagomo, F. & Macdonald, D.W. (2007) The impact of sport-hunting on the population dynamics of an African lion population in a protected area. *Biological Conservation*, **134**(4), 548–558.

Loveridge, A.J., Wang, S.W., Frank, L.G. & Seidensticker, J. (2010) People and wild felids: conservation of cats and management of conflicts. In: *Biology and Conservation of Wild Felids* (eds D.W. Macdonald & A.J. Loveridge). Oxford University Press, Oxford.

Macdonald, D.W. (1981) Dwindling resources and the social behaviour of Capybaras (*Hydrochoerus hydrochaeris*) (Mammalia). *Journal of Zoology*, **194**(3), 371–391.

Macdonald, D.W., Collins, N. M. & Wrangham, R. (2007) Principles, practice and priorities: the quest for alignment. In: *Key Topics in Conservation Biology* (eds D.W. Macdonald & K. Service), pp. 273–292. Blackwell Publishing, Oxford.

Macdonald, D.W., Loveridge, A. J. & Rabinowitz, A. (2010) Felid futures: crossing disciplines, borders, and generations. In: *Biology and Conservation of Wild Felids* (eds D.W. Macdonald & A.J. Loveridge), pp. 599–649. Oxford University Press, Oxford.

Macdonald, D.W., Johnson P.J., Albrechtsen, L., *et al.* (2011) Association of body mass with price of bushmeat in Nigeria and Cameroon. *Conservation Biology*, **25**(6), 1220–1228.

Macdonald, D.W., Herrera, E.A., Ferraz, K.M.P.M.B. & Moreira, J.R. (2012) The Capybara paradigm: from sociality to sustainability. In: *Capybara: Biology, Use and Conservation of a Valuable Neotropical Resource* (eds J.R. Moreira, K.M.P. Ferraz, E.A. Herrera & D.W. Macdonald). Springer, New York.

McMurray, C.A. (2008) Wildlife trafficking: U.S. efforts to tackle a global crisis. *Natural Resources and Environment*, **23**, 16.

Mills, J.A. (1994) Asian dedication to the use of bear bile as medicine. Proceedings of the International Symposium on the Trade of Bear Parts for Medicinal Use, University of Washington, Seattle.

Muth, R.M. & Bowe, J.F. (1998) Illegal harvest of renewable natural resources in North America: toward a typology of the motivations for poaching. *Society and Natural Resources*, **11**(1), 9–24.

Naughton, P. (2009) Thousands stranded by ports blockade by French fishermen. *The Times*, London.

Nowell, K. (2000) *Far from a Cure: the Tiger Trade Revisited*. TRAFFIC, Cambridge. www.trafficj.org/publication/00_Far_from_Cure.pdf

Ojasti, J. (1973) *Estudio Biológico del Chigüire o Capibara*. FONAIAP, Caracas, Venezuela.

Ojasti, J. (1991) Human exploitation of capybara. In: *Neotropical Wildlife Use and Conservation* (eds J. Robinson & K.H. Redford), pp. 236–252. University of Chicago Press, Chicago.

Oldfield, S. (ed.) (2003) *The Trade in Wildlife: Regulation for Conservation*. Earthscan Publications, Sterling, VA.

Outshoorn, J. (2005) The political debates on prostiution and trafficking of women. *Social Politics: International Studies in Gender*, **12**(1), 141–155.

Packer, C. & Pusey, A.E. (1983) Adaptations of female lions to infanticide by incoming males. *American Naturalist*, **121**(5), 716–728.

Parker, C. & Cranford, M. (2010) *The Little Biodiversity Finance Book: A Guide to Proactive Investment in Natural Capital (PINC)*. Global Canopy Project, John Krebs Field Station, Oxford.

Pearce, D.W. & Hecht, S. (2007) What is biodiversity worth? Economics as a problem and a solution. In: *Key Topics in Conservation Biology* (eds D.W. Macdonald & K. Service). Blackwell Publishing, Oxford.

Pearce, F. (1996) Greenpeace: storm-tossed on the high seas. *Green Globe Year Book*. www.fni.no/ybiced/96_07_pearce.pdf

Prescott, G.W., Johnson, P.J., Loveridge, A.J. & Macdonald, D.W. (2011) Does change in IUCN status affect demand for African bovid trophies? *Animal Conservation*, **15**(3), 248–252.

Press Association (2011) Government attacks EU fishing rules. *The Guardian*, London. www.guardian.co.uk/environment/2011/mar/01/government-attacks-eu-fishing-rules

Rivalan, P., Delmas, V., Angulo, E., *et al.* (2007) Can bans stimulate wildlife trade? *Nature*, **447**(7144), 529–530.

Ross, J.P. (1998) *Crocodiles: Status Survey and Conservation Action Plan*. IUCN Crocodile Specialist Group, Gland, Switzerland.

Roumasset, J. & Thaw, M.M. (2003) *The Economics of Prohibition: Price, Consumption and Enforcement Expenditures during Alcohol Prohibition*. Hawaii Reporter, Honolulu.

Salinger, P. (2002) Kennedy, Cuba and Cigars. *Cigar Aficionado*, M. Shanken Communications Inc, New York.

Searle, R., Colby S. & Milway, K.S. (2004) *Moving Eco-certification Mainstream*. Bridgespan Group, Boston, MA.

Shoumatoff, A. (2011) Agony and ivory. *Vanity Fair*, **August**, 120–135.

Stiles, D. (2004) The ivory trade and elephant conservation. *Environmental Conservation*, **31**(04), 309–321.

Stokes, E. J. (2010) Improving effectiveness of protection efforts in tiger source sites: developing a framework for law enforcement monitoring using MIST. *Integrative Zoology*, **5**(4), 363–377.

Sukhdev, P., Wittmer, H., Schroter-Schlaack, C., *et al.* (2010) *The Economics of Ecosystems and Biodiversity. Mainstreaming the Economics of Nature: A Synthesis of the Approach, Conclusions and Recommendations of TEEB*. TEEB, Bonn.

Sumaila, R.U., Dalkmann H., Ayres, R., *et al.* (2010) *Green Economy Report: A Preview*. United Nations Environment Programme, Chatelaine, Switzerland. www.unep.org/greeneconomy/Portals/88/documents/ger/GER_synthesis_en.pdf

Sutinen, J.G. & Kuperan, K. (1999) A socioeconomic theory of regulatory compliance in fisheries. *International Journal of Social Economics*, **26**, 174–193.

Suuronen, P. & Sard, F. (2007) The role of technical measures in European fisheries management and how to make them work better. *ICES Journal of Marine Science: Journal du Conseil*, **64**(4), 751–756.

Swanson, T. & Barbier, E.B. (1992) *Economics for the Wilds: Wild Life, Wildlands, Diversity and Development*. Earthscan Publications, London.

Thorbjarnarson, J. (1999) Crocodile tears and skins: international trade, economic constraints, and limits to the sustainable use of crocodilians. *Conservation Biology*, **13**(3), 465–470.

Thornton, M. (1991) Alcohol prohibition was a failure. *Policy Analysis*, **157**. www.cato.org/pubs/pas/pa-157.html

Treves, A. & Jones, S.M. (2009) Strategic tradeoffs for wildlife-friendly eco-labels. *Frontiers in Ecology and the Environment*, **8**(9), 491–498.

Tuyttens, F.A.M., Delahay, R.J., Macdonald, D.W., Cheeseman, C.L., Long, B. & Donnelly, C.A. (2000) Spatial perturbation caused by a badger (*Meles meles*) culling operation: implications for the function of territoriality and the control of bovine tuberculosis (*Mycobacterium bovis*). *Journal of Animal Ecology*, **69**(5), 815–828.

Tyler, T.R. (2006) *Why People Obey the Law*. Princeton University Press, New Jersey.

UNEP (2010) *Green Economy*. Nairobi, Kenya.

US Department of State (2007) *Trafficking in Persons Report*. USDS, Washington, D.C.

Wasser, S., Poole, J., Lee, P., *et al.* (2010) Elephants, ivory, and trade. *Science*, **327**(5971), 1331–1332.

WCED (1987) *UN Report of the World Commission on Environment and Development: Our Common Future*. Transmitted to the General Assembly as an Annex to document A/42/427. Development and International Co-operation: Environment, United Nations, New York.

Whitman, K., Starfield, A.M., Quadling, H.S. & Packer, C. (2004) Sustainable trophy hunting of African lions. *Nature*, **428**(6979), 175–178.

Wright, S.J. & Muller-Landau, H.C. (2006) The future of tropical forest species. *Biotropica*, **38**(3), 287–301.

WSPA (2002) *The Bear Bile Business*. WSPA, London.

WSPA (2005) *Finding Herbal Alternatives to Bear Bile*. WSPA, London.

Wyler, L.S. & Sheikh, P.A. (2009) *International Illegal Wildlife Trade: Threats and US Policy*. Congressional Research Service, Washington, D.C.

Xiuxiang, M., Caiquan, Z., Jinchu, H., *et al.* (2006) Musk deer farming in China. *Animal Science*, **82**(01), 1–6.

Yang, Q., Meng, X., Xia, L. & Feng, Z. (2003) Conservation status and causes of decline of musk deer (*Moschus* spp.) in China. *Biological Conservation*, **109**(3), 333–342.

6

Leadership and listening: inspiration for conservation mission and advocacy

Andrew Gosler[1], Shonil Bhagwat[2], Stuart Harrop[3], Mark Bonta[4] and Sonia Tidemann[5]

[1]Edward Grey Institute of Field Ornithology *and* Institute of Human Sciences, University of Oxford, Oxford, UK
[2]School of Geography and the Environment, University of Oxford, Oxford, UK
[3]Durrell Institute of Conservation and Ecology, University of Kent, Canterbury, UK
[4]Center for Community and Economic Development, Delta State University, Cleveland, MS, USA
[5]Batchelor Institute of Indigenous Tertiary Education, Batchelor, NT, Australia

The vision of the Convention on Biodiversity is of a world 'living in harmony with nature where by 2050, biodiversity is valued, conserved, restored and wisely used, maintaining ecosystem services, sustaining a healthy planet and delivering benefits essential for all people'.

(CBD UNEP 2010)

Introduction: conservation biology as mission-driven science

When Soulé & Wilcox (1980) stated that conservation biology is a mission-driven discipline, the idea of applying such an openly persuasive term to a supposedly objective science seemed surprising. What was this 'mission' they were talking about? Thirty years later, the term is much more widely understood. It generally refers to the passionate conviction, common to anyone engaged in conservation work, that unless the world's leaders can be persuaded to understand and apply the principles of conservation biology, the earth may no longer be able to support life as we know it. The present status of this major theme as a self-evident truth is underlined in the preface to the first volume of *Key Topics in Conservation Biology* (Macdonald & Service 2007). Yet the concept of mission is densely packed with implications (see also Meine et al. 2006), for while it needs to be separated from its cultural baggage and any negative connotations, the insights gained from its further consideration may hold great potential for real conservation action.

Key Topics in Conservation Biology 2, First Edition. Edited by David W. Macdonald and Katherine J. Willis.

As conservation biologists, trained in the formal sciences that constitute ecology, we may feel frustrated, even betrayed, by the diverse ways in which the superficial, media-led culture of the West makes frequent, and often inappropriate, reference to ecological concepts. Such misappropriation of environmental science has contributed to the development of a cultural construct of ecology, which has been used to market everything from washing powder to cars and corporate bonds, and given rise, as the angst-ridden response of environmentalists, to the notion of 'greenwash' (Tokar 1997). Committed environmentalists may feel indignant, even angry, at the western media's wanton misunderstanding of, or missed opportunities to present, ecological principles and their conservation significance to the public. For example, as well as its huge importance for biodiversity – containing 30% of the world's soft corals and some 1500 species of fish – the Great Barrier Reef of north eastern Australia is a globally significant carbon sink (Kinsey & Hopley 1991). So it was disappointing when a BBC environment correspondent, describing the issue of soil run-off from the Queensland floods of 2011 onto the Great Barrier Reef, focused on the reef's local significance as a tourist attraction, albeit because of its beauty and biodiversity (Richard Black, BBC News, 14 January 2011), but failed to mention its global ecological significance.

In much the same casual way, the world of conservation has appropriated the language of biblical faith traditions. For example, when western conservationists refer to mission, crisis, wilderness, stewardship, apocalypse, apostle, Eden, creatures or Creation, they are (wittingly or unwittingly) using biblical references as a form of cultural shorthand. They are assuming, in effect, that their audience will understand these concepts because both the writer or speaker and the audience are embedded in the same western cultural milieu, which includes fundamental aspects of the biblical tradition. Furthermore, the recognition that conservation must be aligned with global issues such as social justice, poverty alleviation and community building as advocated by the Convention on Biodiversity (CBD UNEP 2002; Macdonald et al. 2007) naturally leads the conservation community onto the turf (or into the hearts) of traditional faith communities for whom these are central ethical issues (Palmer & Finlay 2003, Sluka et al. 2011).

These insights raise important questions about who and where are the natural allies of conservation in the wider world. Since the 1980s, conservationists have increasingly engaged with the commercial sector and with multinational corporations, and have adopted much of the culture of these agencies. However, while that engagement is essential and has yielded significant benefits, concerns about greenwash are rarely far from the surface (for example, consider the transition of BP from green champion to environmental pariah within a decade). Here we argue for a greater engagement with the cultural traditions of faith communities (i.e. the local or global communities associated with a religion), and with the wisdom of long, thoughtful and non-commercial experience of the world that is embedded within them. With more than 5 billion followers around the world (Bhagwat & Palmer 2009), these faith communities also represent significant agencies.

Aims of this chapter

The urgency of the present environmental crisis and the need to forge new alliances have been stated already (e.g. Macdonald et al. 2007). We assume that there is wide recognition of the shift in conservation management practice over the last 100 years from a top-down, managed (or classic) perspective to a ground-up (neopopulist) perspective requiring mass participation and consideration of lifestyle (Blaikie & Jeanrenaud 1996). In reality, conservation action requires a combination of approaches. So, for example, while top-down conservation emphasizes the significance of regulation and of establishing reserves, the ground-up perspective recognizes the essential part hitherto played by amateur naturalists in Citizen Science delivery of essential data (Greenwood 2007) and by traditional communities who have been working for centuries to evolve a viable relationship with their immediate natural environment (Harrop 2007).

We have two broad aims in this chapter. The first is to encourage conservation practitioners to take seriously their duty to engage sensitively with diverse communities (from local to global) and with their leaders (non-political as well as political), and to temper the sense of urgency with the cultural awareness necessary to negotiate effectively. Our second aim is to emphasize that conservation is not the sole preserve of the professional conservation biologist. Indeed, the need to maintain a stable balance between the diversity of life and human activities has been recognized by others quite independently, and has been built into the foundations of custom, cultures and belief throughout the world. Different cultures might express the need for environmental responsibility and ethical action in different ways, but wherever the conservation message is delivered sensitively, it should be culturally acceptable because an evolved sense of natural justice is a universal human trait (de Waal 2006, 2010).

Pro-conservation ethics, originally developed in small but diverse local traditional cultures, has reached billions of people through the ethical teachings of their own cultural and faith traditions (Bhagwat et al. 2011a). Many of the adherents to these traditions are already sensitive to the issues of poverty, to the need for education and to the dangers of greed and materialism. They also have some (often much) understanding of what constitutes appropriate (here meaning 'life-sustaining') conduct between humans, and between people, nature and the earth (Taylor 2005; Hodson & Hodson 2008). They may be unfamiliar with the technicalities of conservation science, and the value it places upon concepts like biodiversity, environmental science and economics, because they place greater value on their own, sometimes rather different and much more ancient, concepts of how the natural world works. The same applies to many ordinary people in the western world who are not trained in science and do not think in its terms, but nevertheless have a passion to conserve the natural world and have access to long-evolved cultural knowledge that is imbued with conservation principles. Their contributions to conservation on the ground, and to conservation strategy, are crucial and must never be underestimated.

Therefore, our second aim emphasizes that, as contemporary conservation scientists, we are 'cultural pirates' in the sense that many of the concepts that seem to be newly devised, such as sustainability and intergenerational equity, in fact derive from ancient and hard-won knowledge. Ultimately, while environmental science is important, we argue that in order to achieve the goals of global biodiversity conservation, conservationists need to engage the collaboration of committed people, whether or not they have a scientific training.

Public sensitivity to the message of conservation

While it is acknowledged increasingly that the biosphere (including human civilization in anything like its present form) cannot be sustained purely through the conservation of a reserve here and a national park there (Berkes 2007), conservationists have usually been slow to recognize the sociopolitical and cultural means that already exist to achieve global public engagement in conservation issues. Yet participatory democracy has the power to bring about rapid change. For example, in the face of a 500,000-strong petition, in 2011 the UK government backed away from a key policy to sell a substantial number of nationally owned forests into private hands (www.bbc.co.uk/news/uk-13350917). Surely there could be few greater demonstrations, in the UK at least, of its people's concern to preserve forests both for their extrinsic and intrinsic values, and to retain existing rights of access to enjoy and appreciate them.

Two key lessons follow. First, while the wider public's motivation for conservation might differ subtly from that of professional conservation biologists (Wilson 1984), it is nonetheless genuine and heartfelt. Great things can be achieved by concerned local people informed by scientists willing to offer guidance to groups who have strong motivations but who lack the knowledge to focus their concern most effectively. Second, as all the major NGOs such as WWF know well, if the public is on your side, in a democracy you can move mountains, or even stop others from moving them! Furthermore, the response of the general public to Fairtrade™ products demonstrates that a wide sense of responsibility to the global community exists. The Fairtrade Foundation was established in 1992 by a consortium consisting of CAFOD, Christian Aid, Oxfam, Traidcraft plc, the World Development Movement and the National Federation of Women's Institutes. Fairtrade™ sales increased from £16.7 million in 1999 to £799 million in 2009 in the UK alone (Fairtrade Foundation 2011), even increasing market share through the economic recession of 2008. Partnerships between conservationists and indigenous peoples established to promote Fairtrade™ products (e.g. forest coffee) have also been successful.

The concept of mission and the sense of vocation in conservation

The origin of the concept of mission that has entered conservation's lexicon lies deeply rooted within Christianity (where it is known as The Great Commission – *Matthew* 28:16–20). It is chiefly concerned with conveying a specific message to the world (advocacy). The message is intended to be transformative: that by changing the outlook of people, most specifically to have concern for the welfare of others in all one's actions, it will transform their modes of action in the world for the better. Those who feel drawn to this message, to engage with it and to the mission itself, experience a sense of being 'called' and hence of *vocation*. Theologically speaking, one way to recognize a vocation is when the work a person most wants to do is also the work that the world most needs to have done. An essential motivation for carrying the message is that it gives hope to people worldwide. But the Christian message also comes with a warning: if humanity does not develop its collective concern for each other and the biosphere (*Creation*), the result will be catastrophic (*apocalypse*).

Conservation biologists are convinced that the future of the world depends on a wider knowledge and practical application of environmental science, but they must be aware that unrestrained missionary zeal can harm their ability to engage effectively with others. The environmental message is urgent, and those engaged with it feel a sense of vocation to bring it to a wider world, and frustration at those 'so blind they *will* not see'. The literature of conservation biology often points to the consequences of humanity's corporate failure to heed this message and suggests that the failure to engage with it might cause or permit the very collapse of the ecosystems on which human civilization has always depended (Ehrlich & Ehrlich 1991). But there is also a sense of hope, because the delight that so many find in the natural world, and the many successes of conservation that have preserved threatened species by involving wide public engagement, are signs that the message of conservation can be truly transformative, both at the personal and at broader levels (Mabey 2005; Louv 2006).

Unpacking the baggage...

Through the history of empire building, especially in the 19th century, both conservation and Christian mission share a somewhat tainted image for many. Fairly or unfairly, it is an image of western supremacy and of the imposition of western interests over those of indigenous people. The history of Yellowstone, the world's first national park, exemplifies why. Established in 1872 'for the benefit and enjoyment of the people', promotion of the park in the eastern USA relied on the myth that the territory was true *wilderness* and essentially uninhabited because the Indians were afraid to go there. In fact, in 1872 Yellowstone was inhabited by thousands of Indians of the Shoshone, Bannock, Nez Percé, Flathead, Crow and Cheyenne tribes. Ignorant of the region's 11,000 years of Indian habitation, and defining their activities as poaching, the US government expelled the tribes from the Yellowstone region to reservations over the next 20 years (Keller & Turek 1999). Conflict with indigenous people was similarly associated with the establishment of other national parks in North America, and in the British Empire, for example in East Africa (Beinart & Hughes 2007). Indeed, this type of conflict persists today in many biodiversity-rich areas of the globe (see Chapter 7).

Through its association with the imperial subjugation of indigenous people, the history of religious mission is similarly tarnished. For while missionaries often engaged sensitively with indigenous people, even vigorously defending their interests (e.g. Jonathan Edwards' defence of the Housatonic Indians in 1750s Massachusetts), their work is overshadowed by later attempts by empire builders to 'civilize' indigenous peoples through so-called 'Christianization'. While this may have been well meant, it contributed to the decline of aspects of indigenous culture and even extirpation of the people themselves (see for example the role of George Augustus Robinson in leading Tasmanian aborigines to the brink of the destruction of their culture and society; Plomley 1987, 2008).

Despite these tainted impressions, advocates for conservation can learn from the experience of those trained in a western culture who have engaged, with due humility, with another culture. Vincent J. Donovan (1926 – 2000), a Christian missionary working with the Maasai in East Africa in the 1960s, described how his understanding of his own mission was itself transformed by his engagement with people whose life experience differed greatly from his own (Donovan 1982). His book, *Christianity Rediscovered*, is an important illustration of how both parties can grow in understanding through a process of mutually respectful engagement.

The sensitivity exemplified by Donovan is essential to the mission of conservation. For example, 'sacred' sites are areas in indigenous or other rural community territories that are held in special respect. They are often of great interest to conservationists because they are effectively long-term protected areas richly endowed with a diversity of rare species (Bhagwat 2009). The local reasons for preserving these sites are likely to be obscure to western observers, but the conservation motive may still mirror to some degree the ethic of a western conservationist (Byers et al. 2001), even if they are protected for wholly unrelated reasons such as a respect for, or fear of, the spirits of ancestors believed to exist in the area. The unique local tradition will be enforced by community access restrictions, perhaps embodied in customary rules handed down from generation to generation.

Conservation science can do much good by directly negotiating partnership agreements with local people that include provisions for strengthening local traditions and the prestige of local leaders, in order to support species at risk. For example, very considerable progress has been made along these lines in New Zealand, where such joint ventures are strongly supported both by academic research and by national legislation (Moller et al. 2009). In some cases this might be achieved by reference to an adopted religious tradition such as Islam or Christianity teaching the importance of stewardship (see below). Many local traditions are based on ecological knowledge passed down through generations. Such knowledge not only forms an important aspect of indigenous people's worldview, but is also often relevant to conservationists' concerns (Posey 2001). It is this traditional ecological knowledge (TEK) that we turn to in the next section.

Traditional ecological knowledge and the problem of anthropocentricity

It is not only for ethical reasons that the extinction of indigenous cultures is of great concern to conservationists. It is of concern also for pragmatic reasons because it often accompanies a loss of knowledge and experience of intimate – and actually or potentially sustainable – engagement with nature. Moreover, the process of cultural extinction is comparable with, and frequently accompanies, biodiversity loss whose cause is often common to both processes (Maffi 2001). In the same way that the pharmaceutical value of ethnobotanical knowledge is well understood (Chivian & Bernstein 2008), humanity cannot afford to lose knowledge that might help to engage more sustainably with the rest of the biosphere.

The issue of direct (as opposed to virtual) engagement with nature has great currency for conservation. In 2008, the earth's human population passed a significant milestone: the majority of people now live in some form of urban environment rather than in a rural, natural or semi-natural one. The shift from rural to urban accompanies a shift in personal space from one with many natural elements and fewer artificial to one with many artificial and few natural elements. This shift is accompanied by a decline in knowledge of the natural world (Balmford et al. 2002; Louv 2006) and increasing fear of it (Louv 2006), both of which are to be expected from the context-dependent nature of human learning and knowledge.

The shift to a more anthropogenic environment also accompanies a shift to an anthropocentric worldview. It is a view expressed in sentiments that prioritize issues perceived to concern humans (e.g. economy, health, education, sport and leisure) ahead of what are purportedly non-human issues, notably including nature and conservation. However, in reality, the long-term persistence and integrity of major ecosystems are essential for the very existence of humanity: without ecology, there is no economy (Millennium Ecosystem Assessment 2005). Therefore, our goal must be a holistic appreciation of the interactions of all the issues: environment (including biodiversity and climate), economics, health, education and leisure (including sport), etc. (see Chapter XXX [biophilia chapter]).

These shifts in perception have become apparent largely through anthropological studies that allow comparison between modern societies and indigenous hunter-gatherer people, and especially consideration of what has become known as traditional ecological knowledge or TEK (see, for example, Atran & Medin 2008, Tidemann & Gosler 2010 for reviews). A number of general findings from these studies are of interest to conservationists. These include the facts that many traditional peoples classify organisms in a box-within-box hierarchy schema consistent with that of systematic biology; traditional, non-western taxonomies often mirror scientific taxonomies (see Maffi 2001 and Atran & Medin 2008 for examples); and furthermore, people commonly name all taxa in their environment: their approach is not simply utilitarian (Desfayes 1998; Maffi 2001; Atran & Medin 2008; Yoon 2010; and *contra* Birkhead 2008; Mynott 2009). Most important is the perception of indigenous hunter-gatherer people themselves that they exist as part of the natural community in which they live, and engage with it in a reciprocal relationship of mutual interdependence (Ingold 1996; Gosler et al. 2010) rather than as impartial observers or in alienation from nature, as is the perception in most modern western societies. For an account of the significance of TEK to conservation see Bonta (2010), who recommends that local, traditional knowledge, whether 'indigenous' or not, should be incorporated into conservation projects at the research, planning, programme design and implementation stages, as much as possible an equal partner to western, scientific approaches.

'Conservation dialogues' between diverse, concerned actors representing these approaches can be challenging, but may also result in synergies built from the trust instilled by mutual respect for differing ways of conceiving and classifying nature. Conservation dialogues are best held at the community level, since communities existed before conservationists became interested in TEK, and those same communities must endure after TEK projects lose their funding (see Manriquez 2001). Community-based conservation projects where TEK and global science (*sensu* Sillitoe 2006) engage each other also open a space for the incorporation of cultural traditions in a complex mixture of secular and religious attitudes and missions toward the common goal.

Many peoples adhering to traditional precepts have perceived (and still do) an original quality of 'humanity' or 'subjectivity', which pervades both animal and human societies (de Castro 1998). Such views may help preserve nature even when they are completely inconsistent with the western view of how nature works. For instance, to understand why the Inuit people persist in low-level traditional whale hunting, and to gauge their level of respect for the whales and thus their rationale for conserving them, it is necessary to appreciate the Inuit cosmological view (though we need not participate in it) whereby whales derive their existence from the body parts of a mythical human and thus possess sacred qualities akin to humanity (Harrop 2011). Science can make a contribution to this issue by estimating the sustainable level of cultural harvest, for example within a context of the role of humans in stewardship (see below), whilst refraining from comment on the reasons (however apparently illogical to westerners) why cultural harvest should or should not be permitted.

Although science provides the western conservationist with a mathematically coherent understanding of the physical unity of, and interdependence inherent within, ecosystems, we must still acknowledge that numerous and complex links between humans and nature are also perceived by people in cultural and spiritual contexts. For example, despite over 200 years of non-Aboriginal occupation of Australia, conservationists there still tend to operate within a limited framework of western environmental knowledge. Promotion of 'alternative' knowledge is slow, and much of it seems to contravene that of western science.

> 'Jutja (Merten's Water Monitor), Danbukarr (Spangled Perch) and Gurruk (Fresh-water Mussel) all live in the same water and belong to the same family. When Gurruk opens its shell, it is sometimes eaten by fish such as Danbukarr. At night, Gurruk opens his shell and flies up out of the water to the tops of paperbarks to eat the flowers or to the woolly butts (*Eucalyptus miniata*) to eat the orange blossoms. Before first light, he drops back down into the water. When you sleep by the river, you can hear them plopping back into the water.'
>
> (Wynjorroc et al. 2001)

Such stories are imbued with symbolic and culturally specific metaphorical meaning. They mean more than the mere biological, or literal, rendering of the story. Failing to recognize this, modern conservationists have a hard time 'listening' to what a culture with around 60,000 years (Thorne et al. 1999) of experience in Australia is saying, and so dismiss it simply because, on first glance, it doesn't make sense (but see Gosler et al. 2010). In fact, if we translate this story from metaphorical into literal truth, we find that Aborigines are systematic zoologists who classify species as related if they occupy the same habitat within their language region (see also Bradley et al. 2010), and who recognize that Gurruk's filter feeding at night results in an accumulation of orange stamens about themselves, and that the dynamism of the ecological system described can be experienced directly if close at hand.

Sometimes the anthropocentricity of western scientists and conservationists is challenged by indigenous knowledge. For example, victims and survivors of the 2009 fires that ravaged south eastern Australia might reel at Aboriginal attitudes towards fire in northern Australia:

> Fire is nothing, just clean up.
> When you burn,
> New grass coming up,
> That mean good animal soon ...
> (Neidjie et al. 1987)

Big Bill Neidje was referring to the long-practised Aboriginal burning of the landscape. Despite its description as firestick farming by Rhys Jones in 1969, this management practice has been ill understood by non-Aboriginals until recently. A return to indigenous burning practices in western Arnhem Land has reduced the frequency of unplanned fires and especially the incidence of destructive late dry season fires (Whitehead et al. 2009), so reducing greenhouse gas emissions from fire (Russell-Smith et al. 2009), influencing biodiversity positively (Woinarski et al. 2009) and reaffirming the valued centrality of fire use to indigenous life (Ritchie 2009).

The recognition that western biologists need to embrace TEK is reflected in the creation of a number of conservation NGOs (see, for example, www.arcadiafund.org.uk, www.ejfoundation.org and www.natureandculture.org) linking the conservation of biodiversity and indigenous cultures.

Religion and conservation

In the rest of the chapter we move from the relative micro scale of engaging with indigenous cultures to the relative macro scale of engaging with global faith communities. The benefits for conservation of engaging with the latter differ from those of the former largely in the opportunity that this dialogue offers the conservation mission to engage with billions, rather than thousands of people. The essential approaches are common to both, however, and resonances will be seen between them.

Synergies in the language of conservation and faith traditions: the concept of stewardship

As with many indigenous cultures (Posey 2001), the Abrahamic traditions centre their environmental ethic on the usufructary concept of stewardship. By this, they mean the upkeep and management of nature as the responsibility of humans, who also hold nature as heritage in trust for future generations (Osland 1999).

The discipline of conservation biology is based on what looks like a similar premise, valid in itself but better described by the modern term 'management' because the religious concept of stewardship is hotly debated (Berry 2006). Nevertheless, conservation biologists Power & Chapin (2009) have proposed 'planetary stewardship' as a framework for harnessing science and society together to reduce rapid anthropogenic damage to the biosphere. Even though people from different backgrounds might have very different motivations for participating, they can still agree on the concept of inter- and intragenerational equity that underpins contemporary sustainability strategies and policies.

The nature of leadership for conservation advocacy

Religious leaders can be important agents in effectively communicating issues in conservation that may not always be accessible or understandable to lay members of their communities. There are two reasons for this. First, faith leaders often command large congregations, and their word is received with much more respect than is that of secular leaders. People often have a much more trusting relationship with their religious leaders than they have with political leaders. A second reason is that faith group leaders are familiar with their group's vocabulary and know what words evoke positive reactions in their congregations. For example, phrases such as 'the creation' and 'creation care' are helpful to communicate a conservation message when working with Abrahamic religions, at least when their terms have approximate equivalents to those used in science. Because these phrases and concepts have been used, often for centuries, they permeate a community's culture in a way that newly created institutional concepts such as 'biodiversity' and 'sustainable use' could never hope, in their short lifetimes, to emulate (Wilson 2006). While there may be a more or less subtle distinction between 'caring for creation' and 'conserving biodiversity', people of good will can still agree on the outcomes.

Modernity's disconnection from nature

Although some religious people see environmentalism as blasphemous – akin to animism or nature worship – or completely irrelevant (e.g. extreme fundamentalist Christians may see no point in conserving a natural world which is due to be destroyed at the end of time), there is increasing interest from religious groups in reconnecting with nature. For example, Paul Gorman is founder and director of the National Religious Partnership for the Environment, which is leading a faith-based environmental movement in the USA attempting to reconnect people, particularly young children, with nature.

Religious and conservation activists therefore share much, both in their means and their aims, although their reasons may differ. Some conservationists are already willing to find common ground on the boundaries between faith and conservation to engage more effectively with faith communities whilst also working out any disagreements. In return, faith communities, active at the local scale, can take a leading role in conservation on the ground. There are abundant signs that this is already happening: see below, for example, on the growth of the A Rocha movement (www.arocha.org/int-en/index.html).

Faith communities helping conservation

Misali is an island in the Zanzibar archipelago off the coast of Tanzania. Well known as a nesting site for green turtle (*Cheloniamydas* L) and hawksbill turtle (*Eretmochelysimbricata* L), the coral reef surrounding this island is also a lucrative fishing ground (Bourjea et al. 2008). The predominantly Muslim fishing communities of neighbouring Pemba Island rely on Misali's fish stocks for their livelihoods. The rapidly depleting fish stocks have forced these fishermen to use dynamite or cyanide fishing and bottom trawling. Such destructive methods are known to remove up to 25% of an area's seabed life on a single run (Worm et al. 2006). Recent educational programmes started by the Islamic Foundation for Ecology and Environmental Sciences (IFEES), however, have inspired the local fishermen to use more sustainable methods of fishing to care for Misali's coral reef and its life forms. In addition to conservation of coral reefs, this programme is also helping to ensure that local fishermen earn their livelihood in a sustainable manner. The IFEES has now proposed to declare Misali a 'hima'. This is in effect a conservation zone and a system of protected area that originated in the early days of Muslim culture. The hima would be established according to Shariah law, which is applied in certain circumstances on the island and is thus binding within the local law (IFEES 2006).

Indonesia has the world's largest Muslim population and contains 10% of the world's remaining rainforests, although these crucial and highly biodiverse forests are receding rapidly. A current project in Sumatra, funded by the UK's Darwin Initiative, is seeking to harness Islamic faith principles that concern nature preservation and which echo and yet predate many contemporary conservation concepts, including the use of protected areas and sustainability. The project is seeking to test whether local people are more likely to understand, and be willing to comply with, Muslim principles and teachings than with remote scientific or institutional directives, and thereby work to stem the relentless tide of rainforest destruction. These faith principles, such as *Khalifah* (which corresponds roughly with the biblical principle of 'stewardship'), are set out in the *Al-Qur'an* (Khalid 2002) and their application in Shariah law, and in local religious teachings where this does not apply (as in Sumatra), has resulted in the development of many concepts with the potential to conserve nature and halt deforestation. Graphic teachings using direct imagery in everyday language, such as the belief that each tree is praying to Allah, may have a powerful impact on the ethical perspectives of local people. The project is working with local communities, Islamic leaders, teachers, scientists and institutional/governmental and non-governmental partners to establish local awareness of these tenets of Islam through local education and religious teaching systems, to facilitate their integration with forest management systems and to measure the impact of the resultant changed behaviour on the quality of these crucial ecosystems. An important finding is that the tenets of Islam appear to integrate well with local Sumatran forest management principles that evolved through the earlier pantheistic religions that dominated the area prior to the arrival of new cultures (Harrop et al. 2011; IFEES 2010).

In South and Central America, two Roman Catholic priests have been especially active in forest protection. In Brazil, Father Edilberto Sena founded the Amazon Defence Front to protect the Amazon forest against commercial development in the region (BBC 2009). Since 2001, he has run a radio station that reaches at least 500,000 people in the Amazon and promotes a model of agriculture that is not dominated by large agribusinesses and provides sustainable livelihoods to local people. In Honduras, Father José Andrés Tamayo Cortez has directed the Environmental Movement of Olancho (MAO). This coalition of subsistence farmers, community and religious leaders is defending forests against uncontrolled commercial logging that threatens both forest ecosystems and water supplies for local communities. In 2003 he was awarded the Honduras National Human Rights Award, and in 2005 the Goldman Environmental Award (GEA). Linking environmental and human ethics, Father Tamayo is quoted as saying: 'Natural resources and life itself are human rights; therefore, to destroy God's creation is to attack human life; our last remaining option is to defend life with our own life' (GEA citation).

These examples also point to the role of faith communities in standing up against corruption and criminal activities, which often lie behind environmental damage, and from which vulnerable people may need protection.

The move toward environmental conservation also extends beyond the Abrahamic faiths to the eastern religions. One example among many is a case from Cambodia. The Association of Buddhists for the Environment (ABE), founded in 2005, involves monks in environmental activities such as community forestry. By undertaking tree ordination ceremonies and promoting protection of sacred forests, the network introduces links between Buddhism and conservation to the local communities. In collaboration with Conservation International, this network also supports the Green Pagodas project where local monks are trained to communicate the conservation message through Buddhist practices and rituals. In addition to forest conservation, this community forestry initiative also provides livelihoods to local residents (ABE 2011).

The Chipko movement is another significant grassroots movement protecting forests, this time in India and Nepal, adhering to the Hindu tradition of *satyagraha* or non-violent protest for the sake of truth and purity. Since 1973 this movement, named after its practice of hugging trees to prevent their being felled, has been led largely by peasant women. Their aim has been to protect local forests and their human inhabitants from contract logging and other development schemes that threatened to be socially and environmentally destructive. Their methods are always non-violent. Chipko participants express a reverence for nature and a perception of trees as sacred; they draw heavily on Hindu tradition. It has been argued that the significance of Chipko as a women's movement reflects the long-standing association of nature and 'the feminine principle' in Indian religious traditions (James 2000).

These examples show clearly that religions are helping conservation on the ground. But this is especially so when faith encourages the recognition that humans are responsible for their actions with respect to non-human as well as human inhabitants of the earth in a spirit of responsible stewardship. By doing so, they are also making conservation culturally acceptable whilst providing alternative, sustainable sources of income to people whose livelihoods have come under increasing pressure from the depletion of resources. Furthermore, they demonstrate that religion-based conservation initiatives can begin at the grassroots and in culturally appropriate contexts, making them culturally more acceptable locally than are many of the top-down approaches in conservationists' 'tool-kits' (Woodhams 2009).

The above case studies come from Latin America, Africa, South and South-east Asia, where many biodiversity hot spots are located, and they encompass terrestrial, freshwater and marine ecosystems. The major, or global, faith traditions – Christianity, Islam, Hinduism, Buddhism, etc. – are seen to be engaged in these conservation activities despite differences in their origins and underlying philosophies. Even though the few examples given here may seem anecdotal, quantitative analyses have suggested that an overwhelming majority of countries included within Conservation International's biodiversity hotspots have a high proportion of population affiliated to these faiths and likely to be influenced by teaching from their own leaders on matters relating to conservation (Bhagwat & Palmer 2009; Bhagwat et al. 2011b). The numerous examples that now exist (see www.arcworld.org) suggest that the above are not isolated case studies but examples that are replicated many times across the world. Together, the global faith communities have the potential to speak to more than 5 billion people through their own leadership and community structures.

Conservation and faith communities

But why work with faith communities? While some recent texts (e.g. Johns 2009) have failed to recognize the significance of such co-operation, the potential benefits for conservation of engaging with faith communities have been noted many times (e.g. Palmer & Finlay 2003; Gottlieb 2006; Bhagwat & Palmer 2009; Bhagwat et al. 2011a,b). The relevance of religion to conservation advocacy was outlined above. Two further issues must be recognized that help to harmonize considerations of faith with the concerns of conservation. On the one hand, a connection with nature is widely recognized as key to human spiritual development (see, for example, entries in Taylor 2005). Hence the connection between spirituality and nature has been especially strongly developed in indigenous forest-dwellers (Ingold 1996; Sponsel 2001; Tidemann & Gosler 2010). On the other hand, a full understanding of what it means to say that life and living organisms have intrinsic value is a concept already well explored in western metaphysical philosophy (Deane-Drummond 2008). A perception of the intrinsic value of life is entrenched within the human psyche, and so is a core motivational driver for conservation action, but its metaphysical definition means that it cannot be used in strategic conservation planning, i.e. to prioritize concern for one species or ecosystem over another (Gosler 2009). For this reason, Justus et al. (2009) promoted the notion of instrumental, rather than intrinsic value. Nevertheless, the concept of intrinsic value lies behind the key statement of the World Charter for Nature UN General Resolution A/RES/37/7, 28 October 1982, which declares that: 'Every form of life is unique, warranting respect regardless of its worth to man'.

Increasingly, conservationists are becoming aware that human valuation of nature can be expressed through spiritual awareness and the establishment and protection of sacred natural sites (Bhagwat 2009; Verschuuren *et al.* 2010). For example, the Alliance of Religions and Conservation (ARC; www.arcworld.org) acknowledges that, while religiously motivated practice can sometimes be harmful to conservation interests, this is comparatively rare. Such cases are best addressed through mutually respectful engagement with the faith community to discuss points of difference (e.g. Gutiérrez & Buchy 2011) in the light of fundamentals such as stewardship and compassion for all beings.

Because a desire for right action drives much religious observance, any realization that a practice might raise ethical concerns is likely to be taken seriously. But faith communities are, understandably, resistant to criticism from without, so experience shows that the swiftest change follows when new perceptions emerge from within the community itself. This approach is vital to productive engagement with faith communities, and is well illustrated by the rapid development over the last 40 years of environmental awareness, concern and action for conservation within the Christian community in the West.

Whilst many faiths could be used to illustrate the key points of synergy, the greening of Christianity (Gottlieb 2006) is particularly relevant for three reasons. First, the Judeo-Christian heart of western culture has been uniquely blamed for the environmental crisis (e.g. White 1967). Second, the conservation message needs to be heard by that western culture whose model of capitalist economics is now the principal driver of global environmental change. That culture is, at least nominally and culturally, Christian, a fact borne out by the observation that in the UK (one of the most secular western countries with just 2 million [3.2%] regular church-goers) a considerable majority of people still list Christianity as their religion on the decadal census (Figure 6.1). Third, this narrative is significant because the strong resonance between the objectives of conservation and Christianity noted earlier suggests a synergy of purpose, the nurturing of which could represent a natural and fruitful progression for both communities (King 2001; Hodson 2010; Sluka et al. 2011).

Two further, practical points might be noted: first, in many countries the Church represents one of the most significant social influences, and its contribution to developing healthy human communities can be substantial (see, for example, Jarvis et al. 2010); second, traditional churches are often significant land-owners (as is the Church of England) so that if appropriate management of all the Church's landholdings, taking due care of wildlife interests, were to be widely observed from churchyard to glebe land and beyond, the results might be significant for many wild populations (see Chapter 13). Globally, there are numerous examples of individual churches with active environmental groups managing their churchyards and other land for wildlife conservation (for example, see www.caringforgodsacre.org.uk/ or www.sthilda.ca/environmental.html).

In a landmark paper, the historian Lynn Townsend White Jr denounced Christianity for its part in the developing global environmental crisis (White 1967). The essence of his critique was that the faith encouraged anthropocentricity. However, being a Christian himself, White suggested that St Francis of Assisi should be recognized as the patron saint of ecology to encourage greater consideration of ecological concerns. Rather than turning within itself and battening down the hatches for another onslaught from the secular world, many

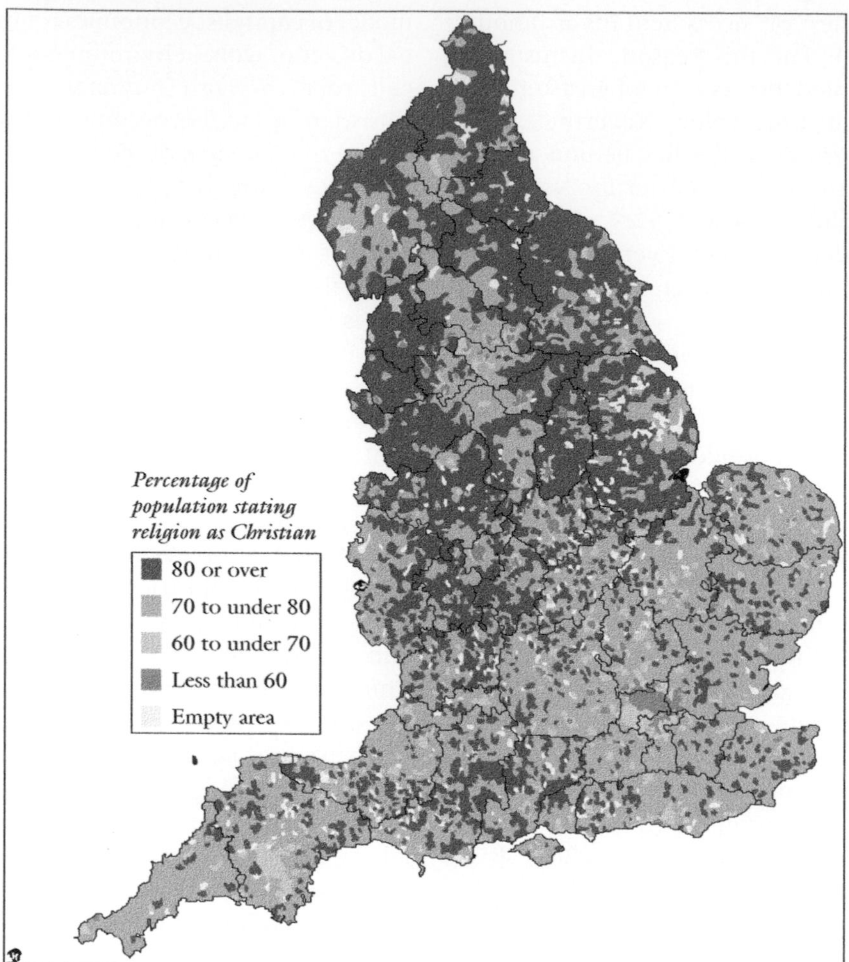

Figure 6.1 Map of England showing by Church of England parishes the percentage of people declaring religion as Christian in the 2001 census. The strongest responses (above 70%) typically come from rural communities with less artificial environments. Data are from census key statistics (2001), Office National Statistics (2003), reconfigured by Archbishops' Council, Research and Statistics. The figure is reproduced here with the kind permission of the Department of Research and Statistics, Archbishops' Council.

Christians took his critique very seriously (Berry 2000; Sponsel 2001), not least because serious discussion of the proper nature of human stewardship of the earth was already ongoing in the World Council of Churches (WCC) by the time of White's paper (King 2001). Over the next decades, the Christian community came to reconsider its interpretation of scripture, in particular that the fundamental commandment, supported by rigorous theological exegesis, was to care for Creation, not to abuse it (Hillel 2007; Hodson & Hodson 2008; Deane-Drummond 2008). In other words, the recognition that there was some truth within White's charge led to contrition, and environmental responses by churches began to burgeon across the world (Gottlieb 2006). In the Anglican Communion, this concern is reflected by the addition in 1990 of the fifth mark of mission: *to strive to safeguard the integrity of creation and sustain and renew the life of the Earth*. For a deeper exploration of these developments, see King (2001).

An exemplary expression of this new concern was the founding in 1980 of A Rocha (www.arocha.org), a Christian organization engaging specifically in environmental mission. From its small beginnings in Portugal (Harris 2000), A Rocha has grown in 30 years into a 'community' of projects across 19 countries (Harris 2008). Although A Rocha is concerned chiefly with issues of biodiversity loss, its associates help members to reduce their own environmental footprint (Bookless 2008; Valerio 2008). Whilst there are still many who consider environmental concern to be a side issue to faith, they are decreasing as a proportion of overall adherents to Christianity. Church leaders preach frequently on the importance of environmental awareness and stewardship (as do leaders of other faith communities, most notably the Dalai Lama, who has long promoted the need for human harmony with nature), and A Rocha is but one of many examples of the expression of the new environmental concern that is developing within the Christian faith community. One of the strongest expressions of the changing perception within Christianity was the 2008 publication of *The Green Bible* (2008, Collins, London). This New Revised Standard Version of the Bible carries substantial introductory commentary by secular environmental and religious leaders alike. Printed on recycled paper, this edition has printed in green ink the copious texts relevant to conservation advocacy.

Conservation and the emotional human bond with nature

> 'If I speak in the tongues of men or of angels, but have not love, I am but a resounding gong or a clanging cymbal.'
>
> (*1 Corinthians* 13:1)

Citing Gould, Macdonald et al. (2007, p.279) stated 'we cannot win this battle to save species and environments without forging an emotional bond between ourselves and nature as well, for we will not fight to save what we do not love'. We argue here that the emotional bond that connects humans with nature already exists, that indeed it is this love of nature (what Wilson (1984) termed *biophilia* (see Chapter 9) that motivates us to become involved in conservation in the first place. Indeed, how often have we heard conservationists refer to their work as a labour of love? So the issue of communicating the conservation message to the world has, essentially, to be one that ignites the natural human motivation to conserve for the love of nature.

When conservationists have tried to make their point in the languages of western science, or of economics, or of fear, they have found that none of these is suitable to convey a message based on love and respect, in the senses in which Leopold used those words in the quotation at the head of this chapter. Science notoriously finds little room within its lexicon for love; economics (e.g. Sukhdev et al. 2010) makes space for a love of nature only when it generates a profit; and fear attempts to drive change through the apocalyptic images of climate change and biodiversity loss. It is said that fear drives out love (Jampolsky 2004). Our task is to communicate to a wider public who understand love, but not the lexicon of science or economics, and who seek to escape fear.

The essential issue is one of alignment (Macdonald et al. 2007), by which conservation moves into the public arena and is seen to resonate with the broader issues of sustainability that concern people more generally. This is a matter of skilful and appropriate communication, and of understanding the cultures with which we are engaging. If we misunderstand this by attempting to impose our views without sensitivity to that need for resonance, we shall fail to make that engagement. This entails respect for other people, their cultures and their concerns (Gosler et al. 2010), and the realization that conservation is part of a global drive toward ecological sustainability (CBD UNEP 2002).

Acknowledgements

We are grateful to Stephen Awoyemi, Carolyn (Kim) King, Andy Lester, Barbara Mearns, Tom Moorhouse, John Paull and the editors for encouragement and constructive comments on earlier drafts of this chapter.

> *We abuse land because we regard it as a commodity belonging to us. When we see land as a community to which we belong, we may begin to use it with love and respect.*
>
> **(Leopold 1949)**

References

ABE (Association of Buddhists for the Environment) (2011) www.sanghanetwork.org

Atran, S. & Medin, D. (2008) *The Native Mind and the Cultural Construction of Nature*. MIT Press, Cambridge, MA.

Balmford, A., Clegg, L., Coulson, T. & Taylor, J. (2002) Why conservationists should heed Pokémon. *Science*, **295**(5564), 2367b.

BBC (2009) The Amazon's most ardent protector. http://news.bbc.co.uk/1/hi/world/americas/7915568.stm

Beinart, W. & Hughes, L. (2007) *Environment and Empire*. Oxford University Press, Oxford.

Berkes, F. (2007) Community-based conservation in a globalized world. *Proceeding of the National Academy of Sciences*, **39**, 15188–15193.

Berry, R.J. (2000) *The Care of Creation: Focusing Concern and Action*. Inter-Varsity Press, Leicester.

Berry, R.J. (ed.) (2006) *Environmental Stewardship: Critical Perspectives – Past and Present*. T. & T. Clark Ltd, London.

Bhagwat, S.A. (2009) Ecosystem services and sacred natural sites: reconciling material and non-material values in nature conservation. *Environmental Values*, **18**, 417–427.

Bhagwat, S.A. & Palmer, M. (2009) Conservation: the world's religions can help. *Nature*, **461**(7260), 37.

Bhagwat, S.A., Dudley, N. & Harrop, S. (2011a) Religious following in biodiversity hotspots: challenges and opportunities for conservation and development. *Conservation Letters*, **4**, 234–240.

Bhagwat, S.A., Ormsby, A.A. & Rutte, C. (2011b) The role of religion in linking conservation and development: challenges and opportunities. *Journal for the Study of Religion, Nature and Culture*, **5**, 39–60.

Birkhead, T. (2008) *The Wisdom of Birds: An Illustrated History of Ornithology*. Bloomsbury, London.

Blaikie, P. & Jeanrenaud, S. (1996) *Biodiversity and Human Welfare*. UN Research Institute for Social Development, New York.

Bonta, M. (2010) Ethno-ornithology and biological conservation. In: *Ethno-Ornithology: Birds, Indigenous Peoples, Culture and Society* (eds S. Tidemann & A.G. Gosler). Earthscan, London.

Bookless, D. (2008) *Planetwise: Dare to Care for God's World*. Inter-Varsity Press, Leicester.

Bourjea, J., Nel, R., Jiddawi, N.S., Koonjul, M.S. & Bianchi, G. (2008) Sea turtle bycatch in the West Indian Ocean: review, recommendations and research priorities. *Western Indian Ocean Journal of Marine Science*, **7**, 137–150.

Bradley, J. and Yanyuwa families (2010) *Singing Saltwater Country*, Allen and Unwin, Sydney.

Byers, A.B., Cunliffe, R.N. & Hudak, A.T. (2001) Linking the conservation of culture and nature: a case study of sacred forests in Zimbabwe. *Human Ecology*, **29**, 187–218.

CBD UNEP (2002) Report on the Sixth Meeting of the Conference of the Parties to the Convention on Biological Diversity (UNEP/CBD/COP/6/20/Part 2) Strategic Plan Decision VI/26. www.biodiv.org/doc/meetings/cop/cop-06/official/cop-06-20-part2-en.pdf

CBD UNEP (2010) *CBD Strategic Plan for Biodiversity 2011–2020 and the Aichi Targets: 'Living in Harmony with Nature'*. www.cbd.int/doc/strategic-plan/2011-2020/Aichi-Targets-EN.pdf

Chivian, E. & Bernstein, A. (2008) *Sustaining Life: How Human Health Depends on Biodiversity*. Oxford University Press, New York.

Deane-Drummond, C. (2008) *Eco-theology*. Darton, Longman and Todd, London.

de Castro, E.V. (1998) Cosmological deixis and Amerindian perspectivism. *Journal of the Royal Anthropological Institute*, **4**, 469–488.

Desfayes, M. (1998) *A Thesaurus of Bird Names: Etymology of European Lexis Through Paradigms*. Vol 1. Musees Cantonaux du Valais, Switzerland.

de Waal, F. (2006) *Primates and Philosophers: How Morality Evolved*. Princeton University Press, Princeton.

de Waal, F. (2010) *The Age of Empathy: Nature's Lessons for a Kinder Society*. Three Rivers Press, New York.

Donovan, V.J. (1982) *Christianity Rediscovered: An Epistle from the Masai*. Orbis Books, New York.

Ehrlich, P.R. & Ehrlich, A.H. (1991) *Healing the Planet*. Addison-Wesley, New York.

Fairtrade Foundation (2011) www.fairtrade.org.uk/what_is_fairtrade/facts_and_figures.aspx

Gosler, A.G. (2009) Surprise and the value of life. In: *Real Scientists, Real Faith* (ed. R.J. Berry). Lion Hudson, Oxford.

Gosler, A., Buehler, D. & Castillo, A. (2010) The broader significance of ethno-ornithology. In: *Ethno-Ornithology: Birds, Indigenous Peoples, Culture and Society* (eds S. Tidemann & A.G. Gosler). Earthscan, London.

Gottlieb, R.S. (2006) *A Greener Faith: Religious Environmentalism and Our Planet's Future*. Oxford University Press, Oxford.

Greenwood, J.J.D. (2007) Citizens, science and bird conservation. *Journal of Ornithology*, **148**(Suppl 1), S77–S124.

Gutiérrez, R.A. & Buchy, P. (2011) The merit release birds: Buddhist ritual and implications in the H5N1 virus contamination cycle. *BMC Proceedings*, **5**(Suppl 1), 64.

Harris, P. (2000) *Under the Bright Wings*. Regent College Publishing, Vancouver.

Harris, P. (2008) *Kingfisher's Fire: A Story of Hope for God's Earth*. Monarch Books, Oxford.

Harrop, S.R. (2007) Traditional agricultural landscapes as protected areas in international law and policy. *Agriculture, Ecosystems and Environment*, **121**, 296–307.

Harrop, S.R. (2011) Impressions: whales and human relationships in myth, tradition, and law. In: *Whales and Dolphins Cognition, Culture, Conservation and Human Perceptions* (eds P. Brakes & M.P. Simmonds). Earthscan, London.

Harrop, S.R. McKay, J. & Dinata, Y. (2011) *Darwin Initiative Annual Report*. http://darwin.defra.gov.uk/project/17009/

Hillel, D. (2007) *Natural History of the Bible: An Environmental Exploration of the Hebrew Scriptures*. Columbia University Press, New York.

Hodson, M.J. & Hodson, M.R. (2008) *Cherishing the Earth: How to Care for God's Creation*. Lion Hudson, Oxford.

Hodson, M.R. (2010) Storm clouds and mission: creation care and environmental crisis. In: *Holistic Mission. God's Plan for God's People* (eds B.E. Woolnough and W. Ma). Regnum International, Oxford.

IFEES (Islamic Foundation for Ecology and Environmental Sciences) (2006) Inspiring Change. ECO Islam: Voice of the Islamic Foundation for Ecology and Environmental Sciences. www.ifees.org.uk

IFEES (Islamic Foundation for Ecology and Environmental Sciences) (2010) Islamic Belief and Sumatran Forest Management. www.ifees.org.uk

Ingold, T. (1996) Hunting and gathering as ways of perceiving the environment. In: *Redefining Nature* (eds R. Ellen & K. Fukui). Berg, Oxford.

James, G.A. (2000) Ethical and religious dimensions of Chipko resistance. In: *Hinduism and Ecology: The Intersection of Earth, Sky, and Water* (eds C.K. Chapple & M.E. Tucker). Harvard University Press, Cambridge, MA.

Jampolsky, G.G. (2004) *Love is Letting Go of Fear*. Celestial Arts Publishing, Berkeley, CA.

Jarvis, D., Porter, F., Lambie, H., Broughton, K. & Farnell, R. (2010) *Building Better Neighbourhoods: The Contribution of Faith Communities to Oxfordshire Life*. Coventry University for Oxfordshire Stronger Communities Alliance, Coventry.

Johns, D. (2009) *A New Conservation Politics: Power, Organization Building, and Effectiveness*. Wiley-Blackwell, Oxford.

Jones, R. (1969) Firestick farming. *Australian Natural History*, **16**, 224–231.

Justus, J., Colyvan, M., Regan, H. & Maguire, L. (2009) Buying into conservation: intrinsic versus instrumental value. *Trends in Ecology and Evolution*, **24**, 187–191.

Keller, R.H. & Turek, M.F. (1999) *American Indians and National Parks*. University of Arizona Press, Tucson, AZ.

Khalid, M.K. (2002) Islam and the environment, In: *Encyclopedia of Global Environmental Change*, Volume 5 (eds T. Munn & P. Timmeran). Wiley, New York.

King, C.M. (2001) Ecotheology: a marriage between secular ecological science and rational, compassionate faith. *Ecotheology*, **10**, 40–69.

Kinsey, D.W. & Hopley, D. (1991) The significance of coral reefs as global carbon sinks – response to Greenhouse. *Palaeogeography, Palaeoclimatology, Palaeoecology*, **89**, 363–377.

Leopold, A. (1949) *A Sand County Almanac*. Oxford University Press, New York.

Louv, R. (2006) *Last Child in the Woods: Saving Our Children from Nature-deficit Disorder*. Algonquin Books, New York.

Mabey, R. (2005) *Nature Cure*. Chatto and Windus, London.

Macdonald, D.W. & Service, K. (eds) (2007) *Key Topics in Conservation Biology*. Blackwell, Oxford.

Macdonald, D.W., Collins, N.M. & Wrangham, R. (2007) Principles, practice and priorities: the quest for 'alignment'. In: *Key Topics in Conservation Biology* (eds D.W. Macdonald & K. Service). Blackwell, Oxford.

Maffi, L. (2001) Introduction: on the interdependence of biological and cultural diversity. In: *Biocultural Diversity: Linking Language, Knowledge and the Environment* (ed. L. Maffi). Smithsonian Institution Press, Washington, DC.

Manriquez, L.F. (2001) Silent no more: Californian Indians reclaim their culture – and they invite you to listen. In: *Biocultural Diversity: Linking Language, Knowledge and the Environment* (ed. L. Maffi). Smithsonian Institution Press, Washington, DC.

Meine, C., Soule, M.E. & Noss, R.F. (2006) A mission-driven discipline: the growth of conservation biology. *Conservation Biology*, **20**, 631–651.

Millennium Ecosystem Assessment (2005) *Ecosystems and Human Well-Being: Synthesis*. Island Press, Washington, DC.

Moller, H., O'Blyver, P., Bragg, C., *et al.* & Rakiura Titi Islands Administering Body (2009) Guidelines for cross-cultural Participatory Action Research partnerships: a case study of a customary seabird harvest in New Zealand. *New Zealand Journal of Zoology*, **36**, 211–241.

Mynott, J. (2009) *Birdscapes: Birds in Our Imagination and Experience*. Princeton University Press, Princeton, NJ.

Neidjie, B., Davis, S. & Fox, A. (1987) *Kakadu Man: Big Bill Neidjie*. Mybrood P/L, New South Wales.

Osland, J.O. (1999) The stewardship of human and natural resources. In: *All Creation is Groaning: An Interdisciplinary Vision for Life in a Sacred Universe* (eds C.J. Dempsey & R.A. Butkus). Liturgical Press, Collegeville, MN.

Palmer, M. & Finlay, V. (2003) *Faith in Conservation: New Approaches to Religions and the Environment*. World Bank, Washington, DC.

Plomley, N.J.B. (1987) *Weep in Silence: A History of the Flinders Island Aboriginal Settlement, with the Flinders Island Journal of George Augustus Robinson, 1835–1839*, Blubber Head Press, Hobart, Tasmania.

Plomley, N.J.B. (ed.) (2008) *Friendly Mission: The Tasmanian Journals and Papers of George Augustus Robinson, 1829–1834*, Queen Victoria Museum and Art Gallery and Quintus Publishing, Launceston, Tasmania.

Posey, D. (2001) Biological and cultural diversity. In: *Biocultural Diversity: Linking Language, Knowledge and the Environment* (ed. L. Maffi). Smithsonian Institution Press, Washington, DC.

Power, M.E. & Chapin, F.S. (2009) Planetary stewardship. *Frontiers in Ecology and the Environment*, **7**, 399.

Ritchie, D. (2009) Things fall apart: the end of an era of systematic Indigenous fire management. In: *Culture, Ecology and Economy of Fire Management in North Australian Savannas: Rekindling the Wurrk Tradition* (eds J. Russell-Smith, P. Whitehead & P. Cooke). CSIRO Publishing, Collingwood, Australia.

Russell-Smith, J., Murphy, B.P., Meyer, M.C.P., *et al.* (2009) Improving estimates of savanna burning emissions for greenhouse accounting in northern Australia: limitations, challenges and applications. In: *Culture, Ecology and Economy of Fire Management in North Australian Savannas: Rekindling the Wurrk Tradition* (eds J. Russell-Smith, P. Whitehead & P. Cooke). CSIRO Publishing, Collingwood, Australia.

Sillitoe, P. (2006) *Local Science Vs Global Science: Approaches to Indigenous Knowledge in International Development*. Berghahn Books, Oxford.

Sluka, R.D., Martin Kaonga, M., Weatherley, J., Anand, V., Bosu, D. & Jackson, C. (2011) Christians, biodiversity conservation and poverty alleviation: a potential synergy? *Biodiversity*, **12**, 108–115.

Soulé, M.E. & Wilcox, B.A. (eds) (1980) *Conservation Biology: An Evolutionary-Ecological Perspective*. Sinauer, Sunderland.

Sponsel, L.E. (2001) Do anthropologists need religion, and vice versa? Adventures and dangers in spiritual Ecology. In: *New Directions in Anthropology and Environment* (ed. C.L. Crumley). Altimara Press, Oxford.

Sukhdev, P., Wittmer, H., Schroter-Schlaack, C., *et al.* (2010) *The Economics of Ecosystems and Biodiversity. Mainstreaming the Economics of Nature: A Synthesis of the Approach, Conclusions and Recommendations of TEEB*. TEEB, Bonn.

Taylor, B.R. (2005) *Encyclopedia of Religion and Nature*. Continuum Books, London.

Thorne, A., Grün, R., Mortimer, G., *et al.* (1999) Australia's oldest human remains: age of the Lake Mungo 3 skeleton. *Journal of Human Evolution*, **36**, 591–612.

Tidemann, S. & Gosler, A. (2010) *Ethno-Ornithology: Birds, Indigenous Peoples, Culture and Society*. Earthscan, London.

Tokar, B. (1997) *Earth for Sale: Reclaiming Ecology in the Age of Corporate Greenwash*. South End Press, Cambridge, MA.

Valerio, R. (2008) *'L' is for Lifestyle: Christian Living That Doesn't Cost the Earth*. Inter-Varsity Press, Leicester.

Verschuuren, B., Wild, R., McNeely, J. & Oviedo, G. (eds) (2010) *Sacred Natural Sites: Conserving Nature and Culture*. Earthscan, London.

White, L.T. (1967) The historical roots of our ecologic crisis. *Science*, **155**, 1203–1207.

Whitehead, P.J., Purdon, P., Cooke, P.M., Russell-Smith, J. & Sutton, S. (2009) The West Arnhem Land Fire Abatement (WALFA) project: the institutional environment and its implications. In: *Culture, Ecology and Economy of Fire Management in North Australian Savannas: Rekindling the Wurrk Tradition* (eds J. Russell-Smith, P. Whitehead & P. Cooke). CSIRO Publishing, Collingwood, Australia.

Wilson, E.O. (1984) *Biophilia*. Harvard University Press, Cambridge, MA.

Wilson, E.O. (2006) *The Creation: An Appeal to Save Life on Earth*. Norton, New York

Woinarski, J.C.Z., Russell-Smith, R., Andersen, A.N. & Brennan, K. (2009) Fire management and biodiversity of the western Arnhem Land Plateau. In: *Culture, Ecology and Economy of Fire Management in North Australian Savannas: Rekindling the Wurrk Tradition* (eds J. Russell-Smith, P. Whitehead & P. Cooke). CSIRO Publishing, Collingwood, Australia.

Woodhams, D.C. (2009) Converting the religious: putting amphibian conservation in context. *Bioscience*, **59**, 463–464.

Worm, B., Barbier, E.B., Beaumont, N., *et al.* (2006) Impacts of biodiversity loss on ocean ecosystem services. *Science*, **314**, 787–790.

Wynjorroc, P., Long, H., Long, P., *et al.* (2001) *Plants, Animals and People: Ethnoecology of the Jawoyn People*. Batchelor Institute Press, Batchelor, Australia.

Yoon, C.K. (2010) *Naming Nature: The Clash Between Instinct and Science*. Norton, New York.

7

The human dimension in addressing conflict with large carnivores

Amy Dickman[1], Silvio Marchini[2] and Michael Manfredo[3]

[1]Wildlife Conservation Research Unit, Department of Zoology, Recanati-Kaplan Centre, University of Oxford, Oxford, UK
[2]Instituto Pró-Carnívoros Av. Horácio Neto, Atibaia, Brazil
[3]Human Dimensions of Natural Resources, Colorado State University, Fort Collins, CO, USA

'With public sentiment, nothing can fail. Without it, nothing can succeed.'

— **Abraham Lincoln**

Introduction

Human–large carnivore conflict is a crucially important conservation issue. Large carnivores have undergone striking declines in both geographic range and population size: lions (*Panthera leo*) have disappeared from over 80% of their historic range and declined by 30–50% in just two decades (Ray et al. 2005; IUCN 2006), while tigers (*Panthera tigris*) have experienced a 93% range contraction and 98% population decline over the past 200 years (Seidensticker et al. 2010). In East Africa, resident populations of cheetahs (*Acinonyx jubatus*) and African wild dogs (*Lycaon pictus*) are now found in only 6% and 7% of their respective historic ranges (IUCN 2007), and wild dogs have been extirpated from at least 25 of the 39 countries where they used to occur (Woodroffe et al. 1997). Human persecution has been a major driver of these declines (Ray et al. 2005) but, ironically, the size and configuration of existing protected areas mean that for many carnivores, most of their remaining range falls outside the reserve network (Nowell & Jackson 1996). Therefore, for many major large carnivores, maintaining viable populations across much of their range depends upon developing effective conservation strategies on human-dominated land.

However, achieving harmonious human–carnivore co-existence is notoriously problematic, as these species can impose significant costs on local communities, mainly through livestock depredation (Sillero-Zubiri & Laurenson 2001; Loveridge et al. 2010). Such depredation can be

Key Topics in Conservation Biology 2, First Edition. Edited by David W. Macdonald and Katherine J. Willis.

devastating; for instance, tiger and leopard (*Panthera pardus*) depredation around Bhutan's Jigme Singye Wangchuck National Park cost villagers over two-thirds of their cash income (Wang & Macdonald 2006). Although depredation often accounts for less stock loss than disease, theft or accidents (Dar et al. 2009; Loveridge et al. 2010), even low levels can impose intolerable costs on poor households; in India, depredation of only 1.6 head of livestock cost households 50% of their per capita income (Bagchi & Mishra 2006). In particular, surplus killing, where predators kill multiple animals in one attack, can severely affect stock owners and engender intense hostility towards carnivores; this was seen in the western US, where wolves (*Canis lupus*) killed up to 98 sheep per attack, reducing local tolerance for them (Muhly & Musiani 2009). Impacts are exacerbated if the stock concerned is particularly valuable, or has cultural or emotional value as well as financial significance (Sillero-Zubiri & Laurenson 2001). People traditionally respond to such conflict by killing carnivores; for instance, fears of livestock depredation led to the removal of over 6700 cheetahs from Namibian farmlands from 1980 to 1991 (CITES 1992). Such killing has been a major driver of widespread large carnivore extirpation, and is a critical threat facing remaining populations (Frank et al. 2006).

Clearly, there is a pressing need to develop effective conflict mitigation strategies, and this has been the focus of extensive research (Loveridge et al. 2010). Researchers have assessed characteristics of 'problem' carnivores, and have found that the propensity to cause conflict is often linked to physical impediments which make them less able to kill normal prey (Loveridge et al. 2010); a lion which killed 35 people in Tanzania from 2002 to 2004 had dental problems which could have affected its hunting ability (Baldus 2006), while the infamous man-eaters of Tsavo, which killed at least 28 people in 1898, also had significant tooth damage (Peterhans et al. 2001). Furthermore, in Namibia, stock-raiding cheetahs were more likely than average to have physical problems (Marker et al. 2003a). Scientists have also characterized habitat features linked to conflict; for instance, in Brazil, cattle depredation occurred around 100 m closer to forest than other mortalities (Azevedo & Murray 2007). Such data, gleaned from the natural sciences, can help target conflict mitigation; in likely conflict hot spots, people can avoid grazing vulnerable stock, improve livestock enclosures or employ livestock guardians. Such approaches can significantly reduce conflict; Namibian farmers who employed at least one method of livestock protection, such as guardian animals, suffered 85% less leopard depredation than others (Stein et al. 2010).

However, there is not a simple, consistent relationship between stock loss and negativity towards large carnivores. In Brazil, levels of livestock depredation did not significantly affect local ranchers' attitudes towards jaguars (Conforti & de Azevedo 2003; Marchini 2010), whereas in Namibia, an average of 14 cheetahs per year were still removed from farmland even where they were not thought to cause depredation (Marker et al. 2003b). Moreover, antagonism towards carnivores can persist for many years; in part of the northern Ethiopian highlands, 19% of respondents were negative towards leopards, and over 70% said leopards killed livestock, even though they no longer occurred locally (Yirga et al. 2011). Conversely, people are often more positive towards carnivores than might be expected; although 91% of people living close to South Africa's Kruger National Park were fearful of carnivores, and 81% had suffered depredation, only 36% expressed negative views towards them, and almost everyone (96%) said it was important to protect predators from extinction (Lagendijk & Gusset 2008).

Hence, attitudes towards carnivores are not merely determined by any direct costs imposed, but are the product of a dynamic and complex web of individual, societal and cultural factors. As humans are the common thread in the highly variable arena of human–wildlife conflict, and the course and resolution of conflict are determined by the thoughts and actions of the people involved, understanding the human dimension is the most crucial prerequisite for developing effective mitigation (Manfredo & Dayer 2004; Loveridge et al. 2010).

The complexity of the human dimension

For decades, conservation practitioners have been told about the importance of considering the human dimensions of natural resource management. They have listened, and have demonstrated how invaluable this understanding can be to conservation success. For instance, a detailed anthropological study of Maasai motivations for lion killing, which revealed the cultural value of such killing amongst young men, led to the creation of the 'Lion Guardians' scheme, where young warriors become central to lion monitoring and conservation (Hazzah et al. 2009). This has been very successful: a 2007 analysis showed no lions had been killed in the study area since inception of the scheme, compared to at least 12 on neighbouring ranches (Hazzah & Dolrenry 2007).

However, the human dimension is extremely complex; easy-to-follow rules of thumb are rare, while humans often seem to ignore scientific information and act in inconsistent, seemingly unpredictable ways. Furthermore, there are multiple layers for examining human response to carnivores. It is at the most basic, species level that the greatest commonality in human response to large carnivores is found. Predators helped shape the evolution of hominids (by preying upon them) for millions of years, and remnants of prey response to large carnivores persist in modern human behaviour (Kruuk 2002). Still, most of the variability in human response to carnivores is found in what we learn and how we are socialized, as there are many layers of societal influences that bear upon human action and thought toward carnivores. First, there is the individual, who will have a unique accumulation of experience and learning that shapes their thought about predators. Most research into human dimensions of carnivore conservation has focused on the individual, examining psychological concepts such as attitudes (evaluations that precede behaviour), motivations (behaviour-directing goals), subjective norms (perceptions of what others approve or disapprove of), or values (fundamental ideals that transcend groups of attitudes). These concepts are useful in understanding and predicting behaviour, and can readily be used in practice; for example, if a high percentage of people surveyed support carnivore reintroduction, such action might be approved. This was the case in Yellowstone National Park; in a tri-state survey, 56% of people in Idaho were positive towards wolf reintroduction (while 27% were negative), while in Montana, 44% were positive and 40% negative, and in Wyoming, 49% were positive and 35% were negative (Varley & Brewster 1992). Despite huge controversy, the support for reintroduction was sufficient that wolves were brought back to Yellowstone in March 1995 (Noecker 1997; Phillips et al. 2004).

However, there are obvious limitations to an attitudinal study like this. The attitudes of citizenry are likely to have much more utility and meaning in a more democratic country compared to a more hierarchical one. The organization, structure and systems of a society will strongly influence (1) how carnivores are thought of and treated and (2) how such information would be used in practice. At the societal level, numerous factors are important: the social groups that operate, the networks of communication, the institutions, power differentials, and the political, economic, legal and class structures that shape action. At an even broader level, factors such as demography, available technology and modes of economy should be considered. Demography is particularly critical as population distribution and population growth strongly influence current and future trends of human encroachment into wildlands, reducing habitat and prey base for carnivores and increasing the likelihood of human–carnivore contact.

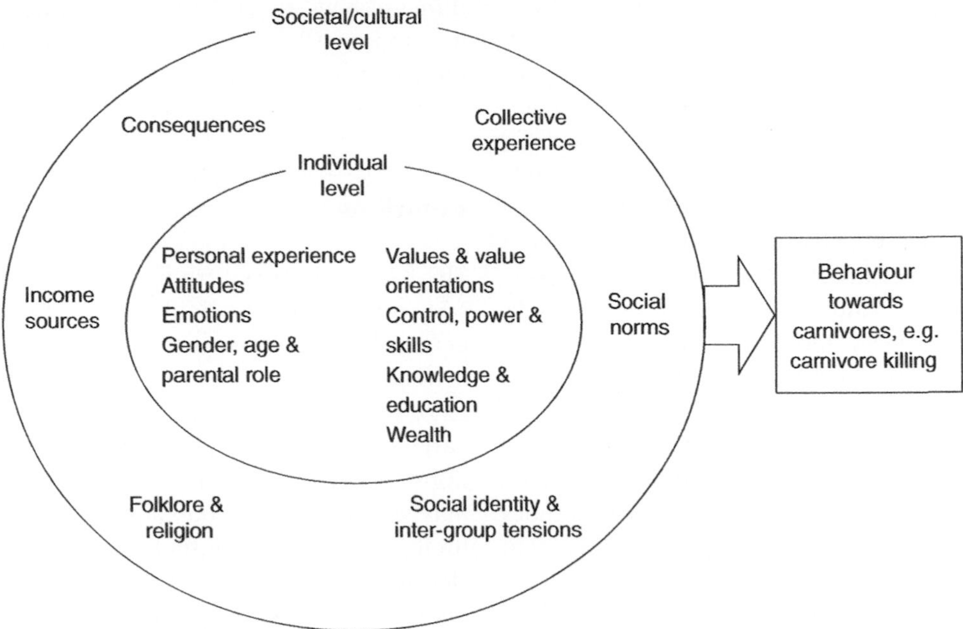

Figure 7.1 Individual and societal/cultural level factors which are likely to shape views towards carnivores and any damage caused, and therefore behaviour towards carnivores.

Ultimately, it is important to remember that the behaviour of an individual is woven into an intricate web of personal, societal and cultural context. With that said, we will turn to reviewing likely determinants of human hostility towards carnivores, both at the micro-level (the individual) and the macro-level (the society or culture) (Figure 7.1).

Individual level

Personal experience

Personal experience of carnivore damage undoubtedly plays an important role in determining hostility towards them. In Kenya, 75% of Maasai respondents who said they would kill lions had suffered depredation amounting to over 10% of their stock losses (Hazzah et al. 2009), while the 82% of ranchers in Brazil's Pantanal who suffered jaguar depredation considered them a greater threat than did other people (Zimmerman et al. 2005). There can also be subtle but important distinctions in exactly what engenders hostility towards carnivores; in Pakistan, 3% of study households suffered depredation, losing a mean of 1.7 stock per attack and accounting for 19.8% of household financial losses (Dar et al. 2009). Leopards caused over 90% of attacks, and 45% of people supported leopard killing but this view did not seem to be linked to the number of livestock lost to leopards, but rather the degree of financial loss incurred (Dar et al. 2009).

Moreover, as seen in areas as diverse as Brazil (Marchini 2010) and South Africa (Lagendijk & Gusset 2008), people who have personally experienced depredation are not always significantly more hostile towards carnivores than people who have not. In India's Spiti region, 41% of respondents in one community reported snow leopard depredation, and 45% were strongly antagonistic towards the cats, whereas in a different community where slightly more people (43%) suffered losses, only 30% were very hostile (Bagchi & Mishra 2006). Therefore, views towards carnivores are not predicated by personal costs alone, but are affected by numerous additional factors.

Attitudes

Although it is human actions, such as killing carnivores, that directly affect conservation, investigations of human–wildlife conflict have focused mainly on attitudes (Karlsson 2007; Manfredo 2008), as these are easier and quicker to measure than the often secretive, comparatively infrequent actions involved in carnivore killing. Furthermore, many conflict studies necessarily rely on self-reported attitudes and actions, but reporting is heavily influenced by social norms, perceived expectations of certain responses, and unwillingness to report behaviour considered unacceptable or illegal. Even when these limitations are explicitly considered, it is often assumed that attitudes can be used as a simple indicator of actions that will be taken. However, the situation is usually much more complex; numerous factors affect people's attitudes towards carnivores and also whether those attitudes are translated into actions, such as killing carnivores (Hazzah et al. 2009).

In social and cognitive psychology, extensive work has been done on the relationship between attitudes and behaviours. The theory of planned behaviour (Ajzen 1985) and the theory of reasoned action (Ajzen & Fishbein 1980; Fishbein & Manfredo 1992) propose that human behaviours are determined by a combination of individual attitudes, social norms and the extent of perceived control over personal behaviour. Furthermore, attitudes towards a behaviour are based on beliefs about the outcomes of that behaviour. Crucially, these theories recognize that attitudes will not predict behaviour unless they are measured with corresponding levels of specificity: attitudes about objects (such as carnivores) will not necessarily predict behaviours (such as killing carnivores). In order for attitudes to predict behaviour, the attitude and behaviour must correspond on four levels of specificity: action, target, context and time. For example, in order to predict ranchers' intention to kill jaguars in Brazil (Marchini & Macdonald 2012), one would have to assess ranchers' attitudes towards the specific behaviour of killing (action) the jaguar (target) on their properties (context) in the near future (time) (Cavalcanti et al. 2010). Attitude specificity has arguably been one of the most significant refinements in improving the applicability of the attitude concept (Manfredo 2008).

Emotions

The 20th-century tradition of the social sciences focused on deliberative thought and generally avoided emotion, but theorists are increasingly recognizing the importance of emotion in human action and decision making. Large carnivores elicit strong, conflicting emotions (Kruuk 2002; Dickman 2008); people may simultaneously like, admire, fear and hate them, and such conflicting emotions were clearly stated by 9% of Maasai respondents interviewed in Kenya and Tanzania (Goldman et al. 2010). In southern Amazonia, Pantanal and São Paulo city, 77% of children and adolescents (ages 10–18) describe the jaguar as either fearful or attractive, and 11% as both fearful and attractive (Marchini & Macdonald 2012). In Tanzania's Tarangire-Manyara ecosystem, young warriors were the most positive towards lions, with 77% liking them (compared to 47% across all respondents); the most common reason for liking them was the celebrations associated with lion hunting (Goldman et al. 2010).

However, the preponderance of attention so far has been given to negative emotions, particularly fear (Manfredo 2008), with people who are fearful of carnivores usually being more antagonistic towards them (Roskaft et al. 2007). The power of carnivores to elicit fear in humans should not be underestimated: as noted earlier, it is likely to have evolved as a heritable antipredator response and humans' inherent, deep-seated fear is probably a key driver of hostility towards wildlife (Berg 2001). Ancient mythology is awash with terrifying representations of predators: for instance, the famously bloodthirsty Egyptian goddess Sehkmet, who was associated with war, disease and death, was represented in the shape of a lion; the Sphinx was a lion with the face of a woman, who sadistically tricked and ate people; and the Babylonian monster, thought to be 630 miles in length and adept at devouring people, was named 'Labbu', meaning 'lion' (Quammen 2004). The concepts of these terrifying carnivorous monsters all came from societies living alongside lions, and fear remains widespread today wherever people still live with predators, even if the real risk of carnivore attack is low. Furthermore, people are far less prepared to accept risks associated with long-established and deep-seated intrinsic dread and fear, such as the possibility of being killed by a man-eating carnivore, compared to those that they can consider relatively calmly, such as the possibility of being killed in a car accident. A 2002 study in Norway revealed that 48% of respondents were very afraid of wolves, with an additional 40% somewhat afraid, despite the fact that only one documented wolf attack has ever occurred in Norway, in 1800 (Linnell et al. 2003).

Gender, age and parental role

People differ in their attitudes and emotions towards carnivores, but some trends have emerged. Women tend to view carnivores in a more negative light than do men (Zinn & Pierce 2002); in British Columbia, 69–84% of women in different age classes viewed mountain lions as dangerous to children, compared to only 45–54% of men (Campbell & Lancaster 2010). Women's fear of dangerous wildlife is often justified; 50% of Indian respondents said that women were disproportionately affected by conflict with wildlife, compared to only 4.3% who said men were more affected. Women's duties often put them in most danger of attack, they may be less physically capable of escaping, and they are also the ones who tend to make sacrifices, for instance if there is less food available following an attack (Ogra 2008).

However, although they tend to be more fearful and negative, women are often less inclined to support carnivore killing. In British Columbia, only 4–8% of women supported shooting mountain lions, compared to 10–15% of men, with similar patterns for black bears (*Ursus americanus*) (Campbell & Lancaster 2010). In Lithuania, 73% of women rated all local large carnivores as dangerous, compared to only 44% of men, but only 11.5% of women were in favour of carnivore populations diminishing, compared to 16.6% of men (Balčiauskiene & Balčiauskas 2001). These differences may be linked to social norms and cultural expectations, as discussed further below.

Parental role can also have an influence; in both Norway and Colorado, people with children at home were significantly more fearful of carnivores than those without (Bjerke et al. 2001; Zinn & Pierce 2002). Age is also important, with younger respondents often expressing relatively positive attitudes towards carnivores (Roskaft et al. 2007). In British Columbia, 18–25 year olds were significantly more tolerant of carnivores than 61–83 year olds; within the young group, 62% of men and 61% of women said that black bears should be tolerated, compared to only 19% and 12% respectively for the older group, and the same pattern was seen regarding wolves (Campbell & Lancaster 2010).

Values and value orientations

People tend to have many attitudes (defined as a favourable or unfavourable disposition towards an action) (Manfredo & Dayer 2004), and these can form quickly; for instance, people could have a hostile attitude towards carnivore reintroduction, based upon hearing recent evidence of conflict in other locations. However, attitudes do not exist in isolation but are influenced by values – these are fundamental, broad-based beliefs about desirable goals and modes of conduct, with examples of values including honesty, self-respect, independence and obedience (Manfredo & Dayer 2004). People usually have few values, which are acquired early in life, are slow to form and are highly resistant to change (Manfredo & Dayer 2004). These values affect the formation of attitudes, which in turn direct behaviour (Manfredo 2008).

Values towards wildlife can be thought of as being 'oriented' somewhere along a continuum from a 'utilitarian' standpoint, which endorses human domination of wildlife, to a 'mutualist' one, where animals have rights similar to those of humans (Manfredo et al. 2009). Such 'value orientations' are dynamic; a comprehensive study across 19 US states revealed that the urbanization of respondents was negatively correlated with 'dominionistic' views and was positively associated with 'mutualistic' views instead (Manfredo et al. 2009). As countries become better educated, wealthier and more urbanized, their view of wildlife appears to shifts from food source or threat to that of companion (Teel et al. 2007). Importantly, people's values towards wildlife affect how they judge the acceptability of wildlife management actions, i.e., unlike utilitarians, mutualists are unlikely to support killing wildlife (Manfredo et al. 2009). Understanding wildlife value orientations of stakeholders can help conservation professionals anticipate attitudes and therefore potential conflict.

Control, power and skills

People's hostility towards hazards is strongly shaped by how much control they perceive they have over that hazard, how predictable it is, its novelty and its impacts, with particular antipathy felt towards uncontrollable, unpredictable, novel risks with severe impacts (Starr 1969). Unfortunately, all these factors usually characterize carnivore attacks, so people are likely to respond more antagonistically to them than towards other potential dangers. Hostility towards carnivores is intensified if people feel powerless and vulnerable; for instance, they are poor, have few livestock, no alternative sources of income and limited opportunities for social reciprocity amongst their community (Naughton-Treves & Treves 2005).

Control and power are also highly significant in determining the actions people are likely and able to take in response to conflict. Perceived behavioural control is important here, which is the extent to which someone views their behaviour to be under their volitional control (Ajzen 1985). People who live in an authoritarian environment, perceiving low behavioural control over their actions, are likely to act differently from people in an independent environment who perceive high behavioural control (Douglas & Wildavsky 1982). For instance, ranchers in remote locations who have the skills to kill predators covertly have high behavioural control, and often implement the desired lethal control, even if there are theoretical legal barriers – the infamous 'shoot, shovel and shut up' approach. This has been observed with large property owners in the Brazilian Pantanal, among whom jaguar killing is relatively common (Cavalcanti et al. 2010). However, relatively powerless, poor livestock owners often have low behavioural control and may be unable to kill carnivores even if they want to, particularly if there is strong, locally imposed protection for the carnivores concerned. For example, 43% of the small landowners on the Amazon deforestation frontier favour jaguar killing but do not actually do it, as they believe they lack the necessary skills and courage (Marchini 2010).

Knowledge and education

Education can be valuable for raising awareness of conservation and creating advocacy for carnivores (Macdonald & Sillero-Zubiri 2004), and within groups of stakeholders, the more knowledgeable people are about carnivores, the more positively they tend to view them (Ericsson & Heberlein 2003). Moreover, knowledgeable people are also more likely to behave in a way that lessens the chance of conflict arising in the first place, such as not approaching wildlife too closely, avoiding habituating carnivores to food, and taking extra care when moving through carnivore habitat (Herrero & Fleck 1990; Conover 2002). Misinformation can be linked to higher conflict; for instance, 55% of respondents in Chile believed (incorrectly) that pumas had been released into the area by the authorities, so they felt antagonistic that the authorities did not understand the risks of living alongside pumas and were unwilling to call them in response to a depredation (Murphy & Macdonald 2010).

Knowledge can also affect perceptions of conflict: an experiment in which Brazilian ranchers were shown photographs of dead livestock and asked to ascribe the cause of death revealed that respondents who had little knowledge of, and negative attitudes towards, jaguars assigned an average of 2.9 photographs to jaguar depredation, compared to 0.7 amongst more knowledgeable ranchers (Marchini 2010). Therefore, attitudes and knowledge may affect the conclusions a rancher draws from livestock loss. Furthermore, providing information on jaguars to school children had significant impacts: it raised their knowledge scores from an average of 3.95 to 7.08, improved their attitudes (from an initial score of 2.82 to 4.52), and made them more negative to jaguar killing (changing from a score of −0.50 to −1.66) (Marchini & Macdonald 2012). Crucially, educating children also positively influenced their fathers' perceptions, on average improving their fathers' attitudinal scores from 1.43 to 2.57 (Marchini & Macdonald 2012). Therefore, if local people are hostile but uninformed about carnivores, then investing in conservation education can clearly be valuable for reducing conflict.

However, while conservation education is frequently suggested as an important mitigation strategy, and can be useful, the reality of using education to influence attitudes, values and actions is complicated. People's values regarding wildlife tend to be established early in life, and are usually fairly resistant to change (Bright & Manfredo 1996; Bruskotter et al. 2007), so education by carnivore conservationists may make little difference to long-held attitudes. For example, an education initiative about wild dogs in South Africa resulted in no obvious long-term positive impact, and local people actually became more negative towards wild dogs over time; in 2000, 43% of locals said they were largely positive towards the dogs, with 16% negative, but by 2003, this had changed to 16% positive and 22% negative (Gusset et al. 2008). Furthermore, educational efforts may even have a polarizing effect: in Brazil, an 'elaboration' exercise, where children discussed their views towards jaguars, resulted in children with negative views actually having them reinforced, with their attitudinal scores declining significantly, from −1.80 to −4.00 on a scale ranging from −10 (highly unfavourable to jaguars) to 10 (highly favourable) (Marchini & Macdonald 2012).

Wealth

Wealth, as a component of modernization in developed countries, can influence shifts in wildlife value orientations (see above). Furthermore, financial costs imposed by carnivores can, unsurprisingly, be a significant driver of negative attitudes (Dar et al. 2009). Increased wealth also acts as a buffer against depredation being catastrophic, making the household less vulnerable to the potentially severe risks of carnivore presence. Wealth allows people to lessen environmental risks by having increased access to capital or labour, enabling the use of more efficient protection strategies such as increasing herder numbers or building sturdy enclosures (Naughton-Treves & Treves 2005). The poorest people, who tend to be most reliant upon biodiversity-rich areas (Loveridge et al. 2010), are therefore at risk of 'compounding vulnerability', as they are least able to either absorb the impact of losses or protect against them (Naughton-Treves 1997). Greater affluence has also been linked to more positive views of conservation (Infield 1988), although this relationship is not clear-cut; sources of income are also important, as discussed below.

Societal/cultural level

Collective experiences

Although people's views are derived in part from their own individual experiences, they are also influenced by their peers, elders, friends, family, teachers and local media, so there is a collective cultural element to attitudes which must be considered (Hunter 2000). Numerous carnivore conflict studies have demonstrated this, as people frequently view carnivores as problematic, even if they have never personally experienced any damage caused by them. For instance, in Tanzania, 63% of respondents who stated that lions were problematic had never themselves experienced any form of lion attack (Dickman 2008). In southern Amazonia, small-property owners often believed that jaguar depredation was more serious on neighbouring ranches than on their own, suggesting that their perceptions of jaguar conflict were shaped primarily by what is heard from other people, rather than personal experience (Marchini 2010). These discrepancies between personal experiences and perceptions reveal the importance of a collective belief in the threat posed by certain species, and the widespread, long-lasting influence that costs imposed on individuals can have across the larger community.

The perceived threat posed by large carnivores can also become exaggerated, because they can impose catastrophic costs which attract intense attention. News of carnivore attacks, particularly on humans, often spreads far beyond the initial attack location, generating widespread and long-lasting fear (Knight 2000a). For example, puma (*Puma concolor*) attacks are rare; from 1890 to 1990 there were only nine fatal attacks in North America and 44 non-fatal attacks (Beier 1991) yet 65% of people surveyed in Montana said they had read or heard of someone being attacked by a puma (Murphy & Macdonald 2010). In Hokkaido, Japan, in 1915, a bear killed seven people, some of them one night and the rest when it returned the following evening to resume the attack at a vigil for the original victims (Knight 2000a). This incident was widely and dramatically reported, and made an important contribution to the widespread perception of bears as bloodthirsty killers, which still persists across Japan almost a century later (Knight 2000a).

Social norms

Norms are 'ought' statements or rules that direct people's behaviour (Manfredo 2008). Descriptive norms reflect an individual's perception of whether other people perform the behaviour in question (Cialdini et al. 1990). Descriptive norms describe what is typical or normal, and motivate action by indicating what is likely to be effective, adaptive and appropriate (White et al. 2009). For instance, in the Pantanal, 70% of ranchers believe that their neighbours often killed jaguars, encouraging them to do the same thing (Marchini 2010). In contrast, social injunctive norms reflect an individual's perceptions of whether significant others approve of the individual performing the behaviour – in other words, social pressure. Social injunctive norms motivate action by offering potential social rewards or punishments for either engaging or not engaging in a particular behaviour. Marchini (2010) found that social motivations are important determinants of the intention to kill jaguars in the Pantanal, where 25% of ranchers justified their approval of jaguar killing on the grounds of tradition. These ranchers often refer, with apparent pride, to the '*Pantaneiro* culture' and the conviction that jaguar hunting has been passed from generation to generation as an element of that culture.

Social identity and intergroup tensions

Social identity is the component of an individual's concept of himself that is derived from his knowledge of group membership, and the value and emotion attached to that membership (Tajfel 1981). According to the social identity theory, people define and evaluate themselves in terms of distinct social categories (e.g. rancher, Maasai). By allocating himself a particular social identity, an individual is encouraged to accentuate both the similarities between himself and other group members and the differences between himself and people outside the group (Fielding et al. 2008). A social identity approach assumes that if a certain behaviour, for example killing carnivores, is central to a social identity, then such killing will be influenced by the norms of that social group rather than by the expectations and desires of generalized 'others'. This is particularly evident in traditional pastoralist groups such as the Maasai, for whom killing lions is central to their culture – young warriors are expected to kill lions and are celebrated when they do so (Hazzah et al. 2009).

Tensions between different social groups, which often have very different values, norms and orientations, can exacerbate human–carnivore conflict. For example, intense debate about wolf recovery in North America was fuelled by underlying cultural conflicts, such as urban versus rural values and the imposition of national government on local decision making (Primm & Clark 1996; Nie 2004). Such tensions are common amongst relatively small, marginalized rural populations, which often perceive that the costs imposed by carnivores are ignored by distant, pro-conservation authorities, and resent external limitations on actions that can be taken in response to conflict (Naughton-Treves & Treves 2005). As people tend to be far more willing to tolerate risks that are undertaken voluntarily, rather than those imposed upon them externally (Starr 1969), imposed co-existence with protected areas inhabited by carnivores can be a particularly important component of conflict. In Kenya, Maasai warriors (*morani*) killed 10 of the 19 lions in Nairobi National Park after they lost over 100 livestock to depredation, to demonstrate their anger that the government did not recognize the cultural value of their cattle and failed to do more to prevent depredation (Anonymous 2003). The government immediately launched a manhunt for the lion killers, deploying over 50 armed agents and a police helicopter in a bid to protect the remaining lions (Nyamwaro et al. 2006). Ultimately, no-one was arrested or charged, but the heavy-handed response reinforced local perceptions that the government valued lions more than the Maasai. In Sweden, traditional Saami pastoralists are frequently enraged at wolf depredation upon their reindeer (*Rangifer tarandus*), and the perceived inadequacy of legal wolf control or compensation (Lindquist 2000). In 1995, the Saami dumped a heap of bloodied reindeer carcasses in a central Stockholm square in a dramatic protest against the government and their perceived favouring of wolf conservation above Saami livelihoods (Lindquist 2000).

Tensions intensifying human–carnivore conflict can also arise between different users of the same landscape. For instance, in Tanzania, Hehe people reported particular hostility towards spotted hyaenas (*Crocuta crocuta*), as they felt that the neighbouring Gogo tribe bewitched hyaenas so that they would kill the Hehe's stock and drag it back to the Gogo, allowing them to eat meat without killing their own livestock (Dickman 2008). Similarly, in the Mueda region of Mozambique, locals believe that some people can create bewitched 'people-lions', in order to kill their rivals and gain from them (West 2001). It is important to understand such dynamics because solving the 'real' problem, perceived from the outside as reducing attacks by carnivores on people or livestock, may not significantly change people's attitudes if carnivores are still regarded as representing the hostile embodiment of a spirit sent by another ethnic group.

Conflicts between social groups over carnivores can be aggravated by the urban–rural divide. Urban residents are frequently more pro-carnivore conservation than people in rural areas; a meta-analysis of worldwide attitudes to wolves showed that 61% of city-dwellers favoured wolves and wolf reintroductions, compared to 45% of rural residents and only 35% of ranchers and farmers (Williams et al. 2002). In 2009, over half the world's population was urban, and urban areas are predicted to grow disproportionately over the next four decades, (United Nations 2009) so in most countries, ranchers and farmers are a dwindling minority group. As a result, they may associate carnivore conservation with the urban values increasingly imposed on them, and might view continued carnivore killing as part of their resistance to this and their struggle to preserve their rural heritage.

Folklore and religion

Awareness of the danger posed by carnivores and the need to control them are often culturally institutionalized through language, stories, rituals and games. For instance, children in livestock-centric Spanish communities often play games re-enacting the hunting of wolves to protect their stock, while in Japan the word for 'bear' can be used as a synonym for fear itself (Knight 2000a,b). In stories and folklore, carnivores are often cast as the central menacing, devious characters (Lindquist 2000; Marchini & Macdonald 2012), which can help entrench lasting negative perceptions of carnivores.

Furthermore, carnivores are often associated with witchcraft and magic (as seen with the 'people-lions' above), and such beliefs are deep-seated and pervasive in many cultures. Various ethnic groups in Tanzania, including the Warangi and the Sukuma, traditionally believe that invisible, naked witches ride upon the backs of hyaenas, to enable the witch to travel long distances quickly (Dunham 2006). Not all folklore and myths about carnivores are negative; in North America, the disappearance of bears in the autumn and their reappearance in the spring led to beliefs that they had two-way access to the afterlife and could travel between worlds, generating deep respect (Schwartz et al. 2003).

Religious beliefs can also significantly affect people's values, cultural norms and actions regarding carnivores. In Kenya, where 48% of respondents belonged to an evangelical church, 35% of the evangelists said that they would kill lions, compared to only 14% of non-evangelists, and this emerged as a significant predictor of conflict, independent of depredation history (Hazzah et al. 2009). Conversely, cultural or religious beliefs may foster a benign attitude towards dangerous animals; in Manang, Nepal, local Buddhists are particularly tolerant of depredation by snow leopards, as they believe the cats are sacred, so killing them is considered a grave sin (Ale 1998).

Such beliefs affect not only views about carnivores but also attitudes towards any losses. In Tanzania, Muslim livestock keepers were particularly incensed by depredation because the meat was not *halal*, so they could receive no benefit at all from the dead animals. Similarly, in traditional Saami beliefs, reindeer killed by wolves are viewed as contaminated and inedible, exacerbating the waste (Lindquist 2000). Conversely, in some Buddhist communities, snow leopard depredation is regarded as a curse from the 'mountain god' in response to forbidden human behaviour, so the cats themselves are rarely held accountable (Ale 1998). There is wide variation in these cultural norms and values between (and often within) groups, but understanding such variations can help identify the reasons behind heterogeneity in conflict and responses, and guide local mitigation strategies.

Income sources

Sources of income affect conflict with carnivores; where people are solely reliant upon livestock, they have few reasons to tolerate carnivores and their attendant risks. Those with income from other sources, especially those linked to wildlife and conservation, often have more positive attitudes towards wildlife (Sillero-Zubiri & Laurenson 2001) and higher tolerance of wildlife-related losses (Naughton-Treves & Treves 2005). However, conservation-related revenue often fails to reach those people most affected by wildlife; in Tanzania's Longido district, adjacent to the world-famous Serengeti National Park, less than 3% of households received any wildlife income, with wildlife generating an average of only US$1.20 annually per household (Homewood & Thompson 2010). Moreover, even where local people receive substantial financial benefits from wildlife-related sources, improved attitudes towards tourism are not necessarily matched with increased positivity towards wildlife. Indonesia's Komodo National Park generates over $1,000,000 from tourism annually, but local people dependent upon tourism for part of their income were actually less positive towards conservation than others (Walpole & Goodwin 2001).

Consequences: rewards and penalties

All the above factors are likely to affect views of carnivores, and each may affect what actions someone desires in response to carnivore damage. However, the consequences of any action, such as carnivore killing, also play a major role in determining whether it is really taken or not. The consequences of action can be positive (rewards) or negative (penalties). In societies which have long co-existed with the threat of carnivore damage, there are often important social rewards connected with carnivore killing. In traditional pastoralist societies, killing a lion is often viewed with such respect by the community that it has traditionally been a rite of passage for young men (Hazzah et al. 2009), who are rewarded with dancing, feasting and the attentions of young women. Within groups such as the Sukuma and Barabaig, lion killers may even be rewarded with gifts of cattle, which are extremely important assets. In Japan, bear killers are seen as protecting rural villagers, and are celebrated within rural societies as heroic, selfless 'men of courage' (Knight 2000a). These social accolades are important positive consequences of carnivore killing, and can make lethal control much more likely, even if it is technically prohibited (Marchini 2010).

Negative consequences can also affect how likely someone is to kill carnivores. The consequences of carnivore killing can be extremely harsh, especially in protected areas. For instance, in Assam's Kaziranga National Park, Park authorities regularly exercise shoot-to-kill policies for poachers (two such killings occurred in December 2010), and this may deter people from taking action. There may also be social and cultural penalties to carnivore killing, as for Nepalese Buddhists, who see snow leopard killing as sinful and fear its consequences (Ale 1998). Potential carnivore hunters may also be concerned for their physical safety; despite their famed bravery, over a quarter of Tanzanian pastoralists reported that the reason they did not engage in traditional lion hunts was the fear of being injured or killed by the lion. Similarly, in Brazil, 43% of the small landowners who did not kill jaguars said that it was because of their fear of the cats (Marchini 2010).

Ultimately, action will be determined not only by the personal attitudes of those affected by carnivore damage, but by the broader rewards and penalties associated with carnivore killing. The challenge for modern conservationists is to develop effective strategies which not only reduce the levels of carnivore damage, but also provide tangible rewards to local communities sufficient to offset any irreducible costs of conflict (Macdonald 2000). Such rewards are often economic, and can be termed 'payments to encourage co-existence' (Dickman et al. 2011). Conservation methods may require innovative 'carrots' to incentivize conservation, backed up by the 'stick' of legal penalties for activities such as illegal wildlife killing (Macdonald in press).

Using an understanding of the human dimension to guide conflict mitigation

Understanding both individual and societal/cultural determinants of conflict with carnivores is critical for mitigation. Information at the individual level helps to understand people's behaviours towards carnivores and to evaluate prevention and conflict mitigation measures (Manfredo & Dayer 2004). Meanwhile, information at the societal/cultural level can reveal what is generalizable in human–carnivore conflict and what is context specific, to predict which proven interventions may be transferable elsewhere. The critical skill is to assess the relative importance of the individual versus societal/cultural factors that motivate or deter the killing of carnivores, to guide conservation efforts.

The role of information in changing behaviour is complex. Education and information might be used to influence individual factors such as attitudes, emotions, values, etc. in order to deter carnivore killing. While information alone will not motivate someone to adopt a new behaviour (e.g. to tolerate carnivores) (Stern 2000), a lack of information can be a barrier to changing behaviour (Schultz 2002). Education and communication may help change the values that people place on carnivores, reducing disproportionate fears and perceptions of carnivore impact, and empowering people to prevent depredation (Macdonald & Sillero-Zubiri 2004). The use of role models, case studies and examples of co-existence with carnivores, combined with information about the negative consequences of killing them, can help to create or redefine a social norm, creating the sense that the community condemns rather than condones or glamorizes carnivore killing.

Arranging for conspicuous, respected group members or community institutions (e.g. co-operatives and rural schools) to promote tolerance of carnivores can influence other group members (Marchini & Macdonald 2012). Communicating information about the behaviour and practices of group members via informal social networks is another possible strategy. Conservation communicators should draw upon the exceptional social and cultural significance of carnivores (Macdonald in press) and on the current societal shift in values from utilitarianism toward mutualism (Manfredo et al. 2003), to promote protecting, rather than killing, carnivores.

Conclusions

Most human–carnivore conflict studies have focused narrowly on the immediate losses caused by depredation, and efforts to increase people's tolerance of predators have focused largely on economic incentives (e.g. monetary compensation for livestock loss) and legal prohibitions and sanctions (e.g. establishment of protected areas). The role and importance of personal, social and cultural factors have been far less considered.

However, carnivore killing results from the interplay between individual influences and social and cultural motivations, so effective strategies for preventing it must be based on these aspects as well as on economic and legal considerations. Conservationists need to be aware of the complex, dynamic nature of such conflicts, and develop solutions that are not externally imposed but which are locally driven, participatory and culturally sensitive, in order to make carnivore killing personally, socially and culturally unacceptable.

"It is time to... look to the very roots of motivation and understand why, in what circumstances and on which occasions, we cherish and protect life."

E.O. Wilson

References

Ajzen, I. (1985) From intentions to actions: a theory of planned behavior. In: *Action–Control: From Cognition to Behaviour* (eds J. Kuhl & J. Beckman), pp. 11–39. Springer, Heidelberg, Germany.

Ajzen, I. & Fishbein, M. (1980) *Understanding Attitudes and Predicting Social Behavior*. Prentice-Hall, New Jersey.

Ale, S. (1998) Culture and conservation: the snow leopard in Nepal. *International Snow Leopard Trust Newsletter*, **16**, 10.

Anonymous (2003) Maasai kill half the lions in Nairobi National Park. *Cat News*, **39**, 5.

Azevedo, F.C.C. & Murray, D.L. (2007) Evaluation of potential factors predisposing livestock to predation by jaguars. *Journal of Wildlife Management*, **71**, 2379–2386.

Bagchi, S. & Mishra, C. (2006) Living with large carnivores: predation on livestock by the snow leopard (*Uncia uncia*). *Journal of Zoology*, **268**, 217–224.

Balčiauskiene, L. & Balčiauskas, L. (2001) *Threat Perception of the Large Carnivores: Are There Sexual Differences?* Institute of Ecology, Vilnius, Lithuania.

Baldus, R. D. (2006) A man-eating lion (*Panthera leo*) from Tanzania with a toothache. *European Journal of Wildlife Research*, **52**, 59–62.

Beier, P. (1991) Cougar attacks on humans in the United States and Canada. *Wildlife Society Bulletin*, **19**, 403–412.

Berg, K.A. (2001) Historical attitudes and images and the implications on carnivore survival. *Endangered Species Update*, **18**, 186–189.

Bjerke, T., Kaltenborn, B. & Thrane, C. (2001) Sociodemographic correlates of fear-related attitudes toward the wolf (*Canis lupus*). A survey in southeastern Norway. *Fauna Norvegica*, **21**, 25–33.

Bright, A.D. & Manfredo, M.J. (1996) A conceptual model of attitudes toward natural resource issues: a case study of wolf reintroduction. *Human Dimensions of Wildlife*, **1**, 1–21.

Bruskotter, J.T., Schmidt, R.H. & Teel, T.L. (2007) Are attitudes toward wolves changing? A case study in Utah. *Biological Conservation*, **139**, 211–218.

Campbell, M. & Lancaster, B.L. (2010) Public attitudes toward black bears (*Ursus americanus*) and cougars (*Puma concolor*) on Vancouver Island. *Society and Animals*, **18**, 40–57.

Cavalcanti, S.M.C., Marchini, S., Zimmerman, A., Gese, E.M. & Macdonald, D.W. (2010) Jaguars, livestock and people in Brazil: realities and perceptions behind the conflict. In: *Biology and Conservation of Wild Felids* (eds D.W. Macdonald & A. Loveridge), pp. 383–402. Oxford University Press, Oxford.

Cialdini, R.B., Reno, R.R. & Kallgren, C.A. (1990) A focus theory of normative conduct: recycling the concept of norms to reduce littering in public places. *Journal of Personality and Social Psychology*, **58**, 1015.

CITES (1992) *Quotas for Trade in Specimens of Cheetah*. Eighth Meeting of the Convention of International Trade in Endangered Species of Wild Fauna and Flora.

Conforti, V.A. & de Azevedo, F.C.C. (2003) Local perceptions of jaguars (*Panthera onca*) and pumas (*Puma concolor*) in the Iguacu National Park area, South Brazil. *Biological Conservation*, **111**, 215–221.

Conover, M. (2002) *Resolving Human–Wildlife Conflicts: The Science of Wildlife Damage Management*. CRC Press, Boca Raton, FL.

Dar, N.I., Minhas, R.A., Zaman, Q. & Linkie, M. (2009) Predicting the patterns, perceptions and causes of human–carnivore conflict in and around Machiara National Park, Pakistan. *Biological Conservation*, **142**, 2076–2082.

Dickman, A.J. (2008) Investigating key determinants of conflict between people and wildlife, particularly large carnivores, around Ruaha National Park, Tanzania. PhD thesis, Biological Anthropology, University College London, London.

Dickman, A.J., Macdonald, E.A. & Macdonald, D.W. (2011) A review of financial instruments to pay for predator conservation and encourage human–carnivore coexistence. *Proceedings of the National Academy of Sciences USA*, August 22 (epub ahead of print).

Douglas, M. & Wildavsky, A. (1982) *Risk and Culture*. University of California Press, Berkeley.

Dunham, M. (2006) The hyena: witch's auxiliary or nature's fool? Witchcraft and animal lore amongst the Valengi of Tanzania. In: *Le Symbolism des Animaux* (eds E. Dounias, E. Motte-Florac, M. Mesnil & M. Dunham), pp. 1–10. Institut de Recherche pour le Developpement, Marseille, France.

Ericsson, G. & Heberlein, T.A. (2003) Attitudes of hunters, locals, and the general public in Sweden now that the wolves are back. *Biological Conservation*, **111**, 149–159.

Fielding, K.S., Terry, D.J., Masser, B.M. & Hogg, M.A. (2008) Integrating social identity theory and the theory of planned behaviour to explain decisions to engage in sustainable agricultural practices. *British Journal of Social Psychology*, **47**, 23–48.

Fishbein, M. & Manfredo, M.J. (1992) Applying Theory of Reasoned Action in understanding and influencing recreation behavior. In: *Influencing Behavior: Theory and Applications in Recreation and Tourism* (ed. M.J. Manfredo), pp. 29–50. Sagamore Press, Champaign, IL.

Frank, L., Hemson G., Kushnir, H. & Packer, C. (2006) *Lions, Conflict and Conservation in Eastern and Southern Africa*. Eastern and Southern African Lion Conservation Workshop, Johannesburg, South Africa.

Goldman, M.J., Roque De Pinho, J. & Perry, J. (2010) Maintaining complex relations with large cats: Maasai and lions in Kenya and Tanzania. *Human Dimensions of Wildlife*, **15**, 332–346.

Gusset, M., Maddock, A.H., Gunther, G.J., *et al.* 2008. Conflicting human interests over the reintroduction of endangered wild dogs in South Africa. *Biodiversity and Conservation*, **17**, 83–101.

Hazzah, L. & Dolrenry, S. (2007) Coexisting with predators. *Seminar*, **577**, 21–27.

Hazzah, L., Borgerhoff Mulder, M. & Frank, L. (2009) Lions and warriors: social factors underlying declining African lion populations and the effect of incentive-based management in Kenya. *Biological Conservation*, **142**, 2428–2437.

Herrero, S. & Fleck, S. (1990) Injury to people inflicted by black, grizzly or polar bears: recent trends and new insights. *Bears: Their Biology and Management*, **8**, 25–32.

Homewood, K. & Thompson, D.M. (2010) Social and economic challenges for conservation in East African rangelands: land use, livelihoods and wildlife change in Maasailand. In: *Wild Rangelands: Conserving Wildlife While Maintaining Livestock in Semi-Arid Ecosystems* (eds J. du Toit, R. Kock & J. Deutsch), pp. 340–366. Wiley-Blackwell, Oxford.

Hunter, L.M. (2000) A comparison of the environmental attitudes, concern, and behaviours of native-born and foreign-born residents. *Population and Environment*, **21**, 565–580.

Infield, M. (1988) Attitudes of a rural community towards conservation and a local conservation area in Natal, South Africa. *Biological Conservation*, **45**, 21–46.

IUCN (2006) *Regional Conservation Strategy for the Lion Panthera leo in Eastern and Southern Africa*. IUCN SSC Cat Specialist Group, Gland, Switzerland.

IUCN (2007) *Regional Conservation Strategy for the Cheetah and African Wild Dog in Eastern Africa*. IUCN SSC, Gland, Switzerland.

Karlsson, J. (2007) *Management of Wolf and Lynx Conflicts with Human Interests*. Swedish University of Agricultural Sciences, Uppsala, Sweden.

Knight, J. (2000a) Culling demons: the problem of bears in Japan. In: *Natural Enemies: People–Wildlife Conflicts in Anthropological Perspective* (ed. J. Knight). Routledge, London.

Knight, J. (ed.) (2000b) *Natural Enemies: People–Wildlife Conflicts in Anthropological Perspective*. Routledge, London.

Kruuk, H. (2002) *Hunter and Hunted: Relationships between Carnivores and People*. Cambridge University Press, Cambridge.

Lagendijk, D.D.G. & Gusset, M. (2008) Human–carnivore coexistence on communal land bordering the Greater Kruger Area, South Africa. *Environmental Management*, **42**, 971–976.

Lindquist, G. (2000) The wolf, the Saami and the urban shaman: predator symbolism in Sweden. In: *Natural Enemies: People–Wildlife Conflicts in Anthropological Perspective* (ed. J. Knight), pp. 170–188. Routledge, London.

Linnell, J.D.C., Solberg, E.J., Brainerd, S., *et al.* (2003) Is the fear of wolves justified? A Fennoscandian perspective. *Acta Zoologica Lituanica*, **13**, 34–40.

Loveridge, A.J., Wang, S.W., Frank, L.G. & Seidensticker, J. (2010) People and wild felids: conservation of cats and management of conflicts. In: *Biology and Conservation of Wild Felids* (eds D.W. Macdonald & A.J. Loveridge), pp. 161–195. Oxford University Press, Oxford.

Macdonald, D.W. (2000) Bartering biodiversity: what are the options? In: *Economic Policy: Objectives, Instruments and Implementation* (ed. D. Helm), pp. 142–171. Oxford University Press, Oxford.

Macdonald, D.W. (in press) From ethology to biodiversity: case studies of wildlife conservation. *Nova Acta Leopoldina N.F.*, **111**(380).

Macdonald, D.W. & Sillero-Zubiri, C. (2004) Conservation. In: *Biology and Conservation of Wild Canids* (eds D.W. Macdonald & C. Sillero-Zubiri), pp. 353–372. Oxford University Press, Oxford, U.K.

Manfredo, M.J. (2008) *Who Cares about Wildlife? Social Science Concepts for Exploring Human–Wildlife Relationships and Conservation Issues*. Springer, New York.

Manfredo, M.J. & Dayer, A. A. (2004) Concepts for exploring the social aspects of human–wildlife conflict in a global context. *Human Dimensions of Wildlife*, **9**, 1–20.

Manfredo, M.J., Teel, T.L. & Bright, A.D. (2003) Why are public values toward wildlife changing? *Human Dimensions of Wildlife*, **8**, 287–306.

Manfredo, M.J., Teel, T.L. & Henry, K.L. (2009) Linking society and environment: a multilevel model of shifting wildlife value orientations in the Western United States. *Social Science Quarterly*, **90**, 407–427.

Marchini, S. (2010) Human dimensions of the conflicts between people and jaguars (*Panthera onca*) in Brazil. PhD dissertation, University of Oxford, UK.

Marchini, S. & Macdonald, D.W. (2012) Predicting ranchers' intention to kill jaguars: case studies in Amazonia and Pantanal. *Biological Conservation*, **147**, 213–221.

Marker, L.L., Dickman, A.J., Mills, M.G.L. & Macdonald, D.W. (2003a) Aspects of the management of cheetahs, *Acinonyx jubatus jubatus*, trapped on Namibian farmlands. *Biological Conservation*, **114**, 401–412.

Marker, L.L., Mills, M.G.L. & Macdonald, D.W. (2003b) Factors influencing perceptions and tolerance toward cheetahs (*Acinonyx jubatus*) on Namibian farmlands. *Conservation Biology*, **17**, 1–9.

Muhly, T.B. & Musiani, M. (2009) Livestock depredation by wolves and the ranching economy in the Northwestern US. *Ecological Economics*, **68**, 2439–2450.

Murphy, T. & Macdonald, D.W. (2010) Pumas and people: lessons in the landscape of tolerance from a widely distributed felid. In: *Biology and Conservation of Wild Felids* (eds D.W. Macdonald & A.J. Loveridge), pp. 431–452. Oxford University Press, Oxford.

Naughton-Treves, L. (1997) Farming the forest edge: vulnerable places and people around Kibale National Park, Uganda. *Geographical Review*, **87**, 27–46.

Naughton-Treves, L. & Treves, A. (2005) Socio-ecological factors shaping local support for wildlife: crop-raiding by elephants and other wildlife in Africa. In: *People and Wildlife: Conflict or Coexistence?* (eds R. Woodroffe, S. Thirgood & A. Rabinowitz), pp. 252–277. Cambridge University Press, Cambridge.

Nie, M. (2004) State Wildlife governance and wildlife conservation. In: *People and Predators: From Conflict to Coexistence* (eds N. Fascione, A. Delach & M.E. Smith), pp. 197–218. Island Press, Washington, D.C.

Noecker, R.J. (1997) *Reintroduction of Wolves*. Congressional Research Service, Washington, D.C.

Nowell, K. & Jackson, P. (1996) *Wild Cats: Status Survey and Conservation Action Plan*. Burlington Press, Cambridge, MA.

Nyamwaro, S.O., Murilla, G.A., Mochabo, M.O.K. & Wanjala, K.B. (2006) *Conflict minimising Strategies on Natural Resource Management and Use: The Case for Managing and Coping with Conflicts Between Wildlife and Agro-Pastoral Production Resources in Transmara District, Kenya*. Policy Research Conference on Pastoralism and Poverty Reduction in East Africa, Nairobi.

Ogra, M.V. (2008) Human–wildlife conflict and gender in protected area borderlands: a case study of costs, perceptions, and vulnerabilities from Uttarakhand (Uttaranchal), India. *Geoforum*, **39**, 1408–1422.

Peterhans, K., Julian, C. & Gnoske, T.P. (2001) The science of 'man-eating*' among lions (*Panthera leo*) with a reconstruction of the natural history of the 'Man-Eaters of Tsavo'. *Journal of East African Natural History*, **90**, 1–40.

Phillips, M., Bangs, E., Mech, L., Kelly, B. & Fazio, B. (2004) Grey wolves – Yellowstone: extermination and recovery of red and grey wolf in the contiguous United States. In: *The Biology and Conservation of Wild Canids* (eds D.W. Macdonald & C. Sillero-Zubiri), pp. 297–309. Oxford University Press, Oxford.

Primm, S.A. & Clark, T.W. (1996). Making sense of the policy process for carnivore conservation. *Conservation Biology*, **10**, 1036–1045.

Quammen, D. (2004) *Monster of God: The Man-Eating Predator in the Jungles of History and the Mind*. W.W. Norton, New York.

Ray, J., Hunter, L. & Zigouris, J. (2005) *Setting Conservation and Research Priorities for Larger African Carnivores*. Wildlife Conservation Society, New York.

Roskaft, E., Handel, B., Bjerke, T. & Kaltenborn, B.P. 2007. Human attitudes towards large carnivores in Norway. *Wildlife Biology*, **13**, 172.

Schultz, P.W. (2002) Knowledge, information, and household recycling: examining the knowledge-deficit model of behavior change. In: *New Tools for Environmental Protection: Education, Information and Voluntary Measures*. (eds T. Dietz & P.C. Stern), pp. 67–82. National Academy Press, Washington, D.C.

Schwartz, C.C., Swenson, J.E. & Miller, S.D. (2003) Large carnivores, moose and humans: a changing paradigm of predator management in the 21st century. *Alces*, **39**, 41–63.

Seidensticker, J., Lumpkin, S. & Shrestha, M. (2010) The status and recovery of Amur tigers in comparison to other tiger subspecies. Paper presented at The Amur Tiger in Northeast Asia: Planning for the 21st Century Conference, Vladivostock.

Sillero-Zubiri, C. & Laurenson, M.K. (2001) Interactions between carnivores and local communities: conflict or co-existence? In: *Carnivore Conservation* (eds J.L. Gittleman, S.M. Funk, D.W. Macdonald & R.K. Wayne), pp. 282–312. Cambridge University Press, Cambridge.

Starr, C. (1969) Social benefits vs. technological risks. *Science*, **165**, 1232–1238.

Stein, A., Fuller, T., Damery, D., Sievert, L. & Marker, L. (2010) Farm management and economic analyses of leopard conservation in north central Namibia. *Animal Conservation*, **13**, 419–427.

Stern, P.C. (2000) New environmental theories: toward a coherent theory of environmentally significant behavior. *Journal of Social Issues*, **56**, 407–424.

Tajfel, H. (1981) *Human Groups and Social Categories: Studies in Social Psychology*. Cambridge University Press, Cambridge.

Teel, T.L., Manfredo, M.J. & Stinchfield, H.M. (2007) The need and theoretical basis for exploring wildlife value orientations cross-culturally. *Human Dimensions of Wildlife*, **12**(5), 297–305.

United Nations (2009) *World Urbanization Prospects: The 2009 Revision*. Department of Economic and Social Affairs, New York.

Varley, J.D. & Brewster, W.G. (1992) *Wolves for Yellowstone? A Report to the United States Congress*. National Parks Service, Yellowstone National Park, Wyoming.

Walpole, M.J. & Goodwin, H.J. (2001) Local attitudes towards conservation and tourism around Komodo National Park, Indonesia. *Environmental Conservation*, **28**, 160–166.

Wang, S.W. & Macdonald, D.W. (2006) Livestock predation by carnivores in Jigme Singye Wangchuck National Park, Bhutan. *Biological Conservation*, **129**, 558–565.

West, H. (2001) Sorcery of construction and socialist modernisation: ways of understanding power in postcolonial Mozambique. *American Ethnologist*, **28**, 119–150.

White, K.M., Smith, J.R., Terry, D.J., Greenslade, J.H. & McKimmie, B.M. (2009) Social influence in the theory of planned behaviour: the role of descriptive, injunctive, and in group norms. *British Journal of Social Psychology*, **48**, 135–158.

Williams, C.K., Ericsson, G. & Heberlein, T.A. (2002) A quantitative summary of attitudes toward wolves and their reintroduction (1972–2000). *Wildlife Society Bulletin*, **30**(20), 1–10.

Woodroffe, R., Ginsberg J. & Macdonald, D.W. (1997) *The African Wild Dog: Status Survey and Conservation Action Plan*. IUCN/SSC Canid Specialist Group, Gland, Switzerland.

Yirga, G., Bauer, H., Worasi, Y. & Asmelash, S. (2011) Farmers' perception of leopard (*Panthera pardus*) conservation in a human dominated landscape in northern Ethiopian highlands. *International Journal of Biodiversity and Conservation*, **3**, 160–166.

Zimmerman, A., Walpole, M.J. & Leader-Williams, N. (2005) Cattle ranchers' attitudes to conflicts with jaguar *Panthera once* in the Pantanal of Brazil. *Oryx*, **39**, 406–412.

Zinn, H.C. & Pierce, C.L. (2002) Values, gender, and concern about potentially dangerous wildlife. *Environment and Behavior*, **34**, 239–256.

8

Citizen science and nature conservation

Jonathan Silvertown[1], Christina D. Buesching[2], Susan K. Jacobson[3] and Tony Rebelo[4]

[1]Department of Environment, Earth and Ecosystems, Faculty of Science, Open University, Milton Keynes, UK
[2]Wildlife Conservation Research Unit, Department of Zoology, Recanati-Kaplan Centre, University of Oxford, Oxford, UK
[3]Department of Wildlife Ecology and Conservation, University of Florida Gainesville FL, USA
[4]Threatened Species Research Unit, South African National Biodiversity Institute, Claremont, South Africa

If you think in terms of a year, plant a seed; if in terms of ten years, plant trees; if in terms of 100 years, teach the people.

Confucius

Introduction

The Convention on Biological Diversity (CBD 1993) stresses the '3 Cs': Conservation, Commerce and Community (also termed the '3 Es': Environment, Economy and Equity). Legislation in itself, however, cannot be the entire solution. A shift in our values and attitudes toward the environment is required to bring about changes in behaviour at all levels Agenda 21 of the CBD thus advocates specific criteria for tackling anthropogenic impacts and attitudes to address the human dimension in conservation.

The general public and corporations are both stakeholders in the biosphere and can be involved in nature conservation through citizen science (Irwin 1995) and corporate social responsibility (CSR: Henningfield et al. 2006). This involvement can take many forms, ranging from volunteer-based environmental clean-up days and implementation of species conservation measures based on established scientific knowledge, to public education in parks and museums, to collecting data on species distribution and abundance for monitoring purposes or for the purpose of testing scientific hypotheses (Leslie et al. 2004). Aside from the immediate benefits, long-term volunteer experience has also been linked to strong advocacy for the environment (Ryan et al. 2001), furthering environmental stewardship.

Key Topics in Conservation Biology 2, First Edition. Edited by David W. Macdonald and Katherine J. Willis.

While there are many advantages to using volunteers as citizen scientists (Toms & Newson 2006), there are limitations, too. Professional scientists often doubt the reliability of scientific data collected by amateur volunteers. Based on arguments from the House of Representatives asserting that 'volunteers are incompetent and biased', some public monitoring programmes in the United States are reverting solely to the use of professional scientists despite an absence of evidence corroborating these doubts (e.g. the US National Biological Survey: www.nap.edu/openbook.php?record_id=2243).

In addition, volunteers are not really a 'free' labour source. Financial and human resources are required to recruit, train and retain volunteers, and to recognize their accomplishments (Jacobson 2009). It is critical that projects relying on the help of citizen scientists minimize the costs and maximize the benefits of their volunteer programmes by understanding volunteer motivations and perceptions.

This chapter considers four key issues that need to be taken into account when working with volunteers: recruitment, motivation, training and deployment, and data validation. We then discuss the circumstances in which a citizen science approach should offer benefits for nature conservation projects.

Recruiting and retaining volunteers

The willingness to volunteer pervades all sectors of society. However, many organizations report a strong age and/ or gender bias in their volunteers, thus making targeted recruitment a necessity. The most suitable method of volunteer recruitment depends upon how many volunteers are required, the desired geographical reach of the project, the project logistics, whether any particular prior skills (e.g. competence in species identification) or physical abilities (e.g. scuba-diving) are needed for participation, and what kind of budget is available for publicity. The spectrum of volunteering opportunities for citizen scientists ranges from intensive and well-structured short- or medium-term residential projects (e.g. www.opwall.com, www.earthwatch.org, www.conservationvip.org) to long-term or open-ended projects, where large numbers of volunteers are asked to send in any records of specific observations they collect to a central data collection agency (e.g. National Biodiversity Network: www.opalexplorenature.org/nbn).

Whereas in the first case scenario, small teams of volunteers usually work under the direct supervision of one or several professional scientists from whom they receive intensive personalized field training, in the second scenario the relationship between citizen scientist and professional researcher remains largely anonymous, and training is usually restricted to written instructions and web-based material (e.g. the Mini Mammal Monitoring survey organized by the Mammal Society in the UK: www.mammal.org.uk; the Great Backyard Bird Count in the US: www.birdsource.org/gbbc). Some surveys utilize specifically designed software applications to facilitate easy and reliable data collection by volunteers, such as the iPhone app for recording and submitting records of different species of mammals killed on Britain's roads to the People's Trust for Endangered Species (http://itunes.apple.com/gb/app/mammals-on-roads/id446109227?mt=8). Such mobile applications can now be produced using customizable freeware (Aanensen et al. 2009).

The longer-running projects conducted, for example, by the British Trust for Ornithology in the UK and the Cornell Lab of Ornithology in the US demonstrate that many volunteers can reach a professional level of expertise (Greenwood 2007). Social networking offers the opportunity to exploit the full range of volunteer expertise by using more experienced and knowledgeable volunteers to guide and train less experienced ones. This is the principle behind iSpot (http://ispot.org.uk), which involves members of more than 90 specialist natural history societies, most of them run by volunteers, to help beginners and

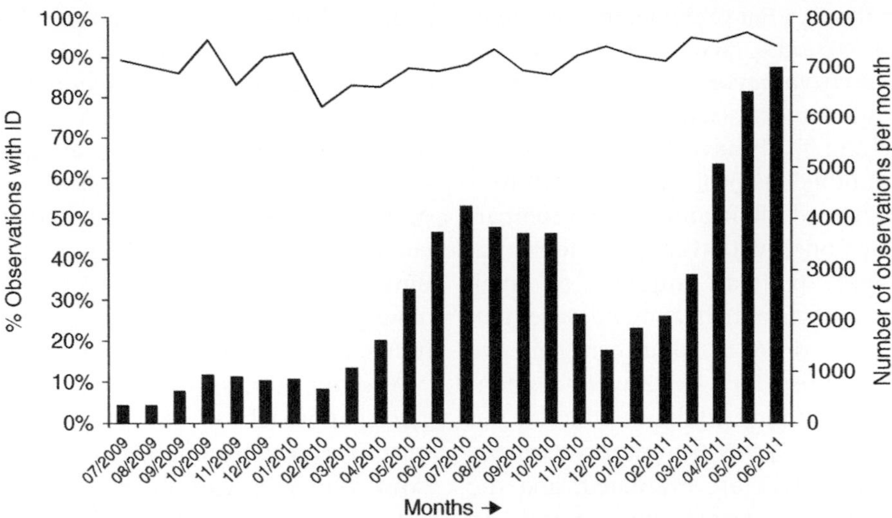

Figure 8.1 The number of observations submitted to the iSpot website during its first 2 years of operation (columns) and the percentage that were named by the social network (line).

non-specialists identify photographs of species that they submit to the website. Once validated, these data become a valuable addition to national biodiversity records. In its first 2 years of operation in the UK, iSpot volunteers made more than 66,000 observations of over 5000 species. Members of the Amateur Entomologists Society are one of the more active expert groups and between them contributed in excess of 2000 identifications to iSpot.

Because iSpot is new, its full potential is still to be realized, but the early signs are that it is highly scalable. The number of observations submitted doubled from 3500 to 7000 per month between June 2010 and June 2011, and the iSpot community proved more than capable of keeping pace with the task of supplying valid determinations (Figure 8.1). Ninety-six percent of observations have received a name, 77% at species level. These include two species of insect never recorded in the UK before, including the first record of the Euonymus leaf notcher moth, discovered by a 6-year-old girl. In a subsample of 2931 species observed on iSpot, 10% had a conservation listing, including 160 Biodiversity Action Plan priority species, 118 species on the UK Red List, and 102 nationally rare/scarce species. iSpot has recently been adopted by the South African National Biodiversity Institute. In its first few months of operation in South Africa, iSpot has reported an endemic species previously thought to be extinct and located an alien species that is a potential threat to endemic plants on Table Mountain.

The goal of an increasing number of conservation organizations is 'to help unite volunteers with scientific projects in need of voluntary assistance' (Mackney & Spring 2000). In some cases, individuals volunteer their time and labour, in others they are also expected to contribute to the cost of the research (Coghlan 2005). Web-based advertisements from (non-) governmental and professional organizations, research institutions or individual scientists offering volunteer placements are widespread. The Earthwatch Institute (www.earthwatch.org), for example, promotes a number of scientific projects in need of volunteer helpers to their members, who book a volunteering holiday package including food, accommodation, transport and insurance with the explicit expectation that they will work alongside professional scientists to collect important data to further scientific knowledge. Some Earthwatch projects can accommodate 10–15 volunteers per team, offer between five and 10 expeditions per year and run for several years. For example, the Mammals of Wytham Woods (www.earthwatch.org/aboutus/research/valuevol/) and the Mammals of Nova Scotia Projects (www.earthwatch.org/exped/buesching.html) have thus far benefitted from the combined help of close to 1000 residential volunteers, building up detailed comparative databases on a wide variety of terrestrial mammal species with an emphasis on small mammals and effects of habitat management on forest ecology. Here, 6–12 volunteers join two professional scientists for 1- or 2-week residential terms to survey for field signs and to lay out and check trapping grids. Simultaneously, the volunteers themselves are evaluated, and the veracity of their data tested and compared to those of the scientists (Newman et al. 2003). Volunteers pay their own travel as well as all expenses occurred on the project in addition to the staff costs and overheads incurred at Earthwatch offices, although the project scientists' salaries' are not covered by the volunteer contribution.

Even residential projects supported by the help of comparatively small volunteer teams of 10–15 people can utilize modern web-based communication tools to increase their outreach and conservation impact, as illustrated by Earthwatch's Life from the Field Teacher programmes, where sponsored teachers skype live from the project with their own (and other teachers') classrooms, regularly reaching thousands of students from the Mammals of Nova Scotia Project.

At the other end of the spectrum are large-scale, mostly web-based citizen science projects, such as the Evolution MegaLab (http://evolutionmegalab.org), directed at sampling a maximum number of populations of banded snails (*Cepaea* spp.) throughout their native geographical range (Silvertown et al. 2011). This project aimed to reach as many potential volunteers as possible throughout the whole of Europe (Silvertown et al. 2011). No specific skills were required for participation, and extensive institutional resources could be deployed in promotion. The project operated in 14 languages in 15 European countries and reached 5–10 million people through widespread publicity in the press, television and radio. To take the UK as an example, of the estimated 5 million who heard about the project, only about 1% visited the website, and of these only about one in 10 registered (Figure 8.2). The Swiss partners, in contrast, used a different approach by recruiting volunteers specifically from among members of Bird Life Switzerland, but achieved a similar rate of participation per head of national population (0.006%) to that obtained in the UK (0.005%).

Another successful example for such a targeted approach is the Protea Atlas Project (http://protea.worldonline.co.za) which, aside from collecting scientific data about the distribution and survival strategies of the 370 different species in the Proteaceae in South Africa, also aimed to encourage amateur involvement in botany and to stimulate public awareness about South African conservation issues. Volunteers were recruited by organizers giving dozens of talks and visiting 42 of the annual flower shows in the western and southern Cape throughout the duration of the project. Help with protea identification was given at the shows, and information was obtained from visitors with local knowledge about species' localities and common names. Posters advocating collection of local distribution records mapping local proteas were designed for each flower show, and later donated to local libraries, information centres and museums. Media coverage heightened awareness of the project, but recruited few volunteers.

During the 10-year duration of this project, 1455 people approached the management team, although 18% of these did not express any further interest in the project. A further 52% ordered identification and sampling kits, but did not send in any data. Some 478 volunteers (30%) sent in data, with 97 of them sending in more than 50 localities and 12 contributors submitting more than 1000 localities. The top 10 volunteers collected 52% of the data. Throughout the project, a high profile was maintained by giving publicity to newly discovered range extensions, new taxa, local threats

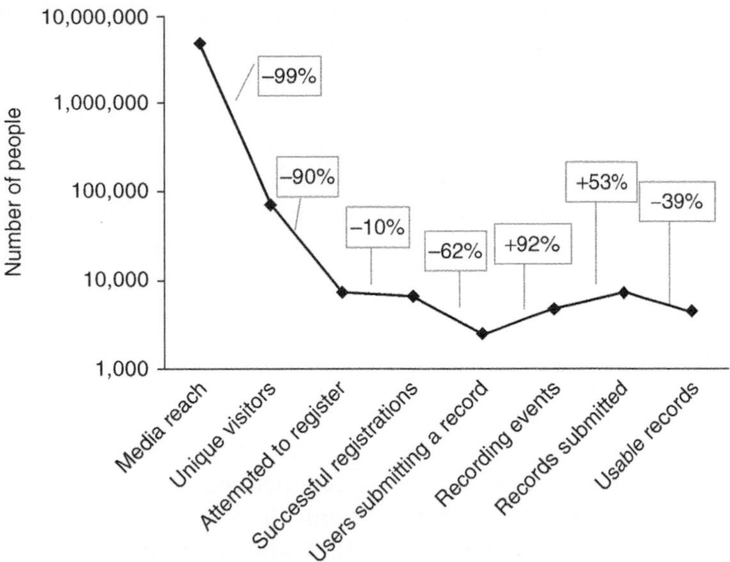

Figure 8.2 A recruitment curve for the Evolution MegaLab showing the numbers with an opportunity to see publicity about the project in the UK press and other media through to the number actually participating and the number of samples submitted. Percentages show the loss or gain in numbers between each stage. From Worthington et al. (2011).

identified during atlassing and by organizing outings and courses in atlassing and plant identification. The organizers estimated that with atlassing teams, over 1000 people contributed to the Protea Atlas Project, excluding ancillaries who joined walks, hikes and trips in which mapping was a focus of the excursion. A substantial number of conservation officers, including 90 novices and 22 experts, contributed data to the atlas.

Volunteer motivation

Researchers and habitat managers have long recognized that understanding volunteer motivation is a valuable component of volunteer management (Cnaan & Goldberg-Glen 1991; Harrison 1995; Omoto & Snyder 1995). Yet little research has been conducted on factors affecting recruitment and retention in citizen science and conservation. Motivated volunteers serve significantly longer than volunteers who do not have their needs met through service (Jacobson et al. 2012). A successful recruitment and retention plan can minimize many of the common challenges that supervisors face, such as attracting good volunteers who can work the appropriate hours, and who bring unique and valuable perspectives, expertise or training to a programme (Hager & Brudney 2005). The recruitment message should therefore be tailored to the audience being sought, and should not only address how volunteers can help meet the specific needs of the organization but must also emphasize the benefits to the volunteers themselves, and how their motivations will be satisfied. To maximize their appeal, some recruitment drives such as the Mini Mammal Monitoring Project (www.mammal.org.uk) offer different methodological approaches (in this case from field sign surveys to life trapping) to appeal to beginners and experienced volunteers alike on a sliding time scale (from one hour to several days), whilst emphasizing the urgent need for voluntary participation as well as the potential for social involvement in regional groups.

People are motivated by various needs. Psychologist Abraham Maslow developed a simple hierarchy of people's needs which translate into general motivations. His theory posits that once people fulfil their physiological needs for food and health as well as safety and security, they progress to addressing personal drives for a sense of social belonging, self-esteem and ultimately self-actualization. Volunteer opportunities can satisfy many of these same drives for social affiliation, personal achievement and esteem.

A number of studies have examined the personal and social functions that are served by initiating and sustaining a particular helping behaviour, such as volunteerism. Research on the motivations of 569 volunteers for Florida's Fish and Wildlife Conservation Commission (FWC) used a functional framework to measure the relative importance of seven primary motivations. Based on a web-based questionnaire, volunteers ranked a series of 30 items that corresponded to seven primary motivations. Each item was scored from 1 (very unimportant) to 7 (very important) and resulted in the following average scores for each motivation category: (6.3) help the environment, (5.2) have opportunities for learning, (5.0) enhance their use of the environment, (5.0) express their personal values, (4.8) be involved in effective projects and esteem, (4.8) engage in social interactions, and (3.4) further their career goals (Jacobson et al. 2012). An earlier study of volunteers for five environmental organizations in Colorado found similar motivations, with helping the environment and related items, such as assisting an environmental cause, seeing improvements in the environment and preserving natural areas for future generations, being among the highest motivational factors (Bruyere & Rappe 2007). Other studies have found that volunteers are motivated by similar factors, such as 'gathering new experiences' (Pearce & Lee 2005), 'being adventurous' (Moscardo et al. 1996), 'learning about the environment' (Manfredo et al. 1996), 'supporting an organization/scientific project' (Bonjean et al. 1994), 'escaping their daily routine' or 'experiencing an entirely new direction in life' (Buesching & Slade 2012).

Motivations of volunteers can vary widely among individuals and can change with time. Age-related differences were seen in the Florida Wildlife Agency study. Younger volunteers were more interested in opportunities offering career development (Jacobson et al. 2012). Different expectations were associated with national and cultural identity as well as societal class and educational background among volunteers in the Earthwatch programme (Buesching & Slade 2012), which has approximately 20,000 members globally and places more than 4000 volunteers per year. The majority of these volunteers come from only three countries: Britain, the USA and Australia, and more than half of all residential conservation volunteer organizations are based in the UK (Coghlan 2005). The largest proportion of volunteers on the Mammal Monitoring Projects in the UK as well as in Nova Scotia came from the UK (ca. 78% in the UK compared to 30% in NS) and the US (ca. 8% in the UK compared to 54% in the US), followed by Australia (ca. 6% in both countries) and Japan (ca. 5% in both countries), whilst all volunteers from other European countries (Germany, France, Holland, Switzerland) contributed only ca. 3% on both projects.

Understanding such national differences in participation and in what motivates volunteers presents a challenge. Differences in cultural expectations relating to division of labour, recreation, career development and financial decisions are all involved (Buesching & Slade 2012). As in some countries, such as the USA, all costs related to volunteering are tax-deductable, residential environmental volunteering programmes can be a cost-effective alternative to conventional eco-holiday travel.

Of a sample of 611 volunteers from the Earthwatch Institute participating in the two residential Mammal Monitoring Projects, approximately 25% were motivated foremost by the opportunity for interacting with other like-minded people and enjoying the social aspects of their activities or participating in a new activity, especially after reaching a turning point (e.g. divorce or unemployment) in their personal life (ca. 25%). However, volunteer recruitment and project design usually focus on the scientific importance of the work. Recruitment and volunteer retention might thus be greatly enhanced on many projects if these aspects are emphasized when designing a programme. In addition, committed volunteers are often more strongly motivated by social considerations, e.g. at environmental stewardship programmes in Michigan, volunteers who were oriented to volunteering for social reasons, and for whom project organization played an important role, were more committed to their work (Ryan et al. 2001). Items such as seeing familiar faces and having fun were important predictors of commitment to their volunteer programme. Creating time for volunteers to socialize during their work activities is thus important for the retention of long-term volunteers. For web-based research activities, regular feedback from organizers, group postings, social networking and other activities may satisfy social drives.

For people who choose to volunteer for achievement opportunities, emphasizing the chance to learn new skills through participation in a project will help to keep them satisfied. On the Nova Scotia Mammal Monitoring Project, approximately 25% of volunteers signed up with the explicit hope and expectation of learning new skills, such as mammal trapping and field sign identification, relevant to their career (e.g. biology students, environmental consultants), and/or to obtain references by the projects scientists for future job or college applications. More experienced and knowledgeable volunteers for Christmas Bird Counts are given the opportunity to share their birding expertise with new volunteers, and receive recognition from their peers through local and regional tallies and reports. In 2010, during the 110th Christmas Bird Count in North America, and more recently Latin America, 60,753 observers counted a total of 55,951,707 birds. Bird count participants can follow feeder observations on websites and see final tallies published in local newsletters and national journals. Extensive media coverage of compiled data provided reinforcement for volunteers. In 2007, when Audubon released their 'Common Birds in Decline' report (Butcher & Niven 2007), more than 700 articles appeared in print and extensive coverage of the bird counts occurred on radio and TV (Bancroft 2007).

Volunteers seeking a feeling of esteem or power might benefit from working independently and having control over a part of the project. Researchers have found that more proactive volunteer activities, such as native plant and stream restoration tasks, are more likely to result in greater frequency and strength of commitment of volunteers than simple manual tasks, such as clean-up activities (Ryan et al. 2001). These activities allowed participants to develop their skills in identifying native plant or aquatic species. Similarly, a content analysis of volunteer newsletters found that the tangible results of ecological restoration work are an important factor in motivating volunteers (Schroeder 2000). In short-term residential projects, volunteer motivation and data quality are both enhanced if volunteers participate in the analysis of data during their stay (Newman et al. 2003; Foster-Smith & Evans 2003; Buesching et al. submitted). Many citizen science projects require supporting activities, such as data input or fence maintenance, and it is important to link these activities to the ultimate objectives of habitat management and to conservation outcomes in order to show volunteers the importance of their completion (Buesching & Slade 2012).

Training of volunteers

Appropriate training is considered increasingly as the most important factor affecting volunteer performance (Brandon et al. 2003; Newman et al. 2003; Foster-Smith & Evans 2003; Blackburn & Frank 2010; Danielsen et al. 2005). Whilst many volunteer-based surveys are conducted by mailing written instructions or web-based training manuals to participants

(Janzen 2004; Cohn 2008), often using field guides, video and an online quiz to train volunteers (Silvertown et al. 2011), several studies have emphasized the importance of hands-on practical training, which needs to include the actual volunteer tasks (e.g. spotting field signs) in a natural setting (Newman et al. 2003; Foster-Smith & Evans 2003).

For example, volunteers for the Protea Atlas Project were given intensive training, either by atlas staff or by other volunteers. The project team led volunteers on numerous field trips to nature reserves, unexplored areas, hot-spot areas, and to farms at the invitation of landowners. In addition, regional co-ordinators also ran weekend and camping trips to areas within their region. Generic skills such as map reading, species identification, estimating population sizes, interpreting the geology, landscape and vegetation, filling in forms, and standardized monitoring techniques all need to be considered in training volunteers. An obvious error to avoid is to focus volunteer training purely on the project-specific techniques, while neglecting support skills (such as map reading or correctly using GPS) that are just as important to the project (Buesching et al. submitted).

For many monitoring tasks, 2–3 hours of practical field training by a professional scientist has been shown to be sufficient (Newman et al. 2003; Foster-Smith & Evans 2003). For example, if provided only with written instructions, drawings of field signs and a map and compass to survey a woodland area for badger setts and latrines, volunteers found only 10% of badger setts and none of the latrines known to be in the area. In contrast, volunteers having been shown one sett and one latrine were able to find 90% of all main setts, 67% of smaller outlying setts, and 56% of latrines (Newman et al. 2003).

Training has to start at the basics, without the assumption of prior volunteer knowledge, to ensure methodological coherency in data records (Macdonald et al. 2002; Newman et al. 2003; Buesching & Slade 2012). Often, complete novices can be better suited than experienced volunteers to studies where methodological consistency is crucial, because prior knowledge implicitly produces methodological preconceptions (Newman et al. 2003). For example, when conducting standing crop faecal pellet count surveys in experimental plots with known numbers of droppings, novice volunteers found 75% (no previous experience), whilst experienced volunteers (>15 plots surveyed) found only 67% (professional field biologists consistently found 74% in this study; Newman et al. 2003). Frequent supervision, especially during the initial training period (Newman et al. 2003), with follow-up spot-checks and intensive training sessions concentrating on specific issues, minimizes observer errors and enhances volunteer performance significantly (Foster-Smith & Evans 2003). The inclusion of some theoretical background and context, detailing why the particular research project in question is important, has been shown to enhance volunteer motivation and comprehension significantly (Martinich et al. 2006), leading to more consistent results (Mumby et al. 1995; Newman et al. 2003). Performance is considerably improved if, during follow-up sessions, the consequences of incorrect task performance are explained, too, rather than just reiterating the correct techniques (Cook & Berrenberg 1981; Newman et al. 2003).

Data validation and analysis

Despite scepticism about the validity of volunteer data, many researchers rely on the use of volunteers in scientific data collection. Scientists expect to collect data, reliable and accurate enough to test hypotheses, and by using volunteers they expect to carry out more work than they could on their own (Crall et al. 2010; Devictor et al. 2010). However, studies generally show that novice volunteers take significantly more time to accomplish a task compared with professional biologists, although with increasing practice they usually achieve similar speed and efficiency (Newman et al. 2003; Foster-Smith & Evans 2003; Buesching & Newman 2005;

Buesching et al. submitted). Even such straightforward tasks as laying out random 10 × 10 m survey quadrants in forests for standing crop faecal counts takes a group of six inexperienced volunteers three times longer than one experienced researcher (Buesching et al. submitted).

It is important that volunteer data should be of equivalent quality to those collected by professionals, or that their quality is consistent and can thus be validated, as shown, for example, in visual badger counts at setts, where volunteers consistently counted 60% of all badgers that were known to live at each sett from long-term trapping records (Newman et al. 2003). Pilot studies (for example, marine: Darwall & Dulvey 1996; Foster-Smith & Evans 2003; woodland mammals: Newman et al. 2003; Buesching & Newman 2005; Buesching et al. submitted) indicate that, with appropriate training, volunteers are capable of mastering many monitoring techniques. Individual variation between volunteers is, however, considerable (Ericsson & Wallin 1999; Barrett et al. 2002; Genet & Sargent 2003) and influenced by individual-specific characteristics, such as gender, fitness and enjoyment/boredom (Buesching & Newman 2005; Buesching & Slade 2012).

Some volunteering programmes afford the opportunity to evaluate the effects of different individual-specific characteristics on volunteer performance, and thus formulate generalized guidelines for researchers to employ new volunteers on their likely strengths and weaknesses. Analyses based on a sample size of 750 volunteers from 65 teams on the UK and Nova Scotia Mammal Monitoring Projects showed that, when evaluated by one male and one female observer on performance of a variety of different tasks on a scale of 1 (= unreliable work/data quality) to 5 (= very efficient/data comparable to professional standards), men scored higher generalized averages than women (Newman et al. 2003; Buesching & Slade 2012; Buesching et al. submitted). The standard deviation among women, however, was much higher than among men (Newman et al. 2003; Buesching & Newman 2005) and in addition, women improved significantly more with training and experience than men (Buesching et al. submitted).

To ensure data quality, it proved important that the tasks be well within the volunteers' physical abilities, as fitness has been shown to be correlated with data veracity, while age (between 11 and 86), if corrected for fitness, had little or no effect on volunteer abilities (Newman et al. 2003; Buesching et al. submitted). Nevertheless, in general, different age classes show different task-specific strengths. Older people are generally more patient and are thus usually well suited to fine-scale monitoring at a leisurely pace, e.g. species lists, bird calls, direct observations (Newman et al. 2003; Buesching et al. submitted) while younger volunteers are usually more energetic and are thus predisposed to more active tasks, e.g. transect walking through difficult terrain (Buesching et al. submitted). Some volunteers also have certain task-specific disadvantages, e.g. red/green colour blindness prevents easy discrimination of animal droppings (especially deer and hare) against a green background (e.g. moss) (Buesching et al. submitted).

Aside from these individual differences, volunteers in general have a lower 'boredom threshold' for repetitive tasks than professionals, resulting in carelessness and thus considerably lower data reliability (Mumby et al. 1995; Martinich et al. 2006; Buesching & Slade 2012). Similarly, some volunteers may consider a task as too demanding to complete, for example due to bad weather or terrain, resulting in loss of or incomplete data, which may violate the sampling methodology (Cook & Berrenberg 1981; Basinger 1998; Buesching & Slade 2012). In addition, many novice volunteers display a tendency to disregard non-events (e.g. the absence of field signs) or undervalue common events (e.g. repeated findings of field signs of the same, common, species), in favour of seeking out the rare and 'more exciting' observations, resulting in considerable data bias (Macdonald et al. 1998; Brandon et al. 2003; Genet & Sargent 2003; Foster et al. 2003; Danielsen et al. 2005; Buesching & Slade 2012; Buesching et al. submitted).

Data validation on non-residential long-term citizen science projects, however, proves more difficult. The web-based Evolution MegaLab survey, for example, used a number of indirect methods to validate the data submitted (Silvertown et al. 2011), of which one was a web-based quiz testing the volunteer's snail species identification skills. However, the original intention of using individual quiz scores to weight the reliability of the data submitted by each volunteer had to be abandoned in order to maximize data collection. Only about 20% of the users who submitted data also participated in the quiz. While this low participation rate made the validation of individual volunteer data impossible, the quiz results nevertheless proved useful in determining the possible extent of species misidentification (Silvertown et al. 2011). For example, the quiz showed that users had difficulty telling juveniles of the species *Cepaea nemoralis* from adults of *C. hortensis*, which resulted in the researchers' decision to include only adult *C. nemoralis* in the analyses (Silvertown et al. 2011).

The potential to use volunteers to validate data should not be overlooked. Not all interested volunteers can do field work, and many field volunteers also enjoy the social camaraderie of office work. Coupled with computer checks to detect possible errors in identification, coding and methodology, a history of volunteers' previous corrections and proforma responses to common problems, data checkers can play a major role in detecting errors of identification. For example, InstantWILD is a wildlife monitoring project that involves the general public in identifying photos of species taken by camera traps. Cameras are placed in remote locations in countries such as Mongolia, Sri Lanka and Kenya and when an animal triggers the camera a photo is automatically taken and instantly sent to both a website (www.edgeofexistence.org/instantwild/) and an InstantWILD cellphone application. The general public then has the opportunity to identify the species using a basic field guide. In cases of consensus the species is allocated to an appropriate bin (such as lion or elephant), otherwise the images are binned for further evaluation by experienced volunteers and professional scientists. During the first week following the launch of the InstantWILD app, it was downloaded over 60,000 times.

Office volunteers can also help to develop methodologies as well as training field volunteers, most especially new recruits who may be too shy to admit their inexperience. During data collection, the Protea Atlas Project staff provided constant feedback (based on automated reports of inconsistencies and cross-referenced identifications with herbarium and other atlas data) to the atlassers on what errors were made when filling in site record sheets and noted any potential misidentification/out-of-range observations for validation and further attention. This practice improved the quality of the data received by the project, with the end-result that after 10 years of data collection, an extremely comprehensive and reliable data set for proteas in South Africa now exists.

Conclusions

The advantages of using citizen science volunteers in conservation research are clear (Toms et al. 1999): they can provide an inexpensive and potentially large labour force (Bruyere & Rappe 2007; Pfeffer & Wagenet 2007), they usually contribute at least indirectly to the costs of the research, and the volunteers themselves gain fulfillment and knowledge (Sharpe & Conrad 2006). Local involvement also contributes to Agenda 21 targets and can improve management responses (Danielsen et al. 2010). For example, the Protea Atlas Project discovered eight species new to science, rediscovered two species previously thought to be extinct, confirmed two suspected extinctions and established the extinction of two additional species, advancing our knowledge of species' geographic ranges considerably. Increased distributional ranges were found for over 33% of species (some increased by over 100 km), and significant populations were added within known areas of distribution for at least 33% more. This project revealed that the centre of richness for the Cape Floral Kingdom was not the Kogelberg, as previously thought, but in the Western Riviersonderend Mountains. The atlas data have been used, and continue to be used, in numerous scientific studies (Thuiller et al. 2004; Bomhard et al. 2005; Midgley et al. 2006; Manne et al. 2007; Latimer et al. 2009) as well as by conservation authorities for assessing development applications and designing integrated development plans (Rebelo et al. 2011).

Conversely, citizen science brings with it a unique set of challenges and potential drawbacks (Buesching & Newman 2005; Dickinson et al. 2010): training and supervising volunteers takes scientists' time away from professional research, data are prone to higher intra- and interobserver variability, and some tasks continue to require professional involvement to conform with statutory regulations relevant to the working environment, e.g. animal handling licensing procedures and welfare considerations, as well as health and safety requirements (Buesching & Slade 2012).

While many of the challenges of maximizing team success and participant enjoyment are well-researched social science paradigms (Cook & Berrenberg 1981), they have not yet been recognized by the biological sciences well enough to develop more optimal team and project management. Recruiting agencies often use 'conservation output and delivery' as project selection criteria, but then evaluate the ongoing success of the project in terms of 'volunteer enjoyment' and marketability; thus, function and gratification are not necessarily linked intrinsically (e.g. Macdonald et al. 2002). For scientists it is important to recognize that volunteers participate in projects partly also for social reasons, in order to feel part of a group of like-minded people (Coghlan 2005), rather than participating exclusively to benefit scientific understanding. Anecdotal reports show that by fostering 'team spirit' (e.g. by follow-up reports on the progress of the project), volunteers are more likely to return (Miles et al. 1998). By definition, volunteers give up their free time to help without monetary reward (Campbell & Smith 2006). Volunteerism is therefore exemplary of prototypic planned helping (Clary et al. 1996, 1998), which calls for planning, sorting priorities and matching personal capabilities and interests with the type of intervention (Brown 1999).

To optimize the data quality and quantity provided by volunteers, researchers must understand which factors affect volunteer performance most (Newman et al. 2003; Buesching & Newman 2005; Buesching & Slade 2012), and then find ways to optimize and mitigate these factors, as appropriate, by allocating tasks to the best suited individuals (Mackney & Spring 2000), and by offering a varied programme to avoid monotony (Mumby et al. 1995; Martinich et al. 2006; Buesching & Slade 2012). Data collection protocols need to be designed to minimize interobserver errors (Basinger 1998), and methods need to be easy to understand and to perform without requiring special government licences. At the same time, training techniques have to be optimized by the scientists (Cook & Berrenberg 1981; Newman et al. 2003).

In Table 8.1 we offer some considerations to help anyone planning a project to decide whether using citizen science is a viable option. Factors to take into account in this context involve not only quantifiables, such as the ratio of fixed to variable costs, but also the value of engagement with volunteers *per se*: educational outreach and project publicity, the availability of volunteers, the suitability of the tasks required and the ease with which the data collected may be validated.

However, in addition to the obvious benefits of citizen science outlined above, there is a wide variety of more subtle benefits of involving the public in environmental research projects. Experiences from the Evolution MegaLab and the Protea Atlas Project show that working with volunteers is not always a cheap option. However, both projects aimed not only to collect scientific

Table 8.1 When is a citizen science approach appropriate?

Issues to be considered	Suitability of a citizen science (CS) approach
Project objectives	
Education/outreach	Appropriately designed, CS can be an effective tool if learning objectives and audiences are identified clearly
Project duration/ legacy	Since CS incurs set-up costs, it is more cost-effective if the project is of long duration and/ or a legacy effect is desirable. Resources for long-term monitoring and data maintenance must be in place
Accessibility of the scientific objectives	The scientific rationale behind the project must be easily explainable to a lay audience in order to recruit the volunteers to participate using CS
Project budget	
Overall cost	If low cost is an over-riding factor, CS may not be the cheapest option. Even if volunteers pay to participate (e.g. Earthwatch volunteers), they will require scientists' time and supervision, which will incur costs. Recognition of service must match volunteer expectations
Fixed costs	Staff time needs to be devoted to managing volunteers, which could be a cost burden on a small project or one that fails to recruit enough volunteers
Project design	
Health and safety	Major safety issues may make CS unviable
Geographical area to be sampled	The larger the area to be sampled, the more worthwhile it will be to invest in a CS approach
Sampling protocol	A robust and easy-to-apply protocol for CS volunteers must be devised
Data validation	Mechanisms must be in place to validate the data
Volunteers	
Recruitment	Projects involving charismatic species or habitats attract volunteers, as do opportunities to help the environment, be part of a well-organized programme, learn something new, join a social group or affiliate with people with similar values. Is there a readily available pool of volunteers from which to recruit and obvious channels for recruiting them? Recruitment materials should address the many reasons why citizens volunteer
Skills, training and supervision	Are special skills required by volunteers? Is appropriate training provided? Can staff and/or experienced volunteers provide ongoing training and supervision? Is the research project organized effectively?
Feedback/ recognition	How will feedback to volunteers be provided? Can a volunteer's personal contribution be tracked and recognized? Rewarding volunteers through recognition is how an organization expresses thanks for donated time, energy and expertise. Recognition should be frequent and meet the expectations of the volunteers
Time commitment	How much of a time commitment is required from volunteers? Is there more than one role/ level of involvement for volunteers? For long-term projects, how will interest and affiliation be maintained?

data but also to educate the public about local and national conservation issues and raise awareness for the project objectives. In 2010, the year after its official end, more than 1000 new records were submitted online to the MegaLab database, showing that a citizen science approach leaves a valuable and unique legacy in terms of public education and outreach. More than 10 years after the Protea Atlas Project finished, volunteers recruited then are making significant contributions to iSpot and to Custodians of Rare and Endangered Wildflowers (CREW) in South Africa, a citizen science project monitoring threatened species (www.sanbi.org/programmes/threats/

custodians-rare-and-endangered-wildflowers-crew-programme).

Although most people committing to volunteering are likely to belong to an already environmentally aware subsector of the population (Coghlan 2005), a surprisingly large number of volunteers (on the Mammals of Nova Scotia Project almost 50%) are motivated by curiosity, tourism motives or because they want to make a new start in life (e.g. after divorce, job redundancy, etc.). Many organizations also offer a number of fully sponsored places either to employees of specific companies or to teachers, students or similar. In addition, opportunities arise to recruit volunteers from unconventional sources, such as resettlement programmes after prison or drug/alcohol rehabilitation courses involving people without any prior conservation experience/interest (Newman et al. 2003). Such volunteers are thus likely to be somewhat uncertain how to behave correctly in this (new) situation, and will orientate their behaviour based on their fellow volunteers and team leaders (Abrams et al. 1990). In all cases, social alignment with other participants on the project (and the scientists) could be expected to lead to a more environmentally responsible attitude and behaviour of volunteers on the project. These small voluntary changes in behaviour can translate into an altered self-image reflecting greater environmental awareness and an altered perception of these volunteers by their social environment (Schlenker et al. 1994). As Wilson (1986) emphasized, fostering biophilia through understanding and first-hand experience of the natural world is likely to result in heightened compassion and raised environmental awareness (see also Chapter 9).

Advocacy for nature conservation has been linked to long-term volunteering (Ryan et al. 2001), thus demanding that we make citizen science programmes as effective as possible for our science, citizens and the environment. The development of mobile apps for monitoring could bring large numbers of new volunteers into conservation and could become a major source of data in the future.

Acknowledgements

We are grateful to Martin Harvey for data on iSpot, and to Jonathan Baillie for data on InstantWILD. We thank J. Stuart Carlton, Martha Monroe and the Florida Fish and Wildlife Commission for research assistance and support, and Chris Newman for his continued involvement with the Mammals of NS Project.

References

Aanensen, D.M., Huntley, D.M., Feil, E J., Al-Own, F.A. & Spratt, B.G. (2009) EpiCollect: linking smartphones to web applications for epidemiology, ecology and community data collection. *PLoS One*, **4**, e6968.

Abrams, D., Wetherell, M., Cochrane, S., Hogg, M. A. & Turner, J.C. (1990) Knowing what to think by knowing who you are. *British Journal of Social Psychology*, **29**, 97–119.

Bancroft, G.T. (2007) Citizen scientists make a difference. *American Birds*, **64**, 9–11.

Barrett, N., Edgar, G. & Morton, A. (2002) Monitoring of Tasmanian inshore reef ecosystems. An assessment of the potential for volunteer monitoring programs and a summary of changes within the Maria Island Marine Reserve from 1992–2001. *Tasmanian Aquaculture and Fisheries Institute Technical Report Series*, 53.

Basinger, J. (1998) To scientists who use paying volunteers in fieldwork, the benefits outweigh the bother. *Chronicle of Higher Education*, **44**, A14–A15.

Blackburn, S. & Frank, L. (2010) *Assessment of Guide Reporting & Preliminary Results of Lion Monitoring*. Mara Predator Project.

Bomhard, B., Richardson, D.M., Donaldson, J.S. *et al.* (2005) Potential impacts of future land use and climate change on the Red List status of the Proteaceae in the Cape Floristic Region, South Africa. *Global Change Biology*, **11**, 1452–1468.

Bonjean, C.M., Markham, W.T. & Macken, P.O. (1994) Measuring self-expression in volunteer organisations: a theory-based questionnaire. *Journal of Applied Behavioural Science*, **30**, 487–514.

Brandon, A., Spyreas, G., Molano-Flores, B., Carroll, C. & Ellis, J. (2003) Can volunteers provide reliable data for forest vegetation surveys? *Natural Areas Journal*, **23**, 254–261.

Brown, E. (1999) Assessing the value of volunteer activity. *Non-profit & Voluntary Sector Quarterly*, 28.

Bruyere, B. & Rappe, S. (2007) Identifying the motivations of environmental volunteers. *Journal of Environmental Planning and Management*, **50**, 503–516.

Buesching, C.D. & Newman, C. (2005) Volunteers in ecological research: amateur ecological monitors – the benefits and challenges of using volunteers. *Bulletin of the Ecological Society of America*, **36**, 20–22.

Buesching, C.D. & Slade, E. (2012) Citizen science in Wytham Woods: collecting scientific data using amateur volunteers. In: *Wildlife Conservation on Farmland* (eds D.W. Macdonald & R.E. Feber). Oxford University Press, Oxford.

Buesching, C.D., Newman, C. & Macdonald, D.W. (submitted) Deer volunteers are dear volunteers: the use of amateurs in estimating deer population sizes. *Oryx*.

Butcher, G.S. & Niven, D.K. (2007) Combining data from the Christmas bird count and the breeding bird survey to determine the continental status and trends of North America birds. National Audubon Society, Ivyland, PA.

Campbell, L.M. & Smith, C. (2006) What makes them pay? Values of volunteer tourists working for sea turtle conservation. *Environmental Management*, **38**, 84–98.

CBD (1993) Convention on Biological Diversity. www.cbd.int/convention/text/

Clary, E.G., Snyder, M. & Stukas, A.A. (1996) Volunteers's motivations: findings from a national survey. *Non-profit and Voluntary Sector Quarterly*, **25**, 485–505.

Clary, E.G., Snyder, M., Ridge, R.D., *et al.* (1998) Understanding and assessing the motivations of volunteers: a functional approach. *Journal of Personality and Social Psychology*, **74**, 1516–1530.

Cnaan, R.A. & Goldberg-Glen, R.S. (1991) Measuring motivation to volunteer in human services. *Journal of Applied and Behavioural Science*, **27**, 269–284.

Coghlan, A. (2005) *Towards an Understanding of the Volunteer Tourism Experience*. School of Business, James Cook University, Townsville, Australia.

Cohn, J.P. (2008) Citizen science: can volunteers do real research? *Bioscience*, **58**, 192–197.

Cook, S.W. & Berrenberg, J.L. (1981) Approaches to encouraging conservation behavior – a review and conceptual framework. *Journal of Social Issues*, **37**, 73–107.

Crall, A.W., Gregory, G.J., Jarnevich, C.S., Stohlgren, T.J., Waller, D.M. & Graham, J. (2010) Improving and integrating data on invasive species collected by citizen scientists. *Biological Invasions*, **12**, 3419–3428.

Danielsen, F., Burgess, N.D. & Balmford, A. (2005) Monitoring matters: examining the potential of locally-based approaches. *Biodiversity and Conservation*, **14**, 2507–2542.

Danielsen, F., Burgess, N.D., Jensen, P.M. & Pirhofer-Walzl, K. (2010) Environmental monitoring: the scale and speed of implementation varies according to the degree of people's involvement. *Journal of Applied Ecology*, **47**, 1166–1168.

Darwall, W.R.T. & Dulvey, N.K. (1996) An evaluation of the suitability of non-specialist volunteer researchers for coral reef surveys, Mafia Islands, Tanzania – a case study. *Biological Conservation*, **78**, 223–231.

Devictor, V., Whittaker, R.J. & Beltrame, C. (2010) Beyond scarcity: citizen science programmes as useful tools for conservation biogeography. *Diversity and Distributions*, **16**, 354–362.

Dickinson, J.L., Zuckerberg, B. & Bonter, D.N. (2010) Citizen science as an ecological research tool: challenges and benefits. *Annual Review of Ecology, Evolution and Systematics*, **41**, 149–172.

Ericsson, G. & Wallin, K. (1999) Hunter observations as an index of moose Alces alces population parameters. *Wildlife Biology*, **5**, 177–185.

Foster, D., Swanson, F., Aber, J., *et al.* (2003) The importance of land-use legacies to ecology and conservation. *Bioscience*, **53**, 77–88.

Foster-Smith, J. & Evans, S.M. (2003) The value of marine ecological data collected by volunteers. *Biological Conservation*, **113**, 199–213.

Genet, K.S. & Sargent, L.G. (2003) Evaluation of methods and data quality from a volunteer-based amphibian call survey. *Wildlife Society Bulletin*, **31**, 703–714.

Greenwood, J.J.D. (2007) Citizens, science and bird conservation. *Journal of Ornithology*, **148**, 77–124.

Hager, M.A. & Brudney, J. L. (2005) Net benefits: weighing the challenges and benefits of volunteers. *Journal of Volunteer Administration*, **23**, 26–31.

Harrison, D.A. (1995) Volunteer motivation and attendance decisions: competitive theory testing in multiple samples from a homeless shelter. *Journal of Applied Psychology*, **80**, 371–385.

Henningfield, J., Pohl, M. & Tolhurst, N. (2006) *ICCA Handbook of Corporate Social Responsibility*. John Wiley, Chichester.

Irwin, A. (1995) *Citizen Science: A Study of People, Expertise and Sustainable Development*. Routledge, London.

Jacobson, S.K. (2009) *Communication Skills for Conservation Professionals*. Island Press, Washington, D.C.

Jacobson, S.K., Carlton, J.S. & Monroe, M.C. (2012) Motivation and satisfaction of volunteers at a Florida natural resource agency. *Journal of Park and Recreation Administration*, **30**(1).

Janzen, D.H. (2004) Setting up tropical biodiversity for conservation through non-damaging use: participation by parataxonomists. *Journal of Applied Ecology*, **41**, 181–187.

Latimer, A.M., Silander, J.A., Rebelo, A.G. & Midgley, G.F. (2009) Experimental biogeography: the role of environmental gradients in high geographic diversity in Cape Proteaceae. *Oecologia*, **160**, 151–162.

Leslie, L.L., Velez, C.E. & Bonar, S.A. (2004) Utilizing volunteers on fisheries projects: benefits, challenges, and management techniques. *Fisheries*, **29**, 10–14.

Macdonald, D.W., Mace, G. & Rushton, S. (1998) *Proposals for Future Monitoring of British Mammals*. DETR/JNCC, London. www.wildcru.org/

Macdonald, D.W., Newman, C., Buesching, C.D. & Tattersall, F.T. (2002) *Volunteer Mammal Monitors: Measuring the Professionalism of Amateurs. ENSUS 2002: Marine Science and Technology for Environmental Sustainability*. University of Newcastle Publishing, Newcastle.

Mackney, P. & Spring, N. (2000) *The Earthwatch Perspective on Volunteers and Data Validity*. Earthwatch Institute, Oxford.

Manfredo, M. J., Driver, B. J. & Tarrant, M. A. (1996) Measuring leisure motivation: a metaanalysis of the recreation experience preference scales. *Journal of Leisure Research*, **28**(3), 188–213.

Manne, L.L., Williams, P.H., Midgley, G.F., Thuiller, W., Rebelo, T. & Hannah, L. (2007) Spatial and temporal variation in species-area relationships in the Fynbos biological hotspot. *Ecography*, **30**, 852–861.

Martinich, J.A., Solarz, S.L. & Lyons, J.R. (2006) Preparing students for conservation careers through project-based learning. *Conservation Biology*, **20**, 1579–1583.

Midgley, G.F., Hughes, G.O., Thuiller, W. & Rebelo, A.G. (2006) Migration rate limitations on climate change-induced range shifts in Cape Proteaceae. *Diversity and Distributions*, **12**, 555–562.

Miles, I., Sullivan, W.C. & Kuo, F.E. (1998) Ecological restoration volunteers: the benefits of participation. *Urban Ecosystems*, **2**, 27–41.

Moscardo, G., Morrison, A.M., Pearce, P.L., Lang, C. & O'LEARY, J.T. (1996) Understanding vacation destination choice through travel motivation and activities. *Journal of Vacation Marketing*, **2**, 109–122.

Mumby, P.J., Harborne, A.R., Raines, P.S. & Ridley, J.M. (1995) A critical assessment of data derived from coral cay conservation volunteers. *Bulletin of Marine Science*, **56**, 737–751.

Newman, C., Buesching, C.D. & Macdonald, D.W. (2003) Validating mammal monitoring methods and assessing the performance of volunteers in wildlife conservation – "Sed quis custodiet ipsos custodies?". *Biological Conservation*, **113**, 189–197.

Omoto, A.M. & Snyder, M. (1995) Sustained helping without obligation: motivation, longevity of service and perceived attitude change among AIDS volunteers. *Journal of Personality and Social Psychology*, **68**, 671–689.

Pearce, P.L. & Lee, U. (2005) Developing the travel career approach to tourist motivation. *Journal of Travel Research*, **43**, 226–237.

Pfeffer, M.J. & Wagenet, L.P. (2007) Volunteer environmental monitoring, knowledge creation and citizen-scientist interaction. In: *The Sage Handbook of Environment and Society* (eds Pretty, J., Benton, T., Guivant, J. *et al.*). Sage, London.

Rebelo, A.G., Holmes, P.M., Dorse, C. & Wood, J. (2011) Impacts of urbanization in a biodiversity hotspot: conservation challenges in Metropolitan Cape Town. *South African Journal of Botany*, **77**, 20–35.

Ryan, R., Kaplan, F. & Grese, R. (2001) Predicting volunteer commitment in environmental stewardship programmes. *Journal of Environmental Planning and Management*, **44**, 629–648.

Schlenker, B.R., Dlugolecki, D.W. & Doherty, K. (1994) The impact of self-presentations on self-appraisals and behaviour. The power of public commitment. *Personality and Social Psychology Bulletin*, **20**, 20–33.

Schroeder, H.W. (2000) The restoration experience: volunteers' motives, values, and concepts of nature. In: *Restoring Nature: Perspectives from the*

Social Sciences and Humanities (eds Gobster, P.H. and Hull, R.B). Island Press, Washington, DC.

Sharpe, A. & Conrad, C. (2006) Community based ecological monitoring in Nova Scotia: challenges and opportunities. *Environmental Monitoring and Assessment*, **113**, 395–409.

Silvertown, J., Cook, L., Cameron, R., *et al.* (2011) Citizen science reveals unexpected continental-scale evolutionary change in a model organism. *PLoS One*, **6**, 8.

Thuiller, W., Lavorel, S., Midgley, G., Lavergne, S. & Rebelo, T. (2004) Relating plant traits and species distributions along bioclimatic gradients for 88 Leucadendron taxa. *Ecology*, **85**, 1688–1699.

Toms, M.P. & Newson, S.E. (2006) Volunteer surveys as a means of inferring trends in garden mammal populations. *Mammal Review*, **36**, 309–317.

Toms, M.P., Siriwardena, G.M. & Greenwood, J.J.D. (1999) *Developing a Mammal Monitoring Programme for the UK*. BTO Research Report. British Trust for Ornithology, Thetford, Norfolk.

Wilson, E.O. (1986) *Biophilia*. Harvard University Press, Cambridge, MA.

Worthington, J.P., Silvertown, J., Cook, L.M., *et al.* (2011) Evolution MegaLab: a case study in citizen science methods. *Methods in Ecology and Evolution*, **3**(2), 303–309.

9

Nature as a source of health and well-being: is this an ecosystem service that could pay for conserving biodiversity?

Joelene Hughes[1], Jules Pretty[2] and David W. Macdonald[3]

[1]Wildlife Conservation Research Unit, Department of Zoology, Recanati-Kaplan Centre, University of Oxford, Oxford, UK
[2]Department of Biological Sciences, University of Essex, Colchester, UK
[3]Wildlife Conservation Research Unit, Department of Zoology, Recanati-Kaplan Centre, Tubney House, University of Oxford, UK

'One of the first conditions of happiness is that the link between Man and Nature shall not be broken.'

(Leo Tolstoy)

In 2008, for the first time, more than half of the global human population was living in urban areas (UN 2008, 2009). The majority of people in more developed regions (Europe, Northern America, Australia/New Zealand and Japan) have lived in urban places since the mid 20th century, while those in less developed countries are expected to pass the 50% urbanite milestone within the next 10 years (UN 2008, 2009). As the human population has increased, so has the proportion living urban lives, and, in developed countries, even those living in rural areas may be increasingly disengaged from nature. The environmental challenges relating to the urbanization of humanity are manifold – water supply, pollution control, disease transmission and waste disposal among the more conspicuous. This chapter, however, concerns a more insidious impact: a possible detrimental effect on health and well-being for people divorced from nature. If divorce from nature is bad for your health, restoring that relationship may be therapeutic. It may be possible to identify the curative or prophylactic elements of the environment, thereby creating a powerful

Key Topics in Conservation Biology 2, First Edition. Edited by David W. Macdonald and Katherine J. Willis.

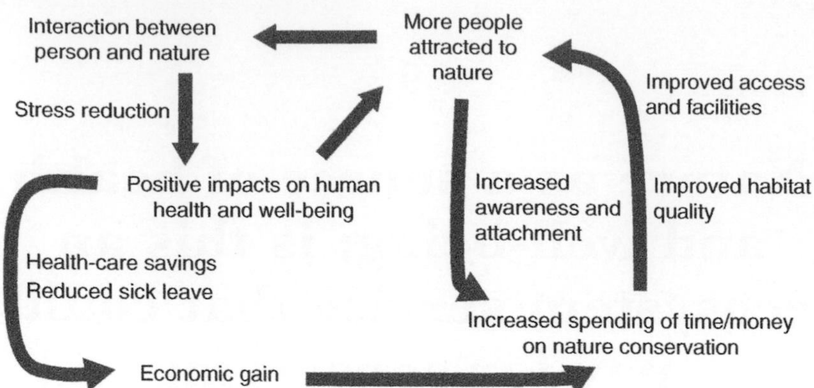

Figure 9.1 A virtuous circle. Framework of how a beneficial relationship between nature and human health and well-being may provide health, economic and biodiversity benefits (Hughes et al. submitted).

lever for their conservation, perhaps with wider benefits for biodiversity conservation.

The provision of ecosystem services is an increasingly recognized, and valued, benefit of nature (Sukhdev et al. 2010; UK National Ecosystem Assessment 2011), conspicuously supplying, for example, water purification, nutrients and medicines (Millennium Ecosystem Assessment 2005). Less tangible, although perhaps intuitively appealing, is the hypothesis that estrangement from nature is detrimental to mental and physical health and well-being. Nature (a hazy category, broadly considered here as the natural world including plants, animals and features of the earth itself) is arguably a critical component of a hypothesized restorative power of, for example, wilderness, rural landscapes or urban greenspaces (parks, nature reserves or gardens). Urbanization, and other forces for disengaging people from nature, for example the increasing technology for indoor entertainment, focuses attention on the possible existence of a valuable ecosystem service: the provision of benefits for health.

But how might this potential relationship between health and nature link to biodiversity conservation? Until recently, human well-being and health might have seemed removed from the concerns of conservationists. However, Hughes et al. (submitted) hypothesize that a 'virtuous circle' may exist, linking health benefits with benefits for conservation (Figure 9.1). This virtuous circle encompasses two key topics, which we present as hypotheses: first, that engagement with nature delivers health benefits (the ecosystem service hypothesis) and second, that these benefits are sufficient to lever directly the conservation of health-giving aspects of the environment, perhaps bundled together, as an indirect benefit, with co-benefits to wider biodiversity (the conservation leverage hypothesis). Here, we evaluate critically the evidence for these two hypotheses.

The ecosystem service hypothesis: health and well-being

The term 'health' is generally taken to incorporate physical, mental, emotional, social and spiritual health, lifestyle and functionality. The World Health Organization's (WHO) definition of health is still the most widely cited and states that: 'health is a state of complete physical, mental and social (individual) well-being, and not merely the absence of disease or infirmity' (WHO 1948). Many sources interpret 'well-being' in a broader context, and in the UK DEFRA (2007) has developed the following meaning of well-being within a policy context.

> 'Well-being is a positive physical, social and mental state; it is not just the absence of pain, discomfort and incapacity. It requires that basic needs are met, that individuals have a sense of purpose, that they feel able to achieve important personal goals and participate in society. It is enhanced by conditions that include supportive personal relationships, strong and inclusive communities, good health, financial and personal security, rewarding employment, and a healthy and attractive environment.'
>
> (Air Quality Strategy)

Globally, there are indications that some aspects of health and well-being are declining. For example, mental health disorders related to stress are increasing (Prince et al. 2007) while since the 1980s, obesity has more than doubled around the world (WHO 2011). Illness and diseases resulting from poor health and well-being burden the British economy: stress, depression and anxiety are the second most reported causes of work-related ill health, affecting 400,000 people and costing an estimated 10.8 million workdays in 2010/2011 (HSE 2011a,b). Coping behaviours, for example increasing smoking and alcohol consumption, will further affect individual health and the amount of money required for health service provision.

These declines in health have been concurrent with numerous factors (population density, family sizes, longevity and nutrition, to name a few) but also increasing urbanization and a waning interest in experiencing nature first hand. For example, in the USA computer activities have shown the biggest rise in hobby popularity since 1995, pushing fishing and gardening out of the top five (HarrisInteractive 2007). Similarly, a national survey of 4941 people in the UK showed only 10% listing sports or outdoor activities as a main occupation in a day, while 80% watch TV, videos, DVDs or listen to the radio/music (ONS 2005). So, in addition to other lifestyle factors, is some of the increase in ill health partly because people are disengaging from nature?

Sedentary occupations are associated with diminished mental or physical health which can be improved with physical activity (Blair & Morris 2009) but over and above this there are indications that 'green exercise' – exercising in the natural environment – may be more effective than comparable indoor activities, with greater improvements in mood and self-esteem alongside greater reduction in anger and depression (Pretty et al. 2005; Barton & Pretty 2010; Bowler et al. 2010b; Thompson Coon et al. 2011). There are various, interrelated ways in which interacting with the natural environment is hypothesized to benefit human health (Figure 9.2): natural green spaces may indirectly provide a range of benefits by providing an attractive space facilitating regular physical activity (CDC 1996; Department of Health 2004; Foresight 2007; Sandercock et al. 2010), social bonding (Coley et al. 1997), family or community cohesion (Kuo et al. 1998) and enhancing an individual's well-being. But perhaps most intriguing is the hypothesis that individuals may gain health benefits directly and specifically from experiencing nature.

The history behind the concept

It has long been clear in art and literature that some people believed in, and advocated, nature's health-giving potency (Pilgrim & Pretty 2010). From Grecian healing groves to 19th-century asylums, Victorian seaside resorts to Frederick Law Olmsted's (1822–1903) transformation of urban planning through the foundation of landscape architecture, the restorative effects of rural environments and (at least the benign face of) wilderness have been appreciated (Philo 1987; Mian et al. 2005). However, whether these benefits were the presence of a positive or the absence of a negative (breathing fresh air versus smog, landscapes akin to Constable's haywain versus open sewers and slum parasites) is not always clear (Philo 1987).

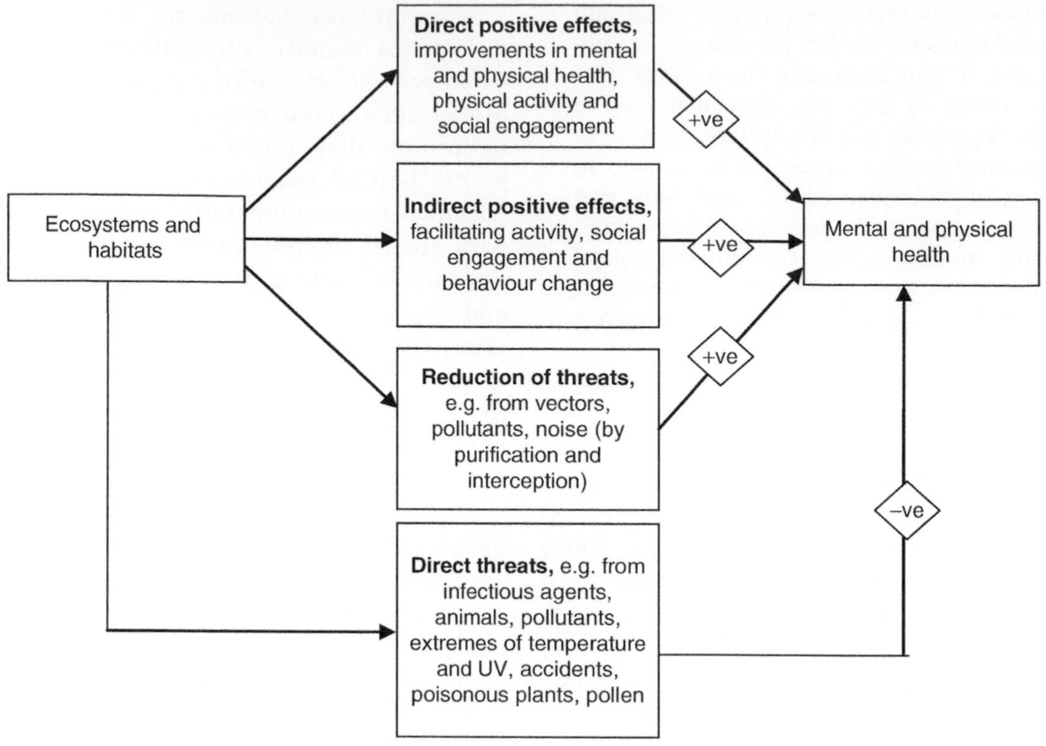

Figure 9.2 Health benefits and threats from ecosystems (Pretty et al. 2011).

Most theoretical and experimental research on this topic occurred in only the latter half of the 20th century. One aspect of this supposed connection with nature has been referred to as 'biophilia'. The term is primarily attributed to the psychoanalyst Erich Fromm, who used biophilia as the antonym of necrophilia, the love of death, to describe the love of life (Fromm 1964). The concept of biophilia in relation to 'nature' was introduced by the biologist E.O. Wilson (1984; Kellert & Wilson 1993) and defined as 'the innately emotional affiliation of human beings to other living organisms'. This definition was intended to encompass both the love and fear of nature (in which usage, to be pedantic, it renders redundant the coining of an opposite term biophobia, as is often done). The defining proposition of biophilia is that the affiliation with, and thus well-being benefits of, nature is innate.

Akin to biophilia, a psycho-evolutionary theory was proposed by behavioural scientist Roger S. Ulrich (Ulrich 1983). He proposed that many environmental qualities required to aid recovery from a stressor are found in nature. While recognizing that both psychological and physiological components of stress are important, Ulrich did not disentangle the effects of short- or long-term stress. He hypothesized that humans have a precognitive response to natural scenes and that this is a consequence of our evolution in the natural world. Humans have precognitive responses (the fight-or-flight response) to negative stimuli that may plausibly have arisen as adaptations through natural selection, for example, humans and naive macaques (*Macaca fuscata*) respond to snakes quicker than to non-threatening stimuli (Shibasaki & Kawai 2009). Ulrich hypothesized that humans may conversely have precognitive responses to positive features of the natural environment, and have not had time to evolve comparable responses to urban environments. Therefore, Ulrich argues, non-natural components of the urban environment will not have the same restorative potential as do natural environments.

A second theory seeking to explain how and why human health and well-being may be related specifically to an experience of nature is attention restoration theory (ART), proposed by the psychologists Rachel and Stephen Kaplan (Kaplan & Kaplan 1989). ART grew from the idea of restoring an individual's capacity for 'directed attention' after suffering mental fatigue, which the authors consider distinct from stress. Kaplan & Kaplan define stress as the build-up to an anticipated threatening or harmful event (Kaplan & Kaplan 1989, p.178). Mental fatigue may be a consequence of stress but, importantly, also results from periods of hard work or protracted effort, diminishing the ability to focus on uninteresting tasks. ART describes four environmental attributes that promote restoration: being away (from the source of stress); extent, the total immersion in the new situation; fascination; and compatibility between a person's requirements and environment. Importantly, this theory differs from Ulrich's in acknowledging that these properties may be available in various built environments or situations without a nature component, but Kaplan and Kaplan argue they are particularly available in natural environments, which thus have special value to people.

The hypothesized beneficial relationship between the experience of nature and human health or well-being could act as a new preventative or curative health intervention, with a 'dose' of nature prescribed (Barton & Pretty 2010). Accurate assessment of the costs and benefits of this resource requires detailed knowledge of the mechanism through which nature may act, and the scale of effect on an individual lifetime or at the population level. However, detailed information on the mechanisms or pathways through which the environment may affect human well-being is sparse. Although human health and well-being have both psychological and physiological components, in a systematic review of literature looking specifically for studies examining the effect of a nature experience on stress, only 24 had taken any physiological measurements (Hughes et al. submitted). The measurements are often blood pressure, skin conductance and heart rate, possibly selected in part because of their simplicity of recording. However, these are all measures that can be affected by a number of different stimuli (for example, presence of other people, room temperature, a developing illness, anxiety) and these confounding factors are not always controlled in the published experimental studies.

A potential pathway of effect is indicated by the focus of much research, based around the psycho-evolutionary and ART theories, on the response of humans to either long- or short-term stressors. Stress at various stages throughout an individual's life has significant implications for mental health, potentially a precursor for illnesses such as depression and schizophrenia (Fumagalli et al. 2007). The causes and results of stress are themselves complex: there is no single, medically accepted definition of stress, it is considered a combination of symptoms and can be acute or chronic, useful or harmful. Stress hormones can have a significant negative impact on cell development and generation in various tissues in mammals (Mirescu & Gould 2006), whilst lower cortisol levels have been associated with more positive mood, better health and well-being (Dockray & Steptoe 2010). These chemicals may, therefore, be an important component in the physiological pathway between nature and improved health. Logically, if the environment causes stress hormone levels to increase, or doesn't promote recovery to baseline levels after a stress event, cell development could be retarded.

Evidence that the environment can have a significant effect on stress and mammalian brain development can be seen from laboratory studies on environmental enrichment (Fox et al. 2006). Environmental enrichment usually involves a combination of social stimulation and inanimate objects that has historically varied between laboratories and species (Sztainberg & Chen 2010). It is difficult, therefore, to determine which specific environmental stimuli are affecting the study animals. Nonetheless, after 40 days in an environmentally enriched cage, 12 mice showed more neurogenesis in the hippocampus and quicker learning in the Morris water maze than their 12 littermates housed in standard conditions (Kempermann et al. 1997). In another seminal work on rats with Huntington's disease, van Dellen et al. (2000) demonstrated that environmental enrichment significantly delayed the onset of behavioural symptoms compared to those in normal laboratory environments. The rats' agility was tested on a weekly basis by seeing if they could balance and turn on a suspended wooden rod. By the end of the experiment only one of the Huntington's rats from the enriched environment failed the task compared to all of the rats kept in standard conditions. Normal laboratory conditions are often a depauperate environment and these experiments demonstrate the enormous influence the environment has on the mammalian brain. This provokes the intriguing question of to what extent humans, obviously having evolved in a natural world, are adapted to urban environments, and to what extent, along with the many advantages it may bring, urbanization is disadvantageous to our physiology.

There are accepted physiological and psychological tests that can be used to evaluate stress convincingly. For example, levels of cortisol respond to psychological and chemical stressors and can be measured through salivary samples (Kudielka et al. 2009). Another simple test that has been demonstrated to be sensitive to changes in stress levels is the Leucocyte Coping Capacity (LCC) test (McLaren et al. 2003). Leucocyte activity has been shown to alter measurably in mammals (badgers, water voles and humans) in response to short-term stress (Montes et al. 2004; Gelling et al. 2010; Shelton-Rayner et al. 2011) and could prove a useful technique with which to measure responses to experiencing nature. For example, the LCC has already been used to quantify the stresses associated with aspects of motor car driving in both stationary and dynamic driving environments (Shelton-Rayner et al. 2010, 2011), and this technique could be adapted to measure stress, or lack of it, associated with engagement with nature. Through using these techniques, it may be possible to elucidate in fine detail the reaction of the body to stress in natural environments as oppose to the urban.

Is engagement with nature health giving?

The quest for a general answer as to whether there is evidence of a health benefit from nature which could be quantified as an ecosystem service is complicated. The connection is intuitively appealing, with the attendant risk that evidence for it is potentially assessed less critically by advocates eager to promote the health benefits of green space. Considering the mesh of confounding factors, critical evaluation is essential ('Do you know what they call "alternative medicine" that's been proved to work? Medicine'; Minchin 2010).

At the broadest scale, some of the most intriguing indications of a connection have come from population and landscape scale studies that combine national census health information with satellite-derived environment data. In the past decade, a number of studies have examined the correlative relationship between health and neighbourhood green space. One of the earliest, carried out in The Netherlands and controlling for socio-economic and demographic variables that may affect choice of living area, found that people living in greener environments appeared to have better physical and mental health (Vries et al. 2003; Maas et al. 2006). Similarly in the UK, research using the early 1990s British Household Panel Survey showed that the rural population has slightly better mental health than the non-rural sample (Weich et al. 2006), a relationship also observed in other countries (Peen et al. 2007). In Denmark, Nielsen & Hansen (2007) found that living in greener areas was correlated with lower stress and obesity levels while a relationship between mortality, cardiovascular disease and green space was observed in England (Mitchell & Popham 2008). Interestingly, the relationship observed by Mitchell & Popham was not seen in a similar study in New Zealand (Richardson et al. 2010), perhaps because New Zealanders generally live at lower density and reputedly enjoy a more outdoors lifestyle than do the British. In Australia, a large-scale study showed that although exercise is a better predictor of physical health than is access to green space, neighbourhood greenness is positively correlated with mental health (Sugiyama et al. 2008). In Sweden, self-reported incidences of stress were negatively correlated with the number of visits to urban green spaces (Grahn & Stigsdotter 2003). Also in Sweden, intriguingly, there is a negative correlation between July temperature and the number of antidepressants dispensed, possibly because people go outside less when the weather is colder, which may inhibit recovery from chronic stress (Hartig et al. 2007). A number of factors may confound such correlations but at least they offer preliminary support for the proposition that a relationship may exist between health and experience of green space.

At the community level, a natural experiment in which people had been randomly assigned to housing blocks, some with trees and green space and others without, enabled Kuo and colleagues to demonstrate that living in the more leafy conditions was associated with beneficial effects on moods and emotions, crime and community cohesion (Kuo 2001; Kuo & Sullivan 2001). However, for individuals, the evidence, while less substantial, is intriguing. In a widely cited paper, Ulrich (1984) reported on the recovery rates and nurses' reports for 46 cholecystectomy patients over nine summers at a Pennsylvanian hospital. All patients were in similarly situated rooms but through the window some had a view of a wall while the others saw trees. Patients with the tree view spent less time in hospital, prompted fewer negative comments from nurses and required fewer painkillers. However, as the author notes, there was no control over the patient room allocation, so nurses may have subconsciously assigned pleasanter patients nicer rooms (with tree views) or those requiring more nurse care nearer their station and the recovery room (in rooms with brick views). Furthermore, results might alternatively indicate that the view of the wall was damagingly monotonous, resulting in understimulation, rather than demonstrating an inherent therapeutic quality of trees.

More recently, experiments where people are exposed to a 'nature treatment' indicate that there may be beneficial effects, although little is known of how long they last. In one of the first experiments to record any physiological measurements, Ulrich et al. (1991) exposed 120 subjects to a 'stressful' movie, before playing them colour and sound video tapes of one of six urban or natural settings, measuring the subjects' heart period, muscle tension, skin conductance and pulse transit time (which correlates with systolic blood pressure). The 'restorative' power of the natural setting was reflected across the physiological measurements, with changes in muscle tension, skin conductance and pulse transit time moving towards baseline levels significantly faster in subjects watching nature scenes. Hartig et al. (2003) found some restorative benefits of viewing trees, where the rate of decline in diastolic blood pressure was significantly increased while walking in a natural compared to an urban environment (30 minutes after a walk systolic blood pressure was approximately 6 mmHg lower, a significant difference, for those walking in a natural environment in comparison with an urban one), indicating reduced stress.

Similarly, Laumann et al. (2003) demonstrated that viewing video simulations of nature led to a significant reduction in heart rate (increased interbeat interval) from the baseline level, that did not occur while watching a video of urban scenery. Pretty et al. (2005) showed that viewing slides of pleasant rural landscapes increases the benefits of exercise on systolic, diastolic and mean arterial blood pressure, with a highly significant drop after exercising. In comparison, people viewing slides of unpleasant rural, pleasant urban, unpleasant urban scenes or no slides showed non-significant changes in blood pressure after exercise. Viewing urban scenes, both unpleasant and pleasant, showed a tendency to increase blood pressure relative to the exercise-only control. Thus, viewing photographs of nature can have a relaxing effect on autonomic functions (the unconscious regulation of internal bodily activity), decreasing heart rate and blood pressure measurements. Furthermore, in Japan, research on *Shinrin-yoku* – the habit of taking in the forest atmosphere or forest bathing/walking – shows that salivary cortisol concentration, pulse rate and blood pressure are all significantly lower after viewing and walking in a forest compared to viewing and walking in a city (Park et al. 2010). Other research where people viewed nature has provided evidence of improved attention and concentration (Tennessen & Cimprich 1995; Lohr et al. 1996; Taylor et al. 2002; Laumann et al. 2003; Berman et al. 2008), mood (Hartig et al. 1991, 1996; Ulrich et al. 1991; van den Berg et al. 2003) and psychological well-being (Kaplan & Kaplan 1989; Hartig et al. 1991; Kaplan 1992, 1995).

Alongside the experimental research, anecdotal information is provided by case studies, interviews and questionnaires. Pets, institution gardens and horticultural activities are increasingly being used in therapies in the US, Europe and a number of other countries around the world (e.g. Kearney & Winterbottom 2005; Pachana et al. 2005) and much research has been conducted into the positive influences of wilderness experiences, for example for holidaymakers (Hartig et al. 1991) or as a therapeutic intervention for adolescents (Becker 2010). Access to nature during youth is considered important insofar as childhood experiences of nature may predict nature contact during adulthood (Ward Thompson et al. 2008) and nearby nature is considered important for the mental well-being of children (Kaplan & Kaplan 1989; Thomas & Thompson 2004; Ward Thompson et al. 2008).

Wells (2000) conducted a study with children of low-income urban families that were moving house, rating characteristics of the house, for example the naturalness of views from windows, and measuring the child's cognitive functioning through a questionnaire for the mother. The increase in naturalness of the surroundings experienced before and after the move was significantly positively correlated to an increase in cognitive functioning. However, these findings should be treated with caution because these families may have been able to select their preferred homes, and the move may have had other influences on family life not measured in the experiment. Cause and effect can be difficult to disentangle and decipher (Wells 2000). Wells & Evans (2003) found that children with more access to nature, as judged by the views from their homes, were generally less stressed than those in urban habitats with less access to green space. However, the issue of causality still remains unresolved, insofar as it is unclear whether (a) having contact with nature aids the development of stress-coping mechanisms for later life, (b) nearby nature provides the opportunity for stress recovery and replenishes attentional fatigue, (c) green space provides the opportunity for social contact, or (d) whether it is a combination of many factors.

Despite the variety of research, conclusions about the specific relationship between nature, human health and well-being remain elusive. Within this research base, definitions of 'nature' range from wilderness or rural landscapes to pot plants, pets or pictures; measurements recorded extend from self-reported questionnaires to physiological tests; and studies may be anecdotal, theoretical or empirical. The diversity of stimuli that have been used to represent 'nature' has made it difficult to compare studies or, in some cases, to distinguish a specific effect of the nature component beyond that of any relaxing experience. Many psychological, social and cultural factors complicate any relationship between people, gardens or pets, which cannot be characterized simply as a human–nature interaction (e.g. Westgarth et al. 2010). Furthermore, even though open nature scenes may provide a restorative environment for some, other peaceful environments, such as monasteries, may be no less restorative (Ouellette et al. 2005). Working with humans can involve the biases of self-selection (people who volunteer for country walks or exercise tasks may do so because they enjoy it, those that don't may not volunteer). An individual's background, for example, socio-economic factors, learning, family and peers, can lead to multiple confounding factors affecting responses to nature. Herzog et al. (2002) presented 630 students with a scenario about undertaking a project that required considerable effort, and asked them to identify what choices would be made for themselves or a best friend when they had become fatigued. The authors asked them to rate a variety of environments or activities they might choose, or recommend to their friend, to restore themselves. Answers were varied, as would be expected, but nature-related activities (e.g. gardening, picnic, boating, etc.) were perceived as more restorative than doing chores or taking drugs, although ranked as less restorative than exercise, entertainment (e.g. watching a film) or grooming activities (e.g. having a bath or haircut). This raises intriguing questions about the generality of any response across society. Will people who have learned to relax, for example, by listening to music with friends or working out at the gym, produce the same response as those that already enjoy nature? The relative benefits may be further confounded by subject knowledge or experience. The outdoors may be perceived as appealing wilderness to some, whereas others see only threat.

Hitherto, a problem with many individual-based studies has been the neglect of proper comparator groups and controls (Bowler et al. 2010b). For example, as with many similarly designed experiments, Chang et al. (2008) used images of nature to elucidate whether they elicited different psychological and physiological responses to a blank screen which was presented as the control. Without comparative 'non-nature' scenes to test the response to viewing pictures of anything versus nothing, they cannot draw conclusions as to whether images of nature have different effects to those of non-nature. Chang et al. (2008) used images containing the four components of ART (being away, fascination, compatibility and coherence) and tested two hypotheses: (1) that the different components of the images will lead to different scores on the Perceived Restorativeness Scale (PRS, as defined in Purcell et al. 2001), and (2) that there will be improved physiological responses, as measured through facial electromyography (EMG), brainwave activity (EEG) and blood volume pulse (BVP), when 110 participants viewed nature scenes compared to the blank image control. The different images did produce different PRS ratings of ART components, and the physiological measures differed significantly between image and non-image conditions ($p \leq 0.05$ in all cases with all measures increasing from non-viewing to viewing except BVP, which decreased). However, there were no statistically significant differences in the physiological measures between the ART components, except for EMG for which there were significant differences between being away versus fascination; being away versus compatibility; coherence versus fascination; coherence versus compatibility. The authors conclude that these results might indicate a restorative value to natural environments. This might be so, if the physiological changes do translate into better well-being, but the results do not permit the conclusion that this effect is greater for pictures of natural environments than it could be for pictures of anything except a blank screen. Many studies seeking to demonstrate a link between nature and well-being have not provided the evidence to rule out alternative, not necessarily exclusive, hypotheses (for example, people living in towns not only experience urban views, but may also face higher levels of pollution).

Despite these caveats, there is enough evidence of positive psychological and physiological reactions to experiencing nature to justify more investigation (Health Council of the Netherlands and Dutch Advisory Council for Research on Spatial Planning 2004; Bowler et al. 2010a, b; Lee & Maheswaran 2011; Thompson Coon et al. 2011). A priority is to discover which features and characteristics of nature are essential to the delivery of this ecosystem service and to characterize the attributes of the 'dose–response' relationship. It might be that some prominent landscape features stimulate a beneficial effect; for example, there is evidence that people prefer images of landscapes containing water (Kaplan & Kaplan 1989). Following this, White et al. (2010) asked 40 people to rank their preferences for 120 photos which variously depicted built, green and aquatic environments. Built environments with some water scored similarly in the participants' preferences to green environments without water. Purely built environments were least preferred and green-aquatic environments the most preferred. There is evidence that the naturalness of an environment may affect the health gain from experiencing it: people watching pleasant rural scenes had healthier scores on physiological measures than did those viewing pleasant or unpleasant urban scenes while exercising (Pretty et al. 2005); a walk in a forest was reported to have improved moods and reduced blood pressure, adrenalin and serum cortisol levels amongst a small sample of elderly participants in comparison to a walk in a rural agricultural environment in Japan (Kondo et al. 2008), and Fuller et al. (2007) found a positive correlation between measures of psychological well-being and increasing species richness of plants and birds in urban greenspaces. However, Parsons et al. (1998) showed that a simulated drive with a golf course as the roadside environment led to a significantly greater reduction in the magnitude of skin conductance – an indicator of stress – and quicker recovery from a stressor than did urban, mixed or forest environments.

So, although evidence suggests that nature is good for your health, the detail remains tantalising. Is a single plant in an office as good as several plants or even time sitting in a forest? Is there a relationship between the quantity and complexity of nature and the effect of the experience? What types of wildlife might be important, and how relevant is, for example, familiarity with the species, its taxon or the threat it might pose? If a minimal quantity of non-specific nature can suffice (a single plant of any species or a monoculture field), then this health service may not be a strong case for conservation (see the Biodiversity Leverage Hypothesis, below). However if a complex natural environment produces measurably greater health gains than does a pot plant, then the many reasons for maintaining these environments, and their biodiversity components, are strengthened.

In summary, improvements in mood and self-esteem have been recorded in studies comparing activities carried out in, or in view of, nature with other more built environments (Hartig et al. 1991, 1996, 2003; Ulrich et al. 1991; van den Berg et al. 2003; Pretty et al. 2005; Morita et al. 2007; Bowler et al. 2010b). There is evidence of reduction in stress measurements (Ulrich et al. 1991; Parsons et al. 1998; Hartig et al. 2003; Park et al. 2010) and a potential for cognitive benefits (Lohr et al. 1996; Pretty et al. 2005; Berman et al. 2008) specifically associated with exercising in nature. However, the specific components of nature stimulating these changes, the significance of results from these short-term experiments for the long-term health of the individual, and the mechanism or pathway through which they may act still have to be ascertained.

The Biodiversity Leverage Hypothesis: health, well-being and biodiversity conservation

The Biodiversity Leverage Hypothesis proposes that the health benefits of engagement with the natural environment are sufficient to lever, directly, conservation of health-giving aspects of the environment; they may also catalyse indirect co-benefits to wider biodiversity. Therefore, the capacity of the natural environment to act as a preventative or curative health resource *could* lead both directly and indirectly to gains for biodiversity. Direct gains for biodiversity could occur if demonstrable health-giving benefits of engagement with nature encouraged investment in this 'medicine'. Indirect benefits to biodiversity could arise in two ways; from linked co-benefits and from enhancing the value people attribute to nature as a result of increased engagement with it.

Securing direct ecosystem benefits

Are the health benefits of engagement with the natural environment sufficient to add to the powerful list of reasons that make it important to invest in environmental conservation in general and biodiversity conservation in particular? In this chapter, we have considered stress as one possible example of how direct gains may be accrued from interaction with the nature. Stress, which is relevant to the realms of medicine, health, economics and politics, incurs a high economic cost in many countries through its effect on people's health both at work and at home (Kalia 2002). A report by the European Agency for Safety and Health at Work (2009) recorded that stress-related illnesses cost France an estimated €830–1656 million in 2000, that psychological disorders (37% of them depression) cost Germany €3000 million in 2001 and in 2005–6 work-related stress, depression and anxiety cost the UK over £530 million. Reductions in the prevalence of stress-related illnesses could reduce these costs and reap economic rewards.

Further, commercial benefits may accrue from improved workforce performance insofar as experiencing nature may improve accomplishment in tasks, enhancing productivity. Rita Berto (2005) conducted an experiment in Italy in which 32 undergraduates took the Sustained Attention to Response Test, a stressor under which the subjects respond to numbers on a screen and to which they may make a correct, an incorrect or no response. Subjects then watched a series of 25 slides showing either nature, geometric patterns or urban scenes before repeating the test. The group that watched nature scenes showed significantly improved responses in target detection (means: before, 1.40, SD 0.71; after, 1.86, SD 0.89), reaction times (means in ms: before, 313.71, SD 38.36; after, 267.38, SD 73.78) and correct responses (means: before, 11.68, SD 5.28; after, 13.62, SD 5.37), with the only other significant change being a reduction in the number of errors by the group that viewed non-nature scenes (means: before, 3.25, SD 6.22; after, 1.62, SD 4.96).

In addition to altering physiology, the environment may also influence behaviour that promotes health. Notably, a self-reported questionnaire study in Bristol, UK, showed that the proximity of green space to their home was related to whether people met government-recommended weekly exercise targets. Increasing distance to formal (well-managed and structured) green spaces was linked to a statistically significant decline in visits to green space, with people living furthest – over 2250 m – from formal green space being 36% less likely to visit it than those living under 830 m away, and 26% less likely to achieve the recommended physical activity level of at least five sessions of 30 minutes of exercise per week (Coombes et al. 2010). Thus, not only did people with less access to green space, unsurprisingly, visit green spaces less, but they also exercised less.

There is mounting evidence suggesting that engagement with the natural environment may be healthy, and in ways that may have significant benefits to a nation's economy. If the effects of interaction with nature (or parts thereof) are demonstrated to have financial benefits, then this merits investment to safeguard or enhance these benefits. However, in the context of biodiversity conservation, a crucial question is to what extent any health benefits from engagement with the natural environment can be attributed to particular species or natural communities (and if so, which components of them). This question has not yet been answered, and inevitably therefore, neither has the obvious sequel of whether any such health benefit thus provides an additional motive for investing in the conservation of these species. Indeed, the evidence so far does not preclude the possibility that any health-giving properties of nature may be provided as fully by an urban park as they are by pristine wilderness.

Securing indirect biodiversity benefits

Even if the ecosystem service of health gains provided by the natural environment do not lever direct investment in wildlife conservation, there are two ways in which they might lead indirectly to biodiversity benefits.

'PLEIOTROPIC' CO-BENEFITS With increasing awareness of the links between all aspects of policy and politics, it is important to optimize solutions and maximize gains (Macdonald et al. 2006). In the context of wildlife conservation, it is advantageous if paying for one outcome can also deliver others. This is essentially the notion behind the 'hot-spots' approach to prioritizing areas for conservation – where investment in protecting a place for one species may maximize the numbers of others also protected (Mace et al. 2006). By analogy, Valenzuela-Galvan et al. (2008) use this concept of complementarity to quantify how to maximize the conservation gains for several species of carnivore by investing in given locations in North America. They showed that conserving just $18\times40{,}000\,km^2$ grid cells could deliver protection for all the 47 species considered. Similarly, looking for efficiencies in the spending of conservation funds (Macdonald et al. 2012) shows how conserving carefully chosen sites for threatened felids may deliver, perhaps at minimal extra cost, shared conservation benefits for threatened primates, and vice versa (e.g. some 1° grid cells in Africa fall within the geographic distributions of 16 species of threatened felids and primates).

This concept of bundled co-benefits has been most explicit in the hope that investment to reduce carbon pollution by protecting forests could deliver gains for biodiversity, for example through REDD+(Reduced Emissions from Deforestation and Degradation). Collins et al. (2011) make explicit that there are two categories of species in this context, and their classification is directly transferable to the issue of health benefits levering biodiversity gains. First, there are those species whose conservation inevitably gains as a co-benefit of protecting the environment for another motive (e.g. to minimize greenhouse gas emissions or to maximize delivery of health gains) and, by analogy with linked genetic traits, Collins et al. (2011) term these 'pleiotropic' species. Second, there are those species for which merely protecting their habitat is not sufficient to secure their conservation, and will require further, targeted funding but benefit from the habitat protection. Thus, in the context of health benefits, even without demonstrating a biodiversity component to the health benefits, investment in the natural environment to secure those benefits may indirectly conserve 'pleiotropic' species and cut the costs of conserving others.

ENGAGEMENT DIVIDEND If health benefits are provided by engagement with the natural environment, these may provide an important catalyst for individual behavioural change, encouraging the adoption of healthier and, in a virtuous circle, more environmentally sensitive lifestyles (Wells & Lekies 2006; Pretty et al. 2009). If health benefits cause people to connect more with nature, this can deliver an engagement dividend to biodiversity conservation. As the burgeoning human population puts pressure on resources, exacerbating conflict between biodiversity and human priorities, ecological awareness is increasingly necessary to motivate societal enthusiasm for conservation in the face of tough choices. Interaction with nature is important because if people do not experience the natural environment, particularly as children, they may become emotionally disconnected from it and value it less (Ward Thompson et al. 2008). This disconnection has potential negative consequences for people's awareness of nature, their concern for green spaces and the efficacy of biodiversity conservation (Miller 2005).

The biodiversity that people are aware of, or regard as appropriate in a particular place, is often related to individual experience or is learned from previous generations. As people become more remote from nature, amnesia about the state of the natural environment may occur, as predicted by the shifting baseline hypothesis (Pauly 1995). Consequently, impoverished levels of biodiversity in an area may be judged acceptable if nobody remembers an earlier, biodiversity-rich period. Evidence of just such generational and personal amnesia in baselines has been demonstrated in the UK by assessment of people's perceptions of common bird species. When asking 50 people to identify the three most common bird species 20 years ago and now, personal amnesia was evidenced, as 36% thought things had remained unchanged, expressing the opinion that the same species of bird were the most common in both 1994 and 2006, although in reality the ranking of bird abundance had changed. Furthermore, generational amnesia was apparent insofar as those aged over 60 generally reported two or three changes in the three common species and those under 41 reported only one or two changes (Papworth et al. 2009).

Additionally, in many countries there is heavy reliance on protected areas to fulfil conservation obligations. These protected areas may be visited by only a small sector of the population; for example, in the UK a survey of visitors to 13 protected areas in Yorkshire showed that there was a highly significant bias towards men, only 0.008% of respondents described themselves as black or from ethnic minorities, there was a skewed age structure with people aged from 56–65 visiting more frequently and visitors were from the more affluent sections of society (Booth et al. 2010). If evidence-based health advice encourages people to experience nature, they will be less disconnected from it, more likely to value it and thereby an engagement dividend will be delivered indirectly by health concerns to wildlife conservation. Insofar as wildlife conservation will benefit from this engagement dividend (whether motivated by health or otherwise), it is important that more types of people are engaged. Since concern for personal well-being is ubiquitous, health may be a potent force for convincing people to engage with nature, and in turn causing them to value wildlife more highly.

Lifecycle economic analysis

The connection between human health, well-being and nature has been proposed as providing an ethical reason for conserving biodiversity (Simaika & Samways 2010). The demonstration of health benefits from engagement with nature may lever financial support for conservation and thereby lead to people experiencing a higher quality of nature (Hughes et al. submitted). To the extent that particular components of biodiversity are demonstrated to deliver health benefits, their conservation would be a priority in this particular context. However, as always, potential benefits need to be evaluated in the light of potential costs. For example, there may be trade-offs to be considered with increased use of delicate habitats. Experiments on a single trampling period of low to high impact (25, 75, 200 or 500 passes) through vegetation in five different forest and grassland communities showed that after the higher impact events, all communities had lost cover even 2 years later (Roovers et al. 2004). Indeed, recreational use can change nutrient states, erode soils, spread pathogens and disperse seeds and alternative plant communities (Pickering et al. 2010). Increasing the public's requirement for access to nature will come with financial and development costs; for example, pathways, car parks, amenities and maintenance will be necessary. In a full lifecycle analysis, the positive outcomes for health of engagement with nature can then be balanced against any negative outcomes, for example a greater risk of slips and falls (Frumkin 2001).

Reuniting people with some shadow of the world in which we evolved, thereby improving their health and increasing their appreciation for nature, may add to the list of important ecosystem services from which humanity benefits. While the absolutely rigorous experiment of the impact of experiencing nature on health and well-being is yet to be performed (if possible), the evidence indicates that nature has important roles. The extent to which these gains will leverage the conservation of wildlife, directly or indirectly, is a key topic for the 21st century.

Acknowledgements

JH was previously supported by a grant from NERC. We would like to thank the participants of meetings and workshops for stimulating discussions: Andrew Rowan, Gareth Edwards-Jones, Michael Depledge, Nicholas Rawlins, Nigel Winser, Philip Lowe, Rubina Mian, Stephen Kellert, William Bird, Philip Riordan, Harry Burns, Katherine Irvine, Katy Cooper, Inigo Montes, Clive Hambler, Angela Clow, Lynne Crowe, Joanne Barton, Joe Hinds, Andrew Pullin, Mark Stanley-Price, Nigel Winser. We are grateful for helpful comments from Tom Moorhouse and Kathy Willis and for suggestions from Isobel Macdonald.

The part can never be well unless the whole is well.

Plato

References

Barton, J., & Pretty J. (2010) What is the best dose of nature and green exercise for improving mental health? A multi-study analysis. *Environmental Science and Technology*, **44**, 3947–3955.

Becker, S.P. (2010) Wilderness therapy: ethical considerations for mental health professionals. *Child and Youth Care Forum*, **39**, 47–61.

Berman, M.G., Jonides, J. & Kaplan, S. (2008) The cognitive benefits of interacting with nature. *Psychological Science*, **19**, 1207–1212.

Berto, R. (2005) Exposure to restorative environments helps restore attentional capacity. *Journal of Environmental Psychology*, **25**, 249–259.

Blair, S.N. & Morris, J.N. (2009) Healthy hearts – and the universal benefits of being physically active: physical activity and health. *Annals of Epidemiology*, **19**, 253–256.

Booth, J.E., Gaston, K.J. & Armsworth, P.R. (2010) Who benefits from recreational use of protected areas? *Ecology and Society*, **15**, 19.

Bowler, D., Buyung-Ali, L., Knight, T.M. & Pullin, A.S. (2010a) *The Importance of Nature for Health: Is There A Specific Benefit of Contact with Green Space?* CEE review 08-003 (SR40). Environmental Evidence: www.environmentalevidence.org/SR40.html

Bowler, D., Buyung-Ali, L., Knight, T.M. & Pullin, A.S. (2010b) A systematic review of evidence for the added benefits to health of exposure to natural environments. *BMC Public Health*, **10**, 10.

CDC (1996) *Physical Activity and Health. A Report of the Surgeon General*. US Department of Health and Human Services, Centers for Disease Control and Prevention, National Centre for Chronic Disease Prevention and Health Promotion, Washington, D.C.

Chang, C.Y., Hammitt, W.E., Chen, P.K., Machnik, L. & Su, W.C. (2008) Psychophysiological responses and restorative values of natural environments in Taiwan. *Landscape and Urban Planning*, **85**, 79–84.

Coley, R.L., Kuo, F.E. & Sullivan, W.C. (1997) Where does community grow? The social context created by nature in urban public housing. *Environment and Behaviour*, **29**, 468–494.

Collins, M.B., Milner-Gulland, E.J., Macdonald, E.A. & Macdonald, D.W. (2011) Pleiotropy and charisma determine winners and losers in the REDD+game: all biodiversity is not equal. *Tropical Conservation Science*, **4**, 261–266.

Coombes, E., Jones, A.P. & Hillsdon, M. (2010) The relationship of physical activity and overweight to objectively measured green space accessibility and use. *Social Science and Medicine*, **70**, 816–822.

DEFRA (2007) *Sustainable Development Indicators in Your Pocket 2007: An Update of the UK Government Strategy Indicators*. Department for Environment, Food and Rural Affairs, London.

Department of Health (2004) *Choosing Health: Making Healthy Choices Easier*. Department of Health, London.

Dockray, S. & Steptoe, A. (2010) Positive affect and psychobiological processes. *Neuroscience and Biobehavioral Reviews*, **35**, 69–75.

European Agency for Safety and Health at Work (2009) *OSH in Figures: Stress at Work – Facts and Figures*. Office for Official Publications of the European Communities, Luxembourg.

Foresight (2007) *Tackling Obesities – Future Choices*. Government Office of Science, London.

Fox, C., Merali, Z. & Harrison, C. (2006) Therapeutic and protective effect of environmental enrichment against psychogenic and neurogenic stress. *Behavioural Brain Research*, **175**, 1–8.

Fromm, E. (1964) *The Heart of Man: Its Genius For Good and Evil*. Harper and Row, New York.

Frumkin, P. (2001) Beyond toxicity – human health and the natural environment. *American Journal of Preventive Medicine*, **20**, 234–240.

Fuller, R.A., Irvine, K.N., Devine-Wright, P., Warren, P.H. & Gaston, K.J. (2007) Psychological benefits of greenspace increase with biodiversity. *Biology Letters*, **3**, 390–394.

Fumagalli, F., Molteni, R., Racagni, G. & Riva, M.A. (2007) Stress during development: impact on neuroplasticity and relevance to psychopathology. *Progress in Neurobiology*, **81**, 197–217.

Gelling, M., Montes, I., Moorhouse, T.P. & Macdonald, D.W. (2010) Captive housing during water vole (*Arvicola terrestris*) reintroduction: does short-term social stress impact on animal welfare? PLoS *ONE*, **5**(3), e9791.

Grahn, P. & Stigsdotter, U.A. (2003) Landscape planning and stress. *Urban Forestry and Urban Greening*, **2**, 1–18.

HarrisInteractive (2007) Reading and TV watching still favorite activities, but both have seen drops. In: The Harris Poll® #115, November 15.

Hartig, T., Mang, M. & Evans, G.W. (1991) Restorative effects of natural environment experiences. *Environment and Behavior*, **23**, 3–26.

Hartig, T., Book, A., Garvill, J., Olsson, T. & Garling, T. (1996) Environmental influences on psychological restoration. *Scandinavian Journal of Psychology*, **37**, 378–393.

Hartig, T., Evans, G.W., Jamner, L.D., Davis, D.S. & Garling, T. (2003) Tracking restoration in natural and urban field settings. *Journal of Environmental Psychology*, **23**, 109–123.

Hartig, T., Catalano, R. & Ong, M. (2007) Cold summer weather, constrained restoration, and the use of antidepressants in Sweden. *Journal of Environmental Psychology*, **27**, 107–116.

Health Council of the Netherlands and Dutch Advisory Council for Research on Spatial Planning (2004) *Nature and Health. The Influence of Nature on Social, Psychological and Physical Well-Being*. Health Council of the Netherlands and RMNO, The Hague.

Herzog, T.R., Chen, H.C. & Primeau, J.S. (2002) Perception of the restorative potential of natural and other settings. *Journal of Environmental Psychology*, **22**, 295–306.

HSE (2011a) *Prevalence – For People Working in the Last 12 Months (Table SWIT3W12)*. Health and Safety Executive, London.

HSE (2011b) *Working Days Lost*. Health and Safety Executive, London.

Hughes, J., Barton, J., Bird, W., *et al.* (submitted) Nature as a resource for health and well-being: the missing ecosystem service.

Kalia, M. (2002) Assessing the economic impact of stress – the modern day hidden epidemic. *Metabolism – Clinical and Experimental*, **51**, 49–53.

Kaplan, R. (1992) The psychological benefits of nearby nature. In: *The Role of Horticulture in Human Well-being and Social Development: A National Symposium* (ed. D. Relf), pp. 125–133. Timber Press, Portland, OR.

Kaplan, R. & Kaplan, S. (1989) *The Experience of Nature: A Psychological Perspective*. Cambridge University Press, Cambridge.

Kaplan, S. (1995) The restorative benefits of nature: toward an integrative framework. *Journal of Environmental Psychology*, **15**, 169–182.

Kearney, A.R. & Winterbottom, D. (2005) Nearby nature and long-term care facility residents: benefits and design recommendations. *Journal of Housing for the Elderly*, **19**, 7.

Kellert, S. & Wilson, E.O. (1993) *The Biophilia Hypothesis*. Island Press, Washington, D.C.

Kempermann, G., Kuhn, H.G. & Gage, F.H. (1997) More hippocampal neurons in adult mice living in an enriched environment. *Nature*, **386**, 493–495.

Kondo, T., Takeda, A., Takeda, N., Shimomura, Y., Yatagai, M. & Kobayashi, I. (2008) A physiopsychological research on Shinrin-yoku. *Journal of the Japanese Association of Physical Medicine, Balneology and Climatology*, **71**, 131.

Kudielka, B.M., Hellhammer, D.H. & Wust, S. (2009) Why do we respond so differently? Reviewing determinants of human salivary cortisol responses to challenge. *Psychoneuroendocrinology*, **34**, 2–18.

Kuo, F.E. (2001) Coping with poverty – impacts of environment and attention in the inner city. *Environment and Behavior*, **33**, 5–34.

Kuo, F.E. & Sullivan, W.C. (2001) Environment and crime in the inner city – does vegetation reduce crime? *Environment and Behavior*, **33**, 343–367.

Kuo, F.E., Sullivan, W.C., Coley, R.L. & Brunson, L. (1998) Fertile ground for community: inner-city neighbourhood common spaces. *American Journal of Community Psychology*, **26**, 823–851.

Laumann, K., Garling, T. & Stormark, K.M. (2003) Selective attention and heart rate responses to natural and urban environments. *Journal of Environmental Psychology*, **23**, 125–134.

Lee, A.C.K. & Maheswaran, R. (2011) The health benefits of urban green spaces: a review of the evidence. *Journal of Public Health*, **33**, 212–222.

Lohr, V.I., Pearson-Mims, C.H. & Goodwin, G.K. (1996) Interior plants may improve worker productivity and reduce stress in a windowless environment. *Journal of Environmental Horticulture*, **14**, 97–100.

Maas, J., Verheij, R.A., Groenewegen, P.P., de Vries, S. & Spreeuwenberg, P. (2006) Green space, urbanity, and health: how strong is the relation? *Journal of Epidemiology and Community Health*, **60**, 587–592.

Macdonald, D.W., Collins, N.M. & Wrangham, R. (2006) Principles, practice and priorities: the quest for 'alignment'. In: *Key Topics in Conservation Biology*. (eds D.W. Macdonald & K. Service), pp. 273–292. Blackwell Publishing, Oxford.

Macdonald, D.W, Burnham, D., Hinks, A. & Wrangham, R. (2012) A problem shared is a problem reduced: seeking efficiency in the conservation of felids and primates. *Folia Primatologica*, **83**(3–5), special issue 'Primate–Predator Interactions'.

Mace, G.M., Possingham, H.P. & Leader-Williams, N. (2006) Prioritizing choices in conservation. In: *Key Topics in Conservation Biology* (eds D.W. Macdonald & K. Service), pp. 17–34. Blackwell Publishing, Oxford.

McLaren, G.W., Macdonald, D.W., Georgiou, C., Mathews, F., Newman, C. & Mian, R. (2003) Leukocyte coping capacity: a novel technique for measuring the stress response in vertebrates. *Experimental Physiology*, **88**, 541–546.

Mian, R., McLaren, G.W. & Macdonald, D.W. (2005) Of stress, mice and men: a radical approach to old problems. In: *Stress and Health. New Research*. (ed. K.V. Oxington), pp. 61–80. Nova Publishers, Hauppauge, NY.

Millennium Ecosystem Assessment (2005) *Ecosystems and Human Well-Being: Health Synthesis. A Report of the Millennium Ecosystem Assessment*. World Resources Institute, Washington, D.C.

Miller, J.R. (2005) Biodiversity conservation and the extinction of experience. *Trends in Ecology and Evolution*, **20**, 430–434.

Minchin, T. (2010) *Storm*: www.timminchin.com

Mirescu, C. & Gould, E. (2006) Stress and adult neurogenesis. *Hippocampus*, **16**, 233–238.

Mitchell, R. & Popham, F. (2008) Effect of exposure to natural environment on health inequalities: an observational population study. *Lancet*, **372**, 1655–1660.

Montes. I., McLaren, G.W., Macdonald, D.W. & Mian, R. (2004) The effect of transport stress on neutrophil activation in wild badgers (*Meles meles*). *Animal Welfare*, **13**, 355–359.

Morita, E., Fukuda, S., Nagano, J., *et al.* (2007) Psychological effects of forest environments on healthy adults: Shinrin-yoku (forest-air bathing, walking) as a possible method of stress reduction. *Public Health*, **121**, 54–63.

Nielsen, T.S. & Hansen, K.B. (2007) Do green areas affect health? Results from a Danish survey on the use of green areas and health indicators. *Health and Place*, **13**, 839–850.

ONS (2005) *Main and Secondary Activities with Rates of Participation, 2005*. Office for National Statistics, London.

Ouellette, P., Kaplan, R. & Kaplan, S. (2005) The monastery as a restorative environment. *Journal of Environmental Psychology*, **25**, 175–188.

Pachana, N.A., Ford, J.H., Andrew, B. & Dobson, A.J. (2005) Relations between companion animals and self-reported health in older women: cause, effect or artifact? *International Journal of Behavioral Medicine*, **12**, 103.

Papworth, S.K., Rist, J., Coad, L. & Milner-Gulland, E.J. (2009) Evidence for shifting baseline syndrome in conservation. *Conservation Letters*, **2**, 93–100.

Park, B.J., Tsunetsugu, Y., Kasetani, T., Kagawa, T. & Miyazaki, Y. (2010) The physiological effects of Shinrin-yoku (taking in the forest atmosphere or forest bathing): evidence from field experiments in 24 forests across Japan. *Environmental Health and Preventive Medicine*, **15**, 18–26.

Parsons, R., Tassinary, L.G., Ulrich, R.S., Hebl, M.R. & Grossman-Alexander, M. (1998) The view from the road: implications for stress recovery and immunization. *Journal of Environmental Psychology*, **18**, 113–140.

Pauly, D. (1995) Anecdotes and the shifting base-line syndrome of fisheries. *Trends in Ecology and Evolution*, **10**, 430.

Peen, J., Dekker, J., Schoevers, R.A., ten Have, M., de Graaf, R. & Beekman, A.T. (2007) Is the prevalence of psychiatric disorders associated with urbanization? *Social Psychiatry and Psychiatric Epidemiology*, **42**, 984–989.

Philo, C. (1987) "Fit localities for an asylum": the historical geography of the nineteenth-century "mad-business" in England as viewed through the pages of the Asylum Journal. *Journal of Historical Geography*, **13**, 398–415.

Pickering, C.M., Hill, W., Newsome, D. & Leung, Y-F. (2010) Comparing hiking, mountain biking and horse riding impacts on vegetation and soils in Australia and the United States of America. *Journal of Environmental Management*, **91**, 551–562.

Pilgrim, S. & Pretty, J. (2010) *Nature and Culture*. Earthscan, London.

Pretty, J., Peacock, J., Sellens, M. & Griffin. M. (2005) The mental and physical health outcomes of green exercise. *International Journal of Environmental Health Research*, **15**, 319–337.

Pretty, J., Angus, C., Bain, M., *et al.* (2009) *Nature, Childhood, Health and Life Pathways*. University of Essex Occasional Paper, pp. 1–30.

Prince, M., Patel, V., Saxena, S., *et al.* (2007) No health without mental health. Global Mental Health Series. *Lancet*, **370**, 859–877.

Purcell, T., Peron, E. & Berto, R. (2001) Why do preferences differ between scene types? *Environment and Behavior*, **33**, 93–106.

Richardson, E., Pearce, J., Mitchell, R., Day, P. & Kingham, S. (2010) The association between green space and cause-specific mortality in urban New Zealand: an ecological analysis of green space utility. *BMC Public Health*, **10**, 14.

Roovers, P., Verheyen, K., Hermy, M. & Gulinck, H. (2004) Experimental trampling and vegetation recovery in some forest and heathland communities. *Applied Vegetation Science*, **7**, 111–118.

Sandercock, G., Voss, C., McConnell, D. & Rayner, P. (2010) Ten year secular declines in the cardiorespiratory fitness of affluent English children are largely independent of changes in body mass index. *Archives of Disease in Childhood*, **95**, 46–47.

Shelton-Rayner, G.K., Macdonald, D.W., Chandler, S., Robertson, D. & Mian, R. (2010) Leukocyte reactivity as an objective means of quantifying mental loading during ergonomic evaluation. *Cellular Immunology*, **263**, 22–30.

Shelton-Rayner, G.K., Mian, R., Chandler, S., Robertson, D. & Macdonald, D.W. (2011) Quantifying transient psychological stress using a novel technique: changes to PMA-induced leukocyte production of ROS in vitro. *International Journal of Occupational Safety and Ergonomics*, **17**, 3–13.

Shibasaki, M. & Kawai, N. (2009) Rapid detection of snakes by Japanese monkeys (*Macaca fuscata*): an evolutionarily predisposed visual system. *Journal of Comparative Psychology*, **123**, 131–135.

Simaika, J.P. & Samways, M.J. (2010) Biophilia as a universal ethic for conserving biodiversity. *Conservation Biology*, **24**, 903–906.

Sugiyama, T., Leslie, E., Giles-Corti, B. & Owen, N. (2008) Associations of neighbourhood greenness with physical and mental health: do walking, social coherence and local social interaction explain the relationships? *Journal of Epidemiology and Community Health*, **62**, e9.

Sukhdev, P., Wittmer, H., Schroter-Schlaack, C., *et al.* (2010) *The Economics of Ecosystems and Biodiversity. Mainstreaming the Economics of Nature: A Synthesis of the Approach, Conclusions and Recommendations of TEEB*. TEEB, Bonn.

Sztainberg, Y. & Chen, A. (2010) An environmental enrichment model for mice. *Nature Protocols*, **5**, 1535–1539.

Taylor, A.F., Kuo, F.E. & Sullivan, W.C. (2002) Views of nature and self-discipline: evidence from inner city children. *Journal of Environmental Psychology*, **22**, 49–64.

Tennessen, C.M. & Cimprich, B. (1995) Views to nature: effects on attention. *Journal of Environmental Psychology*, **15**, 77–85.

Thomas, G. & Thompson, G. (2004) *A Child's Place: Why Environment Matters to Children*. Green Alliance/Demos Report, London.

Thompson Coon, J., Boddy, K., Stein, K., Whear, R., Barton, J. & Depledge, M.H. (2011) Does participating in physical activity in outdoor natural environments have a greater effect on physical and mental wellbeing than physical activity indoors? A systematic review. *Environmental Science and Technology*, **45**, 1761–1772.

UK National Ecosystem Assessment (2011) *The UK National Ecosystem Assessment Technical Report*. UNEP–WCMC, Cambridge.

Ulrich, R.S. (1983) Aesthetic and affective response to natural environment. In: *Human Behavior and Environment*, vol 6. (eds I. Altman & J.F. Wohlwill). Plenum Press, New York.

Ulrich, R.S. (1984) View through a window may influence recovery from surgery. *Science*, **224**, 420–421.

Ulrich, R.S., Simons, R.F., Losito, B.D., Fiorito, E., Miles, M.A. & Zelson, M. (1991) Stress recovery during exposure to natural and urban environments. *Journal of Environmental Psychology*, **11**, 201–230.

UN (2008) *World Population Prospects: The 2008 Revision*. Population Division of the Department of Economic and Social Affairs of the United Nations Secretariat, New York.

UN (2009) *World Urbanization Prospects: The 2009 Revision*. Population Division of the Department of Economic and Social Affairs of the United Nations Secretariat, New York.

Valenzuela-Galvan, D., Arita, H.T. & Macdonald, D.W. (2008) Conservation priorities for carnivores considering protected natural areas and human population density. *Biodiversity and Conservation*, **17**, 539–558.

Van Dellen, A., Blakemore, C., Deacon, R., York, D. & Hannan, A.J. (2000) Delaying the onset of Huntington's in mice. *Nature*, **404**, 721–722.

Van den Berg, A.E., Koole, S.L. & van der Wulp, N.Y. (2003) Environmental preference and restoration: (how) are they related? *Journal of Environmental Psychology*, **23**, 135–146.

Vries, S.D., Verheij, R.A., Groenewegen, P.P. & Spreeuwenberg, P. (2003) Natural environments – healthy environments? An exploratory analysis of the relationship between greenspace and health. *Environment and Planning*, **35**, 1717–1731.

Ward Thompson, C., Aspinall, P. & Montarzino, A. (2008) The childhood factor: adult visits to green places and the significance of childhood experience. *Environment and Behaviour*, **40**, 111–143.

Weich, S., Twigg, L. & Lewis, G. (2006) Rural/non-rural differences in rates of common mental disorders in Britain – prospective multilevel cohort study. *British Journal of Psychiatry*, **188**, 51–57.

Wells, N.M. (2000) At home with nature – effects of "greenness" on children's cognitive functioning. *Environment and Behavior*, **32**, 775–795.

Wells. N.M. & Evans, G.W. (2003) Nearby nature – a buffer of life stress among rural children. *Environment and Behavior*, **35**, 311–330.

Wells, N.M. & Lekies, K.S. (2006) Nature and the life course: pathways from adulthood nature experience to adult environmentalism. *Children, Youth and Environments*, **16**, 1–24.

Westgarth, C., Heron, J., Ness, A.R., *et al.* (2010) Family pet ownership during childhood: findings from a UK birth cohort and implications for public health research. *International Journal of Environmental Research and Public Health*, **7**, 3704–3729.

White, M., Smith, A., Humphryes, K., Pahl, S., Snelling, D. & Depledge, M. (2010) Blue space: the importance of water for preference, affect, and restorativeness ratings of natural and built scenes. *Journal of Environmental Psychology*, **30**, 482–493.

WHO (1948) *Preamble to the Constitution of the World Health Organization as Adopted by the International Health Conference*. World Health Organization, New York.

WHO (2011) *Obesity and Overweight*. Factsheet No. 311. World Health Organization, Copenhagen.

Wilson, E.O. (1984) *Biophilia*. Harvard University Press, Cambridge, MA.

II

Habitat case studies

10

Ocean conservation: current challenges and future opportunities

Alex D. Rogers[1], Dan Laffoley[2], Nick Polunin[3] and Derek P. Tittensor[4]

[1]Department of Zoology, University of Oxford, Oxford, UK
[2]International Union for Conservation of Nature, Gland, Switzerland
[3]School of Marine Science and Technology, Newcastle University, Newcastle, UK
[4]United Nations Environment Programme, World Conservation Monitoring Centre/ Microsoft Research, Cambridge, UK

'The progress of rivers to the ocean is not so rapid as that of man to error.'

(Voltaire)

Introduction

Forty years ago the image of the earth from the Apollo photographs changed forever humankind's perspective of the beauty, fragility and finite nature of the world we live on. However, the subsequent approach to the conservation and management of the world's ocean has not lived up to these new perspectives. Human actions have depleted species, altered ecosystems and even started to change the very chemistry of the blue water that makes the world different from any other we know of. This chapter describes the wealth and variety of marine life, the challenges it faces and the scale of human impacts so far. It also sets out a perspective on current conservation and management approaches, and the significant changes needed to these approaches in the future.

In a biological context, the vital statistics of marine ecosystems are remarkable. They constitute over 90% of the earth's habitable volume (Angel 1993). Thirty-two of the 33 animal phyla (as of 1994) exist in the sea, and 21 of those are exclusively marine (May 1994). Sampling just 6.3 m^2 of coral reef habitat produces 525 species of crustaceans alone (Plaisance et al. 2011). A recent study suggests that the oceans hold around 2.2 million eukaryotic species (±0.18 million SE), of which 91% remain to be discovered (Mora et al. 2011). Even of marine animals 2 m or over in length,

Key Topics in Conservation Biology 2, First Edition. Edited by David W. Macdonald and Katherine J. Willis.

there may yet be a dozen or more species waiting to be found (Solow & Smith 2005).

In 1990, around 38% of the global human population lived within 100 km of the coast (Small & Cohen 2004). Almost 2 billion more people have been added to the planet in the intervening two decades, with coastal populations growing disproportionately fast. At a broad scale, areas of high human impacts are more likely to be hot spots of marine biodiversity (Tittensor et al. 2010). Coastal ecosystems such as rocky reefs and the continental shelf are among the most affected by human activities (Halpern et al. 2008). As we will show, there are major changes taking place but we are unable to track these precisely because knowledge of marine systems remains poor. While over 500 sets of long-term data were available over the 29% of the earth covered by land, for the Intergovernmental Panel on Climate Change (IPCC) process, a key intergovernmental tool for assessing the impacts of climate change, there were only about 25 for the 71% comprising the oceans (Richardson & Poloczanska 2008). Chronic undersampling is the norm in the deep ocean (Webb et al. 2010), yet even remote abyssal plains are experiencing human impact (Ramirez-Llodra et al. 2011), for example from litter, pollution or 'ghost fishing' through lost or discarded nets.

The marine realm can have strikingly different patterns of ecosystem structure to terrestrial systems. The oceans account for just under half of global net primary production (NPP) (Field et al. 1998), yet turnover rates for phytoplankton communities can be orders of magnitude faster than those of forests or grasslands (Cebrian 1999), with the consequence that the estimated standing stock of plant biomass is a fraction of that on land (Polis 1999). Herbivore to producer biomass ratios also tend to be higher, particularly in pelagic systems (Cebrian et al. 2009), and may even reach the point where heterotroph biomasses exceed those of autotrophs, resulting in an inverted biomass pyramid (Gasol et al. 1997). Some of these differences may be a consequence of the generally higher nutritional content (nitrogen and phosphorus) of marine producers, particularly phytoplankton (Cebrian et al. 2009).

Spatial patterns of marine NPP are very different from those on land, with the areas of highest productivity being at high latitudes rather than at the equator (Huston & Wolverton 2009). Temperature is a better cross-taxon predictor of marine species richness than NPP (Tittensor et al. 2010). Observations of the global patterns of marine biodiversity suggest that coastal taxa tend to follow similar patterns to most terrestrial systems, with diversity generally peaking towards the equator, while oceanic species are more likely to have mid-latitudinal peaks in diversity (Tittensor et al. 2010). Unfortunately, why oceanic taxa show this contrast to the latitudinal gradient of terrestrial systems is not understood, but along with differences in environmental characteristics, life-history factors such as dispersal may have an effect. Knowledge of large-scale biodiversity patterns is limited to a select few taxa, such as marine mammals and corals, and analyses have shown that inventories of fish biodiversity (one of the best known) become increasingly patchy at finer resolutions (Mora et al. 2008). This combination of a different ecosystem and food web structure and massive under sampling can present conservation professionals with a distinctive challenge relative to terrestrial systems.

The threats to marine biodiversity

Fisheries exploitation

Relatively few marine species, mainly range-restricted mammals and birds, are known to have gone extinct, but the difficulty of observing extinctions in the oceans and the poor knowledge of diversity may hide the true scale of losses (Dulvy et al. 2009). Although global extinctions are hard to quantify, many species have undergone rapid and substantial declines in abundance and range, sometimes resulting in local extirpations (e.g. Worm & Tittensor 2011).

The main cause of extinction has been overexploitation, including that of fish, turtles, seabirds and marine mammals (Dulvy et al. 2009).

Fisheries biologists have argued that such declines are exaggerated. For example, in the Pacific they have estimated that the major tuna fisheries are depleted by 10–70% rather than the 90% losses estimated by ecologists (Hampton et al. 2005; Hilborn 2007). They cite the use of catch per unit effort data which are an imperfect proxy for abundance. In addition, they claim that fishing may only affect part of the geographic range of a species, fisheries data can be inaccurate and changes in technology, market forces and fisher behaviour are unaccounted for (e.g. Hilborn 2007). However, modelling of past stock biomass (e.g. Rosenberg et al. 2005), time-series data (e.g. Pinsky et al. 2011), direct observations such as by divers (e.g. Ward-Paige et al. 2010), extinction risk assessments (e.g. Collette et al. 2011), catch data (e.g. Baum & Blanchard 2010) and meta-analyses (e.g. Worm et al. 2009) show that the populations of numerous target and by-catch species have been severely reduced, sometimes by 90% or more since the advent of large-scale industrialized fisheries around the 1950s.

Reported global marine fisheries landings were 79.9 million tonnes in 2009 (FAO 2010; Figures 10.1,10. 2) but by-catch of these capture fisheries (as distinct from farmed fisheries) may be up to 38.5 million tonnes/year (Davies et al. 2009). Fish catch that was previously discarded has increasingly been landed as new markets are found for 'trash species' to compensate for declining catches of the prime target species of

Figure 10.1 Trawlers moored in Cape Town, South Africa, in November 2011. Crew members on the decks of the vessels demonstrate the scale of modern industrial fishing vessels. The development of modern 'factory' fishing vessels has allowed an explosion in fishing effort throughout the world's oceans since the 1950s (©AD Rogers).

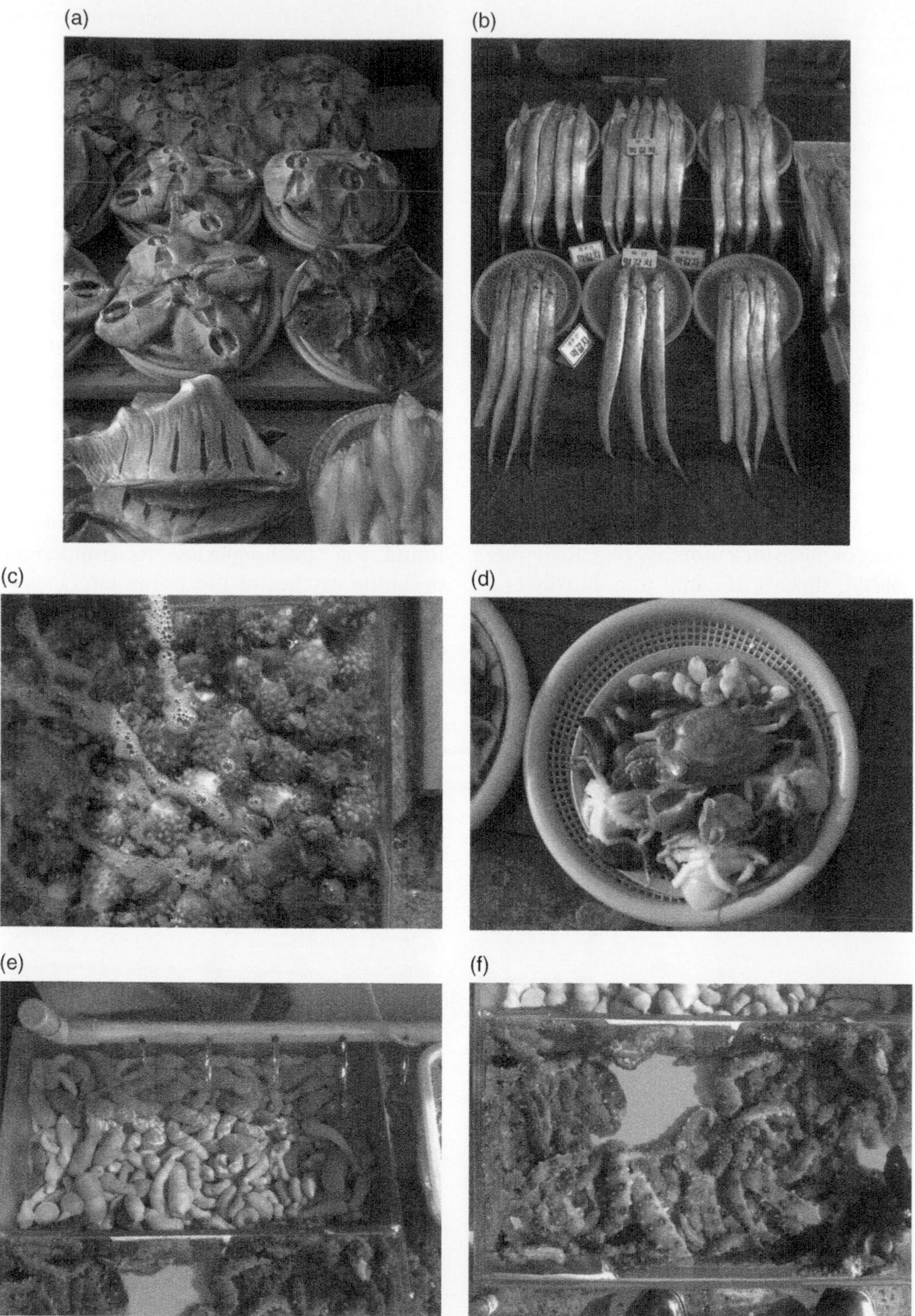

Figure 10.2 Photographs showing the variety of marine animals harvested for human consumption in a fish market in South Korea, recorded in one morning. (a) Rays (Rajidae) and flatfish. (b) Largehead hairtail (*Trichiurus lepturus*) now overexploited in the NW Pacific (FAO 2010). (c) Ascidiacea. (d) Molluscs (Bivalvia, Gastropoda, Cephalopoda), crustaceans (Portunidae, Penaeidae). (e) Echiura. (f) Holothuria or *bêche de mer* (©AD Rogers).

such fisheries. This has the effect of maintaining fisheries which would have otherwise have declined or closed through lack of profitability. Small-scale fisheries in developing countries have catches similar in magnitude to those of large-scale fisheries and are much more important in terms of numbers of fishers (11–12 million versus 1–2 million). Recreational fisheries also have a significant impact, with a global catch of >10 million tonnes/year (Crowder et al. 2008).

Fishing also has significant ecosystems implications through other mechanisms. Depletion of target and by-catch species can lead to trophic cascades, whereby prey items are released from top-down predation pressure, and can have knock-on effects on other organisms in the ecosystem. For example, on the eastern seaboard of the USA, many large shark populations have been driven to low levels, resulting in an increase of small sharks and rays and a trophic cascade down to bay scallops, prey items of these smaller elasmobranchs for which the enhanced predation pressure was sufficient to end a 100-year-old scallop fishery (Myers et al. 2007). Acting in isolation or in conjunction with other environmental changes resulting from natural variability or human impacts, the serial fishing of successive trophic levels of the food web (Pauly et al. 1998) can make marine ecosystems prone to phase shifts (marked changes in the structure and composition of an ecosystem), reversal of which may be slow or improbable. Thus in the Black Sea, targeting of upper trophic level predators and then zooplanktivores, together with eutrophication, led to the dominance of an invasive gelatinous zooplanktivore that contributed to the collapse of the fishery for anchovy (*Engraulis encrasicolus;* Daskalov et al. 2007). Overfishing of grazers on coral reefs has contributed to a widespread phase shift from hard coral to algal domination in the Caribbean and elsewhere (Bellwood et al. 2004).

Habitat destruction can be another consequence of exploitation. For example, blast fishing using explosives to stun fish on coral reefs occurs in at least 40 countries and generates unstable rubble fields which are unfavourable to coral recruitment (Fox et al. 2003). Affected coral reefs may show no recovery after 25 years or more (Raymundo et al. 2007). Blast fishing can be very profitable in the short term but it effectively mines fisheries resources and can displace traditional fishers who may go on to impact other species (Wells 2009). The live fish trade employs poisons (e.g. cyanide) in coral reef fisheries that are toxic to corals (Jones et al. 1999), adding to the reef degradation caused by fishers breaking up the coral matrix to extract stunned fish.

Industrial-scale bottom fisheries further damage seabed habitat, affecting species diversity and composition, with mobile gears such as dredges and trawls being the most destructive. Recovery is likely to be faster if the ecosystem is adapted to a high level of natural disturbance (Jennings et al. 2001) but some organisms, such as deep-sea coral communities, which may take thousands of years to develop (Roberts et al. 2009), are particularly vulnerable (Collie et al. 2000). Another consequence of overfishing can be impacts on the genetic diversity of species, probably through the elimination of subpopulations or distinct spawning stocks (Smedbol & Stephenson 2001). Fishing can also be a selective force on heritable traits. For example, the earlier maturation of northern cod is thought to be the result of size-selective fishing (Olsen et al. 2005). Such changes may render fish less able to maintain recruitment in the face of environmental variation.

Pollution

Pollutants from ocean-based activities (e.g. oil exploration, shipping), the land (e.g. agrochemicals, pharmaceuticals) or the atmosphere (e.g. heavy metals) impact marine ecosystems through acute mortality or sublethal physiological effects. In the open ocean, small quantities of oil are probably readily dispersed and degraded (GESAMP 2009) and in some cases recovery from even significant oil spills can be swift (e.g. the *Braer* spill of 85,000 tonnes which was dispersed relatively quickly by rough weather).

Catastrophic releases, however, can be a greater threat, and recovery of biotas may take a decade or more. When marine animals frequently contact the sea surface, they can suffer from oiling of fur or feathers and die from hypothermia, smothering, drowning, ingestion of oil or inhalation of toxic fumes (Peterson et al. 2003). On beaches, macroalgae and invertebrates can be poisoned by oil or oil dispersants, smothered or displaced by jet washers used to clean beaches. Sublethal toxicity of oil constituents, particularly polycyclic aromatic hydrocarbons (PAHs), can negatively impact populations by reducing recruitment, longevity and recovery of affected species, including fish, birds and sea mammals. Ecosystem cascades may occur; in the *Exxon Valdez* spill this occurred through the elimination of species important in structuring the intertidal environment, notably the kelp, *Fucus gardneri*, along with grazers and predators (gastropod snails). This promoted blooms of ephemeral algae and the settlement of barnacles and recovery of the intertidal ecosystem was inhibited for many years by feedback effects (e.g. *Fucus* spores tend to settle under the kelp canopy as it protects them from dessication; Peterson et al. 2003). Similar cascade effects were observed in the *Torrey Canyon* spill mainly through the elimination of patellid limpets, important grazers on rocky shores in the affected area (Crowe et al. 2000). The *Exxon Valdez* spill also disrupted family groups of cetaceans through high levels of mortality, leading to the disintegration of pods or, through the loss of adult females, leading to suppressed reproduction within pods (Peterson et al. 2003).

Impacts of eutrophication resulting from releases of large quantities of nutrients, mainly from run-off of agricultural fertilizers but also from sewage discharge, include the generation of oxygen-depleted areas, formation of harmful algal blooms (HABs) and increased occurrence of micro-organisms pathogenic to marine life and humans. The deposition of organic material which promotes microbial growth and respiration, if accompanied by stratification of the water column, can lead to oxygen depletion which may then progress to sporadic transient hypoxia, lasting days to weeks, and then to seasonal hypoxia, generally occurring over the warm summer months, with regular mass mortality events of marine organisms (Diaz & Rosenberg 2008). When the oxygen concentration falls below 2 mL/L, benthic infauna (organisms living in the substratum) exhibit stress-related behaviour, such as abandonment of burrows, and by 0.5 mL/L significant mortality occurs. For many species, including fish and crustaceans, even oxygen concentrations well above 2 mL/L may be lethal (Vaquer-Sunyer & Duarte 2008). Where such conditions last for several years and nutrients continue to accumulate in the system, the hypoxic zone expands, anoxia may occur and microbial communities release H_2S (Diaz & Rosenberg 2008). Chronic hypoxia and anoxia reduce secondary production and can eliminate most benthic animals, with flows of energy through the ecosystem and ecosystem services such as nitrogen cycling becoming dominated by micro-organisms (Diaz & Rosenberg 2008). The number of oxygen-depleted 'dead' zones has approximately doubled every decade since the 1960s, being reported from over 400 marine ecosystems by the end of the 20th century, their occurrence positively correlated with human population centres and nutrient input (Diaz & Rosenberg 2008). Where water exchange is restricted (e.g. inland seas, estuaries), the effects of eutrophication are especially prominent.

Harmful algal blooms are caused by at least 200 species of microalgae and a ciliate (Landsberg 2002). The effects can include acute toxicity causing mass mortality of marine life. This can occur with ingestion of toxic organisms, contact with extracellular toxins, cell-to-cell contact (e.g. trapping of some diatoms in the gills of fish) or through the release of toxins when HAB cells lyse. Exposure to toxins from HABs can also be associated with trophic transfer, for example through bio-accumulation, bioconversion or biomagnification of toxins; such indirect exposure may have important impacts on animal health (Landsberg 2002). An example of this is the neurotoxin domoic acid, produced by algae such as the diatom *Pseudonitzschia* which, off the coast of California, has been observed to be transferred up the food chain from zooplankton to zooplanktivorous fish, such as anchovies (e.g. *Anchoa mitchilli, Engraulis mordax*), and then to their predators, including birds, such as brown pelicans (*Pelecanus occidentalis*), Brandt's cormorant (*Phalacrocorax penicillatus*) and California sea lions (*Zalophus californianus*) which accumulate the toxin and suffer mass mortality events (Landsberg 2002).

Many pollutants can have chronic effects on organism health, including pesticides and herbicides, heavy metals, flame retardants and constituents of oil such as PAHs. Persistent organic pollutants (POPs) are a major concern; they are generally stable hydrophobic and lipophilic compounds with low-to-moderate volatility which can be transported long distances in the atmosphere and often accumulate in colder latitudes (Noël et al. 2009; Bossart et al. 2010). Many are subject to bio-accumulation and biomagnification, and their effects can include anaemia, endocrine disruption, immunosuppression, development of tumours, and death. Large-scale mortality in pinnipeds and cetaceans from outbreaks of diseases such as phocine distemper and morobilivirus has also been linked with high concentrations of POPs (e.g. Aguilar & Borrell 1994). Such disease outbreaks, leading to high mortality of marine predators, can result from human introduction of primary hosts, such as cats and dogs, for pathogens such as *Leptospira, Toxoplasma* and *Sarcocystis* (Bossart et al. 2010). Emerging environmental contaminants not covered by current regulations that may pose a risk to marine species include pharmaceuticals, personal care products (PCPs), steroids, hormones, perfluorinated compounds (PFCs) and nanomaterials (La Farré et al. 2008).

Approximately 260 million tonnes of plastics are manufactured annually (Thompson et al. 2009). Plastic debris tends to be buoyant and trap air and as a result it floats. Thus it can constitute 50–80% of the debris on shorelines and up to 3,520,000 items/km^2 can occur on the ocean surface (Thompson et al. 2009). Particles also sink, in some European seas reaching densities of 10,000 items per km^2. Larger floating debris may act as a vector for invasive species as they can settle on such items as larvae and grow into mature adults whilst being transported long distances to reproduce and spread (Thompson et al. 2009). Including lost fishing gear, the main quantifiable threat from plastic debris is entanglement and ingestion by marine animals. More than 260 species have been recorded as being affected by plastic debris, with effects ranging from impairment of movement and feeding, reduced reproductive output, lacerations and ulcers to death. In the 1980s, it is possible that tens of thousands of marine mammals were killed as a result of entanglement in debris in the North Pacific alone (Derraick 2002). Many species including seabirds, turtles, cetaceans, sirenians and fish swallow plastic debris because it resembles their natural food (Derraick 2002).

Microplastic particles (<5 mm to 333 μm or less), containing organic contaminants that are added during manufacture or absorbed, can be consumed by small marine invertebrates (Thompson et al. 2009). Additives including phthalates and bisphenol A act as endocrine disruptors and affect reproduction and development in crustaceans and molluscs (Thompson et al. 2009) whilst absorbed chemicals include a range of POPs. The ingestion of plastic particles may provide a direct but unquantified route for these chemicals into the food chain (Thompson et al. 2009).

Climate change and ocean acidification

Rising sea surface temperatures and ocean acidification resulting from anthropogenic CO_2 input into the atmosphere are increasing threats. The effects of sea surface warming on marine ecosystems are multifaceted, changing the physical structure, stratification, oxygen content and primary productivity of the ocean, and the distribution and phenology of marine species (Reid et al. 2009). Coral reefs, in particular, are sensitive to warming and since the late 1970s have been subject to an increasing frequency and scale of mass coral bleaching. This is a phenomenon whereby corals lose or expel their symbiotic photosynthetic algae (zooxanthellae), which lend them their colour, when exposed to anomalously high temperatures (Hoegh-Guldberg 1999). In 1998 a mass coral bleaching event driven by an El Niño (a quasi-periodic oscillation of climatic temperatures in the tropical eastern Pacific that influences global weather patterns) led to anomalously high temperatures across large areas of the Indian and Pacific oceans, severely damaging 16% of the world's tropical coral reefs (Wilkinson 2004). Corals show differential susceptibility to temperature-induced mass bleaching depending on the species, with fast-growing branched forms, such as *Acropora*, being highly vulnerable. The strain of zooxanthella hosted by a coral can also confer different levels of resistance to bleaching. Although corals and their symbionts therefore may have some capacity to adapt to higher temperatures, this is not likely to be sufficient to reduce the impact of mass bleaching events, which are becoming more frequent as CO_2 levels in the atmosphere rise and global sea surface temperatures increase. About a third of reef-forming coral species are now considered threatened with extinction, climate change being the most important factor (Carpenter et al. 2008). Coral reefs are among the most species-rich habitats on the planet, and the loss of coral cover threatens many associated species (Plaisance et al. 2011).

Sea surface warming is also an issue in polar and subpolar regions, threatening species through declines in sea ice habitat and indirect effects on keystone species such as Antarctic krill (Trathan et al. 2007). Models based on the environmental envelopes of species suggest elevated local extinction rates in tropical (mean 4% local extinction rate) and subpolar (mean 7%) regions by 2050, and greater invasion of higher latitude ecosystems (Cheung et al. 2009; see also Kaschner et al. 2011 for marine mammals). Unpredictable synergies with other human impacts may also occur; for example, recent observations have suggested that expanding oxygen minimum zones (probably influenced by global warming) have vertically compressed available habitat for billfishes and tunas, rendering them more susceptible to surface fishing gear (Stramma et al. 2012).

As well as increased ocean temperatures, another effect of anthropogenic CO_2 emissions is ocean acidification. The oceans absorbed about a third of anthropogenic CO_2 emitted between the years 1850 to 2000, equivalent to about 140 peta grams of carbon (peta gram $= 10^{15}$ g; Fabry et al. 2008). This CO_2 has dissolved in seawater to produce carbonic acid which has reduced seawater pH by about 0.1 units to date (from 8.21 to 8.1) and is expected to continue to decrease a further 0.3–0.4 pH units by the end of the century if emissions continue to increase at present rates. A decrease of 0.1 pH units may sound small but this is a logarithmic scale and represents an increase of 30% in the concentration of hydrogen ions (H^+; Fabry et al. 2008). This process results in a significant shift in carbonate speciation in seawater with carbonate used to build skeletons by marine organisms, being converted to inaccessible bicarbonate (Fabry et al. 2008; Doney et al. 2009). For organisms that secrete calcium carbonate skeletons, including coralline algae, planktonic foraminiferans, corals, most molluscs and echinoderms, the calcification rate of animals and calcium carbonate saturation state of seawater are positively correlated. Observations of declines in the calcification rates of massive coral colonies from the Great Barrier Reef by 21% between 1988 and 2003 likely reflect the effects of ocean acidification but other environmental changes are also likely to have exerted an influence (Doney et al. 2009).

In a very different environment, the Southern Ocean, the skeletons of the planktonic foraminiferan (shelled amoeba) *Globigerina bulloides* were found to be ~30–35% lower than those buried in sediment in the past, probably also reflecting the impacts of ocean acidification (Moy et al. 2009). However, the impacts of acidification vary among groups of organisms; with decreasing pH and calcium carbonate saturation state, calcification declines in some cases and increases in others (Doney et al. 2009). In nitrogen-fixing cyanobacteria and seagrasses, elevated CO_2 may enhance photosynthesis, carbon and nitrogen fixation and growth (Doney et al. 2009). This is supported by observations of natural CO_2 vents in the Mediterranean and tropical Indo-Pacific where there has been a decline or disappearance of calcareous marine organisms with increasing CO_2 levels, although seagrass cover increases (Hall-Spencer et al. 2008; Fabricius et al. 2011). Other aspects of the biology of organisms other than growth rates or rates of skeletal accretion may also be affected by acidification. For example, the larvae of echinoids and molluscs seem particularly sensitive to ocean acidification and fertilization success, occurrence of normal larval development and larval size may decline with increased CO_2 levels (Fabry et al. 2008).

At the ecosystem level, the impacts of ocean acidification are likely to be significant. On coral reefs, the overall rates of calcium carbonate accretion on coral reefs are sensitive to calcium carbonate saturation levels through reduced calcification of corals and coralline algae (Hoegh-Guldberg et al. 2007). There are also complex feedback effects caused by the widespread impact of acidification on reef systems which may act through trophic interactions or which affect the life history of coral reef species. An example of this is the dependence of coral larvae on calcareous algae as an importance substratum for settlement; impact on the algae may feed back as reduced coral recruitment (Hoegh-Guldberg et al. 2007). At high latitudes, important components of pelagic food webs, such as foraminiferans and krill, may be vulnerable to ocean acidification with significant implications for predators that feed on them (e.g. Kawaguchi et al. 2010). Interactive effects of elevated temperatures and decreased pH are expected to be complex (e.g. Rodolfo-Metalpa et al. 2011), as are the interactions between climate change effects and other anthropogenic impacts.

Habitat loss

In addition to habitat impacts through fisheries exploitation and climate change, habitat loss is a particular threat where coral reefs, mangrove forests, seagrass beds, kelp forests and salt marshes have been degraded, with a concomitant increase in the extinction risk to the key structural species of these habitats (Steneck et al. 2002; Carpenter et al. 2008; Alvarez-Filip et al. 2009; Waycott et al. 2009; Polidoro et al. 2010). Degradation has often occurred through coastal development or clearing of habitat for alternative uses, such as aquaculture, which accounts for half of the global loss of mangrove forest (Polidoro et al. 2010). As half the world's population lives within 200 km of the coast and the population is growing, the direct impacts of coastal development are expected to accelerate.

How can the seas be conserved – and where are the successes?

Concerns over human impacts on the marine environment and biodiversity are not new. The problems seen today have often been recognized for years and resulted in the development of initiatives, and calls for action, at global, regional and national levels. Many include the marine environment as part of overall maintenance of biodiversity and incorporate the basic principles which should be addressed in relation to human impacts, the marine environment and conservation (Laffoley 2000).

Primary events in marine conservation include the United Nations Conference on the Human Environment (Stockholm 1972) and the adoption of the World Charter for Nature by the United Nations General Assembly in 1982 (Resolution 37/7) which mark, respectively, the beginning and end of the first decade of world awareness of threats to the natural environment and the importance of remedial action.

The Stockholm Declaration and the World Charter for Nature can be considered as soft law instruments of major importance for the development of international environmental law. Three of the general principles contained in the Charter are of particular relevance to conservation in the sea.

- The genetic viability of the earth shall not be compromised; the population levels of all life forms, wild and domesticated, must be at least sufficient for their survival, and to this end necessary habitats shall be safeguarded (Principle 2).
- All areas of the earth, both land and seas, shall be subject to these principles of conservation; special protection shall be given to unique areas, to representative samples of all the different types of ecosystems and to the habitats of rare and endangered species (Principle 3).
- Ecosystems and organisms, as well as the land, marine and atmospheric resources that are utilized by man, shall be managed to achieve and maintain optimum sustainable productivity, but not in such a way as to endanger the integrity of those other ecosystems or species with which they co-exist (Principle 4).

These principles have provided a framework for the development of conservation treaties. Four global sectoral conservation conventions covering wetlands of international importance (Ramsar Convention), sites of universal value (World Heritage Convention), trade in endangered species (CITES; Convention on International Trade in Endangered Species) and the conservation of migratory species (Bonn Convention) were concluded during the period 1971–1979. They were followed by a number of regional instruments.

It rapidly became clear, however, that such sectoral and regional approaches were not sufficient to cope with the depletion of biological diversity worldwide. The need for a global convention covering all aspects of conservation and sustainable use of biological diversity became increasingly clear to conservationists in the early 1980s. The General Assembly of the International Union for Conservation of Nature (IUCN) in 1982 adopted a resolution calling for the conclusion of such a treaty. This was the first step in a lengthy process. Eleven years later the Convention on Biological Diversity (CBD) was signed at the Earth Summit at Rio de Janeiro, on June 5, 1992 (Quarrie 1992).

Recommendations for a programme of actions to implement this Convention with respect to marine and coastal biodiversity were subsequently made by the Subsidiary Body on Scientific, Technical and Technological Advice (SBSTTA) around the five thematic areas of:

- integrated marine and coastal area management
- marine and coastal protected areas
- sustainable use of marine and coastal living resources
- mariculture
- alien species.

These recommendations, which signatories to the CBD are required to follow, became part of the Jakarta Mandate, established at the second Conference of the Parties meeting in Jakarta in 1995.

At a European level, the Convention on the Conservation of European Wildlife and Natural Habitats (Berne Convention), concluded at Berne in September 1979 under the auspices of the Council of Europe, lists protected species, including some marine species, and requires its parties to prevent the disappearance of endangered natural habitats. The European Union accordingly adopted the Birds Directive in 1979 and the Habitats Directive in 1992, to implement the Berne Convention, establishing Special Protection Areas and Special Areas of Conservation across Europe to tackle the continuing losses of European biodiversity, including in the sea, to human activities (Laffoley 2000).

Along with the development of global conservation conventions, the international community has initiated actions addressing the management of human exploitation of marine biotic resources. Several centuries of uncontrolled harvesting of cetaceans led to the near extinction of several species and efforts to regulate whaling resulted in the establishment of the International Whaling Commission (IWC; see http://iwcoffice.org/commission/convention.htm) in 1946. In the following decades efforts to regulate levels of exploitation of cetaceans by the IWC were generally unsuccessful, mainly because of resistance to reducing quotas by whaling nations and a lack of compliance with new regulations (Evans 1987). In the 1950s stocks of blue whales (*Balaenoptera musculus*) collapsed, followed by fin whales (*Balaenoptera physalus*) in the 1960s and sei whales (*Balaenoptera borealis*) in the late 1960s (Evans 1987). International whaling fleets then turned to previously less exploited species, notably the smallest of all rorquals, the minke whale (*Balaenoptera acutorostrata*; Evans 1987). In 1975 there was an attempt to bring the effort in whale harvesting to a level where populations could recover sufficiently to produce maximum sustainable yield, known as the New Management Procedure (Evans 1987).

Throughout the 1970s public pressure to ban whaling and disagreements on the science in setting appropriate levels of harvesting eventually lead to a majority decision in 1982 to initiate a moratorium on whale hunting which was finally instigated in the winter of 1985–1986 (Evans 1987). Since this time a loophole that allows scientific whaling has allowed some states, notably Japan and Norway, to continue. Some whale species have shown a significant recovery in population sizes following the whaling moratorium (e.g. humpback whales (*Megaptera novaeangliae*) in many areas of the southern hemisphere). Other species have failed to recover and therefore remain endangered (e.g. North Pacific right whale (*Eubalaena glacialis*), north west Pacific grey whale (*Eschrichtius robustus*)), or there is insufficient scientific information for an assessment of current population size (e.g. southern hemisphere fin whales).

Following on from the initiation of the IWC, efforts to regulate human exploitation of living marine resources continued with the 1982 United Nations Convention on Law of the Sea (UNCLOS; www.un.org/Depts/los/convention_agreements/texts/unclos/closindx.htm). Article 61 (2) of UNCLOS stipulates that coastal states shall ensure, through proper conservation and management measures, that the maintenance of the living resources of their exclusive economic zone (EEZ) is not endangered through overexploitation. This is reinforced in the same Article through a statement that proper conservation and management measures should be designed to maintain or restore populations of harvested species at levels which can produce the maximum sustainable yield. Article 61 (4) addresses non-target species, Article 192 further emphasizes that states have an obligation to protect and preserve the marine environment, and Article 194 (5) requires that states take measures necessary to protect and preserve rare or fragile ecosystems as well as the habitat of depleted, threatened or endangered species and other forms of marine life. Many of these stipulations are reinforced in the 1995 UN Fish Stocks Agreement (UNFSA) which applies mainly to straddling fish stocks (those crossing international jurisdictional boundaries) and highly migratory fish stocks of the high seas. Both UNCLOS and the UNFSA place obligations on states to base conservation and management measures on the best scientific evidence.

Through UNCLOS and subsequent agreements, the management of fisheries has fallen largely to states and regional fisheries management organizations (RFMOs), the former focusing on the EEZ of coastal states (waters within 200 nautical miles of the coast) and the latter focusing on large areas of the ocean, including the high seas, or on migratory species such as tuna. The success of both in terms of management of fisheries is varied but the overall picture is one of failure. For example, the European

Commission estimates that for stocks with sufficient data for assessment, 78.5% are exploited unsustainably, with 43% outside of safe biological limits (EC 2011). Statistics on the trends of the state of fish stocks globally show an increase in the numbers of fully exploited and overexploited or collapsed stocks with a decreasing number of stocks which still have potential for increased exploitation (FAO 2010). This has been accompanied by an expansion of fisheries both across latitudes (Swartz et al. 2010) and into greater depths (Morato et al. 2006) with impacts on ecosystems arising through the removal of fish biomass and other large-scale environmental impacts of fishing.

The reasons for failure to comply with international law and agreements are many but several factors stand out. Non-transparent and non-accountable systems of governance and decision making in fisheries management have prevented the adoption of scientific advice in setting sustainable harvest levels for many fish stocks (e.g. Mora et al. 2009). This has occurred because short-term social pressures and commercial interests have superseded long-term sustainability of natural resources. Even where the majority of stakeholders in fisheries have tried to control levels of exploitation, opt-out clauses or poor systems of decision making can allow a minority to stall progress in management. Non-compliance with fisheries regulations is a continuing problem that arises through poor monitoring, control and surveillance and a lack of effective mechanisms for prosecution and punishment of non-compliant vessels (e.g. Mora et al. 2009; Pitcher et al. 2009). This is exacerbated by the exemption of fishing vessels from many rules that apply to other shipping (e.g. an exemption to the requirement to display an International Maritime Organization ship identification number on the hulls of vessels engaged solely in fishing) and a lack of integration of fisheries management with management of other human activities in the ocean.

Many fisheries remain open access, leading to an uncontrolled number of vessels (Mora et al. 2009). Subsidies that increase the fishing power of fleets or maintain profitability in what would otherwise be unprofitable fisheries are also significant contributory factors to overfishing. There are also significant issues with inadequate scientific advice for many fisheries because of a lack of investment in fisheries research or the prevention of access to data of sufficient resolution to estimate levels of mortality on target and non-target species (e.g. Mora et al. 2009).

Examples of good practice in international regimes for fisheries management can be found. The Convention for Conservation of Antarctic Marine Living Resources (CCAMLR) was initiated in 1982 following the successive overexploitation of whales and fin fish and the subsequent development of industrial fishing for Antarctic krill (*Euphausia superba*). Given the key role in Antarctic ecosystems played by the latter species, there was concern that exploitation would have severe repercussions on many Antarctic predators, including seals, whales and seabirds. Thus CCAMLR was established with the primary objective of the conservation of Antarctica's marine living resources. Achievement of this objective was established on the basis of several principles including:

- prevention of decrease in the size of any harvested population to levels below those which ensure its stable recruitment
- maintenance of the ecological relationships between harvested, dependent and related populations of Antarctic marine living resources
- prevention of changes or minimization of the risk of changes in the marine ecosystem which are not potentially reversible over two or three decades (Miller 2009).

The convention has also incorporated principles of precautionary and ecosystem-based management. The underlying principles of CCAMLR in allowing the exploitation of marine biotic resources whilst maintaining the viability of such populations over the long term and preserving the structure and function of the

Antarctic ecosystem has led to innovative approaches to the holistic, scientific and ecosystem-based management of fishing. Thus requirements of reporting and observation of fishing activities, management of levels of harvesting and the initiation of technical measures to prevent ecosystem-wide impacts of fishing activities, such as prevention of seabird bycatch from longlines and monitoring of encounters with vulnerable marine ecosystems, have meant that CCAMLR has largely met its conservation objectives (CCAMLR Review Panel 2008). As such, these may be viewed as examples of good practice that should be adopted widely in fisheries management, although issues still remain with the management of fisheries in the Southern Ocean and Antarctic coastal seas.

Future adoption of broader principles of conservation

Ecosystem-based management

The successes of CCAMLR demonstrate that ecosystem-based frameworks with a focus on conservation and long-term sustainability are needed to address the vast scales and scope of ocean management. Ecosystem-based management (EBM) derives from natural science roots, particularly ecology, but engages with planners and social scientists (Christie 2011). It includes notions of working at an ecosystem level, addressing multiple scales, including humans into management plans, and learning about the system through management practice ('adaptive management'). EBM is a response to the interconnectedness within and among ecosystems and among anthropogenic impacts on the environment. Detractors of EBM highlight the imprecision in the concept, the lack of underlying principles, the data demands and complexity of decision-making tools and the profound change needed in the institutions involved. However, it can be argued that significant progress has been made in all such areas; EBM is a desirable goal and should be considered a work in progress (Murawski 2007).

With origins in planning and applied social sciences, another important framework is integrated coastal management (ICM) which aims to overcome the pitfalls of management approaches focused on single types of activity such as fishing (Cicin-Sain & Knecht 1998). It focuses on environmental governance processes, and seeks to build and sustain collaboration among institutions but clearly requires engagement with the applied natural sciences (Christie 2011). A generic emerging framework that is applicable to marine systems is the social-ecological system (SES). Defined at a range of spatial, temporal and organizational scales, this is a system of interacting biophysical and social factors with the objective of providing a context and set of ideas within which to systematically understand how people interact, or might in future interact, with the environment (Readman et al. 2004). The SES framework is expected to help resolve important questions such as the conditions needed for users to reverse the trend towards accelerated overuse of resources and benefit from time and effort invested in conserving resources (Ostrom 2007). Apart from some analyses of certain traditional coastal societies, coastal fisheries probably provide some of the best examples of users organizing as a group to manage their resources (Wamukota et al. 2012).

Marine protected areas and marine spatial planning

Furthered by the World Summit on Sustainable Development (WSSD) and other international (e.g. Convention on Biological Diversity) and regional agreements (e.g. NE Atlantic OSPAR), there has been a worldwide move to develop marine protected areas (MPAs). Where strict protection has been achieved, biodiversity benefits have sometimes followed, demonstrating significant recovery of species – especially exploited ones – and often, subsequently, habitats. However, even for coral reefs, where the science base is relatively extensive, considerable uncertainty remains about the form and magnitude of wider benefits that may accrue from setting up MPAs (Sale et al. 2005). This remains a significant issue in poor countries where subsistence fishing is important and where there is a long history of MPA and fisheries management failures (McClanahan 1999). There are benefits from tourism and protection of habitat, such as divers perceiving MPAs as an opportunity to observe untrammelled underwater nature (Williams & Polunin 2000). However, MPAs are not resilient to wider effects and threats other than fishery exploitation, such as nutrient enrichment from land run-off and climate-driven impacts such as coral bleaching or ocean acidification, and many have failed by not being large enough or having good management implemented (Graham et al. 2011).

Closing any area to fishing is likely to help protect fragile habitats (e.g. deep-water corals), and sedentary (e.g. sponges) and site-attached species (e.g. scallops), though the degree of benefit relates to the degree to which management regulates impacts on such features. The evidence base for beneficial fishery impacts of MPAs in other ecosystems where species are mobile over considerable distances, such as temperate and tropical continental shelves, is very weak. Making MPAs large enough to investigate such potential benefits is a real problem. In the UK, for example, the 2009 Marine & Coastal Access Act initiated a major move towards what are termed marine conservation zones but following stakeholder consultation, the areas to be fully protected from fishing are generally very limited. This often results from political and social pressure to 'squeeze' biodiversity actions into small spaces in the sea. Frequently political pressures on sites agreed for conservation reduce management to little more than the status quo, rather than to the strict protection that provides a range of benefits by supporting biodiversity recovery. These effects may be seen by those with vested commercial interests to have advantages by allowing their activities to continue unimpeded over the vast majority of the ocean and all too often within MPAs, but ultimately in sustainability terms it damages the aims of such conservation actions and erodes most ecological benefits that can support a range of future, long-term economic interests.

Small areas are subject to future uncertainties such as population distributions which are evidently changing with increasing sea surface temperatures (Dulvy et al. 2008). Another uncertainty lies in the variable survivorship and dispersal of egg and larval stages, the average pelagic larval duration being especially great in fish and crustaceans (Bradbury et al. 2008). The combination of the often small sizes of MPAs and the uncertainties of pelagic dispersal has given rise to much theorizing and modelling of networks of MPAs. In general, MPA science has seriously lagged behind MPA advocacy; although the number of empirical studies has increased, there remains a big gap in MPA science of cold temperate ecosystems (Figure 10.3). Empirical studies need to be strengthened and an element of 'adaptive management', where policy makers, managers and scientists can learn from monitoring of a system following protection, is needed to make up this shortfall.

The scarcity of sound information about how marine systems work means that considerable uncertainty surrounds the understanding of anthropogenic impacts and the optimal means to monitor and manage them. This uncertainty is one of the drivers of change in the science and governance of marine resources. Top-down governance, where central government bodies were responsible for the science, planning and enforcement of management tactics, has tended to give way to a growing element of co-management where the strengths of government actors (e.g. legislation, science) are combined with the strengths of local bodies (e.g. surveillance, compliance), and the burden of risk in management decision making is shared. The evidence

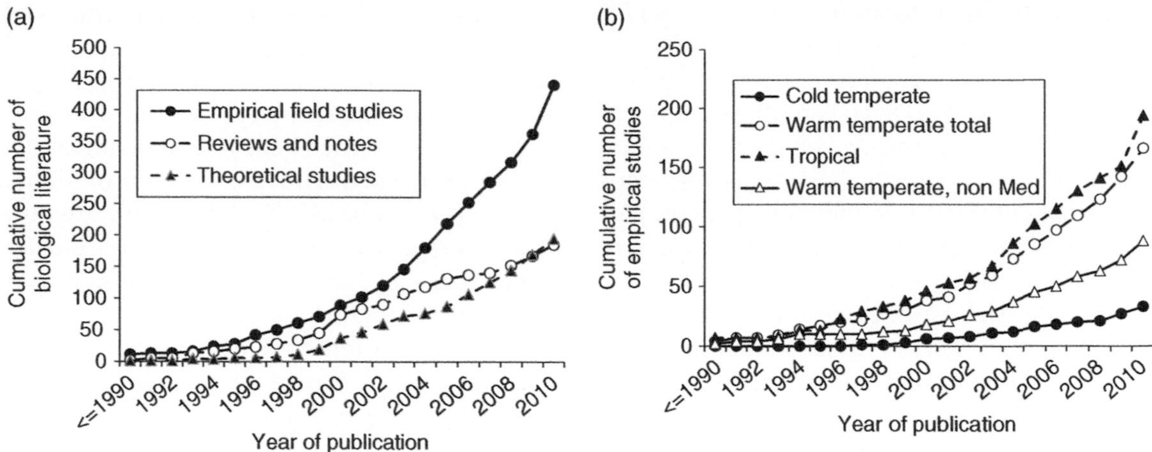

Figure 10.3 (a) Graph plotting the cumulative number of scientific publications on marine protected areas versus year. (b) Cumulative number of empirical studies on marine protected areas versus year broken down into ecosystem type (Caveen et al. 2012).

for co-management being a successful approach to resolving marine environmental problems is abundant but as in the case of MPAs, the data lack resolution and those which have come into the public domain likely favour positive views of the co-management concept (Wamukota et al. 2012). In addition, while government agencies may see some devolution of management decision making and fact finding as a democratic move, as well as a means of sharing risk of management errors, a likely additional motivation is fiscal; government savings may be achieved at the expense of effective management and leadership.

Trends in marine conservation science

Technological advances are providing access to marine ecosystems and information on species that would previously have been inconceivable. The ability to use remote sensing to track annual and interannual changes in marine ecosystems allows the detection of long-term change in temperature and patterns of productivity in the marine environment that underpin the distribution of marine species and the communities to which they belong. Increasingly sophisticated direct intervention technologies such as remotely operated vehicles (ROVs), autonomous underwater vehicles (AUVs) and instrument platforms are enabling direct observation of ecosystems *in situ* as well as the behaviour of the species therein – and the detection of human impacts in the remotest regions of the ocean. Decreasing sizes and falling prices have permitted ever wider deployment of electronic tagging devices (e.g. Block et al. 2011) to monitor animal movement. These types of data can begin to be used to ascertain the effectiveness of MPAs and other management options for highly mobile pelagic organisms, as well as for exploring the consequences of changing oceanographic conditions (Stramma et al. 2012).

Another technological development is the construction and compilation of large marine biological databases such as those associated with fisheries (e.g. the RAM legacy stock assessment database: http://ramlegacy.marinebiodiversity.ca/ram-legacy-stock-assessment-database), marine taxonomy and classification (the World Register of Marine Species: www.marinespecies.org/) and distribution (the Ocean

Biogeographical Information System: www.iobis.org; containing 212,000 valid species and almost 32 million records respectively at the time of writing). Such databases have been valuable in the development of macroecological approaches for answering broad-scale questions of marine conservation and management, particularly with regard to providing baselines and quantifying the effects of fishing (e.g. Worm et al. 2006). The use of historical, anecdotal and opportunistic data is another approach to addressing problems of data paucity (e.g. Luiz & Edwards 2011), as is survey data from volunteer or recreational divers (Ward-Paige et al. 2010).

Other methodological developments include the growing importance of phylogenetic approaches, for quantifying intra-species diversity, identifying cryptic species (Zemlak et al. 2009) and tracing evolutionary dynamics. The combination of phylogenetic and fossil data is likely to yield insights in the future, particularly with regard to quantifying rates of present-day extinction based on traits and taxonomies, and vulnerabilities relative to long-term background rates. The use of species distribution models (SDMs), a technique that broadly relates the distribution of a particular taxon to environmental conditions, is also becoming common in the oceans. They have been deployed to answer a variety of questions, from the potential global distribution of cold-water stony corals (e.g. Tittensor et al. 2009) to exploring the future consequences of climate change and ocean acidification on patterns of diversity and distribution (e.g. Cheung et al. 2009). Species distribution models are open to criticism, particularly when predicting either spatially or temporally in unexplored environments, as they do not take into account the potential plasticity of species responses, nor changing biotic interactions. The next generation of SDMs is likely to include at least some of these important factors, though there tends to be a lag between methods being used terrestrially and then applied in the oceans.

To move towards an integrated future of marine management and conservation at the largest scales, where ecological, climate and socio-economic models are all coupled, process-based (rather than phenomenological) models are needed. The development of multitrophic process-based models is an area in which marine science arguably leads terrestrial. These models have achieved considerable prominence in aiding management and conservation (e.g. Christensen & Walters 2004; Fulton et al. 2011). The marine community has now become engaged in the development of 'end-to-end' models of entire communities from the plankton to the top predators (Fulton 2011). One gap that remains is the development of process-based models focused not on fisheries or biogeochemical cycling, but on the ecosystem *per se*, i.e. the full spectrum of diversity, services and resilience of the marine environment, and its links to the terrestrial realm. These and other long-term modelling endeavours will require large-scale, standardized biological and ecological sampling of the ocean to properly constrain and parameterize processes. Bringing ecological observation programmes and sampling stations nearer to the density and standard of the physical observations used for generating global models of climate will be essential.

Final remarks

Although marine conservation science has developed theory, concepts and principles, the empirical understanding and evidence base which managers and policy makers seek remain limited, due to the vast scales and diversity of systems and approaches involved. This can result in conservation action being taken in the absence of a rigorous evidence base, a source of frustration given the limited resources to address the uncertainties and gaps. Managers and policy makers are thus vulnerable to conflicting pressures (e.g. environmentalist versus commercial interests) while conserving their fiscal resources and the social cohesion of coastal communities (e.g. electorate). Notwithstanding such conflicts, current trends in the status of the oceans are largely negative and in many cases the degradation and the threats to marine species are increasing. Without a significant shift in global approaches to management of the exploitation of the biotic and abiotic resources of marine ecosystems and other human activities impacting the oceans, a further acceleration of decline and extinction of species is inevitable. Change requires scaling up of protection, implementing or increasing the effectiveness of MPA management, and prioritization of policies for the sustainable use, conservation and restoration of marine living resources that moves beyond sectoral management and national self-interest. Such a transition will be difficult but is essential in a world of increasing population and per-capita demand for resources.

References

Aguilar, A. & Borrell, A. (1994) Abnormally high polychlorinated biphenyl levels in striped dolphins (*Steneua coeruleoalba*) affected by the 1990–1992 Mediterranean epizootic. *Science of the Total Environment*, **154**, 237–247.

Alvarez-Filip, L., Dulvy, N.K., Gill, J.A., *et al.* (2009) Flattening of Caribbean coral reefs: region-wide declines in architectural complexity. *Proceedings of the Royal Society of London B*, doi:10.1098/rspb.2009.0339.

Angel, M.V. (1993) Biodiversity of the pelagic ocean. *Conservation Biology*, **7**, 760–772.

Baum, J.K. & Blanchard, W. (2010) Inferring shark population trends from generalized linear mixed models of pelagic longline catch and effort data. *Fisheries Research*, **102**, 229–239.

Bellwood, D.R., Hughes, T.P., Folke, C. & Nystrom, M. (2004) Confronting the coral reef crisis. *Nature*, **429**, 827–833.

Block, B.A., Jonsen, I.D., Jorgensen, S.J., *et al.* (2011) Tracking apex marine predator movements in a dynamic ocean. *Nature*, **475**, 86–90.

Bossart, G.D. (2010) Marine mammals as sentinel species for oceans and human health. *Veterinary Pathology*, **48**, 676–690.

Bradbury, I.R., Laurel, B., Snelgrove, P.V.R., *et al.* (2008) Global patterns in marine dispersal estimates: the influence of geography, taxonomic category and life history. *Proceedings of the Royal Society B*, **275**, 1803–1809.

Carpenter, K.E., Abrar, M., Aeby, G, *et al.* (2008) One third of reef-building corals face elevated extinction risk from climate change and local impacts. *Science*, **321**, 560–563.

Caveen, A.J., Sweeting, C.J., Willis, T.J. & Polunin, N.V.C. (2012) Are the scientific foundations of temperate marine reserves too warm and hard? *Environmental Conservation*, **39**(3), 199–203.

CCAMLR Review Panel (2008) *CCAMLR Performance Review Panel Report*. www.CCAMLR.org

Cebrian, J. (1999) Patterns in the fate of production in plant communities. *American Naturalist*, **154**, 449–468.

Cebrian, J., Shurin, J.B., Borer, E.T., *et al.* (2009) Producer nutritional quality controls ecosystem trophic structure. *PLoS One* **4**, e4929.

Cheung, W.W.L., Close, C., Kearney, K., *et al.* (2009) Projections of global marine biodiversity impacts under climate change scenarios. *Fish and Fisheries*, **10**, 235–251.

Christensen, V. & Walters, C.J. (2004) Ecopath with Ecosim: methods, capabilities and limitations. *Ecological Modelling*, **172**, 109–139.

Christie, P. (2011) Creating space for interdisciplinary marine and coastal research: five dilemmas and suggested resolutions. *Environmental Conservation*, **38**, 172–186.

Cicin-Sain, B. & Knecht, R. (1998) *Integrated Coastal and Ocean Management: Concepts and Practices*. Island Press, Washington, D.C.

Collette, B.B., Carpenter, K.E., Polidoro, B.A., *et al.* (2011) High value and long life – double jeopardy for tunas and billfishes. *Science*, **333**, 291–292.

Collie, J.S., Hall, S.J., Kaiser, M.J. & Poiner, I.R. (2000) A quantitative analysis of fishing impacts on shelf-sea benthos. *Journal of Animal Ecology*, **69**, 785–798.

Crowder, L.B., Hazen, E.L., Avissar, N., *et al.* (2008) The impacts of fisheries on marine ecosystems and the transition to ecosystem-based management. *Annual Review of Ecology, Evolution and Systematics*, **39**, 259–278.

Crowe, T.P., Thompson, R.C., Bray, S. & Hawkins, S.J. (2000) Impacts of anthropogenic stress on rocky intertidal communities. *Journal of Aquatic Ecosystem Stress and Recovery*, **7**, 273–297.

Daskalov, G.M., Grishin, A.N., Rodionov, S. & Mihneva, V. (2007) Trophic cascades triggered by overfishing reveal possible mechanisms of ecosystem regime shifts. *Proceedings of the National Academy of Science*, **104**, 10518–10523.

Davies, R.W.D., Cripps, S.J., Nickson, A. & Porter, G. (2009) Defining and estimating global marine fisheries by-catch. *Marine Policy*, **33**, 661–672.

Derraick, J.G.B. (2002) The pollution of the marine environment by plastic debris: a review. *Marine Pollution Bulletin*, **44**, 842–852.

Diaz, R.J. & Rosenberg, R. (2008) Spreading dead zones and consequences for marine ecosystems. *Science*, **321**, 926–929.

Doney, S.C., Fabry, V.J., Feely, R.A. & Kleypas, J.A. (2009) Ocean acidification: the other CO_2 problem. *Annual Review of Marine Science*, **1**, 169–192.

Dulvy, N.K., Rogers, S.I., Jennings, S., *et al.* (2008) Climate change and deepening of the North Sea fish assemblage: a biotic indicator of warming seas. *Journal of Applied Ecology*, **45**, 1029–1039.

Dulvy, N.K., Pinnegar, J.K. & Reynolds, J.D. (2009) Holocene extinctions in the sea. In: *Holocene Extinctions* (ed. S. Turvey). Oxford University Press, Oxford.

EC (2011) *Impact Assessment: Accompanying Commission Proposal for a Regulation of the European Parliament and of the Council on the Common Fisheries Policy [repealing Regulation (EC) N°2371/2002]*. Commission Staff Working Paper SEC(2011) 891. European Commission, Brussels.

Evans, P.G.H. (1987) *The Natural History of Whales and Dolphins*. Christopher Helm, Bromley, Kent.

Fabricius, K.E., Langdon, C., Uthicke, S., *et al.* (2011) Losers and winners in coral reefs acclimatized to elevated carbon dioxide concentrations. *Nature Climate Change*, **1**, 165–169.

Fabry, V.J., Seibel, B.A., Feely, R.A. & Orr, J.C. (2008) Impacts of ocean acidification on marine fauna and ecosystem processes. *ICES Journal of Marine Science*, **65**, 414–432.

FAO (2010) *The State of World Fisheries and Aquaculture*. Food and Agricultural Organization of the United Nations, Rome.

Field, C.B., Behrenfeld, M.J., Randerson, J.T. & Falkowski, P. (1998) Primary production of the biosphere; integrating terrestrial and oceanic components. *Science*, **281**, 237– 240.

Fox, H.E., Pet, J.S., Dahuri, R. & Caldwell, R.L. (2003) Recovery in rubble fields: long term impacts of blast fishing. *Marine Pollution Bulletin*, **46**, 1024–1031.

Fulton, E.A. (2011) Approaches to end-to-end ecosystem models. *Journal of Marine Systems*, **81**, 171–183.

Fulton, E.A., Link, J.S., Kaplan, I.C., *et al.* (2011) Lessons in modelling and management of marine ecosystems: the Atlantic experience. *Fish and Fisheries*, **12**, 171–188.

Gasol, J.M., del Giorgio, P.A. & Duarte, C.M. (1997) Biomass distribution in marine planktonic communities. *Limnology and Oceanography*, **42**, 1353–1363.

GESAMP (2009) *Pollution in the Open Ocean: A Review of Assessments and Related Studies*. Reports and Studies GESAMP No. 79. GESAMP, London.

Graham, N.A.J., Ainsworth, T.D., Baird, A.H., *et al.* (2011) From microbes to people: tractable benefits of no-take areas for coral reefs. *Oceanography and Marine Biology*, **49**, 105–136.

Hall-Spencer, J.M., Rodolfo-Metalpa, R., Martin S., *et al.* (2008) Volcanic carbon dioxide vents show ecosystem effects of ocean acidification. *Nature*, **454**, 96–99.

Halpern, B.S., Walbridge, S., Selkoe, K.A., *et al.* (2008) A global map of human impacts on marine ecosystems. *Science*, **319**, 948–952.

Hampton, J., Sibert, J.R., Kleiber, P., *et al.* (2005) Decline of Pacific tuna populations exaggerated? *Nature*, **434**, E1–E2.

Hilborn, R. (2007) Moving to sustainability by learning from successful fisheries. *Ambio*, **36**, 296–303.

Hoegh-Guldberg, O. (1999) Climate change, coral bleaching and the future of the world's coral reefs. *Marine and Freshwater Research*, **50**, 839–866.

Hoegh-Guldberg, O., Mumby, P.J. & Hooten, A.J. (2007) Coral reefs under rapid climate change and ocean acidification. *Science*, **318**, 1737–1742.

Huston, M.A. & Wolverton, S. (2009) The global distribution of net primary production: resolving the paradox. *Ecological Monographs*, **79**, 343–377.

Jennings, S., Kaiser, M.J. & Reynolds, J.D. (2001) *Marine Fisheries Ecology*. Blackwell Science, Oxford.

Jones, R.J., Kildea, T. & Hoegh-Guldberg, O. (1999) PAM chlorophyll fluorometry: a new *in situ* technique for stress assessment in scleractinian corals, used to examine the effects of cyanide from cyanide fishing. *Marine Pollution Bulletin*, **38**, 864–874.

Kaschner, K., Tittensor, D.P., Ready, J., Gerrodette, T. & Worm, B. (2011) Current and future patterns of global marine mammal biodiversity. *PLoS ONE*, **6**, e19653.

Kawaguchi, S., Kurihara, H., King, R., *et al.* (2010) Will krill fare well under Southern Ocean acidification? *Biology Letters*, doi:10.1098/rsbl.2010.0777.

La Farré, M., Pérez, S., Kantiani, L. & Barcelo, D. (2008) Fate and toxicity of emerging pollutants, their metabolites and transformation products in the aquatic environment. *Trends in Analytical Chemistry*, **27**, 991–1007.

Laffoley, D.A. (2000) *Historical Perspective and Selective Review of the Literature on Human Impacts on the UK's Marine Environment*. English Nature Research Report 391. Prepared by English Nature for the DETR Working Group on the Review of Marine Nature Conservation. English Nature, Peterborough.

Landsberg, J.H. (2002) The effects of harmful algal blooms on aquatic organisms. *Reviews in Fisheries Science*, **10**, 113–390.

Luiz, O.J. & Edwards, A.J. (2011) Extinction of a shark population in the archipelago of Saint Paul's Rocks (equatorial Atlantic) inferred from the historical record. *Biological Conservation*, **144**(12), 2873–2881.

May, R.M. (1994) Biological diversity – differences between land and sea. *Philosophical Transactions of the Royal Society B*, **343**, 105–111.

McClanahan, T. (1999) Is there a future for coral reef parks in poor tropical countries? *Coral Reefs*, **18**, 321–325.

Miller, D. (2009) Sustainable management in the Southern Ocean: CCAMLR Science. *Science Diplomacy*, 103–121. www.atsummit50.aq/media/book-16.pdf

Mora, C., Tittensor, D.P. & Myers, R.A. (2008) The completeness of taxonomic inventories for describing the global diversity and distribution of marine fishes. *Proceedings of the Royal Society B*, **275**, 149–155.

Mora, C., Myers, R.A., Coll, M., *et al.* (2009) Management effectiveness of the world's marine fisheries. *PLoS Biology*, **7**(6), e1000131.

Mora, C., Tittensor, D.P., Adl, S., *et al.* (2011) How many species are there on earth and in the ocean? *PLoS Biology*, **9**, e1001127.

Morato, T., Watson, R., Pitcher, T.J. & Pauly, D. (2006) Fishing down deep. *Fish and Fisheries*, **7**, 24–34.

Moy, A.D., Howard, W.R. & Bray, S.G. (2009) Reduced calcification in modern Southern Ocean planktonic foraminifera. *Nature Geoscience*, **2**, 276–280.

Murawski, S. (2007) Ten myths concerning ecosystem approaches to marine resource management. *Marine Policy*, **31**, 681–690.

Myers, R.A., Baum, J.K., Shepherd, T.D., *et al.* (2007) Cascading effects of the loss of apex predatory sharks from a coastal ocean. *Science*, **315**, 1846–1850.

Noël, M., Barrett-Lennard, L., Guinet, C., *et al.* (2009) Persistent organic pollutants (POPs) in killer whales (*Orcinus orca*) from the Crozet Archipelago, southern Indian Ocean. *Marine Environmental Research*, **68**, 196–202.

Olsen, E.M., Lilly, G.R., Heino, M., *et al.* (2005) Assessing changes in age and size at maturation in collapsing populations of Atlantic cod (*Gadus morhua*). *Canadian Journal of Fisheries and Aquatic Sciences*, **62**, 811–823.

Ostrom, E. (2007) A diagnostic approach for going beyond panaceas. *Proceedings of the National Academy of Sciences USA*, **104**, 15181–15187.

Pauly, D., Christensen, V., Dalsgaard, J., *et al.* (1998) Fishing down marine foodwebs. *Science*, **279**, 860–863.

Peterson, C.H., Rice, S.D. & Short, J.W. (2003) Long-term ecosystem response to the Exxon Valdez oil spill. *Science*, **302**, 2082–2086.

Pinsky, M.L., Jensen, O.P., Ricard, D. & Palumbi, S.R. (2011) Unexpected patterns of fisheries collapse in the world's oceans. *Proceedings of the National Academy of Science USA*, doi/10.1073/pnas.1015313108.

Pitcher, T., Kalikoski, D., Pramod, G. & Shot, K. (2009) Not honouring the code. *Nature*, **457**, 658–659.

Plaisance, L., Caley, M.J., Brainard, R. & Knowlton, N. (2011) The diversity of coral reefs: what are we missing? *PLoS ONE*, **6**, e25025.

Polidoro, B.A., Carpenter, K.E., Collins, L., *et al.* (2010) The loss of species: mangrove extinction risk and geographic areas of global concern. *PLoS ONE*, **5**, e10095.

Polis, G.A. (1999) Why are parts of the world green? Multiple factors control productivity and the distribution of biomass. *Oikos*, **86**, 3–15.

Quarrie, J. (ed.) (1992) *Earth Summit '92. The United Nations Conference on Environment and Development. Rio de Janerio 1992*. Regency Press, London.

Ramirez-Llodra, E., Tyler, P.A., Baker, M.C., *et al.* (2011) Man and the last great wilderness: human impact on the deep sea. *PLoS ONE*, **6**, e22588.

Raymundo, L.J., Maypa, A.P., Gomez, E.D. & Cadiz, P. (2007) Can dynamite-blasted reefs recover? A novel, low-tech approach to stimulating natural recovery in fish and coral populations. *Marine Pollution Bulletin*, **54**, 1009–1019.

Readman, C.L., Grove, J.M. & Kuby, L.H. (2004) Integrating social science into the Long-Term Ecological Research (LTER) Network: social dimensions of ecological change and ecological dimensions of social change. *Ecosystems*, **7**, 161–171.

Reid, P.C., Fischer, A.C., Lewis-Brown, E., *et al.* (2009) Impacts of the oceans on climate change. *Advances in Marine Biology*, **56**, 1–150.

Richardson, A.J. & Poloczanska, E.S. (2008) Under-resourced, under threat. *Science*, **320**, 1294–1295.

Roberts, J.M., Wheeler, A.J., Freiwald, A. & Cairns, S.D. (2009) *Cold-Water Corals*. Cambridge University Press, Cambridge.

Rodolfo-Metalpa, R., Houlbrèque, F., Tambutté, É., *et al.* (2011) Coral and mollusc resistance to ocean acidification adversely affected by warming. *Nature Climate Change*, **1**, 308–312.

Rosenberg, A.A., Bolster, W.J., Alexander, K.E., *et al.* (2005) The history of ocean resources: modelling cod biomass using historical records. *Frontiers in Ecology and the Environment*, **3**, 78–84.

Sale, P.F., Cowen, R.K., Danilowicz, B.S., *et al.* (2005) Critical science gaps impede use of no-take fishery reserves. *Trends in Ecology and Evolution*, **20**, 74–80.

Small, C. & Cohen, J.E. (2004) Continental physiography, climate and the global distribution of human populations. *Current Anthropology*, **45**, 269–288.

Smedbol, R.K. & Stephenson, R. (2001) The importance of managing within-species diversity in cod and herring fisheries of the north-western Atlantic. *Journal of Fish Biology*, **59**(Suppl A), 109–128.

Solow, A.R. & Smith, W.K. (2005) On estimating the number of species from the discovery record. *Philosophical Transactions of the Royal Society*, **272**, 285–287.

Steneck, R.S., Graham, M.H., Bourque, B.J., *et al.* (2002) Kelp forest ecosystems: biodiversity, stability, resilience and future. *Environmental Conservation*, **29**, 436–459.

Stramma, L., Prince, E.D., Schmidtko, S., *et al.* (2012) Expansion of oxygen minimum zones may reduce available habitat for tropical pelagic fishes. *Nature Climate Change*, **2**, 33–37.

Swartz, W., Sala, E., Tracey, S., *et al.* (2010) The spatial expansion and ecological footprint of fisheries (1950 to present). *PLoS ONE*, **5**, e15143.

Thompson, R.C., Moore, C.J., vom Saal, F.S. & Swan, S.H. (2009) Plastics, the environment and human health: current consensus and future trends. *Philosophical Transactions of the Royal Society of London B*, **364**, 2153–2166.

Tittensor, D.P., Baco-Taylor, A.R., Brewin, P., *et al.* (2009) Predicting global habitat suitability for stony corals on seamounts. *Journal of Biogeography*, **36**, 1111–1128.

Tittensor, D.P., Mora, C., Jetz, W., *et al.* (2010) Global patterns and predictors of marine biodiversity across taxa. *Nature*, **466**, 1098–1101.

Trathan, P.N., Forcada, J. & Murphy, E.J. (2007) Environmental forcing and Southern Ocean marine predator populations: effects of climate change and variability. *Philosophical Transactions of the Royal Society of London B*, **362**, 2351–2365.

Vaquer-Sunyer, R. & Duarte, C.M. (2008) Thresholds of hypoxia for marine biodiversity. *Proceedings of the National Academy of Science USA*, **105**, 15452–15457.

Wamukota, A.W., Cinner, J.E. & McClanahan, T.R. (2012) Co-management of coral reef fisheries: a critical evaluation of the literature. *Marine Policy*, **36**, 481–488.

Ward-Paige, C.A., Mora, C., Lotze, H.K., *et al.* (2010) Large-scale absence of sharks on reefs in the greater Caribbean: a footprint of human pressures. *PLoS ONE*, **5**, e11968.

Waycott, M., Duarte, C.M., Carruthers, T.J.B., *et al.* (2009) Accelerating loss of seagrasses across the globe threatens coastal ecosystems. *Proceedings of the National Academy of Science USA*, **106**, 12377–12381.

Webb, T.J., Vanden Berghe, E. & O'Dor, R. (2010) Biodiversity's big wet secret: the global distribution of marine biological records reveals chronic under-exploration of the deep pelagic ocean. *PLoS ONE*, **5**, e10223.

Wells, S. (2009) Dynamite fishing in northern Tanzania – pervasive, problematic and yet preventable. *Marine Pollution Bulletin*, **58**, 20–23.

Wilkinson, C. (2004) Executive summary. In: *Status of Coral Reefs of the World, Vol 1*. (ed. C. Wilkinson), pp. 7–66. Australian Institute of Marine Science, Townsville, Australia.

Williams, I.D. & Polunin, N.V.C. (2000) Differences between protected and unprotected Caribbean reefs in attributes preferred by dive tourists. *Environmental Conservation*, **27**, 382–391.

Worm, B. & Tittensor, D.P. (2011) Range contraction in large pelagic predators. *Proceedings of the National Academy of Sciences USA*, **108**, 11942–11947.

Worm, B., Barbier, E.B., Beaumont, N., *et al.* (2006) Impacts of biodiversity loss on ocean ecosystem services. *Science*, **314**, 787–790.

Worm, B., Hilborn, R., Baum, J.K., *et al.* (2009) Rebuilding global fisheries. *Science*, **325**, 578–585.

Zemlak, T.S., Ward, R.D., Connell, A.D., *et al.* (2009) DNA barcoding reveals overlooked marine fishes. *Molecular Ecology Resources*, **9**(S1), 237–242.

11

Lost in muddy waters: freshwater biodiversity

Nic Pacini[1], David M. Harper[2], Peter Henderson[3] and Tom Le Quesne[4]

[1]University of Calabria, Cosenza, Italy
[2]Department of Biology, University of Leicester, Leicester, UK
[3]Pisces Conservation, Lymington, UK
[4]WWF-UK, Godalming, UK

"We never know the worth of water till the well is dry."

— Thomas Fuller, Gnomologia, 1732

Introduction

Freshwaters are a vital part of the Earth's natural capital. They deliver essential services to society and hold many species of high economic value to mankind. Without them, industry, agriculture and energy generation would be impossible. They are sacred to many religions and cultures. Their management poses particular challenges due to the many essential services for which they are managed and the complexity of their ecosystems. They tend to be overexploited, polluted and invaded by alien species. The health of freshwater ecosystems has suffered greater declines than any other ecosystem type. This chapter provides an overview of the status of freshwater biodiversity in a number of selected river basins, discusses some of the key pressures on freshwater ecosystems and illustrates possible conservation strategies. It does this by discussing the factors affecting the number of fish species in the largest lake and river basins, using fish as an indicator of aquatic ecosystem biodiversity because they are more easily understood and more extensively documented than most microscopic organisms due to their economic importance.

The extent of freshwater ecosystems and their biodiversity

Freshwater ecosystems are physically varied because of geological history, resulting in many different combinations of size, latitudinal and altitudinal location, hydrology (discharge), morphometry (channel or basin shape), geology and chemistry (Moss 2010). A comprehensive

Key Topics in Conservation Biology 2, First Edition. Edited by David W. Macdonald and Katherine J. Willis.

list of ecosystem types would number several hundred, from small temporary rivulets in upper catchments to the largest rivers tens of kilometres wide in their lower reaches; from small but widespread temporary pools to the great continental lakes and to extensive shallow wetlands. Estuaries and shallow coastal water add another ecosystem dimension, because of their variable salinity gradient. Aquatic organisms are often restricted to a relatively narrow range of these river or lake habitats; in addition, a large number of terrestrial and semi-aquatic organisms are supported by riparian ecotones and wetlands (e.g. Pacini & Harper 2008).

Freshwaters play a critical role in maintaining key terrestrial, as well as aquatic, ecosystem processes due to their widespread presence across continents, cumulative large surface, accessibility by many species and ecotonal character. They can regulate the dynamics of terrestrial populations by providing essential water, food and shelter, particularly in drought (Western 1975; Redfern et al. 2003). For example, the distribution and temporal dynamics of wetlands in the Serengeti-Mara ecosystem of East Africa regulate the dynamics of the largest herbivore migrations (wildebeest, zebra, buffalo and gazelles; Gereta & Wolanski 2008). Water-dependent predators, such as crocodiles, regulate the abundance of migrating wildebeest, zebra, impala and other ungulates, that cross key rivers.

Freshwaters hold a disproportionate share of the earth's biodiversity in relation to their area (IUCN 2011a). The surface covered by rivers and lakes was estimated at the end of the 20th century as 0.8% of the earth (Gleick 1996). Small and shallow temporarily fluctuating wetlands exceed the surface of the world's large lakes and more accurate estimates, based on fine-resolution GIS, have significantly increased the extent covered by lentic waters, now estimated as 3% of the surface of the continents (Downing 2009). When brackish and estuarine wetlands are included, this is 8.6% of the earth's land surface (Zedler & Kercher 2005).

About 126,000 described species rely directly on freshwater habitats – plants, insects, molluscs, fishes, reptiles and mammals (Balian et al. 2008); 18,235 are strictly aquatic vertebrates, of which about 13,000 fish (15,000 when adding brackish waters; Vié et al. 2008). Thus, 38% of recognized vertebrate species (mostly fishes and amphibians; Balian et al. 2010) live in about 3% of the earth's land surface. Freshwater fishes represent 45% of all known fishes on earth and as much as 25% of all known vertebrate species (Balian et al. 2008). This number of extant fish species is probably a large underestimate, due to only partial knowledge of species-rich tropical regions; about 300 new fish species are described every year (Stiassny 1999) and the number of described amphibian species nearly doubled between 1985 and 2006 (Frost et al. 2006), and is still rising. It is now considered that, even in temperate latitudes such as in Europe, fish communities have non-interbreeding populations of many species, showing basin-specific adaptive genetic differences (Keller et al. 2011). The ongoing genetic characterization of European fish populations indicates that several of these may be considered as separate species in the near future (Ole Seehausen, personal communication).

Invertebrate groups, in particular molluscs, insects and crustaceans, remain understudied, e.g. only 6000 species of freshwater molluscs are described out of an estimated 16,000 potential extant species and only 16.5% were assessed for the IUCN Red List in 2008. While higher latitude habitats tend to have a low number of taxa, most freshwater life is concentrated in the intertropical belt. Here, where biodiversity is higher than in any other biome on the planet, tropical freshwaters are 'ultra-hot spots' (IUCN 2011a).

Patterns of diversity

The distribution of freshwater species on the earth's surface is uneven. The pattern of higher species richness at lower latitudes, common also to terrestrial and marine biomes, is generally followed in freshwaters by molluscs, insects, crustaceans, amphibians, fishes and birds (Lévêque 1997; Willig et al. 2003; Mittelbach et al. 2007). Notable exceptions are some aquatic insects, such as invertebrate shredder detritivores that, as a functional group, are more speciose in temperate regions (Boyero et al. 2011). Stoneflies are represented in tropical Africa by one perlid genus (*Neoperla*) with about 50 species (Zwick 2007) and seven genera of *Notonemura*, mainly in the Cape Region. Only 57 stonefly genera occur in the neotropics, 41 in the Oriental regions and 46 in Australia compared with 108 genera in the Palearctic region and 102 in the Nearctic (Fochetti & de Figueroa 2008). Fishes are distinctly more diverse in the tropics. For example, in the Congo River there are 858 species, whereas there are only 250 in the Mississippi, two catchments of equivalent size but different latitude (6°00′S and 29°9′N respectively). Lake Malawi (30,800 km^2) has 800+ species, of which 99% are endemic (Weyl et al. 2010), whereas Lake Baikal in Siberia, with a larger water volume but a similar surface (31,722 km^2), has 27 endemics out of 55 native and 61 extant fish species. The North American Great Lakes, far larger (244,000 km^2), currently hold 148 native fishes out of the 169 natives that predated European colonization (Mandrak & Cudmore 2011).

The 'flood pulse' (Junk 2001) is an important hydrological process structuring the biodiversity and functioning of large rivers. Floods reconnect isolated water bodies and provide an intermediate disturbance regime that reduces short-term stability but perpetuates natural selection processes. Highly dynamic (large difference between baseflow and flood peak) and long-lasting floods are a characteristic of the Amazon basin, of most Sahelian rivers in tropical Africa, of monsoonal rivers in South East Asia and, through snow-melt, of some of the larger European, Asian and North American rivers. Floods provide an ecological rejuvenation mechanism through habitat re-creation and through the intermittent reconnection of temporarily disconnected water bodies that, over time, enhances diversity. Fishes tend to swim upwards towards the rising floodwaters as well as migrating to upstream tributaries and across the floodplain. In the Amazon, other aquatic vertebrates, such as manatees, follow a similar migration from the main river towards the inundated varzéa lakes during the flood, and back to a large river during receding water levels (Arraut et al. 2010). Internal biological clocks are reset by floods to regulate fruiting, reproduction and migration of plants and of aquatic as well as terrestrial vertebrates. Seasonal shifts in habitat use, subject to a significant interannual variation, facilitate the co-existence of a large number of species (Chapman 2001). Temperature coupling in the tropics, when floods occur during high water temperatures to stimulate biological processes such as reproduction, is key to these biological triggers.

The Amazon is the freshwater system on which the flood pulse concept was formulated and illustrates its effect upon biodiversity through speciation. It has 20% of all the freshwater flowing into the oceans of the world and a total discharge greater than the next 10 largest rivers combined. At over 7×10^6 km^2, the catchment covers more than 40% of South America, including Brazil, Venezuela, Colombia, Ecuador, Peru and Bolivia. The river flows through lowlands for much of its course and thus meanders across a wide floodplain, forming a great mosaic of lakes, ponds and channels. Land has been cut and relaid through recent erosions and depositions of the rivers, to maintain an intricate mosaic of forest, scrub, marshes, lakes and channels. Most habitats flood every year, some by up to 11 metres depth. Floating communities do not submerge but are still subject to mixing and disruption (Henderson et al. 1998); no old water bodies develop; most lakes and channels are less than 1000 years old.

The fish fauna of the Amazon is the richest freshwater fish assemblage known. The likely number of species is at least 2500, out of 4035 species from 705 genera present in the neotropical region (Lévêque et al. 2008) and perhaps as many as 8000 species extant in central and southern America combined (Winemiller et al. 2008). Open waters, where only the swiftest of swimmers are able to survive, support large predators such as dolphins and catfish. Many taxa belong to essentially marine families that have become adapted to life in well-oxygenated open water; the herring, *Pellona castelnaeana*, forms shoals just like its marine relative. Many floodplain species enter the rivers at low water either to take refuge or to breed.

The great rivers tend to support a similar fish fauna for long stretches; Goulding et al. (1988) recorded a set of about 450 fish species along a 1200 km stretch of the Negro River between Manaus and San Gabriel, Brazil. The rivers also support a benthic fish fauna that is still far from known; that of whitewater rivers, draining geologically recent uplands and characterized by high levels of nutrients, sediments and conductivity, such as the Rio Solimoes, seems considerably less diverse than that of nutrient-poor blackwater rivers, like the Rio Negro. These drain lowlands, underlain by ancient geological formations, are poor in minerals, exceptionally acidic (pH 3.8–4.9) and support little *in situ* primary production (Henderson & Walker 1986). The fauna of both is similar in the dominance of catfish and knife fish. Caecilians (eel-like amphibians) are also a common component of deeper rivers. Most of the energy input into these stream ecosystems comes from leaves, fruits and animals such as insects that drop from the surrounding forest (Goulding 1980, 1989). As the water is acidic, fungi are favoured over bacteria as the main decomposer group. Leaves and fruits that enter streams are rapidly invaded by fungi, which in turn support insects and prawns. Submerged litter banks are an important habitat sheltering small insects and crustaceans that are food for larger predatory insects and small fish, themselves prey to large dragonfly larvae. The top predators are larger fish and large reptiles (*Anaconda*) or birds. Forest streams hold an abundant open water fauna of small, often brilliantly coloured, fish that feed on insects and other small animals, which accidentally fall into the water. These fish in turn are food to kingfishers and large predatory fish such as cichlids and electric eel. In a study of the leaf litter community food web of the blackwater river Taruma-Mirim near Manaus, Brazil, Henderson & Walker (1986) described over 20 species of small fish collected in a single litter bank. These fish show striking adaptations in form and behaviour to their benthic lifestyle, but almost all are predators on invertebrates living within the litter.

That so many species, feeding on similar foods, can co-exist within a single food web supported primarily by the fungal degradation of forest inputs raises challenging problems with regard to species co-existence under competition. Walker et al. (1991) concluded from a radiolabelling experiment that fish within the leaf litter were remarkably static, normally remaining with a 1 m radius for a number of weeks. Henderson & Walker (1990) found that litter bank fish all live within distinct subregions of the habitat and suggested that it is this subdivision of space that allows high fish diversity to be supported.

Although the habitat subdivisions lead to a static nature of many Amazonian fish, some undertake lengthy spawning migrations from near the mouth of the river to headwater streams, a distance of over 2000 km. These are the longest freshwater migrations known. Catfish such as *Brachyplatystoma* spp. undertake the most impressive migrations of this type, from the brackish waters of the estuary to small headwater streams. The advantages are presumably abundant oxygen, clean gravel spawning substrates where the eggs will not be suffocated with silt, and abundant food supported by the release of nutrients from Andean rocks.

Africa has a higher number of archaeic fish families (Denticipitidae, Distichodontidae, Pantodontidae, Phractolemidae, Kneriidae, Mormyridae and Gymnarchidae; Lévêque 1997) than other tropical continents. The overall number of described species is about 3000 (900 are cichlids), subdivided into 390 genera and 48 families (89 families together with brackish species, Lévêque et al. 2008). More than 858 species have been identified in the Congo River, but as many as 1250 are estimated to live within the basin – that includes

Lake Tanganyika, Lake Kivu and the Malagarasi River (IUCN 2011b). The Congo is characterized by a broad slow-flowing clear water section akin to an 'elongated lake' where more than 80% of the species are endemic and still poorly known (Winemiller et al. 2008). Some species have evolved within the floodplain, where several anabantids, polypterids and cyprinids developed air-film sucking behaviours and striking adaptations; several others evolved to feed on allochthonous material including leaves, flowers, seeds, invertebrates and detritus (Matthes 1961; Chapman & Chapman 2003) in extensive forests covering up to one-third of the basin. The lower river reaches include a 350 km stretch of rapids holding 129 rheophilic species, of which 34 are endemic.

The Nile has a significantly lower fish diversity (128 species) due to the aridity of its basin; this despite its length (6825 km) and the fact that it crosses strikingly different climatic ecoregions. Far richer in fish species are the Nile basin lakes (Victoria, Albert, George, Kyoga, Kivu, Edward and Tana) to the extent that when these are included, the overall fish diversity of the Nile Basin is 800+ species (Witte et al. 2009). More than 800 species are endemic to Lake Malawi, a renowned hot spot of insect, mollusc, amphibian and reptile as well as fish diversity. Some 187 fish species are reported from the Zambesi River, although this is almost certainly an underestimate; a recent fish survey in the upper basin found 120 species, of which 25 were new to science, including five endemics specialized for life at the headwaters (Tweddle et al. 2004).

Asian freshwaters are poorly known, species rich and under a great level of threat because of high human development (Dudgeon 2011). The Indo-Chinese peninsula had 930 described fish species from 87 families by the 1980s (Kottelat 1989) but this estimate had risen to between 1200 and 1700 by the end of the century (Dudgeon 2000); most of these are very poorly known. Some 781 fish species were identified in the Mekong (Vaidyanathan 2011), with scientists describing several new species every year, to the extent that current estimates for this basin are close to 1300, making the Mekong the second or third richest river basin worldwide and the one with the highest diversity per kilometre (Dudgeon 2011). Charismatic endemic species include the giant stingray *Himantura chaophraya*, the world's largest catfish *Pangasianodon gigas*, the giant barb *Catlocarpio siamensis* and the 'freshwater shark' *Wallago attu* (silurid), ranking high among the world's largest freshwater species and all migratory. Beside fish, eight crocodilians live in the region, out of 23 worldwide, and 900 amphibians (Braatz et al. 1992). Freshwater turtles and terrapins are more diverse here than elsewhere, as well as birds, crustaceans (freshwater crabs and shrimps) and several insects, among which are dragonflies (>660 species in Indonesia alone; Braatz et al. 1992) and caddisflies (>50,000 species in the Oriental region; Schmid 1984).

Europe, as an example from temperate regions, has a far lower number of fish species, being at higher latitude and with a history of more intensive human settlement. The 164 km Biebrza River, in north-eastern Poland, with an average annual flow of 30 m^3/s, is a fine example of a middle-sized lowland river that has preserved an almost entirely natural character through a high level of protection of riverside swamps and peatlands. High species richness (for Europe; 39 fish species occur in the 7062 km^2 catchment) is explained by the diverse array of well-preserved lateral floodplain wetlands with numerous oxbow lakes at various stages of succession across a low-gradient floodplain. These habitats provide rich feeding grounds and important winter refuges. The water is highly oxygenated, mildly alkaline (pH 7.1–8.3) and of high quality, although high in dissolved organic matter. Figure 11.1 illustrates how different species distribute themselves across the Biebrza floodplain and change order of dominance according to the frequency of inundation. The natural river regime is key to ensuring the survival of the different fish assemblages. The intermittent connectivity between the main channel and water bodies that are remnants of former

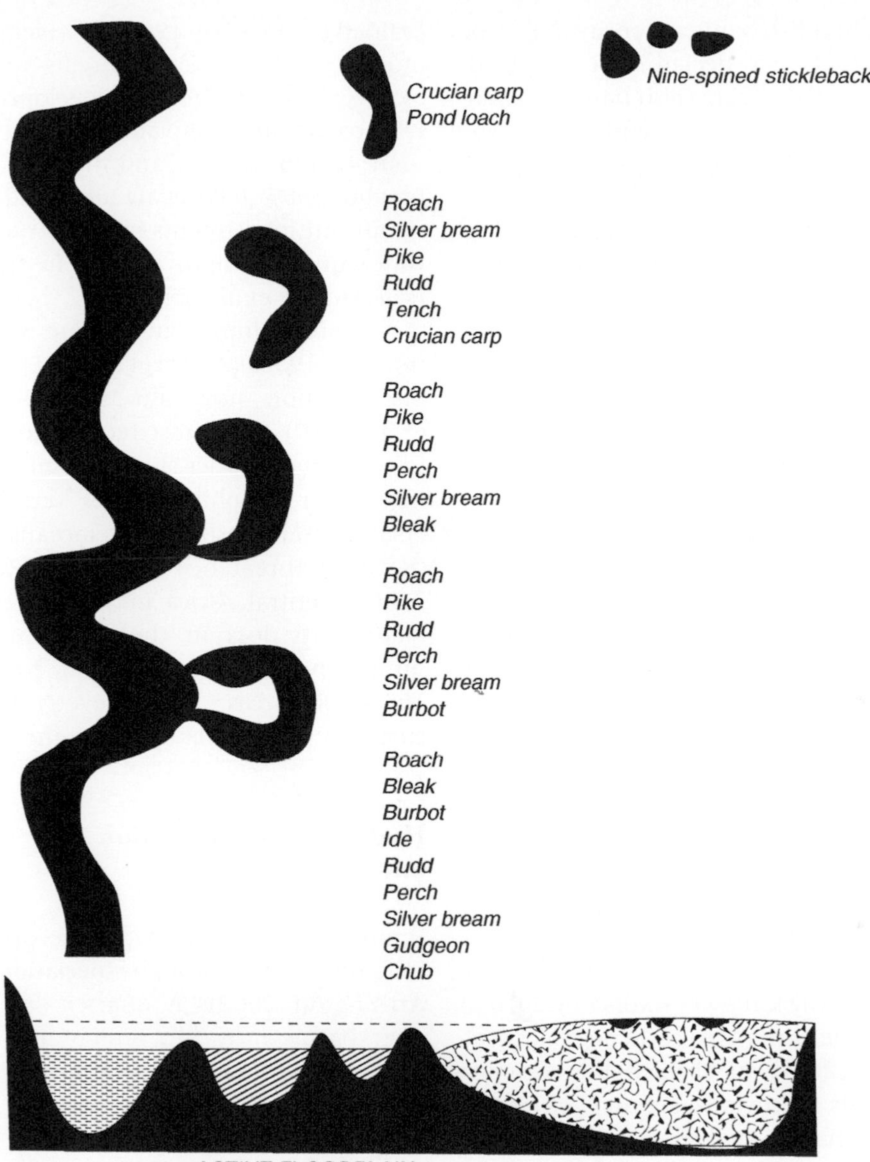

Figure 11.1 Fish species ordered by dominance in the Biebrza Floodplain according to water body type and to the frequency of reconnection to the main channel (redrawn from Witkowski A. and Wiśniewolski W. 2005); the box at the bottom represents a cross section of the floodplain highlighting (from left to right): the main river channel, 2 oxbow lakes that become reconnected during intermediate spates when water level in the main channel rises, and finally -to the far right- wetlands that are now disconnected from the main river, representing remnants of ancient channel migration.

side arms is still an important ecological mechanism supporting the co-existence of diverse arrays of taxa in the Biebrza, despite the lack of temperature coupling, as occurs in the tropical flood pulse. This also applies to many large river basins draining temperate and subarctic plains in Eurasia (Fashchevsky et al. 2008). Prolonged spring and summer floods favour the spawning of species (e.g. pike, ide, roach, rudd, tench and crucian carp) on flooded meadows that gives rise to the highest fish biomass density in Europe (3500 individuals per hectare). Beside floodplain specialists, the Biebrza also harbours rheophilous species such as nase and barbel. The lower Rhine and Meuse rivers are similar, with 70% of the basin species found exclusively in the floodplain lakes, demonstrating that even in temperate Europe, floodplain processes are essential to the maintenance of river biodiversity (van den Brink et al. 1996).

The current decline of freshwater quality and biodiversity

Inland aquatic ecosystems are rapidly deteriorating worldwide; they are the global ecosystem type that has undergone the most significant modifications (Dudgeon et al. 2000; Sala et al. 2000; Carpenter et al. 2011); 46% of freshwater mammals, 35% of amphibians, 38% of freshwater turtles, 32% of freshwater crabs and 36% of fishes that are not data deficient are considered threatened or extinct; in Africa 21% of known aquatic species are threatened compared to 12% of birds or 19% of mammals (Revenga et al. 2005; Strayer & Dudgeon 2010; Dudgeon 2011; Gray 2011; IUCN 2011a). Between 4% and 22% of global freshwater fish may be extinct by 2070 in about 30% of the world's rivers (Xenopoulos et al. 2005). WWF (2011) highlighted this alarming decline using its Freshwater Living Planet Index (LPI), a metric that tracks changes in 2750 populations of 714 species of fish, birds, reptiles, amphibians and mammals found in temperate and tropical freshwater ecosystems. The tropical freshwater LPI has declined by almost 70%, the largest fall among all components of the overall LPI.

The same causes of biodiversity decline in the terrestrial and marine environments also affect freshwaters: habitat change (particularly hydrological modification), overexploitation (fisheries), pollution, alien species and climate change. Habitat change is most pervasive; perhaps 50% of the world's wetlands – the freshwater ecosystem that was formerly most extensive – had been drained by the early 1990s (Barbier et al. 1997), mostly for agriculture. Threats to freshwater biodiversity are higher in developed countries; 28% of African freshwater fishes are threatened , but 40% in Europe and the USA (Kottelat & Freyhof 2007; Jelks et al. 2008) and this despite significant investments in infrastructure and in water quality improvements in these regions (Vörösmarty et al. 2010).

A global amphibian decline was reported from the 1950s (Alford et al. 2001) and a similar phenomenon now affects reptiles (Reading et al. 2010). The most recent large vertebrate to become extinct is the *baiji* or Yangtze freshwater dolphin *(Lipotes vexillifer;* Turvey et al. 2007); freshwater cetaceans are among the most threatened mammals (Reeves et al. 2000). Central Africa may experience greater biodiversity loss in the near future due to human population rise (2.5% for the period 1950–2005; UN 2010), even though it is currently under relatively low threat.

Hydrological alteration and fragmentation

Hydrological patterns were the primary cause of biodiversity through speciation (Bunn & Arthington 2002). A quarter of the planet's river basins now run temporarily dry before reaching the oceans due to abstractions (Molden et al. 2007); almost two-thirds of large rivers are fragmented by dams and 66% of global discharge is affected (Nilsson et al. 2005). The current number of large dams exceeds 50,000 (Berga et al. 2006). Over 472 million people (400 million in Asia) live below the 7000 largest dams, in 120 large rivers in 70 countries and along impacted river reaches (Richter et al. 2010). Worldwide, livelihoods of around 500 million people are degraded because dams 'flatten' seasonal flows that formerly supported high biodiversity and established patterns of fisheries, flood recession agriculture, dry-season grazing and other vital ecosystem services (Hamerlynck & Duvail 2003; Dudgeon 2010).

Some 2200 large dams are under construction and planned for the next decade in South America (Richter et al. 2010). Eight dams exist in the upper Mekong basin in China, but another 11 are planned downstream, which will cause an estimated reduction in flow by as much as 70%, to the extent that many threatened species will be prevented from reaching their upstream spawning sites (Gray 2011). In addition to these, 71 new dams are to be built on Mekong tributaries by 2030. Development schemes of this extent will undoubtedly drive vulnerable species to extinction and also significantly reduce fishery catches and protein availability for South East Asia's rural poor. Some 55% of what is now one of the world's last relatively untamed large rivers will be soon converted into a chain of slow-flowing reservoirs (Vaidyanathan 2011), seriously disrupting the basin's flood pulse.

Overexploitation

Meeting human freshwater needs is an increasing challenge due to rising population numbers which mean that agriculture continues to expand. Fish are intensively exploited as a protein source. Two million tons of fish and other aquatic animals are consumed annually in the lower Mekong basin alone, with 1.5 million tons originating from natural wetlands and 240,000 tons from reservoirs. The Tonle Sap fishery alone on the Mekong system provides 230,000 tons per year (ILEC 2005). Biodiversity is underpinning most of these ecosystem services in ways that it is not yet easy to quantify (Woodward 2009; Dudgeon 2010; Gessner et al. 2010). Overfishing is a major and widespread threat to fish community composition as larger individuals and species are preferentially selected and removed from the system, giving way to smaller, faster growing species which have shorter generation times but lower market appeal. This is a phenomenon that has been known from Africa (Welcomme 2003) and the Laurentian Great Lakes (Regier & Henderson 1973) for several decades.

Alien species

Alien species are an insidious threat to native aquatic species conservation because they have been in place, for so long, in so many places (Cambray 2003), and affect a wide range of species from zooplankton to mammals (Cucherousset & Olden 2011), leading in 77% of cases to loss of native species (Ross 1991). Well-known examples are the introduction of European salmonids by British authorities throughout their colonies (McDowall 2006), the introduction of Nile perch into Lake Victoria, and of black, silver and bighead carp into the Mississippi and the Laurentian Great Lakes (Stokstad 2010). The ecology of the Great Lakes is dominated by non-native species, following recent invasion by the Zebra mussel (*Dreissena*), the round goby and the sea lamprey (Carpenter et al. 2011); currently a complete isolation of the Great Lakes from the Mississippi basin is under consideration. Asian carp species are close to becoming cosmopolitan; their impact on standing waters reduces zooplankton biomass (as well as reducing competing zooplanktivorous native fish) and therefore hinders zooplankton-driven control of phytoplankton growth. In parts of the Mississippi, carp have caused a 90% decline in zooplankton and outcompeted the gizzard shad and the bigmouth buffalo (Stokstad 2010). Current control measures include the development of a selective fish poison and of an active pheromone.

Alien species in rivers may push native species into restricted ranges, often in isolated headwater stretches; in New Zealand, local galaxid fishes are only present in headwaters where they avoid predation by introduced European trout because these cannot pass steep waterfalls (McDowall 2006). In Japan, white-spotted charr (*Salvelinus leucomaenis*) preferred pools to riffles when rainbow trout were scarce but moved from pools to riffles upstream as trout density in the pools increased (Morita et al. 2004). In the western United States, isolating the upper reaches of a river is considered a promising, albeit short-term strategy for protecting cutthroat trout (*Oncorhynchus clarki*) from contact with non-native salmonids (*Salvelinus fontinalis*) (Hildebrand & Kershner 2000; Shepard et al. 2005). Isolation of small trout populations, however, risks genetic diversity erosion, possibly causing further extinctions; cutthroat trout have lost 99% of their original populations and 11 out of 13 subspecies became extinct during the 20th century.

Lake Naivasha, Kenya, is a classic example of the impact of aliens, both deliberate and accidental, upon a lake ecosystem (Gherardi et al. 2011). Complete changes in the ecosystem structure, with aliens the dominant species of the three main levels of the food web, have led to a highly simplified but unstable aquatic system of low diversity. Despite this, the ecosystem services of the lake – particularly provisioning (fishery, water supply), supporting (waste purification) and cultural (recreational, ecotourism) – are still considerable (Harper et al. 2011), albeit now limited by human overexploitation.

Pollution

Water quality deterioration is ubiquitous, not just where humans have settled, but contaminant chemicals such as pesticides and PCBs have been taken to every corner of the earth by winds, water currents and illegal transport of toxic waste, such that waters believed to be pristine contain traces (Bettinetti et al. 2011; Breivik et al. 2011). Human population increase and the intensification of agriculture – livestock rearing, intensive cropping, land reclamation – are causing widespread water pollution, salinization (Martin-Queller et al. 2010), water overabstraction (Gleick et al. 2009) and disruption of global biogeochemical cycles (Canfield et al. 2010; MacDonald et al. 2011). Global fertilizer use has increased by more than 500% over the last 50 years, mainly in China, India, the USA and Western Europe (Foley et al. 2011).

The overall state of freshwater pollution in the northern hemisphere has improved since the 19th century, when there was widespread industrial development; aggregate global indices of freshwater quality show an overall stable trend over the past 2–3 decades and even a marginal improvement in Asia since 1970 (Butchart et al. 2010). However, increased quantities of nitrogen and phosphorus over the same period have caused progressive eutrophication. The causes and consequences of increased nutrient fluxes in waters – human wastes providing most phosphorus, agricultural fertilizer most nitrogen – are well known, causing loss of species at all levels of aquatic food webs and dominance by cyanobacterial phytoplankton (lakes) or filamentous algae (rivers) (Harper 1992). Nitrogen and phosphorus also cause algal blooms in shallow coastal habitats, with anoxia following algal decay and severe impacts on coastal wetlands and marine habitats (dead zones; Rabalais et al. 2002; Diaz & Rosenberg 2008). Widespread nitrogen pollution also leads to formation of nitrous oxide, which contributes to climate forcing (Rockström *et al.* 2009).

Nutrient management still remains a scientific challenge (Larsen et al. 2007), despite phosphorus in surface water in Europe reduced by about 77% between 1985 to 2000 (EU 2003), by lowered phosphorus in detergents and technical advances in urban phosphorus removal. The problem of agriculture and small point sources from villages and dwellings makes almost every lowland watercourse in populated countries eutrophic from its source (Demars & Harper 2003). New pollution comes from the many synthetic and natural trace contaminants, such as synthetic oestrogens. Their toxic effects at local as well as global scale are likely to increase, particularly when present as mixtures (Schwarzenbach et al. 2010).

Pollution, like alien species introductions, reduces species range if it does not totally eliminate them. Salmon, for example, are released and survive in the upper Thames, but are unable to make a return after they move out to sea because of pollution of the lower river (Pope 2009). Pollution and hunting have confined the giant otter (*Pteronura brasiliensis*), naturally found in floodplain wetlands, to remote upstream localities in the Amazon, Orinoco and La Plata basins (Schenck & Staib 1998).

Climate change

The intensive use of water resources and the consequent desiccation of river basins contribute to climate forcing by enhancing greenhouse gas releases from drained soils and by reducing the cooling effect from evapotranspiration (Kravčík et al. 2007). Shifts in the timing of freshwater flows may be the most profound of the impacts of climate change on freshwaters (Le Quesne et al. 2010a), while sea level rise could significantly affect fresh as well as brackish coastal wetlands (Nicholls et al. 1999). Glacier melting in the Himalayas and subsequent changes in floods stress life histories adapted to the snowmelt pulse, lower agricultural productivity, reduce connectivity within floodplains and scour fish by flood peak increase: all are projected to become worse (Xu et al. 2009). Increasing droughts will combine with water overabstraction to increase average temperatures; this is likely to reduce oxygen content, affect biogeochemical cycles, alter microbial metabolism regulating nutrients, reduce productivity in deep lakes through stratification (e.g. Lake Tanganyika; O'Reilly et al. 2003), and consequently to affect fish communities and decrease carbon storage (Schindler 2001; Mohseni et al. 2003; Seitzinger et al. 2010; Carpenter et al. 2011; Stanley et al. 2012). Such impacts on the hydrological cycle will be among the most pervasive impacts of global climate change but deriving predictions of these changes at scales that are meaningful to decision makers remains a significant challenge.

How can societies protect what remains and restore what has been lost?

Freshwater presents one of the greatest conservation challenges, as demand for water will continue to grow at a rapid pace with increasing conflict between water users threatening the maintenance of healthy freshwater ecosystems. In addition, freshwater biodiversity is difficult to manage because of its inherent complexity; habitat templates have evolved to respond to a change in base conditions (e.g. light, nutrients, temperature, flow increase), but no model based on physical parameters can predict a precise species succession when conditions change (Reynolds & Elliott 2012). At the core of the freshwater conservation challenge is the need to respond to changes to water quality and quantity, as well as fragmentation and habitat conversion. This requires changes to the way in which core water management and planning functions are undertaken because of the social and economic importance of the services provided by freshwater systems, but it is done typically by ministries of water, energy and agriculture with little traditional interest in biodiversity conservation.

Managing water flow

An increasing understanding of the impacts of damming and diversions has led to water plans and policies that take account of 'environmental flows', defined by the Brisbane Declaration as: 'the quantity, timing, and quality of water flows required to sustain freshwater and estuarine ecosystems and the human livelihoods, and well-being that depend on these ecosystems' (Le Quesne et al. 2010b). A large number of indicators of hydrological alteration (IHA) were defined during the 1990s, based on 32 hydrological parameters quantifying impacts at various temporal and spatial scales (Richter et al. 1998). Hydrological regime preferences for single species or communities have been developed in a number of studies dealing with the impact of dams and referring to riparian vegetation (Jansson et al. 2000) or migratory fish (Gore & Mead 2008). Controlled artificial flood releases have been planned and operated to moderate dam impacts and to enhance conservation potential through floodplain restoration in the Senegal basin (Hamerlynck & Duvail 2003) and in the Phongolo basin in southern Africa (Acreman et al. 2000). As experience grew in the US, the Nature Conservancy, in co-operation with the Army

Corps of Engineers, issued a conceptual framework for modifying dam operations to restore river health and the survival of aquatic and riparian wildlife in the US (Richter & Thomas 2007), and now there are more than 850 successful river restoration projects (Nature Conservancy 2011). Environmental flows have gained broad recognition elsewhere (Le Quesne et al. 2010b). An environmental 'reserve' was included in the South African 1998 Water Act, and the recognition of limits to water diversion is part of the 2010 Chinese National Water Strategy. However, at the World Water Forum 2009 in Istanbul, the African Ministers' Council on Water (AMCOW) issued a Regional Paper calling for an unprecedented increase in infrastructural development to boost hydropower and water distribution services and to double Africa's irrigation potential by 2025. New infrastructure projects will continue to have significant implications for environmental flows.

Reinvestment in existing structures and in reoperation is the most promising option to increase returns provided by ecosystem services, including biodiversity, and to reduce the need to develop new sites (e.g. the Senegal; Hamerlynck & Duvail 2003). A significant number of large dams developed in Africa in the 1970s need to be restored and upgraded (Richter & Thomas 2007). A Hydropower Sustainability Assessment (Anonymous 2011) recognizes the need for hydropower projects to incorporate environmental flows into their design and operations but, as with environmental flow policies more broadly, the challenge now lies in implementation.

Basin planning and large-scale freshwater conservation planning

An area of increasing interest is the development of laws and planning policies to conserve freshwater systems on a larger scale. Margules & Pressey (2000) advocated the establishment of a complementary set of sites in every ecoregion to protect a representative portion of biodiversity at the broad geographical region scale. Complementarity partially replaces traditional conservation criteria such as naturalness, species richness and rarity due to its perceived higher efficiency and it is based on the gain in representativeness realized when a new site is added to the existing set of protected areas. An outright freshwater-centred conceptual framework proposed by Abell et al. (2007; Figure 11.2) consists of *freshwater focal areas* containing key conservation elements of interest and managed under restrictive use, complemented by *critical management zones*, i.e. portions of the basin located upstream where specific uses need to be carefully managed not to cause damage to the focal area, and finally *catchment management zones* to include the whole catchment upstream, ruled according to sound integrated river basin management principles. This proposal underlines the need to define systematic conservation strategies that are freshwater specific and that go beyond classic categories and definitions transferred from terrestrial conservation practices.

Nel et al. (2011) propose a conceptual framework specific to freshwater biodiversity conservation, consisting of 'persistence' as an additional relevant key criterion. Persistence principles include high ecological integrity, effective multi-dimensional connectivity to neighbouring aquatic ecosystems, thriving populations in core areas and consideration of any additional natural processes that may support effective conservation. These qualities greatly enhance the efficiency and applicability of implementation measures, and will help mainstream conservation action and enhance chances of success.

Alongside these scientific developments, there have been practical attempts to define larger-scale laws, policies and plans for water management that convert these principles into reality (Meng et al. 2011). Examples of laws that recognize some freshwater systems as having particular value are the EU Habitats Directive, the US National Wild and Scenic

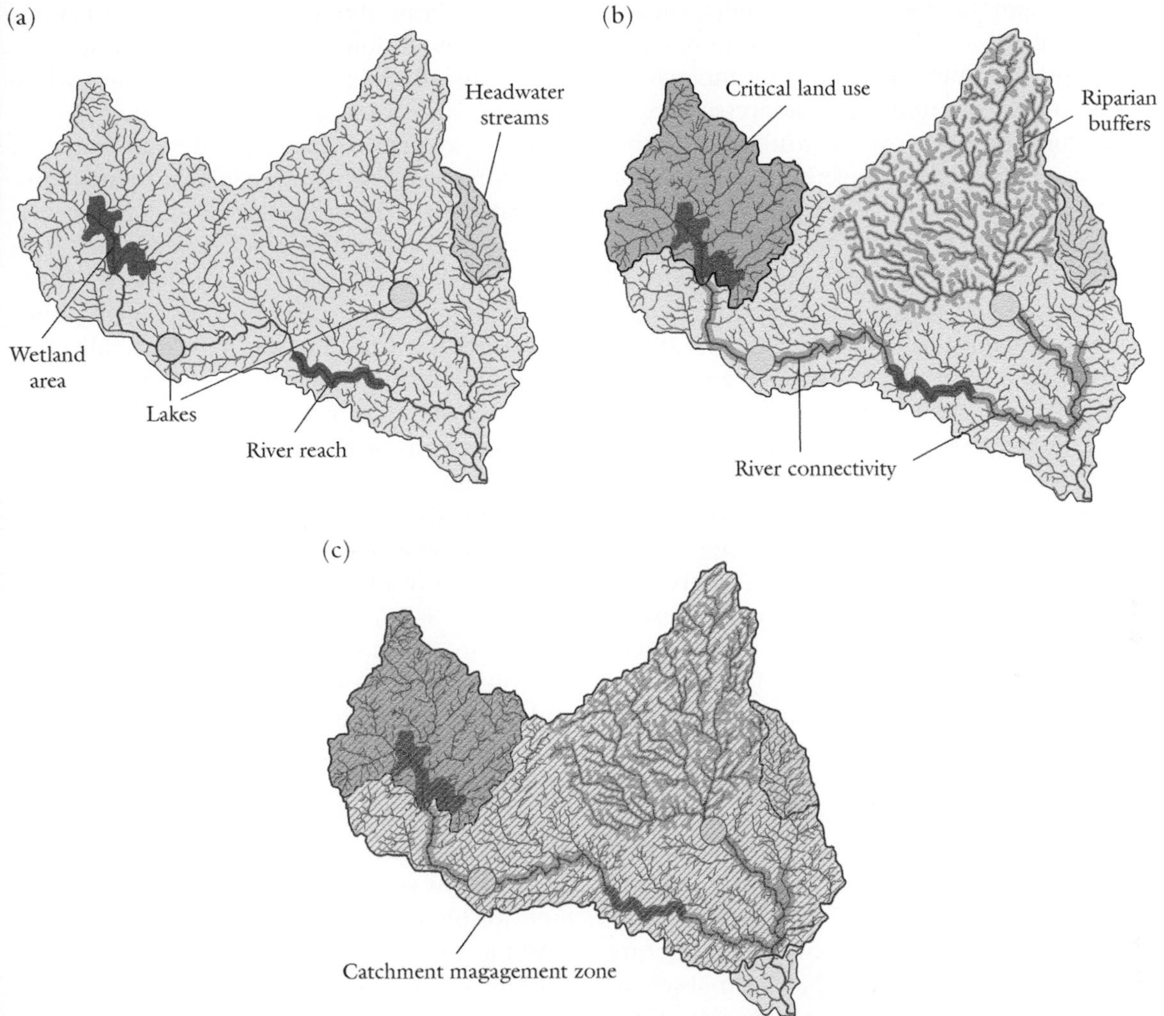

Figure 11.2 Freshwater protected area zones. (a) Freshwater focal areas, such as particular river reaches, lakes, headwater streams or wetlands supporting focal species, populations or communities. (b) Critical management zones, such as river reaches connecting key habitats or upstream riparian areas, whose integrity will be essential to the function of freshwater focal areas. (c) A catchment management zone, covering the entire catchment upstream of the most downstream freshwater focal area or critical management zone, and within which integrated catchment management principles would be applied. After Abell et al. (2007).

Rivers Act, South African Water Act and Kenya's modelled on South Africa's. The EU Water Framework Directive requires the restoration of 'good ecological status' in water bodies by 2015 (Art 4 §1a), requiring justification for any less stringent objectives; it remains to be seen whether implementation can meet the ambitious objectives set out in the policy.

The role of scientific research

There remain substantial obstacles to sound application of freshwater conservation knowledge. The incomplete understanding of the value of biodiversity to ecosystem functioning (Dudgeon 2010) and the inability to predict impacts of large dams and other major drivers on freshwater biodiversity loss under climate change (Pereira et al. 2010) pose practical challenges. Despite a significant recent increase in the number of publications addressing freshwater conservation, several authors feel that there has been little engagement in conservation practices (Strayer & Dudgeon 2010) and advocate the redirection of the more traditional 'science of discovery' into a 'science of engagement' that would be more responsive to society's language and needs (Rogers 2008). The existing gap between 'planning' and 'doing' calls for practicable approaches that should include analysis of costs, opportunities and constraints associated with proposed intervention measures, alongside ecological assessments. Further requirements include the application of spatially explicit and systematic conservation assessments, but with emphasis on implementation procedures and adaptive management rather than mere description of biodiversity patterns (Knight et al. 2008). Targeted monitoring should be implemented using diagnostic research methods, leading the way towards the enhancement of knowledge about the functioning of specific ecosystems and laying the basis of practical intervention measures (Gibbons et al. 2011).

The relationship between ecosystem structure and function in freshwaters has been observed in many case studies but never fully quantified. In the case of Lake Victoria, for example, the haplochromine cichlid collapse following the introduction of Nile perch (*Lates niloticus*) was accompanied by water quality degradation, but concomitant drivers such as human demographic increase and shoreline clearance prevented the establishment of a clear cause-and-effect linkage between biodiversity and habitat quality (Balirwa et al. 2003). Given the complexity of real-world situations, the linkage between biodiversity and ecosystem functioning has often been modelled through laboratory experiments. Harmon et al. (2009) showed that initially identical mesocosms diverged when three different stickleback predator species were introduced, all sharing a recent common ancestor (from about 10,000 years ago) but a different evolutionary history. Benthic specialists produced an increase in zooplankton while limnetic specialists produced an increase in benthic invertebrates. This demonstrated a clear linkage between fish community structure mediated by adaptive radiation and changes in ecosystem functioning. A recent experiment in 150 artificial streams showed a positive linear relationship between number of algal species and water purification capacity (Cardinale et al. 2011). In Germany, a network of 'biodiversity exploratories' has been established for long-term and large-scale multidisciplinary research on functional biodiversity research and ecosystem services, underlining a current trend in ecological research (Fischer et al. 2010). A recent review of biodiversity-ecosystem functioning (B-EF) research in freshwaters (Dudgeon 2010) highlighted challenges posed by issues of scale, of partial species redundancy, and an uneven understanding of key ecosystem processes. Some hydrological modelling (Poff et al. 2010) and understanding of detritivorous food chains (Gessner et al. 2010) have produced applicable tools, but B-EF research still has a long way to go before delivering practical tools for biodiversity conservation.

While the strategies outlined above address options for the preservation of single water bodies, several success stories have emerged from integrated conservation policies targeting broader objectives. A number of new approaches to this have been tried (Bayon & Jenkins 2010). Several governments (Mexico, USA, South Africa, China, Costa Rica) have utilized water surcharges to fund watershed protection by preserving natural forest (Stanton et al. 2010); similar payment schemes for preventing environmental degradation are voluntarily adopted by industries extracting water (Nestlé with its Vittel spring in France) as well as oil and minerals (Bayon & Jenkins 2010). In the USA, a mitigation scheme for preventing the overexploitation of inland aquatic resources requires developers to compensate for damage by restoring a wetland or a stream in another location that is comparable to the one affected by the business activity (Madsen et al. 2010). Further integration of ecological, social and economic skills is urgently required to conserve freshwater biodiversity.

Scientific disciplines dealing with the study of freshwater ecosystems need to become increasingly focused on advocacy and communication activities directed at the mobilization of civil society. Scientific discourse needs to be effectively directed at a wide audience, including policy makers and legislators, planners, land managers, water supply and sewage treatment managers. Standard mechanisms evaluating the quality of scientific research drive research activities towards innovative concepts and discoveries, by-passing the practical applicability of what is being proposed. Gibbons et al. (2011) stress that these criteria that make a study of high academic impact are not necessarily those that ensure impact on solving specific conservation problems. Few journals are specifically designed to communicate with managers and conservationists and vice versa. The market, as well as the current academic system, are not favourable to funding, training and recruiting applied ecologists to tackle pressing conservation problems. The role of the conservation scientist must evolve. Changes are needed to evaluate the effective impact of research (Gibbons et al. 2011) and successful practical research initiatives, such as the UK-based, but international, Ecosystem Services for Poverty Alleviation and the Darwin Initiative plus equivalent national programmes elsewhere, need considerable extension. Conservation societies and pressure groups are advocating for a 'green development' mechanism, analogous to the Kyoto Protocol's Clean Development Mechanism, to create scenarios that would be more favourable to the involvement of private investors in conservation action (James & Vorhles 2010).

Finally, recent evidence underlines that not only do 'biodiversity projects' yield conservation benefits but that 'ecosystem services projects' have a chance to deliver significant improvements in conservation status (Goldman et al. 2008). The most promising 'win–win' solutions can be developed by supporting multiple ecosystem service benefits for local people that motivate and reinforce the protection of diverse and sustainable habitats (Nelson et al. 2009; Nelson & Daily 2010). In Puget Sound, Washington (USA), long-term ecosystem restoration plans aim at six overarching goals: species and food webs, habitats, water quality, water quantity, human health and human well-being (Tallis et al. 2010). The sole option for a sustainable future is when progress towards these objectives proceeds in a parallel fashion.

"On a clear sunny day, the Potamogeton, flourishing at great depth amid the transparent waters, animated by numerous members of the insect and finny races, present a delightful spectacle, the long stems of the white and blue water lilies may be traced from their floating flowers to the root......One feels in such a place estranged for a time from the cares and vicissitudes of the world and the charms of nature penetrate, with their refining influences, the deepest recesses of the heart, denying to human language the power to give them full expression"

Forbes, W. (1848)
The Flora of Forfarshire.

References

Abell, R., Allan, J.D. & Lehner, B. (2007) Unlocking the potential of protected areas for freshwaters. *Biological Conservation*, **134**, 48–63.

Acreman, M.C., Farquharson, F.A.K., McCartney, M.P., *et al.* (2000) *Managed Flood Releases from Reservoirs: Issues and Guidance*. Report to DFID and the World Commission on Dams. Centre for Ecology and Hydrology, Wallingford.

Alford, R.A., Dixon, P.M. & Pechmann, J.H.K. (2001) Global amphibian population declines. *Nature*, **412**, 499–500.

Anonymous (2011) *Hydropower Sustainability Assessment Protocol*. The Hydropower Sustainability Assessment Forum. http://hydrosustainability.org/Document–Library.aspx

Arraut, E.M., Marmontel, M., Mantovani, J.E., Novo, E.M.L.M., Macdonald, D.W. & Kenward, R.E. (2010) The lesser of two evils: seasonal migrations of Amazonian manatees in the Western Amazon. *Journal of Zoology*, **280**, 247–256.

Balian, E.V., Segers, H., Lévêque, C. & Martens, K. (2008) The Freshwater Animal Diversity Assessment: an overview of the results. *Hydrobiologia*, **595**, 627–637.

Balian, E., Harrison, I., Barber-James, H., *et al.* (2010) A wealth of life: species diversity in freshwater systems. In: *Freshwater: The Essence of Life* (eds R.A. Mittermeier, T.A. Farrell, I.J. Harrison, A.J. Upgren & T.M. Brooks), pp. 53–89. CEMEX and ILCP, Arlington, VA.

Balirwa J.S., Chapman C.A., Chapman L.J., *et al.* (2003) Biodiversity and fishery sustainability in the Lake Victoria Basin: an unexpected marriage? *BioScience*, **53**(8), 703–715.

Barbier, E.B., Acreman, M.C. & Knowler, D. (1997). *Economic Valuation of Wetlands: A Guide for Policy Makers and Planners*. Ramsar Convention Bureau, Gland, Switzerland.

Bayon, R. & Jenkins, M. (2010) The business of biodiversity. *Nature*, **466**, 184–185.

Berga, L., Buil, J.M., Bofill, E., *et al.* (2006) *Dams and Reservoirs, Societies and Environment in the 21st Century*. Proceedings of the International Symposium on Dams in Societies of the 21st Century, Barcelona, Spain.

Bettinetti, R., Quadroni, S., Crosa, G., *et al.* (2011) A preliminary evaluation of the DDT contamination of sediments in Lakes Natron and Bogoria (Eastern Rift Valley, Africa). *Ambio*, **40**, 341–350.

Boyero, L., Pearson, R., Dudgeon, D., *et al.* (2011) Global distribution of a key trophic guild contrasts with common latitudinal diversity patterns. *Ecology*, **92**(9), 1839–1848.

Braatz, S., Davis, G., Shen, S. & Rees, C. (eds) (1992) *Conserving Biological Diversity: A Strategy for Protected Areas in the Asia–Pacific Region*. World Bank Technical Paper 193. World Bank Publications, Washington, D.C.

Breivik, K., Gioia, R., Chakraborty, P., Zhang, G. & Jones, K.C. (2011) Are reductions in industrial organic contaminants emissions in rich countries achieved partly by export of toxic wastes? *Environmental Science and Technology*, **45**, 9154–9160.

Bunn, S.E. & Arthington, A.H. (2002) Basic principles and ecological consequences of altered flow regimes for aquatic biodiversity. *Environmental Management*, **30**, 492–507.

Butchart S.H., Walpole M., Collen B., *et al.* (2010) Global biodiversity: indicators of recent declines. *Science*, **328**, 1164–1168.

Cambray, J.A. (2003) The global impact of alien trout species – a review with reference to their impact in South Africa. *African Journal of Aquatic Science*, **28**, 61–67.

Canfield, D.E., Glazer, A.N. & Falkowski, P.G. (2010) The evolution and future of earth's nitrogen cycle. *Science*, **330**, 192–196.

Cardinale, B.J., Matulich, K.L., Hooper, D.U., *et al.* (2011) The functional role of producer diversity in ecosystems. *American Journal of Botany*, **98**(10), 572–592.

Carpenter, S.R., Stanley, E.H. & Vander Zanden, M.J. (2011) State of the world's freshwater ecosystems: physical, chemical, and biological changes. *Annual Review of Environment and Resources*, **36**, 4.1–4.25.

Chapman, L.J. (2001) Fishes of the African rain forests. In: *African Rain Forest Ecology and Conservation* (eds W. Weber, L.J.T. White, A. Vedder & L. Naughton-Treves), pp. 263–290. Yale University, Yale, CT.

Chapman, L.J. & Chapman, C.A. (2003) Fishes of the African rain forests, emerging and potential threats to a little-known fauna. In: *Conservation, Ecology and Management of African Fresh Waters* (eds T.L. Crisman, L.J. Chapman, C.A. Chapman & L. Kaufman), pp. 176–209. University Press of Florida, Gainsville, FL.

Cucherousset, J. & Olden, J.D. (2011) Ecological impacts of non-native freshwater fishes. *Fisheries*, **36**, 215–230.

Demars, B.O.L. & Harper, D.M. (2003) *Assessment of the Impact of Nutrient Removal on Eutrophic Rivers*, Report No. P2–127/TR. Environment Agency, Bristol.

Diaz, R.J. & Rosenberg, R. (2008) Spreading dead zones and consequences for marine ecosystems. *Science*, **321**, 926–929.

Downing, J.A. (2009) Global limnology: up-scaling aquatic services and processes to planet Earth. *Verhandlungen der Internationale Vereinigung für Theoretische und Angewandte Limnologie*, **30**, 1149–1166.

Dudgeon, D. (2000) The ecology of tropical Asian rivers and streams in relation to biodiversity conservation. *Annual Review of Ecology & Systematics*, **31**, 239–263.

Dudgeon D. (2010) Prospects for sustaining freshwater biodiversity in the 21st century: linking ecosystem structure and function. *Current Opinion in Environmental Sustainability*, **2**(5–6), 422–430.

Dudgeon D. (2011) Asian river fishes in the Anthropocene: threats and conservation challenges in an era of rapid environmental change. *Freshwater Biology*, **79**, 1487–1524.

Dudgeon, D., Choowaew, S. & Ho S.-C. 2000. River conservation in South-East Asia. In: *Global Perspectives in River Conservation: Science, Policy and Practice*. (eds P.J. Boon, B.R. Davies & G.E. Petts), pp. 281–310. Wiley-Blackwell, Chichester. www.cabdirect.org/abstracts/20001915171.html;jsessionid=9EF6CE7AE7651597384079274 2F91639

EU (2003) *WRc Report on the Impact on the Environment (Reduction in Eutrophication) that would Result from Substituting Phosphates in Household Detergents*. Health and Consumer Protection Directorate-General, European Commission, Brussels.

Fashchevsky, B., Timchenko, V. & Oksiyuk, O. (2008) Ecohydrological management of impounded large rivers in the former Soviet Union. In: *Ecohydrology: Processes, Models and Case Studies* (eds D.M. Harper, M. Zalewski & N. Pacini), pp. 247–275. CABI International, Wallingford.

Fischer, M., Bossford, O., Gockel, S., *et al.* (2010) Implementing large-scale and long-term functional biodiversity research: the biodiversity exploratories. *Basic and Applied Ecology*, **11**(6), 473–485.

Fochetti, R. & de Figueroa, J.M.T. (2008) Global diversity of stoneflies (Plecoptera; Insecta) in freshwater. *Hydrobiologia*, **595**, 365–377.

Foley, J.A., Ramankutty, N., Brauman, K., *et al.* (2011) Solutions for a cultivated planet. *Nature*, **478**, 337–342.

Frost, D.R., Grant T., Faivovich, J., *et al.* (2006) The amphibia tree of life. *Bulletin of the American Museum of Natural History*, **297**, 1–370.

Gereta, E. & Wolanski, E. (2008) Ecohydrology driving a tropical savannah ecosystem. In: *Ecohydrology: Processes, Models and Case Studies* (eds D.M. Harper, M. Zalewski & N. Pacini), pp. 171–186. CABI International, Wallingford.

Gessner, M.O., Swan, C.M., Dang, C.K., *et al.* (2010) Diversity meets decomposition. *Trends in Ecology and Evolution*, **25**, 372–380.

Gherardi, F., Britton, J.R., Mavuti, K.M., *et al.* (2011) A review of allodiversity in Lake Naivasha, Kenya: developing conservation actions to protect East African lakes from alien species impacts. *Biological Conservation*, **144**(11), 2585–2596.

Gibbons, D.W., Wilson, J.D. & Green, R.E. (2011) Using conservation science to solve conservation problems. *Journal of Applied Ecology*, **48**, 505–508.

Gleick, P.H. (1996) Water resources. In: *Encyclopedia of Climate and Weather* (ed. S.H. Schneider), pp. 817–823. Oxford University Press, New York.

Gleick, P.H., Cooley, H. & Morikawa, M. (2009) The world's water 2008–2009. In: *The Biennial Report on Freshwater Resources* (eds P.H. Gleick, H. Cooley, M.J. Cohen, M. Morikawa, J. Morrison & M. Palaniappan), pp. 202–210. Island Press, Washington, D.C.

Goldman, R.L., Tallis, H., Kareiva P. & Daily, G.C. (2008) Field evidence that ecosystem service projects support biodiversity and diversify options. *Proceedings of the National Academy of Sciences of the USA*, **105**(27), 9445–9448.

Gore, J. & Mead, J. (2008) Ecohydrological modelling in lotic systems; benefits and risks. In: Harper, D.M., Zalewski M. and Pacini N. (2009) *Ecohydrology: Process, Models and Case Studies* (eds D.M. Harper, M. Zalewski & N. Pacini). CABI International, Wallingford.

Goulding, M. (1980) *The Fishes and the Forest*. University of California Press, Berkeley, CA.

Goulding, M. (1989) *Amazon: The Flooded Forest*. BBC Books, London.

Goulding, M., Carvalho, M.L. & Ferreira, E.G. (1988): *Rio Negro: Rich Life in Poor Water*. SPB Academic Publishing, The Hague.

Gray, R. (2011) Third of freshwater fish threatened with extinction. *The Telegraph*, 30th July 2011. www.telegraph.co.uk/earth/wildlife/8672417/Third-of-freshwater-fish-threatened-with-extinction.html

Hamerlynck, O. & Duvail, S. (2003) *The Rehabilitation of the Delta of the Senegal River in Mauritania*. IUCN, Gland, Switzerland.

Harmon, L.J., Matthews, B., Des Roches, S., Chase, J.M., Shurin, J.B. & Schluter, D. (2009) Evolutionary diversification in stickleback affects ecosystem functioning. *Nature*, **458**, 1167–1170.

Harper, D. (1992) *Eutrophication of Freshwater: Principles, Problems and Restoration*. Chapman and Hall, London.

Harper, D.M., Morrison, E.H.J., Macharia, M.M., Mavuti, K.M. & Upton, C. (2011) Lake Naivasha, Kenya: ecology, society and future. *Freshwater Reviews*, **4**, 89–114.

Henderson, P.A. & Walker, I. (1986) On the leaf-litter community of the Amazonian blackwater stream Tarumazinho. *Journal of Tropical Ecology*, **2**, 1–17.

Henderson, P.A. & Walker, I. (1990) Spatial organisation and population density of the fish community of the litter banks within a central Amazonian blackwater stream. *Journal of Fish Biology*, **37**, 401–411.

Henderson, P.A., Hamilton, W.D. & Crampton, W.G.R. (1998) Evolution and diversity in the Amazonian floodplain communities. In: *Dynamics of Tropical Communities* (eds, D.M. Newbury, H.H.T. Prins & N.D. Brown), pp. 385–419. British Ecological Society, Cambridge.

Hildebrand, R.E. & Kershner, J.L. (2000) Conserving inland cutthroat trout in small streams: how much stream is enough? *North American Journal of Fisheries Management*, **20**, 513–520.

ILEC (2005) *Managing Lakes and Their Basins for Sustainable Use: A Report for Lake Basin Managers and Stakeholders*. International Lake Environmental Committee Foundation, Kusatsu, Japan.

IUCN (2011a) *An Analysis of the Status and Distribution of Freshwater Species Throughout Mainland Africa* (eds W. Darwall, K. Smith, D. Allen, R. Holland, I. Harrison & E. Brooks). IUCN, Gland, Switzerland.

IUCN (2011b) *The Status and Distribution of Freshwater Biodiversity in Central Africa* (eds E.G.E. Brooks, D.J. Allen & W.R.T. Darwall). IUCN, Gland, Switzerland.

James, A.N. & Vorhles, F. (2010) Green development credits to foster global biodiversity. *Nature*, **465**, 869.

Jansson, R., Nilsson, C. & Renöfält, B. (2000) Fragmentation of riparian floras in rivers with multiple dams. *Ecology*, **81**(4), 899–903.

Jelks, H.L., Walsh, S.J., Burkhead, N.M., *et al.* (2008) Conservation status of imperilled North American freshwater and diadromous fishes. *Fisheries*, **33**, 372–407.

Junk, W.J. (2001) The flood-pulse concept of large rivers: learning from the tropics. *Verhandlungen der Internationale Vereinigung für Theoretische und Angewandte Limnologie*, **27**, 3950–3953.

Keller, I., Taverna, A. & Seehausen, O. (2011) Evidence of neutral and adaptive genetic divergence between European trout populations sampled along altitudinal gradients. *Molecular Ecology*, **20**, 1888–1904.

Knight, A.T., Cowling, R.M., Rouget, M., Balmford, A., Lombard, A.T. & Campbell, B.M. (2008) Knowing but not doing: selecting priority conservation areas and the research-implementation gap. *Conservation Biology*, **22**(3), 610–617.

Kottelat, M. (1989) Zoogeography of the fishes from Indochinese inland waters with an annotated checklist. *Bulletin of the Zoological Museum, University of Amsterdam*, **12**, 1–56.

Kottelat, M. & Freyhof, J. (2007) *Handbook of European Freshwater Fishes*. Publications Kottelat, Cornol, Switzerland.

Kravčík, M., Pokorný, J., Kohutiar, J., Kovác, M. & Tóth, E. (2007) *Water for the Recovery of the Climate – A New Water Paradigm*. Krupa Print, Žilina, Slovakia.

Larsen, T.A., Maurer, M., Udert, K.M. & Lienert, J. (2007) Nutrient cycles and resource management: implications for the choice of wastewater treatment technology. *Water Science and Technology*, **56**, 229–237.

Le Quesne, T., Matthews, J. & von der Heyden, C. (2010a) *Flowing Forward: Freshwater Ecosystem Adaptation to Climate Change in Water Resources Management and Biodiversity Conservation*, World Bank Water Working Notes 28. World Bank, Washington, D.C.

Le Quesne, T., Kendy, E. & Weston, D. (2010b) *The Implementation Challenge: Taking Stock of Government Policies to Protect and Restore Environmental Flows*. WWF-UK and The Nature Conservancy, Arlington, VA.

Lévêque, C. (1997) *Biodiversity Dynamics and Conservation: The Freshwater Fish of Tropical Africa*, Cambridge University Press, Cambridge.

Lévêque C., Oberdorff T., Paugy D., Stiassny M.L.J. & Tedesco, P.A. (2008) Global diversity of fish (Pisces) in freshwater. *Hydrobiologia*, **595**, 545–567.

MacDonald, G.K., Bennett, E.M., Potter, P.A. & Ramankutty, N. (2011) Agronomic phosphorus imbalances across the world's croplands. *Proceedings of the National Academy of Sciences USA*, **108**, 3086–3091.

Madsen, B., Carroll, N. & Moore Brands, K. (2010) *State of Biodiversity Markets Report: Offset and Compensation Programs Worldwide* (*Ecosystems Marketplace*). www.forest-trends.org/publication_details.php?publicationiD=2388

Mandrak, N.E. & Cudmore, B. (2011) The fall of native fishes and the rise of non-native fishes in the Great Lakes Basin. *Aquatic Ecosystem Health & Management*, **13**(3), 255–268.

Martín-Queller, E., Moreno-Mateos, D., Pedrocchi, C., Cervantes, J. & Martínez, G. (2010) Impacts of intensive agricultural irrigation and livestock farming on a semi-arid Mediterranean catchment. *Environmental Monitoring and Assessment*, **167**(1–4), 423–435.

Matthes, H. (1961). Feeding habits of some Central African freshwater fishes. *Nature*, **192**, 78–80.

Margules, C.R. & Pressey, R.L. (2000) Systematic conservation planning. *Nature*, **405**, 243–253.

McDowall, R.M. (2006) Crying wolf, crying foul, or crying shame: alien salmonids and a biodiversity crisis in the southern cool-temperate galaxioid fishes? *Reviews in Fish Biology and Fisheries*, **16**, 233–422.

Meng, J., Klauschen A., Antonelli F. & Thieme M. (2011) *Rivers for Life: The Case for Conservation Priorities in the Face of Water Infrastructure Development*. WWF Deutschland, Berlin.

Mittelbach, G.G., Schemske, D., Cornell, H., *et al.* (2007) Evolution and the latitudinal diversity gradient: speciation, extinction and biogeography. *Ecology Letters*, **10**, 315–331.

Mohseni, O., Stefan, H.G. & Eaton, J.G. (2003) Global warming and potential changes in fish habitat in U.S. streams. *Climate Change*, **59**, 389–409.

Molden, D., Frenken K., Barker R., *et al.* (2007) Trends in water and agricultural development. In: *Water for Food, Water for Life: A Comprehensive Assessment of Water Management in Agriculture* (ed. D. Molden), pp. 57–89. Earthscan, London.

Morita, K., Tsuboi, J. & Matsuda, H. (2004) The impact of exotic trout on native charr in a Japanese stream. *Journal of Applied Ecology*, **41**, 962–972.

Moss, B. (2010) Climate change, nutrient pollution and the bargain of Dr Faustus. *Freshwater Biology*, **55**(Suppl. 1), 175–187.

Nature Conservancy (2011) *Protecting Lakes and Rivers for People*. www.nature.org/ourinitiatives/habitats/riverslakes/index.htm

Nel, J.L., Reyers, B., Roux, D.J., Impson, N.D. & Cowling, R.M. (2011) Designing a conservation area network that supports the representation and persistence of freshwater biodiversity. *Freshwater Biology*, **56**, 106–124.

Nelson, E.J. & Daily, G.C. (2010) Modelling ecosystem services in terrestrial systems. *F1000 Biology Reports*, **2**, 53.

Nelson, E.J., Mendoza, G., Regetz, J., *et al.* (2009) Modeling multiple ecosystem services, biodiversity conservation, commodity production, and tradeoffs at landscape scales. *Frontiers in Ecology and the Environment*, **7**, 4–11.

Nicholls, R.J., Hoozemans, F.M.J. & Marchand, M. (1999) Increasing flood risk and wetland losses due to global sea-level rise: regional and global analyses. *Global Environmental Change*, **9**, S69–87.

Nilsson, C., Reidy, C.A., Dynesius, M. & Revenga, C. (2005) Fragmentation and flow regulation of the world's large river systems. *Science*, **308**, 405–408.

O'Reilly, C.M., Alin, S.R., Plisnier, P.D., Cohen, A.S. & McKee, B.A. (2003) Climate change decreases aquatic ecosystem productivity of Lake Tanganyika, Africa. *Nature*, **424**, 766–768.

Pacini, N. & Harper, D.M. (2008). Aquatic, semi-aquatic and riparian vertebrates. In: *Tropical Stream Ecology* (ed. D. Dudgeon), pp. 147–197. Academic Press/Elsevier, San Diego, CA.

Pereira, H.M., Leadley, P.W., Proença V., *et al.* (2010) Scenarios for global biodiversity in the 21st century. *Science*, **330**, 1496–1501.

Poff, N.L., Richter, B.D., Arthington, A.H., *et al.* (2010) The ecological limits of hydrologic alteration (ELOHA): a new framework for developing regional environmental flow standards. *Freshwater Biology*, **55**, 147–170.

Pope, F. (2009) Why salmon chose to return to the Seine but not the Thames. *The Times*, August 12.

Rabalais, N.N., Turner, R.E. & Wiseman, W.J. Jr. (2002) Gulf of Mexico hypoxia, a.k.a. "the dead zone". *Annual Review of Ecology and Systematics*, **33**, 235–263.

Reading, C.J., Luiselli, L.M., Akani, G.C., *et al.* (2010) Are snake populations in widespread decline? *Biology Letters*, **6**, 777–780.

Redfern, J.V., Grant, R., Biggs, H. & Getz, W.M. (2003) Surface-water constraints on herbivore foraging in the Kruger National Park, South Africa. *Ecology*, **84**, 2092–2107.

Reeves, R.R., Smith, B.D. & Toshio Kasuya, T. (2000) *Biology and Conservation of Freshwater Cetaceans in Asia*. International Union for Conservation of Nature and Natural Resources, Species Survival Commission, Gland, Switzerland.

Regier, H.A. & Henderson, H.F. (1973) Towards a broad ecological model of fish communities and fisheries. *Transactions of the American Fisheries Society*, **1**, 56–72.

Revenga, C., Campbell, I., Abell, R., de Villiers, P. & Bryer, M. (2005) Prospects for monitoring freshwater ecosystems towards the 2010 targets. *Philosophical Transactions of the Royal Society B: Biological Sciences*, **360**, 397–413.

Reynolds, C.S. & Elliott, J.A. (2012) Complexity and emergent properties in aquatic ecosystems: predictability of ecosystem responses. *Freshwater Biology*, **57**(s1), 74–90.

Richter, B.D. & Thomas, G.A. (2007) Restoring environmental flows by modifying dam operations. *Ecology and Society*, **12**(1), 12. www.ecologyandsociety.org/vol12/iss1/art12/

Richter, B.D., Baumgartner, J.V., Wigington, R. & Braun, D. (1998) A spatial assessment of hydrologic alteration within a river network. *Regulated Rivers*, **14**, 329–340.

Richter, B.D., Postel, S., Revenga, C., *et al.* (2010) Lost in development's shadow: the downstream human consequences of dams. *Water Alternatives*, **3**(2), 14–42.

Rockström, J., Steffen, W., Noone, K., *et al.* (2009) A safe operating space for humanity. *Nature*, **461**, 472–475.

Rogers, K.H. (2008) Limnology and the post-normal imperative: an African perspective. *Verhandlungen der Internationalen Vereinigung fuer theoretische und angewandte Limnologie*, **30**, 171–185.

Ross, S.T. (1991) Mechanisms structuring stream fish assemblages – are there lessons from introduced species? *Environmental Biology of Fishes*, **30**, 359–368.

Sala, O.E., Chapin, F.S., Armesto, J.J., *et al.* (2000). Global biodiversity scenarios for the year 2100. *Science*, **287**, 1770–1774.

Schenck, C. & Staib, E. (1998) Status, habitat use and conservation of the giant otter in Peru. In: *Behaviour and Ecology of Riparian Mammals* (eds N. Dunstone & M.L. Gorman), pp. 359–370. Symposia of the Zoological Society of London 71. Cambridge University Press, Cambridge.

Schindler, D.W. (2001) The cumulative effects of climate warming and other human stresses on Canadian freshwaters in the new millennium. *Canadian Journal of Fisheries and Aquatic Science*, **58**, 18–29.

Schmid, F. (1984) Essai d'évaluation de la faune mondiale des Trichoptères. In: *Proceedings of the Fourth International Symposium on Trichoptera*. Clemson University, South Carolina.

Schwarzenbach, R.P., Egli, T., Hofstetter, T.B., von Gunten, U. & Wehrli, B. (2010) Global water pollution and human health. *Annual Review of Environment and Resources*, **35**, 109–36.

Seitzinger, S.P, Mayorga, E., Bouwman, A.F., Kroeze, C. & Beusen, A.H.W. (2010) Global river nutrient export: a scenario analysis of past and future trends. *Global Biogeochemal Cycles*, **24**, GB0A08.

Shepard, B.B., May, B.E. & Urie, W. (2005) Status and conservation of westslope cutthroat trout within the western United States. *North American Journal of Fisheries Management*, **25**, 1426–1440.

Stanley, E.H., Powers, S.M., Lottig, N.R., Buffam, I. & Crawford, J.T. (2012) Contemporary changes in dissolved organic carbon in human-dominated rivers: is there a role for DOC management? *Freshwater Biology*, DOI:10.1111/j.1365-2427.2011.02613.x.

Stanton, T., Echavarria, M., Hamilton, K. & Ott, C. (2010) State of watershed payments: an emerging marketplace. Ecosystem Marketplace. www.foresttrends.org/publication_details.php?publicationiD=2438

Stiassny, M.L.J. (1999) The medium is the message: freshwater biodiversity in peril. In: *The Living Planet in Crisis: Biodiversity Science and Policy* (eds J. Cracraft & F. Griffo), pp. 53–71. Columbia University Press, New York.

Stokstad, E. (2010) Biologists rush to protect Great Lakes from onslaught of carp. *Science*, **327**, 932.

Strayer, D.L. & Dudgeon, D. (2010) Freshwater biodiversity conservation: recent progress and future challenges. *Journal of the North American Benthological Society*, **29**, 344–358.

Tallis, H., Levin, P.S., Ruckelshaus, M., *et al.* (2010) The many faces of ecosystem-based management: making the process work today in real places. *Marine Policy*, **34**, 340–348.

Turvey, S.T., Pitman, R. L., Taylor, B.L., *et al.* (2007) First human-caused extinction of a cetacean species. *Biology Letters*, **3**, 537–540.

Tweddle, D., Skelton, P.H., van der Waal, B.C.W., Bills, I.R., Chilala, A. & Lekoko, O.T. (2004) *Aquatic Biodiversity Surveys for the "Four Corners" Transboundary Natural Resources Management Area*. Report for the African Wildlife Foundation.

United Nations. (2010) *World Population Prospects: The 2008 Revision*. Population Division of the Department of Economic and Social Affairs of the United Nations Secretariat. http://esa.un.org/unpp

Vaidyanathan, G. (2011) Remaking the Mekong. *Nature*, **478**, 305–307.

Van den Brink, F.W.B., van der Velde, G., Buijse, A.D. & Klink, A.G. (1996) Biodiversity in the Lower Rhine and Meuse river-flood-plains: its significance for ecological river management. *Netherlands Journal of Aquatic Ecology*, **30**(2–3), 129–149.

Vié, J.C., Hilton-Taylor, C. & Stuart, S.N. (2008) *Wildlife in a Changing World: An Analysis of the 2008 IUCN Red List of Threatened Species*. IUCN, Gland, Switzerland.

Vörösmarty, C.J., McIntyre, P.B., Gessner, M.O., *et al.* (2010) Global threats to human water security and river biodiversity. *Nature*, **467**, 555–561.

Walker, I., Henderson, P.A. & Sterry, P. (1991) On the patterns of biomass transfer in the benthic fauna of an Amazonian black-water river, as evidenced by 32P label experiment. *Hydrobiologia*, **215**, 153–162.

Welcomme, R. (2003) River fisheries in Africa: their past, present, and future. In: *Conservation Ecology and Management of African Fresh Waters* (eds T.L Crisman, L.J. Chapman, C.A. Chapman & L. Kaufman), pp. 145–175. University of Florida, Gainesville, FL.

Western, D. (1975) Water availability and its influence on the structure and dynamics of a savannah large mammal community. *African Journal of Ecology*, **13**(3–4), 265–268.

Weyl, O.L.F., Ribbink, A.J. & Tweddle, D. (2010) Lake Malawi: fishes, fisheries, biodiversity, health and habitat. *Aquatic Ecosystem Health and Management*, **13**(3), 241–254.

Willig, M.R., Kaufman, D.M. & Stevens, R.D. (2003) Latitudinal gradients of biodiversity: pattern, process, scale, and synthesis. *Annual Review of Ecology and Systematics*, **34**, 273–309.

Winemiller, K.O., Agostinho, A.A. & Pellegrini Caramaschi, E. (2008) Fish ecology in tropical streams. In: *Tropical Stream Ecology* (ed. D. Dudgeon), pp. 107–146. Academic Press, San Diego, CA.

Witkowski, A. & Wiśniewolski, W. (2005) Ryby i minogi Bierbzy, jej starorzeczy i dopływów. In: *Przyroda Biebrzankiego Parku Narodowego* (eds A. Dyrcz & C. Werpachowski), pp. 247–255. Biebrza National Park editions, Osowiec Twierdza.

Witte, F., van Oijen, M.J.P. & Sibbing, F.A. (2009) Fish fauna of the Nile. In: *The Nile* (ed. H.J. Dumont), pp. 647–675. Monographiae Biologicae 89. Springer, New York.

Woodward, G. (2009) Biodiversity, ecosystem functioning and food webs in fresh waters: assembling the jigsaw puzzle. *Freshwater Biology*, **54**(10), 2171–2187.

WWF. (2011) *Living Planet Report 2010: Biodiversity, Biocapacity and Development*. WWF International, Gland, Switzerland.

Xenopoulos, M.A., Lodge, D.M., Alcamo, J., Märker, M., Schulze, K. & van Vuuren, D. P. (2005) Scenarios of freshwater fish extinctions from climate change and water withdrawal *Global Change Biology*, **11**, 1557.

Xu, J., Grumbine, R.E., Shrestha, A., *et al.* (2009) The melting Himalayas: cascading effects of climate change on water, biodiversity, and livelihoods. *Conservation Biology*, **23**(3), 520–530.

Zedler, J.B. & Kercher, S. (2005) Wetland resources: status, trends, ecosystem services, and restorability. *Annual Review of Environment and Resources*, **30**, 39–74.

Zwick, P. (2007) Neoperla (Plecoptera, Perlidae) emerging from a mountain stream in central Africa. *International Review of Hydrobiology*, **61**(5), 683–697.

12

Habitat case studies: islands

Carolyn King[1], Mark Lomolino[2], Gary Roemer[3] and Brendan Godley[4]

[1]Department of Biological Sciences, University of Waikato, Hamilton, New Zealand
[2]Department of Environmental and Forest Biology, SUNY College of Environmental Science and Forestry, Syracuse, New York, USA
[3]Department of Fish, Wildlife & Conservation Ecology, New Mexico State University, Las Cruces, NM, USA
[4]Centre for Ecology and Conservation, University of Exeter, Exeter, UK

Islands have earned the ironic distinction as the most fertile crucibles and most fatal pitfalls of evolution.
William Stolzenburg, *Rat Island* (2011)

Introduction

Many important issues in conservation biology are more clearly visible on islands than on continental mainlands. Islands are natural laboratories for demonstrating the dynamic nature of life and the consequences both of natural selection and of the human conquest of planet earth (Cook et al. 2006). In recent years, islands have also been places where large-scale experiments in the conservation of endemic species and the management of invasive species have led to huge practical advances in conservation biology. The literature on conservation on islands is enormous, and in this chapter we summarize a few of the key topics that concern island managers worldwide, with exemplary case studies from the islands shown in Figure 12.1.[1]

Conservation biogeography

The view that the patterns and principles of biogeography could serve as a 'cornerstone of conservation biology' is fundamental to the new discipline of conservation biogeography (Lomolino 2004, 2006; Whittaker et al. 2005; Richardson & Whittaker 2010). A brief summary of the key historical developments in island theory helps to illustrate the extent of its applications to endangered species conservation

[1] We use the term 'island' to mean an area of land completely surrounded by water. For discussions of marine habitats, see Chapter 9, and for mainland fragments, see Chapter 21. We consider primarily the land-breeding vertebrate faunas of islands, but only because there are fewer consistent and comparable databases available on plants and invertebrates.

Key Topics in Conservation Biology 2, First Edition. Edited by David W. Macdonald and Katherine J. Willis.

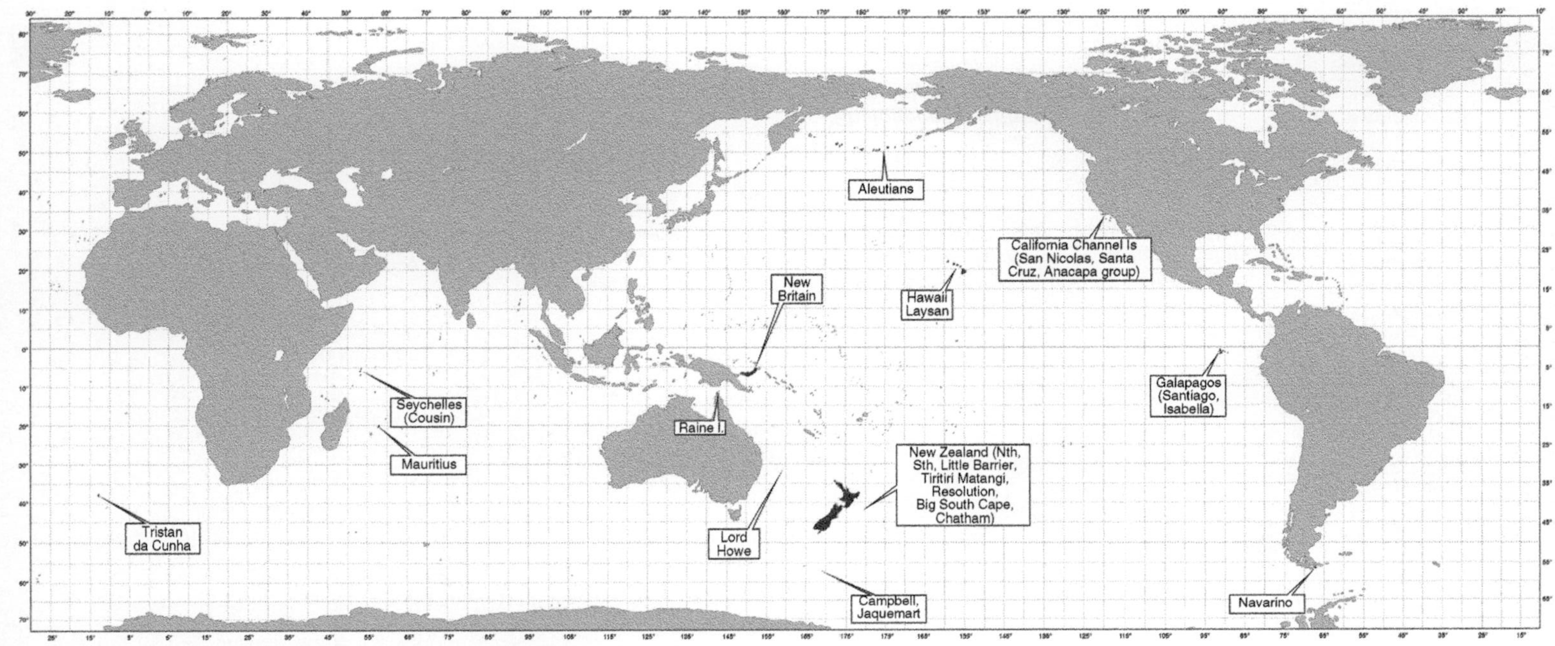

Figure 12.1 World distribution of islands mentioned in the text.

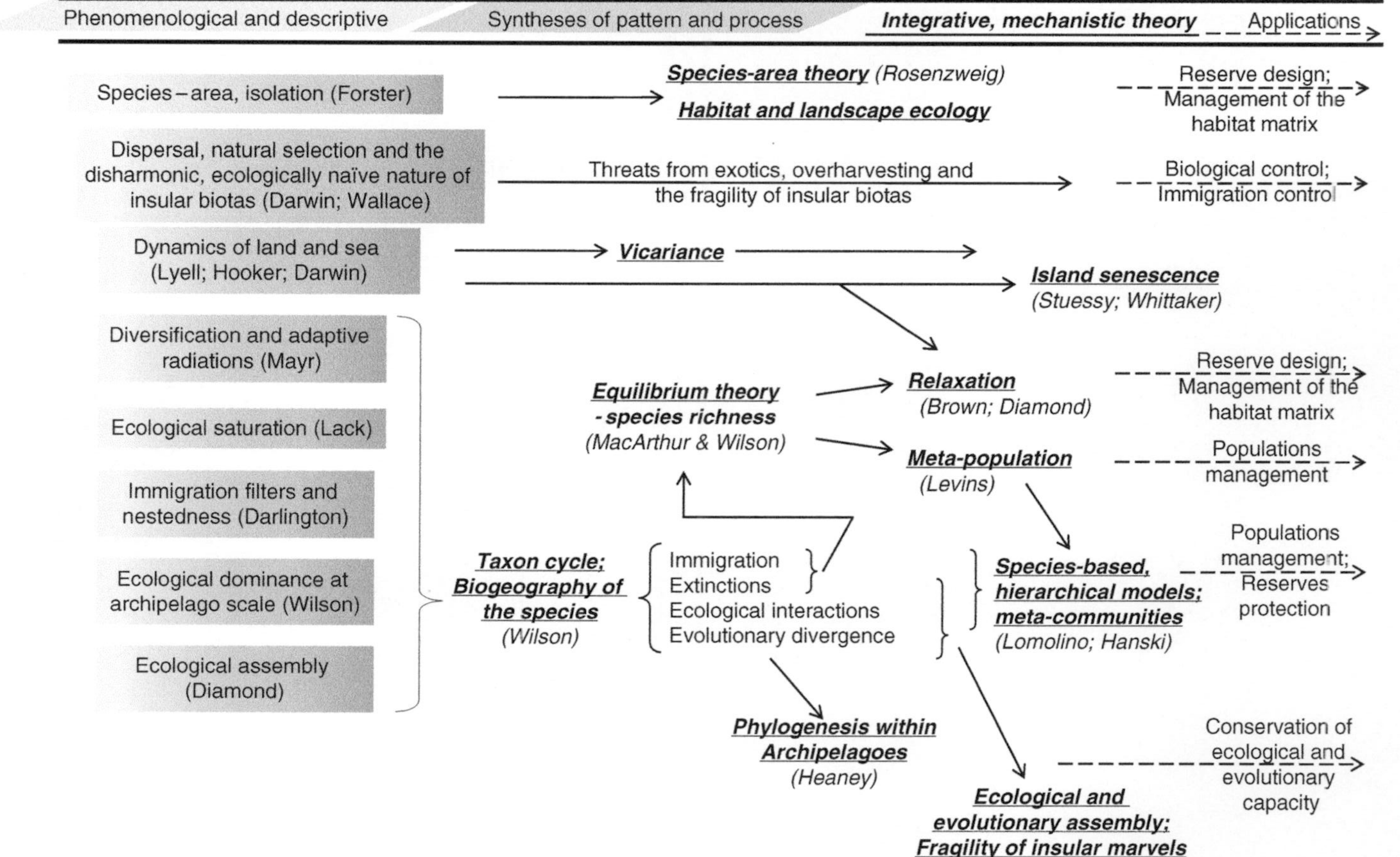

Figure 12.2 Summary of the historical development and influence of theoretical and applied island biogeography. The most fundamental contributions are listed on the left in rough chronological order, and suggest linkages from largely descriptive (light shaded boxes) or synthetic works on pattern and process (darker shading), to more integrative, mechanistic theories (bold, underlined italics) and their applications (dashed arrows). For summaries of the theories mentioned, see Box 12.1. Figure drawn by Mark Lomolino.

Box 12.1 Brief summary of some key theories and models of island biogeography

General definitions

Vicariance biogeography: a set of explanations for distributions of biotas that attribute disjunctions (broad gaps in species occurrences) to the splitting of areas (such as that caused by continental drift) rather than long-distance dispersal (Hooker 1866).

Taxon cycle theory: a comprehensive explanation of the predictable ecological and evolutionary changes in insular populations following their colonization of isolated islands (Wilson 1959, 1961).

Equilibrium theory of island biogeography: an explanation for patterns in species number among islands based on the assumptions that a relatively stable equilibrium is achieved between the processes of immigration of new species and extinctions among those already present (MacArthur & Wilson 1963, 1967).

Faunal relaxation model: explains exceptions to the equilibrium theory, which arise where immigrations and extinctions proceed on such different time scales that an equilibrium number of species is unlikely (e.g. for some biotas on land bridge islands following climate driven changes in sea level) (Brown 1971; Diamond 1972).

Model of dynamic disequilibrium and phylogenesis: an explanation for diversity of insular biotas based on the assertion that phylogenesis (evolutionary diversification) interacts in a complex manner with island area and isolation and, in turn, with extinction and immigration, and should therefore be modelled as a separate process driving biotic dynamics of insular biotas (Heaney 2000).

Demographic/ecological models

Metapopulation/metacommunity models: applications of genetic and demographic models for predicting distributions and persistence of species populations among patchily distributed habitats or true islands, based on the assumption that populations are linked by dispersal among patches (islands) (Hanski 2005, 2010).

Species-based, hierarchical model: an explanation for patterns in species distributions, diversity and species composition among islands based on the assertion that these patterns result from predictable patterns of variation in immigration abilities, resource requirements and interspecific interactions among species (Lomolino 2000).

Island senescence models

A model of ontogeny of insular floras: an explanation for the sequence of predictable changes in species diversity and genetic diversity based on how anagenesis and cladogenesis should vary with characteristics of the species, heterogeneity of habitats and age of the islands (Stuessy 2007).

Geological (general) dynamic model: a comprehensive explanation for changes in species diversity on particular islands as geological processes generate changes in their area, elevation and carrying capacity and, in turn, changes in the fundamental processes of immigration, extinction and speciation (Whittaker et al. 2008).

(Figure 12.2). We emphasize that any distinctions we make between empirical, conceptual and applied approaches are arbitrary, because even the earliest descriptive accounts also included comparisons with similar patterns, and speculations on the underlying causes. As Poincaré (1952) pointed out, 'Science is built upon facts much in the same way that a house is built with bricks; but the mere collection of facts is no more a science than a pile of bricks is a house'. None of the biogeographers we feature here, or those who contributed to the development of biogeography and its descendant disciplines, were simply dutiful collectors and cataloguers of facts.

History and development

The earliest foundations of island biogeography were laid by Johann Reinhold Forster, a naturalist on Cook's HMS *Resolution*, 1772–1775. He described two of the most general patterns in nature – that isolated systems have fewer species than those on the mainland, and that diversity on islands increases with their size (Forster 1778). His explanations for these patterns – the tendency for larger islands to provide more resources and a greater diversity of habitats – established lines of research that would eventually lead to species-area theory, habitat ecology and management, landscape ecology, and the theory of the assembly of biological communities and regional biotas (Box 12.1; see also Figure 12.2).

Darwin and Wallace both pointed to the remarkable products of evolution in isolation, including birds that had lost the power of flight, and to these and countless other species that had become ecologically naive in the absence of mainland competitors and predators (Darwin 1839, 1860; Wallace, 1876). Lyell and Hooker recognized the similarity of plant assemblages in regions now isolated by southern hemisphere oceans (those now known to have formerly comprised the ancient supercontinent of Gondwanaland). Rejecting long-distance dispersal over water, they invoked geological dynamics – the emergence and subsequent submergence of extensive land bridges that once connected isolated continents and oceanic islands (Lyell 1834; Hooker 1866). While their proposed mechanism – dispersal across now submerged oceanic land bridges – was incorrect, their dynamic view of ancient life inspired later models of historical biogeography including vicariance, island senescence and the relaxation of insular biotas (Brown 1971; Diamond 1972). The same dynamics can be applied to predict the declines in biological communities, as previously expansive and continuous systems are reduced and fragmented ('insularized') in the face of environmental changes (see Chapter 21).

These observations and early theories nurtured the insightful and synthetic contributions of the early to middle 20th century by Ernst Mayr (1942, 1954), David Lack (1970), Philip Darlington (1957), Edward O. Wilson (1959, 1961) and Jared M. Diamond (1975a). These authors shared the view that diversification and assembly of insular biotas were strongly influenced by ecological interactions and differences among species in niches, ecological and competitive dominance, and the dispersal capacities of potential colonists. The taxon cycle theory of E.O. Wilson (1959, 1961) telescoped the theories of invasions and ecological dominance among continental biotas to invasions and ecological shifts of particular lineages along 'radiation zones' in archipelagos, and then down to within-island scales (Lomolino & Brown 2009). Wilson's idea of a taxon cycle uses each of the fundamental processes influencing assembly of insular biotas – immigration, extinction, evolution and ecological interactions – to produce a theory hypothesizing that isolated archipelagos were initially depauperate and disharmonic (unbalanced, or lacking taxa that typically dominate mainland communities, e.g., mammalian carnivores), then were colonized by a small subset of species that soon became entrained in a cycle of increased ecological and evolutionary specialization, which ultimately ended in extinctions of their descendant lineages. These extinctions were driven by continued colonizations and ecological divergence of new lineages as they or their descendants eventually out-competed more ancient and now overspecialized species. Wilson's concepts of the ecological dominance of invaders and their threat to native, insular biotas shared much with Darwin and Wallace's earlier observations on the fragility of island life, and foreshadowed the development and continued intensification of programmes to eradicate exotic, insular populations of invasive pests (discussed below).

Wilson's subsequent collaboration with the mathematical ecologist Robert MacArthur reversed the trend toward increasingly more integrative theories in favour of those that put a premium on generalization by simplicity. MacArthur & Wilson's (1963, 1967) equilibrium theory of island biogeography, while retaining the taxon cycle's assumption of recurrent immigration and extinctions, succeeded in explaining general species-isolation and species-area relationships without referring to ecological interactions, differences among species or evolution. Their equilibrium model saw the diversity of insular communities as a balance between recurrent immigrations and extinctions, which were direct functions of island isolation and area, respectively. Community ecologists and conservation biologists applied MacArthur & Wilson's model and, in particular, its assumption of an equilibrium between immigrations and extinctions, to develop mathematical and graphical models for predicting faunal collapse and the prospects of preservation of biotic diversity under alternative scenarios of habitat loss and fragmentation (Harris 1984; Rosenzweig 1995). Best known and most controversial among these applications were a set of alternative, geometric designs for nature reserves (Diamond 1975b; Wilson & Willis 1975). Debates over these alternative prescriptions were often contentious (see Chapter 21) but had little positive and lasting effect on conservation biology.

Recent advances

Ironically, the most enduring and significant works influencing the equilibrium theory on island biogeography and the conservation of isolated communities were those that attempted to reintegrate fundamental processes distilled out of Wilson's taxon cycle theory during the conceptual development of the equilibrium model. These included the models that added non-equilibrial conditions generated by long-term dynamics of landscapes and seascapes, e.g. the relaxation models of Brown (1971) and Diamond (1972); the island senescence models of Stuessy (2007), Whittaker and his colleagues (Whittaker et al. 2008), and other models that emphasized demographic, genetic and ecological differences, and interactions among species, e.g. metapopulation models and species-based and metacommunity models, including those of Lomolino (2000) and Hanski and his colleagues (Hanski and Gilpin, 1997; Hanski, 2005), all summarized in Box 12.1.

Still lacking from these more recent models, but we think essential for understanding the natural character of insular biotas, is a reintegration of evolutionary processes in island theory. Evolutionary divergence has been at the core of historical biogeography at least since Hooker's anticipation of vicariance biogeography, and Darwin's dispersalist and centre of origin theories. Insights from historical biogeography, however, had less influence on the early development of conservation biology than did practical experience in wildlife management (Caughley 1994).

Exemplary case studies in conservation phylogeography such as those by Heaney (2000, 2004, 2007) have demonstrated the importance of phylogenesis to both conceptual and applied island biogeography. If we are to conserve the natural character of insular biotas, it seems essential that conservation programmes be guided by empirical patterns and their underlying processes, i.e. the influence of ecological interactions (among natives as well as exotics) on the distributions, abundances, niches and evolution of native species. The marvels and perils of life on isolated, initially depauperate oceanic islands are the products of diversification, adaptive radiations and evolutionary interactions within a small subset of the mainland biota. The absence or paucity of most species that usually dominate mainland communities, and the relatively intense competition from conspecifics and from an unbalanced assemblage of other long-distance colonists drive character release and displacement and, in turn, convergence on the size, growth form, and trophic and metabolic strategies of absent species. The ultimate consequence is the evolution of insular anomalies including giant shrews and rodents, dwarfed proboscideans and ungulates, flightless and often giant birds, lizards that become 'supergeneralists' as pollinators and seed dispersers, and the evolution of woodiness and tree-like stature in otherwise herbaceous lineages of plants, all described in recent books (Whittaker & Fernandez-Palacios 2007; Lomolino et al. 2010). Different rules apply to the equally fascinating assembly and ecology of marine species using islands as breeding platforms (Box 12.2).

Conservation management on islands

Caughley (1994) proposed a fundamental distinction between two main threads of conservation biology, distinguished by, among other things, their origins, their theoretical significance

Box 12.2 Significance of islands to marine vertebrates

Islands, often naturally lacking terrestrial predators, provide many significant breeding sites for air-breathing marine vertebrates, e.g. pinnipeds, marine turtles and seabirds. The biodiversity importance of these breeding aggregations is threefold. First, single breeding colonies can support great hosts of animals gathered from across vast areas of ocean, e.g. Raine Island, Australia, may host >100,000 breeding green turtles (*Chelonia mydas*), possibly as many as live in the whole Atlantic (Limpus et al. 2003). Second, isolation of breeding sites has lead to important endemism even among wide-ranging species, e.g. the Tristan da Cunha archipelago hosts at least four endemic species of seabirds (the Tristan albatross *Diomedea dabbenena*, yellow-nosed albatross *Thalassarche chlororhynchus*, sooty albatross *Phoebetria fusca* and great shearwater *Puffinus gravis*), all considered globally threatened (Hilton & Cuthbert 2010). Third, the large numbers of marine vertebrates can play important ecosystem engineering roles at their breeding sites, e.g. seabirds acting as conduits of nutrients from marine to terrestrial ecosystems (Polis & Hurd 1996) or affecting soil dynamics via burrowing (Mulder & Keall 2001).

Concentration and isolation define some of the major conservation concerns for these groups. The seasonally predictable distribution of numerous large vertebrates has permitted high levels of exploitation over the last 500 years. Key examples of overexploitation include sealing to near-extinction of fur seals in the southern oceans (Richards 2003), and overfishing extensive enough to cause significant shifts in food webs (Jackson et al. 2001). More than 90% of recent avian extinctions have been of island species, mainly due to predation by invasive mammals (Atkinson 1985; Steadman 1995). Seabirds are especially vulnerable (Hilton & Cuthbert 2010) so invasive species eradication efforts on islands (detailed below) are often undertaken, at least partly, for seabird restoration. Predicted climate change has the potential to affect breeding marine vertebrates in various ways, including loss of key coastal habitats or whole breeding islands through sea level rise (Baker et al. 2006) or through disruption of thermally determined sex ratios (Hawkes et al. 2009). A key consideration is whether animals breeding on islands have the potential to adapt by shifting to nearby, more suitable habitats.

and their technical methods. The *small-population paradigm* is concerned with the effects of small numbers on the chances of extinction, whereas the *declining-population paradigm* is concerned with identifying the reasons why a population becomes small. The dangers of being a small population are intrinsic, whereas for declining populations the focus is less on the distressed population than on the agents of decline. Both are relevant to the theme of protecting island biota, in combination with a special branch of theoretical biology focused specifically on islands.

Both paradigms assume that the main aim of conservation biology is to prevent extinctions, but they have developed more or less independently. Ideas about declining populations are based on decades of practical wildlife management experience. If the causes of a particular decline can be identified in time, experienced field managers can often mitigate or reverse them without reference to, or in the absence of, any useful theory. Ideas about small populations, such as risk of population extinction, population viability and extinction vortices, have their origins in both demographic and population genetic theory, and are often helpful for predicting which populations are or could be especially vulnerable to less manageable stochastic processes. The best rescue remedies, for example the one developed for the island fox (*Urocyon littoralis*) (Bakker et al. 2009), use a combination of both but with an element of defiance that refuses to give up on cases that should in theory be hopeless. When the Mauritius kestrel (*Falco punctatus*) got down to four individuals in 1974, the ICBP was ready to give up on it but a determined captive breeding programme led by Carl Jones succeeded so well that the species' 'Endangered' label was removed in 1994 (Quammen 1996).

Declining populations

The declining- and small-population paradigms are not independent, since a declining population often becomes a small population, and then a locally extinct one, but declining populations may have more options than small ones. For example, the decline of the endemic Galapagos rail (*Laterallus spilonotus*) on Santiago was rapidly reversed by eradication of feral goats and pigs (Donlan et al. 2007). By contrast, the last 122 kakapo (*Strigops habroptilus*), a large, slow-breeding parrot endemic to New Zealand (Fidler et al. 2008), are vulnerable to environmental stochasticity, genetic drift and inbreeding depression, which are less easy to manage.

Diamond (1989) identified four main causes of decline to extinction, all of which are often illustrated on islands: overkill, habitat destruction and fragmentation, impacts of introduced species, and chains of extinction.

Overkill rapidly removed many large, slow-breeding species from remote islands as soon as human explorers, traders and colonists found them. In early postglacial times, nine species of moa (Aves: Dinornithiformes) inhabited the two main islands of New Zealand, which totals about the same area as Britain (Bunce et al. 2009); all were lost after Polynesian colonists arrived in about 1280 AD (Wilmshurst et al. 2008), probably within a century (Holdaway & Jacomb 2000). Similarly, 36 species of giant tortoise have been harvested to extinction on islands around the globe since the Pleistocene (Hansen et al. 2010).

Habitat destruction can range from 100% loss of lowland coastal areas, caused by the rapid post-Pleistocene rise in sea level, to the widespread contemporary logging of tropical forests on islands supporting many endemic species. Diamond (1984) showed that the species most affected by island drowning were those on the smallest islands, those with the largest body size, and those with the most carnivorous or specialist food habits – all characteristics associated with small population size. On New Britain, the largest island of the Bismarck Archipelago (Papua New Guinea), 12% of the original forest was lost in only 11 years (1989–2000), and continued losses are now jeopardizing 21 species of threatened or near-threatened forest birds (Buchanan et al. 2008).

Invasive species often prey on naive native faunas (Hilton & Cuthbert 2010), but do not have to be predatory to be disastrous (Macdonald et al. 2007). Exotic herbivores can reach huge numbers capable of replacing an original forest with grassland, making extinctions of resident forest-adapted endemics inevitable. For example, on Laysan Island (370 ha) in Hawaii, European rabbits (*Oryctolagus cuniculus*) introduced before 1904 destroyed virtually all vegetation by 1923, and reduced five endemic taxa of birds to remnants, from which three never recovered (Caughley & Gunn 1996).

Chains of extinction are observed when a change in status of one species makes life impossible for a different species. Rabbits can kill island populations of seabirds by supporting large populations of feral cats (Courchamp et al. 2000). Abundant exotic prey, feral pigs (*Sus scrofa*), enabled native golden eagles (*Aquila chrysaetos*) to colonize the California Channel islands, where their predation indirectly threatened an endemic island fox (Box 12.3).

Small populations

The concept of a minimum viable population rests on two separate but linked ideas, one derived from genetics and the other from population dynamics, both agreeing that small populations are always at much greater risk of extinction than larger ones. A widely quoted general rule for animals predicts that at least 50 *breeding* individuals are needed to stave off inbreeding depression, and 500 for long-term security against extinction by genetic drift (Franklin 1980). Total populations need to be larger. Inbreeding coefficients estimated for 182 non-endemic and 28 endemic island populations confirmed that breeding groups on islands are usually more inbred than their mainland relatives (Frankham 2005). In small populations, mating between close relatives is frequent, which causes H, the proportion of heterozygous loci in an average individual, to drop to dangerous levels.

Box 12.3 Recent innovations in island restoration

Methodological advances in island restoration, including unique financial instruments, strategic plans and the application of existing and new technologies, have allowed successful removal of entire suites of invasive vertebrates from larger and larger islands. Many of these approaches, first pioneered in New Zealand (Bellingham et al. 2010), are exemplified by recent eradications of invasive mammals in South and North America, as follows.

Project Isabella

Project Isabella is perhaps the most comprehensive invasive mammal eradication programme conducted to date. The goal was to remove invasive ungulates from Santiago Island (58,465 ha), plus the northern portion of Isabella Island (Galapagos). First, >18,000 pigs were removed from Santiago over a 30-year period. At this stage, the goats were ignored, because their intense browsing kept habitats open, facilitating pig removal (Cruz et al. 2005, 2009). Completion of this stage of the plan was followed by the development of a comprehensive strategy for a large-scale, rapid goat eradication campaign, financed by donations from numerous sources totalling US$6.1 million, administered by the Global Environment Facility to the Ecuadorian government. A systematic design used a mixture of methods (ground hunting with and without hunting dogs, and aerial hunting from helicopters), and the efficiency of each method was constantly assessed. The island was divided into hunting zones that varied in size, depending on the removal method (Lavoie et al. 2007) (Figure 12.3). GPS and GIS tools allowed close daily monitoring and confirmation of each stage of the work.

Importantly, this adaptive management strategy took into account the behavioural reactions of both the target animals and the human operators. First, animal behaviour was exploited by the use of the Judas goat, a sterilized individual (male or female) fitted with a radio collar which when released will always search for a herd to join. It is then relocated, every goat not collared is killed and the Judas left to find new companions. Later refinements of this method included the creation of the even more lethal Mata Hari females, 'sexed up' with reproductive hormones to prolong oestrus (Campbell et al. 2007). Mata Hari goats were particularly adept at leading hunters to males, resulting in a shortage of breeding males that reduced female reproductive output. This effect was critical during the crucial latter part of the eradication effort, when density-dependent increases in production could compensate for hunter-induced mortality.

Second, the influence of human behaviour on eradication efficiency was minimized by a keen understanding of the operators' response to one method of ground hunting, the *rastrillo*. Akin to a deer hunting drive, *rastrillo* involves placing hunters 100–150 m apart in a line and systematically covering a hunting block. This method was initially very effective, but prolonged use caused hunters to become disillusioned and less keen to succeed. A switch back to free-range hunting instilled a competitive air among the hunters that then increased efficiency (Cruz et al. 2009).

The net result was that >79,000 goats were removed in less than 4.5 years (Cruz et al. 2009). This campaign was the largest and most 'savvy' approach ever devised to eradicate large ungulates from an archipelago of enormous biological and social importance.

California Channel Islands

On Santa Cruz Island (25,000 ha), the largest of the Channel Islands off the coast of southern California, USA, foraging and soil disturbance by pigs directly endangered native plants, and indirectly endangered the endemic island fox (*Urocyon littoralis*) by attracting a shared predator, the golden eagle (*Aquila chrysaetos*). Eagle predation had a disproportionate effect on the foxes, driving them toward extinction (Roemer et al. 2002). By the deliberate and sequential deployment of different methods of accelerating intensity, 5036 pigs were removed in ~14 months (Parkes et al. 2010). Innovations in this and related programmes included sampling designed to detect when the last animal was removed, and modelling that yielded a framework linking population viability analysis with risk assessment to allow flexible management tactics (Morrison et al. 2007; Bakker et al. 2009).

Removing invasive rodents is 2–3 times more expensive per unit area covered than removing large ungulates (Martins et al. 2006). So another first for the California Channel Islands was the eradication of black rats (*Rattus rattus*) from Anacapa Island (actually a group of three small islands) despite the presence of an endemic rodent species (Howald et al. 2009). The chosen tool was aerial spreading of Brodifacoum, an anticoagulant rodenticide that would kill all rodents and probably cause secondary poisoning of raptors and other species. The operation proceeded in three phases: (1) conservation practitioners first captured and confined peregrine falcons

(Continued next page)

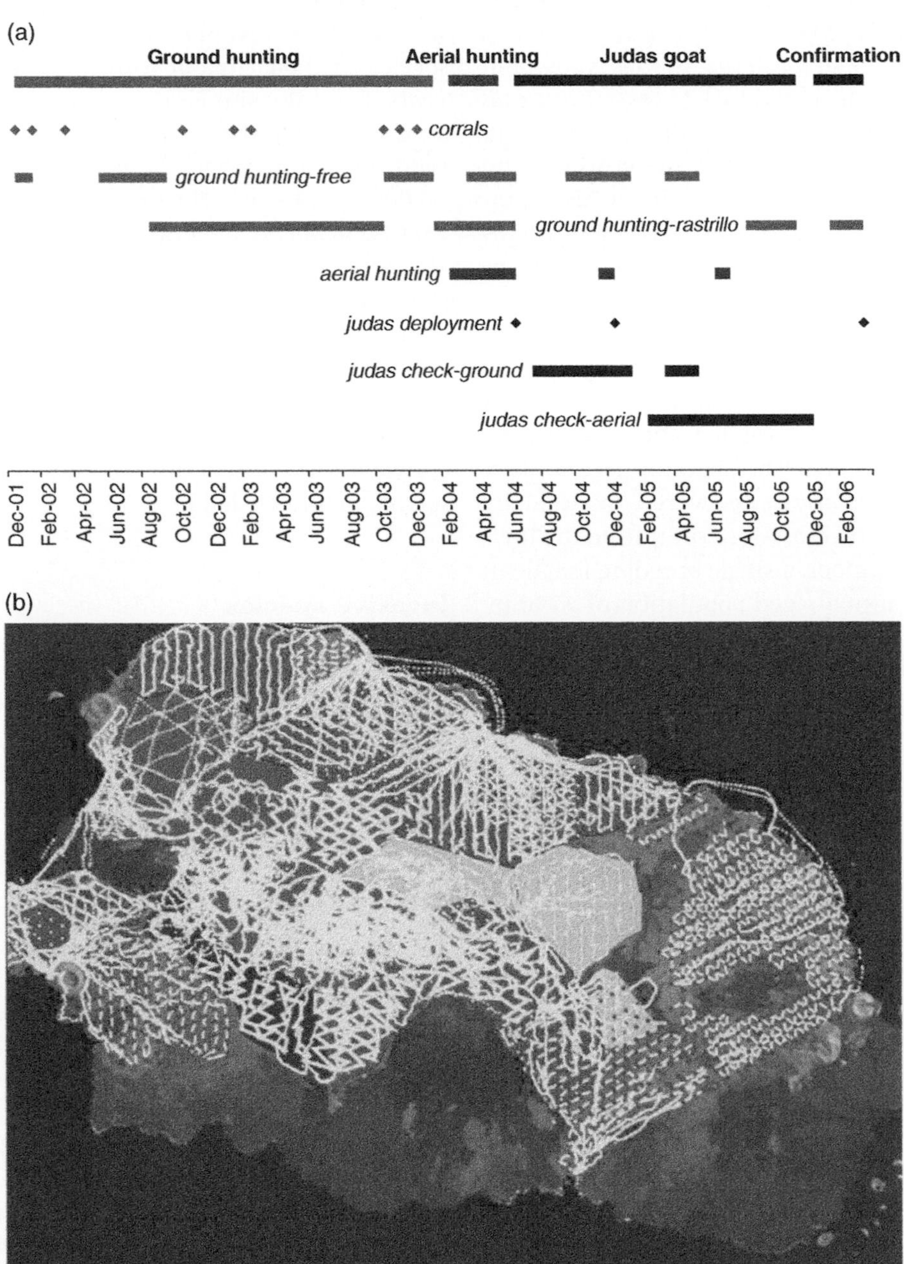

Figure 12.3 (a) The schedule of techniques adopted to eradicate feral goats from Santiago Island, Galapagos Islands, Ecuador. (b) An example of the coverage of Santiago Island by ground-hunting over the course of the eradication campaign. The southern and eastern portions of the island were too steep to be hunted on foot, so these areas were patrolled by helicopter. Figures courtesy of C.J. Donlan.

(*Falco peregrinus*), a listed species, and created a conservation strategy to collect and captive-breed the endemic Anacapa deer mouse (*Peromyscus maniculatus anacapae*); (2) the toxin was spread by helicopter; and (3) monitoring to verify environmental degradation of the remaining toxin determined when it would be safe to release the peregines and reintroduce the deer mice across the three islets (Pergams et al. 2000; Howald et al. 2009).

Low *H* is not always fatal, because genetic diversity can be maintained in other ways, as illustrated on many of the islands shown in Figure 12.1. The individual foxes (*U. littoralis dickeyi*) sampled on San Nicolas Island were virtually identical in DNA fingerprints and microsatellite profiles, but the major histocompatibilty complex, the region involved in microparasite defence, has maintained genetic variation owing, in part, to intense selection (Goldstein et al. 1999; Aguilar et al. 2004). In practice, emergency rescues of tiny and inbred island populations often do succeed. The case of the Mauritius kestrel quoted above is not the only such example. The Chatham Island black robin (*Petroica traversi*) recovered (assisted by translocation and cross-fostering) from the lowest possible number, a single breeding female in 1976, to an unmanaged population of >250 by 2003, because the risk of extinction from total loss of habitat was much more immediate than the risk of inbreeding depression (Jamieson et al. 2006). After rats reached the last refuge of another endemic species, the South Island saddleback (*Philesturnus carunculatus carunculatus*) on Big South Cape Island about 1962, the last remnant group of 36 birds was whisked out of danger in the nick of time. Later genetic management and translocations restored the stock to c.1200 birds spread among 15 island populations (Hooson & Jamieson 2004). The North Island subspecies of the saddleback (*P. c. rufusater*), which was also secured from extinction by a long history of repeated translocations from a single surviving group, provides one of the best documented examples of how serial bottlenecks can reduce genetic diversity in small founding populations (Lambert et al. 2005).

Analyses of minimum viable populations (MVP) examine small populations in terms of population dynamics, to estimate the expected time to extinction of an isolated group with specified characteristics. Almost all islands are too small to maintain any species of larger-bodied resident vertebrates in numbers large enough to provide for adaptation in response to future environmental changes (Soulé 1980; Frankel & Soulé 1981), and most islands are too small even to prevent extinction of large-bodied species in their current form for the next 100 years. Even the simple preservation of unmanaged viable populations of moderate-sized birds requires astonishingly large areas, e.g. about 10,000 ha for the North Island brown kiwi (*Apteryx mantelli*), a flightless bird of 2–3 kg body weight (Basse & McLennan 2003). Conversely, for invertebrates on islands, the potential for future evolution is more likely to be limited by habitat deterioration (New 2008). The stark difference in potential for successful preservation and prospects of future evolution between the vertebrates and invertebrates is seldom acknowledged.

Invasive species

Past a certain point, declining and small populations of vertebrates become linked by the extinction vortex, summarized by Caughley (1994) as a five-step process: 1. declining mate choice, hence more close-relative matings; 2. reduced heterozygosity in the offspring; 3. more lethal recessive alleles exposed to selection; 4. reduced fecundity and increased mortality; 5. reduced numbers and further restriction in mate choice. On islands, this often irreversible sequence of events can be set off by various events, often a combination of human-induced habitat destruction and the arrival of an invasive species.

New arrivals are a natural and ongoing part of the dynamic equilibrium controlling numbers of species on islands. Not all new species arriving on an island become hugely abundant invasives (Macdonald et al. 2007), and not all invasive species become damaging pests, but those that do may change species compositions and simplify community dynamics. They are the most common cause of the declining species syndrome. The chances of establishment of a new population vary; for ungulates, they are directly proportional to propagule size (Forsyth & Duncan 2001).

Pest eradication

To achieve at least temporary eradication of any population, three conditions are absolutely non-negotiable: (1) all individuals, or at least all of one sex, must be at risk of removal; (2) they must be removed faster than they can be replaced; and (3) any future arrivals must be prevented (Parkes & Murphy 2003). Failure to meet these conditions turns an eradication attempt into an exercise in sustained harvesting. The last few pest individuals are always the hardest and the most expensive to find, so success often depends on switching among standard technologies, e.g. from poisoning to introducing disease to shooting to trapping (Bloomer & Bester 1992), or by developing a new technique (see Box 12.3).

The third condition for eradication is easier to meet on islands than elsewhere, which is why successful island eradications are becoming routine, even of small pests on large islands. Removing Norway rats from 11,216 ha on Campbell Island (see Figure 12.1), once regarded as impossible, was achieved by aerial toxic baiting within 3 weeks in 2001 (Towns & Broome 2003; Bellingham et al. 2010). The marriage of strategy, technology and innovation has permitted huge advances in speed and efficiency, best illustrated by Project Isabella (see Box 12.3). Such advances make land area no longer limiting to pest eradication on uninhabited islands, but human residents unsympathetic to the programme or with domestic conspecifics can provide refuges which make eradication much more challenging (Ratcliffe et al. 2009).

The huge benefits of a well-planned and complete island eradication are becoming commonplace, e.g. the recovery of Cook's petrels (*Pterodroma cookii*) after the removal of cats and Pacific rats (*Rattus exulans*) from Little Barrier Island (Rayner et al. 2007). Additional and totally unexpected benefits can surprise everyone. Less than 5 years after Norway rats disappeared from Campbell Island, a previously unknown remnant colony of snipe, an undescribed species of *Coenocorypha* which had survived the rat invasion camped out on tiny (19 ha) Jacquemart Island offshore, recolonized the main island (Baker et al. 2010).

Successful eradication depends primarily on isolation from reinvasion. A systematic survey of 83 studies in which islands were defined by total area, not isolation of the area cleared of pests (Smith et al. 2010), concluded that post-breeding population sizes of birds were not improved on islands up to 2000 km^2. Such large islands are no different from mainlands if they are not treated as a single unit and if the cleared area is not protected from reinvasion. Much smaller islands than this have been vital in conservation provided the predators never got back to them. Fortunately, a theoretical model suggests that the benefits per unit of expenditure are higher if directed towards a larger number of relatively small islands (Brooke et al. 2007), despite the challenges and limitations of protecting small populations discussed above.

Mesopredator release, an increase in numbers of a previously less common predator from further down the food chain (Russell et al. 2009), can add indirect costs to a successful island eradication by inducing complex interactions among the remaining pests, including goats, pigs, cats, rats or rabbits, and mice – all often distributed to islands around the world by early explorers. For example, removal of cats from Little Barrier Island *reduced* the breeding success of Cook's petrels, owing to the consequent booming numbers of Pacific rats (*Rattus exulans*). This effect was reversed only when the rats were also removed (Rayner et al. 2007).

Species regarded as pests by conventional conservation management may be seen differently by indigenous people. *Rattus exulans* is valued by some Maori groups as a *taonga* (tribal treasure), so proposals to remove it from Little Barrier Island were preceded by translocation of some individuals to an alternative island (Ruscoe & Murphy 2005). Intense public debate focussed on conflicts between value judgements and science have so far prevented eradication of

North American beaver (*Castor canadensis*) from Navarino Island, Chile (Haider & Jax 2007). Large carnivores provide many more examples of the same problem (see Chapter 7).

Translocations and reintroductions

Island archipelagos are ideal locations in which to establish metapopulations of an endangered species, often by multiple transfers or assisted recolonizations from one or a few remnant sources to new locations, within or outside their former range. Archipelagos offer an escape from the risk of total loss and, in a world where large conservation islands are very scarce, they are often the best available option. The disadvantage of establishing several small populations is balanced by the great advantage of increasing the breeding stock and maximizing the genetic diversity of the species (Grueber & Jamieson 2008). Among 116 reintroductions analysed by Fischer & Lindenmayer (2000), success was associated with using a wild rather than a captive source population, releasing large propagules (>100 individuals) and removing the cause of the original population decline (see Chapter 22).

Source populations must not be overharvested, and the new location must not have the same limitations as the old one. The Seychelles warbler *Acrocephalus sechellensis* thrived after translocation from Cousin Island, where insects were few, to two islands with abundant insects (Komdeur 1994). The earliest translocations of kakapo to Resolution Island in 1894 (Hill & Hill 1987) failed at the time because stoats (*Mustela erminea*) swam the narrow channel (525 m) separating Resolution from the South Island. When a well-planned new programme of stoat eradication began in 2008 (Elliott et al. 2010), it removed 258 stoats in the first 2 weeks (New Zealand Department of Conservation, unpublished), and is ongoing.

Critically endangered island species retrieved from the brink of extinction by captive breeding, and later reintroduced to their native habitats, include, besides the Mauritius kestrel, the Lord Howe Island woodhen (*Tricholimnas sylvestris*) and the Hawaiian goose or nene (*Branta sandvicensis*) (Caughley & Gunn 1996). Conservation authorities now manage island translocations to maximize the genetic health of reintroduced populations, and monitor the results at least until the new populations stabilize (Armstrong et al. 2002; Taylor et al. 2005; Cardoso et al. 2009).

Island restoration

Restoration programmes attempt to return an island to a state approximating its condition at some point in the past. The focus is the whole island ecosystem rather than any particular species, which means that restoration programmes on severely damaged islands have to be long term, often starting with replanting of vegetation native to the area. On Tiritiri Matangi Island, near Auckland, >280,000 trees were planted over 9 years (1984–93), mostly by volunteers, and increasing numbers of native bird species are being translocated there as the forest matures. The island has developed as an open sanctuary, demonstrating that community participation in conservation is not only possible but strongly desirable, and is now being imitated elsewhere (Miller et al. 1994; Ogden & Gilbert 2009). Spontaneous recovery of native vegetation, including species not seen for years, is also possible if a sufficiently diverse seed bank survives and re-establishment is not prevented by lack of pollinators or dispersal agents (Weerasinghe et al. 2008).

Even without replanting, the restoration possibilities opened up by large-scale pest eradications are often spectacular (Towns et al. 1997, 2006; Towns & Broome 2003; Bellingham et al. 2010). Costs can be roughly estimated in advance if the island area (which accounts for 72% of the variation in expenses) and species to be eradicated are known. Costs have declined over time but increase with remoteness, so date and distance to the nearest main airport (a measure of remoteness) can be used to help refine the first estimate (Martins et al. 2006).

Successful restoration should ideally aim to re-establish the original ecological and evolutionary processes of an island. For example, once rid of introduced Arctic foxes (*Vulpes lagopus*), several of the Aleutian Islands' seabird colonies were revived, and their guano fertilized and restored native vegetation communities that had been radically altered during their absence (Maron et al. 2006). But what if the native island fauna had been driven to extinction? How would these important processes be restored? Guidelines have now been developed on how to use extant taxa as ecological substitutes to replace extinct taxa and reinstate lost ecological and evolutionary processes (Hansen et al. 2010). For example, giant tortoises (e.g. *Geochelone* spp.) have been the targets of successful 'rewilding' efforts of islands (see Chapter 22).

Conclusions

Insular species are often ecologically naive to exotic competitors, predators and parasites, or imperilled by the loss of native pollinators, seed dispersers and other mutualists. Thus, in order to conserve not just the marvellous diversity but also the natural characters of island biota, the challenge is to manage fragile, insular assemblages successfully. That means not only attempting to minimize intense habitat destruction and alteration, such as logging, agricultural conversion and urbanization, but also reducing the threat from ecologically novel, exotic species while maintaining or replacing natural, ecological conditions that shaped the evolution of island life.

Facts and theories are different things, not rungs in a hierarchy of increasing certainty.

Stephen Jay Gould *Hen's Teeth and Horse's Toes* (1994)

References

Aguilar, A., Roemer, G., Debenham, M., Binns, M., Garcelon, D. & Wayne, R.K. (2004) High MHC diversity maintained by balancing selection in an otherwise genetically monomorphic mammal. *Proceedings of the National Academy of Sciences USA*, **101**, 3490–3494.

Armstrong, D.P., Davidson, R.S., Dimond, W.J., *et al.* (2002) Population dynamics of reintroduced forest birds on New Zealand islands. *Journal of Biogeography*, **29**, 609–621.

Atkinson, I.A.E. (1985) The spread of commensal species of *Rattus* to oceanic islands and their effects on island avifaunas. In: *Conservation of Island Birds* (ed. P.J. Moors), pp. 35–81. Technical Publication 3. International Council for Bird Preservation, Cambridge.

Baker, A.J., Miskelly, C.M. & Haddrath, O. (2010) Species limits and population differentiation in New Zealand snipes (Scolopacidae: Coenocorypha). *Conservation Genetics*, **11**, 1363–1374.

Baker, J.D., Littnan, C.L. & Johnston, D.W. (2006) Potential effects of sea level rise on the terrestrial habitats of endangered and endemic megafauna in the Northwestern Hawaiian Islands. *Endangered Species Research*, **2**, 21–30.

Bakker, V.J., Doak, D., Roemer, G.W., *et al.* (2009) Incorporating ecological drivers and uncertainty into a demographic population viability analysis for the island fox. *Ecological Monographs*, **79**, 77–108.

Basse, B. & McLennan, J.A. (2003) Protected areas for kiwi in mainland forests of New Zealand: how large should they be? *New Zealand Journal of Ecology*, **27**, 95–105.

Bellingham, P.J., Towns, D.R., Cameron, E.K., *et al.* (2010) New Zealand island restoration: seabirds, predators, and the importance of history. *New Zealand Journal of Ecology*, **34**, 115–136.

Bloomer, J.P. & Bester, M.N. (1992) Control of feral cats on sub-Antarctic Marion Island, Indian Ocean. *Biological Conservation*, **60**, 211–219.

Brooke, M.D., Hilton, G.M. & Martins, T.L.F. (2007) Prioritizing the world's islands for vertebrate-eradication programmes. *Animal Conservation*, **10**, 380–390.

Brown, J.H. (1971) Mammals on mountain tops: nonequilibrium insular biogeography. *American Naturalist*, **105**, 467–478.

Buchanan, G.M., Butchart, S.H.M., Dutson, G., *et al.* (2008) Using remote sensing to inform conservation status assessment: estimates of recent deforestation rates on New Britain and the impacts upon endemic birds. *Biological Conservation*, **141**, 56–66.

Bunce, M., Worthy, T.H., Phillips, M.J., *et al.* 2009. The evolutionary history of the extinct ratite moa and New Zealand Neogene paleogeography. *Proceedings of the National Academy of Sciences USA*, **106**, 20646–20651.

Campbell, K.J., Baxter, G.S., Murray, P.J., Coblentz, B.E. & Donlan, C.J. (2007) Development of a prolonged estrus effect for use in Judas goats. *Applied Animal Behaviour Science*, **102**, 12–23.

Cardoso, M.J., Eldridge, M.D., Oakwood, M., Rankmore, B., Sherwin, W.B. & Firestone, K.B. (2009) Effects of founder events on the genetic variation of translocated island populations: implications for conservation management of the northern quoll. *Conservation Genetics*, **10**, 1719–1733.

Caughley, G. (1994). Directions in conservation biology. *Journal of Animal Ecology*, **63**, 215–244.

Caughley, G. & Gunn, A. (1996) *Conservation Biology in Theory and Practice*. Blackwell Science, Cambridge, MA.

Cook, J.A., Dawson, N.G. & MacDonald, S.O. (2006) Conservation of highly fragmented systems: The north temperate Alexander Archipelago. *Biological Conservation*, **133**, 1–15.

Courchamp, F., Langlais, M. & Sugihara, G. (2000) Rabbits killing birds: modelling the hyperpredation process. *Journal of Animal Ecology*, **69**, 154–164.

Cruz, F., Donlan, C.J., Campbell, K. & Carrion, V. (2005) Conservation action in the Galapagos: feral pig (*Sus scrofa*) eradication from Santiago Island. *Biological Conservation*, **121**, 473–478.

Cruz, F., Carrion, V., Cambell, K.J., Lavoie, C. & Donlan, C.J. (2009) Bio-economics of large-scale eradication of feral goats from Santiago Island, Galapagos. *Journal of Wildlife Management*, **73**, 191–200.

Darlington, P.J. (1957) *Zoogeography: The Geographical Distribution of Animals*. John Wiley & Sons, New York.

Darwin, C. (1839) *Journal of the Researches into the Geology and Natural History of Various Countries Visited by H.M.S. Beagle, under the Command of Captain Fitzroy, R.N. from 1832 to 1836*. Henry Colburn, London.

Darwin, C. (1860) *The Voyage of the Beagle*. New Jersey, Doubleday.

Diamond, J. (1972) Biogeographic kinetics: estimation of relaxation times for avifaunas of southwest Pacific islands. *Proceedings of the National Academy of Sciences USA*, **69**, 3199–3203.

Diamond, J. (1975a) Assembly of species communities. In: *Ecology and Evolution of Communities* (eds M.L. Cody & J. Diamond), pp. 342–444. Belknap Press, Cambridge, MA.

Diamond, J. (1975b) The island dilemma: lessons of modern biogeographic studies for the design of natural reserves. *Biological Conservation*, **7**, 129–146.

Diamond, J. (1984) Historic extinctions: a Rosetta stone for understanding prehistoric extinctions. In: *Quaternary Extinctions: A Prehistoric Revolution* (eds P.S. Martin & R.G. Klein), pp. 824–862. University of Arizona Press, Tucson, AZ.

Diamond, J. (1989) Overview of recent extinctions. In: *Conservation for the Twenty-first Century* (eds D. Western & M. Pearl), pp. 37–41. Oxford University Press, New York.

Donlan, C.J., Campbell, K., Cabrera, W., Lavoie, C., Carrion, V. & Cruz, F. (2007) Recovery of the Galapagos rail (*Laterallus spilonotus*) following the removal of invasive mammals. *Biological Conservation*, **138**, 520–524.

Elliott, G.P., Williams, M., Edmonds, H. & Crouchley, D. (2010) Stoat invasion, eradication and re-invasion of islands in Fiordland. *New Zealand Journal of Zoology*, **37**, 1–12.

Fidler, A.E., Lawrence, S.B. & McNatty, K.P. (2008) An hypothesis to explain the linkage between kakapo (*Strigops habroptilus*) breeding and the mast fruiting of their food trees. *Wildlife Research*, **35**, 1–7.

Fischer, J. & Lindemayer, D.B. (2000) An assessment of the published results of animal relocations. *Biological Conservation*, **96**, 1–11.

Forster, J.W. (1778) *Observations Made during a Voyage Round the World, on Physical Geography, Natural History and Ethic Philosophy*. G. Robinson, London.

Forsyth, D.M. & Duncan, R.P. (2001) Propagule size and the relative success of exotic ungulate and bird introductions in New Zealand. *American Naturalist*, **157**, 583–595.

Frankel, O.H. & Soulé, M.E. (1981) *Conservation and Evolution*. Cambridge University Press, Cambridge.

Frankham, R. (2005) Genetics and extinction. *Biological Conservation*, **126**, 131–140.

Franklin, I.R. (1980) Evolutionary change in small populations. In: *Conservation Biology: An*

Evolutionary-Ecological Perspective (eds M.E. Soulé & B.A. Wilcox), pp. 135–149. Sinauer Associates Inc, Sunderland, MA.

Goldstein, D.B., Roemer, G.W., Smith, D.A., Reich, D.E., Bergman, A. & Wayne, R. K. (1999) The use of microsatellite variation to infer patterns of migration, population structure and demographic history: an evaluation of methods in a natural model system. *Genetics*, **151**, 797–801.

Grueber, C.E. & Jamieson, I.G. (2008) Quantifying and managing the loss of genetic variation in a free-ranging population of takahe through the use of pedigrees. *Conservation Genetics*, **9**, 645–651.

Haider, S. & Jax, K. (2007) The application of environmental ethics in biological conservation: a case study from the southernmost tip of the Americas. *Biodiversity and Conservation*, **16**, 2559–2573.

Hansen, D.M., Donlan, C.J., Griffiths, C.J. & Campbell, K.J. (2010) Ecological history and latent conservation potential: large and giant tortoises as a model for taxon substitution. *Ecography*, **33**, 272–284.

Hanski, I. (2005) *Metapopulation Ecology*. Oxford University Press, Oxford.

Hanski, I. (2010) The theories of island biogeography and metapopulation dynamics. In: *The Theory of Island Biogeography Revisited* (eds J. Losos & R.E. Ricklefs), pp. 186–213. Princeton University Press, Princeton, NJ.

Hanski, I. & Gilpin, M.E. (1997) *Metapopulation Biology: Ecology, Genetics and Evolution*. Academic Press, San Diego, CA.

Harris, L.D. (1984) *The Fragmented Forest: Island Biogeography Theory and the Preservation of Biotic Diversity*. University of Chicago Press, Chicago, IL.

Hawkes, L.A., Broderick, A.C., Godfrey, M.H. & Godley, B.J. (2009) Climate change and marine turtles. *Endangered Species Research*, **7**, 137–154.

Heaney, L.R. (2000) Dynamic disequilibrium: a longterm, large-scale perspective on the equilibrium model of island biogeography. *Global Ecology and Biogeography*, **9**, 59–74.

Heaney, L.R. (2004) Conservation biogeography in oceanic archipelagoes. In: *Frontiers of Biogeography: New Directions in the Geography of Nature* (eds M.V. Lomolino & L.R. Heaney), pp. 345–360. Cambridge University Press, Cambridge.

Heaney, L.R. (2007) Is a new paradigm emerging for oceanic island biogeography? *Journal of Biogeography*, **34**, 753–757.

Hill, S. & Hill, J. (1987) *Richard Henry of Resolution Island*. John McIndoe Ltd, Dunedin.

Hilton, G.M. & Cuthbert, R.J. (2010) The catastrophic impact of invasive mammalian predators on birds of the UK Overseas territories: a review and synthesis. *Ibis*, **152**, 443–458.

Holdaway, R.N. & Jacomb, C. (2000) Rapid extinction of the moas (Aves: Dinornithiformes): model, test, and implications. *Science*, **287**, 2250–2254.

Hooker, J.D. (1866) *Lecture on Insular Floras*. British Association for the Advancement of Science, August 27, 1866, Nottingham.

Hooson, S. & Jamieson, I.G. (2004) Variation in breeding success among reintroduced island populations of South Island Saddlebacks *Philesturnus carunculatus carunculatus*. *Ibis*, **146**, 417–426.

Howald, G., Donlan, C.J., Faulkner, K.R., *et al.* (2009) Eradication of black rats *Rattus rattus* from Anacapa Island. *Oryx*, **44**, 30–40.

Jackson, J.B.C., Kirby, M.X., Berger, W.H., *et al.* (2001) Historical overfishing and the recent collapse of coastal ecosystems. *Science*, **293**, 629–638.

Jamieson, I.G., Wallis, G.P. & Briskie, J.V. (2006) Inbreeding and endangered species management: is New Zealand out of step with the rest of the world? *Conservation Biology*, **20**, 38–47.

Komdeur, J. (1994) Conserving the Seychelles warbler, *Acrocephalus sechellensis*, by translocation from Cousin Island to the Islands of Aride and Cousine. *Biological Conservation*, **67**, 143–152.

Lack, D. (1970) Island birds. *Biotropica*, **2**, 29–31.

Lambert, D.M., King, T., Shepherd, L.D., Livingston, A., Anderson, S. & Craig, J.L. (2005) Serial population bottlenecks and genetic variation: translocated populations of the New Zealand Saddleback (*Philesturnus carunculatus rufusater*). *Conservation Genetics*, **6**, 1–14.

Lavoie, C., Donlan, C.J., Campbell, K.J., Cruz, F. & Carrion, G.V. (2007) Geographic tools for eradication programs of insular non-native mammals. *Biological Invasions*, **9**, 139–148.

Limpus, C.J., Miller, J.D., Parmenter, C.J. & Limpus, D.J. (2003) The green turtle, *Chelonia mydas*, population of Raine Island and the northern Great Barrier Reef: 1843–2001. *Memoirs of the Queensland Museum*, **49**, 349–440.

Lomolino, M.V. (2000) A species-based theory of insular zoogeography. *Global Ecology and Biogeography*, **9**, 39–58.

Lomolino, M.V. (2004) Conservation biogeography. In: *Frontiers of Biogeography* (eds M.V. Lomolino & L.R. Heaney), pp. 293–296. Sinauer Associates Inc, Sunderland, MA.

Lomolino, M.V. (2006) Space, time, and conservation biogeography. In: *The Endangered Species Act at Thirty: Conserving Biodiversity in Human-Dominated Landscapes* (eds J.M. Scott, D.D. Goble & F.W. Davis). Island Press, London.

Lomolino, M.V. & Brown, J.H. (2009) The reticulating phylogeny of island biogeography theory. *Quarterly Review of Biology*, **84**, 357–90.

Lomolino, M.V., Riddle, B.R., Whittaker, R.J. & Brown, J.H. (2010) *Biogeography*, 4th edn. Sinauer Associates Inc, Sunderland, MA.

Lyell, C. (1834) *Principles of Geology, Being an Attempt to Explain the Former Changes of the Earth's Surface, by Reference to Causes Now in Operation*, 3rd edn. John Murray, London.

MacArthur, R.H. & Wilson, E.O. (1963) An equilibrium theory of insular zoogeography. *Evolution*, **17**, 373–387.

MacArthur, R.H. & Wilson, E.O. (1967) *The Theory of Island Biogeography*. Princeton University Press, Princeton. NJ.

Macdonald, D.W., King, C.M. & Strachan, R. (2007) Introduced species and the line between biodiversity conservation and naturalistic eugenics. In: *Key Topics in Conservation Biology* (eds D.W. Macdonald & K. Service), pp. 186–205. Blackwell Publishing, Oxford.

Maron, J.L., Estes, J.A., Croll, D.A., Danner, E.M., Elmendorf, S.C. & Buckelew, S.L. (2006) An introduced predator alters Aleutian Island plant communities by thwarting nutrient subsidies. *Ecological Monographs*, **76**, 3–24.

Martins, T.L., Brooke, M.D., Hilton, G.M., Farnsworth, S., Gould, J. & Pain, D.J. (2006) Costing eradications of alien mammals from islands. *Animal Conservation*, **9**, 439–444.

Mayr, E. (1942) *Systematics and the Origin of Species, from the Viewpoint of a Zoologist*. Columbia University Press, New York.

Mayr, E. (1954) Changes in genetic environment and evolution. In: *Evolution as a Process* (eds J. Huxley, A.C. Hardy & E.B. Ford), pp. 157–180. Allen and Unwin, London.

Miller, C.J., Craig, J.L. & Mitchell, N.D. (1994) Ark 2020 – a conservation vision for Rangitoto and Motutapu Islands. *Journal of the Royal Society of New Zealand*, **24**, 65–90.

Morrison, S.A., Macdonald, N., Walker, K., Lozier, L. & Shaw, M.R. (2007) Facing the dilemma at eradication's end: uncertainty of absences and the Lazarus effect. *Frontiers in Ecology and the Environment*, **5**, 271–276.

Mulder, C.P. & Keall, S.N. (2001) Burrowing seabirds and reptiles: impacts on seeds, seedlings and soils in an island forest in New Zealand. *Oecologia*, **127**, 350–360.

New, T.R. (2008) Insect conservation on islands: setting the scene and defining the needs. *Journal of Insect Conservation*, **12**, 197–204.

Ogden, J. & Gilbert, J. (2009) Prospects for the eradication of rats from a large inhabited island: community based ecosystem studies on Great Barrier Island, New Zealand. *Biological Invasions*, **11**, 1705–1717.

Parkes, J. & Murphy, E.C. (2003) Management of introduced mammals in New Zealand. *New Zealand Journal of Zoology*, **30**, 335–359.

Parkes, J., Ramsey, D.S., Macdonald, N., *et al.* (2010) Rapid eradication of feral pigs (*Sus scrofa*) from Santa Cruz Island, California. *Biological Conservation*, **143**, 634–641.

Pergams, O.R., Lacy, R.C. & Ashley, M.V. (2000) Conservation and management of Anacapa Island deer mice. *Conservation Biology*, **14**, 819–832.

Poincaré, H. (1952) *Science and Hypothesis*. Dover Publications, New York.

Polis, G.A. & Hurd, S.D. (1996) Linking marine and terrestrial food webs: allochthonous input from the ocean supports high secondary productivity on small islands and coastal land communities. *American Naturalist*, **147**, 396–423.

Quammen, D. (1996) *The Song of the Dodo*. Touchstone, New York.

Ratcliffe, N., Bell, M., Pelembe, T., *et al.* (2009) The eradication of feral cats from Ascension Island and its subsequent recolonization by seabirds. *Oryx*, **44**, 20–29.

Rayner, M.J., Hauber, M.E., Imber, M.J., Stamp, R.K. & Clout, M.N. (2007) Spatial heterogeneity of mesopredator release within an oceanic island system. *Proceedings of the National Academy of Sciences USA*, **104**, 20862–20865.

Richards, R. (2003) New market evidence on the depletion of southern fur seals: 1788–1833. *New Zealand Journal of Zoology*, **30**, 1–9.

Richardson, D.M. & Whittaker, R.J. (2010) Conservation biogeography – foundations, concepts and challenges. *Diversity and Distributions*, **16**, 313–320.

Roemer, G.W., Donlan, C.J. & Courchamp, F. (2002) Golden eagles, feral pigs, and insular carnivores: how exotic species turn native predators into prey. *Proceedings of the National Academy of Sciences USA*, **99**, 791–796.

Rosenzweig, M.L. (1995) *Species Diversity in Space and Time*. Cambridge University Press, New York.

Ruscoe, W.A. & Murphy, E.C. (2005) Kiore. In: *The Handbook of New Zealand Mammals*, 2nd edn. (ed. C.M. King), pp. 159–174. Oxford University Press, Melbourne.

Russell, J.C., Lecomte, V., Dumont, Y. & Le Corre, M. (2009) Intraguild predation and mesopredator release effect on long-lived prey. *Ecological Modelling*, **220**, 1098–1104.

Smith, R.K., Pullin, A.S., Stewart, G.B. & Sutherland, W.J. (2010) Effectiveness of predator removal for enhancing bird populations. *Conservation Biology*, **24**, 820–829.

Soulé, M.E. (1980) Thresholds for survival: maintaining fitness and evolutionary potential. In: *Conservation Biology: An Evolutionary–Ecological Perspective* (eds M.E Soulé & B.A. Wilcox), pp. 151–169. Sinauer Associates Inc, Sunderland, MA.

Steadman, D.W. (1995) Prehistoric extinctions of Pacific Island birds: biodiversity meets zooarchaeology. *Science*, **267**, 1123–1131.

Stuessy, T.F. (2007) Evolution of specific and genetic diversity during ontogeny of island floras: the importance of understanding process for interpreting island biogeographic patterns. In: *Biogeography in a Changing World* (eds M.C. Ebach & R.S. Tangney), pp. 117–134. CRC Press, New York.

Taylor, S.S., Jamieson, I.G. & Armstrong, D.P. (2005) Successful island reintroductions of New Zealand robins and saddlebacks with small numbers of founders. *Animal Conservation*, **8**, 415–420.

Towns, D.R. & Broome, K.G. (2003) From small Maria to massive Campbell: forty years of rat eradications from New Zealand islands. *New Zealand Journal of Zoology*, **30**, 377–398.

Towns, D.R., Simberloff, D. & Atkinson, I.A. (1997) Restoration of New Zealand islands: redressing the effects of introduced species. *Pacific Conservation Biology*, **3**, 99–124.

Towns, D.R., Atkinson, I.A. & Daugherty, C.H. (2006) Have the harmful effects of introduced rats on islands been exaggerated? *Biological Invasions*, **8**, 863–891.

Wallace, A.R. (1876) *The Geographical Distribution of Animals*. Macmillan, London.

Weerasinghe, U.R., Akiko, S., Palitha, J. & Seiki, T. (2008) The role of the soil seed bank in vegetation recovery on an oceanic island severely damaged by introduced goats. *Applied Vegetation Science*, **11**, 355–364.

Whittaker, R.J. & Fernandez-Palacios, J.M. (2007) *Island Biogeography: Ecology, Evolution, and Conservation*, 2nd edn. Oxford University Press, Oxford.

Whittaker, R.J., Araujo, M.B., Jepson, P., Ladle, R.J., Watson, J.E. & Willis, K.J. (2005) Conservation biogeography: assessment and prospect. *Diversity and Distributions*, **11**, 3–23.

Whittaker, R.J., Triantis, K.A. & Ladle, R.J. (2008) A general dynamic theory of oceanic island biogeography. *Journal of Biogeography*, **35**, 977–994.

Wilmshurst, J.M., Anderson, A.J., Higham, T.F. & Worthy, T.H. (2008) Dating the prehistoric dispersal of Polynesians to New Zealand using the commensal Pacific rat. *Proceedings of the National Academy of Sciences USA*, **105**, 7676–7680.

Wilson, E.O. (1959) Adaptive shift and dispersal in a tropical ant fauna. *Evolution*, **13**, 122–144.

Wilson, E.O. (1961) The nature of the taxon cycle in the Melanesian ant fauna. *American Naturalist*, **95**, 169–193.

Wilson, E.O. & Willis, E.O. (1975) Applied biogeography. In: *Ecology and Evolution of Communities* (eds M.L. Cody & J. Diamond), pp. 522–534. Harvard University Press, Cambridge, MA.

13

Conservation of tropical forests: maintaining ecological integrity and resilience

Owen T. Lewis[1], Robert M. Ewers[2], Margaret D. Lowman[3] and Yadvinder Malhi[4]

[1]Department of Zoology, University of Oxford, Oxford, UK
[2]Division of Biology, Imperial College London, Ascot, UK
[3]Nature Research Center, North Carolina Museum of Natural Sciences *and* North Carolina State University, Raleigh, NC, USA
[4]Environmental Change Institute, School of Geography and the Environment, University of Oxford, Oxford, UK

'When we try to pick out anything by itself, we find it hitched to everything else in the Universe.'
—John Muir, My First Summer in the Sierra (1911)

Introduction

Viewed from space, the earth's tropical forests form a narrow green belt around the equator, with three significant blocks: in northern South America, West and Central Africa, and the peninsula and islands of South East Asia and Australasia. The total area has been reduced by about 50% since the beginning of the 20th century (FAO 2001) but around 1200 million ha remains, or approximately 5% of the earth's land surface. This might sound a lot, but deforestation and degradation of tropical forests continue at a high rate worldwide (Curran et al. 2004; Laurance & Peres 2006) and has become a *cause célèbre* for conservationists.

There are several good reasons why we should worry about modifying tropical forests and reducing their area. These ecosystems are a key element of global cycles of water and carbon, and changes to them are likely to have repercussions on a global scale (Lewis 2006; Lewis et al. 2009). More locally, tropical forests provide a suite of ecosystem services for human populations living in and near them. These include harvestable resources of timber, firewood and bushmeat, as well as less immediately

Key Topics in Conservation Biology 2, First Edition. Edited by David W. Macdonald and Katherine J. Willis.

obvious services like erosion control and stabilization of water supplies (Gardner et al. 2009). From a biodiversity conservation perspective, concern about deforestation and modification is motivated by a strong desire to protect the extraordinary concentration of biological diversity within tropical forests. These habitats are the most species rich on earth and may contain up to 75% of all terrestrial species. Ultimately, the ecosystem services provided by tropical forests depend on the persistence of their component species.

Conservation of tropical forest biodiversity is a straightforward proposition, at least in theory. Sophisticated ecological models, intensive single-species conservation efforts and a nuanced understanding of the ecological processes structuring tropical forests seem unnecessary. We already know the most effective way to conserve this ecosystem: stop destroying and modifying it! In practice, things are not that simple. Only a small fraction of the world's tropical forests, about 10%, is currently protected within parks or reserves (Brooks et al. 2009), and the demands of growing populations in tropical countries make it unlikely that this area will increase much in future. Meanwhile, 'external' pressures on existing protected areas, many of which might be characterized as 'paper parks' (Brandon et al. 1998), will continue to increase.

In this chapter, while acknowledging the critical importance of maintaining large, core areas of tropical forests as free as possible from human interference (Gardner et al. 2009), we address the need for tropical forest conservation efforts in the wider tropical landscape, beyond the boundaries of strictly protected areas. We highlight the need to understand the resilience of tropical forests to anthropogenic perturbations, focusing on ecosystem-level processes, particularly food web changes, ecological cascades, and alterations to ecosystem functions. We review empirical evidence for the resilience of tropical forests to different anthropogenic drivers, consider what humans can do to maximize resilience at various scales, and suggest that it may be possible to maintain tropical forest biodiversity by working within the bounds of 'natural' disturbances. We suggest that conservation efforts in the wider tropical landscape may increasingly need to retain functioning and resilient ecosystems, rather than biodiversity *per se*.

Practical approaches for achieving resilience will vary. While much of the news about conservation in tropical forests is negative, there are 'good news' stories from around the tropics. What approaches for conservation of tropical forests are working, and might these be applied more widely? We focus on three situations where, for varying reasons, there is cause to be optimistic and where we believe that significant practical progress is being made towards establishing stable, resilient tropical forests both within and beyond the borders of protected areas.

Destruction versus degradation: ecosystem-level consequences

One common misconception is that human impacts on tropical forests are all-or-nothing: forest is either present or absent. A second common misconception is the romantic notion that only 'virgin' forests are of any value from a conservation perspective (Perfecto & Vandermeer 2008). If tropical forests are clear-felled for plantations or agriculture, the habitat that replaces the forest will indeed support very little of the original biodiversity. Habitat destruction of this sort is a major factor in some parts of the tropics, notably forests in South East Asia converted to oil palm plantations (Sodhi et al. 2004). However, it has been suggested that the total forested area in some tropical regions (particularly Latin America) may actually be increasing (Wright & Muller-Landau 2006), although these calculations are controversial (Laurance 2007). Most would agree that the trend globally is a shift from relatively unmodified forest to modified and 'secondary' forests of various sorts. These forest fragments are embedded in a matrix of agro-ecosystems, which themselves may have considerable conservation potential (Perfecto & Vandermeer 2008).

Given that it will only be possible to protect a small fraction of the earth's tropical forests entirely from human impacts, it might be argued that the real battleground for conservation lies in ensuring that the inevitable harm that people will cause to forests is minimized. Since widespread exploitation of tropical forests appears unavoidable, can we plan this exploitation and manage the habitats that replace natural forests in a way that minimizes the repercussions for biodiversity and associated ecosystem services? In the context of logging, for example, this might involve ensuring that disturbances are within the bounds of those that a forest might experience naturally: so-called 'ecological forestry' (Hunter 1999). Indeed, such disturbances will inevitably be beneficial for some disturbance-adapted species, including commercially important trees such as mahogany (*Swietenia macrophylla* King) (Brown et al. 2003).

While humans have modified tropical habitats significantly for thousands of years (Heckenberger et al. 2008), current human perturbations to tropical forests differ in terms of their scale and intensity. Furthermore, they coincide with other, escalating drivers of global change including climate change, fragmentation, invasive species and overexploitation (reviewed by Laurance & Peres 2006). The extent to which ecological communities are likely to recover to their natural state over the long term is therefore a matter of debate and ongoing research.

It is increasingly recognized that we need to take a more inclusive, ecosystem-level perspective on ecological responses to perturbations. There are at least three reasons why a wider community- and ecosystem-level perspective is important. First, the intricate interconnectedness of ecological networks means that species that are not affected directly by perturbations can, in the long run, suffer through trophic cascades and other indirect effects. Second, changes in richness or composition may not give a full picture of the *functional* consequences of losses or changes to biodiversity. Third, responses to human disturbance may be long delayed, such that tipping points or thresholds of resilience may be crossed, perhaps before the full impacts of human actions are recognized. Understanding these aspects of community and ecosystem responses to tropical forest modification may be key to understanding the extent to which tropical forests can be modified without jeopardizing their biodiversity.

Trophic cascades and food webs

Ecological communities are intimately connected through networks of interactions, both positive (e.g. those between mutualists such as plants and their pollinators) and negative (e.g. competitive interactions, and trophic interactions involving predators and prey, hosts and parasites). Where these interactions are specialized and obligate, local or global extinction of one partner can lead to 'co-extinction' of the other (Koh et al. 2004). More subtly, changes to one part of the network may have repercussions elsewhere. Pace et al. (1999) suggest that trophic cascades may be intrinsically less likely to occur in high-diversity systems like tropical forests. However, there are some clear examples, notably the top-down trophic cascades generated on tropical islands lacking top predators, where high herbivore densities severely restrict plant regeneration (Terborgh et al. 2001). Keystone species do not always occupy the tops of food chains, and it seems likely that similar cascades follow the extinction of species or groups of species at lower trophic levels. For example, local extinction or reduced abundance of a single species in a network of hosts and parasitoids can have widespread repercussions for the abundance of other species, even if these are not directly linked to the impacted species (Morris et al. 2004).

A further possibility is that the ecological processes that help to structure and maintain diversity will be disrupted (Lewis & Gripenberg 2008). 'Mobile links' such as pollinators and seed dispersers play key roles in the dynamics of plant populations. Their decline or loss therefore has the potential to reverberate through food webs (Gardner et al. 2009). Diversity-enhancing processes may also be disrupted by anthropogenic disturbance. For example, the Janzen–Connell mechanism (where specialized natural enemies such as seed predators inhibit regeneration near conspecifics, helping to maintain high plant diversity) may be weakened if seed predator populations are depleted by hunting or habitat modification (Dirzo & Miranda 1991; Bagchi et al. 2011). Since plants form the basis of all food webs, and plant diversity appears to be the main driver of diversity at higher trophic levels (Novotny et al. 2006), such effects can cascade up to affect the diversity of the wider ecological community and, ultimately, plant diversity will be reduced.

Functional changes

Ecosystem functions include the physical, chemical and biological processes or attributes that contribute to the persistence of an ecosystem (Loreau et al. 2002). Notable examples include decomposition, pollination and cycling of elements. Modification of tropical forests is a major concern to conservationists worldwide, because it could lead to shifts in ecosystem functions and ecosystem services (the subset of ecosystem functions that are directly useful for humans). Altered ecosystem functions and services could occur at the global level, for example by changing patterns of rainfall or geochemical cycles (Lewis 2006), but are more easily documented at a local scale. For example, pollination of agricultural crops can be highly dependent on the diversity of insect pollinators, which in turn is sensitive to the management of forested landscapes. In Central Sulawesi, Klein et al. (2003) found that fruit set of coffee was strongly and positively correlated with the diversity of pollinating bees visiting plantations. Bee diversity in turn decreased with distance from the nearest forest, providing a strong incentive for farmers to conserve natural forested habitats in proximity to plantations. Despite such examples, most of the evidence base on the relationship between biodiversity and ecosystem functioning still comes from highly manipulative studies carried out at relatively low levels of diversity, largely in temperate systems. There are very few tropical forest studies that quantify how variations in diversity affect ecosystem functions and services, and how varying levels of anthropogenic impact affect functionally important components of diversity (Lewis 2009).

Of particular interest here are measures of species' sensitivity to different forms and intensities of disturbance ('response traits') and their contributions to ecosystem function ('effect traits': Lavorel & Garnier 2002). Not all species contribute equally to ecosystem functions, and not all species respond similarly to perturbations. Larsen et al. (2005) studied dung beetle assemblages on forest fragments isolated on artificial islands in Lago Guri, a reservoir in eastern Venezuela created in 1986 by the flooding of 4300 km^2 of semi-deciduous tropical forest. Dung beetles use animal dung as a food source and often bury it to provision their offspring. Dung burial by beetles accelerates rates of nutrient cycling, increases plant productivity, and helps seed dispersal and germination. Using dung-baited pitfall traps on 29 of the islands and the adjacent mainland, the researchers found that the smaller islands supported fewer dung beetle species and fewer individuals. They also measured the ecosystem function of dung removal using artificial dung patches of a known mass and volume. Rates of dung removal were lower on islands with low dung beetle richness and abundance. In separate trials they found that large-bodied dung beetle species were particularly important in processing dung. However, these functionally important, large-bodied species were those most likely to go extinct following forest fragmentation: they were absent from the smaller islands. In this case, 'response traits' and 'effect traits' are positively correlated, potentially leading to an accelerating loss of function with loss of species. It should be a priority to determine if such correlations are widespread for other functionally important plant and animal taxa. A further source of uncertainty is that tropical forest modification often leads to the formation of novel species assemblages: interactions among existing sets of species unravel, and new interactions form (Gardner et al. 2009). Inevitably, such compositional changes will have functional consequences, but these are poorly studied.

Resilience

Resilience can be defined as the 'capacity of a system to recover to essentially the same state after a disturbance' (Scheffer 2009). From the perspective of tropical forest conservation, it is important to avoid exploiting tropical forests in a way that exceeds their capacity for resilience. The danger is that changes accumulate past a 'tipping point', beyond which the system enters an alternative stable state and from which recovery to the original state is difficult or impossible (Lenton et al. 2008). It is widely argued that Amazonian forests may be approaching a tipping point where deforestation and a warming climate interact to increase the frequency of severe droughts and forest fires, and reduce overall precipitation to a point where forest dieback cannot be reversed (Malhi et al. 2009).

Whether the high diversity of tropical forest systems makes them intrinsically more or less stable remains an area of considerable debate. In theory, if multiple species can deliver a particular contribution to ecosystem function (i.e. have similar 'effect traits'), and these species respond differently to environmental changes including human perturbations, then collectively the system will be better placed to weather these perturbations, i.e. it will be more resilient (Folke et al. 2004). However, Ehrlich & Pringle (2008) suggest that tropical forests may be less resilient to human impacts, compared with other tropical habitats such as savannas. For example, livestock farming in tropical savannas can largely co-exist with the maintenance of the savanna ecosystem, presumably because it closely mimics the 'natural system' of high densities of wild grazing ungulates.

Other authors point to evidence that tropical forests can be relatively resilient in the long term. For example, Wright & Muller-Landau (2006) suggest that Pleistocene-era fragmentation of forests (particularly in West Africa) and long-term clearance and hunting by indigenous peoples (particularly in Central America) will have acted as an 'extinction filter' (Balmford 1996), making the surviving species relatively resilient to future perturbations. However, it is hard to know for certain, because few tropical forests have escaped all anthropogenic impacts (Willis et al. 2004; Lewis 2006). The Upper Xingu region of the Brazilian Amazon is currently covered by a large swathe of intact forest, but archaeological evidence shows that large parts of this region were densely populated and heavily cultivated between approximately 1250 and 1600 AD (Heckenberger et al. 2008). Similarly, modern-day Belize probably has a smaller human population and a greater area of tropical forest than it did 1000 years ago at the peak of the Maya civilization (Wright et al. 1959). Few biologists would immediately recognize the forests of Belize and the Upper Xingu as secondary regrowth, and they clearly retain relatively high biological diversity.

There is a risk that our expectation of what a diverse, intact and functioning tropical forest ecosystem looks like will be distorted by 'shifting baselines'; past disturbance may have generated patterns in biodiversity that are already substantially altered from the natural state (reviewed by Gardner et al. 2009). For example, Hanski et al. (2007) investigated dung beetle diversity in Madagascar, where approximately 50% of the forested area has been destroyed in the past 50 years and about 10% of the original forest cover now remains. The dung beetle fauna of Madagascar is well known from extensive collecting in the late 19th and early 20th centuries, before major deforestation occurred, providing a rare opportunity to assess the extent to which current levels of biodiversity reflect the baseline situation. In extensive recent survey work, Hanski et al. re-found 29 of the 51 species that had been documented for

Madagascar. It seems likely that many or most of the 'missing' species have gone extinct, because they had small distributions that no longer contain suitable forested habitats. The key message here is that without either excellent historical data or baseline data from 'intact' forest sites to provide a reference, it can be difficult to know what we are missing.

Practical solutions

Having reviewed the rather depressing challenges facing tropical forest conservationists, we now consider some areas for optimism, where co-existence of human activities with tropical forest biodiversity has been demonstrated to work, or where best evidence suggests that it may be feasible for people to co-exist sustainably with forest biodiversity.

Maintaining and restoring biodiversity in secondary forests

While some species are unlikely to persist in human-modified tropical forests, many such forests support very diverse flora and fauna (Dent & Wright 2009). In a study comparing the diversity of 15 taxa among forests of different disturbance levels, Barlow et al. (2007) showed that between 5% and 57% of species are restricted to primary forest. The flipside to this, of course, is that 43–95% of species may persist in modified forests, hinting at their potential value for conservation despite their degraded status. The values presented by Barlow et al. (2007) are more comprehensive than many comparable studies, but probably typical. For example, logging concessions around protected areas in the Democratic Republic of Congo supported very high levels of mammal diversity, including a number of endangered species (Clark et al. 2009), although it is not clear what proportion of those species had formed viable populations or were simply transient individuals. Similarly, secondary forests in Gabon support many of the invertebrate species found in primary forests (Basset et al. 2008). The take-home message from studies such as these is that conservation, and even restoration, of tropical forests need not be an issue that is purely focused on primary forests.

There are, of course, caveats to such a general statement about the importance of degraded forests, and many of these caveats relate to the community and ecosystem perspectives that we emphasized in the preceding section. Most importantly, the presence of any particular species in a modified forest habitat does not necessarily mean that it can maintain a viable population (Gardner et al. 2009). For example, mammal community structure in central African logging concessions changes with distance to protected areas (Clark et al. 2009), strongly indicating that at least some of the individuals detected in logged forests belong to populations that persist in primary forest. Such a spillover effect can lead to misinterpretation of basic forest biodiversity data, as degraded forests may represent unrecognized population sinks for species which persist solely because they have source populations in nearby primary forest. This spatial dependence is important, but rarely quantified. Equally important is to know whether modified habitats and agro-ecosystems have the potential to act as breeding habitat rather than population sinks if we adjust the manner in which they are exploited. For example, while conversion of forest to oil palm invariably leads to extinction of most forest species, riparian strips (areas of forest left uncut bordering existing streams and rivers) are now routinely left within oil palm plantations to prevent erosion from the cleared land leading to heavy sediment loads in the river water, and are often a legal requirement. In Sabah, Malaysia, plantation owners must maintain a 30 m buffer of riparian forest either side of any river greater than 3 m in width. Such habitats may have considerable potential as reservoirs of forest biodiversity, but how their conservation value varies with their width and whether they are significant in themselves, or only because they act as corridors allowing dispersal between remaining forest fragments, remain important questions for guiding conservation policy and practice.

Understanding the spatial relationship between the diversity of degraded forests and their proximity to primary forest leads directly to an important tactic for maximizing biodiversity gains from conservation efforts in degraded forests: the gain per unit effort is likely to be higher if efforts are directed towards degraded forests which persist close to primary forests. The biodiversity value of degraded forests is something that can be improved through careful management and restoration, although the time frames involved are long. The simplest approach is to leave the forest to recover unaided. The likely success of this approach depends, however, on the initial extent of degradation, and while natural regeneration can eventually restore much of the original forest diversity, it is clear that many of these second-growth forests retain only a subset of the original forest species (Lamb et al. 2005; Bhagwat et al. 2008; Chazdon 2008). Moreover, natural recovery is not an option that will work in all cases, and further interventions may be required, of which the most common is to focus on the plant community and to plant or seed a forest with 'missing' tree species (Lamb et al. 2005). The expectation is that animals require the food, shelter and other ecological resources provided by the trees themselves and there is little point in introducing them to a site until after the tree community has been established.

Simply restoring species to a site is unlikely to be sufficient. The new arrivals need to form self-supporting populations, for which they may require the restoration of ecological processes, functions and disturbance regimes such as species interactions, nutrient cycling and hydrological processes (Chazdon 2008; Gardner et al. 2009). To some extent, these processes can be expected to restore themselves as the species come back and begin to form interacting networks. However, other processes, such as natural flooding regimes, might be affected by changes outside the area being restored, highlighting the importance of considering a forest restoration project as being embedded and integrated within a wider landscape context (Lamb et al. 2005; Gardner et al. 2009). Fire regimes represent an important case study. In the humid tropical forests of the Amazon, natural forest fires are a rare event that results in dramatic changes to the structure and composition of tree communities (Barlow & Peres 2008). Human modification to Amazonian landscapes, and human activities in those landscapes, have led to increased fire frequency in 42% of the Brazilian Amazon (Aragão & Shimabukuro 2010). Increased fire frequency compounds the impacts of fire on Amazonian forests, because initial fires alter environmental conditions in a way that increases the probability and intensity of further fires (Cochrane et al. 1999), and because changes in tree communities increase greatly in magnitude when forests are burned multiple times (Barlow & Peres 2008). Restoring degraded forests in this region will require the suppression of anthropogenically increased fire regimes (Aragão & Shimabukuro 2010) to reflect more closely the regimes observed in primary forest.

Local stakeholder engagement and action

The conventional stakeholders of tropical forests are usually large government and conservation NGOs, but the continuing decline of tropical forests globally requires wider engagement and action. Local stakeholders have successfully conserved forests in many tropical regions. For example, islanders in Western Samoa have substituted ecotourism for logging (Cox & Banack 1991; Lowman et al. 2006), and local priests of the Coptic Church have successfully conserved some of Ethiopia's last forest fragments (see below). Increased publicity describing their successes could inspire others and provide models for effective forest conservation solutions elsewhere.

In tropical agro-ecosystems, farmers represent an emerging group of forest conservation stakeholders. Eighteen countries in Africa are currently engaged in trials of fertilizer trees as part of a new agroforestry movement, 'evergreen agriculture' (Garrity et al. 2010). Canopy foliage provides shade and litterfall nutrients for crops grown below the trees. Tropical countries may learn from the experience of Australia, which suffered widespread social problems when deforestation threatened rural livelihoods (Heatwole & Lowman 1987). In this case, clearing for sheep and cattle grazing destroyed more than 90% of the original eucalypt forest cover across much of New South Wales. The resulting loss of insectivorous birds led to outbreaks of Christmas beetles that defoliated and ultimately killed the remaining forest fragments. This forest dieback was only reversed when local farmers instigated planting activities, using local seed sources. A programme called 'A Billion Trees by 2000' was initiated, one farm at a time (Heatwole & Lowman 1987).

In addition to farmers and graziers, religious leaders comprise another successful group of local stakeholders, and land use practices associated with religious observance can be valuable for tropical biodiversity (reviewed in Verschuuren et al. 2010). One notable case study is the Coptic or Christian Orthodox church in Ethiopia (Wassie-Eshete 2007; Jarzen et al. 2010; Lowman 2011). There are over 35,000 church buildings throughout the country, some dating back to 360 AD, each surrounded by a tract of native forest because biodiversity stewardship is fundamental to the church mission (Wassie-Eshete 2007). Loss of these last remaining patches of forest would represent extinction for many native trees, insects, birds, and mammals, since the remaining landscape matrix is arid farmland with little or no forested habitat (but see Wassie-Eshete et al. 2009). Many church forest tree species are listed as threatened on the IUCN Red List (Wassie-Eshete 2007). These forest patches not only preserve biodiversity, but also provide numerous ecosystem services: pollination, native seed stock, shade, spiritual sites, medicines from the plants, and fresh water conservation through sustaining rainfall patterns and underground springs (Jarzen et al. 2010; Lowman 2011). Pressure from subsistence agriculture and demand for firewood threaten these tiny forest fragments, which are embedded in an otherwise brown and arid landscape. Religious leaders are working with an international group of conservation biologists to educate local people about ecosystem services, focusing on insect pollinators as indicator species of forest health and utility (Lowman 2011). Ethiopia has lost more than 95% of its forest cover, but the partnership of religion and science has the capacity to save the remaining 5%, and perhaps ultimately lead to forest restoration (Bongers et al. 2006).

A final example of stakeholders facilitating conservation efforts is through local people engaging in ecotourism as a sustainable income stream. In many tropical regions, the payments derived from logging operations far exceed any economic benefits from conservation (Novotny 2010). In most cases, however, logging provides one-off, non-renewable profits that benefit local people in the short term only. Ecotourism revolving around, for example, canopy access walkways, bird watching, education-based nature tours, spas and holistic medicine (Weaver 2001) sustains 'green businesses' which provide long-term alternative income streams for villagers (Lowman 2009b). For example, over 20 canopy walkways now operate in tropical forests around the world, serving research, education and ecotourism (Lowman 2009a). Canopy walkways range in cost from US$100 to US$3000/m to establish, but then they can generate annual revenues for local stakeholders, as well as providing environmental education opportunities well into the future (Lowman & Bouricius 1995; Lowman 2004). Maintenance costs are minimal in these tropical ecotourism sites, usually because the locals have expertise (and pride) to undertake constant inspection and repair (Lowman 2009a).

For example, in the Sucasari tributary of the Rio Napo in Peru, the world's longest canopy walkway provides employment for over 100 local families as well as educating thousands of western visitors every year about rainforest ecology and conservation (Lowman 2009a). Costing some $250,000 to build, it generates revenue estimated at $1.2 million/year and, most importantly, it provides an economic incentive to conserve the primary forest. Revenues are significantly higher than those that could be achieved by felling the timber, because they have been sustained for over 15 years. In Western Samoa, a canopy access platform was similarly constructed, enabling local villagers on the island of Savaii'i to pay for their new school from ecotourism profits, instead of from selling logs. In a village where there is essentially no cash economy, the metrics are fuzzy but the conservation success is evident (Lowman 2009a).

Protecting a forest the size of a continent: good news from the Brazilian Amazon

One of the main challenges in tropical forest conservation is to work at a sufficiently large scale to ensure survival of viable populations, where individuals are able to move between protected areas through corridors of suitable habitat. This becomes particularly important in the context of global climate change, where some species may be unable to maintain viable populations in the face of warming temperatures or changing moisture supply. Their survival will then depend on their capacity to disperse to cooler or wetter locations.

Although there has been much media coverage of the possibility of climate change-induced 'dieback' of some tropical forest regions, such as in the Amazon basin, a more likely future scenario is one where forests persist under expected climate change, albeit with substantial changes in species composition in response to rising temperatures and changes in atmospheric CO_2 and rainfall regimes (Malhi et al. 2009; Zelazowski et al. 2011). Indeed, protection of sufficiently large areas of intact forest has long been seen as an important tool to help forest species adapt to global climate change, by maintaining the regional rainfall recycling and microclimate cooling services that forests provide (Malhi et al. 2009), as well as maintaining habitat to facilitate future range shifts.

Given the challenges of both adapting to and mitigating climate change, can forest area and a forest matrix be conserved on a sufficiently large scale? In recent years there has been an increased recognition of the role that tropical forest conservation can play in mitigating climate change, and expectation of a major increase in resources available for tropical forest conservation, in particular through the REDD+ (Reduced Emissions from Deforestation and Forest Degradation) mechanism. This has encouraged renewed hope that tropical deforestation can be slowed at global scales. In reality, finance is only part of the solution, and such aspirations run up against the economic and demographic pressures on landowners needing to earn income from food or cash crops such as cocoa, palm oil and beef, as well as the challenges of good governance and sustainable development in tropical forest frontier regions. The opportunities for tropical forest conservation have never been greater, but the challenges are also immense.

Amidst the conflicting reasons for optimism and despair for the future of tropical forests, a compelling and optimistic story emerges from the greatest tropical forest region, the Amazon rainforest of Brazil, which holds two-thirds of the overall Amazon forest. Since the 1980s the deforestation of the Brazilian Amazon has been one of the iconic images of the environmental degradation of the planet, as large areas of primary forest have been converted to cattle ranches, soya fields and small-holder farms, fuelled by government-supported road expansion and settlement schemes, and in many places accompanied by an atmosphere of lawlessness, corruption and poor governance. Over the decade 1996–2005, Brazilian Amazonia had a deforestation rate of 19,500 km^2 per year, about half of total global deforestation.

Then, from July 2005 to July 2010, something remarkable happened. Deforestation rates declined rapidly, dropping to 6450 km^2 per year by 2010, and with many indications that this decline will continue. This is a reduction by 67%, to the lowest levels of deforestation recorded since monitoring began in the 1980s. Such a reduction has led the Brazilian government to declare an intention to reduce deforestation rates by 80% below the 1995–2005 baseline by 2020, and some have suggested it is possible for net deforestation to come to a complete halt by the end of this decade (Nepstad et al. 2009). Such a turnaround is truly remarkable. If Brazil's ambitions can be achieved, it opens the prospect of Brazil achieving an advanced state of economic development with > 70% of its Amazon forest area still intact and supporting native ecosystems. This contrasts markedly with the heavily deforested and altered forest landscapes of North America, Europe and Asia, although the Atlantic Forest and *cerrado* savannas, the other major woody biomes of Brazil, have not fared so well, at least in part because government policy has shifted agricultural activity to these areas.

What are the factors that have driven this reduction in deforestation, and are they sustainable? What lessons do they hold for the future of other tropical forest regions? First, it is important to recognize the nature of deforestation in Brazil. Cattle ranching accounts for 80% of deforestation, with mechanized soya bean agriculture as a second major cause (Nepstad et al. 2009). Small-scale farming causes only a small fraction of deforestation. Hence, Brazilian deforestation is driven by (moderate to high) wealth, national economic integration and global market demand; it is not mainly driven by poverty, local demographic pressure or marginalization, as is the case in many other tropical regions. This level of organization and scale is the reason that deforestation rates are so high, but also means that these processes are more open to pressure for governance, certification and high environmental standards (Nepstad et al. 2009). Some of the initial causes of the reduction in deforestation rates have been economic, as the drop in the price of beef and soya over the period 2004–2006 reduced pressure for new land, but the subsequent rise in these prices coinciding with ongoing decline in deforestation rates suggests that the link between market demand and deforestation seems to have been broken. A number of features explain this dramatic decline.

Technical capacity and information

To manage deforestation, it is important to know where it is and why it is happening. Until recently, most tropical deforestation occurred under conditions of global ignorance. Brazil has led the world in open and sophisticated monitoring of its deforestation by satellite, both through its national space agency INPE and through environmental NGOs. INPE's PRODES system has been providing annual reports of forest loss. This has highlighted hot spots, enabled identification of illegal activities and also, importantly, raised the profile of deforestation. More recently, the DETER (detection of deforestation in real time) system, in parallel with similar initiatives driven by environmental NGOs such as Imazon in Brazil, has allowed monthly or shorter time scale reporting of deforestation activity (albeit at a lower resolution). This has become a powerful enforcement and governance tool, as new hot spots and drivers of deforestation can be acted on before they are a *fait accompli*.

Leadership and governance

Information is only useful if there is a will to use the information, and if it is used to make planning decisions that take forest conservation into account. In this regard, Brazil has shown environmental leadership. Some of this has been by a 'bottom-up' process of consulting stakeholders and planning resource use around new road expansion schemes, but much has also been by 'top-down' processes of enforcement of existing laws, and action on corruption and illegality. Such enforcement becomes much clearer when satellites provide deforestation information, and private land claims are clearly registered and demarcated. Recent examples of governance and enforcement include a federal campaign to identify and imprison illegal operators, including government employees. In 2008, the municipalities responsible for 50% of current deforestation were the focus of another federal campaign to register properties, publicize illegal holdings, cancel lines of credit for illegal landholders, and pressurize buyers of products from deforested lands (Soares et al. 2010).

Protected areas

Where sufficient governance exists, protected areas can be a powerful tool to assist in conservation of forest blocks in the context of regional development. Brazil has been extremely active in this regard, and has expanded its network of protected areas in Amazonia from 1.26 to 1.82 million km^2. This alone is estimated to have contributed 37% of the region's total reduction in deforestation between 2004 and 2006 (Soares et al. 2010).

Scientific institutions and civil society

The fact that controlling Amazonian deforestation has reached such a high profile within the Brazilian government (amidst intense political pressure for maintaining high deforestation from some lobbies) is a credit to the active engagement on these issues by informed Brazilian scientific institutions, and the active engagement by civil society groups, many well informed and with high technical capacity. This has led to Brazilian 'ownership' of the issue of Amazonian deforestation, encouraging dialogue and action. The technical capacity within Brazilian research institutes, governments and civil society is to some extent a product of decades of international scientific collaboration in Amazonia, through which many students and young scientists have been trained and have subsequently risen through the ranks of academia, government and civil society.

The remarkable decline in deforestation in Brazilian Amazonia has lessons for the wider tropics, despite the very different socio-ecological contexts and drivers of deforestation in different regions. Much is possible with leadership, technical capacity, open availability of information and good governance, and very little is possible without these factors. This has lessons for the surge of interest in financing forest conservation through REDD+. There is a need to build technical capacity and solve wider problems of governance and development at appropriate scale if REDD+ is to make a globally meaningful contribution to forest conservation. Even in Brazil, the challenges are ongoing, as a more crowded and wealthier world increases demand for food and biofuels at the expense of natural ecosystems.

Conclusions

Tropical forests are threatened by a suite of co-occurring human impacts. Chief among these threats are deforestation, overexploitation, climate change, habitat fragmentation and degradation, and invasive species. Often their effects will be synergistic; for example, both logging and climate change are likely to increase the frequency of damaging fires, which would be rare in unmodified forests (Barlow & Peres 2008). Ensuring the resilience of tropical forests and the persistence of their biodiversity in the face of this onslaught will require a pragmatic approach that extends well beyond the boundaries of protected areas. Reserves protecting core areas of undisturbed forest will remain the 'gold standard' for tropical forest conservation. We do not wish to downplay their importance: they are likely to be the only strategy able to guarantee the survival of a substantial proportion of tropical forest species. However, humans have already substantially modified much of the tropical landscape, and intense human pressures will inevitably shape its future. Part of the role for protected areas will be as a source of propagules able to colonize nearby human-modified habitats, allowing the natural restoration of species and ecosystem functions following perturbations. Informed by a landscape and ecosystem perspective, managers and scientists need to take advantage of any opportunity to maintain functioning, diverse ecological communities in human-modified tropical landscapes.

There is no single magic solution for tropical forest conservation and conservationists will need to be pragmatic, flexible and adaptable to promote the best solutions in the context of different economic pressures and varying ecological contexts. There is a wide range of possible solutions available in our armoury, some of which we have discussed in this chapter.

Destroying rain forest for economic gain is like burning a Renaissance painting to cook a meal.
— Edward O. Wilson

References

Aragão, L.E.O.C. & Shimabukuro, Y.E. (2010) The incidence of fire in Amazonian forests with implications for REDD. *Science*, **328**, 1275–1278.

Bagchi, R., Philipson, C.D., Slade, E.M., *et al.* (2011) Impacts of logging on density dependent predation of dipterocarp seeds in a Southeast Asian rainforest. *Philosophical Transactions of the Royal Society B-Biological Sciences*, **366**(1582), 3246–3255.

Balmford, A. (1996) Extinction filters and current resilience: the significance of past selection pressures for conservation biology. *Trends in Ecology and Evolution*, **11**, 193–196.

Barlow, J. & Peres, C.A. (2008) Fire-mediated dieback and compositional cascade in an Amazonian forest. *Philosophical Transactions of the Royal Society B: Biological Sciences*, **363**, 1787–1794.

Barlow, J., Mestre, L.A.M., Gardner, T.A. & Peres, C.A. (2007) The value of primary, secondary and plantation forests for Amazonian birds. *Biological Conservation*, **126**, 212–231.

Basset, Y., Missa, O., Alonso, A., *et al.* (2008) Changes in arthropod assemblages along a wide gradient of disturbance in Gabon. *Conservation Biology*, **22**, 1552–1563.

Bhagwat, S., Willis, K.J., Birks, H.J.B. & Whittaker, R.J. (2008) Agroforestry: a refuge for tropical biodiversity? *Trends in Ecology And Evolution*, **23**, 261–267.

Bongers, F., Wassie, A., Sterck, F.J., Bekele, T. & Teketay, D. (2006) Ecological restoration and church forests in northern Ethiopia. *Journal of the Drylands*, 35–44.

Brandon, K., Sanderson, S. & Redford, K. (1998) *Parks in Peril: People, Politics, and Protected Areas*. Island Press, Washington, D.C.

Brooks, T.M., Wright, S.J. & Sheil, D. (2009) Evaluating the success of conservation actions in safeguarding tropical forest biodiversity. *Conservation Biology*, **23**, 1448–1457.

Brown, N., Jennings, S. & Clements, T. (2003) The ecology, silviculture and biogeography of mahogany (*Swietenia Macrophylla*): a critical review of the evidence. *Perspectives in Plant Ecology Evolution and Systematics*, **6**, 37–49.

Chazdon, R.L. (2008) Beyond deforestation: restoring forests and ecosystem services on degraded lands. *Science*, **320**, 1458–1460.

Clark, C.J., Poulsen, J.R., Malonga, R. & Elkan, J.P.W. (2009) Logging concessions can extend the conservation estate for Central African tropical forests. *Conservation Biology*, **23**, 1281–1293.

Cochrane, M.A., Alencar, A., Schulze, M.D., *et al.* (1999) Positive feedbacks in the fire dynamic of closed canopy tropical forests. *Science*, **284**, 1832–1835.

Cox, P.S. & Banack, A. (eds) (1991) *Islands, Plants, and Polynesians: An Introduction to Polynesian Ethnobotany*. Dioscorides Press, Portland, OR.

Curran, L.M., Trigg, S.N., Mcdonald, A.K., *et al.* (2004) Lowland forest loss in protected areas of Indonesian Borneo. *Science*, **303**, 1000–1003.

Dent, D.H. & Wright, S.J. (2009) The future of tropical species in secondary forests: a quantitative review. *Biological Conservation*, **142**, 2833–2843.

Dirzo, R. & Miranda, A. (1991) Altered patterns of herbivory and diversity in the forest understorey: a case study of the possible consequences of contemporary defaunation. In: *Plant-Animal*

Interactions: Evolutionary Ecology in Tropical and Temperate Regions (eds P.W. Price, T.M. Lewinsohn, G.W. Fernandes & W.W. Benson). John Wiley, New York.

Ehrlich, P.R. & Pringle, R. (2008) Where does biodiversity go from here? A grim business-as-usual forecast and a hopeful portfolio of partial solutions. *Proceedings of the National Academy of Sciences USA*, **105**, 11579–11586.

FAO (2001) Global Forest Resources Assessment 2000: main report. Rome.

Folke, C., Carpenter, S., Walker, B., *et al.* (2004) Regime shifts, resilience and biodiversity in ecosystem management. *Annual Review of Ecology, Evolution and Systematics*, **35**, 557–581.

Gardner, T.A., Barlow, J., Chazdon, R.L., *et al.* (2009) Prospects for tropical forest biodiversity in a human-modified world. *Ecology Letters*, **12**, 561–582.

Garrity, D., Akinnifesi, F., Ajayi, O., *et al.* (2010) Evergreen agriculture: a robust approach to sustainable food security in Africa. *Food Security*, **2**, 197–214.

Hanski, I., Koivulehto, H., Cameron, A. & Rahagalala, P. (2007) Deforestation and apparent extinctions of endemic forest beetles in Madagascar. *Biology Letters*, **3**, 344–347.

Heatwole, H. & Lowman, M. (1987) Dieback: death of an Australian landscape. In: *If Atoms Could Talk: Search and Serendipity in Australian Science* (ed. R. Love). Greenhouse Publications, Richmond, Victoria.

Heckenberger, M., Russell, J., Fausto, C., *et al.* (2008) Pre-Columbian urbanism, anthropogenic landscapes, and the future of the Amazon. *Science*, **321**, 1214–1217.

Hunter, M.L. (1999) Biological diversity. In: *Maintaining Biodiversity in Forest Ecosystems* (ed. M.L. Hunter). Cambridge University Press, Cambridge.

Jarzen, D., Jarzen, S.A. & Lowman, M.D. (2010) In and out of Africa. *Palynological Society Newsletter*, **43**, 11–15.

Klein, A.M., Steffan-Dewenter, I. & Tscharntke, T. (2003) Fruit set of Highland coffee increases with the diversity of pollinating bees. *Proceedings of the Royal Society B: Biological Sciences*, **270**, 955–961.

Koh, L.P., Dunn, R.R., Sodhi, N.S., Colwell, R.K., Proctor, H.C. & Smith, V.S. (2004) Species coextinctions and the biodiversity crisis. *Science*, **305**, 1632–1634.

Lamb, D., Erskine, P. & Parrotta, J. (2005) Restoration of degraded tropical forest landscapes. *Science*, **310**, 1628–1632.

Larsen, T.H., Williams, N.M. & Kremen, C. (2005) Extinction order and altered community structure rapidly disrupt ecosystem functioning. *Ecology Letters*, **8**, 538–547.

Laurance, W.F. (2007) Have we overstated the tropical biodiversity crisis? *Trends in Ecology and Evolution*, **22**, 65–70.

Laurance, W.F. & Peres, C.A. (eds) (2006) *Emerging Threats to Tropical Forests*, University of Chicago Press, Chicago.

Lavorel, S. & Garnier, E. (2002) Predicting changes in community composition and ecosystem functioning from plant traits: revisiting the holy grail. *Functional Ecology*, **16**, 545–556.

Lenton, T.M., Held, H., Kriegler, E., *et al.* (2008) Tipping elements in the earth's climate system. *Proceedings of the National Academy of Sciences USA*, **105**, 1786–1793.

Lewis, O.T. (2009) Biodiversity change and ecosystem function in tropical forests. *Basic and Applied Ecology*, **10**, 97–102.

Lewis, O.T. & Gripenberg, S. (2008) Insect seed predators and environmental change. *Journal of Applied Ecology*, **45**, 1593–1599.

Lewis, S L. (2006) Tropical forests and the changing earth system. *Philosophical Transactions of the Royal Society B: Biological Sciences*, **361**, 435–439.

Lewis, S.L., Lloyd, J., Sitch, S., Mitchard, E.T.A. & Laurance, W.F. (2009) Changing ecology of tropical forests: evidence and drivers. *Annual Review of Ecology, Evolution and Systematics*, **40**, 529–549.

Loreau, M., Naeem, S. & Inchausti, P. (eds) (2002) *Biodiversity and Ecosystem Functioning: Synthesis and Perspectives*. Oxford University Press, Oxford.

Lowman, M. (2004) Ecotourism and the treetops. In: *Forest Canopies* (eds M.D. Lowman & H.B. Rinker). Elsevier, San Diego, CA.

Lowman, M. (2009a) Canopy walkways for conservation: a tropical biologist's panacea or fuzzy metrics to justify ecotourism? *Biotropica*, **41**, 545–548.

Lowman, M. (2009b) Biodiversity in tropical forest canopies as a "hook" for science education outreach and conservation. *Journal of Tropical Ecology*, **50**, 125–136.

Lowman, M. (2011) Finding sanctuary – conserving the forests of Ethiopia, one church at a time. *Explorers Journal*, **Winter**, 22–27.

Lowman, M. & Bouricius, B. (1995) The construction of platforms and bridges for forest canopy access. *Selbyana*, **16**, 179–184.

Lowman, M.D., Burgess, E. & Burgess, J. (2006) *It's a Jungle Out There – More Tales from the Treetops*. Yale University Press, New Haven, CT.

Malhi, Y., Aragão, L.E.O.C., Galbraith, D., *et al.* (2009) Exploring the likelihood and mechanism of a climate-change-induced dieback of the Amazon rainforest. *Proceedings of the National Academy of Sciences USA*, **106**, 20610–20615.

Morris, R.J., Lewis, O.T. & Godfray, H.C.J. (2004) Experimental evidence for apparent competition in a tropical forest food web. *Nature*, **428**, 310–313.

Nepstad, D., Soares, B.S., Merry, F., *et al.* (2009) The end of deforestation in the Brazilian Amazon. *Science*, **326**, 1350–1351.

Novotny, V. (2010) Rain forest conservation in a tribal world: why forest dwellers prefer loggers to conservationists. *Biotropica*, **42**, 546–549.

Novotny, V., Drozd, P., Miller, S.E., *et al.* (2006) Why are there so many species of herbivorous insects in tropical rainforests? *Science*, **313**, 1115–1118.

Pace, M.L., Cole, J.J., Carpenter, S.R. & Kitchell, J.F. (1999) Trophic cascades revealed in diverse ecosystems. *Trends in Ecology and Evolution*, **14**, 483–488.

Perfecto, I. & Vandermeer, J. (2008) Biodiversity conservation in tropical agroecosystems: a new paradigm. *Annals of the New York Academy of Science*, **1134**, 173–200.

Scheffer, M. (2009) Alternative stable states and regime shifts in ecosystems. In: *Princeton Guide to Ecology* (ed. S.A. Levin). Princeton University Press, Princeton, NJ.

Soares, B., Moutinho, P., Nepstad, D., *et al.* (2010) Role of Brazilian Amazon protected areas in climate change mitigation. *Proceedings of the National Academy of Sciences USA*, **107**, 10821–10826.

Sodhi, N.S., Koh, L.P., Brook, B.W. & Ng, P.K.L. (2004) Southeast Asian biodiversity: an impending disaster. *Trends in Ecology and Evolution*, **19**, 654–660.

Terborgh, J., Lopez, L., Nunez, P., *et al.* (2001) Ecological meltdown in predator-free forest fragments. *Science*, **294**, 1923–1926.

Verschuuren, B., Wild, R., Mcneely, J. & Oviedo, G. (eds) (2010) *Sacred Natural Sites: Conserving Nature and Culture*. Earthscan, London.

Wassie-Eshete, A. 2007. Ethiopian church forests – opportunities and challenges for restoration. PhD thesis, Wageningen Universiteit.

Wassie-Eshete, A., Sterck, F.J., Teketay, D. & Bongers, F. (2009) Tree regeneration in church forests of Ethiopia: effects of microsites and management. *Biotropica*, **41**, 110–119.

Weaver, D.B. (2001) *The Encyclopedia of Ecotourism*. Cabi Publishing, New York.

Willis, K.J., Gillson, L. & Brncic, T.M. (2004) How "virgin" is virgin rainforest? *Science*, **304**, 402–403.

Wright, A.C.S., Romney, D.H., Arbuckle, R.H. & Vial, V.E. (1959) *Land in British Honduras*. H.M. Stationery Office, London.

Wright, S.J. & Muller-Landau, H.C. (2006) The future of tropical forest species. *Biotropica*, **38**, 287–301.

Zelazowski, P., Malhi, Y., Huntingford, C., Sitch, S. & Fisher, J.B. (2011) Changes in the potential distribution of humid tropical forests on a warmer planet. *Philosophical Transactions of the Royal Society A: Mathematical, Physical and Engineering Sciences*, **369**, 137–160.

III

Taxonomic case studies

14

A global perspective on conserving butterflies and moths and their habitats

Thomas Merckx[1], Blanca Huertas[2], Yves Basset[3] and Jeremy Thomas[4]

[1]Wildlife Conservation Research Unit, Department of Zoology, Recanati-Kaplan Centre, University of Oxford, Oxford, UK
[2]Life Sciences Department, The Natural History Museum, Cromwell Road, London, UK
[3]Smithsonian Tropical Research Institute, Apartado, 0843-03092, Balboa, Ancon, Peru
[4]Department of Zoology, University of Oxford, Oxford, UK

Just living is not enough, said the butterfly, one must have sunshine, freedom and a little flower.
— Hans Christian Andersen

Introduction

Lepidoptera are one of the four major insect orders, and one of the best studied invertebrate groups, containing over 160,000 described species and an estimated equal number of undescribed species, arranged in 124 families (Kristensen et al. 2007). Lepidoptera occupy all except the very coldest terrestrial regions, but the Neotropics and Indoaustralian region have five times more species per unit area than the Palaearctic and Nearctic, and three times more than the Afrotropical region (Heppner 1991). They are scale-winged insects, traditionally divided into three major assemblages: micro-moths, butterflies and macro-moths (Kristensen et al. 2007).

The order represents a mega-diverse radiation of almost exclusively phytophagous insects, probably correlated with the great diversification of flowering plants since the Cretaceous (Menken et al. 2010). They provide many vital and economically important services within terrestrial ecosystems (e.g. nutrient recycling, soil formation, food resources and pollination). The scale of these contributions is illustrated by the estimate that blue tit (*Parus caeruleus*) chicks consume at least 35 billion caterpillars each year in the UK alone (Fox et al. 2006). Lepidoptera also have considerable human significance, both economic and scientific. A growing industry farms pupae for supply to butterfly houses across the world. One moth species has been domesticated in order to provide silk (i.e. *Bombyx mori*

Key Topics in Conservation Biology 2, First Edition. Edited by David W. Macdonald and Katherine J. Willis.

from the wild *B. mandarina*). For scientists, the group offers a model system valuable to studies of biodiversity conservation, ecology, ethology, genetics, (co)evolution and systematics (Samways 1995; Boggs et al. 2003).

Human appreciation of the beauty and vulnerability of (especially) butterflies has grown exponentially in recent decades, particularly in developed nations. For example, among the 40 national biodiversity mapping schemes extant in the UK, more than 2.5 million records were submitted for Lepidoptera thanks to the work of the Centre for Ecology and Hydrology, Rothamsted Research and Butterfly Conservation (BC) UK, with key input from amateur enthusiasts (butterflies 2.2 m; macro-moths 384 k) prior to 2000, roughly double the number received for all other invertebrates combined, or indeed for birds (1.2 m) (Thomas 2005). Despite a burgeoning interest in UK moths that has seen 12.4 million records amassed by over 5000 volunteer recorders in recent years (BC's Moths Count initiative) (see Chapter 8), butterflies are currently (and regrettably) probably the only taxon of terrestrial invertebrates across much of the world for which it is realistic to assess the scale and rates of change in species' ranges or populations (Lewis & Senior 2011). For the same reason, butterflies have been successfully used as charismatic flagships and umbrella species (see Box 14.1) in insect conservation programmes (New 1997; Thomas & Settele 2004; Fleishman et al. 2005; Guiney & Oberhauser 2009).

Long-term change in populations of Lepidoptera

Rates and causes

Before discussing practical conservation of Lepidoptera, it is necessary to consider their known rates and causes of change, and whether these are representative of other insect species (Thomas & Clarke 2004; Fleishman et al. 2005). In the UK, which has the longest history of rigorous recording, butterfly populations have changed dramatically since the 1850s. Some species have increased their range but most have declined, and 7% of British species are extinct. The mean decline in butterflies has been an order of magnitude greater than that of birds or vascular plants (or, where monitored, mammals), whether measured at the scale of single sites, regions or the entire nation (Thomas et al. 2004). Moreover, until the recent application of ecological principles to conservation described below, local extinction rates on nature reserves often exceeded those on commercially managed land, in sharp contrast to the stability achieved for vertebrates and plants (Thomas 1991).

The four methods of assessment available in the UK (Red Data Books (RDBs), species surveys, mapping and population monitoring schemes) are of shorter antiquity elsewhere but it is clear that UK declines are typical of other developed nations, and are exceeded by some (Maes & Van Dyck 2001). In The Netherlands, for example, 24% of 71 butterfly species became extinct during the 20th century while the number of breeding birds increased by 20% (Thomas 1995).

There is some debate as to whether butterflies are indicators of change in other insects: Hambler & Speight (2004) argued that butterflies have suffered higher extinction rates than other invertebrates according to UK RDBs; Thomas & Clarke (2004) attributed this discrepancy to an artefactual underestimate of decline inherent in comparing poor with well-sampled taxa, and showed that butterflies experienced similar extinction rates to other groups when sampling intensity is factored in. Moreover, observed extinction rates in dragonflies, bumblebees and macro-moths have unequivocally been slightly higher than those of UK butterflies (Thomas 2005; Conrad et al. 2006). Whilst clearly unrepresentative of certain species and functional types, because of their popularity, ease of study (e.g. conspicuous, day-active, often identifiable in the field) and patterns of species richness and endemism that mirror those of many other insects, butterflies are increasingly used as indicators of change in other taxa in Europe and to a lesser extent elsewhere.

Butterflies may be useful indicators of habitat change (Ricketts et al. 1999). We distinguish two types of indicator (Thomas et al. 2005): (i) the 'miners' canary' whose decline heralds future losses for less sensitive species; and (ii) taxa which mirror change or predict the presence (Fleishman et al. 2005) of poorly monitored organisms. Across Europe (Erhardt & Thomas 1991; Thomas 2005), butterflies in general are early warning systems for future change in vertebrate (apart from mega-fauna) and vascular plant populations. One family of butterflies, the Lycaenidae, provides ultra-sensitive indicators of coming change in other families, because many of them need two specialized larval resources (foodplants, ants) to co-occur (Thomas et al. 2005). We therefore advocate the application of standardized mapping and population monitoring schemes in nations where such schemes are absent, and the monitoring of other arthropod taxa.

Drivers of change

Of the main drivers of global biodiversity loss, the spread of exotic pest species and direct over-exploitation by humans have had negligible detectable impacts on populations of Lepidoptera (e.g. Collins & Morris 1985). So the banning of trade and collecting – the main measure applied in many nations today and for the first century of conservation practice in the UK – is inappropriate (unless coupled with habitat conservation), because it fails to account for the very different population dynamics and life-history traits of insects compared with vertebrates and many plants (Thomas 1995). This resilience arises because Lepidoptera populations typically have high intrinsic rates of increase wherever the quality of habitat is high, with individual females laying many eggs which subsequently experience high density-dependent mortalities (in unperturbed populations), especially in the later larval stages, allowing numbers to recover quickly if the previous generation of adults was depleted by collectors. Furthermore, collectors are seldom able to remove more than a small proportion of the effective breeding population of adult Lepidoptera per generation, because in most studied species the majority of eggs are laid within 2–5 days of each female's emergence, and for species with discrete generations, the short-lived individuals emerge over a period of 4–8 weeks.

Habitat loss, undoubtedly the prime culprit (Stewart et al. 2007), can broadly be divided into two processes (Thomas 1991): (i) the destruction of primary and species-rich secondary ecosystems by intensive modern agriculture, exotic-species forestry, mining, armed conflict and illicit crops (Dávalos et al. 2011) and, to a lesser extent, urbanization; (ii) the reduced size, increased isolation and degradation in quality of those fragments of potentially inhabitable biotopes that survive. The first process effectively eliminates all populations apart from pests of crops and exploiters of ruderal plants. The second is less clear-cut but equally harmful, especially in developed regions (see below).

Another driver of population reductions is climate change. Its observed impacts on Lepidoptera are relatively minor so far but it is predicted to rival habitat change (with which it interacts) in future decades (van Swaay et al. 2010a). Already in the Holarctic, non-migratory species have shown southern or lowland contractions that exceed their northward or altitudinal shifts in ranges (e.g. Parmesan et al. 1999), whilst similar altitudinal shifts are detectable in moth communities in Borneo (Chen et al. 2009). The net impacts on Lepidoptera of future climate and land use changes are rightly a research priority.

Single-species conservation

Compared with tropical regions, species richness within temperate biomes is generally much lower, both in ecosystems as a whole and among the taxa they support. Lower diversity has undoubtedly made it easier to name and understand a large proportion of temperate

Box 14.1 From single-species to community and landscape conservation: *Maculinea arion* in the UK

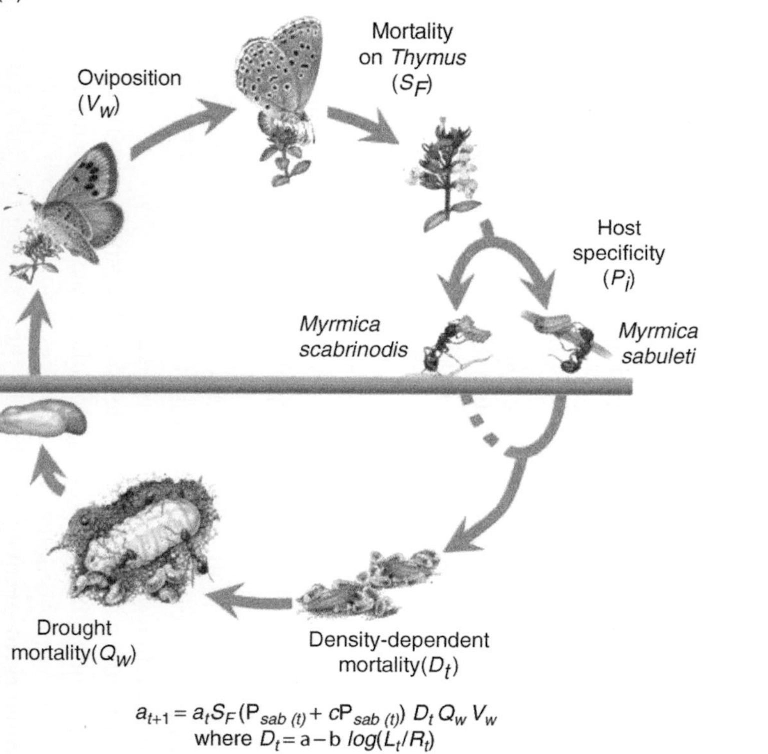

(a) *M. arion* is a globally threatened flagship butterfly with specialized larvae that briefly eat *Thymus* flowers before inhabiting *Myrmica* ant nests, where they prey on ant brood. A 6-year study identified and modelled the parameters driving its population dynamics (see Thomas et al. 2009 for symbols). The key discovery was its host specificity to one ant, *Myrmica sabuleti*, rather than to any *Myrmica* species.

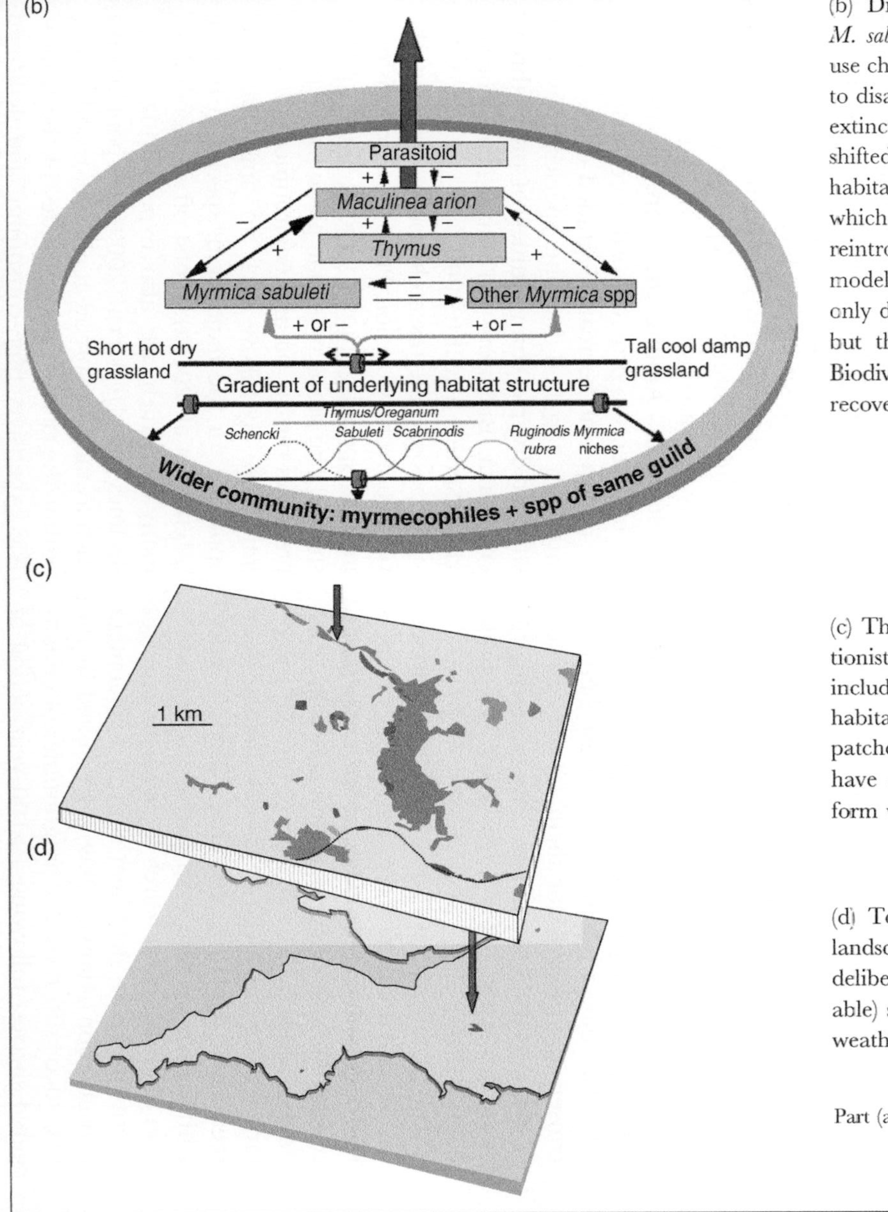

(b) Different *Myrmica* ants occupy different niches within grassland; *M. sabuleti* dominates in warm short swards under UK climates. Land use changes resulted in the abandonment of most sites causing *M. sabuleti* to disappear or be displaced by unsuitable congeners, causing *M. arion's* extinction in the UK in 1979. Targeted conservation management shifted the sward structure (indicated by the position of sleeves along the habitat gradient in the diagram) back to the optimum for *M. sabuleti*, which quickly returned to dominate the turf. *M. arion* was then reintroduced from Sweden and increased rapidly, closely matching model predictions on 21 tested sites (Thomas et al. 2009, 2011). Not only did *M. arion's* immediate community of interacting rarities benefit, but the restoration of a disappearing type of habitat caused other Biodiversity Action Plan organisms belonging to the same guild to recover, including threatened plants, insects and birds.

(c) The collateral benefits from *M. arion* restorations justified conservationists in applying similar targeted management across the landscape, including the restoration of calcareous grassland and the creation of new habitat on railway constructions. *M. arion* has spread to 25 sites (dark patches), forming a meta-population of loosely connected colonies, as have other priority species. Mid-grey indicates woodland plantations form which most sites were restored.

(d) Today, similar programmes are being initiated in the other UK landscapes formerly inhabited by *M. arion*. In all regions, some sites are deliberately managed to create heterogeneous and suboptimal (but suitable) swards for *M. arion*, to buffer the new communities from extreme weather and to prepare for future climate change.

Part (a) adapted with permission from Thomas et al. (2009)

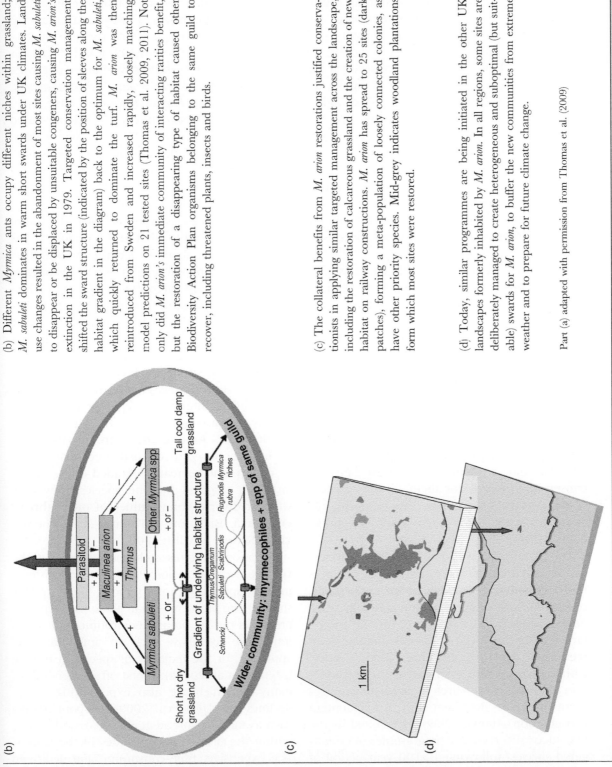

biodiversity, a process that began with the rise of 'natural history' in Europe during the 1700s. Hence, there is much more knowledge on distribution, ecology and life-history of Lepidoptera, and of many more species, in temperate regions than in the intrinsically more diverse tropical regions, where a significant percentage of taxa remains undescribed. Today, many temperate countries have valid Red List assessments of nearly all butterfly species, and in Europe there is even a continental Red List (van Swaay et al. 2010b) whereas tropical countries do not, or rather, only of subsamples (see below). However, only Norway and Finland have national Red Lists for moths, although other developed countries are considering them.

By contrast, tropical birds and mammals have received considerable attention, both because they are more popular and because their combined global species richness is only 5% of that estimated for Lepidoptera. The relatively small numbers of species-specific conservation actions in the tropics mainly focus on these popular taxa, whereas most species-specific conservation research is done within temperate regions. The lack of the ecological knowledge necessary for species-specific actions for tropical Lepidoptera, combined with financial constraints, may explain the contrast between conservation efforts for them (mainly as ecosystem protection) and in the west (mainly as species action plans) (New 2009; Bonebrake et al. 2010).

The current and welcome increase in species-specific research and action plans mainly focuses on regionally threatened habitat specialists, confined to rare biotopes within generally small nature reserves. The prospects for the insects targeted are improved by specific management measures applying evidence-based science. Although now well established, the introduction of ecological research to practical management marked a paradigm shift in Lepidoptera conservation (Hanski 1999). It resulted in the first successful recovery programmes of endangered species, following a >100-year period of failed (simpler) approaches such as the regulation of collecting and the establishment of biotopes as nature reserves without recourse to managing their internal structures or successional dynamics (Thomas 1991). In the UK, it is arguable that five butterfly species (*Satyrium pruni, Polyommatus bellargus, Hesperia comma, Melitaea athalia, Papilio machaon*) have been saved from national extinction by science-based management since the 1970s–90s, whilst three nationally extinct, globally threatened *Maculinea* species have been re-established on specifically managed sites in the UK or The Netherlands (Thomas et al. 2011; Wynhoff et al. 2011).

The single-species approach has therefore had several benefits. First and foremost, it has succeeded for several declining butterflies, whereas biotope protection *per se* historically failed to maintain them, for population extinction rates on nature reserves up to 1980 typically exceeded those on neighbouring land (Thomas 1991). Importantly, success breeds success in conservation, and the demonstrable recovery of (alas rather few) declining iconic species in Europe has greatly increased public and political interest in Lepidoptera, and has hugely increased the flow of funding, leading to wider gains. Among these are multi-million euro projects to protect pristine ecosystems and to restore and recreate degraded ones, which, although targeted for one species, inevitably support diverse communities of other threatened taxa (Bickmore & Thomas 2000; Thomas 2001; Settele & Kühn 2009).

Species-specific research and actions, e.g. the UK Biodiversity Action Plan, now increasingly include specialist moths among their priority species. The problem is that, given the increased pressures on the natural environment, it seems impossible to follow such an 'intensive care' approach for the majority of species in trouble (Merckx et al. 2010a). Worst off are generalists, many of which are also experiencing severe declines (Van Dyck et al. 2009), less popular taxa such as moths (Conrad et al. 2006), and species within the diverse communities of the tropics. The expensive, time-consuming and dedicated approach is obviously desirable to rescue highly threatened iconic species in desperate situations, such as the monarch roosts of central Mexico, and the Queen Alexandra's birdwing (the world's largest and perhaps most endangered butterfly) in Papua New Guinea (Thomas & Settele 2004). There is an argument that such care should be temporary and eventually relaxed as threat levels decrease. Perhaps the strongest arguments for its current continuation are: (i) beneficial umbrella impacts on other species, and (ii) the gaining of the knowledge required for designing efficient biotope-specific management.

We therefore believe there is currently still a need for both species-specific and biotope-specific approaches in temperate and (sub)tropical regions. The focus on biotopes should gradually be increased in temperate regions, including on land outside nature reserves managed for agriculture, silviculture and urban-industrial purposes such as road and railway constructions. At the same time, the species-specific conservation effort for highly endangered iconic tropical species should be considerably increased, especially those with severely restricted ranges. Indeed, a consequence of the lack of population studies in the tropics is the minimal knowledge of foodplants for larvae, and spatial structure and dispersal capabilities of most populations of tropical Lepidoptera (Bonebrake et al. 2010, but see Marini-Filho & Martins 2010).

From single sites to meta-populations: ecological conservation at landscape scales

The development of the modern ecological approach to Lepidoptera conservation included two paradigm shifts: first, the concept that surviving patches of habitat were declining in quality and had reduced (or no) capacity to support a valued species; and later, that irrespective of their quality, the surviving islands of habitat were too few, small and isolated for a meta-population of populations to persist in a landscape. In practice, the two concepts are inextricably entwined, both in their causes and in their consequences for population dynamics. The same socio-economic changes that lead to ecosystem destruction and fragmentation typically alter both successional dynamics and the development of plagioclimaxes within the surviving habitat islands (Thomas et al. 2001).

Apart from special cases, such as monarch butterfly overwintering sites, reduced habitat quality primarily affects fitness of Lepidoptera in the larval stage (Thomas 1991). In a degraded biotope, the larval foodplant typically remains abundant (in some cases increases) but grows in less suitable forms or with altered nutritional value in response to pollution, changed water levels or other attributes such as microclimate (Thomas et al. 2011). For most butterflies studied, the mean density of adults on isolated sites containing optimum larval habitat was around 100 times greater (spatially or temporally) than on those containing the lowest-quality source habitat that had supported a population for at least 10 consecutive generations. In contrast, fluctuations caused by weather and variation in adult resources (e.g. nectar, mating sites), were 10 to 100 times smaller. And only in one studied species (*Celastrina argiolus*) were interactions with an enemy (a specific parasitoid) sufficient to overrule the population variation due to habitat quality (Thomas et al. 2011).

In contrast, isolation of habitat patches affects mainly the adult stage, which is unexpectedly sedentary in most studied species of Lepidoptera, especially in habitat specialists. Increased isolation causes meta-populations to disappear from landscapes as populations die out, having failed either to track the generation of new habitat patches or resources across modern landscapes or to replenish those that go extinct (Hanski 1999). On the spectrum between classic 'blinking light' meta-populations or mainland-island ones, the positions of most populations of 'colonial' species of Lepidoptera have been a matter of some debate. The former structure probably approaches the norm for several Melitaenini and related fritillaries (Schtickzelle & Baguette 2009), but in other species may be a transient phase following habitat fragmentation, presaging the breakdown of the whole system (Harrison 1994).

At the applied level, there was an unwelcome dichotomy during the 1980s–90s in Europe between (i) spending scarce resources solely on improving the (degenerating) internal habitat quality of existing conservation sites for endangered guilds or species of Lepidoptera, rather than adding to the suite of reserves, and (ii) the exact opposite approach. Today an evidence-based consensus recognizes that it is equally important to address both processes. This resulted from a series of field studies from Thomas et al. (2001) onwards, which found that, whilst the density of individuals within a site reflects the quality of habitat within each occupied patch rather than its isolation or size, the chance of a patch being occupied in the first place was correlated strongly, independently and almost equally with both its isolation from neighbours and (again) the quality of the patch. Surprisingly, very few single-species studies (as opposed to whole communities of species) have shown patch size to be a significant predictor of occupancy once account is taken of habitat quality (Thomas et al. 2011). There are three explanations for the contribution of patch quality to meta-population stability in Lepidoptera: (i) for a given patch size, optimum habitat supports populations that are up to 100-fold larger, and hence more likely to persist, than those of low (but just suitable) source habitat; (ii) the former release more emigrants into the matrix to seek new sites, disproportionally so since individual butterflies hatching into high-density populations are more likely to leave their natal patch (Hovestadt *et al.* 2010); (iii) new sites containing high-quality habitat are more likely to be colonized successfully by an immigrant female, since the vulnerable period of low numbers is short (Thomas et al. 2011).

By applying the above principles, European conservationists concerned with Lepidoptera are (at last) matching the successes achieved for many plants and vertebrates (see Box 14.1). Still, there are key questions to be addressed by future research both at patch and landscape scales. To what extent are subsets of species regionally adapted to different biotic interactions or environmental conditions (e.g. local climate)? How responsive are local phenotypes to rapid environmental change, for example in selection for more dispersive forms or types physiologically adapted to a warmer climate or capable of switching foodplants or hosts? Although larval habitat quality and adult dispersal are the two overriding factors so far found to regulate populations and meta-populations in most studied Lepidoptera (Thomas et al. 2001), do other factors (e.g. adult resources) play a key role for populations elsewhere (Dennis 2010) or under tropical climates? How important is the matrix separating patches of breeding habitat in providing resources for adults or facilitating/obstructing their spread (Dennis 2010)?

Advancing towards multi-species conservation

As we argue elsewhere, with only half of the world's estimated Lepidoptera described and a minute proportion well studied, it is impossible to provide targeted conservation programmes for more than a small number of highly valued species, at least outside nations where diversity is exceedingly low (e.g. the UK). Moreover, very few nature reserves are managed primarily for Lepidoptera, although typically the plants, ants and (often more endangered) specific parasitoids with which each interacts benefit directly (e.g. Anton et al. 2007).

We have seen that declines in Lepidoptera are driven primarily by factors that affect all species, rather than by targeted overcollecting (see above). In addition, we have suggested that butterflies can be sensitive predictors of the impacts of environmental change on other organisms, as well as useful representatives of less conspicuous terrestrial insects (Thomas

2005). In theory, therefore, the restoration of optimum conditions for a rapidly declining butterfly will require restoring a type of habitat within existing biotopes, or a network of sites to landscapes, that should benefit the community characteristic of that configuration of habitat patches.

In practice, there are few examples of observed umbrella benefits resulting from single-species conservation of insects, which has caused its efficacy to be doubted (Stewart et al. 2007) despite abundant evidence in support from biotope manipulation for other taxa (e.g. Dunk et al. 2006). This is an inevitable consequence of a lack of resources to monitor collateral changes in most Lepidoptera conservation projects. In the USA, Launer & Murphy (1994) found that protection of all those serpentine soil-based grasslands in central California occupied by the federally protected butterfly *Euphydryas editha bayensis* would also conserve 98% of native spring-flowering plant species (although the percentage fell sharply if only the largest butterfly sites were targeted). In the UK (Thomas et al. 2011), the restoration of scrub habitats designed to increase *Satyrium pruni*, of chalk grassland for *Polyommatus coridon*, and of limestone and acid grasslands for *Maculinea arion* (see Box 14.1) produced rapid and diverse gains across a suite of declining insects, plants and (where studied) birds. In the case of *M. arion*, the beneficiaries included 33% of the declining butterfly species listed in the UK's Biodiversity Action Plan, as well as RDB-listed species of ants, beetles, flies, cockroaches, plants and birds. Most species increased because they were adapted to the warm early-successional grassland structures created for *M. arion* in the UK. Several also increased due to a direct or indirect relationship with ants. For example, alongside another *Viola* species, the UK-RDB *V. lactea* has myrmecochorous seeds that are particularly attractive to *Myrmica* ants, which after eating the eliasomes eject the seeds into sparsely vegetated soil around their nests. The consequences of up to a thousand-fold increase of *Maculinea arion*'s host-ant *Myrmica sabuleti* was a ca. 100-fold increase on three sites of *Viola* and a 10–15-fold increase (and new populations) of three fritillary butterflies whose larvae eat *Viola*, including *Argynnis adippe* and *Boloria euphrosyne*, two of the UK's most endangered insect species (Randle et al. 2005). In this respect the target butterfly, *M. arion*, acted as an indicator for a keystone species of ant, *M. sabuleti*, which dominates its habitat at its scale. Since most Lycaenidae (nearly a third of butterfly species) interact with ants as larvae or pupae, we suspect that they will prove to be an especially useful umbrella taxon for other groups.

Encouraging as these results may be, there is clearly a pressing need for future research and conservation practice to understand the ecological requirements of a spectrum of endangered Lepidoptera with enough precision to ensure the continuity of sufficient habitat for each component within a single heterogeneous ecosystem, for example as co-existing successional stages within woodland, heath and grassland. Equally clearly, this is more likely to succeed in multi-sited landscape-scale projects which, in our view rightly, have been the trend in recent years.

Two multi-species approaches

Landscape restoration

The successful landscape-scale conservation of multi-species assemblages basically boils down to either conserving/protecting 'pristine' landscapes or restoring human-altered landscapes. Although adequate protection of the world's remaining 'wilderness' areas is both more important and also more effective, such areas are dwindling worldwide. Human-altered landscapes keep extending ever larger, so conservation efforts within them are becoming relatively more significant. Overall, landscape-scale restoration approaches within such areas should minimize the fragmentation of specific habitat resources.

We believe it is useful to make a distinction between areas that (i) have a short history of human alteration, e.g. deforested in recent decades, (ii) have a long history, e.g. farmed for several centuries, and (iii) are 'pristine'. We argue that, above the general ecological conservation principles applying to all, successful conservation of the extant biodiversity in each needs a different approach.

The massive recent and ongoing deforestation/ degradation of tropical forests (in several West African countries, forest cover loss exceeds 90%; Safian et al. 2011) has dramatic consequences for the highly diverse Lepidoptera fauna adapted to these climax ecosystems. Agri- and silvicultural mosaics where the percentage of converted original forest exceeds 30%, including selective logging of three or more large trees per hectare, show species compositional shifts with loss of many components of the butterfly community (Brown 1997). Hence, the only way to repair some of the damage is the combination of thorough protection of remaining patches of primary forests, successful regeneration of secondary forests, and targeted reforestation projects within agro-forestry systems and clear-cut areas (Schulze et al. 2004; Safian et al. 2011). Species recolonization rates depend primarily on life-history characteristics, patch size and biological structure, distance to source patches and permeability of the intervening matrix. Still, the often substantial dispersal/recolonization potential of tropical butterfly and moth communities allows them to regenerate several decades after clearance (Hilt & Fiedler 2005; Safian et al. 2011). Obviously, recent conversion by agriculture in areas originally not fully forested need protection and restoration of the original set of biotopes (e.g. Afromontane mosaic landscapes; Tropek & Konvicka 2010).

The situation is more complex in countries, such as most of Europe, with a long history of human alteration. Here, ultimate conservation goals are the subject of much, but not enough, debate. The natural climax forests and sparsely forested pastures kept open by mega-herbivores have been fully replaced, often millennia ago, by so-called semi-natural biotopes, which are essentially different versions of early- to mid-successional natural seral stages or plagioclimaxes, prevented from reverting towards forest. Only scattered fragments of ancient woodland remain, and these have suffered continuous disturbance as human needs for woodland products changed. For example, up to a century ago most European woodland was maintained as coppice or very open coppice-with-standards, to provide fuel and fencing materials, whereas today much woodland has a more uniformly closed shady canopy than is found in ancient forests with no history of disturbance. And since very old trees were no more valuable to humans than middle-aged ones, the rarest type of arboreal habitats in developed countries today are those associated with rotting wood on ancient trees: yet these saproxylic habitats support numerous endangered invertebrate specialists, especially Coleoptera and Diptera, and some moths (e.g. *Parascotia fuliginaria*). Although many species were undoubtedly lost over the centuries during this transition, others have managed to adapt successfully as specialists in these semi-natural biotopes (e.g. heaths, grasslands, hay meadows, marshes and coppiced woodland).

The intensification of agriculture and forestry and of urbanization since the 1950s, and more recently abandonment in response to socio-economic imperatives, has severely decreased these semi-natural biotopes in quantity and quality, and with them their specialist fauna and Lepidoptera. The Red List considers agricultural intensification and abandonment a major threat for almost 30% of European butterfly species (van Swaay et al. 2010b). As a result, most 'conservationists' seek to sustain or restore semi-natural biotopes, and do so by maintaining very specific disturbance regimes (often simply by copying traditional agricultural practices, since most large wild herbivores were excluded centuries ago) (New 2009). Nevertheless, popular management operations to influence vegetation structure, such as burning and grazing/ mowing, need careful planning (mainly to ensure refugia) as they may destroy much of the existing invertebrate fauna if applied too intensively, too infrequently, on too large a scale or at unsuitable times of year (New 2009). Hence, although wrongly applied conservation management may have unintended negative consequences, management is seen as a good thing overall, whereas abandonment of human disturbance is often perceived as a threat.

The situation should not be black and white. Abandonment undoubtedly poses a threat to many specialist species that have become adapted to certain semi-natural biotopes, and is a threat to specialists that have nowhere else to go because natural succession dynamics are currently too disturbed and suitable natural patches are too small and/or isolated. On the other hand, abandonment will most certainly benefit some endangered specialist faunal groups too (e.g. closed-woodland Lepidoptera, saproxylic groups), and may also provide 'rewilding' opportunities in biotopes evolving towards mosaics including mature climax vegetation or natural successional stages such as coastal and river areas, wood gaps and high-altitude areas, although even here, open habitats may be much rarer than in prehuman landscapes, owing to the absence of most former natural herbivores, e.g. bison, aurochs, wild sheep and horses (and, more locally, beaver) in Europe (see Chapter 23).

We believe there should be room for both active management (i.e. restoration of semi-natural biotopes) and passive abandonment (i.e. rewilding), even at small spatial scales (Merckx et al. 2012a). They are complementary, and now is the time to designate areas to one or the other, on a European-wide scale. Such allocations should be done carefully, and for as many taxa as possible, taking into account many variables, such as historic and recent distributional data, international threat status and life-history traits, and they should have clear, quantifiable conservation goals. The information on butterflies (i.e. European Red List: van Swaay et al. 2010b; Prime Butterfly Areas in Europe: van Swaay & Warren 2003) will be a valuable input to such a multi-taxa exercise. The current and possible future distribution and intrinsic properties of nature reserves (e.g. Natura 2000 sites, UK Sites of Special Scientific Interest) and High Nature Value farmland, together with what will be politically and financially achievable through the soon to be revised Common Agricultural Policy, will all be instrumental in producing a road-map, with substantially increased funds and mechanisms for semi-natural biotope restoration, which will clearly delineate sites/areas best managed under each approach.

In addition, restoration of areas usually regarded as 'the matrix' in between (semi-)-natural patches, such as intensive farmland, brownfields and even urbanized areas, is of value too. These areas are currently often rather 'simple' and 'homogeneous' in terms of habitat resources, so restoration may make a relatively large difference to their conservation value (Tscharntke et al. 2005), as well as increasing the value of neighbouring (semi-)natural patches (Dennis 2010). Brownfield sites provide many opportunities for restoration of successional biotopes otherwise not strongly represented locally, and restoration plans should be tailored to focal species and/or generally improve biotopes by assuring a sufficient quantity, quality and spatiotemporal diversity of resources (New 2009; Dennis 2010). Restoration of intensive agricultural areas may be globally important given their huge and growing footprint. Here, the aim should be to reconcile intensive agricultural practices with wider societal benefits, including biodiversity. The question is how to decide which landscape elements to restore, how, and at what spatial scale in order to make farmland less hostile to a broad range of declining 'wider countryside' and rare, localized species (Merckx et al. 2010a).

Agri-environment schemes (AES) can reverse negative biodiversity trends by increasing resource heterogeneity and improving dispersal success (Shreeve & Dennis 2011). However, they must be made more efficient and cost-effective, so that they actually achieve their goals (Kleijn et al. 2006; Settele et al. 2009). One way to achieve this is by implementing specific measures for high-priority species within AES targeted at landscapes where such species occur, as this approach has been shown to benefit specialist butterfly species (Brereton et al. 2011). However, we argue that this species-specific approach must be complemented by a multi-species approach in order to more fully address the steep declines in farmland biodiversity. General AES that are focused on the restoration and implementation of vital landscape elements are key to this multi-species approach. Even simple AES management prescriptions applied to relatively small areas can benefit Lepidoptera populations. For example, the restoration and management of arable field margins has been shown to benefit a range of butterfly species, both on conventionally managed (Feber & Smith 1995; Feber et al. 1996) and organically managed (Feber et al. 2007) farmland. Hedgerow management can have a positive effect on declining species such as the brown hairstreak *Thecla betulae* butterfly (Merckx & Berwaerts 2010). In addition, we have recently discovered that the protection of existing hedgerow trees, and the provision of new ones, is likely to be a highly beneficial conservation tool for populations of moths, and probably many other flying insects too, as hedgerow trees provide a sheltered microclimate and other key habitat resources (Merckx et al. 2009a, 2010b, 2012b). Nevertheless, hedgerow tree and field margin AES options are likely to obtain best results for moth populations where farmers are targeted to join these schemes across the landscape, probably because this results in a landscape-scale joining up of habitat resources, which especially benefits the large proportion of moth species of intermediate mobility that use the agricultural biotope and move through it on a scale larger than the field scale (Merckx et al. 2009a,b, 2010b) (Figure 14.1).

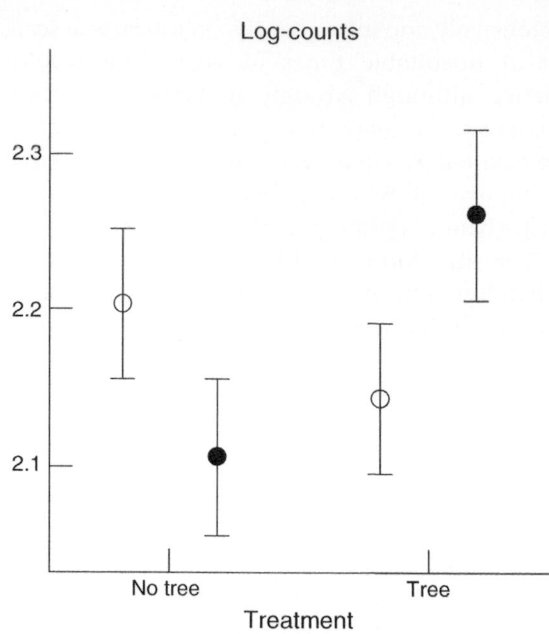

Figure 14.1 Fortnightly individual moth counts (log $N+1$) (with SE) contrasting the effects of presence/absence of a hedgerow tree on moth abundance within areas where farmers had (•) and had not been (○) targeted to apply for agri-environment schemes. Reprinted from Merckx *et al.* (2009a), with permission from Elsevier.

Landscape conservation

Sites with 'no' or little human disturbance, and with high biodiversity and/or threat levels, are candidates for landscape conservation, a very effective strategy for protecting endangered species and biotopes (Bruner et al. 2001). The number of national park declarations, mainly for biodiversity conservation purposes, has increased in recent years. Resources for these national parks are finite, and must be directed to the most important sites. GIS-based prioritization exercises with a multitude of data layers, including forest structure change and sound distributional and modelled occupancy data of threatened species, are a prerequisite.

The first threatened species assessments were published for mammals and birds in the mid-1960s. So far, out of the entire planet's biodiversity, only ca. 45,000 species have been assessed against IUCN criteria, mostly vascular plants and vertebrates, plus <1500 insects including ca. 300 Lepidoptera (Vié et al. 2008). The IUCN is now conducting assessments, using the Sampled Red List Index (SRLI) methodology, of the threat status of samples of 1500 species of the lesser known groups without complete Red Lists, such as many invertebrates. Recent initiatives include IUCN assessments of butterflies in Europe (van Swaay et al. 2010b) and the rest of the world (Afrotropics: Lewis & Senior 2011; Neotropics: Willmott et al. 2011).

Some site-based assessment methods have been developed with particular reference to butterflies. For example, Ackery & Vane-Wright (1984) proposed the concept of *critical fauna analysis*, whereby regions that contain certain local faunas are identified by their endemic species, and if these regions were protected, all species of a particular group would be too. Only a few analyses of 'critical faunas' have been conducted in tropical areas (e.g. Collins & Morris 1985; Hall 1999; Vane-Wright & de Jong 2003; Willmott 2003). Data are assessed to identify optimally efficient, single-site sequences of near-equal priority areas for a group using the complementarity principle (Vane-Wright et al. 1991; Williams 2001) and incorporating other criteria (Margules & Pressey 2000; Araújo et al. 2002). This method is also useful in assessing habitats on a broader geographic scale, where it is possible to detect areas of unusually high significance in understanding evolutionary processes (New 1997).

Van Swaay & Warren (2003) proposed a selection of Prime Butterfly Areas (PBAs) for threatened species in Europe. PBAs are defined as a preliminary selection of areas supporting species meeting three criteria.

- *Biogeography*: European range-restricted species
- *Conservation*: threatened species defined by IUCN criteria
- *Legislation*: species listed in Appendix II of the Bern Convention and/or the EU Habitats and Species Directive.

This initiative identified 431 PBAs (covering 1.8% of Europe). However, not enough is known of threatened species in tropical regions to apply this method there. Threat assessments for butterflies are available but are not yet comprehensive for the group as a whole. Wells et al. (1983) produced the first IUCN invertebrate RDB including some butterflies. The first national assessment for invertebrates was produced for England by Shirt (1987) and assessments of threatened butterflies for Europe by Heath (1981).

Only one family of butterflies, the Papilionidae (i.e. 'swallowtails'), has been assessed on a global scale (Collins & Morris 1985). This family has a pan-global distribution and includes both widespread and habitat-restricted species, making them well suited for conservation studies. They include some of the largest and most spectacular butterfly species, attracting the attention of amateurs and specialists alike, so their taxonomy and distribution are relatively well known. The assessments were made under earlier versions of IUCN criteria and were mostly qualitative; 170 papilionids were considered threatened or near-threatened in this study (Collins & Morris 1985). Later, New & Collins (1991) showed that nearly 14% of papilionid taxa are believed to be threatened or declining, plus 17% with no information to be assessed. The Lycaenidae is another major butterfly group that has been subject to some detailed conservation studies (e.g. New 1993; Thomas et al. 2005). This family is difficult to study due to its high diversity (almost 40% of all described butterfly species) and complex and poorly known taxonomy in many regions. The other major butterfly families (Pieridae, Nymphalidae, Hesperiidae and Riodinidae) have not had a global assessment.

Van Swaay & Warren (1999) provided the first comprehensive regional review of the status of butterflies for Europe (except Turkey and Cyprus). Out of 576 butterfly species assessed, 71 (12% of the total) were categorized as threatened, of which 19 were globally threatened species and 52 regionally threatened. In the tropical regions, hardly any such regional assessments have yet been done, though various assessments have been carried out at a national level, some of which use modified IUCN criteria or subjective assessments for butterflies and a few moth species (e.g. Colombia: Amat-García et al. 2007; Venezuela: Rodríguez & Rojas-Suárez 2008; Brazil: MMA 2008). However, those assessments do not cover the entire butterfly fauna of each country, and variation in the criteria used introduces discrepancies. For example, the high-elevation butterfly *Lymanopoda paramera* was separately assessed twice as it lives in the border region of Colombia and Venezuela. It was assessed as Vulnerable in Venezuela, but as Critically Endangered in Colombia, perhaps because different methodologies were used during the assessments, and/or the difference in extent of each national range. In addition, there are some *ad hoc* assessments of certain recently described species of Andean butterflies (e.g. Hall 1999; Willmott 2003; Huertas et al. 2009; Huertas 2011). However, less than 10% of the Neotropical butterfly fauna (ca. 45% of the world) has been assessed so far (Willmott et al. 2011).

Various site-based conservation assessment initiatives are based mainly on birds. The Important Bird Areas (IBAs) of Grimmett & Jones (1989) was later applied to plants as Important Plant Areas (IPAs) (PlantLife 2004), and then to all groups as Key Biodiversity Areas (KBA) (Langhammer et al. 2007). In order to be listed as a KBA, a site must support at least one globally threatened species, range-restricted species, biome-restricted species or congregation of species. Notably, in the context of invertebrates, the designation of KBAs does not require the identity of all threatened or range-restricted species, just that there is *a* threatened species at the locality. The KBA programme is less useful in tropical regions, where almost all habitats include some threatened species. As a result, a further important initiative is the Alliance for Zero Extinction (AZE) (Ricketts et al. 2005). AZE sites are defined much more restrictively (using three specific criteria) as the most important locality for an Endangered or Critically Endangered species. The Tropical Andean Butterfly Diversity Project (TABDP; www.andeanbutterflies.org) is currently developing a list of KBAs for the Neotropics based on the presence of threatened butterfly species studied in the IUCN SRLI (Box 14.2) (Willmott et al. 2011).

The lack of threat assessments for Lepidoptera and other invertebrates in the tropics has serious consequences. Research and conservation resources may not be adequately targeted or may rely upon passive conservation of other, better known, taxa. The lack of threat assessments may bias funding towards better known groups. Some grant research programmes require supported projects to concentrate on species rated as threatened. Consequently, one way of obtaining funding for Lepidoptera research has been to conduct surveys within multi-taxa studies involving other faunistic groups (e.g. Huertas & Donegan 2006). The NGO ProAves (www.proaves.org) has established various nature reserves and helped in the declaration of a national park in Colombia for threatened bird and butterfly species based on some of the latter studies.

Conclusion

Because of the now global dimension of rapid biodiversity decline, and its detrimental impact on humanity, we need to manage unsustainable land use and massive conversion and degradation of natural habitats. It is our duty to preserve and restore natural areas, not only because of their intrinsic value but also, from a utilitarian point of view, to avoid the functional breakdown of the ecosystems on which we depend. The worldwide ubiquity, abundance, sheer diversity, indicator capacity and both historic and current appeal to scientists and amateur naturalists make the Lepidoptera an excellent group to monitor conservation efforts worldwide, and they complement conservation narrowly focused on birds and mammals alone (not least because butterflies decline faster than birds and plants: Thomas et al. 2004). Here, while we have commented on how approaches to Lepidoptera conservation

Box 14.2 A case study: the Tropical Andean Butterfly Diversity Project (TABDP)

This project (www.andeanbutterflies.org) is a major initiative involving international collaboration among scientists, institutions and students working to establish a foundation for future research on the butterflies of the Andean region, a global biodiversity hotspot. The project was inspired by the BioMap Project (www.biomap.net) and started in 2005 as an international collaboration among institutions based in five Andean countries (Colombia, Peru, Bolivia, Venezuela and Ecuador), the UK and the USA. Threatened species and site assessments require baseline distributional data. This project sought to collect and collate these data, based largely on the resources and information available in museum collections.

The TABDP started from scratch in the capture of the data on the distribution of Neotropical butterfly species. A tailored database was designed to capture data on the locality, identification and other details of specimens in museums. Also, a manual for the use of databases (Willmott & Huertas 2006) and another for butterfly photography (Huertas & Willmott 2006), in both Spanish and English, were published. A database of ca. 200,000 records and photographs of the types of Andean species is now freely available online.

Building the capacity of host countries to conduct research on tropical Andean butterflies and train a new generation of researchers was a primary goal of this project. Around 300 students and professionals from various collections were trained in eight courses in five countries and a first Andean butterfly network, now with ca. 600 members, has been established. Threatened species assessments have been produced for 350 Neotropical butterfly species, in collaboration with the IUCN, using the SRLI methodology. Based on these assessments, the first KBAs based on butterflies have been proposed for South America (Willmott et al. 2011).

The TABDP data capture methods and threat assessment methods can be replicated for any insect group and applied in other conservation initiatives, for example using more species or at national or local levels. Projects and institutions should not spare any effort in improving data sources and providing accessibility. Research programmes and targeted surveys are key sources of data, which should be considered (and funded) as part of the conservation process. Locality information necessary to produce threat assessments can be gathered only with an army of naturalists or parataxonomists (Basset et al. 2004), trained in collating and analysing data effectively. However, as taxonomic expertise is crucial when doing the assessments, more people need to be trained and more resources be facilitated for core taxonomy.

differ between regions and land use types, we stress the importance of adopting a landscape scale allied to a resource-based view, both for single-species and for biotope/community conservation.

Rapid land use change, especially in recent decades, has caused serious declines in butterflies and moths worldwide, despite the recent designation of many new nature reserves (e.g. van Swaay et al. 2010b). It is hence clear that we need to go a lot further, with far greater long-term resources. Society should now start to invest massively on five fronts: (i) protecting and buffering natural areas (i.e. more, better managed and larger reserves), (ii) restoring and managing robust networks of semi-natural biotopes, (iii) rewilding areas where this is appropriate, (iv) improving typical 'matrix' areas, and (v) gathering data for poorly known areas and species concerning distribution, habitat requirements, population changes and taxonomy, which would also benefit such areas by increasing local awareness and by the production of field guides. Such an increased effort will not be in vain, as there is compelling evidence, for vertebrates at least, that conservation efforts can halt and even reverse biodiversity loss, provided there are sufficient resources and the collective will to protect critical habitat resources (Hoffmann et al. 2010). The efforts need to be monitored too. Lepidoptera are uniquely easy to survey for a better understanding of biodiversity change, and as recent experiences in Europe suggest that the challenge could be met, we call for projects to make rigorous population trend estimates in undermonitored regions (see also Pereira et al. 2010). Such projects would be especially welcome in the tropics, where few (if any) Lepidoptera monitoring schemes exist.

References

Ackery, P.R. & Vane-Wright R.I. (1984) *Milkweed Butterflies. Their Cladistics and Biology. Being an Account of the Natural History of the Danainae, A Subfamily of the Lepidoptera, Nymphalidae*. British Museum of Natural History, London.

Amat-García, G., Andrade, M.G. & Amat, E. (2007) *Libro Rojo de los Invertebrados Terrestres de Colombia*. Instituto de Ciencias Naturales, Universidad Nacional de Colombia, Conservación Internacional Colombia, Bogotá, Colombia.

Anton, C., Musche, M. & Settele, J. (2007) Spatial patterns of host exploitation in a larval parasitoid of the Dusky Large Blue *Maculinea nausithous*. *Basic and Applied Ecology*, **8**, 66–74.

Araújo, M.B., Williams, P.H. & Turner, A. (2002) A sequential approach to minimise threats within selected conservation areas. *Biodiversity and Conservation*, **11**, 1011–1024.

Basset, Y., Novotny, V., Miller, S.E., Weiblen, G.D., Missa, O. & Stewart, A.J.A. (2004) Conservation and biological monitoring of tropical forests: the role of parataxonomists. *Journal of Applied Ecology*, **41**, 163–174.

Bickmore, C.J. & Thomas, J.A. (2000) The development of habitat for butterflies in former arable cultivation. *Aspects of Applied Biology*, **58**, 305–312.

Boggs, C.L., Watt, W.B. & Ehrlich, P.R. (2003) *Butterflies: Ecology and Evolution Taking Flight*. University of Chicago Press, Chicago.

Bonebrake, T.C., Ponisio, L.C., Boggs, C.L. & Ehrlich, P.R. (2010) More than just indicators: a review of tropical butterfly ecology and conservation. *Biological Conservation*, **143**, 1831–1841.

Brereton, T., Roy, D.B., Middlebrook, I., Botham, M. & Warren, M. (2011) The development of butterfly indicators in the United Kingdom and assessments in 2010. *Journal of Insect Conservation*, **15**, 139–151.

Brown, K.S. (1997) Diversity, disturbance, and sustainable use of Neotropical forests: insects as indicators for conservation monitoring. *Journal of Insect Conservation*, **1**, 25–42.

Bruner, A.G., Gullison, R.E., Rice, R.E. & da Fonseca, G.A.B. (2001) Effectiveness of parks in protecting tropical biodiversity. *Science*, **291**, 125–128.

Chen, I.C., Shiu. H.J., Benedick. S., *et al.* (2009) Elevation increases in moth assemblages over 42 years on a tropical mountain. *Proceedings of the National Academy of Sciences USA*, **106**, 1479–1483.

Collins, N.M. & Morris, M.G. (1985) Threatened swallowtail butterflies of the world. In: *The IUCN Red Data Book*. IUCN, Gland, Switzerland.

Conrad, K.F., Warren, M.S., Fox, R., Parsons, M.S. & Woiwod, I.P. (2006) Rapid declines of common, widespread British moths provide evidence of an insect biodiversity crisis. *Biological Conservation*, **132**, 279–291.

Dávalos, L., Bejarano, A.C., Hall, M.A., Correa, H.L, Corthals, A. & Espejo, O.J. (2011) Forests and drugs: coca-driven deforestation in tropical biodiversity hotspots. *Environmental Science and Technology*, January 11 (epub ahead of print).

Dennis, R.L.H. (2010) *A Resource-based Habitat View for Conservation – Butterflies in the British Landscape*. Wiley-Blackwell, Chichester.

Dunk, J.R., Zielinkski, W.J. & Walsh, H.H. (2006) Evaluating reserves for species richness and representation in northern California. *Diversity and Distributions*, **12**, 434–442.

Erhardt, E. & Thomas, J.A. (1991) Lepidoptera as indicators of change in semi-natural grasslands of lowland and upland Europe. In: *The Conservation of Insects and Their Habitats* (eds N.M. Collins & J.A. Thomas), pp.213–236. Academic Press, London.

Feber, R.E. & Smith, H. (1995) Butterfly conservation on arable farmland. In: *Ecology and Conservation of Butterflies* (ed. A.S. Pullin). Chapman and Hall, London.

Feber, R.E., Smith, H. & Macdonald, D.W. (1996) The effects of management of uncropped edges of arable fields on butterfly abundance. *Journal of Applied Ecology*, **33**, 1191–1205.

Feber, R.E., Johnson, P.J., Firbank, L.G., Hopkins, A. & Macdonald, D.W. (2007) A comparison of butterfly populations on organically and conventionally managed farmland. *Journal of Zoology* (London), **273**, 30–39.

Fleishman, E., Thomson, J.R., MacNally, R., Murphy, D.D. & Fay, J.P. (2005) Using indicator species to predict species richness of multiple taxonomic groups. *Conservation Biology*, **19**, 1125–1137.

Fox, R., Conrad, K.F., Parsons, M.S., Warren, M.S. & Woiwod, I.P. (2006) *The State of Britain's Larger Moths*. Butterfly Conservation and Rothamsted Research, Wareham, Dorset.

Grimmett, R.F. & Jones, T.A. (1989) *Important Bird Areas in Europe*. ICBP Technical Publication No. 9. ICBP, Cambridge.

Guiney, M.S. & Oberhauser, K.S. (2009) Insects as flagship conservation species. *Terrestrial Arthropod Reviews*, **1**, 111–123.

Hall, J.P.W. (1999) *A Revision of the Genus Theope: Its Systematics and Biology (Lepidoptera: Riodinidae: Nymphidiini)*. Scientific Publishers, Gainesville, FL.

Hambler, C. & Speight, M.R. (2004) Extinction rates and butterflies. *Science*, **305**, 1563.

Hanski, I. (1999) *Metapopulation Ecology*. Oxford University Press, Oxford.

Harrison, S. (1994) Metapopulations and conservation. In: *Large-Scale Ecology and Conservation* (eds P.J. Edwards, N.R. Webb & R.M. May), pp. 111–128. Blackwell Scientific Publishers, Oxford.

Heath, J. (1981). *Threatened Rhopalocera (Butterflies) in Europe*. Nature and Environment Series No. 23. Council of Europe, Strasbourg.

Heppner, J.B. (1991). Faunal regions and the diversity of Lepidoptera. *Tropical Lepidoptera*, **2**, 1.

Hilt, N. & Fiedler, K. (2005) Diversity and composition of Arctiidae moth ensembles along a successional gradient in the Ecuadorian Andes. *Diversity and Distributions*, **11**, 387–398.

Hoffmann, M., Hilton-Taylor, C., Angulo, A. *et al.* (2010) The impact of conservation on the status of the world's vertebrates. *Science*, **330**, 1503–1509.

Hovestadt, T., Kubisch, A. & Poethke, H.J. (2010) Information processing in models for density-dependent emigration: a comparison. *Ecological Modelling*, **221**, 405–410.

Huertas, B.C. & Donegan, T.M. (2006) *Proyecto YARÉ: Investigación Y Evaluación de las Especies Amenazadas de la Serranía de los Yariguíes, Santander, Colombia*. Informe final. Colombian EBA Project Report Series 7. www.proaves.org.

Huertas, B. & Willmott, K.R. (2006) *Manual para la Toma de Fotografías Digitales del Proyecto TABD*. www.andeanbutterflies.org

Huertas, B. (2011) A new species of Satyrinae butterfly from Peru (Nymphalidae: Satyrini: Euptychiina). *Zootaxa* (in press).

Huertas, B., Rios, C. & Le Crom, J.F. (2009) A new species of Splendeuptychia from the Magdalena Valley in Colombia (Lepidoptera: Nymphalidae: Satyrinae). *Zootaxa*, **2014**, 51–58.

Kleijn, D., Baquero. R.A., Clough, Y., *et al.* (2006) Mixed biodiversity benefits of agri-environment schemes in five European countries. *Ecology Letters*, **9**, 243–254.

Kristensen, N.P., Scoble, M.J. & Karsholt, O. (2007) Lepidoptera phylogeny and systematics: the state of inventorying moth and butterfly diversity. *Zootaxa*, **1668**, 699–747.

Langhammer, P.F., Bakarr, M.I., Bennun, L.A. *et al.* (2007) *Identification and Gap Analysis of Key Biodiversity Areas: Targets for Comprehensive Protected Area Systems*. IUCN, Gland.

Launer, A.E. & Murphy, D.D. (1994) Umbrella species and the conservation of habitat fragments: a case of a threatened butterfly and a vanishing grassland ecosystem. *Biological Conservation*, **69** 145–153.

Lewis, O.T. & Senior, M.J.M. (2011) Assessing conservation status and trends for the world's butterflies: the Sampled Red List Index approach. *Journal of Insect Conservation*, **15**, 121–128.

Maes, D. & Van Dyck, H. (2001) Butterfly diversity loss in Flanders (north Belgium): Europe's worst case scenario? *Biological Conservation*, **99**, 263–276.

Margules, C.R. & Pressey, R.L. (2000) Systematic conservation planning. *Nature*, **405**, 243–253.

Marini-Filho, O.J. & Martins, R.P. (2010) Nymphalid butterfly dispersal among forest fragments at Serra da Canastra National Park, Brazil. *Journal of Insect Conservation*, **14**, 401–411.

Menken, S.B.J., Boomsma, J.J. & van Nieukerken, E.J. (2010). Large-scale evolutionary patterns of host plant associations in the Lepidoptera. *Evolution*, **64**, 1098–1119.

Merckx, T. & Berwaerts, K. (2010) What type of hedgerows do Brown hairstreak (*Thecla betulae* L.) butterflies prefer? Implications for European agricultural landscape conservation. *Insect Conservation and Diversity*, **3**, 194–204.

Merckx, T., Feber, R.E., Riordan, P., *et al.* (2009a) Optimizing the biodiversity gain from agri-environment schemes. *Agriculture, Ecosystems and Environment*, **130**, 177–182.

Merckx, T., Feber, R.E., Dulieu, R.L., *et al.* (2009b) Effect of field margins on moths depends on species mobility: field-based evidence for landscape-scale conservation. *Agriculture, Ecosystems and Environment*, **129**, 302–309.

Merckx, T., Feber, R.E., Parsons, M.S., *et al.* (2010a) Habitat preference and mobility of *Polia bombycina*: are non-tailored agri-environment schemes any good for a rare and localised species? *Journal of Insect Conservation*, **14**, 499–510.

Merckx, T., Feber, R.E., Mclaughlan, C., *et al.* (2010b) Shelter benefits less mobile moth species: the field-scale effect of hedgerow trees. *Agriculture, Ecosystems and Environment*, **138**, 147–151.

Merckx, T., Feber, R.E., Hoare, D.J., *et al.* (2012a) Conserving threatened Lepidoptera: towards an effective woodland management policy in landscapes under intense human land-use. *Biological Conservation*, **149**, 32–39.

Merckx, T., Marini, L., Feber, R.E., *et al.* (2012b) Hedgerow trees and extended-width field margins enhance macro-moth diversity: implications for management. *Journal of Applied Ecology*, **49**: 1396–1404.

MMA (Ministerio do Meio Ambiente). (2008) *Livro Vermelho da Fauna Brasileira Ameaçada de Extinção*. Belo Horizonte, MG, Fundação Biodiversitas, Brasília.

New, T.R. (1993) *Conservation Biology of Lycaenidae (Butterflies)*. Occasional Paper of the IUCN Species Survival Commission, No. 8. IUCN, UK.

New, T.R. (1997) *Butterfly Conservation*, 2nd edn, Oxford University Press, Melbourne.

New, T.R. (2009) *Insect Species Conservation*. Cambridge University Press, Cambridge.

New, T.R. & Collins, N.M. (1991) *Swallowtail Butterflies: An Action Plan for Their Conservation*. IUCN, Gland, Switzerland.

Parmesan, C., Rytholm, N., Stefanescu, C., *et al.* (1999) Poleward shifts of species' ranges associated with regional warming. *Nature*, **399**, 579–583.

Pereira, H.M., Belnap, J., Brummitt, N., *et al.* (2010) Global biodiversity monitoring. *Frontiers in Ecology and the Environment*, **8**, 459–460.

PlantLife (2004) *Identifying and Protecting the World's most Important Plant Areas*. Plantlife, Salisbury.

Randle, Z., Simcox, D.J., Schönrogge, K., Wardlaw, J.C. & Thomas, J.A. (2005) *Myrmica* ants as keystone species and *Maculinea arion* as an indicator of rare niches in UK grasslands. In: *Studies in the Ecology & Conservation of Butterflies in Europe 2. Species Ecology Along a European Gradient: Maculinea Butterflies as a Model* (eds J. Settele, E. Kuehn & J.A. Thomas), pp.26–27. Pensoft, Sofia.

Ricketts, T.H., Dinerstein, E., Olson, D.M. & Loucks, C. (1999) Who is where in North America? Patterns of species richness and utility of indicator taxa for conservation. *Bioscience*, **49**, 369–381.

Ricketts, T.H., Dinerstein, E., Boucher, T., *et al.* (2005) Pinpointing and preventing imminent extinctions. *Proceedings of the National Academy of Sciences USA*, **51**, 18497–18501.

Rodríguez, J.P. & Rojas-Suárez, F. (2008) *Libro Rojo de la Fauna Venezolana*, 3rd edn. Provita y Shell Venezuela, S.A., Caracas, Venezuela.

Safian, S., Csontos, G. & Winkler, D. (2011) Butterfly community recovery in degraded rainforest habitats in the Upper Guinean Forest Zone (Kakum forest, Ghana). *Journal of Insect Conservation*, **15**, 351–359.

Samways, M.J. (1995) *Insect Conservation Biology*. Chapman and Hall, London.

Schtickzelle, N. & Baguette, M. (2009) Metapopulation viability analysis: a crystal ball for the conservation of endangered butterflies? In: *Ecology of Butterflies in Europe* (eds J. Settele, T. Shreeve, M. Konvicka & H. Van Dyck), pp339–352. Cambridge University Press, Cambridge.

Schulze, C.H., Waltert, M., Kessler, P.J.A., *et al.* (2004) Biodiversity indicator groups of tropical land-use systems: comparing plants, birds, and insects. *Ecological Applications*, **14**, 1321–1333.

Settele, J. & Kühn, E. (2009) Insect conservation. *Science*, **325**, 41–42.

Settele, J., Dover, J., Dolek, M. & Konvicka, M. (2009) Butterflies of European ecosystems: impact of land use and options for conservation management. In: *Ecology of Butterflies in Europe* (eds J. Settele, T. Shreeve, M. Konvicka & H. Van Dyck), pp353–370. Cambridge University Press, Cambridge.

Shirt, D.B. (1987) *British Red Data Books, 2. Insects*. Nature Conservancy Council, Peterborough.

Shreeve, T.G. & Dennis, R.L.H. (2011) Landscape scale conservation: resources, behaviour, the matrix and opportunities. *Journal of Insect Conservation*, **15**, 179–188.

Stewart, A.J.A., New, T.R. & Lewis, O.T. (2007) *Insect Conservation Biology*. CABI, Wallingford.

Thomas, J.A. (1991) Rare species conservation: case studies of European butterflies. In: *The Scientific Management of Temperate Communities for Conservation* (eds I. Spellerberg, B. Goldsmith & M.G. Morris, M.G), pp149–197. Blackwells, Oxford.

Thomas, J.A. (1995) Why small cold-blooded insects pose different conservation problems to birds in modern landscapes. *Ibis*, **137**, 112–119.

Thomas, J.A. (2001) Can ecologists recreate habitats and restore absent species? In: *Nature, Landscape and People Since the Second World War: A Celebration of the 1949 Act* (eds C. Smout & J. Sheail), pp150–160. Royal Society of Edinburgh, Edinburgh.

Thomas, J.A. (2005) Monitoring change in the abundance and distribution of insects using butterflies and other indicator groups. *Philosophical Transactions of the Royal Society B*, **360**, 339–357.

Thomas, J.A. & Clarke, R.T. (2004) Extinction rates and butterflies. *Science*, **305**, 1563–1564.

Thomas, J.A., & Settele, J. (2004) Butterfly mimics of ants. *Nature*, **432**, 283–284.

Thomas, J.A., Bourn, N.A.D., Clarke, R.T., *et al.* (2001) The quality and isolation of habitat patches both determine where butterflies persist in fragmented landscapes. *Proceedings of the Royal Society of London B*, **268**, 1791–1796.

Thomas, J.A., Telfer, M.G., Roy, D.B., *et al.* (2004) Comparative losses of British butterflies, birds, and plants and the global extinction crisis. *Science*, **303**, 1879–1881.

Thomas, J.A., Clarke, R.T., Randle, Z., *et al.* (2005) *Maculinea* and myrmecophiles as sensitive indicators of grassland butterflies (umbrella species), ants (keystone species) and other invertebrates. In: *Studies in the Ecology and Conservation of Butterflies in Europe 2. Species Ecology Along a European Gradient: Maculinea Butterflies as a Model* (eds J. Settele, E. Kühn & J.A. Thomas), pp28–31.Pensoft, Sofia.

Thomas, J.A., Simcox, D.J. & Clarke, R.T. (2009) Successful conservation of a threatened *Maculinea* butterfly. *Science*, **325**, 80–83.

Thomas, J.A., Simcox, D.J & Hovestadt, T. (2011) Evidence based conservation of butterflies. *Journal of Insect Conservation*, **15**, 241–258.

Tropek, R. & Konvicka, M. (2010) Forest eternal? Endemic butterflies of the Bamenda highlands, Cameroon, avoid close-canopy forest. *African Journal of Ecology*, **48**, 428–437.

Tscharntke, T., Klein, A.M., Kruess, A., Steffan-Dewenter, I. & Thies, C. (2005) Landscape perspectives on agricultural intensification and biodiversity – Ecosystem service management. *Ecology Letters*, **8**, 857–874.

Vane-Wright, R. & de Jong (2003) *The Butterflies of Sulawesi. Annotated Checklist for a Critical Island Fauna*. Backhuys, Leiden.

Vane-Wright, R.I., Humphries, C.J. & Williams, P.H. (1991) What to protect? Systematics and the agony of choice. *Biological Conservation*, **55**, 235–254.

Van Dyck, H., van Strien, A.J., Maes, D. & van Swaay, C.A.M. (2009) Declines in common, widespread butterflies in a landscape under intense human use. *Conservation Biology*, **23**, 957–965.

Vié, J.C., Hilton-Taylor, C. & Stuart, S.N. (2008) *The 2008 Review of the IUCN Red List of Threatened Species*. IUCN Gland, Switzerland.

Van Swaay, C.A.M. & Warren, M.S. (1999) *Red Data Book of European butterflies (Rhopalocera)*. Nature and Environment No. 99, Council of Europe Publishing, Strasbourg.

Van Swaay C.A.M. & Warren, M. (2003) *Prime Butterfly Areas in Europe – Priority Sites for Conservation*. National Reference Centre for Agriculture, Nature and Fisheries. Ministry of Agriculture, Nature Management and Fisheries, Wageningen, The Netherlands.

Van Swaay, C.A.M., Harpke, A., van Strien, A., *et al.* (2010a) *The Impact of Climate Change on Butterfly Communities 1990–2009*. Report VS2010.025, Butterfly Conservation Europe and De Vlinderstichting, Wageningen, The Netherlands.

Van Swaay, C.A.M., Cuttelod, A., Collins, S., *et al.* (2010b) *European Red List of Butterflies*. IUCN and Butterfly Conservation Europe, Gland, Switzerland.

Williams, P.H. (2001) Key sites for conservation: area-selection methods for biodiversity. In: *Conservation in a Changing World: Integrating Processes into Priorities for Action* (eds G.M. Mace, A. Balmford & J.R. Ginsberg). Cambridge University Press, Cambridge.

Willmott, K.R. (2003) *The Genus Adelpha: Its Systematics, Biology and Biogeography (Lepidoptera: Nymphalidae: Limenitidini)*. Scientific Publishers, Gainesville, FL.

Willmott, K.R. & Huertas, B. (2006) *Manual para el Manejo de las Bases de Datos el Proyecto TABD*. www.andeanbutterflies.org.

Willmott, K.R., Lamas, G. & Huertas, B. (2011) *Priorities for Research and Conservation of Tropical Andean Butterflies*. McGuire Center for Lepidoptera and Biodiversity, Gainesville, FL.

Wells, S.M., Pyle, R.M. & Collins, N.M. (1983) *The IUCN Invertebrate Red Data Book*. IUCN, Gland, Switzerland.

Wynhoff, I., van Gestel, R., van Swaay, C. & van Langevelde, F. (2011) Not only the butterflies: managing ants on road verges to benefit *Phengaris (Maculinea)* butterflies. *Journal of Insect Conservation*, **15**, 189–206.

15

Bird conservation in tropical ecosystems: challenges and opportunities

Joseph A. Tobias[1], Çağan H. Şekercioğlu[2] and F. Hernan Vargas[3]

[1]Edward Grey Institute, Department of Zoology, University of Oxford, Oxford, UK
[2]Department of Biology, University of Utah, Salt Lake City, UT, USA
[3]Peregrine Fund, Boise, ID, USA

"What would the world be, once bereft of wet and of wildness?"

Gerard Manley Hopkins

Introduction

Bird conservation is a global mission but most of the key battles are being played out in the tropics. Tropical ecosystems are generally under greater pressure than their temperate counterparts from human population growth, agricultural expansion and a host of related factors. They also support 87% of bird species, many of them highly susceptible to habitat loss or climate change (Şekercioğlu et al. 2012). The concern that many tropical species are therefore destined for extinction has focused much effort towards setting global conservation priorities based on a minimum number of protected areas, often one per species (e.g. Rodrigues et al. 2004; Ricketts et al. 2005). However, this approach is founded on traditional conservation strategies developed in the temperate zone, and the extent to which it can be applied to tropical birds remains unclear.

Here, we summarize the key attributes of tropical ecosystems and implications for bird conservation. First, we outline threats to key tropical environments. Then we argue that tropical species often differ from their temperate-zone counterparts in ways that pose novel challenges for conservation. We conclude that sustainable conservation of tropical birds and the ecosystem services they provide will be achieved only if attention is focused not merely on current snapshots of species distributions and protected areas but on biotic processes and interactions operating at larger spatial and temporal scales. To support these ideas, we consider (i) why and how tropical ecosystems work in different ways from temperate systems, (ii) the shortcomings of standard conservation strategies when applied to the tropics, and (iii) the

Key Topics in Conservation Biology 2, First Edition. Edited by David W. Macdonald and Katherine J. Willis.

outstanding priorities for policy, practice and future research. The strategies we propose have broad relevance for the management of tropical diversity because birds have long been viewed as a model system for assessing conservation priorities, and act as flagships for numerous conservation programmes (Tobias et al. 2005).

Threats to tropical environments

Lowland tropical forests

Tropical forests support vast numbers of species. They also limit soil erosion, reduce floods, contribute to hydrological cycles, help to stabilize the climate, and generally play a major role in human well-being. Nonetheless, people are currently removing tropical forests at a stupendous rate: an average of 1.2% of rainforest area, equivalent to ~15 million hectares, is destroyed annually (Laurance 2008) (see Chapter 13). One view holds that this process is self-reinforcing, and therefore likely to accelerate, because deforestation opens previously remote regions to agriculture and development, and increases the impact of fires and hunting. An alternative view holds that such forecasts are unduly pessimistic because human populations will become increasingly urbanized, and primary forest loss will be offset by regrowth of secondary forests in depopulated areas (Wright & Muller-Landau 2006). Either way, the pressure on old-growth tropical forests is intense, and set to get much worse in future (Laurance 2006).

Over the next 50 years, the human population is expected to reach 9 billion, with an ever greater proportion climbing the socio-economic ladder (Laurance & Peres 2006). Demand for food is predicted to rise by 70–100% (Godfray et al. 2010), and global industrial activity may expand 3–6-fold over the same period (Soh et al. 2006). The scale of projected increases in overall and per capita consumption of food, timber and countless other products will further stimulate industrial drivers of tropical deforestation, such as cattle ranches, soy farms, paper mills, oil palm plantations, and major highways and infrastructure projects, all of which have expanded markedly during recent decades (Smith et al. 1993; Vargas et al. 2006). Biofuel production from food crops is expected to increase by 170% in the next 10 years (Fargione et al. 2010). Crucially, the spiralling demand for crops and commodities is likely to drive further degradation of large areas of remaining tropical forests, regardless of whether human populations decline in rural areas (Vargas et al. 2006; DeFries et al. 2010). All projections carry with them a degree of uncertainty, yet the precautionary principle dictates that we should prepare for a future in which lowland tropical forests are highly fragmented.

Tropical mountains

Tropical mountains offer hope for conservation because montane habitats often remain relatively intact, at least at higher elevations. They play an essential role as refuges for lowland biodiversity that may otherwise disappear in response to climate warming or habitat loss at lower elevations. They are also important hubs of bird diversification and endemism because of their isolation. However, adaptation to these higher elevation refuges is risky: tropical montane endemics have limited capacity to shift their ranges across unsuitable lowland habitats and are therefore particularly vulnerable to climate change (Şekercioğlu et al. 2008). Moreover, this risk is faced by precisely those species that are least affected by habitat loss in highly threatened lowlands (Pimm 2008).

Tropical islands

Tropical regions contain 45,000 islands over a minimum size of 5 ha (Arnberger & Arnberger 2001). Each island supports relatively few species but many of these are important in terms of rarity and uniqueness (see Chapter 12). Birds endemic to islands often have relatively small geographical ranges, having diverged from continental ancestors through isolation. This isolation has often resulted in the evolution of flightlessness, fearlessness and loss of immunity, rendering island birds poorly adapted to novel anthropogenic pressures, including habitat change and the introduction of alien predators and pathogens (Milberg & Tyrberg 1993). Because of these issues, the arrival of humans on islands in the tropical Pacific led to the disappearance of at least 2000 insular bird species (Steadman 1995). It is no surprise, therefore, that 88% of bird extinctions since 1600 occurred on islands, mainly in the tropics (Butchart et al. 2006).

Today, island birds are coming under increasing pressure from human exploitation and invasive species (see Chapter XX [Cross-ref invasives chapter]). For example, it is thought that various species of rat (*Rattus*) have been introduced, accidentally or otherwise, to 90% of the world's islands (Jones et al. 2008; Oppel et al. 2010). Other threats vary case by case. The popularity of some tropical islands as places to live or visit makes them susceptible to rapid coastal development, overfishing or disease. On St Lucia, for example, roughly 40% of habitat occupied by the white-breasted thrasher (*Ramphocinclus brachyurus*), an endangered species with a global population of roughly 1200 individuals, is slated for potential tourist development (Young et al. 2010). Likewise, Bataille et al. (2009) provide evidence that tourism drives the ongoing introduction of avian malaria vectors (e.g. *Culex quinquefasciatus*) from mainland Ecuador to the Galapagos Islands, transported by cruise boats and aeroplanes.

New insights into threats facing tropical avifaunas

The litany of threats outlined above is broadly familiar but its ultimate impact on birds remains debatable. BirdLife International (2000) predicted that 13% of bird species may be extinct or consigned to extinction within 100 years, most of them due to tropical deforestation and hunting. Given the scale of potential impact, the time is ripe for a more detailed assessment of the ecological processes underlying decline and extinction in tropical birds.

Demography and life-history

Tropical birds differ from their temperate-zone counterparts in numerous ways but three factors have a disproportionate impact on the survival of populations. First, a far larger proportion of tropical species are highly sedentary, presumably as a result of climatic (and thus resource) stability. For example, year-round territories are defended by only nine (4.5%) of 193 North American forest bird species, whereas the same figure rises to 379 (51%) of 739 Amazonian bird species (Salisbury et al. 2012). Second, tropical species have lower reproductive output. They tend to lay only two-egg clutches, whereas temperate birds lay 4–13 eggs per clutch (Jetz et al. 2008). Third, their ranges are often patchy and population densities low (Donlan et al. 2007). At one locality in Amazonian Peru, for example, 106 (35%) of 329 resident species occurred at densities of <1 pair/100 ha, with a median abundance (i.e. average population density) across all species of 2.5 pairs/100 ha (Terborgh et al. 1990). This contrasts with the last remaining primeval European forests, where the median abundance of >50 forest bird species is much higher, at 10–30 pairs/100 ha (Wesołowski et al. 2006). Indeed, median abundance of European forest birds often exceeds the maximum estimated population density in Amazonia (*Cercomacra cinerascens* at 20 pairs/100 ha), providing a striking illustration of the relative rarity of tropical birds.

This combination of low dispersal, low reproductive output and low population density suggests that tropical birds are more sensitive to habitat fragmentation or disturbance, and less able to recover after population bottlenecks (Stratford & Robinson 2005; Soh et al. 2006).

Most importantly, it implies that a far greater area of intact habitat is required to protect viable populations. Judging only by the population densities reported above, an area 4–12 times higher on average per species may be required. However, the added constraints of low dispersal and low reproductive output, as well as significant numbers of extreme low-density species, suggest that diverse tropical bird communities can only be conserved in much larger areas.

Dispersal and migration

The term 'sedentary' fails to convey the extreme dispersal limitation of insectivorous birds in tropical forests, many of which are unable or unwilling to cross relatively minor gaps such as roads (Stratford & Robinson 2005; Laurance et al. 2009). In experimental tests, 50% of rainforest understorey species struggled to cross 100 m gaps (Moore et al. 2008), and comparative studies suggest that thousands of tropical bird species face this problem (Stratford & Robinson 2005; Salisbury et al. 2012). This contrasts sharply with the situation in temperate-zone habitats where gap aversion is rare. Amongst European woodland birds, for example, even the wren (*Troglodytes troglodytes*) is able to cross broad (>100 m) gaps in habitat, as long as low vegetation is present.

Unsurprisingly, limited mobility is an overlooked determinant of extinction risk in birds. Two-thirds of the planet's bird species are sedentary, and 74% of sedentary species live exclusively in the tropics. Of sedentary species, 26% are globally threatened or near threatened with extinction, compared to 10% of migratory species (Şekercioğlu 2007). Projections of land bird extinctions expected from the combined effects of climate change and habitat loss also indicate that sedentary bird species are approximately five times more likely to go extinct by 2100 than migratory birds (Şekercioğlu et al. 2008). This partly reflects the disadvantages of a sedentary lifestyle when it comes to tracking moving climate optima. Despite the greater risks to tropical sedentary bird species, they have generally been neglected in comparison to northern hemisphere migratory species that have laws and conventions dedicated to them (e.g. Neotropical Migratory Bird Conservation Act in the USA). For example, between 1990 and 2007, international agencies provided the island of Hispaniola (i.e. the Dominican Republic and Haiti) with more than $1.3 million for migratory bird research, compared with only $300,000 for the study of resident birds, including several rare endemics (Latta & Faaborg 2009).

Although year-round territorial systems make up the largest component of tropical avifaunas, a different set of risks is faced by many frugivores and nectarivores. Telemetry studies show that these birds are typically mobile because they track patchy food resources. Unlike temperate-zone birds, however, they are often unable to cross degraded landscapes. Classic examples include elevational migration in white-ruffed manakins (*Corapipo altera*) (Boyle 2008) and unpredictably complex annual movements over hundreds of kilometres in the three-wattled bellbird (*Procnias tricarunculata*) (Powell & Bjork 2004). In effect, these movements compound the challenges of designing effective protected area networks because isolated reserves will fail to provide sufficient coverage for many species.

Physiological constraints and preferences

Most tropical birds experience lower climatic variability than do their temperate-zone counterparts, both within and between years (Ghalambor et al. 2006). They are therefore thought to have lower thermal plasticity (Stratford & Robinson 2005). The evidence from a geographical study of rufous-collared sparrows (*Zonotrichia capensis*) supports the hypothesis that populations in stable environments are less able to adapt to novel environmental conditions (Cavieres & Sabat 2008). Moreover, there are good reasons to expect tropical birds to be relatively intolerant of temperature fluctuation: in particular, they have lower basal metabolic rates (BMR; Wiersma et al. 2007; McNab 2009), whereas species with higher BMRs are more adaptable to climate change (Bernardo et al. 2007). In general, therefore, tropical birds are likely to be constrained by narrower environmental niches, reduced tolerance of thermal stress and habitat change (Janzen 1967; Stratford & Robinson 2005; Şekercioğlu et al. 2012).

Biotic interactions

In terms of genetic diversity, 80% of the tree of life could be retained even when approximately 95% of species are lost (Nee & May 1997); the only problem is that the resulting ecosystems would not work! Healthy ecosystems depend on associations between hosts and parasites, predators and prey, seeds and seed dispersers, and so on. These networks are the lifeblood of biodiversity. They are particularly important in tropical ecosystems because co-evolutionary associations tend to increase in abundance towards the equator (Schemske et al. 2009). In effect, the complex architecture of food webs and other biotic interactions in tropical ecosystems increases the likelihood of cascading co-extinctions, and cautions against a simplistic view of phylogenetic diversity (PD) as a means to prioritize conservation action (see Chapter 1).

Tropical birds offer good examples of finely tuned interactions, as they have more specialized dietary niches than their temperate-zone counterparts (Belmaker et al. 2012). The most clear-cut cases are nectarivores. For example, some species of hummingbird with highly specialized bills (e.g. sword-billed hummingbird, white-tipped sicklebill, etc.) can only forage on, and pollinate, very few species of flowering plants. Thus, the survival of specialist nectarivores and their associated foodplants is tightly interwoven via co-evolutionary adaptations. Similarly, frugivorous birds dictate the fate of forests. Many tropical trees produce large, lipid-rich fruits adapted for animal dispersal, so the disappearance of frugivores can have serious consequences for forest regeneration, even when the drivers of habitat loss and degradation are controlled.

The impact of bird declines on ecosystem function and services

Ecosystem services are 'the set of ecosystem functions that are useful to humans' (Kremen 2005). In general, increased biodiversity in a particular environment is thought to increase ecosystem efficiency and productivity, and decrease susceptibility to perturbation. Birds play a role as 'mobile link' animals connecting habitats and ecosystems through their movements (Lundberg & Moberg 2003), and providing services such as pollination, seed dispersal, nutrient deposition, pest control and scavenging (Şekercioğlu 2006; Wenny et al. 2011). Thus, the ongoing decline in bird species is likely to have far-reaching ecological consequences, from the spread of disease and loss of agricultural pest control to the extinction of plants dependent on avian pollinators and seed dispersers (Şekercioğlu et al. 2004).

In the tropics, large frugivorous birds are particularly threatened by hunting and habitat fragmentation, which can have significant consequences for shade-tolerant, late successional tree species with large seeds (e.g. Lauraceae, Burseraceae, Sapotaceae). For example, in the East Usambara mountains of Tanzania, the endemic tree *Leptonychia usambarensis* (Sterculiaceae) is dependent on avian seed dispersers such as greenbuls (*Andropadus* spp.), most of which are rare or absent in small forest fragments (Cordeiro & Howe 2003). This results in lower seed removal, shorter dispersal distance, greater seedling aggregation under the parent trees, and reduced recruitment in fragments than in continuous forest (Cordeiro & Howe 2003). Similarly, previous studies have shown that overhunting of large frugivores (e.g. guans, toucans, hornbills) has a detrimental impact on recruitment of bird-dispersed trees in humid forests of Amazonia (Terborgh et al. 2008) and India (Sethi & Howe 2009). In effect, widespread losses of avian frugivores may result in the domination of short-lived pioneer trees, with long-term effects cascading through plant communities (Şekercioğlu et al. 2004; Terborgh et al. 2008). They also mean that the potential for long-distance dispersal is declining for many plant species at precisely the time when flexible range shifts are becoming more important because of land use and climate change.

Meanwhile, the susceptibility of insectivorous birds to habitat degradation and fragmentation may cause a different set of problems. Insectivores often control invertebrate populations and play a significant role in limiting foliage damage in tropical forests (van Bael et al. 2010) and plantations (Greenberg et al. 2000; Kellermann et al. 2008; van Bael et al. 2008; Mooney et al. 2010). This form of control of insect herbivores can be economically important in agricultural regions. For example, bird-mediated predation of the coffee berry borer (*Hypothenemus hampei*) in Jamaican coffee plantations saves farmers $310/ha per year (Johnson et al. 2010).

Finally, tropical scavengers such as vultures provide one of the most important yet under-appreciated ecosystem services. Avian scavengers worldwide comprise the most threatened avian functional group, with about 40% of the species being threatened or near threatened with extinction (Şekercioğlu et al. 2004). Because of the loss of decomposition services provided by vultures, increased disease transmission and consequent health spending are likely. Markandya et al. (2008) suggested that the spiralling Indian dog population, which increased by 7.25 million between 1992 and 2003, could be caused by the 90–99% declines in vultures. Partly based on the associated surge of ~48,000 additional rabies deaths, they estimated that the health costs attributable to vulture declines were US$ 18 billion (Markandya et al. 2008).

What are the implications for biodiversity conservation?

Armed with this information about tropical environments and their native species, we can explore the implications for conservation strategies. The following sections deal with major threats to biodiversity at a global scale, highlighting the type and scale of impact they may have on tropical ecosystems.

Habitat fragmentation

Millions of hectares of tropical forests exist in fragments and many thousands of new fragments are created every year (Sodhi et al. 2011). These exist in a variable matrix of land uses, and consequently the loss of avian diversity is highly site dependent (Sodhi et al. 2011). Nevertheless, some patterns have emerged. Understorey insectivores, for instance, tend to be particularly sensitive to fragmentation (Peh et al. 2005; Barlow et al. 2006; Yong et al. 2011). This may be because other dietary guilds, including frugivores, granivores and nectarivores, are often dispersive, non-territorial and naturally adapted to exploit patchy resources, and therefore better equipped to cross gaps and travel through degraded landscapes. This process is highlighted by studies in Amazonia showing that fragments are primarily inhabited by bird species with good dispersal abilities or high tolerance of the non-forest matrix (Lees & Peres 2008). A key latitudinal difference linked to dispersal ability involves the likelihood of recolonization after local extinction. This is illustrated by the fact that temperate bird species dropping out from forest fragment avifaunas often recolonize within a few years if source populations are found nearby, whereas tropical species rarely return to the community unless the habitat linking the fragment to the source population actually recovers (see Stratford & Robinson 2005).

Another issue influencing the survival of bird populations is elevated predation in forest fragments. For example, in the Eastern Arc mountains of Tanzania, seven common understorey bird species, e.g. forest batis (*Batis mixta*), suffered 2–200 times lower rates of nest success in fragments versus continuous forest (Newmark & Stanley 2011). Similar reductions in reproductive success appear to be widespread in forest fragments, and at least partly linked to higher densities of nest predators, including rodents, raccoons, mongooses and snakes, as well as greater visibility of nests to hawks and other arboreal predators (Chalfoun et al. 2002). Importantly, the density and diversity of nest predators are typically far higher in tropical than temperate regions, again suggesting that fragmentation has a greater negative impact on tropical birds (Stratford & Robinson 2005).

Climate change

Most studies of the impacts of climate change on biodiversity have focused on boreal or temperate regions, perhaps because it is often assumed that the threat of warming increases towards the poles. However, the tropics are a far more likely setting for climate-mediated mass extinctions, for three main reasons. First, the direct impact of global warming will be severe in tropical mountains, where high-elevation climates will shrink or disappear (Ohlemüller et al. 2008). Second, species inhabiting extensive lowlands may find it difficult to cope with further warming or to migrate towards cooler refuges (Wright et al. 2009). And third, the sheer diversity of tropical lineages means that these threats will affect large numbers of bird species. Specifically, if a warming climate threatens taxa restricted to tropical highlands (≥500 m asl), this would amount to approximately 10% of the world's ~10,000 bird species, including the majority of those currently considered to be safeguarded from land use change (Pimm 2008; Harris et al. 2011). In contrast, the melting of polar ice caps attracts much more media attention despite affecting a relatively small number of bird species, most of which have large global ranges.

Previous studies indicate that tropical organisms are as likely as temperate organisms to track moving climates (Pounds et al. 1999; Colwell et al. 2008). Although the full impact of this process under global warming is difficult to predict, it is likely to force many bird species uphill in the montane tropics, reducing their ranges, sometimes entirely (Shoo et al. 2005; Şekercioğlu et al. 2008; Wormworth & Şekercioğlu 2011; Şekercioğlu et al. 2012). For example, the potential habitat of the golden bowerbird (*Prionodura newtonia*), a montane forest species endemic to Australia, would be reduced by 98% if temperatures rose by 3 °C (Hilbert et al. 2004). Likewise, a 4 °C increase would result in almost complete removal of the Pantepui vegetation of Venezuela (Nogue et al. 2009), presumably with devastating effects on 30–40 endemic bird species, e.g. rose-collared piha (*Lipaugus streptophorus*).

This so-called 'escalator to extinction' may be exacerbated by competitive interactions, a problem particularly relevant to tropical birds because of strong interspecific territoriality between elevational replacement species (Jankowski et al. 2010). Finally, warming could expose montane birds to an array of new pathogens, predators and competitors that migrate upslope (Bradshaw et al. 2009; Wright et al. 2009). For these reasons, tropical high-elevation specialists could be among the most endangered species on earth if global temperatures rise (Williams et al. 2003; Ricketts et al. 2005; Şekercioğlu et al. 2008).

Birds of tropical lowlands face a different set of problems. Current projections suggest that, by 2100, 75% of modern lowland tropical forests will experience mean annual temperatures warmer than the warmest forests known today, i.e. >28 °C (Wright et al. 2009). This represents an increase of 2–4 °C from present-day temperatures, well below the upper realistic bound of 6.4 °C (IPCC 2007).

Thus, lowland species may be ill suited to further warming because they are already close to their tolerance limits (Weathers 1997; Colwell et al. 2008; Wright et al. 2009). In conjunction, latitudinal range shifts to cooler environments are unlikely because there is virtually no latitudinal temperature gradient in the tropics (Colwell et al. 2008). In other words, the escape route to cooler climates is closed to many lowland tropical species, particularly those living far from the nearest topographical relief, e.g. central Amazonia (Loarie et al. 2009; Wright et al. 2009). This contrasts with the situation at higher latitudes, where latitudinal climatic gradients are a permanent feature.

Tropical islands are likely to fare little better. Many rare island endemics are restricted to the montane zone where the extent of suitable habitat could recede dramatically in response to warming (Fordham & Brook 2010). Moreover, climatic fluctuations can seriously threaten bird populations. For example, El Niño conditions, characterized by high sea temperature and rainfall in the Equatorial Pacific, are associated with disrupted oceanic food webs. Specifically, they tend to switch off the Cromwell current, thereby reducing fish stocks and causing population crashes in Galapagos penguins (*Spheniscus mendiculus*), an endangered species with a population of around 1500 individuals (Vargas et al. 2006, 2007). They also promote vector-borne diseases (Gilbert & Brindle 2009; Kolivras 2010). El Niños are natural cyclical phenomena but the frequency of severe events has increased sevenfold over the last century; moreover, global warming is predicted to drive further increases in their frequency and intensity (Timmermann et al. 1999). Although these changes may have severe consequences for island birds and ecosystems, they are overshadowed by the threat of warming to atolls and other low-lying oceanic islands. Here, extreme climatic perturbations and sea-level rises may have a catastrophic impact on human welfare and livelihoods, as well as wildlife (Barnett & Adger 2003).

Pathogens and disease

Diversity, virulence and abundance of pathogens are all higher in the tropics. Moreover, since many pathogens are sensitive to temperature, rainfall and humidity, climate warming has the potential to increase pathogen development and survival rates, disease transmission and host susceptibility (Harvell et al. 2002; Garamszegi 2011). This is coupled with the problem of range shifts through introductions, invasions and climate-related expansions and contractions of distributions. Such processes can lead to parasites becoming pathogenic when established in novel environments or in contact with susceptible new hosts. The results for birds can be devastating. In Hawaii, for example, more than half of the native avifauna became extinct due to avian malaria and other introduced pathogens (LaPointe et al. 2010). Upward shifts in mosquito distribution predicted from continued global warming will further reduce the refuge habitat available, with serious implications for some of the surviving Hawaiian endemics: a predicted 2 °C rise will probably eliminate all remaining disease-free forested refugia in Hawaii in the next century (Benning et al. 2002). Montane avifaunas throughout the tropics could be similarly threatened by changes in the distribution of hosts and vectors caused by land use and climate change (Harvell et al. 2002).

Invasive species

Invasive plants, predators and pathogens are present on islands worldwide but they are more likely to cause catastrophic declines in the tropics. This is partly because tropical islands support many endemic species with tiny ranges and populations, and also because they provide suitable environments for invasive species to thrive. Introductions can have major negative impacts on tropical birds, particularly on island endemics that have evolved without predation pressure. For example, the predatory brown tree snake (*Boiga irregularis*) was introduced to the island of Guam shortly after World War II, causing the extirpation or serious decline of 17 of the island's 18 native bird species (Wiles et al. 2003). Most island extinctions have been caused by similar events, with feral cats and rats being the most damaging culprits (see Chapter XX [XREF INVASIVES CHAPTER]). Invasive plants and pathogens can also have strongly deleterious impacts. The red quinine tree (*Cinchona pubescens*), for example, is causing drastic changes to the native plant community of the Galápagos Islands, and is destroying the breeding habitat of the Galápagos rail *Laterallus spilonotus* (Shriver et al. 2010). The main line of defence against invasive species is eradication schemes, which are often costly but sometimes highly effective (e.g. Donlan et al. 2007).

The central role of synergisms

At a local scale, the fate of tropical biodiversity is typically shaped by a suite of factors: biotic, abiotic and socio-economic (Seddon et al. 2000). It is often the inevitable interactions and synergisms between these factors that pose the greatest threat of all (Dobson et al. 2006; Brook et al. 2008; Tylianakis et al. 2008). For example, as habitat declines, hunting pressure often increases, a twin threat responsible for an elevated extinction risk in tropical vertebrates (Laurance & Useche 2009). Likewise, the fragmentation of tropical forest increases the risk of fires, further reducing avian diversity (Barlow et al. 2006).

The importance of synergisms in tropical ecosystems is highlighted by the issue of climate change. While many species coped perfectly well with wide temperature fluctuations and other stressors during their evolutionary history (Balmford 1996), none did so in a heavily human-modified environment with numerous additional barriers to dispersal. This is important because the inherent low dispersal ability of many tropical bird species means that they are often unable to disperse across agricultural landscapes (Gillies et al. 2011). Thus, those species forced to disperse polewards or uphill by climate change may now find that human settlements and agricultural areas stand in the way (Şekercioğlu et al. 2008). Unfortunately, this outcome is likely to be extremely common because mid-elevations are among the most densely populated and heavily cultivated lands in the tropics. People preferentially settle this elevational band because of its pleasant climate and suitability for a range of important cash crops, including coffee, tea and coca. It therefore seems likely that human activity is reducing the ability of many species to track climates, with potentially disastrous effects on long-term survival prospects.

Even discounting the problem of dispersal constraints, climate change may significantly exacerbate a number of threats to tropical birds (Şekercioğlu et al. 2012). Contemporary communities of birds, plants, pathogens and other interacting species will be disassembled by individualistic range shifts, such that novel communities of species will mix in the future (Laurance & Peres 2006; Thuiller 2007). There will be winners and losers in these new interactions, and the outcome across entire networks of species may produce unanticipated effects on ecosystems (Parmesan 2006). Overall, the dramatic pace of contemporary habitat loss, combined with the synergistic effects of future climatic change, overhunting, emerging pathogens and many forms of habitat degradation, could sharply increase the rate of species extinctions (Tylianakis et al. 2008).

Governance, legislation and economics

Many tropical nations are too poorly equipped, financially and institutionally, to cope with the current rate of environmental change (Butchart et al. 2006). This problem is further exacerbated by widespread corruption (Barrett et al. 2001; Lee & Jetz 2008). The triumvirate of poverty, weak institutions and corrupt officials leads to the destruction or mismanagement of protected areas, and the uncontrolled exploitation of natural resources at the whim of global markets (Smith et al. 2003). Rising prices for commodities, such as gold, can unleash a surge of tropical deforestation by making exploitation profitable in regions formerly too remote to be affected (Swenson et al. 2011). Biofuel crops, such as sugar cane, oil palm or soy, grow best at low latitudes, and thus escalating demands are creating an incentive for converting tropical habitats to farmland (Laurance 2006; Laurance & Peres 2006; Brook et al. 2008). These mechanisms can now operate rapidly and with dramatic consequences, not least because infrastructural projects have opened up trade routes, making the exploitation of once remote regions economically viable. For example, the Trans-Oceanica Highway links the Brazilian soya belt of Acre and Mato Grosso with Peruvian ports, such that the expanding markets of China and other Asian countries have direct influence on the deforestation frontiers of southern and western Amazonia (Tobias et al. 2008b).

Knowledge

A final piece in the jigsaw is knowledge, or lack thereof. The quality of information about tropical birds lags far behind that available for the temperate zone in terms of taxonomy (Tobias et al. 2008a), population size (Tobias & Seddon 2002; Seddon & Tobias 2007) and conservation status (Tobias & Brightsmith 2007). Taxonomic revisions tend to reveal that many tropical bird species consist of more than one cryptic species, each with global ranges and populations smaller than previous estimates (Lohman et al. 2010). The scale of taxonomic uncertainty in the tropics suggests that conservation issues need to be re-evaluated, as many analyses are sensitive to estimates of global range (e.g. Rodrigues et al. 2004; Vale et al. 2008). Moreover, latitudinal bias in knowledge is not restricted to taxonomy: for example, only a tiny number of long-term climate datasets originate from the tropics, in contrast to many thousands from Europe or North America (Rosenzweig et al. 2008).

Towards long-term and broad-scale strategies for the conservation of tropical ecosystems

Most conservation efforts and priority-setting exercises focused on tropical systems have attempted to apply strategies developed in the temperate zone. Many rely on species as units of conservation; others emphasize metrics such as phylogenetic diversity (PD), such that conservation may target the maximum amount of 'evolutionary history'. However, the preceding sections outline several reasons why these approaches may fail in the tropics. Up to this point we have dwelt on the underlying problems; the following sections focus on conservation priorities and solutions.

Protected areas and beyond

Between 1994 and 2004, the extinction of 16 bird species was prevented through a combination of habitat protection, control of invasive species and captive breeding (Butchart et al. 2006). For example, wild populations of Seychelles magpie-robin (*Copsychus sechellarum*) increased from approximately 15 to 136 individuals, and Mauritius kestrel (*Falco punctatus*) from four to 500–800 individuals. These targeted schemes, often costing millions of dollars per species, are clearly effective in some cases. However, they only involve <10% of critically endangered bird species, suggesting that conservation efforts and funding need to be greatly expanded.

Aside from financial constraints, it is doubtful whether this fire-fighting approach is viable from the perspective of ecosystem function. Small or fragmented populations will cease to play a significant ecological role, and thus conservation strategies assuming that species will survive in a few isolated reserves are unlikely to result in healthy tropical ecosystems. Only by maintaining large, interconnected populations can we avoid numerous species being officially extant but functionally extinct (Sodhi et al. 2011). Thus, conservation needs to focus not only on protected areas but the broader landscape context in which reserves are embedded. This is especially important in the humid tropics, where the fate of avian diversity within protected areas is often inextricably linked to the surrounding agricultural matrix (Harvey et al. 2008).

Lowland and mid-elevation forests of tropical Asia (Peh et al. 2005), Africa (Norris et al. 2010) and Meso-America (Harvey et al. 2008) are tightly coupled agro-forestry systems, with most habitats supporting rural livelihoods. Amazonia is inexorably going this way. A priority for birds is to promote landscape configurations that connect reserves, maintain a diverse array of habitats and retain high structural and floristic complexity (Koh 2008; Chazdon et al. 2009). In particular, the preservation of forest remnants, including riparian strips, secondary forests and individual trees, helps birds to move between forest patches and even breed in human-dominated landscapes (Şekercioğlu et al. 2007; Martensen et al. 2008; Dent & Wright 2009).

Promoting these features is a huge challenge, particularly in areas suited to mechanized agriculture. Where possible, engagement is required to ensure that farmers are stakeholders in creating landscapes that preserve birds as well as rural livelihoods (Harvey et al. 2008). One successful blueprint involves Payment for Ecosystem Service schemes, such as those operating in Costa Rica, which motivate landowners to maintain forest patches and watershed reserves (Sodhi et al. 2011) (see Chapter 4). These measures need to be expanded where possible, and coupled with outreach and education initiatives, to ensure that we preserve key ecosystem processes as well as populations of rare species.

Conservation priorities: too much pattern, not enough process

The idea that conservationists need to focus on the processes sustaining biodiversity is not new (Smith et al. 1993; Moritz 2002), but it has proved difficult to incorporate into conservation strategies. The approach requires us to see beyond the current pattern of extant species, and to consider longer-term evolutionary perspectives. But what perspectives are these? There is a tendency to assume that by using molecular phylogenies to map conservation priorities, we have somehow captured the evolutionary process but a phylogeny is as much a pattern as anything else (Losos 2011). Instead, the defining attributes of process-based conservation are (i) the preservation of genetic, phenotypic and functional diversity below the species level; (ii) the maintenance of abundance so that lineages continue to perform their ecological functions across their natural ranges; and (iii) the promotion of conditions necessary for long-term population connectivity and dispersal, even under climate and land use change scenarios.

The contrasts between pattern-based and process-based conservation are striking. A pattern-based approach tends to prioritize rare endemics, many of which are only narrowly divergent from their closest relatives, by preserving small populations in isolated reserves. A process-based approach does not target these potentially ephemeral 'twigs' of the tree of life, and attempts instead to sustain the underlying mechanisms that provide the impetus for adaptation and speciation. This can be achieved in different ways. One method is to focus attention on all genetically and phenotypically distinct lineages within species (Moritz 2002); another is simply to maintain the context for selection, rather than protecting phenotypes *per se* (Thomassen et al. 2011). A key practical component is the maintenance of habitat heterogeneity and connectivity within and between reserves, particularly where this captures genetic variation across environmental gradients (Moritz 2002).

Aldo Leopold famously observed that 'to keep every cog and wheel is the first precaution of intelligent tinkering', meaning that all parts of ecosystems should be saved, even though their function may be unclear. Conservationists worldwide have tended to interpret 'parts' as 'species'. However, this misses the point that in all ecosystems the essential cogs and wheels are adaptation, connection and interaction – in other words, not patterns but processes. Incorporating this viewpoint into the bigger picture of conservation thinking is particularly vital in the tropics.

Key measures to preserve healthy tropical avifaunas

With the goal of managing tropical ecosystems and preventing further declines in tropical birds, we propose a list of conservation and research priorities. Although these are designed with birds in mind, similar strategies will greatly improve our chances of preserving a vast proportion of tropical biodiversity in resilient, functional and flexible ecosystems (Hole et al. 2009).

Conservation targets

- Extensive protected areas, i.e. 'megareserves', sustaining viable populations of birds in large areas of intact habitat (Laurance 2005; Peres 2005). These areas are always richer in bird diversity, and more important for bird conservation, than land shared with agricultural productivity (Edwards et al. 2010; Gibson et al. 2011; Phalan et al. 2011).
- Improved yields of tropical crops. This is an important factor in allowing a maximum area of natural habitat to be spared for conservation purposes.
- Improved targeting of clearance and conversion of forests for human land uses. For example, intensive agricultural development should be focused on grasslands or precleared land, rather than forests, as this limits losses in terms of both carbon balance and bird diversity (Danielsen et al. 2009).
- Island restoration. The eradication of invasive mammals, plants and pathogens is key to preserving avifaunas of tropical islands.
- Connectivity between bird populations. This can be achieved by integrating degraded, secondary and fragmented habitats into conservation schemes. Techniques include minimizing forest fragmentation and fragment degradation, preventing fires, cattle incursions and other types of disturbance, reducing the contrast between fragments and the surrounding non-forest matrix, and increasing fragment connectivity via ecological corridors, riparian strips and landscape restoration. Where feasible, efforts should focus on developing and implementing community schemes and government legislation to maintain these landscape features at deforestation frontiers.
- Intact elevational and ecological gradients. Even supposedly complete gradients, such as Manu National Park in Peru, are currently being truncated or interrupted by habitat degradation. One laudable government-sponsored initiative attempts to connect protected areas from the western lowlands of Ecuador, over the Andes and down into eastern Amazonia, and there is an urgent need to replicate this approach elsewhere to safeguard elevational corridors. The logic of preserving elevational transects extends to all habitat mosaics and ecotones, i.e. gradations between core habitat types. In all cases, land-sparing efforts need to consider that environmental gradients may fluctuate spatially over time.
- Strengthened incentives to promote the sustainable economic use of bird-friendly habitats. This includes much greater emphasis on educating people in tropical nations about the key functional roles played by ecosystems in flood protection, sustainable food production and delivery of clean water (Edwards et al. 2010; Şekercioğlu 2012).

Research agenda

- Promote research focused on tropical birds. Bird behaviour and ecology are fundamental to process-based conservation strategies, yet most work on this subject has targeted temperate-zone taxa (Newmark & Stanley 2011).
- Dispersal. Research should clarify the ability of tropical birds to use human-modified landscapes, cross gaps and navigate corridors, as these are critical aspects of avian life-history from the perspective of landscape management. The recent use of tracking technology to quantify dispersal ability and habitat use (e.g. Gillies & St. Clair 2008; Hawes et al. 2008; Hadley & Betts 2009) should be expanded.
- Understanding biotic interactions. These are complex even in temperate-zone bird communities but in the tropics more research is required to understand the long-term implications of interactions with food plants, prey and parasites. A range of modern tools, from field experiments to phylogenies, should be used to explore the role of interactions in maintaining stable tropical ecosystems.
- Improving knowledge about range and abundance. Most tropical regions remain poorly known in terms of the distribution and abundance of bird species, and how these fluctuate over time. Progress requires field surveys and monitoring exercises, ideally coupled with the development of automated monitoring tools, e.g. song identification software.
- Improving knowledge about taxonomy. Detailed phenotypic and molecular studies are required to revise tropical species limits and relationships in many avian families.
- Conservation potential of agricultural landscapes, including logged and secondary forests. Most studies have looked at bird diversity from the perspective of isolated natural habitat, whereas the matrix of human land uses provides a useful framework for a different set of questions. What are the relative impacts of different agricultural practices on conservation outcomes and ecosystem services? To what extent can species with different ecologies survive in or travel through different crops or marginal habitats? These questions require a focus on the biodiversity value of plantations and other agricultural habitats, and management practices that maintain connectivity in natural populations (Harvey et al. 2008).
- Long-term population dynamics. The extent to which degraded habitats serve as a 'safety net' for tropical bird populations needs detailed study. It is clear that these habitats support a wide diversity of bird species and are important for conservation (Edwards et al. 2011), yet studies that only consider presence/absence, or even relative abundance, may provide an overly optimistic scenario if surveys detect temporary transients or if degraded habitats act as 'ecological traps' (Part et al. 2007). Thus, we need an improved understanding of underlying population dynamics in key habitats, with a focus on individual fitness, demography, extinction lags, etc.
- Resolving the land-sharing versus land-sparing debate. Retaining habitat patches in the landscape reduces agricultural yield, so that agricultural production needs to spread over larger areas to meet demand (Edwards et al. 2010). More research is required to identify optimum conservation strategies for balancing the trade-off between larger protected areas and a more connected landscape for birds.
- Climate change. Most studies have focused on the temperate zone, and we need to know much more about the physiological constraints, and climate-tracking potential of tropical birds (Stratford & Robinson 2005; Wormworth & Şekercioğlu 2011). For example, a global review of range shifts in response to climate warming was unable to include a single study on tropical birds (Chen et al. 2011).

Conclusions

When it comes to bird conservation, lessons learnt in temperate systems cannot simply be transferred to the tropics. The current practice of preserving tropical species in small numbers of protected areas embedded in an inhospitable matrix is not sustainable. Instead, conservation efforts should seek to maximize the size of reserves and the connectivity between them. On one hand, land sparing in the form of large pristine blocks of habitat is paramount. On the other hand, we urgently need to harness the potential of agricultural, secondary and degraded habitats as refuges for tropical birds. As part of this process, more attention should focus on developing community-based environmental education, outreach and income generation programmes that support bird conservation initiatives. Only by integrating these approaches, and expanding basic research, can we succeed in preserving tropical avifaunas for the future.

References

Arnberger, H. & Arnberger, E. (2001) *The Tropical Islands of the Indian and Pacific Oceans*. Austrian Academy of Sciences Press, Vienna.

Balmford, A. (1996) Extinction filters and current resilience: the significance of past selection pressures for conservation biology. *Trends in Ecology and Evolution*, **11**, 193–196.

Barlow, J., Peres, C.A., Henriques, L.M., Stouffer, P.C. & Wunderle, J.M. (2006) The responses of understorey birds to forest fragmentation, logging and wildfires: an Amazonian synthesis. *Biological Conservation*, **128**, 182–190.

Barnett, J. & Adger, W.N. (2003) Climate dangers and atoll countries. *Climatic Change*, **61**, 321–337.

Barrett, C.B., Brandon, K., Gibson, C. & Gjertson, H. (2001) Conserving tropical biodiversity amid weak institutions. *Bioscience*, **51**, 497–502.

Bataille, A., Cunningham, A.A., Cedeno, V., *et al.* (2009) Evidence for regular ongoing introductions of mosquito disease vectors into the Galapagos Islands. *Proceedings of the Royal Society B: Biological Sciences*, **276**, 3769–3775.

Belmaker, J., Şekercioğlu, C.H. & Jetz, W. (2012) Global patterns of specialization and coexistence in bird assemblages. *Journal of Biogeography*, **39**, 193–203.

Benning, T.L., LaPointe, D., Atkinson, C.T. & Vitousek, P.M. (2002) Interactions of climate change with biological invasions and land use in the Hawaiian Islands: modeling the fate of endemic birds using a geographic information system. *Proceedings of the National Academy of Sciences USA*, **99**, 14246–14249.

Bernardo, J., Ossola, R.J., Spotila, J. & Crandall, K.A. (2007) Interspecies physiological variation as a tool for cross-species assessments of global warming-induced endangerment: validation of an intrinsic determinant of macroecological and phylogeographic structure. *Biology Letters*, **3**, 695–698.

BirdLife International (2000) Threatened Birds of the World. BirdLife International, Cambridge.

Boyle, W.A. (2008) Partial migration in birds: tests of three hypotheses in a tropical lekking frugivore. *Journal of Animal Ecology*, **77**, 1122–1128.

Bradshaw, C.J., Sodhi, N.S. & Brook, B.W. (2009) Tropical turmoil: a biodiversity tragedy in progress. *Frontiers in Ecology and the Environment*, **7**, 79–87.

Brook, B.W., Sodhi, N.S. & Bradshaw, C.J. (2008) Synergies among extinction drivers under global change. *Trends in Ecology and Evolution*, **23**, 453–460.

Butchart, S.H., Stattersfield, A.J. & Collar, N.J. (2006) How many bird extinctions have we prevented? *Oryx*, **40**, 266–278.

Cavieres, G. & Sabat, P. (2008) Geographic variation in the response to thermal acclimation in rufous-collared sparrows: are physiological flexibility and environmental heterogeneity correlated? *Functional Ecology*, **22**, 509–515.

Chalfoun, A.D., Thompson, F.R. & Ratnaswamy, M.J. (2002) Nest predators and fragmentation: a review and meta-analysis. *Conservation Biology*, **16**, 306–318.

Chazdon, R.L., Harvey, C.A., Komar, O., *et al.* (2009) Beyond reserves: a research agenda for conserving biodiversity in human-modified tropical landscapes. *Biotropica*, **41**, 142–153.

Chen, I.C., Hill, J.K., Ohlemuller, R., Roy, D.B. & Thomas, C.D. (2011) Rapid range shifts of species associated with high levels of climate warming. *Science*, **333**, 1024–1026.

Colwell, R.K., Brehm, G., Cardelus, C.L., Gilman, A.C. & Longino, J.T. (2008) Global warming, elevational range shifts, and lowland biotic attrition in the wet tropics. *Science*, **322**, 258–261.

Cordeiro, N.J. & Howe, H.F. (2003) Forest fragmentation severs mutualism between seed dispersers and an endemic African tree. *Proceedings of the National Academy of Sciences USA*, **100**, 14052–14056.

Danielsen, F., Beukema, H., Burgess, N.D., *et al.* (2009) Biofuel plantations on forested lands: double jeopardy for biodiversity and climate. *Conservation Biology*, **23**, 348–358.

DeFries, R.S., Rudel, T., Uriarte, M. & Hansen, M. (2010) Deforestation driven by urban population growth and agricultural trade in the twenty-first century. *Nature Geoscience*, **3**, 178–181.

Dent, D.H. & Wright, S.J. (2009) The future of tropical species in secondary forests: a quantitative review. *Biological Conservation*, **142**, 2833–2843.

Dobson, A., Lodge, D., Alder, J., *et al.* (2006) Habitat loss, trophic collapse, and the decline of ecosystem services. *Ecology*, **87**, 1915–1924.

Donlan, C.J., Campbell, K., Cabrera, W., Lavoie, C., Carrion, V. & Cruz, F. (2007) Recovery of the Galápagos rail (*Laterallus spilonotus*) following the removal of invasive mammals. *Biological Conservation*, **138**, 520–524.

Edwards, D.P., Hodgson, J.A., Hamer, K.C., *et al.* (2010) Wildlife-friendly oil palm plantations fail to protect biodiversity effectively. *Conservation Letters*, **3**, 236–242.

Edwards, D.P., Larsen, T.H., Docherty, T.D., *et al.* (2011) Degraded lands worth protecting: the biological importance of Southeast Asia's repeatedly logged forests. *Proceedings of the Royal Society B: Biological Sciences*, **278**, 82–90.

Fargione, J.E., Plevin, R.J. & Hill, J.D. (2010) The ecological impact of biofuels. *Annual Review of Ecology, Evolution and Systematics*, **41**, 351–377.

Fordham, D.A. & Brook, B.W. (2010) Why tropical island endemics are acutely susceptible to global change. *Biodiversity and Conservation*, **19**, 329–342.

Garamszegi, L.Z. (2011) Climate change increases the risk of malaria in birds. *Global Change Biology*, **17**, 1751–1759.

Ghalambor, C.K., Huey, R.B., Martin, P.R., Tewksbury, J.J. & Wang, G. (2006) Are mountain passes higher in the tropics? Janzen's hypothesis revisited. *Integrative and Comparative Biology*, **46**, 5–17.

Gibson, L., Lee, T.M., Koh, L.P., *et al.* (2011). Primary forests are irreplaceable for sustaining tropical biodiversity. *Nature*, **478**, 378–381.

Gilbert, M. & Brindle, R. (2009) El Nino and variations in the prevalence of *Plasmodium vivax* and *P. falciparum* in Vanuatu. *Transactions of the Royal Society of Tropical Medicine and Hygiene*, **103**, 1285–1287.

Gillies, C.S. & St Clair, C.C. (2008) Riparian corridors enhance movement of a forest specialist bird in fragmented tropical forest. *Proceedings of the National Academy of Sciences USA*, **105**, 19774–19779.

Gillies, C.S., Beyer, H.L. & St Clair, C.C. (2011) Fine-scale movement decisions of tropical forest birds in a fragmented landscape. *Ecological Applications*, **21**, 944–954.

Godfray, H.C., Beddington, J.R., Crute, I.R., *et al.* (2010) Food security: the challenge of feeding 9 billion people. *Science*, **327**, 812–818.

Greenberg, R., Bichier, P., Angon, A.C., MacVean, C., Perez, R. & Cano, E. (2000) The impact of avian insectivory on arthropods and leaf damage in some Guatemalan coffee plantations. *Ecology*, **81**, 1750–1755.

Hadley, A.S. & Betts, M.G. (2009) Tropical deforestation alters hummingbird movement patterns. *Biology Letters*, **5**, 207–210.

Harris, J.B., Şekercioğlu, C.H., Sodhi, N.S., Fordham, D.A., Paton, D.C. & Brook, B.W. (2011) The tropical frontier in avian climate impact research. *Ibis*, **153**, 877–882.

Harvell, C.D., Mitchell, C.E., Ward, J.R., *et al.* (2002) Climate warming and disease risks for terrestrial and marine biota. *Science*, **296**, 2158–2162.

Harvey, C., Komar, O., Chazdon, R., *et al.* (2008) Integrating agricultural landscapes with biodiversity conservation in the Mesoamerican hotspot. *Conservation Biology*, **22**, 8–15.

Hawes, J., Barlow, J., Gardner, T.A. & Peres, C.A. (2008) The value of forest strips for understorey birds in an Amazonian plantation landscape. *Biological Conservation*, **141**, 2262–2278.

Hilbert, D.W., Bradford, M., Parker, T. & Westcott, D.A. (2004) Golden bowerbird (Prionodura newtonia) habitat in past, present and future climates: predicted extinction of a vertebrate in tropical highlands due to global warming. *Biological Conservation*, **116**, 367–377.

Hole, D.G., Willis, S.G., Pain, D.J., *et al.* (2009) Projected impacts of climate change on a continent-wide protected area network. *Ecology Letters*, **12**, 420–431.

IPCC (2007) *Climate Change 2007: Impacts, Adaptation, and Vulnerability*. Cambridge University Press, Cambridge.

Jankowski, J.E., Robinson, S.K. & Levey, D.J. (2010) Squeezed at the top: interspecific aggression may constrain elevational ranges in tropical birds. *Ecology*, **91**, 1877–1884.

Janzen, D.H. (1967) Why mountain passes are higher in the tropics. *American Naturalist*, **101**, 233–249.

Jetz, W., Şekercioğlu, C.H. & Bohning-Gaese, K. (2008) The worldwide variation in avian clutch size across species and space. *PloS Biology*, **6**, 2650–2657.

Johnson, M.D., Kellermann, J.L. & Stercho, A.M. (2010) Pest reduction services by birds in shade and sun coffee in Jamaica. *Animal Conservation*, **13**, 140–147.

Jones, H.P., Tershy, B. R., Zavaleta, E. S., *et al*. (2008) Severity of the effects of invasive rats on seabirds: a global review. *Conservation Biology*, **22**, 16–26.

Kellermann, J.L., Johnson, M.D., Stercho, A.M. & Hackett, S.C. (2008) Ecological and economic services provided by birds on Jamaican Blue Mountain coffee farms. *Conservation Biology*, **22**, 1177–1185.

Koh, L.P. (2008) Can oil palm plantations be made more hospitable for forest butterflies and birds. *Journal of Applied Ecology*, **45**, 1002–1009.

Kolivras, K.N. (2010) Changes in dengue risk potential in Hawaii, USA, due to climate variability and change. *Climate Research*, **42**, 1–11.

Kremen, C. (2005) Managing ecosystem services: what do we need to know about their ecology? *Ecology Letters*, **8**, 468–479.

LaPointe, D.A., Goff, M.L. & Atkinson, C.T. (2010) Thermal constraints to the sporogonic development and altitudinal distribution of avian malaria *Plasmodium relictum* in Hawaii. *Journal of Parasitology*, **96**, 318–324.

Latta, S.C. & Faaborg, J. (2009) Benefits of studies of overwintering birds for understanding resident bird ecology and promoting development of conservation capacity. *Conservation Biology*, **23**, 286–293.

Laurance, W.F. (2005) When bigger is better: the need for Amazonian mega-reserves. *Trends in Ecology and Evolution*, **20**, 645–648.

Laurance, W.F. (2006) Have we overstated the tropical biodiversity crisis? *Trends in Ecology and Evolution*, **22**, 65–70.

Laurance, W.F. (2008) Can carbon trading save vanishing forests? *Bioscience*, **58**, 286–287.

Laurance, W.F. & Peres, C.A. (2006) *Emerging Threats to Tropical Forests*. University of Chicago Press, Chicago.

Laurance, W.F. & Useche, D.C. (2009) Environmental synergisms and extinctions of tropical species. *Conservation Biology*, **23**, 1427–1437.

Laurance, W.F., Goosem, M. & Laurance, S.G. (2009) Impacts of roads and linear clearings on tropical forests. *Trends in Ecology and Evolution*, **24**, 659–669.

Lee, T.M. & Jetz, W. (2008) Future battlegrounds for conservation under global change. *Proceedings of the Royal Society B: Biological Sciences*, **275**, 1261–1270.

Lees, A.C. & Peres, C.A. (2008) Avian life-history determinants of local extinction risk in a hyper-fragmented Neotropical forest landscape. *Animal Conservation*, **11**, 128–137.

Loarie, S.R., Duffy, P.B., Hamilton, H., Asner, G.P., Field, C.B. & Ackerly, D.D. (2009) The velocity of climate change. *Nature*, **462**, 1052–1055.

Lohman, D.J., Ingram, K.K., Prawiradilaga, D.M., *et al*. (2010) Cryptic genetic diversity in "widespread" Southeast Asian bird species suggests that Philippine avian endemism is gravely underestimated. *Biological Conservation*, **143**, 1885–1890.

Losos, J.B. (2011) Seeing the forest for the trees: the limitations of phylogenies in comparative biology. *American Naturalist*, **177**, 709–727.

Lundberg, J. & Moberg, F. (2003) Mobile link organisms and ecosystem functioning: implications for ecosystem resilience and management. *Ecosystems*, **6**, 87–98.

Markandya, A., Taylor, T., Longo, A., Murty, M.N., Murty, S. & Dhavala, K. (2008) Counting the cost of vulture decline: an appraisal of the human health and other benefits of vultures in India. *Ecological Economics*, **67**, 194–204.

Martensen, A.C., Pimentel, R.G. & Metzger, J.P. (2008) Relative effects of fragment size and connectivity on bird community in the Atlantic Rain Forest: implications for conservation. *Biological Conservation*, **141**, 2184–2192.

McNab, B.K. (2009) Ecological factors affect the level and scaling of avian BMR. *Comparative Biochemistry and Physiology a-Molecular and Integrative Physiology*, **152**, 22–45.

Milberg, P. & Tyrberg, T. (1993) Naive birds and noble savages – a review of man-caused prehistoric extinctions of island birds. *Ecography*, **16**, 229–250.

Mooney, K.A., Gruner, D.S., Barber, N.A., van Bael, S.A., Philpott, S.M. & Greenberg, R. (2010) Interactions among predators and the cascading effects of vertebrate insectivores on arthropod communities and plants. *Proceedings of the National Academy of Sciences USA*, **107**, 7335–7340.

Moore, R.P., Robinson, W.D., Lovette, I.J. & Robinson, T.R. (2008) Experimental evidence for extreme dispersal limitation in tropical forest birds. *Ecology Letters*, **11**, 960–968.

Moritz, C. (2002) Strategies to protect biological diversity and the evolutionary processes that sustain it. *Systematic Biology*, **51**, 238–254.

Nee, S. & May, R.M. (1997) Extinction and the loss of evolutionary history. *Science*, **278**, 692–694.

Newmark, W.D. & Stanley, T.R. (2011) Habitat fragmentation reduces nest survival in an Afrotropical bird community in a biodiversity hotspot. *Proceedings of the National Academy of Sciences USA*, **108**, 11488–11493.

Nogue, S., Rull, V. & Vegas-Vilarrubia, T. (2009) Modeling biodiversity loss by global warming on Pantepui, northern South America: projected upward migration and potential habitat loss. *Climatic Change*, **94**, 77–85.

Norris, K., Asase, A., Collen, B., *et al.* (2010) Biodiversity in a forest-agriculture mosaic – the changing face of West African rainforests. *Biological Conservation*, **143**, 2341–2350.

Ohlemüller, R., Anderson, B.J., Araújo, M.B., *et al.* (2008) The coincidence of climatic and species rarity: high risk to small-range species from climate change. *Biology Letters*, **4**, 568–572.

Oppel, S., Beaven, B.M., Bolton, M., Vickery, J. & Bodey, T.W. (2010) Eradication of invasive mammals on islands inhabited by humans and domestic animals. *Conservation Biology*, **25**, 232–240.

Parmesan, C. (2006) Ecological and evolutionary responses to recent climate change. *Annual Review of Ecology, Evolution and Systematics*, **37**, 637–669.

Part, T., Arlt, D. & Villard, M.A. (2007) Empirical evidence for ecological traps: a two-step model focusing on individual decisions. *Journal of Ornithology*, **148**, S327–S332.

Peh, K.S.H., de Jong, J., Sodhi, N.S., Lim, S.L. & Yap, C.A.M. (2005) Lowland rainforest avifauna and human disturbance: persistence of primary forest birds in selectively logged forest and mixed-rural habitats of southern Peninsular Malaysia. *Biological Conservation*, **123**, 489–505.

Peres, C.A. (2005) Why we need megareserves in Amazonia. *Conservation Biology*, **19**, 728–733.

Phalan, B., Onial, M., Balmford, A. & Green, R.E. (2011) Reconciling food production and biodiversity conservation: land sharing and land sparing compared. *Science*, **333**, 1289–1291.

Pimm, S.L. (2008) Biodiversity: climate change or habitat loss – which will kill more species? *Current Biology*, **18**, 117–119.

Pounds, J.A., Fogden, M.P. & Campbell, J.H. (1999) Biological response to climate change on a tropical mountain. *Nature*, **398**, 611–615.

Powell, G.V. & Bjork, R.D. (2004) Habitat linkages and the conservation of tropical biodiversity as indicated by seasonal migrations of three-wattled bellbirds. *Conservation Biology*, **18**, 500–509.

Ricketts, T.H., Dinerstein, E., Boucher, T., *et al.* (2005) Pinpointing and preventing imminent extinctions. *Proceedings of the National Academy of Sciences USA*, **102**, 18497–18501.

Rodrigues, A.S., Andelman, S.L., Bakarr, M.I., *et al.* (2004) Effectiveness of the global protected area network in representing species diversity. *Nature*, **428**, 640–643.

Rosenzweig, S., Karoly, D., Vicarelli, M., *et al.* (2008) Attributing physical and biological impacts to anthropogenic climate change. *Nature*, **453**, 353–358.

Salisbury, C.L., Seddon, N., Cooney, C.R. & Tobias, J.A. (2012) The latitudinal gradient in dispersal constraints: ecological specialisation drives recent diversification in tropical birds. *Ecology Letters*, **15**(8), 847–855.

Schemske, D.W., Mittelbach, G.G., Cornell, H.V., Sobel, J.M. & Roy, K. (2009) Is there a latitudinal gradient in the importance of biotic interactions? *Annual Review of Ecology, Evolution and Systematics*, **40**, 245–269.

Seddon, N. & Tobias, J.A. (2007) Population size and habitat associations of the Long-tailed Ground-roller *Uratelornis chimaera*. *Bird Conservation International*, **17**, 1–13.

Seddon, N., Tobias, J.A., Yount, J.W., Ramanampamonjy, J.R., Butchart, S.H. & Randrianizahana, H. (2000) Conservation issues and priorities in the Mikea Forest of south-west Madagascar. *Oryx*, **34**, 287–304.

Şekercioğlu, C.H. (2006) Increasing awareness of avian ecological function. *Trends in Ecology and Evolution*, **21**, 464–471.

Şekercioğlu, C.H. (2007) Conservation ecology: area trumps mobility in fragment bird extinctions. *Current Biology*, **17**, R283–R286.

Şekercioğlu, C.H. (2012) Promoting community-based bird monitoring in the tropics: conservation, research, environmental education, capacity-building, and local incomes. *Biological Conservation*, **151**, 69–73.

Şekercioğlu, C.H., Daily, G.C. & Ehrlich, P.R. (2004) Ecosystem consequences of bird declines.

Proceedings of the National Academy of Sciences USA, **101**, 18042–18047.

Şekercioğlu, C.H., Loarie, S.R., Oviedo-Brenes, F., Daily, G.C. & Ehrlich, P.R. (2007) Persistence of forest birds in the Costa Rican agricultural countryside. *Conservation Biology*, **21**, 482–494.

Şekercioğlu, C.H., Schneider, S.H., Fay, J.P. & Loarie, S.R. (2008) Climate change, elevational range shifts, and bird extinctions. *Conservation Biology*, **22**, 140–150.

Şekercioğlu, C.H., Primack, R.B. & Wormworth, J. (2012) The effects of climate change on tropical birds. *Biological Conservation*, **148**, 1–18.

Sethi, P. & Howe, H.F. (2009) Recruitment of hornbill-dispersed trees in hunted and logged forests of the Indian Eastern Himalaya. *Conservation Biology*, **23**, 710–718.

Shoo, L.P., Williams, S.E. & Hero, J.M. (2005) Climate warming and the rainforest birds of the Australian Wet Tropics: using abundance data as a sensitive predictor of change in total population size. *Biological Conservation*, **125**, 335–343.

Shriver, W.G., Gibbs, J.P., Woltz, H.W., Schwarz, N.P. & Pepper, M.A. (2010) Galápagos Rail *Laterallus spilonotus* population change associated with habitat invasion by the Red-barked Quinine Tree *Cinchona pubescens*. *Bird Conservation International*, **21**(2), 221–227.

Smith, R.J., Muir, R.D., Walpole, M.J., Balmford, A. & Leader-Williams, N. (2003) Governance and the loss of biodiversity. *Nature*, **426**, 67–70.

Smith, T.B., Bruford, M.W. & Wayne, R.K. (1993) The preservation of process: the missing element of conservation programs. *Biodiversity Letters*, **1**, 164–167.

Sodhi, N.S., Şekercioğlu, C.H., Barlow, J. & Robinson, S.K. (2011) *Conservation of Tropical Birds*. John Wiley & Sons, Oxford.

Soh, M.C., Sodhi, N.S. & Lim, S.L.H. (2006) High sensitivity of montane bird communities to habitat disturbance in Peninsular Malaysia. *Biological Conservation*, **129**, 149–166.

Steadman, D.W. (1995) Prehistoric extinctions of Pacific island birds: biodiversity meets zooarchaeology. *Science*, **267**, 1123–1131.

Stratford, A.J. & Robinson, W.D. (2005) Gulliver travels to the fragmented tropics: geographic variation in mechanisms of avian extinction. *Frontiers in Ecology and the Environment*, **3**, 85–92.

Swenson, J.J., Carter, C.E., Domec, J.C. & Delgado, C.I. (2011) Gold mining in the Peruvian Amazon: global prices, deforestation, and mercury imports. *PLoS ONE*, **6**, e18875.

Terborgh, J., Robinson, S.K., Parker, T.A., Munn, C.A. & Pierpont, N. (1990) Structure and organization of an Amazonian forest bird community. *Ecological Monographs*, **60**, 213–238.

Terborgh, J., Nunez-Iturri, G., Pitman, N.C., *et al.* (2008) Tree recruitment in an empty forest. *Ecology*, **89**, 1757–1768.

Thomassen, H.A., Fuller, T., Buermann, W., *et al.* (2011) Mapping evolutionary process: a multi-taxa approach to conservation prioritization. *Evolutionary Applications*, **4**, 397–413.

Thuiller, W. (2007) Climate change and the ecologist. *Nature*, **448**, 550–552.

Timmermann, A., Oberhuber, J., Bacher, A., Esch, M., Latif, M. & Roeckner, E. (1999) Increased El Niño frequency in a climate model forced by future greenhouse warming. *Nature*, **398**, 694–697.

Tobias, J.A. & Brightsmith, D.J. (2007) Distribution, ecology and conservation status of the Blue-headed Macaw *Primolius couloni*. *Biological Conservation*, **139**, 126–138.

Tobias, J.A. & Seddon, N. (2002) Estimating population size in the Subdesert Mesite: new methods and implications for conservation. *Biological Conservation*, **108**, 199–212.

Tobias, J.A., Bennun, L. & Stattersfield, A. (2005) *Listening to the Birds*. Island Press, Washington, D.C.

Tobias, J.A., Bates, J.M., Hackett, S.J. & Seddon, N. (2008a) Comment on the latitudinal gradient in recent speciation and extinction rates of birds and mammals. *Science*, **319**, 901.

Tobias, J.A., Lebbin, D.J., Aleixo, A., *et al.* (2008b) Distribution, behavior, and conservation status of the rufous twistwing (*Cnipodectes superrufus*). *Wilson Journal of Ornithology*, **120**, 38–49.

Tylianakis, J.M., Didham, R.K., Bascompte, J. & Wardle, D.A. (2008) Global change and species interactions in terrestrial ecosystems. *Ecology Letters*, **11**, 1351–1363.

Vale, M.M., Cohn-Haft, M., Bergen, S. & Pimm, S.L. (2008) Effects of future infrastructure development on threat status and occurrence of Amazonian birds. *Conservation Biology*, **22**, 1006–1015.

Van Bael, S.A., Philpott, S.M., Greenberg, R., *et al.* (2008) Birds as top predators across natural and managed systems. *Ecology*, **89**, 928–934.

Van Bael, S.A., Brawn, J.D. & Robinson, S.K. (2010) Birds defend trees from herbivores in a Neotropical forest canopy. *Proceedings of the National Academy of Sciences USA*, **100**, 8304–8307.

Vargas, F.H., Harrison, S., Reab, S. & Macdonald, D.W. (2006) Biological effects of El Niño on the Galápagos penguin. *Biological Conservation*, **127**, 107–114.

Vargas, F.H., Lacy, R.C., Johnson, P.J., *et al.* (2007) Modelling the effect of El Niño on the persistence of small populations: the Galápagos penguin as case study. *Biological Conservation*, **137**, 138–148.

Weathers, W.W. (1997) Energetics and thermoregulation by small passerines of the humid, lowland tropics. *The Auk*, **114**: 341–353.

Wenny, D.G., DeVault, T.L., Johnson, M.D., *et al.* (2011) The need to quantify ecosystem services provided by birds. *The Auk*, **128**, 1–14.

Wesołowski, T., Rowinski, P., Mitrus, C. & Czeszczewik, D. (2006) Breeding bird community of a primeval temperate forest (Białowieza National Park Poland) at the beginning of the 21st century. *Acta Ornithologica*, **16**, 55–70.

Wiersma, P., Muñoz-Garcia, A., Walker, A. & Williams, J.B. (2007) Tropical birds have a slow pace of life. *Proceedings of the National Academy of Sciences USA*, **104**, 9340–9345.

Wiles, G.J., Bart, J., Beck, R.E. & Aguon, C.F. (2003) Impacts of the Brown Tree Snake: patterns of decline and species persistence in Guam's avifauna. *Conservation Biology*, **17**, 1350–1360.

Williams, S.E., Bolitho, E. & Fox, S. (2003) Climate change in Australian tropical rainforests: an impending environmental catastrophe. *Proceedings of the Royal Society B: Biological Sciences*, **270**, 1887–1892.

Wormworth, J. & Şekercioğlu, C.H. (2011) *Winged Sentinels: Birds and Climate Change*. Cambridge University Press, Port Melbourne.

Wright, S.J. & Muller-Landau, H.C. (2006) The future of tropical forest species. *Biotropica*, **38**, 287–301.

Wright, S.J., Muller-Landau, H. & Schipper, J. (2009) The future of tropical species on a warmer planet. *Conservation Biology*, **6**, 1418–1426.

Yong, D.L., Qie, L., Sodhi, N.S., *et al.* (2011) Do insectivorous bird communities decline on land-bridge forest islands in Peninsular Malaysia? *Journal of Tropical Ecology*, **27**, 1–14.

Young, R.P., Baptiste, T.J., Dornelly, A., *et al.* (2010) Potential impacts of tourist developments in St Lucia on the Endangered White-breasted Thrasher *Ramphocinclus brachyurus*. *Bird Conservation International*, **20**, 354–364.

16

Conserving large mammals: are they a special case?

David W. Macdonald[1], Luigi Boitani[2], Eric Dinerstein[3], Hervé Fritz[4] and Richard Wrangham[5]

[1]Wildlife Conservation Research Unit, Department of Zoology, Recanati-Kaplan Centre, Tubney House, University of Oxford, UK
[2]Department of Biology & Biotechnologies, Università La Sapienza, Viale Università 32, Rome, Italy
[3]WWF-US, 1250 24th St. NW, Washington, D.C. USA
[4]Laboratoire Biométrie et Bilogie Evolutive, CNRS UMR 5558, Université de Lyon, Villeurbanne Cedex, France
[5]Department of Human Evolutionary Biology, Peabody Museum, Harvard University, Cambridge, MA, USA

'All animals are equal but some animals are more equal than others.'

(*Animal Farm*, George Orwell)

Introduction

Large terrestrial mammals (LTMs) are charismatic and widely valued, and thus potent emblems for conservation. The ecological costs and constraints of being a large mammal in the 21st century create special challenges in their conservation compared to smaller mammalian species. In this chapter we review the conservation problems posed by large body size, among the most important being range collapse – the dramatic reduction in area that LTMs formerly occupied. We probe the reasons why such problems are associated with larger species, and propose several solutions to increase their chances for persistence and in some cases, lead to recovery of the former range.

Conservation and management problems of large mammals

Size and conservation status

Large body size among the earth's 5200 mammal species has become an attribute with added risk in the 21st century. Among mammalian species threatened with extinction, the IUCN Red List clearly illustrates that large

Key Topics in Conservation Biology 2, First Edition. Edited by David W. Macdonald and Katherine J. Willis.

species (mammals >20 kg) are almost twice as likely to be threatened (39% are threatened) than are mammals as a whole (Schipper et al. 2008). Similarly, the mean body weight of those mammals threatened with extinction is approximately 1.4 kg compared to 140 g for those not threatened (Cardillo et al. 2005).

Large-bodied taxa are consistently among the most threatened species, whether carnivores, ungulates or primates, and are at greater risk than smaller species in the same orders (Purvis et al. 2000, 2001; Cardillo et al. 2008; Matthews et al. 2010). A century ago 100,000 tigers may have roamed the forests of Asia but today fewer than 3500 may survive in the wild (Dinerstein et al. 2007; Seidensticker et al. 2010; Wikramanayake et al. 2011). In China, an estimated 37–50 (IUCN 2008) or so tigers survive in a range that has been reduced by about 95% and in 2007, India, which contains the largest number of tigers, more than halved the estimate of its tiger population, from over 3000 to less than 1400 (Chundawat et al. 2011). Most important, a range-wide survey of tiger conservation landscapes in 2005 showed that tiger populations occupy only 7% of their historical range and 40% less area than they did 10 years earlier (Dinerstein et al. 2007). The recency of mammalian carnivore losses is shocking: the thylacine went extinct in the 1930s, the last Barbary lion was killed in Morocco in 1942 (Macdonald et al. 2010) and three of the eight tiger subspecies – the Bali tiger (*Panthera tigris balica*), the Caspian tiger (*P.t. virgata*) and the Javan tiger (*P.t. sondaica*) – have gone extinct in the last 50 years, the latter disappearing as recently as the mid-1970s (Jackson & Nowell 2008), and while in Turkey the last documented Caspian tiger was killed in 1970, there are reports of them being hunted there until the mid-1980s (O.E. Can, personal communication).

Among primates, although some of the most threatened are tiny (tarsiers, tamarins and lorises), all of the great apes (bonobos, chimpanzees, gorillas and orang-utans) are classified by the IUCN as Critically Endangered or Endangered. In Gabon, a stronghold for African apes, the ape population halved between 1983 and 2000 (Walsh et al. 2003). In Ivory Coast, chimpanzee populations shrank by 90% between 1989 and 2007 (Campbell et al. 2008). Orang-utans suffered a 10-fold decline in population size during the 20th century (Nantha & Tisdell 2009).

Among the ungulates, the Pyrenean ibex, widespread in the Middle Ages, was reduced to 10 in the 1990s and the last one died in captivity in January 2000, ironically when a tree fell on it. The Père David's deer (*Elaphurus davidianus*) currently persists only in captivity, as does the scimitar-horned oryx (*Oryx dammah*), with the exception of a small number recently reintroduced to Tunisia. In November 2011, the Vietnamese government and the IUCN officially declared extinct the last remaining Javan rhinoceros in that country, leaving those in Ujung Kulon National Park on the western tip of Java, Indonesia, as the sole remaining population. That species now stands at <50 individuals. Finally, the northern white rhino (*Ceratotherium simum cottoni*), recently considered a distinct species from the more abundant southern white rhino, was also declared extinct from its last refuge in Garamba National Park, in the Democratic Republic of the Congo, in 2005. Both of these species once ranged widely in Asia and Africa, respectively, and they illustrate a key theme of this chapter: many LTM species are experiencing a massive range collapse.

Where do large mammals live, and where are they most at risk?

Nowadays, large mammals are most abundant and diverse in the equatorial belt (Silva et al. 2001). However, a long history of conflict with people means that many occupy only fragments of their former geographical ranges. By one estimate, the last 500 years have witnessed a >50% range loss in 35% of mammals >20 kg, and only about a fifth of the earth's surface (68% in Australasia to only 1% in Indomalaya) still contains its mediaeval complement of such mammals (Morrison et al. 2007; Figure 16.1). The link with human impacts is not straightforward: large remote and undisturbed portions of

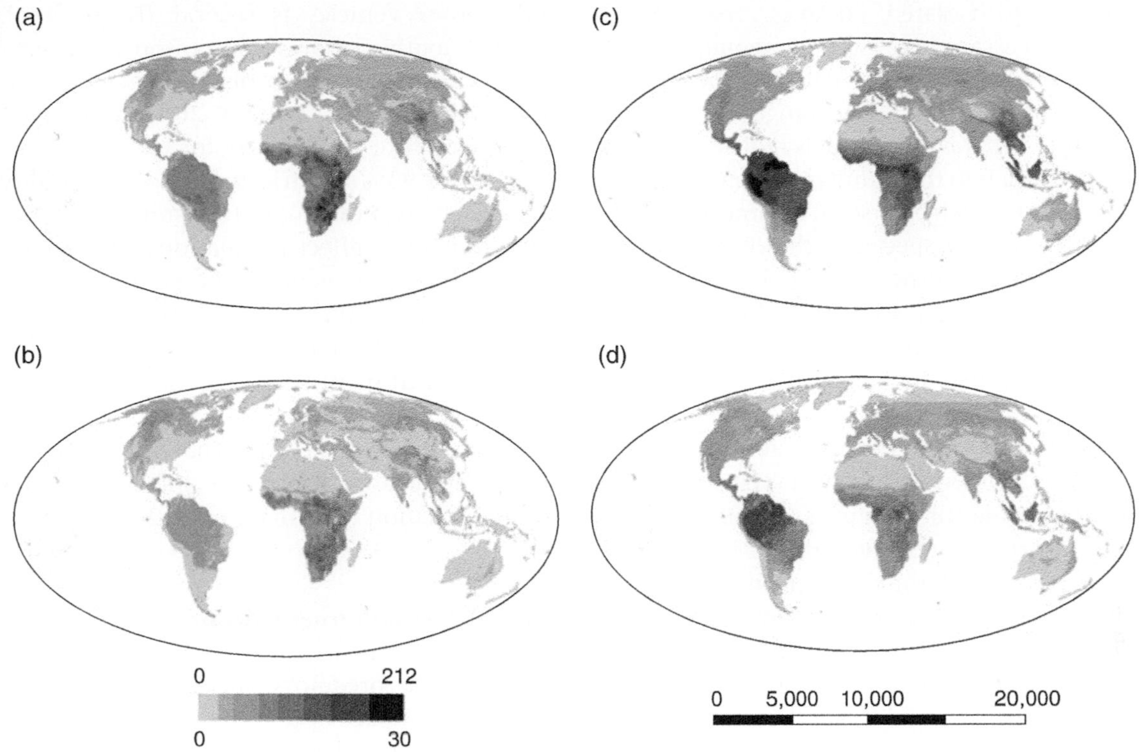

Figure 16.1 Richness of large mammals (>20 kg, 260 species including sea lions, seals and walruses) compared to richness of all mammals using: (a) large mammals ranges, (b) large mammals' distribution models based on habitat suitability, (c) all mammals ranges, (d) all mammals' distribution models based on habitat suitability. Shaded bars indicate min/max number of overlapping species: all mammals 0–212, large mammals 0–30. The most interesting feature is the concentration of large mammals in Africa and the extraordinary reduction of ranges and overlap when habitat suitability models are used as surrogates for species distribution. All mammals figures (c and d) from Rondinini, C., Chiozza, F., di Marco, M., *et al.* (2011) Global habitat suitability models of terrestrial mammals. *Philosophical Transactions of the Royal Society B*, **366**, 2633–2641.

the Nearctic, Neotropical and Palearctic wilderness have lost one or more of their large mammals, while some human-populated areas in the Congo Basin, the Amazon Basin, Australia and areas in Siberia still retain theirs. The greatest numbers of species of large mammals occur in two African sites (30 spp. in Hwange (Zimbabwe) and Serengeti-Mara (Tanzania)), fewest in northern Eurasia and Siberia (seven species). Ten sites in sub-Saharan Africa and only one in the Palearctic realm each conserve more than 25 species. Overall, only 12% of the total area that retains large mammal assemblages is formally protected, although regionally this varies from 9% protected in the Palearctic to 44% in Indomalaya (Morrison et al. 2007).

Macdonald et al. (2010) illustrate range collapse in lions (120–250 kg) of the *Panthera leo* complex, noting an ominous coincidence in the arrival of prehistoric humans and a 96–98% shrinkage in lions' range over c.20,000 years (see also Yamaguchi et al. 2004). In North America, wolves (20–40 kg), despite their versatility and continued IUCN status of Least Concern, have been radically reduced by persecution (thereby conceding to the red fox (c.6 kg) their former title as most widespread wild canid) (Laliberté & Ripple 2004). The African wild dog and the Ethiopian wolf share similar predicaments of range collapse for different reasons (Sillero-Zubiri et al. 2004; Macdonald & Sillero-Zubiri 2004a): both are currently confined, in shrinking numbers, to isolated enclaves, from which neither can escape because of burgeoning human populations surrounding refuges. The addax's (*Addax nasomaculatus*) former range encompassed the whole Sahara, but now it is reduced to about 200 individuals in the Temrit mountains in southern Niger (J. Newby, personal communication).

The five extant species of rhinoceros provide the most dire statistics of decline among large mammals (although, see below, offering hope of reversing range collapse). Black rhino numbers in Africa crashed by more than 96% in just 30 years, with just 2410 surviving in 1995 (Emslie & Brooks 1999); the western black rhino was recently declared extinct (Emslie 2011). Northern white rhino went extinct in 2005. The Sumatran rhino and Javan rhino, having been abundant up to the mid-19th century, are now facing extinction, the Javan rhino's single population mentioned above and only a few small populations of Sumatran rhinos in peninsular Malaysia, Sumatra and Sabah (Dinerstein 2011). The list goes on: of seven equids, four are endangered or worse (Moehlman 2002; Boyd et al. 2008), as are five antelope and 30 of the wild goat subfamily Caprinae (Shackleton 1997). Of Asiatic elephants (*Elephas maximus*), only one population (in the western Ghats of southern India) is larger than 8000 (Williams 2011). All the remaining 12,000 or so Asiatic elephants are in populations smaller than 1000, most smaller than 100.

In short, we are witnessing the demise of a number of LTMs that represent lineages dating from the Eocene, contemporary LTMs that are several million years old, and many large mammals that until recently were mostly widespread and common. What factors have hastened their precipitous decline?

Threats to large mammals

Their size predisposes LTMs to be exploited by people and to conflict with them. The consequences are far-reaching. While intraspecific fighting is the primary cause of death amongst pumas around Los Angeles, California, collision with motor vehicles is second (Beier et al. 2010). Shooting, snares and poison have been for centuries the legal tools to manage carnivores (e.g. wolves; Boitani 2003) and even today they take heavy tolls: for example, they account for 93% of African wild dog mortality on Zimbabwean ranch-land (Rasmussen 1997). These pressures affect populations and behaviour: where humans persecute wild dogs, they become more nocturnal but at a cost of increased competition with lions and hyaenas which are more active at night (Rasmussen & Macdonald 2012; Cozzi et al. in press). Similarly, the risk of being shot increased nocturnal behaviour in large trophy antelopes, thus increasing their risk of predation (Crosmary et al. 2012).

Habitat loss and fragmentation

The largest predators have the greatest demands for space and are most susceptible to habitat loss and fragmentation; tigers are a prime example (Seidensticker et al. 2010). However, opportunistic predators such as wolves, Asian leopards, coyote and pumas are less susceptible to habitat degradation because all of them are habitat generalists and good dispersers (Mech & Boitani 2003; Patil et al. 2011; Murphy et al. submitted), and so even degraded habitats can have high conservation potential. Hearn et al. (submitted) found that Sunda clouded leopards (*Neofelis diardi*) can occur at relatively high densities in well-managed, selectively logged forests in Borneo and concluded that these forests, although not pristine, are important for the conservation of this species. Tigers, too, do well in large contiguous blocks of lightly logged, mineral soil, lowland forests in Sumatra, so long as prey are abundant, ample understorey cover is available, and human disturbances are minimized (Sunarto et al. 2012, in press).

The consequences of fragmentation upon African elephants exemplify many of the more general issues for LTMs. African elephants need copious resources (~150 kg of vegetation and

~100 L of water daily), so their habitat use and ranges are largely dictated by the distribution of surface water (Chamaillé-Jammes et al. 2007) and food resources (Loarie et al. 2009). In arid and semi-arid savannas, where the distribution of these resources is characterized by a high spatial and temporal variability, elephants need to roam widely. However, most of the elephant's range (>60%) is beyond protected areas, and although they cannot tolerate human densities >15 ind/km^2 (Hoare & du Toit 1999), they have no choice but to traverse agricultural landscapes (Fritz et al. 2003). Fences restrict elephants' movements in response to fires and drought, with implications for natural mechanisms of population control, migration and genetic flow, and source-sink population dynamics (Hayward & Kerley 2009; Trimble & van Aarde 2010). This issue argues for removal of fences and the 'megapark' concept (van Aarde & Jackson 2007; see also Chapter 23), where the free movements of elephants could favour self-regulation (Shrader et al. 2010; Young & van Aarde 2010). Zimbabwe's Hwange National Park elephant population has stabilized at a population density of c.2 ind/km^2 and illustrates self-regulation in a managed unfenced system (Chamaillé-Jammes et al. 2008): the interaction between population size and the available foraging range determined by the number of active waterholes during the dry season controls the park population, and density-dependent responses to fluctuations in the carrying capacity of the park occur through dispersal at a larger, regional scale. The induced locally high densities of elephants around water sources may have several implications for the vegetation (Chamaillé-Jammes et al. 2009; Valeix et al. 2011), other large herbivores (Valeix et al. 2007a, 2007b, 2008, 2009a) and ultimately large carnivores (Loveridge et al. 2006).

Thus, designing unfenced 'megaparks', an underlying principle for trans-frontier conservation areas, ultimately needs to address management of human artifacts in the ecosystems. Such unfenced 'megaparks' encompassing protected areas and a mosaic of adjacent land uses can provide sufficient space for the operation of self-regulatory processes, but makes human–elephant conflicts inevitable. Indeed, as human populations increase around protected areas, dispersal and movements are thwarted and conflict heightened (Hoare 2000; Sitati et al. 2003). Unfenced areas may be a vital conservation tool for many LTMs but they increase opportunities for conflict with the human population; sometimes fences may be the best hope for co-existence – certainly the cattle ranchers adjoining Mkgadikgadi National Park (Botswana) called for fencing to separate lions from their stock (Hemson et al. 2009).

Exploitation

An extensive body of literature, principally derived from fisheries, pertains to the science of harvesting wild populations. A central concept for management in fisheries science is the maximum sustainable yield (MSY). This metric is the greatest amount of biomass or individuals that can be removed and replaced by population growth without detriment to a population (Primack 2006). MSY is maximal when a population is at about half its carrying capacity, and increases with the intrinsic growth rate of the population. Beddington & Kirkwood (2007) discuss the problems of simplistic application of MSY to harvest management. One problem is that estimates of current and limiting levels of fish exploitation are far from adequate. For LTMs, where harvests are frequently illegal, patchy and entirely unregulated (see Chapter 5; Moreira et al. 2012), this is even more problematical. Many attempts have been made to develop the production systems based on wild LTMs, mostly through farming and ranching, or even organized cropping operations (Hudson et al. 1989; Macdonald et al. 2012), but there is little evidence that traditional exploitation is staunched, perhaps partly for cultural reasons (Chardonnet et al. 2002; Teel et al. 2007).

FOOD RESOURCES Large mammals are frequently overexploited for food. Tropical forest dwellers relying on bushmeat for protein prefer species that yield the most meat per unit of energy or time spent hunting: that is, generally, large mammals (Jerozolimskia & Peres 2003). The price of bushmeat carcasses is determined principally by size (Macdonald et al. 2011), so there is a strong incentive for hunters to focus on large, conspicuous taxa. Thus, in Asia, 90% of large (>10 kg) mammals are harvested compared with 28% for smaller mammals (the comparison for Africa and the Neoptropics is 80% versus 15% and 64% versus 11%, respectively) (Schipper et al. 2008). Furthermore, bushmeat hunting can have indirect effects, such as the decline of leopard numbers in the Congo Basin through exploitation competition (Henschel et al. 2011).

SPORT HUNTING Hunting of a different sort, for sporting trophies, is also often 'sizeist': Johnson et al. (2010) report that the cost of hunting African cats is approximately proportional to their size. In Zimbabwe, for example, the mean prices advertised in 2004 were approximately $140 for a wildcat, $2400 for a cheetah and $5200 for a lion. Large bovids, however, were relatively cheaper for their size to hunt than were smaller species (possibly because their horns determine their value, and are relatively smaller in larger bovids). A bushbuck licence in Zimbabwe cost around $700 in 2004 compared with $3000 for a buffalo. In short, for felids, big means pricy, to the extent that more than 500 lions are shot for trophies annually, each total package costing about US$50,000–100,000, all in economies where both hard currency and lions are scarce (Loveridge et al. 2007). In the Hwange National Park, Zimbabwe, the legal quota for trophy lions totalled 60 males in 2000. Since only about 22 existed, this could not be fulfilled. Nonetheless,72% of the adult males tagged in the Park by Loveridge et al. (2007) were shot outside it (82% within 1 km), accounting for 44% of the adult male population annually, the average trophy being 5 years old (not yet fully matured). So, even at 15,000 km^2, the park was too small because a vacuum at the boundary drew males from the interior into the firing line. Consequently, lion hunting was suspended in western Zimbabwe for 4 years, during which numbers of males trebled and numbers of coalitions doubled. As the social perturbation triggered by hunting was quelled, coalition take-overs, and thus infanticide, declined, and cub survival to 2 years rose from 41% to 64%. Hunting restarted in 2009, the quota being set at four lions.

The vulnerability of large species to hunting can be exacerbated by an 'anthropogenic Allee effect (AAE)' (Courchamp et al. 2006). That is, an 'extinction vortex' can result if, as the species gets rarer, the last individuals are valued more and more, and extinction follows. Trophy species are vulnerable to this phenomenon if too many licences are issued in response to increased demand, as the cost to hunt an individual is not closely linked to their rarity. An association between level of threat (indexed by IUCN status) and trophy prices has been observed in both bovids (Johnson et al. 2010) and cats (Palazy et al. 2011). Prescott et al. (2012) observed that bovid species that changed IUCN status from less to more vulnerable between 2004 and 2010 increased in price more than did those whose status was unchanged.

TRADE IN BODY PARTS While the body parts of mammals of all sizes are traded for folk medicines – from the os penis of binturong to the perineal glands of musk deer and down to the bones of mole rats (reputedly a less favoured alternative to tiger bone) – populations of larger mammals are, due to their life-history characteristics, likely to be harder hit and slower to recover from such exploitation. This topic is explored by Dutton et al. (see Chapter 5) but unsustainable exploitation for body parts or fur has worsened the predicament of large mammals such as the snow leopard (*P. uncia*)

(Jackson et al. 2010), as it did the Iberian lynx in the 19th century (Ferreras et al. 2010) and threatens the extinction of the Amur tiger (Miquelle et al. 2010). There has been a resurgence in the fashion for wearing tiger furs in Tibet (despite recent criticism from the Dalai Lama), and the burgeoning trade in tiger parts for traditional Chinese medicine (TCM) is a major driver propelling tigers towards extinction. Nijman et al. (2011) document the use of more than 100 primate species in traditional medicines.

Pestilence and persecution

Mammals of all sizes face conflict with people. Among smaller species, a small proportion of rodents exert a disproportionate burden on agriculture, forestry and human health (e.g. Macdonald et al. 2012) and indeed, as invasive species, their spread has led to local extinctions of other vertebrates and plants (see Chapter 12; Macdonald et al. 2007a). When the problematic mammal is small the attempted interventions are generally termed 'pest control' but when the mammal is large, the interactions are referred to as human–wildlife conflict (HWC). This discontinuity in vocabulary may belie inconsistencies not only in approaches but also in values and ethics. Thus regular damage by numerous small, inconspicuous mammals is commonly accepted more readily than occasional damage by large animals. Squirrels damaged 40% of Equatorial Guinea's cacao crop (Smith & Nott 1988). Commensal rodents commonly destroy 8–10% and sometimes 90% of crops (Lund 2013). Vole damage to the eastern USA apple crop cost c.US$50 M, while feral mouse plagues damage c.A$10 M worth of cereals in the state of Victoria, Australia (Wood 2012). Farmers tend to tolerate these smaller competitors as a fact of life, yet view the larger animals with less tolerance despite their lesser impact (see Chapter 7). Furthermore, wider society includes advocates for many larger mammals, indicating a sense of ownership by some group that the farmer can blame ('your cheetah ate my goat' or 'your bears destroyed my apiary'; Can & Macdonald submitted), whereas small mammalian pests may be considered more a blight of nature.

Conflict between people and wildlife can be measured according to several different metrics: the numbers of animals involved and the average impact of the damage they cause; the numbers of people involved and the average costs of the damage they suffer; and their impact on the problem species. For example, tiger attacks on valuable cattle may be much less frequent, but much more costly, than rat damage to grain among subsistence farmers that may own one or two animals, or jackal predation on lambs or game. Among the factors predisposing carnivores to conflict with people, the foremost is their tendency for livestock depredation. This is exacerbated by the depletion of natural prey (livestock provide an alluring alternative), habitat loss (increase in edge effects heightening wildlife–human encounters and creating population sinks; Woodroffe & Ginsberg 1998), and inadequate livestock management (Hoogesteijn 2003; Ogada et al. 2003).

Few vertebrate families rank higher than the Felidae with regard to intensity of HWC when it occurs (Inskip & Zimmermann 2009). Nine are particularly prone to conflict: caracal (*Caracal caracal*), cheetah, Eurasian lynx, jaguar, leopard, lion, puma (*Puma concolor*), snow leopard and tiger, and across the 37 felid species, the severity of conflict is greatest for the largest cats (>50 kg). Thus, snow leopards, jaguars, pumas, tigers and lions can all devastate a small farmer's livelihood (Jackson et al. 2010; Wang & Macdonald 2006; Maclennan et al. 2009; Murphy et al. submitted). For example, Hemson et al. (2009) studied depredation on cattle by the 40 or so lions living in the 4900 km^2 Mkgadikgadi Pans National Park, Botswana. Herders took a relaxed attitude to husbandry (13% of the average herd roamed the park overnight), resulting in high predation rate by lions on livestock (Valeix et al. 2012). As a result, herders were hostile to lions (killing eight in retaliation for one stock-killing episode); 59% of herdsmen claimed to have lost livestock to lions in the previous year, at an annual cost of US$646 each (2002 prices). The cattle herders thought that the benefit derived from lions accrued to somebody else (government or tour operators), and that somebody else should therefore assume responsibility for them. Despite the availability of compensation for cattle killed within corrals, a third of the cattle killed by lions were predated while roaming at night, and herdsmen were unmotivated to improve stockmanship: 61% suggested fencing the park while 16% suggested killing the lions.

In South Africa, black-backed jackals preyed extensively on newborn wild ungulates on game and hunting reserves, consuming up to 31% of individuals in some populations (Klare et al. 2010). On nearby sheep ranches, the consumption of sheep, mostly lambs, was similarly high (Kamler et al. 2012). Such high predation rates on valuable game species and domestic stock potentially cost landowners a considerable amount of money, which may explain why jackals are so heavily persecuted on private lands in South Africa.

Large herbivores similarly damage agriculture. Baboons (*Papio* spp.) and chimpanzees (*Pan troglodytes*) are particularly consistent nuisances to farmers living adjacent to African forests (e.g. Naughton-Treves et al. 1998; Tweheyo et al. 2005). Although group-living monkeys are often the most significant crop-raiding primates, gorillas and even sometimes the mainly arboreal orang-utans (Campbell-Smith et al. 2010) can also wreak important damage (Hockings & Humle 2009). In Europe or North America, deer and wild boar (*Sus scrofa*) affect commercial agriculture and natural habitats (see Côté et al. 2004 for a review on deer). This has raised public debate (Reiter et al. 1999; Goulding & Roper 2002) and conservation dilemmas (Putman & Moore 1998) and stimulated innovative solutions (VerCauteren et al. 2006). Indeed, crops can be a major part (>30%) of the diet of these common ungulates (Schley & Roper 2003) and the costs can be high, e.g. over US$300 million to agriculture and households and US$1.6 billion in deer–vehicle collisions annually (Conover 2002). The same applies for forest production (Gill 1992), with over US$700 million damage to the timber industry in the USA (Conover 2002). Similarly, wild boar and monkies accounted for 60% and 50% of crop damages around Nanda Devi Biosphere Reserve, in India, where losses amounted to 50% of all crops at the vicinity of reserve (Rao et al. 2002). Wild boar is the main pest around the Jigme Singye Wangchuck National Park, in Bhutan, being involved in 97% of the crop damage complaints, followed by barking deer (*Muntiacus muntjak*) (67%), macaques (*Macaca mulata*) (46%) and sambar deer (*Cervus unicolor*) (41%) (Wang et al. 2006a), leading to deteriorating perception of the protected area and of conservation programmes (Wang et al. 2006b).

Translocating LTMs can worsen conflict: the expanding, newly reintroduced rhinoceros population (*Rhinoceros unicornis*) in the Royal Bardia National Park, Nepal, damaged paddy rice and lentils, adding to the existing conflict with wild boar and chital deer (*Axis axis*), resulting in crop losses of up to 47% (for lentils) and 25% (for wheat) around the park (Studsrød & Wegge 1995). Between 1997 and 2008, annual cases of hippo conflict (90% due to crop damage) in Kenya increased to a peak of 937 with a concomitant rise in hippo mortality (Kanga et al. in press). The elephant, both Asian and African, certainly accounts for major damage to crops around protected areas (up to 66% and over US$500 per farmer in Africa, as reviewed by Naughton-Treves et al. 1999), and mitigation continues to be a challenge (Osborn & Parker 2003; Sitati et al. 2005, 2006; Davies et al. 2011).

Beyond the costs to agriculture, tradition and cultural perception have played a role in LTM persecution. Conflict with grey wolves has been rife across North America, Europe and most of Asia. Pastoralists have persecuted them universally, and a bounty system was established by the 6th century BC (Boitani 1995). Wolf persecution has commanded immense resources from farmers, and in the USA this was supported by state and federal forces, culminating in obsessive and irrational investment to kill the few remaining wolves in the lower 48 states. Only a small population remained in northern Minnesota. However, today wolves have recovered in many of the northern states and as we write, a male wolf, dispersing from north east Oregon, has arrived in California for the first time in 88 years. In Europe, wolves were eradicated as early as the 16th century in England and subsequently from most of central Europe. Predation on livestock was one motivation but it was fuelled by disproportionate hatred and fear that they might attack people (Boitani 2003). The Little Red Riding Hood syndrome of fearing wolves has its roots in relatively common attacks up to the 18th century (Fritts et al. 2003; Moriceau 2007) and remains deeply engrained although wolf attacks (outside the context of rabies) have been rare in North America and much of Eurasia for at least two centuries (Boitani 2003; Jhala & Sharma 1997). Even though human deaths to predators may be few, their emotional and political impact is great.

Puncture marks on early hominid skulls (Brain 1969) suggest that the habit of killing humans and human ancestors is long-standing for leopards, and continues in the c.100 attacks annually in India (Athreya et al. 2004); 294 people were killed by tigers between 1984 and 2001 in the Indian Sundarbans, but the recent killing of an adult (inebriated) human by a jaguar in Brazil was unprecedented (de Paula et al. 2008). Nowhere, however, is intolerance of big cats more poignantly understandable than in Tanzania, where between 1990 and 2004 almost 563 people were killed and over 300 more injured by lions (Packer 2005). Large herbivores can be dangerous too, with more than 500 people killed by elephants annually, mostly in India, and hippos considered as amongst the most dangerous species in Africa. Over 27 months from July 2006, Dunham et al. (2010) recorded that in Mozambique 31, 24 and 12 people were killed by elephants, lions and hippos respectively (and 134 by crocodiles). Between 1997 and 2008 annual cases of hippo conflict (90% due to crop damage) in Kenya increased to a peak of 937 with a concomitant rise in deaths caused by hippos (Kanga et al. in press).

Even primates can be dangerous: chimpanzees occasionally prey on human infants (Wrangham et al. 2000). The relative risk to life from large mammals may be exaggerated: according to Herrero et al. (2011), at least 63 people were killed by non-captive black bears in Alaska, Canada and the lower 48 states of USA between 1900 and 2009, while 45,900 people die annually as a result of snakebites in India (the global tally is 100,000 people) (Mohapatra et al. 2011).

Nonetheless, with such fatalities, conserving LTMs can involve high stakes and raise difficult questions about how much a human life is worth. One answer, reported on 12th January 2012 in *The Times of India*, was that friends of a man killed and eaten by a tiger while working his field 500m from the Tipeshwar Wildlife Sanctuary sought compensation of Rs 5 lakh (c.US$10,000) (along with calls to fence the park and increase natural prey therein). The extremes of militaristic interventions are illustrated by President Moi of Kenya empowering the Kenya Wildlife Service to use shoot-on-sight policy against poachers and, in 1998, President Mugabe of Zimbabwe similarly permitted his Parks Department to shoot rhino poachers on sight (Duffy 2010). Translocating villages to make way for protected areas is apparently not inevitably resisted: after three villages had been resettled from their traditional land in Melghat Tiger Reserve during 2001–2 (www.downtoearth.org.in/content/happily-uprooted), residents of other villages in the reserve approached the Maharashtra Forest Department requesting to be moved too. For the 350 families of Amona, Nagartas and Bharukheda, relocation in 2011 brought Rs 10 lakh (c.US$20,000) per adult, free land for a house and better access to markets, education, health facilities, transport, the court and the police. One villager, formerly resident in a thatched hut, was reported to have used his share of the compensation (Rs 40 lakh or c.US$80,000) to purchase a three-storey building, a hectare of farm land and a tractor, according to www.downtoearth.org.in/content/happily-uprooted.

Why are large mammal species particularly vulnerable to extinction?

The spectrum of mammalian weights ranges from Kitti's hog-nosed bat (*Craseonycteris thonglongyai*), weighing only 1.5 g, to the blue whale (*Balaenoptera musculus*) at 150 tonnes (Macdonald 2009a). Space restricts us to only terrestrial mammals, of which elephants weigh a million times more than the smallest shrew (the 2 g Etruscan shrew, *Suncus etruscus*), the remaining c.5200 spanning this body size continuum (Figure 16.2). Of these, why are LTMs (somewhat arbitrarily, those >20 kg) particularly vulnerable to extinction?

For mammalian herbivores, Fritz & Loison (2006) define as large those heavier than 2 kg. On this basis, a 6 tonne elephant is about 3000 times heavier than the smallest 'large herbivorous mammals' (e.g., the rock ringtail possum *Petropseudes dahli*, lesser mouse deer *Tragulus kanchil*, the Royal antelope, *Neotragus pygmaeus*). In 1999, Carbone et al. published a benchmark analysis that provided, for the first time, a biological definition of 'large', phrased in terms of the energetics of hunting behaviour, with reference to one order, the Carnivora. They detected a sharp discontinuity in prey sizes such that large carnivores, heavier than about 14.5 kg, specialized in large prey (see below). Based on this definition, and in contrast to the aforementioned span of large herbivore sizes, the large carnivores differ only 23-fold in size (although, in totality, size variation in the terrestrial Carnivora is 10,000-fold, from polar bear, *Ursus maritimus*, to least weasel, *Mustela nivalis*). Our purpose here is to identify aspects of size that have functional significance for conservation; we will adopt Carbone et al.'s (1999) approach in seeking to identify and explain the biological corollaries of size that are most relevant.

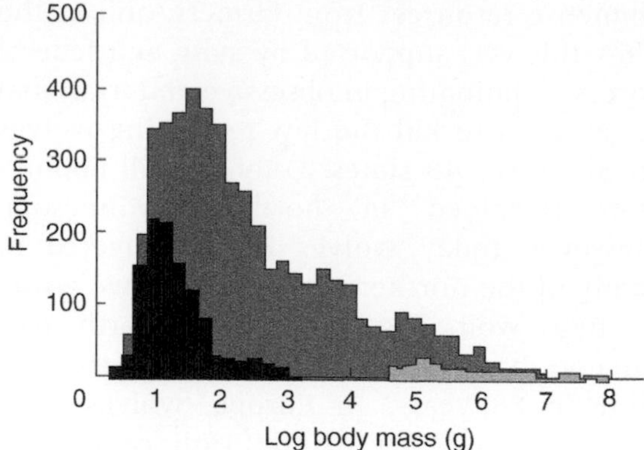

Figure 16.2 The global distribution of mammalian body mass at the Late Quaternary. Patterns are shown separately for volant (dark grey bars, left-hand side of graph), aquatic (grey bars, right-hand side of graph) and terrestrial (light grey) mammals. Note $\log_{10}$ scale. Data are prior to the anthropogenic extinction of megafauna in the Americas at the terminal Pleistocene, which significantly depressed the right mode and led earlier authors to characterize the overall distribution as unimodal. From Smith, F.A. & Lyons, S.K. (2011) How big should a mammal be? Macroecological look at mammalian body size over space and time. *Philosophical Transactions of the Royal Society B*, **366**, 2364–2378.

Diet, home range size and population density

Larger mammals need more food than do smaller ones because they have greater absolute energetic requirements. Smaller mammals, in contrast, have higher energetic demands per unit weight because they lose heat faster due to their relatively larger surface area to volume ratio (Kleiber 1963). This generality, whereby energy requirements scale with body mass to the power of three-quarters, known as Kleiber's Rule, applies in principle to all mammals and affects many aspects of their ecology.

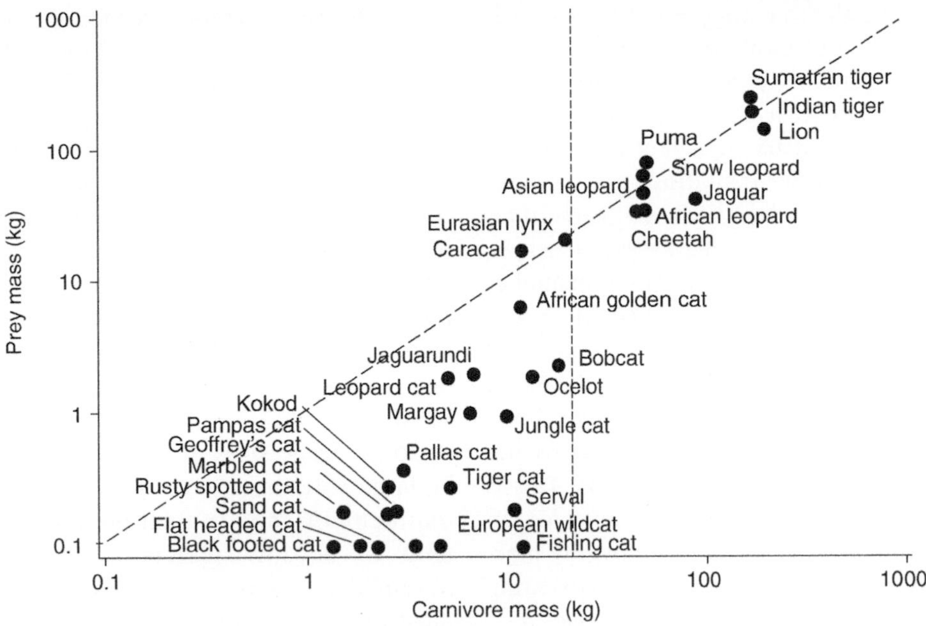

Figure 16.3 Body mass of most common prey compared to mass of felid species. From Macdonald, D.W., Mosser, A. & Gittleman, J.L. (2010) Felid society. In: *Biology and Conservation of Wild Felids* (eds D.W. Macdonald & A.J. Loveridge), pp.125–160. Reproduced by permission of Oxford University Press, Oxford.

Carnivores are particularly revealing because Carbone et al. (1999) discovered that they divide into two size categories. Small or mid-sized carnivores, those of 14.5 kg or less, feed mostly on prey less (often much less) than 45% of their own mass. Large terrestrial carnivores tend to hunt large prey near their own mass (for them prey mass averaged 1.19 times the predator's mass) (Figure 16.3). So, large carnivores over about 20 kg offer a clear example about the special nature of large mammals – they generally cannot live without large prey.

Herbivores are different because energy-rich fruits and buds are inevitably less abundant than are less nutritious leaves. The latter are filled with nutrients, but these are protected by a layer of cellulose and locked within cell walls. Such nutrients are available only to those species whose digestive tracts, body size and energy demands allow for the slower breakdown of the cell walls of grass and leaves. This digestive dilemma means that large vegetarian mammals have different lifestyles from small ones. Indeed, as the energy requirement per kilogram increases with the allometric exponent of 0.75, the digestive tract efficiency increases isometrically. Therefore, larger herbivores can deal better with poorer quality food than can smaller ones (Bell–Jarman principle, from Bell 1970 and Jarman 1974). For example, the 900 kg giant eland (*Taurotragus derbianus*) survives on coarse grasses while the 7 kg duiker (*Cephalophus monticola*) selects buds and shoots; a 180 kg mountain gorilla (*Gorilla gorilla*) eats leaves and stems, while a 50 g tarsier (*Tarsius* spp.) eats only insects (Grow & Gursky-Doyen 2010); and the 50 kg capybara (*Hydrochoerus hydrochaeris*) forages on grass whereas the 20 g bank vole (*Myodes glareolus*, formerly *Clethrionomys glareolus*) consumes seeds and roots.

Does this separation imply a similar energetic discontinuity between small and large vegetarians as there is for carnivores? Broadly, yes. The smallest ungulates, such as the 2 kg lesser mouse deer (*Tragulus kanchil*), have such high energetic demands that they need food that is both nutritious and easily digested (fruits and buds) and is relatively rare. The mouse deer is therefore trapped in a circle of size and diet: because it is small, it has to select high-quality food requiring little fermentation, and because it selects such food, which is more scarce than the abundant coarse forage of grasses and leaves, it has to be small. This same energy trap applies to any vegetarian below 10 kg which appears to be the lower limit to strict folivory amongst mammals (Demment & van Soest 1985). Among primates, a correlation between larger body mass and lower diet quality occurs across the order as a whole (Leonard et al. 2003). This is illustrated among the New World Ceboidea, in which the smaller species (e.g. the pygmy marmosets, *Cebuella pygmaea*) are restricted to eating insects, fruits and gums while larger species (e.g. the southern muriqui, *Brachyteles arachnoids*) are extensively folivorous (Gaulin 1979).

Larger vegetarians have lower energy requirements relative to their mass and can thus survive by eating a greater bulk of less nutritious leaves from which the cellulose can be unlocked by one of two processes. Rumination (chewing the cud predigested in a fore-stomach) is time-consuming but efficient, and is the approach evolved by the Cetartiodactyls, the even-toed ungulates, but because it is time consuming, there is a limit to the energy that can be extracted in this way, which in turn puts a size limit on Cetartiodactyls (amongst giraffes, the largest male may reach 1500 kg). To sustain larger herbivores requires a faster, if less efficient throughput through a hindgut, as typified by Perissodactyls (the horses, tapirs and rhinos) or by elephants. These digestive differences have consequences that link life histories to conservation: hindgut fermenters have slower metabolisms and longer generation time than ruminants, amongst which browsers live faster than grazers (Gaillard et al. 2008), and tend towards specialized diets and habitats with the upshot that they have higher rates of both extinction and speciation (Vrba 1987). For a given body size class, browsers are thus more extinction prone (Janis et al. 2000; review in Fritz & Loison 2006).

Kleiber's relationship between body size and food requirements leads to another allometry: between body size and home range size. This has conservation implications for exclusive territorial species insofar as only a finite number of territories can be squeezed into a given space, and so animals with larger ranges are obviously more likely to overflow the edges of a protected area which are also likely to encompass precariously fewer of them (Woodroffe & Ginsberg 1998).

Home range sizes of large mammals tend to be large (McNab 1963) to meet their resource requirements, and so their population densities tend to be low (Peters & Raelson 1984). Mammalian facts from Macdonald (2009a) illustrate the generalities across the range of body sizes: elephants range up to 21,000 km^2, wolves over 1000 km^2 while naked mole rats never stray from their burrow. Otters at 7 kg may patrol 40 km of river and occur at densities of one per 15 km, while the 200 kg tiger ranges over at least 20 km^2 but at densities as low as 0.52 per 100 km^2 (Wang & Macdonald 2009) (although there is always variation: tigers in alluvial grasslands in Kaziranga National Park and other reserves near large rivers in South Asia live at around 16 per 100 km^2; Jhala et al. 2011; Seidensticker et al. 2010). Big cats all have ranges larger than predicted from their body size and metabolic needs compared with other carnivores, as do African wild dogs compared to other canids, perhaps due to preying on species that form herds that are widely dispersed (Macdonald et al. 2004, 2010).

Larger mammals appear to need disproportionately large areas (perhaps this compensates for loss of resources to neighbours as home range size increases and spatial overlap becomes inevitable; Kelt & van Vuren, 2001; Buskirk 2004; Jetz et al. 2004). However, as ranges grow bigger it becomes harder to defend them or to know them intimately (Kruuk & Macdonald 1985). This together with limits to the capacity for food consumption and travel may explain why, for mammals over 100 kg, the relationship between maximum home range size and body mass becomes asymptopic or negative, and why the largest mammals can sustain higher densities than predicted by body mass (Silva & Downing 1995; Silva et al. 2001).

Risky life histories: predispositions to rarity and conflict

Larger mammals have longer intervals between generations and a lower potential for increasing numbers from one generation to the next. A Norway lemming (*Lemmus lemmus*) at 10–100 g, for instance, can bear its first litter of six young at 14 days. In contrast, the musk oxen (*Ovibos moschatus*), at 200–360 kg, only produces its first single calf after 2 years. If small mammals can produce many more young, why are any mammals big? The answer is that large size confers qualities that can be indispensable assets (Cope's Law; see Peters 1986). Attributes of bulk include ability to survive on poorer food, to store metabolites efficiently to stave off starvation (Miller & Hickling 1990), to travel further and faster and thus exploit widely separated resources, to repel larger predators and to survive colder temperatures, although the fat endurance hypothesis has been criticized as oversimplistic (Dunbrack & Ramsay 1993).

The generally lower productivity rates of larger species (primates, bears, elephants and rhinos; Sibly & Brown 2007; Bielby et al. 2007; Jones 2011) make them susceptible to overhunting and extinction (Price & Gittleman 2007). Nonetheless, ungulates of 200–400 kg have high maximum population growth rates relative to their body size, due to a combination of high adult survival, early reproduction and high fecundity, with cow:calf ratios of, for example, 0.41 for bovids and 0.59 for cervids (Gaillard et al. 2000). This phenomenon explains the explosive recovery of deer in Europe following the eradication of predators and control of hunting.

Risk itself can offer an answer to the question of when size becomes dangerous. Cardillo et al. (2008) estimate that life-history traits conspire to put mammals heavier than 3 kg at greater risk of extinction and these intrinsic factors (e.g. slow reproduction rates) combine with extrinsic risk factors (e.g. small geographic range) to worsen the odds (Davidson et al. 2009). Fritz et al. (2009) point out that the size-related risk of extinction is statistically significant only in tropical regions or in places with a lesser historical impact of agriculture, presumably because many larger species have already gone from areas historically affected by agriculture.

Migrations

Most migratory mammals are large (Peters 1986; Fryxell & Sinclair 1988). Many LTMs that migrate travel long distances between winter and summer range or dry-season and wet-season habitats. Not surprisingly, many of them are hooved and gregarious, including tens of thousands of springbok and, historically, up to 30 million or more bison (Bolger et al. 2008), and barren-ground caribou travelling up to 2500 km from summer calving grounds in open tundra to winter in boreal forest (Berger 2004).

Over the last two centuries familiar blights to large mammals in general (overhunting, habitat loss and human impediments) have increasingly disrupted migrations (Wilcove & Wikelski 2008; Harris et al. 2009). Examples include the overhunting of Mongolian gazelle (*Procapra gutturosa*) in Inner Mongolia and the saiga (*Saiga tatarica*) in Kazakhstan (Wang et al. 1997; Milner-Gulland et al. 2001), and the fences erected for veterinary purposes in the Kalahari mainly for wildebeest and hartebeest (Williamson & Williamson 1984) and in the Kruger-Limpopo region for wildebeest (Whyte & Joubert 1988). Few protected areas can encompass migratory species (Bolger et al. 2008). One example, the 24,000 km^2 Serengeti-Mara ecosystem barely contains the famous migrations of 1.3 million wildebeest (*Connochaetes taurinus*) (about 2 million ungulates in total) (Thirgood et al. 2004). More appropriate in size is the newly declared KAZA Landscape between Angola, Botswana, Namibia, Zambia and Zimbabwe, at 285,000 km^2 (the size of Italy).

Across the continents, few large ungulate migrations remain, most in Africa. There is the Boma-Jonglei ecosystem in south west Sudan, with more than a million white-eared kobs (*Kobus kob leucotis*) and several thousands of elephants and buffaloes, still migrating despite ongoing wars. Populations of zebras and wildebeest travel between the Liuwa Plains National Park in north west Zambia and Kameia National Park in Angola. In Asia, the tiang (*Damaliscus korrigum korrigum*) and the Mongalla gazelle (*Eudorcas albonotata*) still occur in dwindling numbers. The migrations of chiru (*Pantholopos hodgsoni*) on the Tibetan plateau occur largely within protected areas (Harris et al. 2009). In North America, some caribou populations occur within reserves but most cross them. Berger (2004) reports that 105 fences now lie in the route of pronghorn migrations in the Yellowstone ecosystem.

Mass migrations of LTMs are perhaps more endangered than some of the species themselves. Former mass migrations of springbok (*Antidorcas marsupialis*), black wildebeest (*Connochaetes gnou*), blesbok (*Damaliscus dorcas*), kulan (*Equus hemionus*), scimitar horned oryx (*Oryx dammah*) and the extinct quagga (*Equus quagga*) are now only memories, and oil and gas exploration may soon threaten reindeer/caribou (*Rangifer tarandus*) migrations in Canada and Russia (Mahoney & Schaefer 2002; Harris et al. 2009), and the longest pronghorn antelope migration along the US–Canada border in Montana and Saskatchewan (C. Loucks, personal communication).

In summary, life-history corollaries of size not only make larger animals more susceptible to persecution and the loss and fragmentation of habitats (Haskell et al. 2002), but also make them slower to recover (Price & Gittleman 2007). Because fewer can be packed per unit area, their necessarily small populations are at higher risk of extinction than are smaller species (with larger populations) (Brown & Maurer 1989; Marquet & Taper 1998; Diniz-Filho et al. 2005), further compounded by edge effects (Woodroffe & Ginsberg 1998). Larger animals tend to be preferred by subsistence and trophy hunters (Johnson et al. 2010). They are also more conspicuous, potentially more harmful and aggressive, often (and sometimes irrationally) feared, and thus likely to be blamed for conflict and persecuted. Almost all these risks scale allometrically with power laws (Peters 1986), although Carbone et al. (1999) provide an important threshold at 14.5 kg. In short, large mammals are more likely to get into trouble with people, become the butt of their vengeance, and suffer population loss.

Why losing large terrestrial mammals matters

A pragmatic, if subjective, reason why the conservation of large mammals is special is that, almost globally, humans assign them disproportionate value and attention (Macdonald et al. 2007b), dating back to the Lascaux cave paintings of early humans (Curtis 2006). Today, this larger-than-life attribute is reflected in their representation in zoos, where large mammals are displayed despite expense and lack of alignment with captive breeding or 'insurance' arguments (Leader-Williams et al. 2007). While new justifications are found for this function (calling them ambassadors for fundraising and education), the fact is that large mammals attract a wide public worldwide. Charisma has considerable relevance to conservation (Collins et al. 2011), adding to their importance as keystone species (Estes et al. 2011).

Ecological functions

Fritz & Purvis (2010) calculate that the loss of large mammals on the scale projected by Schipper et al. (2008) would dramatically change nature, reducing mammalian variation in weight by 14% per ecoregion (with lesser losses to species richness, by 14 species per region) and phylogenetic diversity (by 283 million years of evolutionary history). The question remains, however, of whether large mammal conservation is especially important for community function (see Chapter 23).

Paine (1966, 1969a,b) was one of the first researchers to draw attention to 'keystone species' when he discovered that removal of a predatory starfish (neither large nor a mammal) led to strong competition at a lower trophic level and, ultimately, to simplification of the food web (Berlow et al. 2004). The appearance or disappearance of a keystone species can lead to demographic changes in other species and ultimately to changes in species richness (Crooks & Soulé 1999; Henke & Bryant 1999; Berger et al. 2001) or ecosystem processes (e.g. the alteration of streams by beavers; Naiman et al. 1988). Thus, keystone species need be neither large nor abundant; a rare species may have a large effect (Tanner et al. 1994). Keystone species are necessary to the maintenance of ecosystem integrity and stability (Emmerson & Raffaelli 2004) and their conservation is a priority (Soulé et al. 2005), but is this particularly relevant to large mammals?

Size relationships are central in structuring trophic linkages within food webs (Owen-Smith & Mills 2008), influencing biomass fluxes within food webs and ultimately determining the relative contribution of top-down and bottom-up limitation processes (Sinclair et al. 2003; Fritz et al. 2011). Larger, often generalist, species often cause major transfer of energy (and nutrients) from one trophic level to the other; for example, the abundance of megaherbivores (>1000 kg; Owen-Smith 1988), and particularly elephants, largely governs the consumption of primary production (Fritz et al. 2011).

Similarly, the largest predators, such as lions, account for the major share of herbivores killed across a wide size range (Owen-Smith & Mills 2008). Predator guilds in which the largest carnivore species represent a larger share of carnivore biomass are likely to exert a stronger top-down impact on herbivores (Fritz et al. 2011). In ecosystems with a large guild of predators, smaller herbivores are subject to predation by a wider size range of predator species (nested-predation effects; see Hopcraft et al. 2010) and their populations are likely to be top-down limited (Sinclair et al. 2003).

The relative importance of bottom-up and top-down processes in structuring communities (abundance, composition, diversity) seems to depend on the distribution of body sizes between and within trophic levels. A herbivore community dominated by very large herbivores is predicted to be bottom-up limited (i.e. limited by the abundance, type and distribution of forage) whereas a herbivore community dominated by small herbivores in a system characterized by a rich guild of carnivores is predicted to be top-down limited (i.e. limited by the distribution and type of predators; Hunter & Price 1992; Power 1992; Fritz et al. 2011). And a carnivore guild dominated by large carnivores is likely to affect more herbivore species.

Elephants are often considered keystone competitors or facilitators and their abundance is likely to impact several species of large herbivores. However, studies investigating the influence of elephants on other large herbivores are few and provide contrasting results. Some have suggested possible competition with browsers (O'Kane et al. 2011), which is supported by a negative correlation between elephant biomass and browser biomass across ecosystems (Fritz et al. 2002), and opposite trends in the Hwange National Park elephant and other browser populations (Valeix et al. 2007a, 2008). Conversely, other studies have suggested facilitation between elephants and browsers. In Chobe National Park, Botswana, the increase in the elephant population has been correlated with an increase in some populations for which medium-term facilitation has been suggested, as elephants may generate more browse resources through coppicing (Rutina et al. 2005; Fornara & du Toit 2007). Further, elephants uproot trees and bushes in savannas, acting as ecosystem engineers modifying not only food availability but also vegetation structure and ultimately visibility (and hence predation risk for other herbivores), and a positive impact on habitat use by other herbivores has been demonstrated in Hwange (Valeix et al. 2011).

Rhinos are also important landscape engineers and can drastically shift vegetation composition and structure through their feeding behaviour. White rhinos can graze grasses to a short stubble and maintain large patches of grazing lawns, which influence not only the fluxes of nutrients and productivity in savanna grasslands (Waldram et al. 2008; Bonnet et al. 2010) but also the rate at which fire spreads in the landscape mosaic of tall (flammable) and short grass (fire break) patches (Archibald et al. 2005; Waldram et al. 2008). Dinerstein & Wemmer (1988) showed how fruit removal and seed dispersal by greater one-horned rhinoceros would rapidly shift the world's tallest grasslands to riverine forest by manuring the seeds of the shade-intolerant common tree *Trewia nudiflora* into grassland latrines. These latrines become outposts of woody vegetation in a sea of elephant grass. Without annual mortality of *Trewia* seedlings by monsoon floods and annual but unpredictable fires, rhino-mediated seed dispersal would lead to the succession from grassland to woodland and then forest within decades. Similarly, Dinerstein (1992) found that the winter browsing of riverine forest tree species by greater one-horned rhinoceros impeded the vertical growth of certain woody species. The conclusion was that the individuals of tree species that compose the riverine forest canopy are those that are unpalatable to rhinos and wild elephant as seedlings and saplings. Palatable browse plants are heavily pruned so that an Asian riverine forest without its herbivorous megafauna would look much different than one where such species are still extant.

Large carnivores can influence the numbers of large and small herbivores and their distribution and behaviour (Berger et al. 2001; Sinclair et al. 2003; Terborgh 1988; Terborgh et al. 2001). The risk of predation by lions influences the spatial distribution and habitat selection (Valeix et al. 2009b), use of waterholes (Valeix et al. 2009c) and vigilance behaviour (Périquet et al. 2010) of their prey. In Yellowstone National Park, USA, similar indirect effects have been demonstrated with wolves and elk (Creel et al. 2005; Liley & Creel 2007). By influencing the distribution and behaviour of herbivores ('ecology of fear' theory, Brown et al. 1999; 'landscape of fear' concept *sensu* Laundré et al. 2010), carnivores may initiate trophic cascades, i.e. 'the progression of indirect effects by predators across successively lower trophic levels' (Estes et al. 2001). Famously, a scarcity of pumas in the Yosemite National Park, California, in the 1920s led to an increase in mule deer (*Odocoileus hemionus*) numbers, and thus a decrease in the recruitment of California black oak (*Quercus kelloggii*) with cascading effects on biodiversity (Ripple & Beschta 2008). This process may have occurred in reverse with the restoration of wolves to Yellowstone National Park in the 1990s. This heightens interest in the cascades that may flow from reintroducing large mammals in ecosystems (e.g. wolves or lynx) in boreal and temperate ecosystems (Beschta & Ripple 2009; see also Chapter 23). In sub-Saharan Africa, reduction in lions and leopards is associated with increased numbers of olive baboon, whose greater contact with humans is associated with higher rates of intestinal parasites in both baboons and humans (Estes et al. 2011).

Some species may have such an important role that they may drive an ecosystem to shift to new characteristics (Grebmeier et al. 2006). Estes et al. (2011) summarize the extensive, and often unanticipated, cascading effects of removing LTMs on marine, terrestrial and freshwater ecosystems worldwide. This 'trophic downgrading' may impact processes as diverse as the dynamics of disease, wildfire, carbon sequestration, invasive species and biogeochemical cycling, as well as the more predictable effects on the intensity of herbivory and thus the abundance and composition of plants.

Large herbivores, and particularly megaherbivores, have major impacts on vegetation when abundant, for example elephants on trees (O'Connor et al. 2007) or white rhinoceros on grass (e.g. Waldram et al. 2008). Increased wildebeest, following eradication of rinderpest, reduced plant biomass and thus wildfires, changing the Serengeti ecosystem from carbon source to carbon sink (Holdo et al. 2009) and the demise of the Pleistocene megaherbivores may have contributed to the reduced atmospheric methane concentration and the resulting 9°C temperature decline that defines the Younger-Dryas period (Smith et al. 2010). Shurin & Seabloom (2005) conclude that the greater the size of a consumer (carnivore or herbivore) *relative* to its prey (animal or vegetable), but not necessarily its absolute size, the bigger its effect on the population of its prey (Borer et al. 2005).

Conservation interventions for large mammals

What are our options for conserving and restoring LTM populations and reversing range collapse? Broadly, there are three: education, which necessitates information (ultimately from research), regulation, which necessitates enforcement, and incentivization, which provides a basis for valuation, such as a market. Of course, these are all interwoven (see Winterbach et al. 2012). On the road to sustainability, there is an iterative process flowing from problem to solution and these solutions typically reduce to mitigation of the impacts of the problematic creature or compensation of aggrieved stakeholders. Later we will give examples, but Macdonald (2000) characterizes the generalities in terms of steps in the Biodiversity Impacts Compensation Scheme (BICS). Imagine a species in conflict; problems can be partitioned between reducible and irreducible elements, and the balance between these will shift as currently intractable elements are rendered reducible by innovation (itself engendered by research). Mitigating reducible problems minimizes current conflict, the residue being the currently irreducible problem. Depending on levels of tolerance, themselves heavily influenced by education (and affected by culture and religion), the problem will be either bearable or not. The unbearable component will require an interim solution (Macdonald & Sillero-Zubiri 2004b). Options are either to control the problematic creature, using one of the mitigation interventions described below, or to compensate the aggrieved stakeholder.

Protected areas, corridors and landscape approaches

More than 100,000 protected areas (PA) cover about 12% of the earth's land surface (www.wdpa.org). Protected areas are especially important to large mammals as they offer core refuges, but they are often insufficient to encompass viable populations. Only three European PAs exceed 1000 km^2 and in the lower 48 states of USA, none exceeds 10,000 km^2. Analysis of 583 time-series revealed an average decline in average population size for large mammals in protected areas of 59% (Craigie et al. 2010); much of the variation in success was associated with wealth (protection had been more effective in South Africa, which is more prosperous than West or East Africa). Options to mitigate this dearth of large PAs are to aggregate adjacent protected areas into larger complexes, and to create corridors that prevent protected areas becoming islands.

Even before the effects on their predicaments unfold, some 13% of globally threatened mammals need conservation action at the landscape scale, and Boyd et al. (2008) argue that because of their functional importance in ecosystems, the need for landscape-scale conservation applies particularly to wide-ranging large carnivores. Many large mammals, especially generalist species, are tolerant of a certain level of human activity and cause limited conflicts: their integration within landscapes only partially used by humans seems a more promising long-term conservation strategy than the expectation of sustainable conservation within few protected areas. Large carnivores in Europe, for example, would not exist if they had to be limited within the small European protected areas but they thrive in many countries in co-existence (with some, but manageable, conflicts) with livestock, hunters and relatively high human densities (Boitani & Ciucci 2009).

There are 188 transboundary complexes, that include 818 protected areas in 112 countries (Besançon & Savy 2005), incorporating 16.8% of the land surface within protected areas (Mittermeier et al. 2005). Examples include the 36,000 km^2 Great Limpopo Trans-frontier Park (GLTP) between South Africa, Mozambique and Zimbabwe and the ambitiously planned Kavango-Zambezi Trans-frontier Park (KAZA-TFCA) which would cover 285,000 km^2 of Angola, Botswana, Namibia, Zambia and Zimbabwe (Braack 2005; Hanks 2005). One idea developed for elephants (van Aarde & Jackson 2007) is the creation of a permeable matrix between protected areas joined by corridors, several home range diameters in width, which may vary from a ribbon of habitat to, probabilistically, valleys through the peaks of risk that together offer a fair hope of safe passage. An ambitious project is the Jaguar Corridor Initiative, launched in 2004 and approved at ministerial level in 2006, as part of the larger Mesoamerican Biological Corridor Program (Rabinowitz & Zeller 2010). Despite high rates of deforestation in Latin America, there are still pathways outside protected areas that enable jaguars to move between populations. The Jaguar Corridor Initiative extends beyond Mesoamerica southwards to Brazil – linkage which resonates with the high levels of gene flow throughout the species' range (Eizirik et al. 2001; Ruiz-Garcia et al. 2006).

Restoring natural processes

If corridors between refuges are impractical, a surrogate is to create 'virtual corridors' to maintain a genetically integrated population. This intervention mimics the functioning of meta-populations in nature and, using this as a metaphor, has been termed meta-population management. Davies-Mostert et al. (2009) moved African wild dogs (*Lycaon pictus*) between fenced reserves in South Africa. Although successful in enriching species assemblages and stimulating ecotourism, they faced challenges with neighbours following break-outs, the ability of prey to sustain wild dog predation, and overcoming stochastic processes that affect small populations and curtail natural population dynamics. The managed meta-population approach might stave off extinction for other large mammal species in fragmented habitats, but it is intensive and expensive. For wild dogs, the ideal will be restoration over a much larger connected area (Davies-Mostert et al. 2009).

The next step is reintroduction (see Chapter 22) and rewilding (see Chapter 23). Plans exist to reintroduce mammals, from beavers to the UK (Macdonald et al. 1995) to Amur leopard in the Russian Far East (Christie 2009) following the classic case of the Arabian oryx in Oman (Stanley Price 1989), and there are refined IUCN guidelines to follow. Vitally, the capacity of human communities to cope with, and benefit from, the reintroduction (Reading & Clark 1996; Hayward & Somers 2009) needs to be considered. Repairing the environment will involve the restoration of habitats, of species that are threatened and locally or globally extinct in the wild, and of entire functioning communities (Macdonald 2009b; Macdonald et al. 2002). Three species of rhino illustrate successes: black rhinos recovered to 4840 in 2011 and the southern white rhino (*Ceratotherium simum simum*) has now crested 20,000, having been the subject of intensive translocation efforts to rebuild from its sole source population around 1900 to more than 430 populations today. Even the greater one-horned rhinoceros numbers have reached a record high not seen since the 1950s, at 2900 (Williams 2011). More importantly, risk of extinction has been reduced by re-establishing numerous populations in all three species and continued translocations are a key part of recovering lost range.

Arguably, large mammals as umbrella species, and their charisma, make them particularly suitable flagships for, and barometers of, the restoration movement. But to be realistic, because large mammals can be dangerous, while easily advocated from the safety of an armchair, restoring large mammals is an ambitious goal and one that carries serious responsibilities.

European lynx provide an interesting case study. Following the last Ice Age, the 20 kg European lynx (*L. lynx*) probably spread north from a southern refugium, perhaps on the Italian peninsula or in eastern France, through the Alps and across land bridges to Scandinavia. Such a journey is impossible for lynxes in modern, environmentally fragmented Europe so lynx (phylogenetically indistinguishable from the extinct Alpine form) were transported from northern Europe, and have been used to repopulate the Alps artificially (Breitenmoser et al. 2010).

A radical further step would be to reintroduce the lynx to Scotland. The last remains of this species in Britain date to c.1600. Hetherington et al. (2008) estimated that the Highlands and the Southern Uplands in Scotland could sustain approximately 400 and 50 lynx, respectively.

What might be the pros and cons of such a reintroduction? In the French Jura, lynx kill 100–400 sheep annually; however, 70% of attacks occur in only 1.5% of the area inhabited by lynx, and of those flocks experiencing damage, most suffer only one or two attacks per year. In the Swiss Alps, 80% of farmers who lost livestock during the years 1979–99 lost fewer than three and 95% of attacks were within 360 m of the forest edge (Stahl et al. 2001, 2002). Although the abundance of roe deer in Scotland might minimize depredation on sheep, it would nonetheless occur. Conservationists might fear for impacts upon populations of capercaillies (*Tetrao urogallus*), wild cats (*Felis silvestris*) and pine martens (*Martes martes*).

Macdonald et al. (2010) use this example to illustrate the centrality of consumer choice in conservation. Lynx occurred in Scotland before, and conditions are now such that they could be there again. They could cause problems to some, delight to others. They would cause financial loss to some, generate revenue for others (e.g. through ecotourism) and it would surely not be beyond the wit of a nation such as Scotland to balance these costs and benefits with suitable economic instruments (see below). The question is simply whether society wants to bring back the lynx, and for those who think the answer should be yes, the road ahead lies in ordering their priorities and honing their advocacy. The answer has some significance further afield because this is a case of a developed country considering the denizens of its own backyard; should the British public be unenthusiastic about having lynx in its midst, this would hamper the British conservationists' case that South Americans, Africans and Asians should welcome, respectively, jaguars, lions and tigers into theirs. The credibility of American scientists promoting LTM conservation in developing nations also requires the recovery of grizzly bears, mountain lions and wolves on the home front.

Amongst large mammals, extreme, controversial but aspirational examples include the Barbary lion (Macdonald et al. 2010) and Caspian tiger (*P. t. virgata*). These Central Asian tigers recently inhabited forests from Xinjiang to Anatolia but scarcely differ genetically from Amur (Siberian) tigers (*P. t. altaica*) (Driscoll et al. 2009). In short, Caspian tigers are not extinct, they just live in the Russian Far East. Thus, the Central Asian tiger range is, in principle, open to reintroductions from Amur stock (Macdonald et al. 2010; Driscoll et al. 2011). Of course, restoring the Caspian tiger would necessitate the most thorough confirmation that there were prey, habitat and a willing human population to support them. But such a bold reintroduction, championed by the government of Kazakhstan and several other former Caspian tiger range states, would help to reverse range collapse in tigers (Dinerstein et al. 2007). Reintroduction would also achieve the central goal of the Global Tiger Summit held in St Petersburg, Russia, in November 2011, the first ever such summit attended by heads of state and devoted to an endangered species. The goal is to double the number of tigers by 2022, the next Year of the Tiger (Global Tiger Initiative 2011). Wikramanayake et al. (2011) show that enough habitat remains to underpin a near-tripling of the wild tiger population. Other reintroductions under consideration include the eastern plains wilderness of Cambodia, and in the Changbaishan region of China bordering North Korea and Russia: in both, efforts are under way to remove snares and rebuild the prey base. There is no reason why such restoration should distract funding from existing tiger priorities (most notably the 6% of tiger range that supports 70% of the world's remaining wild tigers; Walston et al. 2010). Nonetheless, perhaps one day tiger corridors could link Azerbaijan, Armenia and Iran, and sites in Tadjikistan and the swamps of the Illi River in Kazakhstan. This wondrous dream is, for the time being, just that! Considerable efforts are likewise being devoted to the reintroduction of all great apes, but the scale is inevitably local so successes are counted more by individual than by population (e.g. chimpanzee, Goosens et al. 2005; orang-utan, Grundmann 2006).

The idea of reintroduction or translocation provokes a pragmatic question about the role of fenced areas in carnivore conservation. Sandom et al. (2012) point out that fenced areas, intended to keep carnivores and people apart, raise pithy questions such as what level of biological 'naturalness' does an enclosure of a given size provide, and what level of management intervention does such an enclosure necessitate? This dilemma raises the question of the values attributed to areas along the continuum where LTMs live: from zoo cage to enclosure to park to wilderness. These values almost invariably include the hope that the quasi-natural circumstances of the fenced, but still charismatic, carnivores will generate revenue and employment to sustain the enterprise.

Conservation interventions

Mammals damage crops (Naughton-Treves et al. 1998; Schley & Roper 2003) and prey on livestock (Loveridge et al. 2010a,b). Historically, the response has generally been to kill wildlife accused of damage, whether it was proven or perceived (an important distinction, often ignored). Under what conditions is killing problem mammals rational? Three questions must be posed. First, does the farmer suffer extensive, recurrent loss to the pest? Second, does action taken reduce that loss? Finally, is control cost-effective? These issues are rarely addressed.

Control measures have yielded mixed results. For example, culling can reduce elephant numbers but fails to reduce their impacts. Culling may even be counterproductive (Tuyttens et al. 2000), increasing growth rates due to an upturn in birth and immigration rates in response to a reduction in density (Jonker et al. 2008). Some lethal methods of animal control to prevent or reduce such damage raise questions of efficacy, ethics and legality (Baker et al. 2008). The search is on for ingenious, humane solutions to conflict (e.g. aversive conditioning; Baker et al. 2007), one inspiration being to repel elephants by using chillies or bees (the latter also generating about 20 litres of honey worth $300) (Parker & Osborn 2006; King et al. 2007). In general, both traditional methods and innovative ones (e.g chillies or beehives) can be reasonably effective (Fernando et al. 2005; King et al. 2011; Davies et al. 2011). Another, albeit long-term possibility is contraception, which can succeed in reducing elephant numbers when applied to at least 75% of all breeding females, continually for 11–12 years (Fayrer-Hosken et al. 2000), although its cost-effectiveness is under debate (Pimm & van Aarde 2001). The indirect effects on social structure and behaviour remain to be ascertained (Kerley & Shrader 2007), and there is a risk of unwittingly provoking extinction (Bradford & Hobbs 2008).

Another technique is the use of livestock guarding dogs, originally selectively bred by the Romans more than 2000 years ago to protect sheep from wolves, now revived worldwide as a conservation tool. In Slovakia, following protection in 1975, wolf numbers have recovered to about 300, and although wolf depredation of livestock is nationally inconsequential, there are local hot spots of predation. Rigg et al. (2011) reintroduced Slovensky cuvacs, which cut predation by an average of 69%.

Guard dogs have also been tested in Namibia to protect livestock from cheetah. Namibian ranchlands are home to a third of the world's 9000 or so cheetah (*Acinonyx jubatus*) (a century ago there were 10 times as many) (Marker 1998). There, Marker et al. (2003) have sought to reconcile the conflict that arises when cheetah on these 8000 ha farms attack calves, and farmers retaliate by killing cheetahs. This conflict has had a huge impact on Namibia's cheetah population – nearly 7000 were trapped and removed (usually killed) from Namibian farms between 1980 and 1991 (CITES 1992). Marker et al. (2005a,b) imported 10 Anatolian sheep dogs to Namibia, from which 143 more were bred and given to farmers. After having the dog for a year, three-quarters of recipient farmers (73%) reported a large decline in stock losses (Marker et al. 2005a). Before dogs, almost no farmers believed they were free of losses to cheetah and 71% reckoned they lost 10 or more head of stock to predators. After receiving a livestock guarding dog, 65% of farmers recorded no losses to cheetah. This positive outcome partly explains why the average number of cheetah removed annually on problem farms declined from an average of 29.0 to 3.5 per farmer (Marker et al. 2010).

Financial mechanisms, development and poverty

Where mitigation is exhausted, the remaining option within the BICS framework is to incentivize conservation through Payments to Encourage Co-existence (PEC) – the term coined by Dickman et al. (2011) for such financial schemes as 'compensation and insurance', 'revenue sharing' and 'conservation payments', which seek to protect species which are highly valued globally but which have little or even negative value locally – a classic market failure. For example, landowners may be compensated for the opportunity costs incurred in conserving problematic large mammals or for the direct costs of damage (Chapter 4). The objective is to make the conserved population worth more in the bush than would be the profits (or savings) in the bank, yielding interest, from its destruction. Variants of this idea are being explored for the snow leopard in India (Jackson et al. 2010) and the orang-utan and the pygmy elephant in Borneo (Venter et al. 2009), in the latter case using payments for reducing carbon emissions (REDD) to offset the costs of forest conservation (and the lost opportunity costs of conversion to production landscapes, e.g. oil palm).

The ideal PEC would be a framework for translating the global value of, for example, big cats into tangible local benefits large enough to drive conservation 'on the ground' and might meet seven criteria: minimizing conflict, reducing costs, providing additional revenue, avoiding moral hazard, not requiring external revenue, conditionality and reducing poverty. By these criteria, no one (existing) financial mechanism is perfect, but a new framework designed to tick all the required boxes takes the form of a combined PEC in which traditional streams of conservation revenue flow into a central fund which is then distributed to the local community (Dickman et al. 2011;

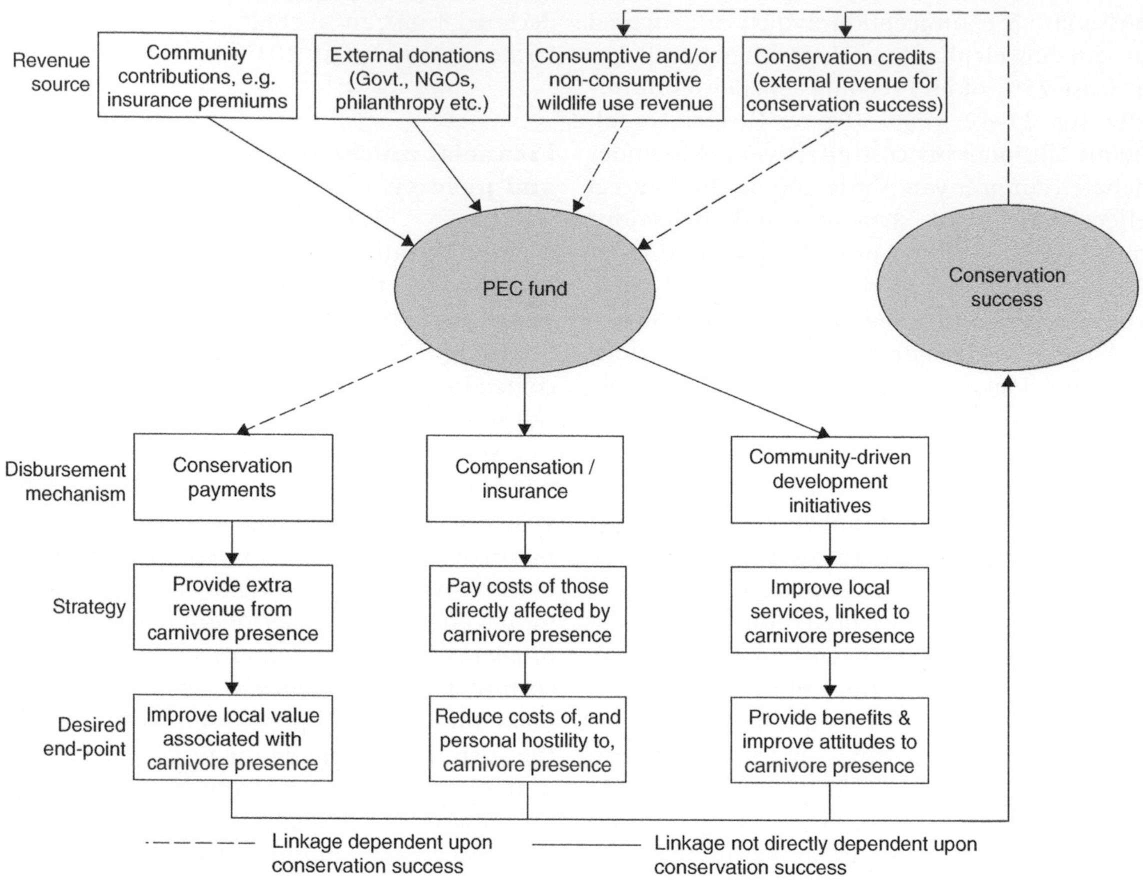

Figure 16.4 Example of how existing PEC strategies could be incorporated under a single scheme to encourage carnivore conservation on human-dominated land. From Dickman et al. 2011, with permission.

Figure 16.4). The ingredients of a large mammal PEC will vary with the direct benefits that might be realized; in the case of dangerous large herbivores, such as elephants, buffaloes and hippos, this could be meat secured during problem animal control. Community-driven development initiatives would provide benefits community-wide, while a compensation scheme would minimize the costs to those most directly disadvantaged. Both these approaches may sway attitudes but neither incentivizes conservation, so additional payments would be made contingent on the delivery of conservation goals (see Chapter 4).

Compensation schemes have the disadvantage of paying people not to do something – for example, not to kill predators. This has often been fruitless (Boitani et al. 2010); there is no evidence that, for example, the payments of 5–7000 euros/wolf/year in France in 2010 reduced poaching or improved acceptance. Another example comes from the Amboseli ecosystem in Kenya, where a privately funded livestock compensation scheme has contributed toward a reduction in the rate of lion killing. Amboseli, Kenya's third most visited national park, nestles amidst a constellation of unfenced ranches, across which roams wildlife, including predators which daily kill livestock. Lions are crucial to Amboseli's US$3.5 million dollar tourist revenue. In 2001–2, mainly in retaliation for livestock losses and partly for traditional reasons, the Maasai community owning the 1200 km^2 Mbirikani ranch speared 24 lions.

Fearing their local annihilation, in 2003 a local NGO drew on American donors to establish the Mbirikani Predator Compensation Fund. So, wider society – in this case the NGO representing the ideological interests of conservation and the financial interests of ecotourism – valued lions sufficiently to pay herdsmen not to kill them (Maclennan et al. 2009; Hazzah et al. 2009). Taking into account the costs of compensation and scientific monitoring, keeping the average lion alive cost US$6124 a year. Since 2007 a new scheme has paid Maasai Lion Guardians to protect lions and there have been no further spearings on Mbirikani but over 60 on neighbouring ranches; in 2011 the guardians prevented 32 hunting parties from killing lions (Hazzah & Dolrenry 2011).

Another approach, being promoted through a partnership of conservation NGOs, governments, the World Bank's Environment Units and its Global Tiger Initiative, known as the Wildlife Premium Mechanism (WPM), seeks to reverse range collapse for a subset of large charismatic terrestrial mammals (and birds) (Dinerstein et al. in press). This performance-based payment scheme would allow stakeholders in lower-income countries to generate revenue streams by recovering and maintaining threatened fauna that serve as flagship or umbrella species for other biodiversity. The hypothesis of the WPM is that investors and philanthropists will be more likely to purchase or contribute to voluntary payment schemes for carbon or other ecosystem services if the forests also support conservation of a subset of endangered, charismatic flagship species, and the local communities benefit financially through conservation stewardship. Four options for applying the premium are being explored among several pilots now under way in Nepal's Terai Arc Landscape (tigers), Kenya's Kasigau elephant corridor linking Tsavo East and Tsavo West, a jaguar corridor in Madre de Dios, Peru, and others in development in Thailand, Madagascar and elsewhere. Three of these options are linked to payment schemes for carbon and other ecosystem services, and one stands alone. Each option presents distinct advantages and challenges, but all can create economic incentives through payment schemes with potential to improve livelihoods for the rural poor who reside in or near areas within the range of these threatened species. The tiger example is a good case in point. Of the 350 or so protected areas in the tigers' range, none is large enough to support a viable population; in fact, 40% of tigers in India live outside reserves. Thus, tiger conservation depends upon metapopulation management and landscape-scale conservation. Interventions at this scale require new ways to incentivize LTM conservation, wherever possible linked to PEC schemes and livelihoods. That even the most intact refuges for LTMs – the KAZA landscape in southern Africa and Madre de Dios in south eastern Peru – are considering WPM initiatives illustrates the priority of linking large mammal conservation to larger development and conservation goals.

Conclusion

Natural selection has no foresight; it cannot predict the future. The traits we see in LTMs today (large body size, low reproductive rates, massive home ranges, migratory or wandering behaviour, consumption of large vertebrate prey or large quantities of vegetation) were selected for ages ago. Some of these traits make their survival precarious in the present, the Anthropocene, where the human footprint has spread rapidly into most corners of the globe, the human population surges past 7 billion, and with the looming spectre of global climate change. Nonetheless, some traits buffer LTMs from extinction. Many of the large herbivores are generalists and can live in different habitats and eat coarse vegetation, features that may help them cope in a changing climate. Tigers and wolves, although conservation-dependent species, breed fast and rebound quickly from poaching if protected. Even some of the slow breeders compensate by long life spans and reproductive periods, giving them the potential to regain former numbers. In fact, one of the greatest successes in conservation history features an LTM – the southern white rhinoceros. This slow-breeding giant herbivore, heavily poached for its horn, numbered as few as 100 individuals at the turn of the last century. Today, there are over 20,000 white rhinoceros translocated into 430 separate populations. Thus, even in developing countries, dramatic population increases and range recovery are possible for LTMs with proper planning, protection and good governance.

We have long known of the exceptional vulnerability of large terrestrial mammals, whose reductions in Australia, the Americas and Madagascar around the time of humans arriving were early signals that species of about our own body size or larger do not co-exist easily with people. In the face of habitat loss, persecution, infectious disease and hunting, these species tend to suffer particularly badly thanks to their low numbers, need for large areas and low reproductive rates. Even in 1871 Charles Darwin predicted that the great apes 'will no doubt be exterminated' (Darwin 1871). He could justifiably have expressed the same fears for large carnivores and ungulates.

But while intense threats and slow population growth continue to jeopardize the future of many LTMs, the data reviewed in this chapter also offer hope. There is a growing sense of how to use knowledge of different species' biology and circumstances to devise effective strategies (see Chapter 21). There is positive experience on which to draw, and there are new tools to deploy (such as ingenious financial instruments, see Chapter 4 and established ones to use more radically, such as reintroduction (see Chapters 22 and 23). And since large species from different taxa experience in many ways parallel problems, there is increasing room for synergy in the approaches towards conservation of multiple large species (for example, Macdonald et al. (2012) illustrate that threatened felids and primates both often face the same threats in the same places, and so curbing a threat to one often has the potential efficiently to protect the other). If humans are to continue sharing the world with other terrestrial species of our size and bigger, we will need a new vision of the future, deeply interdisciplinary knowledge with which to implement it, and the solutions to very tough choices.

Acknowledgements

We are grateful for the expert help of Lauren Harrington in preparing this essay, and for insightful comments from Christos Astaras, Özgün Emre Can, Paul Johnson, Tom Moorhouse, Robert Smith, Marion Valeix, Alexandra Zimmermann and Zinta Zommers.

References

Archibald, S., Bond, W.J., Stock, W.D. & Fairbanks, D.H.K. (2005) Shaping the landscape: Fire–grazer interactions in an African savanna. *Ecological Applications*, **15**, 96–109.

Athreya, V.R., Thakur, S.S., Chaudhuri, S. & Belsare, A.V. (2004) A study of the man–leopard conflict in the Junnar Forest Division, Pune District, Maharashtra. Submitted to the Office of the Chief Wildlife Warden, Nagpur. Maharashtra Forest Department and the Wildlife Protection Society of India, New Delhi. www.ncra.tifr.res.in/~rathreya/JunnarLeopards/report.pdf

Baker, S., Johnson, P., Slater, D., Watkins, R. & Macdonald, D.W. (2007) Learned food aversion with and without an odour cue for protecting untreated baits from wild mammal foraging. *Applied Animal Behaviour Science*, **102**, 410–428.

Baker, S.E., Ellwood, S.A., Slater, D., Watkins, R.W. & Macdonald, D.W. (2008) Food aversion plus odor cue protects crop from wild mammals. *Journal of Wildlife Management*, **72**, 785–791.

Ballard, W.B., Whitman, J.S. & Gardner, C.L. (1987) Ecology of an exploited wolf population in south-central Alaska. *Wildlife Monographs*, **98**, 54.

Beddington, J.R. & Kirkwood, G.P. (2007) Fisheries. In: *Theoretical Ecology: Principles and Applications* (eds R.M. May & A.R. Maclean), pp.148–157. Oxford University Press, Oxford.

Beier, P., Riley, S.P.D. & Sauvajot, R.M. (2010) Mountain lions (*Puma concolor*). In: *Urban Carnivores: Ecology, Conflict, and Conservation* (eds S.D. Gehrt, S.P.D. Riley & B. Cypher), pp. 141–155. Johns Hopkins University Press, Baltimore, MD.

Bell, R.H.V. (1970) The use of the herb layer by grazing ungulates in the Serengeti. In: *Animal Populations in Relation to Their Food Resources* (ed. A. Watson), pp.111–123. Blackwell Scientific, Oxford.

Berger, J. (2004) The last mile: how to sustain long-distance migration in mammals. *Conservation Biology*, **18**, 320–331.

Berger, J., Stacey, P.B., Bellis, L. & Johnson, M.P. (2001) A mammalian predator–prey imbalance: grizzly bear and wolf extinction affect avian neotropical migrants. *Ecological Applications*, **11**, 947–960.

Berlow, E.L., Neutel, A.M., Cohen, J.E., *et al.* (2004) Interaction strengths in food webs: issues and opportunities. *Journal of Animal Ecology*, **73**, 585–598.

Besançon, C. & Savy, C. (2005) Global list of internationally adjoining protected areas and other transboundary initiatives. In: *Transboundary Conservation: A New Vision for Protected Areas* (eds R.A. Mittermeier, C.F. Kormos, C.G. Mittermeier, P. Robles Gil, T. Sandwith & C. Besançon). Chicago University Press, Chicago, IL.

Beschta, R.L. & Ripple, W.J. (2009) Large predators and trophic cascades in terrestrial ecosystems of the western United States. *Biological Conservation*, **142**, 2401–2414.

Bielby, J., Mace, G.M., Bininda-Emonds, O.R.P., *et al.* (2007) The fast–slow continuum in mammalian life history: an empirical reevaluation. *American Naturalist*, **169**, 748–757.

Boitani, L. (1995) Ecological and cultural diversities in the evolution of wolves–humans relationships. In: *Ecology and Conservation of Wolves in a Changing World* (eds L.N. Carbyn, S.H. Fritts & DR. Seip), pp.3–12. Canadian Circumpolar Institute, University of Alberta, Edmonton, Canada.

Boitani, L. (2003) Wolf conservation and recovery. In: *Wolves: Behavior, Ecology and Conservation* (eds L.D. Mech & L. Boitani), pp.317–340. Chicago University Press, Chicago, IL.

Boitani, L. & Ciucci, P. (2009) Species conservation without boundaries – wolf conservation across Europe. In: *A New Era for Wolves and People: Wolf Recovery, Human Attitudes and Policy* (eds M. Musiani, L. Boitani & P. Paquet), pp. 15–39. University of Calgary Press, Calgary, Alberta, Canada.

Boitani, L., Ciucci, P. & Raganella-Pelliccioni, E.R. (2010) Ex-post compensation payments for wolf predation on livestock in Italy: a tool for conservation? *Wildlife Research*, **37**, 722–730.

Bolger, D.T., Newmark, W.D., Morrison, T.A. & Doak, D.F. (2008) The need for integrative approaches to understand and conserve migratory ungulates. *Ecology Letters*, **11**, 63–77.

Bonnet, O., Fritz, H. & Gignoux, J. (2010) Grazing and primary production dynamics define grazing lawns as a flux resource. *Journal of Ecology*, **98**, 908–916.

Borer, E.T., Seabloom, E.W., Shurin, J.B., *et al.* (2005) What determines the strength of a trophic cascade? *Ecology*, **86**, 528–537.

Boyd, C., Brooks, T.M., Butchart, S.H.M., *et al.* (2008) Spatial scale and the conservation of threatened species. *Conservation Letters*, **1**, 37–43.

Braack, L. (2005) The Great Limpopo Transfrontier Park: a benchmark for international conservation. In: *Transboundary Conservation: A New Vision for Protected Areas* (eds R.A. Mittermeier, C.F. Kormos, C.G. Mittermeier, P. Robles Gil, T. Sandwith & C. Besançon). Chicago University Press, Chicago, IL.

Bradford, J.B. & Hobbs, N.T. (2008) Regulating overabundant ungulate populations: an example for elk in Rocky Mountain National Park, Colorado. *Journal of Environmental Management*, **86**, 520–528.

Brain, C.K. (1969) The probable role of leopards as predators of the Swartkrans australopithecines. *South African Archaeological Bulletin*, **24**, 170–171.

Breitenmoser, U., Ryser, A., Molinari-Jobin, A., *et al.* (2010) The changing impact of predation as a source of conflict between hunters and reintroduced lynx in Switzerland. In: *Biology and Conservation of Wild Felids* (eds D.W. Macdonald & A.J. Loveridge), pp.493–506. Oxford University Press, Oxford.

Brown, J.H. & Maurer, B.A. (1989) Macroecology: the division of food and space among species on continents. *Science*, **243**, 1145–1150.

Brown, J.S., Laundré, J.W. & Gurung, M. (1999) The ecology of fear: optimal foraging, game theory, and trophic interactions. *Journal of Mammalogy*, **80**, 385–399.

Buskirk, S.W. (2004) Keeping an eye on the neighbours. *Science*, **306**, 238–239.

Campbell, G., Kuehl, H., Goran Kouame, P. & Boesch, C. (2008) Alarming decline of west African chimpanzees in Cote d'Ivoire. *Current Biology*, **18**, R903–R904.

Campbell-Smith, G., Simanjorang, H.V. P., Leader-Williams, N. & Linkie, M. (2010) Local attitudes and perceptions towards crop-raiding by orangutans (*Pongo abelii*) and other nonhuman primates in northern Sumatra, Indonesia. *American Journal of Primatology*, **72**, 866–876.

Can, Ö.E. & Macdonald, D.W. (submitted) Human–bear conflict management. *BioScience*.

Carbone, C., Mace, G.M., Roberts, S.C. & Macdonald, D.W. (1999) Energetic constraints on the diet of terrestrial carnivores. *Nature*, **402**, 286–288.

Cardillo, M., Mace, G.M., Jones, K.E., *et al.* (2005) Multiple causes of high extinction risk in large mammal species. *Science*, **309**, 1239–1241.

Cardillo, M., Mace, G.M., Gittleman, J.L., Jones, K.E., Bielby, J. & Purvis, A. (2008) The predictability of extinction: biological and external correlates of decline in mammals. *Proceedings of the Royal Society B*, **275**, 1441–1448.

Chamaillé-Jammes, S., Valeix, M. & Fritz, H. (2007) Managing heterogeneity in elephant distribution: interactions between elephant population density and surface-water availability. *Journal of Applied Ecology*, **44**, 625–633.

Chamaillé-Jammes, S., Fritz, H., Valeix, M., Murindagomo, F. & Clobert, J. (2008) Resource variability, aggregation and direct density dependence: the local regulation of an African elephant population. *Journal of Animal Ecology*, **77**, 135–144.

Chamaillé-Jammes, S., Fritz, H. & Madzikanda, H. (2009) Piosphere contribution to landscape heterogeneity: a case-study of remote-sensed woody cover in a high elephant density landscape. *Ecography*, **32**, 871–880.

Chardonnet, P., des Clers, B., Fischer, J., Gerhold, R., Jori, F. & Lamarque, F. (2002) The value of wildlife. *Revue Scientifique et Technique, International Office of Epizootics*, **21**, 115–51.

Christie, S. (2009) Breeding Far Eastern leopards for reintroduction – the zoo program perspective. In: *Reintroduction of Top-Order Predators* (eds M. Hayward & M. Somers). Wiley-Blackwell, Oxford.

Chundawat, R.S., Habib, B., Karanth, U., *et al.* (2011) Panthera tigris. In: *IUCN Red List of Threatened Species. Version 2011.2*. IUCN, Gland, Switzerland.

CITES (1992) *Quotas for trade in specimens of cheetah. Eighth meeting of the Convention of International Trade in Endangered Species of Wild Fauna and Flora*, pp.1–5. DEFRA, Bristol.

Collins, M., MilnerGulland, E.J., Macdonald, E.A. & Macdonald, D.W. (2011) Pleiotropy and charisma determine winners and losers in the REDD+ game: all biodiversity is not equal. *Tropical Conservation Science*, **4**, 261–266.

Conover, M.R. (2002) *Resolving Wildlife Conflicts: The Science of Wildlife Damage Management*. Lewis, Boca Raton, FL.

Côté, S.D., Rooney, T.P., Tremblay, J.P., Dussault, C. & Waller, D.M. (2004) Ecological impact of deer over-abundance. *Annual Review of Ecology and Systematics*, **35**, 113–147.

Courchamp, F., Angulo, E., Rivalan, P., *et al.* (2006) Rarity value and species extinction: the anthropogenic Allee effect. *Plos Biology*, **4**, 2405–2410.

Cozzi, G., Broekhuis F., et al. (In Press). "Fear of the dark or dinner by moonlight? Reduced temporal partitioning among Africa's large carnivores." *Ecology*.

Craigie, I.D., Baillie, J.E.M., Balmford, A., *et al.* (2010) Large mammal population declines in Africa's protected areas. *Biological Conservation*, **143**, 2221–2228.

Creel, S., Winnie, J.A., Maxwell, B., Hamlin, K. & Creel, M. (2005) Elk alter habitat selection as an antipredator response to wolves. *Ecology*, **86**, 3387–3397.

Crooks, K.R. & Soulé, M.E. (1999) Mesopredator release and avifaunal extinctions in a fragmented system. *Nature*, **400**, 563–566.

Crosmary, W.G., Valeix, M., Fritz, H., Madzikanda, H. & Côté, S.D. (2012) African ungulates and their drinking problems: hunting and predation risks constrain access to water. *Animal Behaviour*, **83**, 145–153.

Curtis, G. (2006) *The Cave-Painters: Probing the Mysteries of the World's First Artists*. Knopf, New York.

Darwin, C. (1871) *The Descent of Man, and Selection in Relation to Sex*. John Murray, London.

Davidson, A.D., Hamilton, M.J., Boyer, A.G., Brown, J.H. & Ceballos, G. (2009) Multiple ecological pathways to extinction in mammals. *Proceedings of the National Academy of Sciences USA*, **106**, 10702–10705.

Davies, T.E., Wilson, S., Hazarika, N., *et al.* (2011) Effectiveness of intervention methods against crop-raiding elephants. *Conservation Letters*, **4**, 346–354.

Davies-Mostert, H.T., Mills, G.L. & Macdonald, D.W. (2009) A critical assessment of South Africa's managed metapopulation recovery strategy for African wild dogs and its value as a template for large carnivore conservation elsewhere. In: *Reintroduction of Top-Order Predators* (eds M. Hayward & M. Somers), pp.10–42. Wiley-Blackwell, Oxford.

Demment, M.W. & van Soest, P.J. (1985) A nutritional explanation f body-size patterns of ruminant and nonruminant herbivores. *American Naturalist*, **125**, 641–672.

De Paula, R., Campos Neto, M.F. & Morato, R.G. (2008) First official record of human killed by jaguar in Brazil. *Cat News*, **49**, 31–32.

Dickman, A.J., Macdonald, E.A. & Macdonald, D.W. (2011) A review of financial instruments to pay for predator conservation and encourage human–carnivore coexistence. *Proceedings of the National Academy of Sciences USA*, **108**, 13937–13944.

Dinerstein, E. (1992) Effects of *Rhinoceros unicornis* on riverine forest structure in lowland Nepal. *Ecology*, **73**, 701–704.

Dinerstein, E. (2011) Family Rhinocerotidae (Rhinoceroses). In: *Handbook of the Mammals of the World. Vol. 2. Hoofed Mammals*. (eds D.E. Wilson & R.A. Mittermeier). Lynx Edicions, Barcelona.

Dinerstein, E., Loucks, C., Wikramanayake, E., *et al.* (2007) The fate of wild tigers. *BioScience*, **57**, 508–514.

Dinerstein, E., Varma, K. et al. (2012). "Enhancing conservation, ecosystem services, and local livelihoods through a wildlife premium mechanism." *Conservation Biology*, in press.

Dinerstein, E. & Wemmer, C. (1988) Fruits rhinoceros eat: dispersal of Trewia nudiflora in lowland Nepal. *Ecology*, **69**, 1768–1774.

Diniz-Filho, J.A.F., Carvalho, P., Bini, L.M. & Torres, N.M. (2005) Macroecology, geographic range size–body size relationship and minimum viable population analysis for new world carnivora. *Acta Oecolgica*, **27**, 25–30.

Driscoll, C.A., Yamaguchi, N., Bar-Gal, G.K., *et al.* (2009) Mitochondrial phylogeography illuminates the origin of the extinct Caspian tiger and its relationship to the Amur tiger. *PLoS ONE*, **4**, e4125.

Driscoll, C.A., Luo, S., Macdonald, D., *et al.* (2011) Restoring tigers to the Caspian region. *Science*, **333**, 822.

Duffy, R. (2010) *Nature Crime: How We're Getting Conservation Wrong*. Yale University Press, New Haven, CT.

Dunbrack, R.L. & Ramsay, M.A. (1993) The allometry of mammalian adaptations to seasonal environments: a critique of the fat endurance hypothesis. *Oikos*, **66**, 336–342.

Dunham, K.M., Ghiurghi, A., Cumbi, R. & Urbano, F. (2010) Human–wildlife conflict in Mozambique: a national perspective, with emphasis on wildlife attacks on humans. *Oryx*, **44**, 185–193.

Eizirik, E., Kim, J.H., Menotti-Raymond, M., Crawshaw P. G. Jr, O'Brien, S.J. & Johnson, W.E. (2001) Phylogeography, population history and conservation genetics of jaguars (*Panthera onca*, Mammalia, Felidae). *Molecular Ecology*, **10**, 65–79.

Emmerson, M.C. & Raffaelli, D. (2004) Predator–prey body size, interaction strength and the stability of a real food web. *Journal of Animal Ecology*, **73**, 399–409.

Emslie, R. (2011) Diceros bicornis ssp. longipes. In: *IUCN Red List of Threatened Species. Version 2011.2*. IUCN, Gland, Switzerland.

Emslie, R. & Brooks, M. (1999) *African Rhino. Status Survey and Conservation Action Plan*. IUCN/SSC African Rhino Specialist Group, Gland, Switzerland.

Estes, J.A., Crooks, K. & Holt, R. (2001) Predators, ecological role of. In: *Encyclopedia of Biodiversity* (ed S.A. Levin), pp.857–878. Academic Press, San Diego, CA.

Estes, J.A., Terborgh, J., Brashares, J.S., *et al.* (2011) Trophic downgrading of planet earth. *Science*, **333**, 301–306.

Fayrer-Hosken, R.A., Grobler, D., van Altena, J.J., Bertschinger, H.J. & Kirkpatrick, J.F. (2000) Immunocontraception of African Elephants – a humane method to control elephant populations without behavioural side effects. *Nature*, **407**, 149.

Fernando, P., Wikramanayake, E.D., Weerakoon, D., Jayasinghe, L.K.A., Gunawardene, M. & Janaka, H.K. (2005) Perceptions and patterns of human–elephant conflict in old and new settlements in Sri Lanka: insights for mitigation and management. *Biodiversity and Conservation*, **14**, 2465–2481.

Ferreras, P., Rodriguez, A., Palomares, F. & Delibes, M. (2010) Iberian lynx: the uncertain future of a critically endangered cat. In: *Biology and Conservation of Wild Felids* (eds D.W. Macdonald & A.J. Loveridge), pp.507–520. Oxford University Press, Oxford, UK.

Fornara, D.A. & du Toit, J.T. (2007) Browsing lawns? Responses of Acacia nigrescens to ungulate browsing in an African savanna. *Ecology*, **88**, 200–209.

Fritts, S.H., Stephenson, R.O., Hayes, R.D., and Boitani, L. (2003) Wolves and humans. In: *Wolves: Behavior, Ecology and Conservation* (eds L.D. Mech & L. Boitani), pp.289–316. University of Chicago Press, Chicago, IL.

Fritz, H. & Loison, A. (2006) Large herbivores across biomes. In: *Large Herbivore Ecology, Ecosystem Dynamics and Conservation* (eds K. Danell, R. Bergström, P. Duncan & J. Pastor). Cambridge University Press, Cambridge.

Fritz, H., Duncan, P., Gordon, I.J. & Illius, A.W. (2002) Megaherbivores influence trophic guilds structure in African ungulate communities. *Oecologia*, **131**, 620–625.

Fritz, H., Saïd, S., Renaud, P.C., Mutake, S., Coid, C. & Monicat, F. (2003) The effects of agricultural fields and human settlements on the use of rivers by wildlife in the mid-Zambezi valley, Zimbabwe. *Landscape Ecology*, **18**, 293–302.

Fritz, H., Loreau, M., Chamaillé-Jammes, S., Valeix, M. & Clobert, J. (2011) A food web perspective on large herbivore community limitation. *Ecography*, **34**, 196–202.

Fritz, S.A. & Purvis, A. (2010) Phylogenetic diversity does not capture body size variation at risk in the world's mammals. *Proceedings of the Royal Society B*, **277**, 2435–2441.

Fritz, S.A., Bininda-Edmonds, O.R.P. & Purvis, A. (2009) Geographical variation in predictors of mammalian extinction risk: big is bad, but only in the tropics. *Ecology Letters*, **12**, 538–549.

Fryxell, J.M. & Sinclair, A.R.E. (1988) Causes and consequences of migrations by large herbivores. *Trends in Ecology and Evolution*, **3**, 237–241.

Gaillard, J.M., Festa-Bianchet, M., Yoccoz, N.G., Loison, A. & Toïgo, C. (2000) Temporal variation in fitness components and population dynamics of large herbivores. *Annual Review of Ecology and Systematics*, **31**, 367–393.

Gaillard, J.M., Duncan, P., van Wieren, S.E., Loison, A., Klein, F. & Maillard, D. (2008) Managing large herbivores in theory and practice: is the game the same for browsing and grazing species? In: *The Ecology of Browsing and Grazing. Ecological Studies 195* (eds I.J. Gordon & H.H.T. Prins), pp.293–307. Springer, New York.

Gaulin, S.J.C. (1979) A Jarman–Bell model of primate feeding niches. *Human Ecology*, **7**, 1–20.

Gill, R.M.A. (1992) A review of damage by mammals in north temperate forests: 1. deer. *Forestry*, **65**, 145–169.

Global Tiger Initiative (2011) *Global Tiger Recovery Program 2010–2022*. Global Tiger Initiative Secretariat, World Bank, Washington D.C.

Goossens, B., Setchell, J.M., Tchidongo, E., *et al.* (2005) Survival, interactions with conspecifics and reproduction in 37 chimpanzees released into the wild. *Biological Conservation*, **123**, 461–475.

Goulding, M.J. & Roper, T.J. (2002) Press responses to the presence of free-living wild boar (*Sus scrofa*) in southern England. *Mammal Review*, **32**, 272–282.

Grebmeier, J.M., Overland, J.E., Moore, S.E., *et al.* (2006) A major ecosystem shift in the northern Bering Sea. *Science*, **311**, 1461–1464.

Grow, N. & Gursky-Doyen, S. (2010) Preliminary data on the behavior, ecology, and morphology of pygmy tarsiers (*Tarsius pumilus*). *International Journal of Primatology*, **31**, 1174–1191.

Grundmann, E. (2006). Back to the wild: will reintroduction and rehabilitation help the long-term conservation of orang-utans in Indonesia? *Social Science Information*, **45**, 265–284.

Hanks, J. (2005) Kavango–Zambezi: the four-corners transfrontier conservation area. In: *Transboundary Conservation. A New Vision for Protected Areas* (eds R.A. Mittermeier, C.F. Kormos, C.G. Mittermeier, P. Robles Gil, T. Sandwith & C. Besançon). Chicago University Press, Chicago, IL.

Harris, G., Thirgood, S., Hopcraft, J.G.C., Cromsigt, J.P.G.M. & Berger, J. (2009) Global decline in aggregated migrations of large terrestrial mammals. *Endangered Species Research*, **7**, 55–76.

Haskell, J.P., Ritchie, M.E. & Olff, H. (2002) Fractal geometry predicts varying body size scaling relationships for mammal and bird home ranges. *Nature*, **418**, 527–530.

Hayward, M.W. & Kerley, G.I.H. (2009) Fencing for conservation: restriction of evolutionary potential or a riposte to threatening processes? *Biological Conservation*, **142**, 1–13.

Hayward, M.W. & Somers, M. (2009) *Reintroduction of Top-Order Predators*. Blackwell Publishing, Oxford.

Hazzah, L. & Dolrenry, S. (2011) *Lion Guardians Annual Report 2011*. Nairobi, Kenya. www.lionguardians.org

Hazzah, L., Mulder, M.B. & Frank, L. (2009) Lions and warriors: social factors underlying declining African lion populations and the effect of incentive-based management in Kenya. *Biological Conservation*, **142**, 2428–2437.

Hearn, A.J., Ross, J., Pamin, D., Bernard, H., Hunter, L. & Macdonald, D.W. (submitted) The consequences of forest management for the Sunda clouded leopard in Sabah, Malaysian Borneo. *Journal of Mammalogy*.

Hemson, G., Maclennan, S., Mills, G., Johnson, P. & Macdonald, D. (2009) Community, lions, livestock and money: a spatial and social analysis of attitudes to wildlife and the conservation value of tourism in a human–carnivore conflict in Botswana. *Biological Conservation*, **142**, 2718–2725.

Henke, S.E. & Bryant, F.C. (1999) Effects of coyote removal on the faunal community in western Texas. *Journal of Wildlife Management*, **63**, 1066–1081.

Henschel, P., Hunter, L.T.B., Coad, L., Abernethy, K.A. & Mühlenberg, M. (2011) Leopard prey choice in the Congo Basin rainforest suggests exploitative competition with human bushmeat hunters. *Journal of Zoology*, **285**, 11–20.

Herrero, S., Higgins, A., Cardoza, J.E., Hajduk, L.I. & Smith, T.S. (2011) Fatal attacks by American black bear on people: 1900–2009. *Journal of Wildlife Management*, **75**, 596–603.

Hetherington, D. A., Miller, D.R., Macleod, C.D. & Gorman, M.L. (2008) A potential habitat network for the Eurasian lynx Lynx lynx in Scotland. *Mammal Review*, **38**, 285–303.

Hoare R.E. (2000) African elephants and humans in conflict: an outlook for co-existence. *Oryx*, **34**, 34–38.

Hoare, R.E. & du Toit, J.T. (1999) Coexistence between people and elephants in African savannas. *Conservation Biology*, **13**, 633–639.

Hockings, K. & Humle, T. (2009) *Best Practice Guidelines for the Prevention and Mitigation of Conflict Between Humans and Great Apes*. IUCN/SSC Primate Specialist Group, Gland, Switzerland.

Holdo, R.M., Sinclair, A.R.E., Dobson, A.P., *et al.* (2009) A disease-mediated trophic cascade in the Serengeti and its implications for ecosystem C. *PLoS Biol*, **7**, e1000210.

Hoogesteijn, R. (2003) *Manual on the Problem of Depredation Caused by Jaguars and Pumas on Cattle Ranches* (trans. B. Kuperman). Wildlife Conservation Society and IUCN, Gland, Switzerland.

Hopcraft, J.G.C., Olff, H. & Sinclair, A.R.E. (2010) Herbivores, resources and risks: alternating regulation along primary environmental gradients in savannas. *Trends in Ecology and Evolution*, **25**, 119–128.

Hudson, R.J., Drew, K.R. & Baskin, L.M. (1989) *Wildlife Production Systems: Economic Utilisation of Wild Ungulates*. Cambridge University Press, Cambridge.

Hunter, M.D. & Price, P.W. (1992) Playing chutes and ladders: heterogeneity and the relative roles of bottom-up and top-down forces in natural communities. *Ecology*, **73**, 724–732.

Inskip, C. & Zimmermann, A. (2009) Human–felid conflict: a review of patterns and priorities worldwide. *Oryx*, **43**, 18–34.

IUCN. (2008) *2008 IUCN Red List of Threatened Species*. IUCN Species Survival Commission, Gland, Switzerland.

Jackson, P. & Nowell, K. (2008) Panthera tigris ssp. sondaica. In: *IUCN Red List of Threatened Species. Version 2011.2*. IUCN, Gland, Switzerland.

Jackson, R.M., Mishra, C., McCarthy, T.M. & Ale, S.B. (2010) Snow leopards: conflict and conservation. In: *Biology and Conservation of Wild Felids* (eds D.W. Macdonald & A.J. Loveridge), pp.417–430. Oxford University Press, Oxford.

Janis, C.M., Damuth, J. & Theodor, J.M. (2000) Miocene ungulates and terrestrial primary productivity: where have all the browsers gone? *Proceedings of the National Academy of Sciences USA*, **97**, 7899–7904.

Jarman, P.J. (1974) The social organisation of antelope in relation to their ecology. *Behaviour*, **48**, 215–266.

Jerozolimskia, A. & Peres, C.A. (2003) Bringing home the biggest bacon: a cross-site analysis of the structure of hunter–kill profiles in Neotropical forests. *Biological Conservation*, **111**, 415–425.

Jetz, W., Carbone, C., Fulford, J. & Brown, J.H. (2004) The scaling of animal space use. *Science*, **306**, 266–268.

Jhala, Y.V. & Sharma D.K. (1997) Child lifting by wolves in eastern Uttar Pradesh, India. *Journal of Wildlife Research*, **2**, 94–101.

Jhala, Y.V., Qureshi, Q., Gopal, R. & Sinha, P.R. (2011) *Status of the Tigers, Co-predators, and Prey in India, 2010*. National Tiger Conservation Authority, Government of India, New Delhi, and Wildlife Institute of India, Dehradun.

Johnson, P.J., Kansky, R., Loveridge, A.J. & Macdonald, D.W. (2010) Size, rarity and charisma: valuing African wildlife trophies. *PLos ONE*, **5**, e12866.

Jones, J.H. (2011) Primates and the evolution of long, slow life histories. *Current Biology*, **27**, R708–R717.

Jonker, J., van Aarde, R.J. & Ferreira, S.M. (2008) Temporal trends in elephant Loxodonta africana numbers and densities in northern Botswana: is the elephant really increasing? *Oryx*, **42**, 58–65.

Kamler, J.F., Klare, U. et al. (2012). "Seasonal diet and prey selection of black-backed jackals on a small-livestock farm in South Africa." *African Journal of Ecology*, **50**(3): 299–307.

Kanga, E.M., Ogutu, J.O., Piepho, H.P. & Olff, H. (in press) Human–hippo conflicts in Kenya during 1997–2008: vulnerability of a megaherbivore to anthropogenic land use changes. *Journal of Land Use Science*.

Kelt, D.A. & van Vuren, D.H. (2001) The ecology and macroecology of mammalian home range area. *American Naturalist*, **157**, 637–645.

Kerley, G.I.H. & Shrader, A.M. (2007) Elephant contraception: silver bullet or a potentially bitter pill? *South African Journal of Science*, **103**, 181–182.

King, L.E., Douglas-Hamilton, I. & Vollrath, F. (2007) African elephants run from the sound of disturbed bees. *Current Biology*, **17**, R832–R833.

King, L.E., Douglas-Hamilton, I. & Vollrath, F. (2011) Beehive fences as effective deterrents for crop-raiding elephants: field trials in northern Kenya. *African Journal of Ecology*, **49**, 431–439.

Klare, U., Kamler, J.F., Stenkewitz, U. & Macdonald, D.W. (2010) Diet, prey selection, and predation impact of black-backed jackals in South Africa. *Journal of Wildlife Management*, **74**, 1030–1042.

Kleiber, M. (1963) *The Fire of Life*, John Wiley and Sons, New York.

Kruuk, H. & Macdonald, D. (1985) Group territories of carnivores: empires and enclaves. In: *Behavioural Ecology* (eds R.M. Sibly & R.H. Smith), pp.521–536. Blackwell Scientific Publications, Oxford.

Laliberté, A.S. & Ripple, W.J. (2004) Range contractions of North American carnivores and ungulates. *Bioscience*, **54**, 123–138.

Laundré, J.W., Hernández, L. & Ripple, W.J. (2010) The landscape of fear: ecological implications of being afraid. *Open Ecology Journal*, **3**, 1–7.

Leader-Williams, N., Balmford, A., Linkie, M., *et al.* (2007) Beyond the ark: conservation biologists' views of the achievements of zoos in conservation. In: *Zoos in the 21st Century: Catalysts for Conservation?* (eds A. Zimmerman, M. Hatchwell, L.A. Dickie & C. West), pp.236–254. Cambridge University Press, Cambridge.

Leonard, W.R., Robertson, M.L., Snodgrass, J.J. & Kuzawa, C.W. (2003) Metabolic correlates of hominid brain evolution. *Comparative Biochemistry and Physiology, Part A*, **136**, 5–15.

Liley, S. & Creel, S. (2007) What best explains vigilance in elk: characteristics of prey, predators, or the environment? *Behavioral Ecology and Sociobiology*, **19**, 245–254.

Loarie, S.R., van Aarde, R.J. & Pimm, S.L. (2009) Elephant seasonal vegetation preferences across dry and wet savannas. *Biological Conservation*, **142**, 3099–3107.

Loveridge, A.J., Hunt, J.E., Murindagomo, F. & Macdonald, D.W. (2006) Influence of drought on predation of elephant (*Loxodonta africana*) calves by lions (*Panthera leo*) in an African wooded savannah. *Journal of Zoology*, **270**, 523–530.

Loveridge, A.J., Searle, A.W., Murindagomo, F. & Macdonald, D.W. (2007) The impact of sport-hunting on the population dynamics of an African lion population in a protected area. *Biological Conservation*, **134**, 548–558.

Loveridge, A.J., Wang, S.W., Frank, L.G. & Seidensticker, J. (2010a) People and wild felids: conservation of cats and management of conflicts. In: *Biology and Conservation of Wild Felids* (eds D.W. Macdonald & A.J. Loveridge), pp.161–195. Oxford University Press, Oxford.

Loveridge, A.J., Hemson, G., Davidson, Z. & Macdonald, D.W. (2010b) African lions on the edge: reserve boundaries as 'attractive sinks'. In: *Biology and Conservation of Wild Felids* (eds D.W. Macdonald & A.J. Loveridge), pp.283–304. Oxford University Press, Oxford.

Lund, M. (2013) Commensal rodents. In: *Rodent Pests and Their Control*, 2nd edn (eds A.P. Buckle & R.H. Smith), in press. CABI, Wallingford.

Macdonald, D.W. (2000) Bartering biodiversity: what are the options? In: *Environmental Policy. Objectives, Instruments, and Implementation* (ed. D. Helm), pp.142–171. Oxford University Press, Oxford.

Macdonald, D.W. (2009a) *The New Encylopedia of Mammals*. Oxford University Press, Oxford.

Macdonald, D.W. (2009b) Lessons learnt and plans laid: seven awkward questions for the future of reintroductions. In: *Reintroduction of Top-Order Predators* (eds M.W. Hayward & M.J. Somers), pp.411–448. Blackwell Publishing, Oxford.

Macdonald, D.W. & Sillero-Zubiri, C. (2004a) Dramatis personae. In: *Biology and Conservation of*

Wild Canids (eds D.W. Macdonald & C. Sillero-Zubiri), pp.3–36. Oxford University Press, Oxford.

Macdonald, D.W. & Sillero-Zubiri, C. (2004b) Conservation: from theory to practice, without bluster. In: *Biology and Conservation of Wild Canids* (eds D.W. Macdonald & C. Sillero-Zubiri), pp.353–372. Oxford University Press, Oxford.

Macdonald, D.W., Moorhouse, T.P. & Enck, J.W. (2002) The ecological context: a species population perspective. In: *Handbook of Ecological Restoration* (eds M.R. Perrow & A.J. Davy), pp.47–65. Cambridge University Press, Cambridge.

Macdonald, D.W., Creel, S. & Mills, M.G.L. (2004) Society. In: *Biology and Conservation of Wild Canids* (eds D.W. Macdonald & C. Sillero-Zubiri), pp.85–106. Oxford University Press, Oxford.

Macdonald, D.W., King, C.M. & Strachan, R. (2007a) Introduced species and the line between biodiversity and naturalistic eugenics. In: *Key Topics in Conservation Biology* (eds D.W. Macdonald & K. Service), pp.186–205. Blackwell Publishing, Oxford.

Macdonald, D.W., Collins, N.M. & Wrangham, R. (2007b) Principles, practice and priorities: the quest for 'alignment'. In: *Key Topics in Conservation Biology* (eds D. Macdonald & K. Service), pp.271–289. Blackwell Publishing, Oxford.

Macdonald, D.W., Loveridge, A.J. & Rabinowitz, A. (2010) Felid futures: crossing disciplines, borders, and generations. In: *Biology and Conservation of Wild Felids* (eds D.W. Macdonald & A.J. Loveridge), pp.599–650. Oxford University Press, Oxford.

Macdonald, D.W., Johnson, P.J., Albrechtsen, L., *et al.* (2011) Association of body mass with price of bushmeat in Nigeria and Cameroon. *Conservation Biology*, **25**, 1220–1228.

Macdonald, D.W., Tattersall, F.H., Brown, E.D. & Balharry, D. (1995) Reintroducing the European beaver to Britain: nostalgic meddling or restoring biodiversity? *Mammal Review*, **25**, 161–200.

Macdonald, D.W., Fenn, M.G.P. & Gelling, M. (2012) The natural history of rodents: preadaptations to pestilence. In: *Rodent Pests and Their Control*, 2nd edn (eds A.P. Buckle & R.H. Smith). CABI, Wallingford.

Macdonald, D.W., Herrera, E.A., Ferraz, K.M.P.M.B., and Moreira, J.R. (2012) The capybara paradigm: from sociality to sustainability. pp. 385–408. (In: *Capybara: Biology, Use and Conservation of an Exceptional Neotropical Species*). Eds. Moreira, J.R., Ferraz, K.M.P.M.B., Herrera, E.A. and Macdonald, D.W.. Springer, New York.

Macdonald, D.W., Burnham, D., Hinks, A.E., and Wrangham, R. (2012) A problem shared is a problem reduced: seeking efficiency in the conservation of felids and primates. *Folia Primatologica*, DOI: 1159/00034 2399

Maclennan, S.D., Groom, R.J., Macdonald, D.W. & Frank, L.G. (2009) Evaluation of a compensation scheme to bring about pastoralist tolerance of lions. *Biological Conservation*, **142**, 2419–2427.

Mahoney, S.P. & Schaefer, J.A. (2002) Hydroelectric development and the disruption of migration in caribou. *Biological Conservation*, **107**, 147–153.

Marker, L.L. (1998) Current status of the Cheetah (*Acinonyx jubatus*). In: *Symposium on Cheetahs as Game Ranch Animals* (ed. B.L. Penzhorn), pp.1–17. Wildlife Group of the South African Veterinary Association, Onderstepoort, South Africa.

Marker, L.L., Mills, M.G.L. & Macdonald, D.W. (2003) Factors influencing perceptions and tolerance toward cheetahs (*Acinonyx jubatus*) on Namibian farmland. *Conservation Biology*, **17**, 1–9.

Marker, L.L., Dickman, A.J. & Macdonald, D.W. (2005a) Perceived effectiveness of livestock guarding dogs placed on Namibian Farms. *Rangeland Ecology and Management*, **58**, 329–336.

Marker, L.L., Dickman, A.J. & Macdonald, D.W. (2005b) Survivorship and causes of mortality for livestock gurading dogs on Namibian Rangeland. *Rangeland Ecology and Management*, **58**, 337–343.

Marker, L.L., Dickman, A.J., Mills, M.G.L. & Macdonald, D.W. (2010) Cheetahs and ranchers in Namibia: a case study. In: *The Biology and Conservation of Wild Felids* (eds D.W. Macdonald & A.J. Loveridge), pp.353–372. Oxford University Press, Oxford.

Marquet, P.A. & Taper, M.L. (1998) On size and area: patterns of mammalian body size extremes across landmasses. *Evolutionary Ecology*, **12**, 127–139.

Matthews, L.J., Arnold, C., Machanda, Z. & Nunn, C.L. (2010) Primate extinction risk and historical patterns of speciation and extinction in relation to body mass. *Proceedings of the Royal Society B*, **278**, 1256–1263.

McNab, B.K. (1963) Bioenergetics and the determination of home range size. *American Naturalist*, **97**, 133–140.

Mech, L.D. & Boitani, L. (2003) *Wolves. Behaviour, Ecology and Conservation*. University of Chicago Press, Chicago, IL.

Miller, J.S. & Hickling, G.J. (1990) Fasting endurance and the evolution of mammalian body size. *Functional Ecology*, **4**, 5–12.

Milner-Gulland, E.J., Kholodova, M.V., Bekenov, A.B., *et al.* (2001) Dramatic declines in saiga antelope populations. *Oryx,* **35**, 340–345.

Miquelle, D.G., Goodrich, J.M., Smirnov, E.N., *et al.* (2010) Amur tiger: a case study of living on the edge. In: *Biology and Conservation of Wild Felids* (eds D.W. Macdonald & A.J. Loveridge), pp.325–340. Oxford University Press, Oxford.

Mittermeier, R.A., Kormos, C.F., Mittermeier, C.G. & Robles Gil, P. (2005) *Transboundary conservation. A New Vision for Protected Areas.* University of Chicago Press, Chicago, IL.

Moehlman, P. (2002) *Equids: Zebras, Asses and Horses. Status Survey and Conservation Action Plan.* IUCN/SSC Equid Specialist Group, Gland, Switzerland.

Mohapatra, B., Warrell, D.A., Suraweera, W., *et al.* (2011) Snakebite mortality in India: a nationally representative mortality survey. *PLoS Neglected Tropical Diseases,* **5**, e1018.

Moreira, J.R., Verdade, L.M., Ferraz, K.M.P.M.B. & Macdonald, D.W. (2012) The sustainable management of capybaras. pp 283–302 In: *Capybara. Biology, Use and Conservation of an Exceptional Neotropical Species* (eds J.R. Moreira, K.M.P.M.B. Ferraz, E.A. Herrera & D.W. Macdonald). Springer, New York.

Moriceau, J.M. (2007) *Histoire Du Méchant Loup.* Fayard, Paris.

Morrison, J.C., Sechrest, W., Dinerstein, E., Wilcove, D.S. & Lamoreux, J.F. (2007) Persistence of large mammal faunas as indicators of global human impacts. *Journal of Mammalogy,* **88**, 1363–1380.

Murphy, T., Gálvez, N., Boutin, A., Laker, J., Bonacic, C. & Macdonald, D.W. (submitted) Puma density and activity in a fragmented, human dominated landscape.

Naiman, R.J., Johnston, C.A. & Kelley, J.C. (1988) Alteration of North American streams by beaver. *Bioscience,* **38**, 753–762.

Nantha, H.S. & Tisdell, C. (2009) The orangutan–oil palm conflict: economic constraints and opportunities for conservation. *Biodiversity and Conservation,* **18**, 487–502.

Naughton-Treves, L., Treves, A., Chapman C.A., & Wrangham, R. (1998) Temporal patterns of crop-raiding by primates: linking food availability in croplands and adjacent forest. *Journal of Applied Ecology,* **35**, 596–606.

Naughton-Treves, L., Rose, R. & Treves, A. (1999) *The Social Dimensions of Human–Elephant Conflict in Africa: A Literature Review and Case Studies from Uganda And Cameroon.* A report to the African Elephant Specialist, Human–Elephant Conflict Task Force. IUCN, Gland, Switzerland.

Nijman, V., Nekaris, K.A.I., Donati, G., Bruford, M. & Fa, J. (2011) Primate conservation: measuring and mitigating trade in primates. *Endangered Species Research,* **13**, 159–161.

O'Connor, T.G., Goodman, P.S. & Clegg, B. (2007) A functional hypothesis of the threat of local extirpation of woody plant species by elephant in Africa. *Biological Conservation,* **136**, 329–345.

Ogada, M.O., Woodroffe, R., Oguge, N.O. & Frank, L.G. (2003) Limiting depredation by African carnivores: the role of livestock husbandry. *Conservation Biology,* **17**, 1521–1530.

O'Kane, C.A.J., Duffy, K.J., Page, B.R. & Macdonald, D.W. (2011) Overlap and seasonal shifts in use of woody plant species amongst a guild of savanna browsers. *Journal of Tropical Ecology,* **27**, 249–258.

Osborn, F.V. & Parker, G.E. (2003) Towards an integrated approach for reducing the conflict between elephants and people: a review of current research. *Oryx,* **37**, 1–5.

Owen-Smith, N. (1988) *Megaherbivores. The Influence of Very Large Body Size on Ecology.* Cambridge University Press, Cambridge.

Owen-Smith, N. & Mills, M.G.L. (2008) Predator–prey size relationships in an African large-mammal food web. *Journal of Animal Ecology,* **77**, 173–183.

Packer, C. (2005) Lion attacks on humans in Tanzania. *Nature,* **436**, 927–928.

Paine, R.T. (1966) Food web complexity and species diversity. *American Naturalist,* **100**, 65–75.

Paine, R.T. (1969a) A note on trophic complexity and community stability. *American Naturalist,* **103**, 91–93.

Paine, R.T. (1969b) The Picaster–Tegula interaction: prey patches, predator food preference, and intertidal community structure. *Ecology,* **52**, 1096–1106.

Palazy, L., Bonenfant, C., Gaillard, J.M. & Courchamp, F. (2011) Cat dilemma: too protected to escape trophy hunting? *PLoS ONE,* **6**, e22424.

Parker, G.E. & Osborn, F.V. (2006) Investigating the potential for chilli Capsicum annuum to reduce human–wildlife conflict in Zimbabwe. *Oryx,* **40**, 1–4.

Patil, N., Kumar, N.S., Gopalaswamy, A.M. & Karanth, K.U. (2011) Dispersing tigers make a point. *Oryx,* **45**, 472.

Périquet, S., Valeix, M., Loveridge, A.J., Madzikanda, H., Macdonald, D.W. & Fritz, H. (2010) Individual vigilance of African herbivores while drinking: the role of immediate predation risk and context. *Animal Behaviour*, **79**, 665–671.

Peters, R.H. (1986) *The Ecological Implications of Body Size*, Cambridge University Press, Cambridge.

Peters, R.H. & Raelson, J.V. (1984) Relations between individual size and mammalian population density. *American Naturalist*, **124**, 498–517.

Pimm, S.L. & van Aarde, R.J. (2001) African elephants and contraception. *Nature*, **411**, 766.

Power, M.E. (1992) Top-down and bottom-up forces in food webs: do plants have primacy? *Ecology*, **73**, 733–746.

Prescott, G.W., Johnson P.J., et al. (2012). "Does change in IUCN status affect demand for African bovid trophies?" *Animal Conservation*, **15**(3): 248–252.

Prescott, G.W., Johnson, P.J., Loveridge, A.J. & Macdonald, D.W. (in press) Does change in IUCN status affect demand for African bovid trophies? *Animal Conservation*.

Price, S.A. & Gittleman, J.L. (2007) Hunting to extinction: biology and regional economy influence extinction risk and the impact of hunting on artiodactyls. *Proceedings of the Royal Society B*, **274**, 1845–1851.

Primack, R.B. (2006) *Essentials of Conservation Biology*, 4th edn. Sinauer Associates, Sunderland, MA.

Purvis, A., Gittleman, J.L., Cowlishaw, G. & Mace, G.M. (2000) Predicting extinction risk in declining species. *Proceedings of the Royal Society B*, **267**, 1947–1952.

Purvis, A., Mace, G.M. & Gittleman, J.L. (2001) Past and future carnivore extinctions: a phylogenetic perspective. In: *Carnivore Conservation* (eds J.L. Gittleman, S.M. Funk, D. Macdonald & R.K. Wayne), p.675. Cambridge University Press, Cambridge.

Putman, R.J. & Moore, N.P. (1998) Impact of deer in lowland Britain on agriculture, forestry and conservation habitats. *Mammal Review*, **28**, 141–164.

Rabinowitz, A. & Zeller, K.A. (2010) A range-wide model of landscape connectivity and conservation for the jaguar, *Panthera onca*. *Biological Conservation*, **143**, 939–945.

Rao, K.S., Maikhurib, R.K., Nautiyala, S. & Saxena, K.G. (2002) Crop damage and livestock depredation by wildlife: a case study from Nanda Devi Biosphere Reserve, India. *Journal of Environmental Management*, **66**, 317–327.

Rasmussen, G.S.A. (1997) *Conservation Status of the Painted Hunting Dog Lycaon pictus in Zimbabwe.* Ministry of Environment and Tourism, Department of National Parks and Wildlife Management, Zimbabwe.

Rasmussen, G.S.A. & Macdonald, D.W. (2012). "Masking of the zeitgeber: African wild dogs mitigate persecution by balancing time." *Journal of Zoology*, **286**(3): 232–242.

Rasmussen, G.S.A. & Macdonald, D.W. (in press) Overriding the Zeitgeber: balancing time to mitigate human persecution. *Journal of Zoology*.

Reading, R.P. & Clark, T.W. (1996) Carnivore reintroductions: an interdisciplinary examination. In: *Carnivore Behaviour, Ecology and Evolution, Volume 2* (ed. J.L. Gittleman), pp.296–336. Cornell University Press, New York.

Reiter, D.K., Brunson, M.W. & Schmidt, R.H. (1999) Public attitudes towards wildlife damage management and policy. *Wildlife Society Bulletin*, **27**, 746–758.

Rigg, R., Findo, S., Wechselberger, M., Gorman, M.L., Sillero-Zubiri, C. & Macdonald, D.W. (2011) Mitigating carnivore–livestock conflict in Europe: lessons from Slovakia. *Oryx*, **45**, 272–280.

Ripple, W.J. & Beschta, R.L. (2008) Trophic cascades involving cougar, mule deer, and black oaks in Yosemite National Park. *Biological Conservation*, **141**, 1249–1256.

Ruiz-Garcia, M., Payán, E., Murillo, A. & Alvarez, D. (2006) DNA microsatellite characterisation of the jaguar (*Panthera onca*) in Colombia. *Genes and Genetic Systems*, **81**, 115–127.

Rutina, L.P., Moe, S.R. & Swenson, J.E. (2005) Elephant *Loxodonta africana* driven woodland conversion to shrubland improves dry-season browse availability for impalas *Aepyceros melampus*. *Wildlife Biology*, **11**, 207–213.

Sandom, C., Bull, J., Canney, S. & Macdonald, D.W. (2012) Exploring the value of wolves (*Canis lupus*) in landscape-scale fenced reserves for ecological restoration in the Scottish highlands. In: *Fencing for Conservation*. (eds M.J. Somers & M. Hayward), pp.245–276. Springer, New York.

Schipper, J., Chanson, J.S., Chiozza, F., *et al.* (2008) The status of the world's land and marine mammals: diversity, threat, and knowledge. *Science*, **322**, 225–230.

Schley, L. & Roper, T.J. (2003) Diet of wild boar Sus scrofa in Western Europe, with particular reference to consumption of agricultural crops. *Mammal Review*, **33**, 43–56.

Seidensticker, J., Dinerstein, E., Goyal, S.P., *et al.* (2010) Tiger range collapse and recovery at the base of the Himalayas. In: *Biology and Conservation of Wild Felids* (eds D.W. Macdonald & A.J. Loveridge), pp.305–324. Oxford University Press, Oxford.

Shackleton, D.M. (1997) *Wild Sheep and Goats and Their Relatives: Status Survey and Conservation Action Plan for Caprinae.* IUCN/SSC Caprinae Specialist Group, Gland, Switzerland.

Shrader, A.M., Pimm, S.L. & van Aarde, R.J. (2010) Elephant surival, rainfall and the confounding effects of water provision and fences. *Biodiversity and Conservation*, **19**, 2235–2245.

Shurin, J.B. & Seabloom, E.W. (2005) The strength of trophic cascades across ecosystems: predictions from allometry and energetics. *Journal of Animal Ecology*, **74**, 1029–1038.

Sibly, R.M. & Brown, J.H. (2007) Effects of body size and lifestyle on evolution of mammal life histories. *Proceedings of the National Academy of Sciences USA*, **104**, 17707–17712.

Sillero-Zubiri, C., Hoffmann, M. & Macdonald, D.W. (2004) *Canids: Foxes, Wolves, Jackals and Dogs. Status Survey and Conservation Action Plans.* IUCN/SSC Canid Specialist Group, Gland, Switzerland.

Silva, M., Brimacombe, M. & Downing, J.A. (2001) Effects of body mass, climate, geography, and census area on population density of terrestrial mammals. *Global Ecology and Biogeography*, **10**, 469–485.

Silva, M. & Downing, J.A. (1995) The allometric scaling of density and body mass: a non-linear relationship for terrestrial mammals. *American Naturalist*, **145**, 704–727.

Sinclair, A.R.E., Mduma, S. & Brashares, J.S. (2003) Patterns of predation in a diverse predator–prey system. *Nature*, **425**, 288–290.

Sitati, N.W. & Walpole, M.J. (2006) Assessing farm-based measures for mitigating human–elephant conflict in Transmara district, Kenya. *Oryx*, **40**, 279–286.

Sitati, N.W., Walpole, M.J., Smith, R.J. & Leader-Williams, N. (2003) Predicting spatial aspects of human–elephant conflict. *Journal of Applied Ecology*, **40**, 667–677.

Sitati, N.W., Walpole, M.J. & Leader-Williams, N. (2005) Factors affecting susceptibility of farms to crop raiding by African elephants: using predictive model to mitigate conflict. *Journal of Applied Ecology*, **42**, 1175–1182.

Smith, F.A., Elliott, S.M. & Lyons, S.K. (2010) Methane emissions from extinct megafauna. *Nature Geosciences*, **3**, 374–375.

Smith, R.H. & Nott, H.M.R. (1988) Rodent damage to Cacao in Equatorial Guinea. *FAO Plant Protection Bulletin*, **36**, 119–124.

Soulé, M.E., Estes, J.A., Miller, B. & Honnold, L. (2005) Strongly interactive species: conservation, policy, management and ethics. *BioScience*, **55**, 168–176.

Stahl, P., Vandel, J.M., Herrenschmidt, V. & Migot, P. (2001) Predation on livestock by an expanding reintroduced lynx population: long-term trend and spatial variability. *Journal of Applied Ecology*, **38**, 674–687.

Stahl, P., Vandel, J.M., Ruette, S., Coat, L., Coat, Y. & Balestra, L. (2002) Factors affecting lynx predation on sheep in the French Jura. *Journal of Applied Ecology*, **39**, 204–216.

Stanley Price, M.R. (1989) *Animal Reintroductions: The Arabian Oryx in Oman.* Cambridge University Press, Cambridge.

Studsrød, J.E. & Wegge, P. (1995) Park–people relationships: the case of damage caused by park animals around the Royal Bardia National Park, Nepal. *Environmental Conservation*, **22**, 133–142.

Sunarto, S., Kelly, M.J., Parakkasi, K., Klenzendorf, S., Septayuda, E. & Kurniawan, H. (2012) Tigers need cover: multi-scale occupancy study of the big cat in Sumatran forest and plantation landscapes. *PLoS ONE*, **7**, 14.

Sunarto, S., Kelly, M.J., Klenzendorf, S., *et al.* (in press) Threatened tigers on the equator: multipoint abundance estimates in central Sumatra. *Oryx*.

Tanner, J.E., Hughes, T.P. & Connell, J.H. (1994) Species coexistence, keystone species, and succession: a sensitivity analysis. *Ecology*, **75**, 2204–2219.

Teel, T. L., Manfredo, M.J. & Stinchfield, H.M. (2007) The need and theoretical basis for exploring wildlife value orientations cross-culturally. *Human Dimensions of Wildlife*, **12**, 297–305.

Terborgh, J. (1988) The big things that run the world – a sequel to E.O. Wilson. *Conservation Biology*, **2**, 402–403.

Terborgh, J., Lopez, L., Nuñez, P., *et al.* (2001) Ecological meltdown in predator-free forest fragments. *Science*, **294**, 1923–1926.

Thirgood, S., Mosser, A., Tham, S., *et al.* (2004) Can parks protect migratory ungulates? The case of the Serengeti wildebeest. *Animal Conservation*, **7**, 113–120.

Trimble, M.J. & van Aarde, R.J. (2010) Fences are more than an issue of aesthetics. *BioScience*, **60**, 486.

Tuyttens, F.A.M., Delahay, R.J., Macdonald, D.W., Cheeseman, C.L., Long, B. & Donnelly, C.A. (2000) Spatial perturbation caused by a badger (*Meles meles*) culling operation: implications for the function of territoriality and the control of bovine tuberculosis (*Mycobacterium bovis*). *Journal of Animal Ecology*, **69**, 815–828.

Tweheyo, M., Hill, C.M. & Obua, J. (2005). Patterns of crop raiding by primates around the Budongo Forest Reserve, Uganda. *Wildlife Biology*, **11**, 237–247.

Valeix, M., Fritz, H., Dubois, S., Kanengoni, K., Alleaume, S. & Saïd, S. (2007a) Vegetation structure and ungulate abundance over a period of increasing elephant abundance in Hwange National Park, Zimbabwe. *Journal of Tropical Ecology*, **23**, 87–93.

Valeix, M., Chamaillé-Jammes, S. & Fritz, H. (2007b) Interference competition and temporal niche shifts: elephants and herbivore communities at waterholes. *Oecologia*, **153**, 739–748.

Valeix, M., Fritz, H., Chamaillé-Jammes, S., Bourgarel, M. & Murindagomo, F. (2008) Fluctuations in abundance of large herbivore populations: insights into the influence of dry season rainfall and elephant numbers from long-term data. *Animal Conservation*, **11**, 391–400.

Valeix, M., Fritz, H., Canévet, V., Le Bel, S. & Madzikanda, H. (2009a) Do elephants prevent other African herbivores from using waterholes in the dry season? *Biodiversity and Conservation*, **18**, 569–576.

Valeix, M., Loveridge, A.J., Chamaille-Jammes, S., *et al.* (2009b) Behavioral adjustments of African herbivores to predation risk by lions: spatiotemporal variations influence habitat use. *Ecology*, **90**, 23–30.

Valeix, M., Fritz, H., Loveridge, A.J., *et al.* (2009c) Does the risk of encountering lions influence African herbivore behaviour at waterholes? *Behavioral Ecology and Sociobiology*, **63**, 1483–1494.

Valeix, M., Fritz, H., Sabatier, R., Murindagomo, F., Cumming, D. & Duncan, P. (2011) Elephant-induced structural changes in the vegetation and habitat selection by large herbivores in an African savanna. *Biological Conservation*, **144**, 902–912.

Valeix, M., Hemson, G., Loveridge, A.J., Mills, G. & Macdonald, D.W. (2012) Behavioural adjustments of a large carnivore to access secondary prey in a human-dominated landscape. *Journal of Applied Ecology*, **49**, 73–81.

Van Aarde, R.J. & Jackson, T.P. (2007) Megaparks for metapopulations: addressing the causes of locally high elephant numbers in southern Africa. *Biological Conservation*, **134**, 289–297.

Venter, O., Meijaard, E., Possingham, H.P., *et al.* (2009) Carbon payments as a safeguard for threatened tropical mammals. *Conservation Letters*, **2**, 123.

VerCauteren, K.C., Lavelle, M.J. & Hygnstrom, S. (2006) Fences and deer-damage management: a review of designs and efficacy. *Wildlife Society Bulletin*, **34**, 191–200.

Vrba, E.S. (1987) Ecology in relation to speciation rates. Some case histories of Miocene–recent mammal clades. *Evolutionary Ecology*, **1**, 283–300.

Waldram, M., Bond, W.J. & Stock, W.D. (2008) Ecological engineering by a mega-grazer: white rhino impacts on a South African savanna. *Ecosystems*, **11**, 101–112.

Walsh, P.D., Abernethy, K.A., Bermejo, M., *et al.* (2003) Catastrophic ape decline in western equatorial Africa. *Nature*, **422**, 611–614.

Walston, J., Robinson, J.G., Bennett, E.L., *et al.* (2010) Bringing the tiger back from the brink – the six percent solution. *PLoS Biology*, **8**, e1000485.

Wang, S.W. & Macdonald, D.W. (2006) Livestock predation by carnivores in Jigme Singye Wangchuck National Park, Bhutan. *Biological Conservation*, **129**, 558–565.

Wang, S.W. & Macdonald, D.W. (2009) The use of camera traps for estimating tiger and leopard populations in the high altitude mountains of Bhutan. *Biological Conservation*, **142**, 606–613.

Wang, S.W., Curtis, P.D. & Lassoie, J.P. (2006a) Farmer perceptions of crop damage by wildlife in Jigme Singye Wangchuck National Park, Bhutan. *Wildlife Society Bulletin*, **34**, 359–365.

Wang, S.W., Lassoie, J.P. & Curtis, P.D. (2006b) Farmer attitudes towards conservation in Jigme Singye Wangchuck National Park, Bhutan. *Environmental Conservation*, **33**, 148–156.

Wang, X., Sheng, H., Bi, J. & Li, M. (1997) Recent history and status of the Mongolian gazelle in Inner Mongolia, China. *Oryx*, **31**, 120–126.

Whyte, I.J. & Joubert, S.C.J. (1988) Blue wildebeest population trends in the Kruger National Park and the effects of fencing. *South African Journal of Wildlife Research*, **18**, 78–87.

Wikramanayake, E., Dinerstein, E., Seidensticker, J., *et al.* (2011) A landscape-based conservation strategy to double the wild tiger population. *Conservation Letters*, **4**, 219–227.

Wilcove, D.S. & Wikelski, M. (2008) Going, going, gone: is animal migration disappearing? *PLoS Biology*, **6**, e188.

Williams, A.C. (2011) *WWF Asian Elephant and Rhino Conservation Programme. Financial Years 2012–2014.* WWF–AREAS, Nepal.

Williamson, D. & Williamson, J. (1984) Botswana's fences and the depletion of Kalahari wildlife. *Oryx*, **18**, 218–222.

Winterbach, H.E.K., Winterbach, C.W., Somers, M.J. & Hayward, M.W. (2012) Key factors and related principles in the conservation of large African carnivores. *Mammal Review*. DOI: 10.1111/j.1365-2907.2011.00209.x

Wood, B.J. & Singleton, G.R. (2013) Rodents in agriculture and forestry. In: *Rodent Pests and Their Control*, 2nd edn (eds A.P. Buckle & R.H. Smith), in press. CABI, Wallingford.

Woodroffe, R. & Ginsberg, J.R. (1998) Edge effects and the extinction of populations inside protected areas. *Science*, **280**, 2126–2128.

Wrangham, R.W., Wilson, M.L., Hare, B. & Wolfe, N.D. (2000) Chimpanzee predation and the ecology of pathogen exchange. *Microbial Ecology in Health and Disease*, **12**, 186–188.

Yamaguchi, N., Cooper, A., Werdelin, L. & Macdonald, D.W. (2004) Evolution of the mane and group-living in the lion (*Panthera leo*): a review. *Journal of Zoology*, **263**, 329–342.

Young, K.D. & van Aarde, R.J. (2010) Density as an explanatory variable of movements and calf survival in savanna elephants across southern Africa. *Journal of Animal Ecology*, **79**, 662–673.

17

Plant conservation: the seeds of success

Timothy Walker[1], Stephen A. Harris[2] and Kingsley W. Dixon[3]

[1]University of Oxford Botanic Garden, Oxford, UK
[2]Department of Plant Sciences, University of Oxford, Oxford, UK
[3]Kings Park and Botanic Garden,
The University of Western Australia, West Perth, 6005, Nedlands, Australia

'As he extends the range of his observations, he will meet with more cases of difficulty'
Darwin (1859)

Introduction

The products and ecological services provided by plants are fundamental to the survival of all animal life on earth. In common with many other organisms, individual plant species are threatened with rapid, human-mediated range change and extinction, processes which over the long term are also important in species evolution. Conservation is not a part of nature but a part of the human character and the unique ability of *Homo sapiens* to look forward and to calculate the consequences of actions. The significance of these observations is two-fold. Conservation requires human action if it is to happen, let alone succeed. Furthermore, species conservation is counter-intuitive, since organisms evolve; tomorrow takes care of itself or rather has until now. Although every day has enough problems, tomorrow will be very difficult for humans if vegetation of some sort is not successfully conserved by someone, somewhere.

In May 2010, in his foreword to *Global Biodiversity Outlook 3*, Ban Ki-moon, Secretary-General of the United Nations, wrote:

> 'In 2002, the world's leaders agreed to achieve a significant reduction in the rate of biodiversity loss by 2010 ... the target has not been met ... the principal pressures leading to biodiversity loss are not just constant but are ... intensifying'.

Awareness of the negative effects of humans began hundreds of years ago, although John Muir's work, and the establishment of the national parks in the USA, perhaps mark the start of the modern conservation movement. Much has been achieved since the signing and ratification of the Convention on Biological Diversity in 1992,

Key Topics in Conservation Biology 2, First Edition. Edited by David W. Macdonald and Katherine J. Willis.

although as Ban Ki-moon emphasized, meeting targets is a challenge (Stokstad 2010).

A major challenge to planning plant conservation is to answer the apparently simple question: how many plant species are found on earth? The difficulty is partially due to bickering about taxonomic rank and regional taxonomic inflation and promotion (Scotland & Wortley 2003; Paton et al. 2008). Darwin observed that 'all naturalists know vaguely what they mean when they speak of species' (Darwin 1859), although about 3 years earlier he commented to Joseph Hooker that perhaps we were trying 'to define the undefinable' (Darwin 1856). However, species continue to be used as biological currency. If we accept a species as a group of individuals that share a unique set of characters, that can reproduce, then there are 352,828 named plant species with perhaps another 50,000 waiting to be named. Unexpectedly, the majority of unnamed plant species may have been collected already; they are in herbaria back-rooms (Bebber et al. 2010). This brings us to a major problem – the uneven and inadequate supply of taxonomists working on floras and monographs; the so-called 'taxonomic impediment'. Some countries have been completely surveyed, e.g. the UK. The UK may be small, with a depauperate, stabilized, postglacial flora with few, if any, endemics, but the fact that it is so well catalogued shows that the quantity of skilled labour, not technology, limits completion of the World Flora project.

How many of these species are in trouble? Not all of them; *Poa annua* will be here for many years. The estimates for the percentage of plant species that will be extinct within 50 years vary from 22% to 62% (Pitman & Jørgensen 2002). As with species lists, there are problems of chauvinism; locally rare species are included in lists of vulnerable species when these species are abundant globally. The reverse is also true, with locally abundant but globally rare species appearing in some lists but not others. The compilation of lists is also difficult if the IUCN's nine categories have to be assessed for every taxon. Such assessments may be possible given infinite resources but a simpler heuristic approach is needed that will enable monographers to assign quickly a level of 'threatened ness' (Pennisi 2010a,b).

The enumeration of taxa is one-dimensional; it gives you a place of occurrence, and is only one aspect of conservation. Other important aspects are associated with pollinator and seed biology, whilst the judicial application of genetic analyses provides important insights for plant conservation protocols. For plant conservation to be effective, research findings must be incorporated into practical conservation strategies, such as the Millennium Seedbank Project.

Research in plant conservation science

Pollination science in conservation

Pollination services underpin the capacity of ecosystems for ongoing reproductive capacity, while ensuring genetic diversity is maintained. Despite pollination biology having a long and respected pedigree of scientific enquiry, the discipline is yet to be fully integrated with conservation sciences, particularly with ecological restoration (Dixon 2009). This discrepancy reflects in part the scientific origins of conservation science in ecosystems dominated by wind-pollinated species. In such ecosystems, investigations of pollen flow do not need to understand or integrate biotically mediated pollination. For example, the ecologically young landscapes of northern Europe are dominated by wind pollination syndromes whilst in some of the world's oldest landscapes, such as the biodiverse landscapes of the south west Australian biodiversity hot spot, complex pollination syndromes operate (Table 17.1) (Hopper & Gioia 2004; Phillips et al. 2010).

Thus, in plant conservation, biotically-mediated pollination, either by generalist or specialist pollinators, requires conservation planners to consider plant and animal interactions and ensure that planning all aspects of plant conservation, from plant reintroductions to reserve design, captures the necessary trophic and habitat support

Table 17.1 Comparison of pollination syndromes in a postglacial European flora compared with the flora from a geologically stable environment in south west Australia (Hopper & Gioia 2004; Phillips et al. 2010)

Pollination syndrome	European flora	South west Australian biodiversity hot spot
Wind	25%	8%
Mammal	0%	2%
Bird	0%	15%
Invertebrates	75%	78%*

*Many invertebrate-pollinated plants are also visited by birds and mammals.

for pollinators. In generalist pollination, planning for pollination services may be straightforward, particularly since the European honeybee can be an effective surrogate where native insect pollinators are lacking or depleted (Lomov et al. 2010). However, caution is needed before substituting honeybees for native insect pollinators and, conversely, in ensuring that honeybee eradication (often practised in many natural areas such as southern Australia) does not eliminate or reduce local pollination services.

Where there is a requirement for an obligate specialist pollinator, for example hummingbirds or bats (Fleming & Muchhala 2008) or the many and varied specialist pollinators of orchids (Tremblay et al. 2005), surrogate pollinators may not be available. In these cases, planning for plant conservation may require setting aside areas of native vegetation or ensuring fragments have effective corridors for movement of pollinators (Hadley & Betts 2009). An alternative is ecological restoration with plant species that encourage and attract, sustain and enhance pollinator communities but this will be difficult where the techniques and technology are lacking, particularly for specialist pollinators (Dixon 2009). However, simple conservation measures, such as weed control, reduced grazing, fire suppression or targeted plantings, can preserve pollination capabilities (Sabatino et al. 2010). For example, high fire frequencies in some diverse ecosystems, such as the Mediterranean biomes, can lead to grasses replacing fire-sensitive shrubs that provide nectar or pollen sources.

In the face of global climate change, the general absence of information on conservation of pollination services, particularly at the ecosystem level, and on how to manage and restore pollinators in conservation planning to maintain genetic fitness and reproductive capacity, represent important research areas (Menz et al. 2011).

Seed biology in conservation

Seeds provide a complementary, cost-effective and efficient means for off-site (*ex situ*) conservation of plant species. Seeds of many plants remain viable after long-term storage, e.g. hundreds, potentially a thousand years (Smith et al. 2003; Havens et al. 2006; De-Zhu & Pritchard 2009).

Ex situ *seedbanks*

Ex situ or institutional seedbanks (where seed is stored under controlled environment conditions) of wild species worldwide are gearing up for large-scale operations, from the Millennium Seedbank Project that aims to store 25% of the world's flora over the next decade, through a 100 tonne-capacity wild seed facility in Utah to the Kunming seedbank in south China that aims to store 16,000 species from South East Asia. Globally, wild seedbank programmes, though found only in 200 of the 1600 seedbanks worldwide, represent important adjuncts in conservation planning. Seedbanks are insurance against species extinction but, more importantly, they are 'working' seedbanks that deliver seed for restoration programs (Havens et al. 2006; Nelleman & Corcoran 2010). Delivering 'at-scale' ecological restoration will require seedbanks to reconsider their storage capacity to ensure that tonnes rather than grams of seed are able to be stored, ready for restoration (Merritt & Dixon 2011).

There is limited understanding of the biological constraints and opportunities of seed-based conservation strategies. The seeds most at risk of long-term deterioration are usually those of wild species, where little research has been done on even the most basic of dormancy and storage conditions (Walck & Dixon 2009). For example, seed from highly endemic floras such as the Cape flora of South Africa or the south west Australian biodiversity hot spot have many endemic genera, families and even an order (Dasypogonales for south west Australia) where research on seed dormancy and longevity is unknown.

Seedbank managers face many unknowns in storing and managing wild seed. How long can seed be stored in seedbanks, what pretreatments maximize seed germination, what dormancy states exist, how best to direct seeds to site or nursery propagation are just some of the key questions faced by scientists working on wild seed conservation. This lack of understanding is even more acute in biodiversity hot spots where seed knowledge is limited, both by the bewildering diversity of plants and often by limited research capability. Often research skills are not where the problems are. For example, three-quarters of the world's botanic gardens, with plant conservation science capability, are outside biodiversity hot spots. Twinning hot spot botanic gardens and those with the technological capability for capacity building is one solution to this problem that is being tackled by some of the larger botanic gardens, such as the Royal Botanic Gardens Kew in the United Kingdom.

Soil-stored seed

Soil seedbanks, sometimes referred to as *in situ* seedbanks, are an untapped resource in species and ecosystem conservation planning, but limited in practice by ignorance of the spatial and temporal dynamics of soil seedbanks, key features that determine the chances of reinstating species. Defining the ecology of soil seedbanks is vital for understanding species resilience to habitat alteration and, for invasive species, the ecological liabilities of persistent weed seedbanks. In fire-prone and disturbance-based ecosystems, soil seedbanks are sometimes stimulated to germination in response to karrikinolide, a molecule in smoke that is known to control seed germination (Chiwocha et al. 2009). Effective at remarkably low concentrations, karrikinolides are a new approach to managing spatially specific, on-demand stimulation of soil seedbanks.

One of the most significant challenges in seed science is the capacity to deliver seed-based solutions at the scale needed for ecological restoration. Seed is often the most expensive component in restoration programmes, yet little research is devoted to technological improvements in efficiencies. Whereas the commercial horticulture sector has made remarkable strides in improving seed use efficiency, for wild seed the story is less clear, with scant development of large-scale applications for improving seeding success. If we are to move effective seed science into practical plant conservation, then at-scale technology will be required, built on a solid foundation of seed science research.

Genetics in plant conservation

In the early 20th century, the Soviet geneticist Nicolai Vavilov recognized the importance of genetics for plant conservation when he defined global centres of crop plant diversity based on patterns in morphological traits among crops and their close relatives (Harlan 1992). By the 1960s collecting expeditions were stocking seed banks with crop landraces (diverse genotypes adapted to traditional agricultural regimes); landrace genes were used to fuel the Green Revolution. However, it took nearly 25 years before technological, theoretical and political developments converged to change the focus of genetic conservation from crops to wild plants (Frankel et al. 1995). Today's technologies mean that DNA from any plant species is quickly and easily investigated, and low-cost, highly reproducible genetic markers, scattered across the genome, are readily available. Low-cost computing and data storage mean large data sets are analysed in ever more sophisticated manners. Furthermore, there is now little gap between the time when a theory is developed and when it is applied empirically. For example, Lander et al. (2011) showed that pollen flow (as measured using DNA markers) in *Gomortega keule*, a rare and endangered tree from the mediterranean Chile, depends strongly on types of land use between habitat fragments, following the development of a model for gene movement across heterogeneously fragmented landscapes. Politically, the Convention on Biological Diversity recognized the importance of both *in situ* and *ex situ* genetic resources, whilst recognition of a global biodiversity crisis and the long-term consequences of climate change on species survival have spurred actions by some governments.

Estimating genetic variation in endangered plants, using more or less neutral markers, is commonplace. Co-dominant markers are commonly used, although dominant markers also find application in conservation (Frankham et al. 2010). A co-dominant marker reveals all the alleles at a heterozygous locus; therefore allele frequencies may be estimated. A dominant marker cannot differentiate between dominant homozygotes and heterozygotes, so allele frequencies cannot be estimated unless particular assumptions are made about the population. Low levels of genetic variation are often associated with the effects of small population size and inbreeding, or with a historic bottleneck or founder event. Once genetic variation has been determined, the amounts of differentiation within and among populations may be estimated; high differentiation is associated with low gene flow and vice versa. Comparisons between population pairs are used to estimate genetic distance and overall population similarities. If data are to be useful for species conservation, it is important to ensure adequate sampling of a species' geographic range, of individuals within a population and of genes within a genome (Lowe et al. 2004). For example, Chamberlain (1998) studied genetic diversity and differentiation (at 23 nuclear loci) of the widespread, economically important, leguminous shrub *Calliandra calothyrsus* within and among 17 populations (mean c.17 individuals per population) across its Mesoamerican range. Three, possibly four, genetic groups were identified across the range of *C. calothyrsus*, which became the basis for the conservation of the species' genetic resources and revision of the species' taxonomy.

Genetically, plant conservation efforts are concerned with two issues: maintaining (or increasing) effective population size and minimizing inbreeding effects (Frankel et al. 1995). The effective or genetic size of a population and its census size show complex relationships determined by processes such as census size fluctuations, unequal sex ratios and family sizes, differential fertility, age structure and non-random mating. Consequently, effective population size is usually smaller than the census size so merely counting individuals in an area may severely underestimate a population's long-term genetic viability.

As population size decreases, genetic variation will be reduced as a consequence of genetic drift, and inbreeding is likely to increase. Inbreeding increases the probability that disadvantageous alleles will be exposed to selection, and hence reduces the reproductive success of any individual homozygous for these alleles, under particular environmental conditions.

A special conservation concern is fragmentation of species' ranges, where a habitat is broken into fragments of varying sizes and separations. The genetic consequences of habitat fragmentation for particular species will depend on factors such as the number of population fragments, the spatial patterns of the fragments, the distribution of population sizes, species' dispersal, gene flow between fragments, structure of the interpopulational matrix, time since fragmentation and the species' life history (Fahrig 2003; Holderegger & Wagner 2008).

Successful species conservation depends on being able to define and identify a species unequivocally. DNA markers, combined with phylogenetic analyses, have transformed understanding of plant species relationships and offered the possibility of apparently simple species identification. In plants, species delimitation may also be problematical because of complex patterns of morphological variation, polyploidy and hybridization and introgression. Such complexities blur efforts to gather rapidly primary conservation data (e.g. species range, population size, gene flow patterns) and generate, and implement, species conservation plans, plus the associated legal frameworks. DNA barcoding, using standard, short DNA regions as universal tools has been promoted for ready species identification, at least if laboratory resources are available. In animal studies, the region of choice is the mitochondrial gene *CO1*, since high-quality sequences may be used to distinguish taxa across a wide taxonomic range (Fazekas et al. 2009). For barcoding plants, the mitochondrial genome is useless as substitution rates are too low (Kress et al. 2005). Therefore, attention has been focused on various combinations of chloroplast gene sequences, e.g. the two-locus combination *rbc*L+*mat*K (CBOL Plant Working Group 2009). DNA barcoding appears an attractive idea, although critics emphasize that resources may be diverted from much needed basic taxonomic research (Ebach & Holdrege 2005), uniparental inheritance of chloroplast genes make them useless for hybrid identification, morphologically complex groups may contain many closely related taxa which are difficult to separate using short chloroplast DNA sequences, and barcoding does nothing for the identification of plants in the field, since laboratory resources are needed. Chloroplast genomes are assumed to be maternally transmitted, yet at least 30% of vascular plants have biparental plastid transmission (Harris & Ingram 1991), leading to heteroplasmy and complex patterns of within-population variation (Hansen et al. 2007).

Despite the issues associated with universal plant barcodes, Lahaye *et al.* (2008) have shown that *mat*K barcodes distinguish *Phragmipedium* species from all other Mesoamerican orchids. Since global trade in all *Phragmipedium* species is illegal, smuggled specimens of this otherwise difficult to identify genus can readily be detected (Hollingsworth 2008).

Genetics gives added value to conservation efforts when analysed in the context of other types of data. Practical species conservation programmes have benefited from the integration of genetic and biogeographical data (phylogeography) for the identification of broad-scale patterns of genetic variation, often as the consequences of species recolonization following long-term environmental change (e.g. Petit et al. 2002). Such studies are all the more convincing when multiple species, scattered across the tree of life, are included. Overlapping areas of high phylogeographic diversity may offer justification for the location of *in situ* reserves (Moritz & Faith 1998; Crandall et al. 2000).

Once patterns of genetic variation have been identified across species' ranges, it becomes possible to find from where unlocalized samples are likely to have come. The forensic use of genetic data in plant conservation has yet to reach the sophistication found in animal conservation (Ogden et al. 2009), although much attention has been directed towards the identification of timber. Timber forensics is concerned with identifying processed wood products from trees covered by Convention on International Trade in Endangered Species legislation (e.g. Ogden et al. 2008) and tracking timbers from illegal sources (Dutech et al. 2003). Such investigations are challenging because of the problems of extracting DNA and sampling geo-referenced specimens for building reference databases. Specimens in herbaria and xylaria may lack important meta-data or have been sampled in ways that reduce their usefulness for such research. Preliminary results appear to offer promise but for these approaches to be accepted by courts, much needs to be done to design rigorous testing and verification strategies, and to understand patterns of genetic variation across species' ranges, at appropriate geographical scales.

Population genetic and geographic data have also been used to answer questions focused on microevolutionary processes such as gene flow, bottlenecks and habitat fragmentation. Such data have shown the importance of rare long-distance gene flow events, and emphasized the importance of understanding patterns of landscape permeability when investigating the conservation consequences of fragmentation (Sork & Waits 2010). In *Gomortega keule*, paternity analysis shows that the tree's pollinators may travel up to 6 km between patches across the fragmented landscape. Furthermore, pollen moved from single trees into large sites, indicating that these sites are important as stepping stones between sites. Thus habitat fragmentation, at the scale investigated, has not led to genetic isolation in *Gomortega*, and genetic connectivity *per se* is not a conservation priority (Lander et al. 2010).

However, genetic approaches may be too expensive for practical conservation purposes. For example, '-omics' technologies, which have been of such great interest in plant sciences, are unlikely to have much impact on species conservation efforts (van Straalen & Roelofs 2006; Höglund 2009). Resources for practical plant conservation are very limited, and genetic technologies are often expensive compared to more traditional plant conservation approaches. Consequently, the role of genetics in conservation should be pragmatic, focused on asking the right questions and determining whether genetic solutions are likely to provide the appropriate answers. Biology, rather than technology, should lead genetic conservation research. The discovery of genetic variation in a species is unlikely to be either interesting or useful for conservation in the absence of other data.

Changing plant conservation priorities

Conservation research usually advances much more slowly than the threat to plant populations. Ideally, well-targeted, practical conservation work would begin only once the global species checklist has been completed, relative species densities calculated from distribution maps, each taxon assessed, and priorities ascribed but one runs the risk of condemning some species to extinction. Yet, to start conservation in the absence of an accurate inventory is to rely on intuitive conservation, based on hunches. This type of conservation may be better than nothing *but* it might also be misguided and have to be repaired at a later date. The reason why Target 3 of the Global Strategy for Plant Conservation was included was to avoid hearsay conservation. Large volumes of robust data are needed to enable the 'best' pieces of vegetation for plant conservation to be identified. Kremen et al. (2008) investigated the distribution of 2315 species from six Madagascan taxonomic groups (lemurs, ants, geckos, frogs, butterflies and plants) to identify the optimum 10% of the island needed for their protection; the data were collected over centuries. Among the many results was confirmation that each taxonomic group occupied a different important area. When the different 10% optima were combined, 26.4% of Madagascar would need to be protected, compared with 6.3% of the island that is currently protected; plants were the best conservation surrogate. If only one taxon group can be surveyed, fewer species overall were left exposed if plants were conserved than if, for example, lemurs were chosen as the indicator group.

In some countries the majority of the land is under some form of production. Globally, some 25% of land is classified as productive, but the biodiversity value of such land may not be zero. The cork oak woodlands of the Iberian peninsula support a great variety of species contributing to the Mediterranean Basin's position as a global biodiversity hot spot (Myers & Cowling 1999). Farming activity is popularly blamed for species decline, yet productive farming and conservation may not be mutually exclusive. For example, in South Africa, farmers compete at annual flower shows to see who can find the largest number of different species flowering on their properties, so agriculture does not have to be synonymous with low diversity.

The human population is expected to peak in 2050, at c.9.5 billion, and to feed this population intensive agriculture is expected to increase. In affluent countries consumers can currently choose where they acquire their food and other biological resources. How they make these choices is personal. It is sometimes stated that poverty deprives people of the luxury of choice (Adams et al. 2004) but plant conservation is needed in all countries (Sachs et al. 2009). Ten years ago, the survey of the Ethiopian flora was far from complete but by 2010 it had been completed by Ethiopian botanists. In the mountains of Colombia at La Cocha, near Pasto, the Botanic Garden of the Inheritors of the Earth was created by people under 20 years old who wanted to conserve their biological inheritance, particularly the local crop landraces. The greening of school grounds in the suburbs of South Africa using only native species is another example of people taking responsibility for their local flora.

Plants can be conserved in their habitats, provided ecosystem services are also preserved. The value of these services has been calculated at $35,000,000,000 per annum (Faith 2010). The problem with this type of calculation is that the ecosystem services are irreplaceable and priceless, and thus can be perceived in commercial terms as worthless and valueless. There are many well-rehearsed reasons to explain why it is better to conserve functioning communities (Maxted et al. 1997). If a threatened plant species is conserved in its habitat then its fungal, pollinators and other associated ecosystem partners will also be conserved. It is also believed that more genetic diversity will be conserved compared to *ex situ* approaches. However, the threat will also be conserved. *In situ* conservation will never succeed if the threat is not removed. Furthermore, if the threat is climate change then trapping plants in protected areas is unlikely to provide the necessary protection (Hannah et al. 2002, 2007).

The importance of specific pollinators has recently been questioned. Pollination syndromes are complex (Wilmer 2011) and Ollerton et al. (2007, 2009) have cast doubt on simplistic ideas by showing that apparently generalist flowers may have specific pollinators, whilst seemingly specialized flowers may have many pollinators. The conservation of fungi is a serious problem, and probably the most powerful reason for habitat preservation as a mechanism for *in situ* species conservation. Fungi, particularly basidiomycetes, are important for the formation of mycorrhizae, especially in forest trees. Mycorrhizae are ancient, intimate associations between plant roots and fungi that are important for the movement of nutrients through ecosystems (Remy et al. 1994; van der Heijden et al. 1998). If the 'taxonomic impediment' is a handicap to conservation, then the 'mycological impediment' is becoming crippling as the number of professional mycologists continues to fall.

Seedbanks do not replace other forms of conservation. On the contrary, the increase in knowledge of seed biology, woefully inadequate for much of the 20th century (Baskin & Baskin 2001), is directly the result of research undertaken in seedbanks. In addition, large seedbanks have been successful conduits for funding of collaborative projects with biologists in developing countries (Thompson 2010). Well-designed sampling strategies may mean that the genetic diversity of an *ex situ* seedbank is as diverse as a natural population (Walck & Dixon 2009), although the diversity of the *ex situ* collection compared with that of the same natural population in the future is unknown. It is clear that the value of seedbanks for species conservation and habitat restoration is very high. The re-creation of species-rich meadowland in the UK totally depends upon the supplementation of seed mixes with species from seedbanks.

Orchid conservation as a case study in global issues in conservation

Orchids are an exemplary group for understanding conservation risks in plants and for deriving principles for developing effective conservation solutions. Their diverse and often intricate and obligate internal associations with mycorrhizae, combined with pollinator specialization, means that orchids are effective indicators of environmental change (Cribb et al.

2003). It is therefore not surprising that orchids are classed at the highest level of threat through habitat loss and habitat alteration, overcollection and sensitivity to environmental change (Swarts & Dixon 2009). As a plant group, orchids provide excellent study opportunities and focal species for understanding the consequences of habitat alteration and potential approaches to restoration. In a plant family replete with complex pollinator and mycorrhizal interactions, success with restoring orchid populations will lead to understanding of how to restore the multiple components of ecological networks.

The diversity of the orchid family, with an estimated 26,000 species, and their array of adaptive features have been studied in depth across many continents. Consequently, there is a wealth of information spanning phylogeography and physiology through to studies of taxonomic inflation (e.g. Phillips et al. 2009). However, with the level of threat to orchids increasing globally, the question is just how well orchid science is answering key conservation questions.

Orchid research has been focussed on the fields of taxonomy, phylogeny, mycorrhizal investigations, horticultural science (including propagation) and evolutionary biology. Some of these areas are directly relevant to conservation science, but just how effective is the translation of the research directly related to conservation outcomes in delivering practice-based conservation benefits? Little of the outstanding repertoire of science has helped to develop practical conservation solutions, especially in the area of orchid translocations. Of 1560 published works on orchids appearing between 2005 and 2010, most do not relate to conservation benefits and direct conservation-based questions; there is a handful of outstanding exceptions (Ramsay & Stewart 1998; Smith 2006; Dutra et al. 2009). For example, few studies provide general principles to guide *ex situ* orchid conservation (mycorrhiza banking and seedbanking) or translocation science (the how, when and where of orchid reintroduction). Equally, although often noted as critical for many orchids, the specialized area of pollination is all but ignored in orchid reintroductions, which seldom monitor or facilitate pollinator activity (Koh et al. 2005; Dixon 2009).

The future conservation benefits of orchid research will rely on delivering science that will guide conservation actions. This science must range from breeding analysis, to ensure that the highest quality (outbred, high-viability) seed is produced, through mycorrhiza banking, seedbanking and propagation science, to translocation and the attendant monitoring of plant health, plant resilience and reproductive capacity (Swarts & Dixon 2009). Importantly, enthusiasts and commercial operators are gearing up to establish plant production capabilities for wild orchids that could provide feedstock for future orchid reintroductions. However, ensuring that these programmes remain grounded in sound science is paramount (Seaton 2007).

Conservation of evolutionary processes rather than evolutionary patterns

Conservation policies are generally directed towards the preservation of the temporal endpoints of evolution rather than the processes that gave rise to these endpoints (e.g. Balmford et al. 1998). Consequently, policies usually concern species, and the details of species-based conservation strategies are much discussed (see Chapter 2). Species-based approaches require that species are readily and unambiguously defined. However, in some plant groups, in which processes such as inbreeding, apomixis and hybridization, often at multiple ploidy levels, are prevalent, discrete species may be impossible to delimit (Coyne & Orr 2004). Persistent lineages may have restricted distributions, be recognized as endemic and given taxonomic names. Endemism tends to generate popular interest, and popular interest is an important component of conservation initiatives and funding. Consequently, substantial conservation effort may be directed towards poorly defined taxa, as a consequence of taxonomic inflation.

Within the European whitebeams and rowans (*Sorbus*), apomixis (the production of seed without the genetic involvement of a male parent), combined with hybridization across ploidy levels, has generated diverse arrays of morphological forms. Some authors choose to recognize such forms as 'microspecies', whilst others argue that they are intermediates in the speciation process and recognize them as proper species (Rich et al. 2010). Currently, Rich and colleagues recognize 45 native *Sorbus* taxa in Britain and Ireland, of which 37 are considered endemic (16 are critically endangered). Morphological, cytological and molecular evidence indicates that the relationships among the taxa may be complex. For example, the Scottish endemic, obligate-apomictic, triploid *Sorbus arranensis* has hybridized with the widespread European diploid *S. aucuparia* on at least five separate occasions to produce the Scottish endemic, sexual, tetraploid *S. pseudofennica* (Robertson et al. 2004).

Complex patterns of taxonomic variation, where population processes (e.g. hybridization and apomixis) are important for the generation and maintenance of variation, create problems for traditional, pattern-based conservation approaches. These considerations led Ennos *et al.* (2005) to suggest that conservation practices for such groups should be modified to recognize explicitly the maintenance of evolutionary processes rather than merely the conservation of named taxa. Process-based conservation practice argues for the importance of understanding evolutionary dynamics. In turn, conservation practitioners, and the funders on whom they rely, would have to recognize that evolutionary processes are dynamic, that over the long term, products of evolutionary processes go extinct and new products arise. At any one point in time, according to this controversial view, there would therefore be little value in trying to conserving particular genotypes. Process-, rather than pattern-driven conservation also presents challenges to legislation designed to protect named taxa. Despite the potential difficulties, explicit consideration is being given to the incorporation of process-based approaches into conservation planning (e.g. Klein et al. 2009; Ferrier & Drielsma 2010).

Natural complexities in the genus *Sorbus* are restricted to particular areas of Britain, for example, south west England, the Cheddar and Avon gorges, south west Wales and Arran in Scotland. *Sorbus* conservation and management suggestions made by Rich et al. (2010) implicitly recognize some aspects of process-driven approaches to conservation, despite naming particular genotypes. Rich et al. (2010) emphasize the importance of *in situ* conservation combined with appropriate *ex situ* approaches, but specifically reject the translocation of rare *Sorbus* taxa into other parts of Britain and Ireland since this would detract from understanding the dynamics of their origins. *Sorbus* genotypes are both localized and rare which, combined with an individual's short life span (less than 100 years), means their population sizes are likely to be small. For example, the endemic hybrid *S.* x *houstoniae* is known from only a single tree (Rich et al. 2009) and the endemic *S. rupicoloides* is known from only one site (13 trees) (Houston et al. 2009). The processes of evolutionary change make individual genotypes labile. For example, if rare, self-incompatible apomictic genotypes are to be maintained, they will need access to cross-pollen from other taxa, so common diploids, along with rare apomicts, will need to be conserved. Habitat management is also a critical issue for recruitment of the products of evolutionary processes into the adult population. Thus, habitat clearance and changes in grazing patterns may have serious long-term consequences for a process-driven conservation of *Sorbus* in parts of Britain and Ireland. For example, in early 2009 plans were made to release goats into the Avon Gorge, near Bristol, to open areas and create new habitat for rare grassland species. However, the Gorge is also one of the main sites of *Sorbus* diversity in the UK, so considerable concern was expressed about the impact of grazing on *Sorbus* seedling recruitment, and hence the evolution of the genus. Six feral goats were released in June 2011.

Conclusion

Once a plant conservation strategy has been developed, people must be educated about the strategy. Environmental education is a challenge since it covers activities ranging from 6-year-old children visiting a botanic garden through university students to courses for those returning to study later in life (Clayton & Myers 2009). Raising public awareness of conservation issues, and the need to live sustainably, is rarely questioned yet there is little evidence that educational strategies change attitudes and practical behaviour (Dillon et al. 2006; Heimlich & Ardoin 2008). There is extraordinary resistance among British gardeners to worry about non-native invasive species, plants, animals and diseases outside their gardens. The arrival of a new species in cultivation is celebrated as a great event, whereas if it becomes yet another garden plant escape, it may herald habitat transformation in some non-domestic ecosystem. In some regions, the effect of non-native species is merely controversial (Pearman & Walker 2009; Thomas & Dines 2010), but in others the effects may be dramatic. For example, in South Africa the introduction of *Acacia saligna* from Australia is associated with the lowering of water tables and increasing wildfire burn temperatures (Cronk & Fuller 2001). In *On the origin of species*, Darwin observed that non-native species were a problem in the Cape but comforted himself with the thought that no species had become extinct due to an invasion by a non-native. This appears to be no longer true; at least one, *Leucadendron levisianus*, is now extinct in the Cape (Rebelo 2000).

Over the last 20 years plant conservation has been increasingly justified in terms of conservation of ecosystem services, rather than traditional concerns of plants for human food and medicine (the remit of genetic resource conservation). However, there may be a conflict between ecosystem service conservation and the conservation of genetic resources. One appears based on the conservation of many species in an area, the other on specific species. If ecosystem services are dependent on specific species then strategies must be found that combine conservation at all levels of biodiversity. However, if there is no link between ecosystem services and genetic variation then species and genetic conservation strategies should be considered differently to strategies for ecosystem service conservation.

> *'We shall never, probably, disentangle, the inextricable web but when we have a distinct object in view we may hope to make sure but slow progress'*
>
> **Darwin (1859)**

References

Adams, W.M., Aveling, R., Dickson, B.D., *et al.* (2004) Biodiversity conservation and the eradication of poverty. *Science*, **306**, 1146.

Balmford, A., Mace, G.A. & Ginsberg, J.R. (1998) The challenges of conservation: putting processes on the map. In: *Conservation in a Changing World* (eds G.A. Mace, A. Balmford & J.R. Ginsberg). Cambridge University Press, Cambridge.

Baskin, C.C. & Baskin, J.M. (2001) *Seeds: Ecology, Biogeography, and Evolution of Dormancy and Germination*. Academic Press, San Diego, CA.

Bebber, D.P., Carine, M.A., Wood, J.R., *et al.* (2010) Herbaria are a major frontier for species discovery. *Proceedings of the National Academy of Sciences USA*, **107**, 22169–22171.

CBOL Plant Working Group (2009) A DNA barcode for land plants. *Proceedings of the National Academy of Sciences USA*, **106**, 12794–12797.

Chamberlain, J.R. (1998) Isozyme variation in *Calliandra calothyrsus* (Leguminosae): its implications for species delimitation and conservation. *American Journal of Botany*, **85**, 37–47.

Chiwocha, S.D., Dixon, K.W., Flematti, G.R., *et al.* (2009) Karrikins: a new family of plant growth regulators in smoke. *Plant Science*, **177**, 252–256.

Clayton, S. & Myers, O.G. (2009) *Conservation Psychology: Understanding and Promoting Human Care for Nature*. Cambridge University Press, Cambridge.

Coyne, J.A. & Orr, H.A. (2004) *Speciation*. Sinauer Associates Inc, Sunderland, MA.

Crandall, K.A., Bininda, E.O., Mace, G.M. & Wayne, R.K. (2000) Considering evolutionary processes in conservation biology. *Trends in Ecology and Evolution*, **15**, 290–295.

Cribb, P.J., Kell, S.P., Dixon, K.W. & Barrett, R.L. (2003) *Orchid conservation*: a global perspective. In: Orchid Conservation (eds K.W. Dixon, S.P. Kell, R.L. Barrett & P.J. Cribb). Natural History Publications, Sabah.

Cronk, Q.C. & Fuller, J.L. (2001) *Plant Invaders: The Threat to Natural Ecosystems*. Earthscan, London.

Darwin, C.R. (1856) *Letter 2022. Letter to Sir Joseph Hooker, 24 Dec [1856]*. Darwin Correspondence Project Database.

Darwin, C.R. (1859) *On the Origin of Species by Means of Natural Selection, Or the Preservation of Favoured Races in the Struggle for Life*. John Murray, London.

De-Zhu, L. & Pritchard, H.W. (2009) The science and economics of *ex situ* plant conservation. *Trends in Plant Science*, **14**, 614–621.

Dillon, J., Rickinson, M., Teamey, K., *et al.* (2006) The value of outdoor learning: evidence from research in the UK and elsewhere. *School Science Review*, **87**, 107–111.

Dixon, K.W. (2009) Pollination and restoration. *Science*, **325**, 571–573.

Dutech, C., Maggia, L., Tardy, C., Joly, H.I. & Jarne, P. (2003) Tracking a genetic signal of extinction-recolonization events in a neotropical tree species: *Vouacapoua americana* Aublet in French Guiana. *Evolution*, **57**, 2753–2764.

Dutra, D., Kane, M., Adams, C. & Richardson, L. (2009) Reproductive biology of *Cyrtopodium punctatum in situ*: implications for conservation of an endangered Florida orchid. *Plant Species Biology*, **24**, 92–103.

Ebach, M.C. & Holdrege, C. (2005) DNA barcoding is no substitute for taxonomy. *Nature*, **434**, 697.

Ennos, R. A., French, G.C. & Hollingsworth, P.M. (2005) Conserving taxonomic complexity. *Trends in Ecology and Evolution*, **20**, 164–168.

Fahrig, L. (2003) Effects of habitat fragmentation on biodiversity. *Annual Review of Ecology, Evolution and Systematics*, **34**, 487–515.

Faith, D.P. (2010) Biodiversity transcends services. *Science*, **330**, 1744–1745.

Fazekas, A.J., Kesanakurti, P.R., Burgess, K.S. (2009) Are plant species inherently harder to discriminate than animal species using DNA barcoding markers? *Molecular Ecology Resources*, **9**, 130–139.

Ferrier, S. & Drielsma, M. (2010) Synthesis of pattern and process in biodiversity conservation assessment: a flexible whole-landscape modelling framework. *Diversity and Distributions*, **16**, 386–402.

Fleming, T.H. & Muchhala, N. (2008) Nectar-feeding bird and bat niches in two worlds: pantropical comparisons of vertebrate pollination systems. *Journal of Biogeography*, **35**, 764–780.

Frankel, O.H., Brown, A.H. & Burdon, J.J. (1995) *The Conservation of Plant Biodiversity*. Cambridge University Press, Cambridge.

Frankham, R., Ballou, J.D. & Briscoe, D.A. (2010) *Introduction to Conservation Genetics*. Cambridge University Press, Cambridge.

Hadley, A. & Betts, M. (2009) Tropical deforestation alters hummingbird movement patterns. *Biology Letters*, **5**, 207–210.

Hannah, L., Midgley, G., Lovejoy, T., *et al.* (2002) Conservation of biodiversity in a changing climate. *Conservation Biology*, **16**, 264–268.

Hannah, L., Midgley, G., Andelman, S., *et al.* (2007) Protected area needs in a changing climate. *Frontiers in Ecology and the Environment*, **5**, 131–138.

Hansen, A.K., Escobar, L.K., Gilbert, L.E. & Jansen, R.K. (2007) Paternal, maternal, and biparental inheritance of the chloroplast genome in *Passiflora* (Passifloraceae): implications for phylogenetic studies. *American Journal of Botany*, **94**, 42–46.

Harlan, J.R. (1992) *Crops and Man*. American Society of Agronomy, Madison, WI.

Harris, S.A. & Ingram, R. (1991) Chloroplast DNA and biosystematics: the effects of intraspecific diversity and plastid transmission. *Taxon*, **40**, 393–412.

Havens, K., Vitt, P., Maunder, M., Guerrant, E.O. & Dixon, K.W. (2006) *Ex situ* plant conservation and beyond. *Bioscience*, **56**, 525–531.

Heimlich, J.E. & Ardoin, N.M. (2008) Understanding behavior to understand behavior change: a literature review. *Environmental Education Research*, **14**, 215–237.

Höglund, J. (2009) *Evolutionary Conservation Genetics*. Oxford University Press, Oxford.

Holderegger, R. & Wagner, H.H. (2008) Landscape genetics. *Bioscience*, **58**, 199–207.

Hollingsworth, P.M. (2008) Progress and outstanding questions. *Heredity*, **101**, 1–2.

Hopper, S.D., & Gioia, P. (2004) The Southwest Australian Floristic Region: evolution and conservation of a global diversity hotspot. *Annual Review of Ecology, Evolution and Systematics*, **35**, 623–650.

Houston, L., Robertson, A., Jones, K., Smith, S.C., Hiscock, S.J. & Rich, T.C. (2009) An account of the whitebeams (*Sorbus* L., Rosaceae) of Cheddar Gorge, England, with description of three new species. *Watsonia*, **27**, 283–300.

Klein, C., Wilson, K., Watts, M., *et al.* (2009) Incorporating ecological and evolutionary processes into large-scale conservation planning. *Ecological Applications*, **19**, 206–217.

Koh, L P., Dunn, R.R., Sodhi, N.S., Colwell, R.K., Proctor, H.C. & Smith, V.S. (2005) Species coextinctions and the biodiversity crisis. *Science*, **305**, 1632–1634.

Kremen, C., Cameron, A., Moilanen, A., *et al.* (2008) Aligning conservation priorities across taxa in Madagascar with high-resolution planning tools. *Science*, **320**, 222–226.

Kress, W.J., Wurdack, K.J., Zimmer, E.A., Weigt, L.A. & Janzen, D.H. (2005) Use of DNA barcodes to identify flowering plants. *Proceedings of the National Academy of Sciences USA*, **102**, 8369–8374.

Lahaye, R., van der Bank, M., Bogarin, D., *et al.* (2008) DNA barcoding the floras of biodiversity hotspots. *Proceedings of the National Academy of Sciences USA*, **105**, 2923–2928.

Lander, T.A., Boshier, D.H. & Harris, S.A. (2010) Fragmented but not isolated: contribution of single trees, small patches and long-distance pollen flow to genetic connectivity for *Gomortega keule*, an endangered Chilean tree. *Biological Conservation*, **143**, 2583–2590.

Lander, T.A., Bebber, D.P., Choy, C.T., Harris, S.A. & Boshier, D.H. (2011) The Circe Principle explains how resource-rich land can waylay pollinators in fragmented landscapes. *Current Biology*, **21**, 1–6.

Lomov, B., Keith, D. & Hochuli, D. (2010) Pollination and plant reproductive success in restored urban landscapes dominated by a pervasive exotic pollinator. *Landscape and Urban Planning*, **96**, 232–239.

Lowe, A.J., Harris, S.A. & Ashton, P.A. (2004) *Ecological Genetics: Design, Analysis and Application*, Blackwell Publishing, Oxford.

Maxted, N., Ford-Lloyd, B.V. & Hawkes, J.G. (1997) *Plant Genetic Conservation. The in situ Approach*. Chapman and Hall, London.

Menz, M.H., Phillips, R.D., Winfree, R., *et al.* (2011) Reconnecting plants and pollinators: challenges in restoration of pollination mutualisms. *Trends in Plant Science*, **16**, 4–12.

Merritt, D.J. & Dixon, K.W. (2011) Restoration seed banks – a matter of scale. *Science*, **332**, 424–425.

Moritz, C. & Faith, D.P. (1998) Comparative phylogeography and the identification of genetically divergent areas for conservation. *Molecular Ecology*, **7**, 419–429.

Myers, N. & Cowling, R. (1999) Mediterranean Basin. In: *Hotspots – Earth's Biologically Richest and most Endangered Terrestrial Ecoregions* (eds R.A. Mittermeier, N. Myers & C.G. Mittermeier). CEMEX and Conservation International, Mexico City.

Nelleman, C. & Corcoran, E. (2010) *Dead Planet, Living Planet – Biodiversity and Ecosystem Restoration for Sustainable Development. A Rapid Response Assessment*. GRID-Arendal, United Nations Environment Programme Office, New York.

Ogden, R., McGough, H.N., Cowan, R.S., Chua, L., Groves, M. & McEwing, R. (2008) SNP-based method for the genetic identification of ramin *Gonostylus* spp. timber and products: applied research meeting CITES enforcement needs. *Endangered Species Research*, **9**, 255–261.

Ogden, R., Dawnay, N. & McEwing, R. (2009) Wildlife DNA forensics – bridging the gap between conservation genetics and law enforcement. *Endangered Species Research*, **9**, 179–195.

Ollerton, J., Killick, A., Lamborn, E., Watts, S. & Whiston, M. (2007) Multiple meanings and modes: on the many ways to be a generalist flower. *Taxon*, **56**, 717–728.

Ollerton, J., Alarcon, R., Waser, N.M., *et al.* (2009) A global test of the pollination syndrome hypothesis. *Annals of Botany*, **103**, 1471–1480.

Paton, A.J., Brummitt, N., Govaerts, R., *et al.* (2008) Towards Target 1 of the Global Strategy for Plant Conservation: a working list of all known plant species – progress and prospects. *Taxon*, **57**, 602–611.

Pearman, D. & Walker, K. (2009) Alien plants in Britain – a real or imagined problem? *British Wildlife*, **21**, 22–27, 150.

Pennisi, E. (2010a) Filling gaps in global biodiversity estimates. *Science*, **330**, 24.

Pennisi, E. (2010b) Tending the global garden. *Science*, **330**, 1274–1277.

Petit, R.J., Csaikl, U.M., Bordacs, S., *et al.* (2002) Chloroplast DNA variation in European white oaks: phylogeography and patterns of diversity based on data from over 2600 populations. *Forest Ecology and Management*, **156**, 5–26.

Phillips, R.D., Faast, R., Bower, C.C., Brown, G.R. & Peakall, R. (2009) Implications of pollination by food and sexual deception for pollinator specificity, fruit set, population genetics and conservation of *Caladenia* (Orchidaceae). *Australian Journal of Botany*, **57**, 287–306.

Phillips, R.D., Hopper, S.D. & Dixon, K.W. (2010) Pollination ecology and the possible impacts of environmental change in the southwest Australian biodiversity hotspot. *Philosophical Transactions of the Royal Society B: Biological Sciences*, **365**, 517–528.

Pitman, N.C. & Jørgensen, P.M. (2002) Estimating the size of the world's threatened flora. *Science*, **298**, 989.

Ramsay, M.M. & Stewart, J. (1998) Re-establishment of the Lady's Slipper Orchid (*Cypripedium calceolus* L.) in Britain. *Botanical Journal of the Linnean Society*, **126**, 173–181.

Rebelo, A.G. (2000) *Proteas of the Cape Peninsula*. Protea Atlas Project, National Botanical Institute, Cape Town.

Remy, W., Taylor, T.N., Hass, H. & Kerp, H. (1994) 4 hundred million year old vesicular–arbuscular mycorrhizae. *Proceedings of the National Academy of Sciences USA*, **91**, 11841–11843.

Rich, T.C., Charles, C.A., Houston, L. & Tillotson, A. (2009) The diversity of *Sorbus* L. (Rosaceae) in the Lower Wye Valley. *Watsonia*, **27**, 301–313.

Rich, T.C., Houston, L., Robertson, A. & Proctor, M. (2010) *Whitebeams, Rowans and Service Trees of Britain and Ireland. A Monograph of British and Irish Sorbus L. B.S.B.I. No. 14*. Botanical Society of the British Isles, London.

Robertson, A., Newton, A.C. & Ennos, R.A. (2004) Multiple hybrid origins, genetic diversity and population genetic structure of two endemic *Sorbus* taxa on the Isle of Arran, Scotland. *Molecular Ecology*, **13**, 123–134.

Sabatino, M., Maciera, N. & Aizen, M.A. (2010) Direct effects of habitat area on interaction diversity in pollination webs. *Ecological Applications*, **20**, 1491–1497.

Sachs, J.D., Baillie, J.E., Sutherland, W.J., *et al.* (2009) Biodiversity conservation and the Millennium Development goals. *Science*, **325**, 1502–1503.

Scotland, R.W. & Wortley, A.H. (2003) How many species of seed plant are there? *Taxon*, **52**, 101–104.

Seaton, P.T. (2007) Orchid conservation: where do we go from here? *Lankesteriana*, **7**, 13–16.

Smith, R.D., Dickie, J.B., Linington, S.H., Pritchard, H.W. & Probert, R.J. (2003) *Seed Conservation: Turning Science into Practice*, Kew Publishing, Kew, London.

Smith, Z.F. (2006) *Developing a Reintroduction Program for the Threatened Terrestrial Orchid Diuris Fragrantissima*. School of Resource Management, Faculty of Land and Food Resources, University of Melbourne.

Sork, V.L. & Waits, L. (2010) Contributions of landscape genetics – approaches, insights, and future potential. *Molecular Ecology*, **19**, 3489–3495.

Stokstad, E. (2010) Despite progress, biodiversity declines. *Science*, **329**, 1272.

Swarts, N.D. & Dixon, K.W. (2009) Orchid conservation in the age of extinction. *Annals of Botany*, **104**, 543–556.

Thomas, S. & Dines, T. (2010) Non-native invasive plants in Britain – a real, not imagined, problem. *British Wildlife*, **21**, 177, 225.

Thompson, P. (2010) *Seeds, Sex and Civilization: How the Hidden Life of Plants has Shaped our World*. Thames and Hudson, London.

Tremblay, R.L., Ackerman, J.D., Zimmerman, J.K. & Calvo, R.N. (2005) Variation in sexual reproduction in orchids and its evolutionary consequences: a spasmodic journey to diversification. *Biological Journal of the Linnean Society*, **84**, 1–54.

Van der Heijden, M.G., Klironomos, J.N., Ursic, M., *et al.* (1998) Mycorrhizal fungal diversity determines plant biodiversity, ecosystem variability and productivity. *Nature*, **396**, 69–72.

Van Straalen, N.M. & Roelofs, D. (2006) *An Introduction to Ecological Genetics*. Oxford University Press, Oxford.

Walck, J. & Dixon, K.W. (2009) Time to future-proof plants in storage. *Nature*, **462**, 721.

Wilmer, P. (2011) *Pollination and Floral Ecology*. Princeton University Press, Princeton, NJ.

IV

Safeguarding the future

18

The 'why', 'what' and 'how' of monitoring for conservation

Julia P.G. Jones[1], Gregory P. Asner[2], Stuart H.M. Butchart[3] and K. Ullas Karanth[4]

[1]School of Environment, Natural Resources and Geography, Bangor University, Bangor, UK
[2]Department of Global Ecology, Carnegie Institution for Science, Stanford, CA, USA
[3]BirdLife International, Cambridge, UK
[4]Wildlife Conservation Society, Centre for Wildlife Studies, India

'..monitoring can be an inefficient use of scarce conservation funding, it also can become a form of political and intellectual displacement behavior, or worse, a deliberate delaying tactic'

(James D. Nichols and Byron K. Williams (2006))

Introduction

The importance of monitoring to conservation is universally recognized. It is a core activity of conservation biology (Marsh & Trenham 2008), consuming, for example, more than 10% of the budgets of agencies charged with managing national biodiversity in Australia and the US (McDonald-Madden et al. 2010). There are innumerable books, reports and peer-reviewed journal articles on the subject, and yet an increasing number of voices have been questioning the way in which monitoring is carried out and the resulting value for conservation (Yoccoz et al. 2001; Nichols & Williams 2006; Lindenmayer & Likens 2010). These authors do not doubt that effective monitoring is needed for effective conservation, but have concerns about what Yoccoz et al. (2001) call the 'why', 'what' and 'how' of monitoring. They warn that poorly designed monitoring can be worse than useless, as it may result in poor decision making and divert conservation resources from other activities.

Monitoring can be defined as the process of gathering information about a state variable (such as the population size of a threatened species, habitat condition, forest cover or the distribution of an invasive pest) to assess the state of the system and draw inferences about changes

Key Topics in Conservation Biology 2, First Edition. Edited by David W. Macdonald and Katherine J. Willis.

over time (Yoccoz et al. 2001). It is a diverse activity covering spatial scales ranging from a single locality to the globe. It can use data from field surveys by professional ecologists, observations by citizen scientists (see Chapter 8), commercial catches of harvested species, remote sensing by orbiting satellites, and opportunistic reporting by members of the public. In this chapter we offer a path through the rapidly expanding literature on monitoring in the context of conservation. We illustrate the importance of having a clearly articulated reason to monitor, discuss questions of what to monitor, and describe how analytical and technical advances are changing the landscape of conservation monitoring.

Why monitor?

Reasons for conservationists to be interested in monitoring

The reasons for setting up a monitoring programme have been classified many different ways. We follow Jones et al. (2011) in considering them as lying along a spectrum (Figure 18.1) from knowledge focused (the information collected has no direct link to management actions but allows learning) to action focused (the information collected can be applied directly to management action).

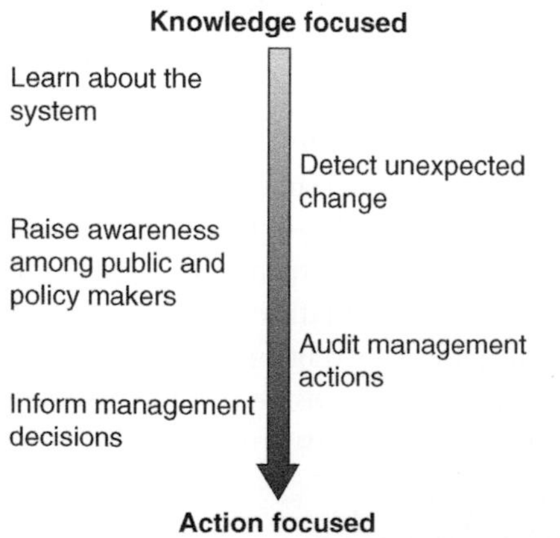

Figure 18.1 The spectrum of reasons to monitor biodiversity. Adapted from Jones et al. 2011.

To learn about the system

Monitoring, over a period of change, can be an important way to improve our understanding of complex interactions in natural systems. For example, the Biological Dynamics of Forest Fragments Project was set up in the Brazilian Amazon to investigate the effect of fragment size and isolation on the composition of animal and plant communities in forest fragments. The long-term monitoring of the fragments over more than 25 years answered many of the original questions, informed the design of protected areas and provided a much deeper understanding of the ecology of forest species (Laurance et al. 2002).

To detect unexpected change

Long-term ecological monitoring can be extremely useful in ways that could not have been foreseen. For example, aggregated data from 153 tropical forest plots, which were established for a variety of different reasons and monitored for more than 20 years, have provided valuable insights into the role of tropical forests as carbon sinks (Phillips et al. 1998).

To raise awareness among the public and policy makers

Monitoring data are also vitally important in raising the profile of conservation issues among the general public. For example, many thousands of Americans are involved every year in the Christmas Bird Counts (Silvertown 2009) which also generate significant media coverage, increasing awareness of biodiversity trends among the general public and policy makers. Similar citizen science bird monitoring schemes now operate in over 160 countries (www.worldbirds.org). The importance of public engagement with the cause of conservation cannot be overestimated, as it creates the necessary political and media pressure on decision makers to ratify international treaties, and to pass and enforce relevant national legislation.

To audit management actions

The possibility of future scrutiny is important in motivating those in positions of power. To give just three examples, it might apply to governments being held to account by NGOs when national biodiversity commitments are broken, government agencies ensuring a private company fulfils any statutory responsibilities with respect to local impact on biodiversity, or a land owner proving they have delivered environmental benefits promised in exchange for payments under an agri-environment scheme. All such auditing requires effective monitoring. Information on the successes and failures of past interventions is also important to improve decision making in the future; there have been suggestions that conservationists have been poor at building up and using an evidence base about the efficacy of potential interventions (Sutherland et al. 2004).

Inform management decisions

One of the most commonly stated objectives for monitoring programmes is to drive management decisions. Knowledge of the state of the system may be required to trigger a given management action (e.g. a manager may want to know if a wild population has fallen below a certain size before instigating captive breeding or reintroduction) or to provide information on which to decide between competing options (such as changing grazing intensity or burning regimes). The principle that monitoring should be part of the management process is well established, although adaptive management, which places an explicit value on learning about the effectiveness of management by monitoring its outcomes, is still not as widely applied as it could be (Nichols & Williams 2006; Lindenmayer & Likens 2009).

Optimizing monitoring

So, there are many valid reasons why monitoring may be valuable but monitoring is costly and in a resource-constrained world, conservation biologists have to make difficult decisions between monitoring and investment in other activities. The cost of monitoring and the value of the improved decision making which monitoring will allow therefore need to be explicitly considered. In rare cases, the best use of conservation resources may be not to carry out monitoring at all. This is more likely to be true if the cost of monitoring is particularly high, if the risk of incorrect inference about a trend is high and if relatively low-cost management options exist which can be implemented without information from monitoring. For example, Field et al. (2004) argue that the relevant authorities in Coffs Harbour, New South Wales, Australia, with an important koala population, should invest in management to safeguard koala (e.g. through protecting habitat) rather than in monitoring to determine trends. This is because robust monitoring is costly and there would remain the risk of inferring that the population was stable when in fact it was declining and therefore failing to intervene when intervention was necessary. This would be potentially a very expensive error for an area economically dependent on koala-based tourism.

Work on optimal monitoring has focused on monitoring to inform very specific decisions (e.g. how long to monitor a recovering fishery in order to set limits for a population model for determining the fraction of the stock to include in a marine reserve; Gerber et al. 2005). Critics have said that it is not possible to know the benefits of wider monitoring programmes in advance, making formal optimization impossible. There are many examples of long-term monitoring data being valuable for a reason unforeseen by the instigators of the monitoring programme (for example, monitoring of badgers in Wytham Wood over 25 years produced data that were later found to be important for understanding the effect of climate change on populations; Macdonald et al. 2010). However, even where the costs and benefits of monitoring cannot be formally estimated, thinking explicitly about why monitoring is needed in a given situation, how costly it is, and how those resources could otherwise be spent should result in a more efficient allocation of limited conservation resources (Salzer & Salafsky 2006).

What to monitor?

Pressure, state, impact or benefit?

The DPSIR framework was developed to describe the interactions between environmental problems, such as biodiversity loss, and society. **D**riving forces (such as population growth or economic development) exert **P**ressures on the environment (e.g. overexploitation or habitat fragmentation) and as a consequence, the **S**tate of the environment changes (e.g. species become threatened with extinction). This leads to **I**mpacts (through the loss of ecosystem services underpinned by biodiversity) which may elicit a societal **R**esponse that feeds back on driving forces, pressures or state (Maxim et al. 2009). The DPSIR framework is now increasingly simplified to a Pressure–State–Response–Benefits framework, wherein drivers are combined with pressures and impacts are renamed 'benefits', to emphasize the direct relevance of biodiversity to decision makers (Butchart et al. 2010; Sparks et al. 2011). This framework is useful for considering what conservationists should monitor.

Pressures

A useful measure at large geographic scales of the underlying general human pressures on the environment through resource consumption is provided by the 'ecological footprint' (Wackernagel et al. 2002) but monitoring of more specific pressures on biodiversity is also important for understanding trends in threats. Direct use and exploitation of species is one of the most immediate pressures. However, there is relatively little systematic monitoring of this, apart from for particular species (e.g. legal international trade in species listed on CITES), products (e.g. ivory, timber), sectors (e.g. fisheries) or markets (e.g. some Asian wildlife markets). Monitoring of the pressures from agriculture, mining and transport is increasingly possible using remote sensing. For example, excellent, high-resolution and rapidly updated (daily) data are now available to monitor fire frequency through the Web Fire Mapper service (http://maps.geog.umd.edu/firms/wms.htm). On the other hand, human impacts on freshwater ecosystems through water management and abstraction are generally not well monitored. For example, although a global map of river fragmentation resulting from dams is available (Nilsson et al. 2005), temporal trends are poorly known. Trends in the arrival and establishment of non-native species are monitored in some countries and at the European scale, but globally our knowledge of the distribution of alien invasive species is patchy (albeit increasing, e.g. through the Global Invasive Species Database and other such compilations) and trends are unquantified (McGeoch et al. 2010). Pollution from a variety of sources is a key pressure on biodiversity, but few are monitored with any degree of robustness at broad scales. Climate change, now well established as a major threat to biodiversity, is quite well monitored at most spatial scales.

State

To assess comprehensively the state of biodiversity would be prohibitively costly: one estimate is that an inventory of all taxa in just one hectare of tropical forest might take 50–500 scientist-years to complete (Lawton et al. 1998). There is therefore considerable interest in finding proxy taxa that can act as indicators for biodiversity more broadly. Lindenmayer & Likens (2010) report that over 55 major taxonomic groups have been proposed as indicators for monitoring programmes, ranging from viruses to flowering plants and virtually all major vertebrate groups, although often it is not clear what they are meant to indicate nor the robustness and representativeness of the trends derived. Monitoring may focus on trends in parameters relating to individual species, such as population density, population size, spatial distribution, survival or recruitment. Such data can be combined into indicators which plot population trends in a defined suite of species (e.g. the Wild Bird Index; Gregory et al. 2005) or a shifting suite of species (the Living Planet Index; Collen et al. 2009), or which estimate how fast species are moving along trajectories towards or away from extinction (the Red List Index; Butchart et al. 2007).

The scale at which such indicators are useful depends on the richness of the data feeding into them. For example, robust national-scale indicators such as the UK's Farmland Bird Indicator (used to indicate the general health of farmland habitats) requires systematic bird population monitoring based on replicated surveys at hundreds of sampling plots, whereas the more opportunistic collation of population time-series that underpin the Living Planet Index means that its strengths lie at global or regional scales (Collen et al. 2009). Monitoring biodiversity at the level of habitats is likely to prove increasingly cost-effective in future, as reliable, repeated, high-resolution satellite imagery becomes more cheaply available. At present, field technology and techniques are advancing so rapidly that newly available, ever-higher resolution datasets may not be comparable with previously published analyses. For studies looking at changes in community structure, the focus of monitoring may be on species richness or some measure of diversity (i.e. combining species richness and the relative abundance of each species), and may focus on taxonomic levels above the species. The choices made will depend on a variety of factors, including the purpose of the monitoring and the resources available.

Responses

Monitoring of the degree of implementation of responses to biodiversity loss is important in order to adapt and target such interventions. Examples of relevant global datasets include the World Database on Protected Areas, which tracks the establishment of new protected areas (Chape et al. 2005), their management effectiveness (Leverington et al. 2010) and the degree to which they cover key sites for biodiversity (Butchart et al. 2010). Policy responses for tackling invasive alien species (McGeoch et al. 2010), the extent of forest areas that are sustainably managed, and the levels of aid provided by developed countries to the developing world for biodiversity conservation are also monitored globally (Butchart et al. 2010; SCBD 2010). At national and local scales, monitoring is patchier overall, but covers a broader suite of responses. For example, the UK government collates data on the proportion of priority species with a biodiversity action plan, the area of land under agri-environment schemes and spending in the UK on biodiversity conservation.

Benefits

Monitoring of trends in the benefits that people derive from biodiversity (ecosystem services) is least well developed. Trends in populations and species that are utilized directly (e.g. for food, medicine or other purposes) have been developed based on the Living Planet Index and Red List Index (Butchart et al. 2010). At a local scale, there are numerous examples of evaluations of a suite of ecosystem services, from tourism revenues (Naidoo et al. 2011) to the contribution that wild food makes to human health (Golden et al. 2011), water flows (Brauman et al. 2007), and pollination (Ricketts et al. 2008). However, it is rarer to find robust long-term monitoring of these benefits over time, and even fewer data are available at a global scale.

Scaling up conservation monitoring from the local to the global

Monitoring effort is unequally distributed, because most effort is focused on large, charismatic species in rich countries (Pereira & Cooper

Box 18.1 Monitoring global biodiversity

Nearly all of the world's governments made a commitment in 2002, through the Convention on Biological Diversity (CBD), 'to achieve by 2010 a significant reduction of the current rate of biodiversity loss at the global, regional and national level as a contribution to poverty alleviation and to the benefit of all life on Earth'. This '2010 target', as it became known, stimulated considerable interest and efforts in monitoring biodiversity trends at the global scale during the latter part of the last decade. The CBD developed a framework of indicators spanning measures of the status of components of biodiversity (population trends, species extinction risk, habitat extent and condition, community composition), sustainable use of biodiversity (extent of sustainable managed resources, ecological footprint), threats to biodiversity (nitrogen deposition, invasive alien species) and ecosystem integrity (water quality, fragmentation), as well as metrics of benefits sharing, traditional knowledge and resourcing of biodiversity conservation. A '2010 Biodiversity Indicators Partnership', encompassing dozens of organizations and agencies, was established to stimulate, co-ordinate and communicate indicator development. By 2010, the results of all these efforts were synthesised in Butchart et al. (2010) and the CBD's flagship report: the *Global Biodiversity Outlook 3* (SCBD 2010). These concluded that measures of the state of biodiversity had continued to decline in recent decades, with no significant deceleration, while pressures had continued to grow. Although responses (policy interventions and actions) had increased, they were not keeping pace with the growing pressures. In short, the world failed to meet the 2010 target.

Encouragingly, and perhaps surprisingly, this failure did not discourage nations from their attempts to tackle the biodiversity crisis. At the 10th meeting of the CBD in Nagoya, Japan, in October 2010, parties to the convention adopted a revised strategic plan for biodiversity, agreed a protocol of sharing access to and benefits from biodiversity surveys (particularly from genetic resources), and committed to developing a means to provide the necessary finances. From a monitoring perspective, the new strategic plan is a considerable improvement. It contains 20 biodiversity targets for 2020, which are much more focused and contain more specific commitments. For example, nations have agreed to increase the terrestrial coverage of protected areas to 17%, halt extinction of known threatened species, restore at least 15% of degraded habitats, and at least halve (and where feasible bring close to zero) the rate of loss of natural habitats.

Monitoring biodiversity is clearly a business that is here to stay. Through the coming months and years, the CBD will define the set of indicators to be used to monitor progress towards these targets, and enhanced efforts will need to be made to establish and expand on-the-ground monitoring schemes, collate and combine existing and historic data, and deliver and communicate robust and informative indicators.

2006). Ecological monitoring where it does occur is a highly decentralized activity in which a given locality or taxon can potentially be covered by a number of different monitoring programmes carried out by a variety of government agencies, local groups and NGOs (Marsh & Trenham 2008). It is important that data collected at a variety of scales by a variety of actors can be integrated in such a way as to allow questions to be answered that might not have been envisaged by those who collected the data originally (Jones 2011) or which are possible only with large data sets covering large spatial-temporal scales. For example, targets for conserving biodiversity are increasingly set globally (Box 18.1) and monitoring that can provide robust inference at the global scale is therefore needed properly to audit progress against these global targets. A prescriptive, top-down global monitoring programme would be prohibitively expensive (Scholes et al. 2008) so the global biodiversity indicators which have been used to measure progress against previous targets have been built up from a wide variety of data sets from around the world (Butchart et al. 2010).

There are a number of challenges to integrating monitoring data from different sources. Once a relevant data set has been identified and located, the data may not be stored in a way which maximizes their usefulness. For example, local place or species names may be used rather than standard co-ordinates and scientific names, and essential information on the sampling protocol used may not be stored with the data, limiting their usefulness. Increasingly, journals publishing ecological research are encouraging authors to archive their data with established databases (Whitlock et al. 2010). Such practice should be widely encouraged, and will be a great resource for future conservationists looking to understand how systems have changed over time. This is particularly important to help avoid the problem of shifting baselines, where knowledge of how things were in the past is lost and the degree of change in a system is underestimated (Papworth et al. 2009). For example, a conservationist scientist may use the population size of a species at the start of her career as a baseline, forgetting that the species had already suffered extreme depletion.

How to monitor?

Statistical approaches

Making the correct inference about a trend is fundamental to the concept of monitoring. Drawing inferences from data is the realm of statistics so an interest in statistics is essential to anyone involved in the design or interpretation of monitoring data.

Statistical power is the probability that an analysis will reject a null hypothesis which is, indeed, false. In the context of ecological monitoring, it is the probability that an analysis will conclude that there is a trend (e.g. the population size of a threatened species is declining), when there is indeed such a trend. Power is positively related to the sample size, the size of the trend and the risk of a false positive, and it is inversely related to the variability in the system (Dytham 2003). This concept is of central importance, as a monitoring programme with low power risks using up conservation resources without providing robust information. For example, Field et al. (2007) show how a simple power analysis carried out on the first 5 years of monitoring data on bird populations in Australia showed that for some rare species, researchers would be unlikely to be able to conclude whether the population size was stable, increasing or decreasing, even after a further 5 years of data collection. Similarly, Sommerville et al. (2011) show that it may not be possible to monitor target species in a biodiversity payment scheme with sufficient power to allow payments to be made based on changes in population or distribution of species, suggesting that monitoring of threats may be more cost-efficient. Up to a point, the more data points (e.g. number of locations, frequency of observations or duration of time series), the more likely that a true trend will be detected. The larger the trend (e.g. a steeper population decline or faster rate of habitat loss), the more likely it is to be detected. The more willing the researchers are to accept the possibility of erroneously concluding there is a trend when there is in fact no trend, the more likely they are to detect a true trend. And finally, the less variability there is in the data (due to natural variability or measurement error), the more likely it is that a true trend will be detected.

The statistical framework upon which power analysis is based (known as null hypothesis testing) was originally developed for analyses of experimental data that were, in principle, unbiased and independent. Such assumptions are rarely met in the real world of monitoring programmes and this, along with the rapid increase in computing power available to most ecologists, is driving a shift towards information-theoretic model selection (Burnham & Anderson 2002; Stephens et al. 2005) and Bayesian model-based approaches. These allow greater flexibility to deal with non-independence of data and other challenges with much greater realism and rigour. For example, hierarchical Bayesian modelling allows researchers to link explicitly the parameter of interest (e.g. animal density) and the observation process involved in the survey, as well as any other ecological or management variables expected to influence the parameter of interest (e.g. hunting pressure, forest type), all within a single modelling framework (Royle & Dorazio 2008) to improve understanding of the influence of potential drivers on system state.

Accounting for detectability

Just because something is present in an area does not mean it will be detected. This fact of 'imperfect detectability' is of central importance in any study monitoring the abundance of a species (Box 18.2), species richness or community composition. For example, if aerial transects recorded much lower numbers per kilometre of elephants in a forested area than in savanna, it would be wrong to conclude that elephant density was higher in the savanna (instead, higher detectability in the savanna may simply mean that a higher proportion of elephants present were detected).

A number of methods have been developed which account for detectability when estimating population size. For example, distance sampling uses the distribution of distances between the surveyor and encounters with the target species to estimate the species' detectability in the habitat (assuming all individuals on the transect line are perfectly detected). This estimate of detectability allows simple counts to be translated into estimates of population size (Buckland et al. 2001). Capture-recapture methods allow estimation of the proportion of individuals in the population that can be captured, and take account of this detectability when estimating population size (Amstrup et al. 2005).

Occupancy modelling is increasingly widely recognized as a valuable approach for monitoring species distribution (Mackenzie & Royle 2005). The basic principle is that a surveyor makes multiple visits to a site and records whether a species was detected or not on each visit. Such replications can even be spatial rather than temporal (Karanth et al. 2011). Using a series of assumptions, the detectability (the probability that the species is detected) and occupancy (the probability that a given location is occupied by the species) can be estimated. Occupancy approaches can be used to monitor trends in species' distributions over time, or even trends in species abundance or evidence of the effectiveness of a pest control campaign. However, despite the value of the approach for large-scale monitoring (Jones 2011), it has not yet been widely taken up by national monitoring schemes (see Box 18.2).

Species richness and the diversity of a community are parameters frequently used in conservation monitoring. However, if simple presence/absence data are used, estimates will be biased because some species are much more readily detectable than others (Yoccoz et al. 2001). Recent papers have investigated the extent of imperfect and variable detectability among species of birds, butterflies and bats, and demonstrated the importance of taking detectability into account when monitoring species richness over time at a site, or making regional comparisons. For example, Kery et al. (2009) showed that failing to account for detectability could result in an overestimate of nearly 50% of the number of butterfly species present in Switzerland.

Free software packages such as DISTANCE (for distance sampling; www.ruwpa.st-and.ac.uk), MARK (for capture-recapture data and occupancy modelling), PRESENCE (for occupancy) and SPECRICH (for species richness) (the latter three all downloadable from www.mbr-pwrc.usgs.gov/software) now make it relatively easy for biologists to fit specialized models, properly accounting for detectability, to their survey data.

The role of remote sensing

The use of remote sensing for ecological monitoring has grown rapidly over the past few decades. With increased access to satellite data, and to software for analysing imagery, many science and conservation organizations are now

Box 18.2 Monitoring one of the world's most charismatic animals: counting India's wild tigers

Monitoring tigers is inherently challenging because they live at low population densities and across a very large area. However, as tigers belong to one of the world's most charismatic species, you might expect that robust monitoring would be in place. Unfortunately, three decades of monitoring wild tigers in India clearly illustrate the challenges of integrating sound conservation monitoring science with policy.

When India's flagship conservation approach, Project Tiger, was launched in 1972, the Indian forestry service developed a monitoring system called the 'pugmark census'. The method assumed that all tigers in India (a country of 3 million km^2 area with ~10% under natural forest cover) could be counted by government staff surveying forests on foot, collecting track prints of tigers and identifying them individually to come up with a total count. The method was implemented faithfully by the hierarchical forest law enforcement machinery and generated 'tiger numbers', which increased from a low of 1800 to ~3500 over the next 30 years. Scientific criticisms of the pugmark census (Karanth et al. 2003) were ignored, probably at least in part because of the common bias against disrupting a long-term data set. Finally in 2004, the pugmark census lost credibility at a time when tiger monitoring was a particularly hot political topic, because tigers continued to be reported from the Sariska Tiger Reserve some years after they probably became extinct there. In 2005, the Tiger Task Force, appointed by the Indian Prime Minister, advised that the pugmark census should be abandoned.

A new 'national tiger estimation' approach was developed by scientists at the Wildlife Institute of India. This was implemented in 2006 at a cost of 120 million rupees (US$2.6 million) and it estimated that approximately 1411 adult tigers remained in the country. The method uses a combination of photographic capture-recapture estimation of tiger densities in a few reserves (Figure 18.2), calibrated against tiger sign encounter rates derived from trail surveys over a larger spatial scale (Jhala et al. 2011). Although this sampling-based approach is a marked improvement over the pugmark census, sampling frequencies and intensities may be insufficient to provide reliable estimations of population size, survival and recruitment in individual populations. Furthermore, it makes extrapolations of tiger densities based on standard regressions rather than occupancy modelling. Despite these deficiencies, a second country-wide exercise has recently been undertaken and has reported 'increased tiger numbers' to 1636, a 16% increase in 4 years.

Getting tiger monitoring right is an important conservation issue, because national monitoring uses up limited resources and so carries a grave responsibility to provide the information needed for management. There remains concern that the national scheme may not be of sufficient temporal resolution to pick up sudden declines or changes in survival rates in key tiger source populations in time for action to be taken (Karanth et al. 2011).

Figure 18.2 A tiger caught in a camera trap. Camera trap mark-recapture is costly but does not disturb the tigers and can be used to estimate population size robustly. © Ullas Karanth/WCS.

using remote sensing to map ecosystem extent (e.g. forest cover) and changes associated with land use such as deforestation (e.g. www.imazon.org.br). Conservationists have long recognized that monitoring ecosystem extent can be used to infer threats to species, using the well-understood relationship between species richness and area. However, recent technical advances have made it possible to go far beyond inferences based on assumed simple relationships, and directly to map and monitor the dynamics of ecological disturbance or degradation, vegetation canopy structure and biomass, and even plant species composition (Chambers et al. 2007; Box 18.3). Here we update and expand on these and other themes to cover the most promising approaches using advanced remote sensing techniques for conservation monitoring.

Ecological disturbance and degradation

A key determinant of species composition and abundance is the type, spatial patterning and temporal frequency of disturbance. Disturbance regimes are abundant in nature, ranging from large size but low-frequency events, such as periodic hurricanes, to small size and high-frequency events such as gap-phase tree turnover in forests. Human-mediated disturbance, such as fire and selective logging, has increased in virtually all ecosystems. Remote sensing can provide valuable information about changes in disturbance. Mainstream satellite sensors, such as Landsat Thematic Mapper, are now routinely analysed for forest disturbances associated with logging and fire at quite fine scale (down to 30 m × 30 m pixels) (Asner et al. 2005; Shearman et al. 2009). This information has, in turn, informed conservation plans and improved environmental law enforcement. There is also progress in using other technologies, such as radar and LIDAR (see below) to monitor ecological disturbance (see Chambers et al. 2007 for a helpful review).

Structure and habitat

There continue to be rapid scientific advances in the use of remote sensing for mapping ecosystem structure and habitat. Here structure refers to the three-dimensional architecture of the system – the size, shape and orientation of canopies, branches and foliage. Of all tested technologies, none has come close to the results obtained with airborne (from aircraft) Light Detection and Ranging (LiDAR), which allows for 3-D imaging of vegetation and terrain. LiDAR has been repeatedly used to map canopy structure to better understand which habitats are occupied by birds in temperate to tropical systems (e.g. see Box 18.3), and this has revolutionized efforts to establish conservation goals based on habitat (Vierling et al. 2008). Recently, a new satellite-based LiDAR was used for habitat suitability analysis (Bergen et al. 2009). Although still in its infancy and not yet globally available, this type of 3-D remote sensing is likely to make major contributions to conservation monitoring in the near future.

Species composition

A holy grail in remote sensing is the identification and mapping of the species composition, richness and abundance of vegetation, but this has proved difficult to achieve. Whether or not individual plant species can be identified within a habitat using remote sensing depends upon factors ranging from technological (e.g. type of remote sensing measurement, spatial and spectral resolution, temporal resolution, etc.) to biological (e.g. phylogeny, chemical traits, structural properties). However, progress has been made, for example in the context of the detection and management of invasive species (Huang & Asner 2009; see Box 18.3). Imaging spectroscopy, also known as hyperspectral imaging, is an emerging technology making an important contribution in this area (see Ustin & Gamon 2010 for a review).

Remote sensing is well recognized for its role in monitoring at large spatial scales where

Box 18.3 Monitoring habitat use by tropical forest species with remote sensing

Tropical forests harbour tall, spatially complex canopies that have proved a challenge to measure and monitor from the ground. Airborne ecological remote sensing can, when combined with field observations and telemetry, drive breakthroughs in understanding and monitoring of forest habitat use by species of conservation concern. Airborne Light Detection and Ranging (LiDAR) is a key technology. Some systems can produce three-dimensional images of forest canopies, thereby allowing for detailed monitoring of habitat in both horizontal and vertical space (Figure 18.3). In Hawaiian rainforests, airborne LiDAR has provided spatially detailed information on how forest fragmentation alters canopy volume and maximum tree height; both were negatively correlated with remaining forest fragment area. In turn, remaining canopy volume and maximum height were positively related to the richness of remaining endangered native Hawaiian bird species (Flaspohler et al. 2010). Although the relationships between species richness and canopy structure were somewhat linear, LiDAR allowed for identification of a minimum forest fragment size (~0.7 ha) beyond which avian species richness drops steeply. In another Hawaiian study, LiDAR was used to map the effects of invasions by multiple exotic plant species on native forest canopy, which indicated changes in the fundamental structure of forests and the habitat they provide (Asner et al. 2008). Combined with bioacoustical and observer counts of bird richness and abundance, Boelman et al. (2007) then found that total avian abundance, and the ratio of native to exotic avifauna, were highest in habitats with the most forest canopy closure and tallest trees. The spatially explicit, co-located data from LiDAR and field studies facilitated a new discovery that native birds provide biotic resistance against invasion by exotic trees by successfully competing for space in the remaining native canopies, thus slowing invasion 'meltdowns' that otherwise radically restructure Hawaiian ecosystems.

These and many other advances, such as a second advanced remote sensing technology called high-fidelity imaging spectroscopy (HiFIS), are changing views on the ecology of spatially complex ecosystems and are helping conservation managers and decision makers to improve their plans for managing lands to ensure biodiversity protection and recovery.

Figure 18.3 Airborne LiDAR data from the Carnegie Airborne Observatory (CAO; http://cao.ciw.edu) indicate the detailed three-dimensional structure of tropical forest canopies such as the one shown here from the Colombian Amazon. These data are produced by emitted laser light at ultra-high pulse rates from an overflying aircraft. The light penetrates the canopy to the ground level, recording interactions with foliage and woody tissues. The measurements are made at resolutions ranging from 0.5 to 1.0 meters, but the laser is sensitive to the presence of canopy materials within each laser beam cross-section, thereby allowing for branch-level detail.

ground-based monitoring would be prohibitively expensive or even impossible. Increasingly, new and emerging techniques will become accessible and more widely used for conservation monitoring at much finer spatial and temporal scales. However, there is a gap between what *can* be remotely sensed and the data currently available to the conservation science and policy community. For example, at a global scale we still lack a system that reports the extent of habitats at relatively frequent intervals (e.g. annually).

Who should monitor?

Many developed countries have formal monitoring programmes that collect systematic information on the state of the environment. For example, the UK's Countryside Survey reports on the status of soil, water and vegetation at approximately 10-year periods, and Biodiversity Monitoring Switzerland surveys plants, butterflies and birds annually. However, such schemes are costly to operate and, because of the limited resources available for conservation, can only ever have a limited focus. Monitoring that harnesses the energies and interest of non-professionals therefore plays a vital role. An advantage of involving non-professionals is that more data can be collected for lower cost. For example, Schmeller et al. (2009) found that volunteer biodiversity monitoring in five European countries resulted in 148,000 person-days/year of effort for a total cost of only €4 million, much less than it would have cost to use paid professionals to produce the same data. Another advantage is that schemes involving non-professionals are a means of promoting public engagement, which has a value to conservation well beyond that of the data collected. Citizen science projects, in which volunteers contribute to data collection or analysis, have a long tradition in natural history and ecology (see Chapter 8). The majority are in high-income countries where people may have more available leisure time to volunteer, although systems for capturing data on bird observations are now operating in most countries of the world, usually in local languages (www.worldbirds.org and http://ebird.org/content/ebird). Non-avian examples include a recent study of tiger distribution in India that generated robust inferences from data collected by volunteer naturalists (Karanth et al. 2011).

Participatory monitoring, where local stakeholders are involved in both monitoring *and* managing natural resources, with some degree of support from external bodies, is a form of non-professional biodiversity monitoring that is increasingly important in both high- and lower-income countries. The idea behind participatory monitoring is that if local stakeholders are involved in monitoring, the information gathered can be more effectively and efficiently used. A recent review of 104 monitoring schemes found that the degree of local involvement was an important predictor of the speed with which data collected were translated into management action (Danielsen et al. 2010). Participatory monitoring schemes were particularly effective at instigating action at the local scale (e.g. a village or a district). Local stakeholders of course may also be involved in monitoring and managing their biodiversity without any external involvement. Such autonomous local monitoring may range from customary management practices by indigenous groups to self-regulating hunting or fishing clubs in high-income countries (Danielsen et al. 2009).

Conclusions

No monitoring programme is ever perfect; the design will always be the result of a compromise between what would be ideal and what is possible given limited financial resources and time. However, effective monitoring must have a clear purpose, a full understanding of 'why' it is being carried out and what the data are needed for. 'What' should be monitored and 'how' the monitoring should be implemented will depend on the objectives, the context and the resources available, but we hope this chapter highlights some of the considerations which are essential if monitoring is to result in robust inference. We also stress that professional ecologists should not necessarily have a monopoly and that many types of monitoring can benefit enormously from the involvement of non-professionals. To make sure monitoring programmes get the 'why', 'what' and 'how' right requires close collaboration between those with a clear idea of what is needed (whether these are decision makers at the local, national or global scale) and conservation scientists (who can provide the technical and practical skills required). Such collaboration is essential if investment of valuable conservation resources in monitoring is to provide the maximum possible conservation benefit.

> *'..most conservation monitoring is, but need not be, a waste of time'*
>
> **(Colin J. Legg, & Laszlo Nagy (2006))**

References

Amstrup, S.C., McDonald, T. & Manly, B. (2005) *Handbook of Capture-Recapture Analysis*. Princeton University Press, New Jersey.

Asner, G.P., Knapp, D.E., Broadbent, E.N., Oliveira, P.J.C., Keller, M. & Silva, J.N. (2005) Selective logging in the Brazilian Amazon. *Science*, **310**, 480–482.

Asner, G.P., Jones, M.O., Martin, R.E., Knapp, D.E. & Hughes, R.F. (2008) Remote sensing of native and invasive species in Hawaiian forests. *Remote Sensing of Environment*, **112**, 1912–1926.

Bergen, K.M., Goetz, S.J., Dubayah, R.O., *et al.* (2009) Remote sensing of vegetation 3-D structure for biodiversity and habitat: review and implications for lidar and radar spaceborne missions. *Journal of Geophysical Research-Biogeosciences*, **114**.

Boelman, N.T., Asner, G.P., Hart, P.J. & Martin, R.E. (2007) Multi-trophic invasion resistance in Hawaii: bioacoustics, field surveys, and airborne remote sensing. *Ecological Applications*, **17**, 2137–2144.

Brauman, K.A., Daily, G.C., Duarte, T.K. & Mooney, H.A. (2007) The nature and value of ecosystem services: an overview highlighting hydrologic services. *Annual Review of Environmental Resources*, **32**, 67–98.

Buckland, S.T., Anderson, D.R., Burnham, K.P., *et al.* (2001) *Introduction to Distance Sampling: Estimating Abundance of Biological Populations*. Oxford University Press, Oxford.

Burnham, K.P. & Anderson, D.R. (2002) *Model Selection and Multimodel Inference: A Practical Information-Theoretic Approach*. Springer-Verlag, New York.

Butchart, S.H.M., Akçakaya, H.R., Chanson, J., *et al.* (2007) Improvements to the Red List Index. *PLoS ONE*, **2(1)**, e140.

Butchart, S.H.M., Walpole, M., Collen, B., *et al.* (2010) Global biodiversity: indicators of recent declines. *Science*, **328**, 1164–1168.

Chambers, J.Q., Asner, G.P., Morton, D.C., *et al.* (2007) Regional ecosystem structure and function: ecological insights from remote sensing of tropical forests. *Trends in Ecology and Evolution*, **22**, 414–423.

Chape, S., Harrison, J., Spalding, M. & Lysenko, I. (2005) Measuring the extent and effectiveness of protected areas as an indicator for meeting global biodiversity targets. *Philosophical Transactions of the Royal Society B*, **360**, 443–455.

Collen, B., Loh, J., Whitmee, S., McRae, L., Amin, R. & Baillie, J.E.M. (2009) Monitoring change in vertebrate abundance: the Living Planet Index. *Conservation Biology*, **23**, 317–327.

Danielsen, F., Burgess, N.D., Balmford, A., *et al.* (2009) Local participation in natural resource monitoring: a characterization of approaches. *Conservation Biology*, **23**, 31–42.

Danielsen, F., Burgess, N.D., Jensen, P.M. & Pirhofer-Walzl, K. (2010) Environmental monitoring: the scale and speed of implementation varies according to the degree of people's involvement. *Journal of Applied Ecology*, **47**, 1166–1168.

Dytham, C. (2003) *Choosing and Using Statistics*. Wiley-Blackwell, Oxford.

Field, S.A., Tyre, A.J., Jonzen, N., Rhodes, J.R. & Possingham, H.P. (2004) Minimizing the cost of environmental management decisions by optimizing statistical thresholds. *Ecology Letters*, **7**, 669–675.

Field, S.A., O'Connor, P.J., Tyre, A.J. & Possingham, H.P. (2007) Making monitoring meaningful. *Austral Ecology*, **32**, 485–491.

Flaspohler, D.J., Giardina, C.P., Asner, G.P., *et al.* (2010) Long-term effects of fragmentation and fragment properties on bird species richness in Hawaiian forests. *Biological Conservation*, **143**, 280–288.

Gerber, L.R., Beger, M., McCarthy, M.A. & Possingham, H.P. (2005) A theory for optimal monitoring of marine reserves. *Ecology Letters*, **8**, 829–837.

Golden, C.D., Fernald, L.C.H., Brashares, J.S., *et al.* (2011) Benefits of wildlife consumption to child nutrition in a biodiversity hotspot. *Proceedings of the National Academy of Sciences USA*, **108**, 19653–19656.

Gregory, R.D., van Strien, A., Vorisek, P., *et al.* (2005) Developing indicators for European birds. *Philosophical Transactions of the Royal Society B*, **360**, 269–288.

Huang, C. & Asner, G.P. (2009) Applications of remote sensing to alien invasive plant studies. *Sensors*, **9**, 4869–4889.

Jhala, Y., Qureshi, Q. & Gopal, R. (2011) Can the abundance of tigers be assessed from their signs? *Journal of Applied Ecology*, **48**, 14–24.

Jones, J.P.G. (2011) Monitoring species distribution and abundance at the landscape scale. *Journal of Applied Ecology*, **48**, 9–13.

Jones, J.P.G., Collen, B., Atkinson, G., *et al.* (2011) The why, what and how of global biodiversity indicators beyond the 2010 Target. *Conservation Biology*, **25**, 450–457.

Karanth, K.U., Nichols, J.D., Seidensticker, J., *et al.* (2003) Science deficiency in conservation practice: the monitoring of tiger populations in India. *Animal Conservation*, **6**, 141–146.

Karanth, K.U., Gopalaswamy, A.M., Vaidyanathan, N., Nichols, J.D. & MacKenzie, D.I. (2011) Monitoring carnivore populations at the landscape scale: occupancy modelling of tigers from sign surveys. *Journal of Applied Ecology*, **48**, 1048–105.

Kery, M., Royle, J.A., Plattner, M. & Dorazio, R.M. (2009) Species richness and occupancy estimation in communities subject to temporary emigration. *Ecology*, **90**, 1279–1290.

Laurance, W.F., Lovejoy, T.E., Vasconcelos, H.L., *et al.* (2002) Ecosystem decay of Amazonian forest fragments: a 22-year investigation. *Conservation Biology*, **16**, 605–618.

Lawton, J., Bignell, D. & Bolton, B. (1998) Biodiversity inventories, indicator taxa and effects of habitat modification in tropical forest. *Nature*, **391**, 72–76.

Leverington, F., Costa, K.L., Pavese, H., Lisle, A. & Hockings, M. (2010) A global analysis of Protected Area management effectiveness. *Environmental Management*, **46**, 685–698.

Lindenmayer, D.B. & Likens, G.E. (2009) Adaptive monitoring: a new paradigm for long-term research and monitoring. *Trends in Ecology and Evolution*, **24**, 482–486.

Lindenmayer, D.B. & Likens, G.E. (2010) The science and application of ecological monitoring. *Biological Conservation*, **143**, 1317–1328.

Macdonald, D.W., Newman, C., Buesching, C.D., *et al.* (2010) Are badgers 'under the weather'? Direct and indirect impacts of climate variation on European badger (Meles meles) population dynamics. *Global Change Biology*, **16**, 2913–2922.

Mackenzie, D.I. & Royle, J.A. (2005) Designing occupancy studies: general advice and allocating survey effort. *Journal of Applied Ecology*, **42**, 1105–1114.

Marsh, D.M. & Trenham, P.C. (2008) Current trends in plant and animal population monitoring. *Conservation Biology*, **22**, 647–655.

Maxim, L., Spangenberg, J.H. & O'Connor, M. (2009) An analysis of risks for biodiversity under the DPSIR framework. *Ecological Economics*, **69**, 12–23.

McDonald-Madden, E., Baxter, P.W.J., Fuller, R.A., *et al.* (2010) Monitoring does not always count. *Trends in Ecology and Evolution*, **25**, 547–550.

McGeoch, M.A., Butchart, S.H.M., Spear, D., *et al.* (2010) Global indicators of biological invasion: species numbers, biodiversity impact and policy responses. *Diversity and Distributions*, **16**, 95–108.

Naidoo, R., Weaver, L.C., Stuart-Hill, G. & Tagg, J. (2011) Effect of biodiversity on economic benefits from communal lands in Namibia. *Journal of Applied Ecology*, **48**, 310–316.

Nichols, J.D. & Williams, B.K. (2006) Monitoring for conservation. *Trends in Ecology and Evolution*, **21**, 668–673.

Nilsson, C., Reidy, C.A., Dynesius, M. & Revenga, C. (2005) Fragmentation and flow regulation of the world's large river systems. *Science*, **308**, 405–408.

Papworth, S.K., Rist, J., Coad, L. & Milner-Gulland, E.J. (2009) Evidence for shifting baseline syndrome in conservation. *Conservation Letters*, **2**, 93–100.

Pereira, H.M. & Cooper, H.D. (2006) Towards the global monitoring of biodiversity change. *Trends in Ecology and Evolution*, **21**, 123–129.

Phillips, O.L., Malhi, Y., Higuchi, N., *et al.* (1998) Changes in the carbon balance of tropical forests:

evidence from long-term plots. *Science*, **282**, 439–442.

Ricketts, T.H., Regetz, J., Steffan-Dewenter, I., *et al.* (2008) Landscape effects on crop pollination services: are there general patterns? *Ecology Letters*, **11**, 499–515.

Royle, J. & Dorazio, R.M. (2008) *Hierarchical Modeling and Inference in Ecology: The Analysis of Data from Populations, Metapopulations, and Communities.* Academic Press, San Diego, CA.

Salzer, D. & Salafsky, N. (2006) Allocating resources between taking action, assessing status, and measuring effectiveness of conservation actions. *Natural Areas Journal*, **26**, 310–316.

SCBD (2010) *Global Biodiversity Outlook-3.* Convention on Biological Diversity, Montreal.

Schmeller, D.S., Henry, P.Y., Julliard, R., *et al.* (2009) Advantages of volunteer-based biodiversity monitoring in Europe. *Conservation Biology*, **23**, 307–316.

Scholes, R.J., Mace, G.M., Turner, W., *et al.* (2008) Ecology – toward a global biodiversity observing system. *Science*, **321**, 1044–1045.

Shearman, P.L., Ash, J., Mackey, B., Bryan, J.E. & Lokes, B. (2009) Forest conversion and degradation in Papua New Guinea 1972–2002. *Biotropica*, **41**, 379–390.

Silvertown, J. (2009) A new dawn for citizen science. *Trends in Ecology and Evolution*, **24**, 467–471.

Sommerville, M.M., Milner-Gulland, E.J. & Jones, J.P.G. (2011) The challenge of monitoring biodiversity in payment for environmental service interventions. *Biological Conservation*, **144**, 2832–2841.

Sparks, T.H., Butchart, S.H.M., Balmford, A., *et al.* (2011) Linked indicator sets for addressing biodiversity loss. *Oryx*, **45**, 411–419.

Stephens, P.A., Buskirk, S.W., Hayward, G.D. & del Rio, C.M. (2005) Information theory and hypothesis testing: a call for pluralism. *Journal of Applied Ecology*, **42**, 4–12.

Sutherland, W.J., Pullin, A.S., Dolman, P.M. & Knight, T.M. (2004) The need for evidence-based conservation. *Trends in Ecology and Evolution*, **19**, 305–308.

Ustin, S.L. & Gamon, J.A. (2010) Remote sensing of plant functional types. *New Phytologist*, **186**, 795–816.

Vierling, K.T., Vierling, L.A., Gould, W.A., Martinuzzi, S. & Clawges, R.M. (2008) Lidar: shedding new light on habitat characterization and modeling. *Frontiers in Ecology and the Environment*, **6**, 90–98.

Wackernagel, M., Schulz, N.B., Deumling, D., *et al.* (2002) Tracking the ecological overshoot of the human economy. *Proceedings of the National Academy of Sciences USA*, **99**, 9266–9271.

Whitlock, M.C., McPeek, M.A., Rausher, M.D., Rieseberg, L. & Moore, A.J. (2010) Data archiving. *American Naturalist*, **175**, 145–146.

Yoccoz, N.G., Nichols, J.D. & Boulinier, T. (2001) Monitoring of biological diversity in space and time. *Trends in Ecology and Evolution*, **16**, 446–453.

19

Effective conservation depends upon understanding human behaviour

Freya A.V. St John,[1] Aidan M. Keane[2] and Eleanor J. Milner-Gulland[3]

[1]Durrell Institute of Conservation and Ecology, School of Anthropology and Conservation, University of Kent, Canterbury, UK
[2]Department of Anthropology, University College London and Institute of Zoology, London, UK
[3]Department of Life Sciences, Imperial College London, Ascot, UK

'I have striven not to laugh at human actions, not to weep at them, nor to hate them, but to understand them.'

Baruch (Benedict de) Spinoza

Introduction

The United Nation's Convention on Biological Diversity (CBD) obliged 193 signatory nations to achieve a significant reduction in the rate of biodiversity loss by 2010 (CBD 2007) but in 2011 biodiversity loss continues unabated. Human actions threatening biodiversity, such as overexploitation, spread of invasive species, pollution and climate change, show no sign of slowing in the near future (Ehrlich & Pringle 2008). The ultimate drivers of biodiversity loss are a growing human population and rising per capita consumption (van Vuuren & Bouwman 2005). How humans meet increasing global demands for food, energy, timber and other goods will in large part determine the fate of current biodiversity. Modifying human actions to minimize their negative impacts upon biodiversity will be ever more important to conservation in an increasingly crowded world.

Humans influence the environment at a range of spatial, temporal and institutional scales. The behaviour of governments and large companies has a huge influence on the environment. However, in this chapter we focus on the small-scale behaviour of individuals and households in poor rural areas of the developing world because much conservation activity is

Key Topics in Conservation Biology 2, First Edition. Edited by David W. Macdonald and Katherine J. Willis.

focused in these biodiversity-rich places which are under threat. Many conservation projects seek to alter human behaviour, for example by restricting resource extraction (Gelcich et al. 2005) or reducing poaching from protected areas (Jachmann 2008). However, often when designing projects inadequate attention is given to potential feedbacks between conservation activities and human behaviour (Nicholson et al. 2009). There is frequently a relatively simplistic narrative, for example: 'if we give people an alternative livelihood it will reduce their need to hunt, which will lead to recovery of the exploited population'. This narrative may be true but the additional income may also allow people to buy better hunting equipment, worsening the situation for some species (Crookes et al. 2005), or it may increase immigration to the area, causing further habitat loss (Oates 1999), or increase bushmeat consumption as people become able to afford it (Wilkie & Godoy 2001).

Conservation practitioners on the ground gain an understanding of the likely outcomes of different approaches as they implement them. For example, the Integrated Conservation and Development (ICD) approach has fallen out of fashion due to the many problems encountered as projects were implemented (Hughes & Flintan 2001). As new approaches to conservation gain popularity (e.g. Payments for Ecosystem Services; Engel et al. 2008, Pagiola 2008) they too tend to be widely and relatively uncritically applied at first, with the danger of the pendulum swinging back away from them as limitations become apparent (Wilshusen et al. 2002). This tendency is destructive, expensive and inefficient. Conservation needs to be able to predict and design for human reactions to conservation interventions, based on an understanding of what drives human behaviour, backed up by well-designed experimental studies of the effectiveness of interventions in different places and circumstances.

We do not need to build a new theory of human behaviour to develop this predictive approach to conservation science; tools are already available in the mainstream social sciences. We therefore start this chapter by providing an overview of the most important and relevant theories of human decision making, and how they relate to environmental behaviour. We then show how this theory can be applied in conservation interventions, both in general and through the use of particular social science research tools. We conclude by suggesting what is needed to facilitate proactive, dynamic conservation in the future that takes account of the adaptive powers that people have to change their behaviour as circumstances change.

Understanding individual behaviour

When making decisions, people are influenced by the potential financial costs and benefits of a given course of action, by social-psychological factors including their attitudes towards an action (Albarracín et al. 2005), and by their understanding of how they are expected to behave within society (Ostrom 2000). Models of human decision making have been offered by many disciplines, including economics, sociology, anthropology and social psychology. Economic models have been applied to household decision making and natural resource management for many years (Rae 1971). Social psychological models which consider characteristics such as individual attitudes and motivations, and the social pressure that leads people to behave in certain ways, are less well used in conservation.

Economic models of behaviour

Traditionally, economic analyses of behaviour have been underpinned by a model of rational choice. Rational decision makers aim to maximize their 'utility' – the level of satisfaction or happiness that they get from consuming a particular amount of a good, be it a beautiful view or a plate of duiker meat. Rational choice theory assumes that people have well-defined preferences, so that when faced with decisions, they choose their most preferred option. By itself, this theory does not impose many restrictions on expected behaviour, as it does not prescribe what a person's preferences should be, but it does impose a consistency assumption; if duiker is preferred to chicken, and chicken is preferred to beans, then duiker should also be preferred to beans. Rational choice becomes a more powerful predictive theory when it is combined with assumptions about preferences, or with data on people's observed choices. For example, a typical assumption is that more consumption is preferred to less consumption. Economic analysis usually focuses on goods and services traded in a market, for which there are quantitative data available. By observing the choices that people make about what to buy, at what prices, economists can infer how much value people place on goods.

The model of humans as rational decision makers has been widely adopted, underpinning the study of consumer choice and demand. Models of rational decision making have also been applied to resource management and conservation. For example, Barrett & Arcese (1998) modelled household decision making in order to investigate the effectiveness of ICD projects in reducing poaching in the Serengeti, Tanzania. In their model, households gained utility by consuming wild meat, crops, bought goods and leisure time, and maximized their utility through choices regarding allocation of labour to hunting and farming. Model results indicated that ICD projects depending upon wildlife harvest as a tool for reducing poaching are unlikely to be sustainable, particularly when game meat is a tradable commodity and environmental shocks reduce the productivity of labour invested in agriculture. Such models have also played an important part in understanding the role of enforcement activities, such as ranger patrols in protected areas, in limiting rule breaking in conservation. For example, Milner-Gulland & Leader-Williams (1992) used a model to show that the extirpation of rhinos from the Luangwa Valley, Zambia, in the 1980s was probably due to opportunistic poaching and driven by the substantial profits to be made from hunting elephants for ivory (then priced at ~ \$50/kg in country). Rhinos were too scarce for much of this period to be profitable targets on their own, even though rhino horn was worth substantially more per kilo than ivory (then priced at ~ \$500/kg in country).

Rational choice is widely used as a basis for hypothesis testing in behavioural ecology as well as in economics. Some studies have found that empirical observations of hunter behaviour fit well with simple models based on ecological theories of rational choice. For example, Rowcliffe et al. (2003) showed that subsistence hunters, such as the Semaq Beri blowpipe hunters of peninsular Malaysia, chose their portfolio of prey items (e.g. white-handed gibbons (*Hylobateslar*) and large squirrels (*Ratufa* spp.)) in order to maximize offtake rates in the same way that non-human animal foragers would, while Alvard (1993) showed that Piro shotgun hunters of Amazonian Peru, targeting species such as collared peccary (*Tayassutajaca*) and agouti (*Dasyproctavariegata*), acted purely to maximize offtakes.

The social, economic, biological and regulatory reasons why people harvest in particular places have received attention in the fisheries literature (Branch et al. 2006). The ecological concept of the ideal free distribution, whereby foraging animals (or fishers) distribute themselves among patches of resources (or fishing grounds), has become a behavioural null hypothesis for explaining the spatial distribution of fishers and their prey. With respect to fisheries, the theory has two main assumptions: that fishers have perfect knowledge of the resource, and that they are free to move between fishing grounds (Gillis 2003). Deviations from the ideal free distribution theory can provide useful information about the relationships between resource user behaviour, regulations and the environment. Swain & Wade (2003) showed that the distribution of commercial fishers in the southern Gulf of St Lawrence was related to the distribution of snow crab (*Chionoecetesopilio*), suggesting that they had knowledge of the resource which they used to decide where to fish. However, regulations restricting access to some fishing grounds, and the higher cost of reaching stocks further from port, inhibited fishers' 'free' movement between foraging grounds. In addition, competition between fishers limited the exchange of information, so fishers did not have perfect knowledge of the resource.

A better understanding of constraints that shape fishers' use of their environment enables the social costs of different conservation interventions to be factored into spatial planning. For example, as seen in the Philippines, positioning no-take zones in areas important to fishers can reduce human well-being or result in non-compliance, causing hostility towards conservation interventions (Majanen 2007). However, despite its widespread use, the convenient assumption that humans act as rational utility maximizers, whose responses to conservation interventions can be predicted through models of trade-offs between costs and benefits, is increasingly being challenged. A growing number of studies, particularly from the fields of psychology and behavioural economics, show that humans display inconsistent preferences and are influenced by supposedly irrelevant aspects of the context and framing of decisions. For example, Bulte et al. (2005) found that Dutch people were willing to pay more to conserve the threatened common seal (*Phocavitulina*) when their decline was presented as a man-made problem than when it was framed as having a natural cause such as a virus.

Social-psychological models of human behaviour

An influential approach to understanding people's behaviour is the theory of reasoned action (Fishbein & Ajzen 1975) and its extension, the theory of planned behaviour (TPB; Ajzen 1991). These theories come from social psychology rather than economics, and highlight the interacting effects of internal and external influences on people's behaviour. They have been used to investigate a broad range of behaviours, including condom use (Albarracín et al. 2001) and alcohol abuse (Schlegel et al. 1992). The approach assumes that people make rational decisions based upon a systematic evaluation of information available to them, evaluating the implications of alternative behaviours before choosing a course of action (Ajzen & Fishbein 1980). Economic models make similar assumptions (Blume & Easley 2008), but social psychologists use quite different predictors. The theory of reasoned action states that behavioural intention, the precursor to performing a behaviour, can be predicted from a person's attitude and subjective norms (what we think others will think of us if we do, or do not do, the behaviour). The TPB extends this model by including a measure of the control that people perceive they have over performing the behaviour (Ajzen 1991). For example, to predict a person's intention to poach a duiker, a series of statements might look as shown in Figure 19.1.

By knowing the relative importance of the different components of the TPB for a specific behaviour, it is possible to design behavioural change interventions targeted towards the most influential beliefs motivating that behaviour (Ajzen & Fishbein 1980; Ajzen 1991). For example, a meta-analysis of studies using the TPB to understand people's condom use, with the aim of persuading people to use condoms consistently to reduce HIV risk, concluded that attitude made a greater contribution to explaining people's intention to use condoms than either perceived behavioural control or subjective norms. Therefore interventions that change the beliefs underlying a person's attitudes towards condom use (e.g. counsellors may explore a person's negative beliefs associated with condom use) would be more successful in influencing condom use than interventions emphasizing subjective norms (e.g. social pressure), or the level of control people perceive they have over condom use (Albarracín et al. 2001).

The TPB has not been widely used within conservation but where it has been used, the results indicate that attitudes, whilst important, reveal only a limited picture of the beliefs underlying human behaviour. Zubair & Garforth (2006) evaluated the reasons behind poor take-up of an on-farm tree-planting scheme, and showed that an inhibiting factor

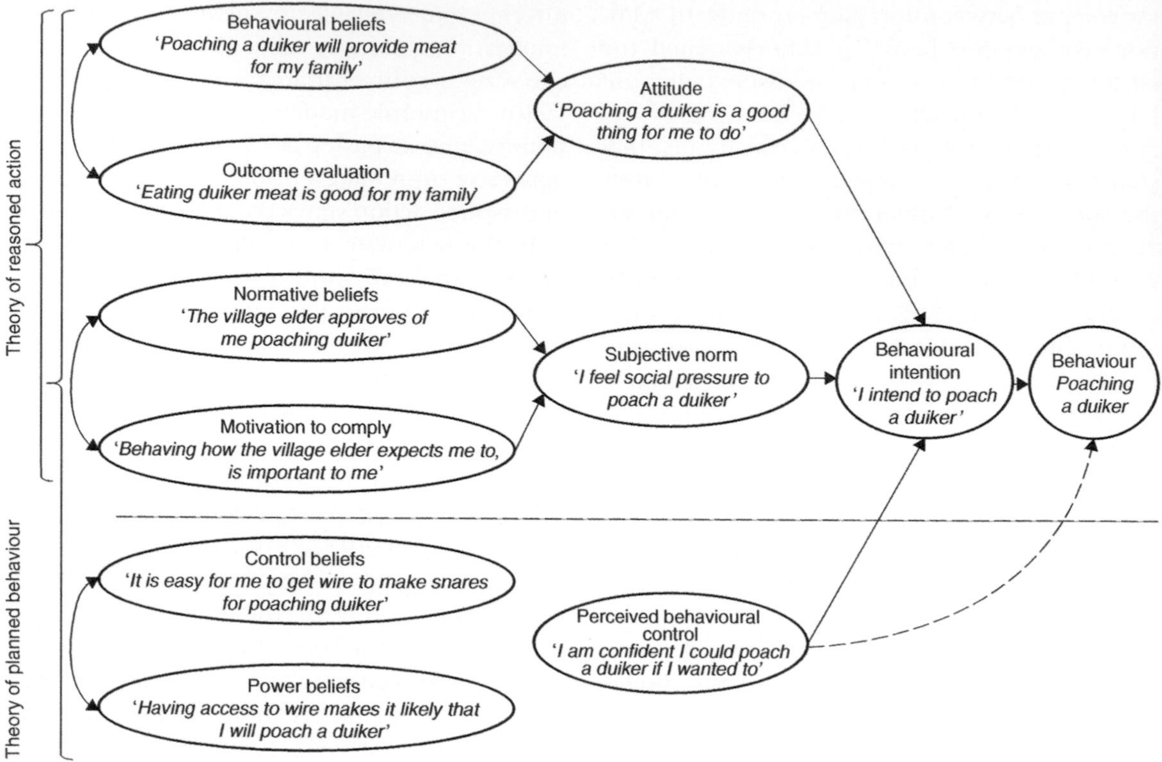

Figure 19.1 The theory of reasoned action and the theory of planned behaviour. All things held equal, the more positive a person's attitude, subjective norm and perceived behavioural control, the greater their behavioural intention and therefore the likelihood that they perform the behaviour. Dotted lines indicates that perceived behavioural control can directly predict behaviour. Adapted from St John et al. (2010b).

was the lack of control that farmers perceived they had due to a lack of tree nurseries and markets. Aipanjiguly et al. (2003) studied a manatee (*Trichechusmanatus*) conservation area and found that the intentions of boaters to abide by speed limits were strongly influenced by their perception of what others would think of them if they did not (subjective norm). Subjective norms were also important to farmers in northern Pantanal, Brazil, deciding whether to hunt jaguars (Cavalcanti et al. 2010; Marchini & Macdonald 2012) and to farmers in Pakistan deciding whether to take part in on-farm tree planting (Zubair & Garforth 2006). All too often, studies measure attitudes to wildlife conservation as proxies for pro-conservation behaviour, even though it has been pointed out on a number of occasions that attitudes alone are a poor predictor of behaviour (e.g. Holmes 2003; Waylen et al. 2009). The TPB provides insights into why this is so.

The importance of other people's views in shaping someone's behavioural intentions directs us to our next topic. Here we argue that whilst conservation at the small scale is about decisions made by individual people, the fact that people exist within societies rather than in a vacuum has a major influence upon how they behave.

Individuals in society

People live embedded within societies, and their interactions are mediated by institutions which shape attitudes, social norms and behaviour (Ostrom 2000; Agrawal 2001). Institutions are the 'rules of the game' for human interactions, and can be formal, based on laws and regulations, or informal, based on social norms or traditional rules (Agrawal 2001). 'Social norms' is a general term describing shared understandings about actions that are obligatory, permitted or forbidden (Ostrom 2000); they are similar to the 'subjective norms' used in the TPB (see Figure 19.1). Failure to observe social norms may result in shame, retribution or social rejection; some actions, which incur the disapproval not only of the community but also of deities, are referred to as taboo (Jones et al. 2008a).

There is a long history of informal institutions governing natural resources by establishing and enforcing social norms and taboos. For example, a millennia-old system of reef lagoon tenure established harvesting rules and no-take zones to safeguard fish stocks in Oceania (Johannes 1978), and in ancient Hawaii, crop and fish harvests from *ahupua'a* (tracts of land containing a watershed zone, upland and coastal farming zones, and fish ponds) were distributed via local institutional structures which observed rules of reciprocity (Costa-Pierce 1987). In Madagascar, taboos originating in respect for ancestors contribute to the protection of threatened species, including the fossa (*Cryptoptoctaferox*), believed to scavenge on the remains of dead ancestors, making eating them strictly taboo, and lemurs of the Indiridae family, protected because they are believed to embody dead ancestors (Jones et al. 2008a). More recently, social norms have been shown to be an important indicator of farmers' intentions to re-enrol in China's Grain-to-Green Programme, a payment for ecosystem services scheme. Social norms were of little importance when conservation payments were at the lowest or highest ends of the scale when none or all farmers re-enrolled. However, when conservation payments were intermediate, farmers based their decision on what neighbouring farmers were doing: if others re-enrolled, they too re-enrolled (Chen et al. 2009).

The study of institutions and their associated rules, norms and taboos is prominent in the natural resources literature. The 'tragedy of the commons' is a vision of a world without institutions, in which there is open access to resources for anyone who wishes to exploit them: 'Ruin is the destination towards which all men rush, each pursuing his own best interest in a society that believes in the freedom of the commons. Freedom in the commons brings ruin to all' (Hardin 1968). Led by Ostrom (1990), researchers have demonstrated that open access is actually relatively rare. Many resources (e.g. forests, fishing areas and rangelands) that lack obvious formal regulation are actually common property resources (CPRs), governed by informal institutions.

The essential problem raised by CPRs is that exploitation by one user often negatively affects others. All users are better off if they can co-operate in managing the commons, but each could potentially gain by taking more than others. To prevent overexploitation, there must be some type of institution, either formal or informal, that detects and punishes non-co-operation (Ostrom 1990). There are many cases in which people have established resource use rules amongst themselves in order to ensure sustainable use. Most successful CPR examples involve repeated interactions among relatively small numbers of users who observe each other's actions, can effectively exclude outsiders, live in stable environmental and social conditions, and have a long history of interactions and trust (Ostrom 1990). CPR governance is fragile and vulnerable to erosion, particularly by external influences that change social norms, reduce the bonds of society or provide alternative worldviews. For example, the drive towards modernization and the introduction of new religions have contributed to the erosion of social norms that traditionally protected streams and forest in Nigeria (Anoliefo et al. 2003).

How do people respond to conservation interventions?

The importance of conditionality

Conservation interventions aimed at changing behaviour can be divided into three broad types: carrots, sticks and distractions (Milner-Gulland & Rowcliffe 2007). Carrots are positive incentives designed to encourage conservation, and may include activities such as providing money (or benefits in kind) to individuals or communities for acting in the appropriate way. Sticks are negative incentives designed to discourage environmentally destructive actions. These include law enforcement to deter activities such as illegal poaching or campaigns that make activities socially costly by altering public opinion (such as those against furs or ivory). Distractions aim to make alternative options more attractive than environmentally destructive ones, for example improving crop yields to reduce the comparative attractiveness of bushmeat hunting.

Most conservation interventions combine carrots, sticks and distractions. For example, ICD projects usually offer alternative sources of income or sustenance to discourage illegal resource extraction from a protected area, as well as developing and enforcing rules for resource use, usually in consultation with the community (Hughes & Flintan 2001). One of the main lessons of over a decade of ICD projects is the importance of conditionality; benefits should be provided *only* when people are actually changing their behaviour. For example, around Kerinci Seblat National Park in Indonesia, where the receipt of development grants under an ICD project was not conditional upon maintaining forest cover, deforestation rates did not differ between villages receiving development grants and those that did not. Therefore, despite considerable financial investment (US$1.5 million), the scheme failed to halt illegal forest clearance (Linkie et al. 2008). Models of ICD interventions in the Serengeti-Mara ecosystem also emphasize the critical importance of linking the transfer of benefits to measurable conservation goals: if the aim is to stop poaching within a protected area, benefits should ideally only be received by people who do not poach, otherwise the scheme creates no incentive to shift effort from poaching to other activities which could contribute to household well-being (Johannesen 2006).

Embedding interventions within institutions

The ICD approach has often failed to achieve conservation objectives, leading some to question whether it is possible to achieve both development and conservation simultaneously in a single, integrated intervention (Berkes 2004). Based on the lessons learnt about the importance of conditionality, payments for ecosystem services (PES) has been proposed as a promising alternative (Pagiola 2008). PES can be described as a voluntary transaction in which a provider of a service (e.g. a community providing clean water by maintaining forest cover on their land) receives a benefit from the buyer of that service (e.g. the government), conditional on that service being provided to an acceptable standard (Engel et al. 2008). The main difference between PES and other conservation interventions is this conditionality.

Large-scale PES schemes have been established in a number of countries with strong institutional frameworks capable of facilitating monitoring, enforcement and outside validation of service providers, including watershed protection schemes in Costa Rica (Pagiola 2008) and agri-environment schemes for conserving farmland biodiversity in the European Union (Dobbs & Pretty 2008). However, many high-priority areas for conservation are located in places with weak institutional frameworks, or which lack clear natural resource and land tenure rights (Agrawal et al. 2008). Given the growing popularity of PES, it is important to understand the best approach to such schemes in the context of weak institutions. Clements et al. (2010) evaluated three PES programmes initiated by the Wildlife Conservation Society in Cambodia in 2002. Situated within the Indo-Burma biodiversity hot spot, Cambodia has one of the highest land use change rates globally (0.5% deforestation per annum, 2000–2005), primarily driven by large-scale developments and smallholder encroachments. Cambodia's current protected area network excludes many areas important for biodiversity conservation, so it is necessary to find ways of conserving biodiversity beyond protected areas.

The schemes were a community-based ecotourism programme, agri-environment payments for wildlife-friendly products, and direct contracts for bird nest protection. The ecotourism and agri-environment programmes were set up with considerable institutional involvement (village committees, protected area authorities, marketing associations and an international NGO) and were implemented upon village land under government-approved land use plans enforced by an elected village committee. In contrast, the nest protection programme involved just two partners; agreements were made between individual nest protectors and the international NGO to protect a resource (nests of globally threatened or near-threatened species) over which the protector had only *de facto* rights, in villages that did not have established land use plans.

The simple institutional structure of the nest protection programme required lower administration and start-up costs. It was also more efficient at retaining funds locally, 71–78% of funds were retained, compared to 24% and 55–60% retained by the ecotourism and agri-environment programmes respectively, as both paid some funds to external agencies. However, fewer people in the village benefited from nest protection payments (5% of families), and those that did benefit received less money ($120/year) than those involved in the ecotourism (10% of families earning $160/year) or agri-environment (5–10% of families earning $160/year) programmes. With respect to habitat protection, the simplicity of the nest protection programme was a weakness; in the absence of institutional support from village committees enforcing land use plans, nest protectors were powerless to prevent the clearance of nesting habitat. In contrast, people involved in either the ecotourism or agri-environment programmes, which were embedded within local institutions, had the capacity to enforce land use rules domestically, turning away in-migrants and minimizing habitat conversion. Critically, whilst more indirect in their approach to conservation, the conservation aims of the two more institutionally complex programmes (to reduce habitat loss and over-harvesting of species) were more widely understood and supported by local people. This case study suggests that grounding conservation interventions within the local institutional structure may make them more sustainable in the long term than interventions focusing exclusively on changing individual behaviour.

The role of enforcement

The conservation and management of natural resources are widely dependent upon negative incentives, principally the making and enforcing of rules which restrict human use. These include rules to restrict the harvest of certain species, and to restrict particular activities within protected areas (Jachmann 2008). However, the existence of rules alone does not change behaviour (Rowcliffe et al. 2004). Enforcement – monitoring adherence to rules and punishing detected infractions – is essential to successful conservation and natural resource management (Keane et al. 2008).

The negative incentive provided by enforcement has two components: the probability of being caught and the penalty incurred when apprehended (e.g. a fine or incarceration). Empirical evidence from fisheries and elsewhere suggests that increasing the perceived probability of detection and capture improves compliance far more than increasing the penalty (Becker 1968; Sutinen & Gauvin 1989). The same effect was observed in the Serengeti National Park, Tanzania; poaching rapidly increased after 1977 during a period of low patrol effort related to a collapse in park budgets; although the penalty remained the same, poaching declined from late 1980s onwards in response to increased patrol effort (Hilborn et al. 2006).

Enforcement is costly and with limited resources available to conservation, it is important to design efficient enforcement strategies. Bio-economic models of poacher behaviour in Luangwa Valley, Zambia, suggested that fines tiered according to the number of trophies in a poacher's possession would create greater deterrence than fixed fines (Milner-Gulland & Leader-Williams 1992; Leader-Williams & Milner-Gulland 1993). Jachmann & Billiouw (1997) demonstrated that intelligence-led law enforcement is more effective than simple patrolling. This is likely to be particularly true because poachers' decision making is influenced by their perceived probability of capture before the event, rather than by the actual probability of capture, which they do not know. In north Sulawesi, the effectiveness of market inspections for the endemic babirusa (*Babyrousacelebensis*) declined over time; when inspectors first patrolled the market, babirusa were not traded again for over a year. However, by the third patrol the halt in sales only lasted for one month, as traders refined their expectations of the likelihood of punishment from high to the true level of virtually zero (Milner-Gulland & Clayton 2002).

Indirect approaches can have ambiguous effects

'Distraction' approaches often involve improving the returns to alternative livelihoods, either ones which are already part of a household's activities (e.g. improving crop yields) or new ones introduced by the conservation intervention (e.g. crafts or honey production). However, empirical evidence suggests that this approach has a number of pitfalls. For example, people may simply add any new alternative to their existing portfolio of activities, rather than it replacing the environmentally damaging activity (Sievanen et al. 2005). If a family already pursues a range of activities, one option is to try to change the balance of costs and benefits associated with these existing activities so that people choose to spend their time on the less environmentally damaging activities (Barrett & Arcese 1998). However, this approach can have counterintuitive results. Damania et al. (2005) modelled Ghanaian hunter-farmers' responses to an increase in the income obtained from selling agricultural crops due to a price premium from an ICDP intervention. Hunter-farmers allocated most labour to agriculture when it was the most profitable activity. However, rather than reducing the pressure on bushmeat species, the model suggested that hunter-farmers would invest some of their increased income in new hunting gear (such as a gun), enabling them to target more economically valuable species. In this way, the overall effect of a rise in crop prices upon the sustainability of bushmeat hunting was ambiguous (Damania et al. 2005). Hill et al. (2012) investigated whether introducing seaweed farming as an alternative livelihood strategy in Danajon Bank, central Philippines, reduced the number of fishers exploiting a declining fishery. The intervention did not reduce fisher numbers across all villages; responses to the intervention differed according to different combinations of economic and non-economic factors (e.g. income patterns and capacity), making intervention outcomes difficult to predict. In particular, even though many individual households did take up seaweed farming and reduce their fishing activities, the aggregate number of fishers increased in some villages due to continuing human population growth.

Disrupting existing institutions is dangerous

So what happens in practice when conservationists enter a situation with the aim of changing people's existing behaviour towards something more environmentally sustainable? The theory we have discussed above will have demonstrated already that people's behaviour is subject to a range of external and internal influences, and that the outcome of interventions can be hard to predict. Indeed, the evidence suggests that conservation interventions which are not well founded in an understanding of existing institutions and cultures are more likely to fail (Waylen et al. 2010). It is critical, therefore, to understand the role of existing institutions (including taboos and social norms), which could be providing at least partial protection to threatened species or habitats, before intervening in a way which may erode them (Jones et al. 2008a).

One potentially important effect of superimposing external rules and values on existing institutions is called 'crowding out' (Vollan 2008). This happens when a person's intrinsic motivation to comply with established social norms is weakened, perhaps because of reduced self-governance (i.e. reduced perceived behavioural control), violated norms of reciprocity, or changed value systems focusing on the monetary rather than intrinsic value of a resource. For example, designation of Ranomafana National Park in Madagascar caused the breakdown of traditional management of pandans (*Pandanus* spp.), a plant used for weaving; the prevailing social norm prohibiting damage to the growing tip of pandan plants became widely disregarded once the resource was no longer the property, or responsibility, of the users (Jones et al. 2008a). Similarly, experimental studies in rural Colombia suggested that external regulations can trigger the crowding out of socially desirable behaviour limiting the collection of firewood to minimize the negative impact of forest degradation on communal water supplies (Cardenas et al. 2000). In this experiment, firewood collecting under open access (in the absence of communication) was compared with that under weak enforcement of an external rule restricting how much time could be spent collecting firewood. Initially the imposition of the external rule reduced the time spent collecting, but as players learnt that the external rule was poorly enforced and fines were low, average firewood collection time increased to rates similar to those observed under open access conditions (Cardenas et al. 2000). This is because individuals realized that the rule was being violated by others; this crowded out their motivation to act in the collective interest. People became more interested in maximizing their own well-being by increasing their own firewood collection, and less concerned about the well-being of others. Allowing people to communicate, however, increased individual motivation to act in the collective interest (Cardenas et al. 2000). Clearly, it is important to investigate the degree to which this effect, observed in an experimental setting, represents real situations.

Financial incentives can also crowd out social norms. Money can change a task from something previously carried out due to a sense of moral obligation into a market interaction whereby people generally become less willing to contribute their effort (Vollan 2008). This crowding-out effect was first reported by Titmuss (1970), who found that more people donated blood under the voluntary UK scheme compared to the United States' incentive-based system. Similarly, in an attempt to make parents collect children promptly from nursery, fines were imposed; however, rather than increasing parents' respect for closing times, the fine was perceived as legitimizing parents' behaviour (Gneezy & Rustichini 2000). These results are cause for concern about the effectiveness of monetizing biodiversity conservation, either through placing a monetary value on wildlife or through offering financial incentives to carry out conservation actions. As yet, there has been little quantitative research on crowding out in the conservation literature, although a number of authors have expressed their concern that the intrinsic values of wildlife are lost if conservation becomes a market transaction (e.g. Oates 1999). Crowding out may on most occasions be outweighed by the beneficial effects of monetary incentives on people's

Box 19.1 The randomized response technique

This box provides an example of how to use the 'forced response' randomized response technique. Respondents are provided with an opaque beaker containing two dice and a set of identical question cards, each of which displays the RRT instructions, for example:

5.

INSTRUCTIONS

Please do not let me see what number they land on

Remember the rules, add together the numbers on the two dice:

2, 3, 4 = say **'Yes'**

5 – 10 answer the question below truthfully **'Yes'** or **'No'**

11, 12 = say **'No'**

QUESTION **In the last 12 months did you kill any leopards?**

Respondents provide their answers by simply saying 'yes' or 'no' out loud to the interviewer. The result of the dice throw is never revealed to the interviewer so it is impossible to distinguish a true response from a 'forced' one. However, by knowing the probability of respondents having to answer the sensitive question truthfully, and the probability of the 'forced' responses, the proportion of the population with the sensitive characteristic can be calculated using the following formula:

$$\pi = \frac{\lambda - \theta}{s}$$

where π is the estimated proportion of the sample who have performed the behaviour, λ is the proportion of all responses in the sample that are 'yes', θ is the probability of the answer being a 'forced yes' (in this case 1/6), and s is the probability of having to answer the sensitive question truthfully (in this case 3/4) (St John et al. 2011).

behaviour, but these studies do suggest that incentives are more complex than would be assumed under assumptions of rational self-interest in a social vacuum.

Exciting research tools for conservation scientists

Researchers in conservation science can draw on a wide range of tools and conceptual models from social science, where much of the theoretical and empirical work has already been done. There have been several recent reviews of different fields of social science aimed at conservation audiences, which draw together lessons from these disparate disciplines and suggest how they can be applied in conservation (e.g. Keane et al. (2008) for law enforcement, St John et al. (2010b) for social psychological theory (see Chapter 7) and Colyvan et al. (2011) for game theory). Here we discuss two particularly promising approaches to collecting empirical data on people's behaviour to give an idea of the types of methods that are available. One estimates levels of rule breaking, and the other helps us understand how people behave in situations when the outcomes of their decisions are contingent on the behaviour of others.

Estimating levels of rule breaking

The first requirement for designing interventions to improve compliance is to know how many people, and of what type, are currently breaking the rules (St John et al. 2010a). However, directly investigating rule breaking is problematic, as rule breakers may not wish to reveal themselves for fear of punishment or social opprobrium (Keane et al. 2008). Face-to-face interviews asking people directly about their behaviour can be a cost-effective method for investigating legal and socially acceptable exploitation of wild species (Jones et al. 2008b). Yet if the exploitation is illegal, or sensitive because it violates social norms, people may answer with what they think is the socially appropriate thing to say, or not answer the question at all; both types of response can compromise the validity of data (King & Bruner 2000).

The randomized response technique (RRT; Box 19.1) reduces the risk that people perceive in providing honest answers to sensitive questions (Warner 1965). RRT has been used across a range of sensitive behaviours including insurance fraud (Bockenholt & van der Heijden 2007), illegal abortion (Silva & Vieira 2009) and illegal resource extraction (Solomon et al. 2007). In studies where the actual level of rule breaking was known, RRT provided more accurate estimates of the prevalence of rule breaking than conventional methods (Lensvelt-Mulders et al. 2005). Further, studies comparing methods for collecting sensitive data reported that RRT returned higher estimates of the sensitive behaviour compared to anonymous self-completed questionnaires (St John et al. 2010a) and face-to-face questionnaires (Solomon et al. 2007; Silva & Vieira 2009). For example, Solomon et al. (2007) reported that 39% of respondents admitted to hunting illegally inside Kibale National Park, Uganda, when asked via RRT, whilst just 1.7% of respondents admitted to this behaviour when asked directly. Similarly 26% of fishers admitted to fishing without a licence when asked via RRT, whilst just 1.6% of fishers admitted to this behaviour through an anonymous self-completed questionnaire (St John et al. 2010a). Solomon et al.'s (2007) application of the method in Uganda suggests that RRT can be adapted for use in areas with low literacy and little training in probability theory, enabling conservationists to obtain more realistic estimates of illegal resource use in such situations.

St John et al. (2011) used RRT to estimate the proportion of farmers killing carnivores, some of which were legally protected; they then adapted the logistic regression model to investigate if non-sensitive characteristics of farmers, such as their attitudes towards killing carnivores, predicted their carnivore-killing behaviour. The study demonstrated a relationship between farmers' attitudes and their carnivore-killing behaviour as reported via RRT; how sensitive farmers perceived carnivore killing to be, and their estimates of the proportion of their peers killing carnivores also related to reported behaviour. Scenarios generated from a mixed effects model indicated that farmers reporting the attitude that leopards should be killed on ranches, who estimated that all their peers kill leopards, and thought that killing leopards was not at all sensitive, were 70% more likely to have reported (via RRT) killing leopards, compared to farmers reporting the opposite in attitudes and perceptions. This study provides evidence that carefully specified attitude statements can be useful indicators of actual behaviour, and provides support for the notion that the false consensus effect, whereby people who do socially undesirable acts provide higher estimates of the behaviour within the population compared to those who do not perform such behaviours, offers potential for identifying people involved in sensitive behaviours (Petroczi et al. 2008). This study paves the way for using RRT to identify non-sensitive indicators of people's involvement in illicit behaviours. Such indicators may be used to identify groups of people most likely to be involved in environmentally harmful behaviours, allowing for effective targeting of interventions aimed at influencing behaviour.

Experimental games

Experimental games, which are controlled experiments designed to study the behaviour of participants in stylized situations, have been used to investigate how people behave in social dilemmas (i.e. situations in which individual interests conflict with those of the group; Heckathorn 1996). Experimental games have been widely applied to the study of CPRs (e.g. Ostrom et al. 1994) and public goods (e.g. Isaac et al. 1994). Results from experimental games show that people often play more co-operatively than would be predicted based on just maximizing their own pay-offs (Fischbacher et al. 2001). This indicates that people have more complex utility functions than might be expected from rational utility theory, which factor in the pay-offs to other players and notions of fairness or equity (Fehr & Schmidt 1999). Individuals willingly impose sanctions on non-co-operative behaviour even at a cost to themselves, in order to improve group outcomes (Poteete et al. 2010).

A large number of experimental games have been developed and played in a wide variety of settings (Camerer 2003), but mostly in laboratories using American or European university students as subjects. For example, Fischbacher et al. (2001) recruited University of Zurich students to complete a computerized public goods game in a lab in order to investigate the nature of co-operative behaviour. However, if the results of these games are to have relevance to real-life problems of resource overuse, they need to be conducted *in situ*, involving local people with direct experience of resource use dilemmas. Henrich et al. (2001) used the ultimatum game with people in 15 non-industrial societies across 12 countries. In the ultimatum game, an individual (the proposer) is assigned an amount of money (e.g. equivalent to one or two days' local salary). The proposer is asked to offer another individual (the respondent) some of this money; the respondent can accept the offer and both players receive their agreed split of money, or the respondent can reject the offer in which case neither player receives any money. If both players attempt to maximize their own pay-off, then a respondent should accept any positive offer, and the proposer should offer the smallest possible amount. However, contrary to this prediction, proposers' mean offers to respondents ranged from 26% to 58% of their own stake. The large variation in proposers' offers suggests that preferences and expectations are affected by locally defined rules and norms of fairness (Henrich et al. 2001).

What makes field-based experimental games so interesting, and a potentially powerful tool for investigating how people may respond to different interventions, is that when playing experimental games, the players apparently seek to solve dilemmas by looking for examples in their everyday life. Indeed, for many of the cultural groups that Henrich et al. (2001) observed, similarities between locally established norms and how players behaved in experimental games were evident. For example, in New Guinea some individuals refused offers that were a high proportion of the proposer's stake because, within their society, recipients of gifts are obliged to reciprocate the gift at some time in the future, which can create considerable anxiety.

Travers et al. (2011) showed how experimental games might be used to predict the effects of different conservation interventions on resource users' behaviour. Individuals from four villages in the Northern Plains landscape of Cambodia played a series of simple common pool resource games in which they had to choose how many fish to extract from a shared pond. This scenario was chosen because it mirrored the sort of collective action problem participants face day to day. Each game was played in groups of 10 for five consecutive rounds, and individuals could harvest up to 10 fish per round. A harvested fish was worth 80 Cambodian riel to the individual who caught it; remaining fish were each worth 12 Cambodian riel to all participants, reflecting their contribution to future harvests. Players were allowed to keep their earnings from the games and the expected earnings amounted to two to three times the average local daily wage. The structure of the game meant that although an individual acting alone could achieve a high pay-off by harvesting the full quota, the maximum pay-off could only be achieved if all the players co-operated to extract no fish.

In order to assess the effects of possible interventions on this system, several variants of the game were played. Each had a different institutional structure designed to reflect a common form of conservation intervention: a fine if extraction was too high; individual payments for the lowest extractors; or communal payments if group extraction was below a specified threshold. The effect of each intervention was measured by comparing it with baseline scenarios in which no intervention was applied. Results suggested that each type of intervention reduced overexploitation. However, the most effective, longest lasting interventions were those encouraging participants to form agreements amongst themselves. Carefully designed games such as these have the potential to become valuable tools for making context-specific predictions about the effects of different conservation interventions.

Conclusion

In poor rural areas where much conservation effort is focused, conservation outcomes are ultimately determined by individual behaviour; therefore, conservation success requires that we understand human decision making at this level. An understanding of the drivers of human behaviour can help us to change people's motivations to alter their livelihood strategies; to structure incentive schemes; to balance patrols and fines in law enforcement; and to predict the effects of individual and collective decision making on ecosystem service provision and demand. In order to do this, we must embrace a range of tools and analytical frameworks from the social sciences that seek to understand and model human behaviour.

To illustrate the importance of understanding human behaviour, we consider three topical areas in which a more sophisticated application of approaches from social science may provide important new insights. First, there is a need to understand *indirect interactions between multiple livelihoods* which are likely to have a substantial impact on the predicted future supply of, and demand for, natural resources. For example, the sustainability of bushmeat hunting in the forests of Central Africa appears to be more strongly determined by changes in employment opportunities and the price of agricultural commodities than by the price or availability of bushmeat itself (Barrett & Arcese 1995). Second, the *impact of changes in biodiversity and natural resources on human well-being* should be measured explicitly if we are to ensure that conservation interventions benefit the people living in areas of high and threatened biodiversity. In particular, how are changes in well-being arising from conservation distributed within society (do the poor and marginalized suffer the majority of the costs?), and how does well-being change over time as people adapt to new circumstances? Third, *resource supply and use should be modelled dynamically*, such as done by Ling & Milner-Gulland (2006), rather than assuming an equilibrium that typically does not exist. For example, if the resource is abundant and household demand is opportunistic, rather than a necessity for well-being, the indirect link between resource supply and demand and household decision making may be weak. However, as resource scarcity increases and/or reliance on resources for well-being increases, the indirect link may strengthen, a process only discerned with dynamic models. See de Merode & Cowlishaw's (2006) investigation of fluctuating bushmeat markets in the Democratic Republic of Congo for an example of this.

Conservation science is undergoing a period of rapid and exciting development, with social science at the forefront. The tools and conceptual frameworks for understanding human behaviour are already available, and there are increasing opportunities for collaboration between conservationists and a range of professionals in other disciplines, including economists, social psychologists and anthropologists. Very soon, it will be unforgivable to carry out second-rate social science in conservation, just as now it is unacceptable to use shoddy methods to monitor animal abundance. To guide policy and management effectively, we need to be able to predict the likely consequences of our interventions on human behaviour. Our aspiration should be to create an interdisciplinary field of predictive conservation science, based on a solid conceptual and empirical foundation, rather than learning by trial and error. This requires conservation research to integrate closely with conservation practice, so that we can learn actively and systematically from experience, using controls, counterfactuals and experimental designs for our interventions. This approach is already standard within development economics (e.g. Barrett et al. 2005; Ravallion 2007).

The current generation of conservation scientists is leading the way in transforming conservation science into a robust discipline. The first step is to move beyond simplistic and static assumptions about how interventions affect individual choices, towards a more nuanced understanding of the complex social landscape within which conservationists operate as just one of many stakeholders.

> *'And, most important, we have to shift our understanding of ourselves as separate individuals, each seeking our own welfare, to an understanding of how we fit into social, biological, and physical environments.'*
>
> **Robert Ornstein**

Acknowledgements

The authors would like to thank Steve Polasky, David Macdonald, Kathy Willis and Tom Moorhouse for helpful comments. EJMG acknowledges the support of a Royal Society Wolfson Research Merit award.

References

Agrawal, A. (2001) Common property institutions and sustainable governance of resources. *World Development*, **29**, 1649–1672.

Agrawal, A., Chhatre, A. & Hardin, R. (2008) Changing governance of the world's forests. *Science*, **320**, 1460–1462.

Aipanjiguly, S., Jacobson, S.K. & Flamm, R. (2003) Conserving manatees: knowledge, attitudes, and intentions of boaters in Tampa Bay, Florida. *Conservation Biology*, **17**, 1098–1105.

Ajzen, I. (1991) The theory of planned behavior. *Organizational Behavior and Human Decision Processes*, **50**, 179–211.

Ajzen, I. & Fishbein, M. (1980) *Understanding Attitudes and Predicting Social Behavior*. Prentice-Hall, Englewood Cliffs, NJ.

Albarracín, D., Johnson, B.T., Fishbein, M. & Muellerleile, P.A. (2001) Theories of reasoned action and planned behavior as models of condom use: a meta-analysis. *Psychological Bulletin*, **127**, 142–161.

Albarracín, D., Johnson, B.T., Zanna, M.P. & Kumkale, T.G. (2005) Attitudes: introduction and scope. In: *The Handbook of Attitudes* (eds D. Albarracín, B.T. Johnson & M.P. Zanna), pp.3–20. Lawrence Erlbaum Associates, Mahwah, NJ.

Alvard, M. (1993) Testing the ecologically noble savage hypothesis: interspecific prey choice by Piro hunters of Amazonian Peru. *Human Ecology*, **21**, 355–387.

Anoliefo, G.O., Isikhuemhen, O.S. & Ochije, N.R. (2003) Environmental implications of the erosion of cultural taboo practices in Awka-South local government area of Anambra State, Nigeria: 1. Forests, trees, and water resource preservation. *Journal of Agricultural and Environmental Ethics*, **16**, 281–296.

Barrett, C.B. & Arcese, P. (1995) Are integrated conservation-development projects (ICDPs) sustainable? On the conservation of large mammals in sub-Saharan Africa. *World Development*, **23**, 1073–1084.

Barrett, C.B. & Arcese, P. (1998) Wildlife harvest in integrated conservation and development projects: linking harvest to household demand, agricultural production, and environmental shocks in the Serengeti. *Land Economics*, **74**, 449–465.

Barrett, C.B., Lee, D.R. & McPeak, J.G. (2005) Institutional arrangements for rural poverty reduction and resource conservation. *World Development*, **33**, 193–197.

Becker, G. (1968) Crime and punishment: an economic approach. *Journal of Political Economy*, **76**, 169–217.

Berkes, F. (2004) Rethinking community-based conservation. *Conservation Biology*, **18**, 621–630.

Blume, L.E. & Easley, D. (2008) Rationality. In: *The New Palgrave Dictionary of Economics* (eds S.N. Durlauf & L.E. Blume). Palgrave Macmillan, Basingstoke.

Bockenholt, U. & van der Heijden, P.G. (2007) Item randomized-response models for measuring noncompliance: risk-return perceptions, social influences, and self-protective responses. *Psychometrika*, **72**, 245–262.

Branch, T.A., Hilborn, R., Haynie, A.C., *et al.* (2006) Fleet dynamics and fishermen behavior: lessons for fisheries managers. *Canadian Journal of Fisheries and Aquatic Sciences*, **63**, 1647–1668.

Bulte, E., Gerking, S., List, J. & de Zeeuw, A. (2005) The effect of varying the causes of environmental problems on stated WTP values: evidence from a field study. *Journal of Environmental Economics and Management*, **49**, 330–342.

Camerer, C.F. (2003) Behavioural studies of strategic thinking in games. *Trends in Cognitive Sciences*, **7**, 225–231.

Cardenas, J.C., Stranlund, J. & Willis, C. (2000) Local environmental control and institutional crowding-out. *World Development*, **28**, 1719–1733.

Cavalcanti, S.M., Marchini, S., Zimmermann, A., Gese, E.M. & Macdonald, D.W. (2010) Jaguars, livestock and people in Brazil: realities and perceptions behind the conflict. In: *The Biology and Conservation of Wild Felids* (eds D.W. Macdonald & A. Loveridge), pp.383–402. Oxford University Press, Oxford.

CBD (2007) *About the 2010 Biodiversity Target*. www.cbd.int/2010-target/about.shtml

Chen, X., Lupi, F., He, G. & Liu, J. (2009) Linking social norms to efficient conservation investment in payments for ecosystem services. *Proceedings of the National Academy of Sciences USA*, **106**, 11812–11817.

Clements, T., John, A., Nielsen, K., An, D., Tan, S. & Milner-Gulland, E.J. (2010) Payments for biodiversity conservation in the context of weak institutions: comparison of three programs from Cambodia. *Ecological Economics*, **69**, 1283–1291.

Colyvan, M., Justus, J. & Regan, H.M. (2011) The conservation game. *Biological Conservation*, **144**, 1246–1253.

Costa-Pierce, B.A. (1987) Aquaculture in ancient Hawaii. *BioScience*, **37**, 320–331.

Crookes, D.J., Ankudey, N. & Milner-Gulland, E.J. (2005) The value of a long-term bushmeat market dataset as an indicator of system dynamics. *Environmental Conservation*, **32**, 333–339.

Damania, R., Milner-Gulland, E.J. & Crookes, D.J. (2005) A bioeconomic analysis of bushmeat hunting. *Proceedings of the Royal Society B: Biological Sciences*, **272**, 259–266.

De Merode, E. & Cowlishaw, G. (2006) Species protection, the changing informal economy, and the politics of access to the bushmeat trade in the Democratic Republic of Congo. *Conservation Biology*, **20**, 1262–1271.

Dobbs, T.L. & Pretty, J. (2008) Case study of agri-environmental payments: the United Kingdom. *Ecological Economics*, **65**, 765–775.

Ehrlich, P.R. & Pringle, R.M. (2008) Where does biodiversity go from here? A grim business-as-usual forecast and a hopeful portfolio of partial solutions. *Proceedings of the National Academy of Sciences USA*, **105**, 11579–11586.

Engel, S., Pagiola, S. & Wunder, S. (2008) Designing payments for environmental services in theory and practice: an overview of the issues. *Ecological Economics*, **65**, 663–674.

Fehr, E. & Schmidt, K.M. (1999) A theory of fairness, competition, and cooperation. *Quarterly Journal of Economics*, **114**, 817–868.

Fischbacher, U., Gächter, S. & Fehr, E. (2001) Are people conditionally cooperative? Evidence from a public goods experiment. *Economics Letters*, **71**, 397–404.

Fishbein, M. & Ajzen, I. (1975) *Belief, Attitude, Intention and Behaviour: An Introduction to Theory and Research*. Addison-Wesley, Reading, MA.

Gelcich, S., Edwards-Jones, G. & Kaiser, M.J. (2005) Importance of attitudinal differences among artisanal fishers toward co-management and conservation of marine resources. *Conservation Biology*, **19**, 865–875.

Gillis, D.M. (2003) Ideal free distributions in fleet dynamics: a behavioral perspective on vessel movement in fisheries analysis. *Canadian Journal of Zoology*, **81**, 177–187.

Gneezy, U. & Rustichini, A. (2000) A fine is a price. *Journal of Legal Studies*, **29**, 1–17.

Hardin, G. (1968) The tragedy of the commons. *Science*, **162**, 1243–1248.

Heckathorn, D.D. (1996) The dynamics and dilemmas of collective action. *American Sociological Review*, **61**, 250–277.

Henrich, J., Boyd, R., Bowles, S., *et al.* (2001) In search of Homo Economicus: behavioral experiments in 15 small-scale societies. *American Economic Review*, **91**(2), 73–78.

Hilborn, R., Arcese, P., Borner, M., *et al.* (2006) Effective enforcement in a conservation area. *Science*, **314**, 1266.

Hill, N.A.O., Rowcliffe, M., Koldeway, H., Milner-Gulland, E.J. (2012) The interaction between seaweed farming as an alternative occupaction and fisher numbers in the Central Philippines. *Conservation Biology*, **26**, 324–334.

Holmes, C.M. (2003) The influence of protected area outreach on conservation attitudes and resource use patterns: a case study from western Tanzania. *Oryx*, **37**, 305–315.

Hughes, R. & Flintan, F. (2001) *Integrating Conservation and Development Experience: A Review and Bibliography of the ICDP Literature*. International Institute for Environment and Development, London.

Isaac, R.M., Walker, J.M. & Williams, A.W. (1994) Group size and the voluntary provision of public goods: experimental evidence utilizing large groups. *Journal of Public Economics*, **54**, 1–36.

Jachmann, H. (2008) Illegal wildlife use and protected area management in Ghana. *Biological Conservation*, **141**, 1906–1918.

Jachmann, H. & Billiouw, M. (1997) Elephant poaching and law enforcement in the Central Luangwa Valley, Zambia. *Journal of Applied Ecology*, **34**, 233–244.

Johannes, R.E. (1978) Traditional marine conservation methods in Oceania and their demise. *Annual Review of Ecology and Systematics*, **9**, 349–364.

Johannesen, A.B. (2006) Designing integrated conservation and development projects (ICDPs): illegal hunting, wildlife conservation, and the welfare of the local people. *Environment and Development Economics*, **11**, 247–267.

Jones, J.P., Andriamarovololona, M.M. & Hockley, N. (2008a) The importance of taboos and social norms to conservation in Madagascar. *Conservation Biology*, **22**, 976–986.

Jones, J.P., Andriamarovololona, M.M., Hockley, N., Gibbons, J.M. & Milner-Gulland, E.J. (2008b) Testing the use of interviews as a tool for monitoring trends in the harvesting of wild species. *Journal of Applied Ecology*, **45**, 1205–1212.

Keane, A., Jones, J.P.G., Edwards-Jones, G. & Milner-Gulland, E.J. (2008) The sleeping policeman: understanding issues of enforcement and compliance in conservation. *Animal Conservation*, **11**, 75–82.

King, M.F. & Bruner, G.C. (2000) Social desirability bias: a neglected aspect of validity testing. *Psychology and Marketing*, **17**, 79–103.

Leader-Williams, N. & Milner-Gulland, E.J. (1993) Policies for the enforcement of wildlife laws: the balance between detection and penalties in Luangwa Valley, Zambia. *Conservation Biology*, **7**, 611–617.

Lensvelt-Mulders, G.J., Hox, J.J., van der Heijden, P.G. & Maas, C.J. (2005) Meta-analysis of randomized response research: thirty-five years of validation. *Sociological Methods Research*, **33**, 319–348.

Ling, S. & Milner-Gulland, E.J. (2006) Assessment of the sustainability of bushmeat hunting based on dynamic bioeconomic models. *Conservation Biology*, **20**, 1294–1299.

Linkie, M., Smith, R.J., Zhu, Y.U., *et al.* (2008) Evaluating biodiversity conservation around a large Sumatran protected area. *Conservation Biology*, **22**, 683–690.

Majanen, T. (2007) Resource use conflicts in Mabini and Tingloy, the Philippines. *Marine Policy*, **31**, 480–487.

Marchini, S. & Macdonald, D.W. (2012) Predicting ranchers' intention to kill jaguars: case studies in Amazonia and Pantanal. *Biological Conservation*, **147**, 213–221.

Milner-Gulland, E.J. & Clayton, L. (2002) The trade in babirusas and wild pigs in North Sulawesi, Indonesia. *Ecological Economics*, **42**, 165–183.

Milner-Gulland, E.J. & Leader-Williams, N. (1992) A model of incentives for the illegal exploitation of black rhinos and elephants: poaching pays in Luangwa Valley, Zambia. *Journal of Applied Ecology*, **29**, 388–401.

Milner-Gulland, E.J. & Rowcliffe, J.M. (2007) *Conservation and Sustainable Use: A Handbook of Techniques*. Oxford University Press, Oxford.

Nicholson, E., Mace, G.M., Armsworth, P.R., *et al.* (2009) Priority research areas for ecosystem services in a changing world. *Journal of Applied Ecology*, **46**, 1139–1144.

Oates, J.F. (1999) *Myth and Reality in the Rainforest: How Conservation Strategies are Failing in West Africa*. University of California Press, Berkeley, CA.

Ostrom, E. (1990) *Governing the Commons: the Evolution of Institutions for Collective Action*. Cambridge University Press, Cambridge.

Ostrom, E. (2000) Collective action and the evolution of social norms. *Journal of Economic Perspectives*, **14**, 137–158.

Ostrom, E., Gardener, R. & Walker, J. (1994) *Rules, Games, and Common-Pool Resources*. University of Michigan Press, Ann Arbor, MI.

Pagiola, S. (2008) Payments for environmental services in Costa Rica. *Ecological Economics*, **65**, 712–724.

Petróczi, A., Mazanov, J., Nepusz, T., Backhouse, S. & Naughton, D. (2008) Comfort in big numbers: does over-estimation of doping prevalence in others indicate self-involvement? *Journal of Occupational Medicine and Toxicology*, **3**, 19.

Poteete, A.R., Janssen, M.A. & Ostrom, E. (2010) *Working Together: Collective Action, the Commons, and Multiple Methods in Practice*. Princeton University Press, Princeton, NJ.

Rae, A.N. (1971) Stochastic programming, utility, and sequential decision problems in farm management. *American Journal of Agricultural Economics*, **53**, 448–460.

Ravallion, M. (2007) Evaluating anti-poverty programs. In: *Handbook of Development Economics*, Vol 4 (eds T.P. Schultz & J. Strauss), pp.3787–3846. Elsevier, Amsterdam.

Rowcliffe, J.M., Cowlishaw, G. & Long, J. (2003) A model of human hunting impacts in multi-prey communities. *Journal of Applied Ecology*, **40**, 872–889.

Rowcliffe, J.M., de Merode, E. & Cowlishaw, G. (2004) Do wildlife laws work? Species protection and the application of a prey choice model to poaching decisions. *Proceedings of the Royal Society B: Biological Sciences*, **271**, 2631–2636.

Schlegel, R.P., Davernas, J.R., Zanna, M.P., DeCourville, N.H. & Manske, S.R. (1992) Problem drinking: a problem for the theory of reasoned action? *Journal of Applied Social Psychology*, **22**, 358–385.

Sievanen, L., Crawford, B., Pollnac, R. & Lowe, C. (2005) Weeding through assumptions of livelihood approaches in ICM: seaweed farming in the Philippines and Indonesia. *Ocean snf Coastal Management*, **48**, 297–313.

Silva, R.S. & Vieira, E.M. (2009) Frequency and characteristics of induced abortion among married and single women in São Paulo, Brazil. *Cadernos de Saúde Pública, Rio de Janeiro*, **25**(1), 179–187.

Solomon, J., Jacobson, S.K., Wald, K.D. & Gavin, M. (2007) Estimating illegal resource use at a Ugandan park with the randomized response technique. *Human Dimensions of Wildlife*, **12**, 75–88.

St John, F.A.V., Edwards-Jones, G., Gibbons, J.M. & Jones, J.P. (2010a) Testing novel methods for assessing rule breaking in conservation. *Biological Conservation*, **143**, 1025–1030.

St John, F.A.V., Edwards-Jones, G. & Jones, J.P. (2010b) Conservation and human behaviour: lessons from social psychology. *Wildlife Research*, **37**, 658–667.

St John, F.A.V., Keane, A.M., Edwards-Jones, G., Jones, L., Yarnell, R.W. & Jones, J.P. (2011) Identifying indicators of illegal behaviour: carnivore killing in human-managed landscapes. *Proceedings of the Royal Society B: Biological Sciences*, **279**(1729), 804–812.

Sutinen, J.G. & Gauvin, J.R. (1989) Assessing compliance with fishery regulations. *Maritimes*, **33**, 10–12.

Swain, D.P. & Wade, E.J. (2003) Spatial distribution of catch and effort in a fishery for snow crab (Chionoecetes opilio): tests of predictions of the ideal free distribution. *Canadian Journal of Fisheries and Aquatic Sciences*, **60**, 897–909.

Titmuss, R.M. (1970) *The Gift Relationship: From Human Blood to Social Policy*. Allen and Unwin, London.

Travers, H., Clements, T., Keane, A. & Milner-Gulland, E.J. (2011) Incentives for cooperation: the effects of institutional controls on common pool resource extraction in Cambodia. *Ecological Economics*, **71**, 151–161.

Van Vuuren, D.P. & Bouwman, L.F. (2005) Exploring past and future changes in the ecological footprint for world regions. *Ecological Economics*, **52**, 43–62.

Vollan, B. (2008) Socio-ecological explanations for crowding-out effects from economic field experiments in southern Africa. *Ecological Economics*, **67**, 560–573.

Warner, S.L. (1965) Randomized response: a survey technique for eliminating evasive answer bias. *Journal of the American Statistical Association*, **60**, 63–69.

Waylen, K.A., McGowan, P.J.K., Pawi Study Group & Milner-Gulland, E.J. (2009) Ecotourism positively affects awareness and attitudes but not conservation behaviours: a case study at Grande Riviere, Trinidad. *Oryx*, **43**, 343–351.

Waylen, K.A., Fischer, A., McGowan, P.J.K., Thirgood, S.J. & Milner-Gulland, E.J. (2010) Effect of local cultural context on the success of community-based conservation interventions. *Conservation Biology*, **24**, 1119–1129.

Wilkie, D.S. & Godoy, R.A. (2001) Income and price elasticities of bushmeat demand in lowland Amerindian societies. *Conservation Biology*, **15**, 761–769.

Wilshusen, P.R., Brechin, S.R., Fortwangler, C.L. & West, P.C. (2002) Reinventing a square wheel: critique of a resurgent "Protection Paradigm" in international biodiversity conservation. *Society and Natural Resources*, **15**, 17–40.

Zubair, M. & Garforth, C. (2006) Farm level tree planting in Pakistan: the role of farmers' perceptions and attitudes. *Agroforestry Systems*, **66**, 217–229.

20

Designing effective solutions to conservation planning problems

Andrew T. Knight[1], Ana S.L. Rodrigues[2], Niels Strange[3], Tom Tew[4] and Kerrie A. Wilson[5]

[1]Division of Ecology and Evolution, Imperial College London, Ascot, UK *and* Department of Botany, Nelson Mandela Metropolitan University, Port Elizabeth, South Africa
[2]Centre d'Ecologie Fonctionnelle et Evolutive, CNRS-CEFE UMR5175, Montpellier, France
[3]Department of Food and Resource Economics, Centre for Macroecology, Evolution and Climate, University of Copenhagen, Frederiksberg, Denmark
[4]The Environment Bank, Stamford, UK
[5]School of Biological Sciences, University of Queensland, St Lucia, QLD, Australia

"... it is important that [conservation] theorists do not overestimate the contribution that conservation theory can make in a field that, whether we like it or not, is driven largely by socio-economic imperatives."

John R. Prendergast (1999)

Introduction

The need for conservation planning

Human well-being depends on utilising natural resources but these activities are often not conducted sustainably, leading to significant, sometimes permanent, declines in natural ecosystems and the goods and services they provide. Habitat destruction, overutilisation, pollution, invasive species and freshwater extraction are the major causes of this decline (Millennium Ecosystem Assessment 2005; Vie et al. 2009). The resources allocated to nature conservation activities globally are inadequate for ensuring the persistence of nature specifically, or sustainable development more generally. For example, the total annual cost of an effective global protected area programme on land and sea was estimated at US$45 billion, dwarfing the estimated $6.5 billion actually spent (Balmford et al. 2002).

Key Topics in Conservation Biology 2, First Edition. Edited by David W. Macdonald and Katherine J. Willis.

Protected areas are widely regarded as the cornerstone of conservation action (Margules & Pressey 2000). Approximately 13% of the world's land area has been formally protected (Coad et al. 2010) but this global network still does not comprehensively protect the world's species and ecosystems (Rodrigues et al. 2004). The species of tropical forests, grasslands and savannas, and Mediterranean forests are particularly under-represented (Ferrier et al. 2004). Most protected areas have been located where conservation interests will avoid conflicts with development activities, rather than through strategic regional- or national-scale protected area network expansion programmes (Pressey 1994). This *ad hoc* approach is neither a cost-efficient nor effective way of utilising conservation resources, leading to an over-representation of ecosystems with low economic value, such as rocky mountain tops, and an under-representation of ecosystems whose natural resources can fuel economic growth, such as wetlands and forests (Pressey & Tully 1994). These challenges to establishing effective protected area networks highlight the importance of complementing them with conservation initiatives on private land (e.g. Pence et al. 2003).

We outline an approach for ensuring the effectiveness of regional- or local-scale conservation planning initiatives. We define 'conservation planning' as a collaborative, social learning-driven activity whose goal is to implement actions that ensure the persistence of nature by integrating the processes of spatial prioritisation and implementation strategy development to achieve effective conservation management (Knight et al. 2006a,b). Many studies in the literature use 'conservation planning' to describe an activity we term 'spatial conservation prioritisation' – systematic analyses of mapped information to identify locations for conservation investment (Wilson et al. 2009a) – but do not include implementation strategies or stakeholder collaboration (Box 20.1) (Knight et al. 2008). The spatial decision-making approach we outline here is a subset of a broader group of decisions, such as land use and species recovery planning, for protecting the environment (Knight et al. 2006a,b; Ferrier & Wintle 2009).

Conservation planning comprises three broad activities: assessment, planning and management (Figure 20.1). Assessment involves the scientific evaluation of factors defining the implementation of effective conservation action. These factors include species, habitats and ecological processes; the pressures upon them; the willingness and capacity of stakeholders to enact conservation; and the costs of implementation. The technical activities of identifying the subset of natural features protected, and unprotected, in a planning region (i.e. gap analysis) and spatial prioritisation are core assessment activities that generate information to assist decision makers to identify where conservation action is most cost-efficiently and effectively implemented (Scott et al. 1993; Margules & Pressey 2000). Planning activities then move toward action by linking assessments to processes for developing an implementation strategy (i.e. how conservation initiatives will be undertaken) in collaboration with stakeholders (see Figure 20.1). Management comprises activities that maintain or enhance the persistence of species, ecosystems and ecological processes and the benefits that flow from them to society. Examples include capacity building, and the operations of formally protected areas, mainstreaming conservation activities into privately owned landscapes, and ecosystem restoration.

Specifically defining the scope and differences between assessment, planning and management is essential for ensuring effective conservation action because some activities (e.g. management) directly conserve nature whereas others (e.g. assessment) do not. This chapter outlines an approach for conducting assessment activities so as to better ensure effective collaborative processes for achieving conservation goals (see Figure 20.1). We outline the evolution of these ideas and set the approach within contemporary conservation planning thinking and practice.

Box 20.1 Conservation planning glossary box

Collaboration: the process of pooling appreciations and tangible resources (e.g. funding, labour) by two or more stakeholders to find solutions to a set of problems that neither can solve individually (Wondolleck & Yaffee 2000). Not to be confused with simply 'informing' or 'involving' stakeholders in conservation planning initiatives, which are less empowering for those involved.

Conceptual framework: a representation of the world in words and/or figures that provides context and helps people think about planning phenomena so as to order knowledge and reveal patterns from which models and theories can be developed, tested and improved (Rapoport 1985). Provides the theory behind conservation planning which complements the practice of applying an operational model so as to facilitate learning and adaptation (Knight et al. 2006a).

Conservation planning: a transdisciplinary social process informed by a spatial conservation prioritisation whose goals are to develop collaborative strategies that ensure the persistence of functional social-ecological systems, and to continuously improve their effectiveness through social learning (Salafsky et al. 2002; Knight et al. 2006a). A term often conflated with spatial conservation prioritisation (Knight et al. 2006b).

Conservation planning products: pragmatic, tangible outputs (such as maps of priority areas) whose purpose is to facilitate decision making by specified implementers that result from the redesigning and interpretation of the results of spatial conservation prioritisation analyses. Contrast with spatial prioritisation outputs (Pierce et al. 2005; Knight et al. 2006b).

Implementation strategy: a common plan of action for partners involved in a conservation planning initiative which details the underlying principles, directions and tasks for translating conservation planning products into action. It provides a common vision and responsibilities for stakeholder organisations, assistance for identifying synergies among organisations, and guidance on how implementation activities should be prioritised and executed (Knight et al. 2011).

Landscape management model: an optimal mix of complementary instruments, incentives and institutions collaboratively designed by stakeholders, and contextualised for local and regional opportunities and constraints, which describes how implementation of conservation action will manifest. It forms one of the products of problem orientation, and links spatial prioritisations to implementation strategies (Knight et al. 2006a). Examples include conservation corridors and biosphere reserves.

Operational model: a simplified conceptualisation of a process for implementing conservation action at priority conservation areas that guides and assists understanding of how these processes function, embody best practice, and provide an entity that can be refined as approaches evolve to improve the practice of conservation (Knight et al. 2006a).

Problem formulation: the act of defining a spatial conservation prioritisation problem inclusive of an explicitly stated objective, prescribed actions, the contribution of actions towards achieving the objective, and the costs of and constraints upon implementation (Moilanen et al. 2009b). It is typically represented mathematically and operationalised through computer-based resource allocation software.

Problem orientation: the analytical process of clarifying goals, describing trends, analysing conditions, projecting developments, and inventing, evaluating and selecting alternative futures regarding a policy (conservation) problem (Clark 2002). Also referred to as social assessment (Cowling & Wilhelm-Rechmann 2007) or scoping (Pressey & Bottrill 2009).

Spatial conservation prioritisation: a suite of techniques, typically operationalised through computer software, for conducting systematic analyses of mapped information to identify locations for investing resources to achieve conservation goals (Wilson et al. 2009b). Forms a subset of the activities comprising a conservation planning process (Knight et al. 2006a,b).

A short history of conservation planning

Spatial conservation prioritisation approaches emerged in the early 1980s in response to the limited funding and *ad hoc* way in which protected areas had traditionally been established. Studies focused on the strategic spatially explicit allocation of conservation resources so as to ensure that all natural features were sampled in formally protected areas whilst maximising the cost-efficiency of conservation activities (Kirkpatrick 1983; Margules et al. 1988; Pressey & Nicholls 1989). The principles of efficiency and comprehensiveness (sometimes termed representativeness) (Table 20.1) were integrated into these spatial prioritisations through the principle of complementarity – the

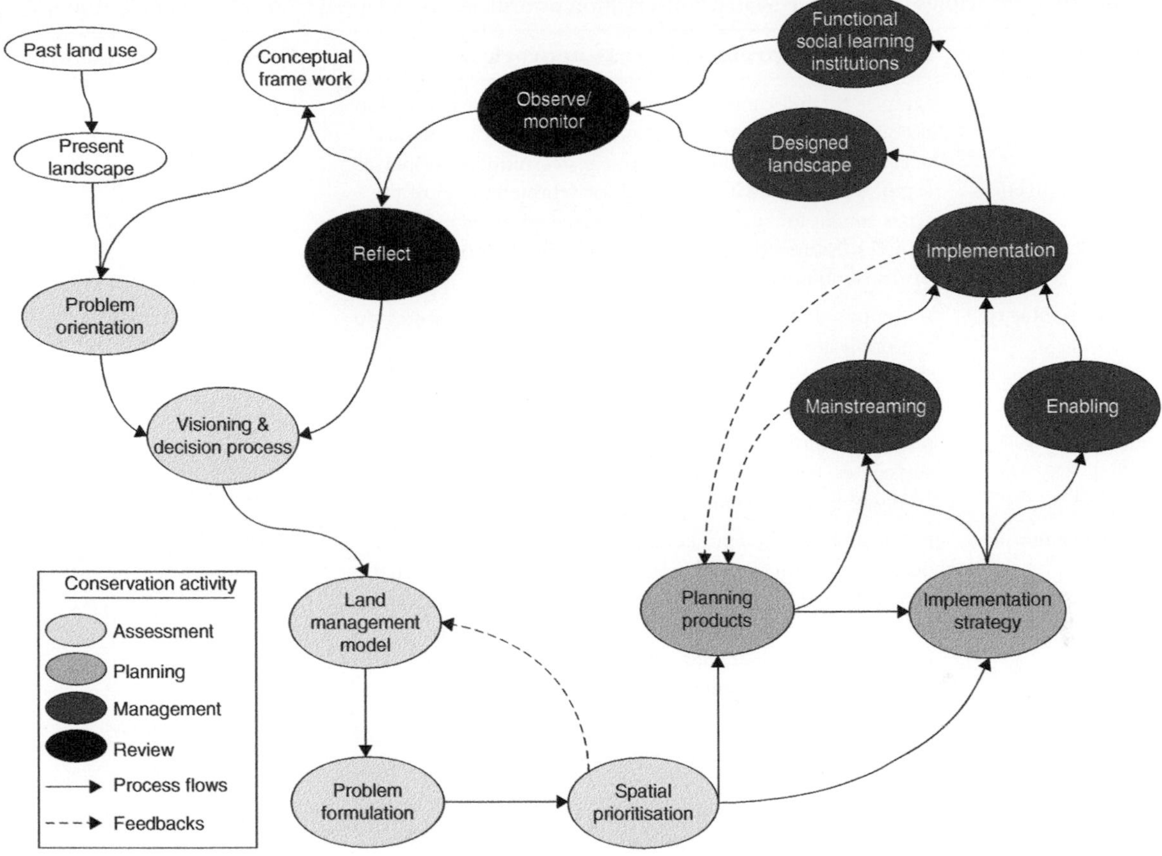

Figure 20.1 An operational model for conservation planning (adapted from Knight et al. 2006a), highlighting the importance and context of problem orientation and problem formulation.

idea that individual areas within a protected area network should complement, rather than duplicate, one another in terms of the natural features they protect (Pressey et al. 1993; Justus & Sarkar 2002). Kirkpatrick et al. (1980) first outlined the benefits of complementarity in Tasmania where a network of seven complementary candidate protected areas were identified, which have subsequently all been gazetted (Pressey 2002). This simple but at the time revolutionary idea helped to ensure that protected area networks were identified that *collectively* maximised their cost-efficiency and effectiveness.

The principle of complementarity remains core to spatial prioritisation theory and practice. Methods for its implementation have evolved rapidly since the development of the first simple algorithms used to identify complementary candidate protected areas for a planning region (e.g. Margules et al. 1988; Pressey & Nicholls 1989). Sophisticated software has been developed to assist planners to map and analyse increasingly complex and realistic conservation scenarios. This evolution has included development of techniques that advanced beyond simply comprehensively sampling species or ecosystems to ensuring their long-term persistence by incorporating connectivity (McDonnell et al. 2002), climate change (Araújo et al. 2004) and ecological processes (Cowling et al. 1999; Rouget et al. 2003; Pressey et al. 2007; Klein et al. 2009). The inclusion of data on the vulnerability of natural features has assisted planners to schedule their actions through time to avoid losing species (e.g. Pressey & Taffs 2001; Noss et al. 2002). Socio-economic considerations, for example implementation costs (Ando et al. 1998; Pence et al. 2003; Naidoo et al. 2006) have also been included

Table 20.1 Principles for effective spatial conservation prioritization. Adapted from Wilson et al. (2009a)

Principle	A candidate protected areas network...	Group aims to ensure...
Comprehensiveness	samples the full range of natural features we seek to include, accounting for composition (e.g. species, genes), structure (e.g. habitat types) and function (e.g. recruitment, dispersal).	Inclusion of features
Representativeness	samples features that are typical (or representative) of the types sought for inclusion in the protected areas network.	
Surrogacy	samples features that act as proxy for those whose existence or distribution is unknown.	
Complementarity	is composed of individual planning units whose features do not duplicate (i.e. complement) existing protected areas or that have been prioritised for future investment.	Comprehensiveness and efficiency
Adequacy and persistence	ensures the existence and evolution of all features in perpetuity through the careful choice of targets and surrogates, consideration of ecological processes, spatial configuration and connectivity, and feature replication.	Viability in perpetuity
Cost-efficiency	is comprehensive, representative, and adequate for the least possible cost, thereby increasing defensibility and optimal expenditure. Costs include acquisition costs (for purchasing property rights), management costs, transaction costs (those associated with identifying and negotiating with land managers) or opportunity costs (the value of foregone use).	Maximum return on investment
Threat	is designed to address, and be resilient to, the pressures that will potentially lead to the decline of the natural features being protected.	Minimisation of loss
Vulnerability	is selected, and whose implementation is scheduled, to avoid or minimise: (i) exposure to (e.g. the probability of a threatening process impacting an area); (ii) intensity of (i.e. magnitude, frequency, and duration); and (iii) impact from (i.e. the response of features to the threat) activities or processes that will produce a decline in conservation values.	
Irreplaceability	includes all sites required for ensuring that all conservation goals are achieved efficiently, or alternatively, the degree to which conservation goals are compromised if a specific site is unavailable for inclusion.	Protection of high-value, or important, features
Replacement cost	reduces the implications of economic or biological differences between the optimal solution and the feasible solution created by either (i) the forced exclusion of desirable sites, or (ii) the forced inclusion of a poor-quality site.	
Flexibility	is implemented to account for emerging opportunities, or conversely respond to lost opportunities, presented where multiple areas provide conservation opportunities to protect the same feature or set of features.	Avoidance of conflict

to improve cost-efficiency. The integration of people's and organisations' willingness and capacity to effectively collaborate and implement conservation action has also recently improved our understanding of opportunities for feasibly implementing proposed actions (e.g. Knight & Cowling 2007; Knight et al. 2010, 2011b; Raymond & Brown 2011).

Techniques have also evolved in complexity, moving beyond both the simple protected/

non-protected dichotomy and a single objective (e.g. transfer of forestry concessions to formally protected areas; Kirkpatrick 1983) to multiple-objective analyses (e.g. Ciarleglio et al. 2010) that allocate multiple conservation instruments or zones (Wilson et al. 2007; Watts et al. 2009) that facilitate trade-offs and synergies between objectives. Dynamic analyses now also promote increasingly cost-efficient and effective solutions by explicitly accounting for changes in land use over time (e.g. Costello & Polasky 2004; Wilson et al. 2007), for example, by using spatially explicit modelling to predict changes in ecosystem services, biodiversity conservation and commodity production levels (Nelson et al. 2009). These advances in the use of data and software have been complemented with improved techniques for establishing quantitative targets for natural features to better ensure their persistence (Cowling et al. 1999; Rodrigues et al. 2000; Desmet & Cowling 2004; Rondinini & Chiozza 2010).

The types of software available to assist conservation planners' decision making have evolved and diversified. For example, C-Plan offers interactive irreplaceability analysis to facilitate real-time stakeholder negotiations (Pressey et al. 2009), Marxan allows users to rapidly analyse multiple zonings and their costs to identify efficient near-optimal networks of candidate protected areas (Watts et al. 2009), Zonation can trade off species connectivity requirements to improve their persistence (Moilanen et al. 2009a), whilst ConsNet facilitates multi-criteria analyses that allow integration of sociopolitical criteria into spatial prioritisation (Ciarleglio et al. 2010). All provide conservation planners with rapidly advancing support for decision making. The use of such sophisticated spatial prioritisation software perhaps gives the impression that decision-making processes have become automated, reducing the need for expert input. However, many implicit and explicit decisions are required of individuals and groups that complement spatial prioritisation software to ensure effective decision making (Cowling et al. 2003). Firstly, a conservation context must be accurately and holistically understood (i.e. problem orientation), so that the drivers of the decline of species, ecosystems and the processes that sustain them, and the opportunities and constraints on implementing effective action are known. This understanding also allows spatial prioritisations to be appropriately designed and situated. Secondly, spatial prioritisation problems must be accurately and precisely defined (i.e. problem formulation) to promote the delivery of cost-efficient and effective solutions. We discuss these two types of conservation planning problems below.

Most land use decisions affect stakeholders, necessitating their involvement in conservation decision-making processes (Clark 2002; Knight et al. 2006a; Smith et al. 2009). Stakeholders demand, and have a right to, technically sound, scientifically defensible decisions (Noss et al. 1997) but as contexts become more complex, fewer stakeholders have the technical expertise required to either meaningfully contribute to, or provide effective critique of, these decisions. This tension between technical competence and genuine collaborative process creates a compelling dynamism between a narrow politic of scientific expertise and a broad politic of inclusion (Daniels & Walker 2001). Bridging the gaps between science, policy and implementation therefore requires finding ways to simultaneously increase the quality of technical expertise whilst including stakeholders in decision-making processes. This necessitates ensuring collaboration between conservation practitioners, government officials, local communities and researchers (Wondolleck & Yaffee 2000; Cowling & Pressey 2003; Smith et al. 2009). For example, conservation planning initiatives in the Succulent Karoo, Cape Floristic Region and Maputaland-Pondoland-Albany 'hotspots' of South Africa were commonly founded upon a collaborative process that fused scientific, experiential and traditional or local knowledge (e.g. Gelderblom et al. 2003; Knight et al. 2011a). Effective conservation planning thus embodies a policy process integrating science and society (Clark 2002) whose goal is social learning and adaptive management (Holling 1978; Salafsky

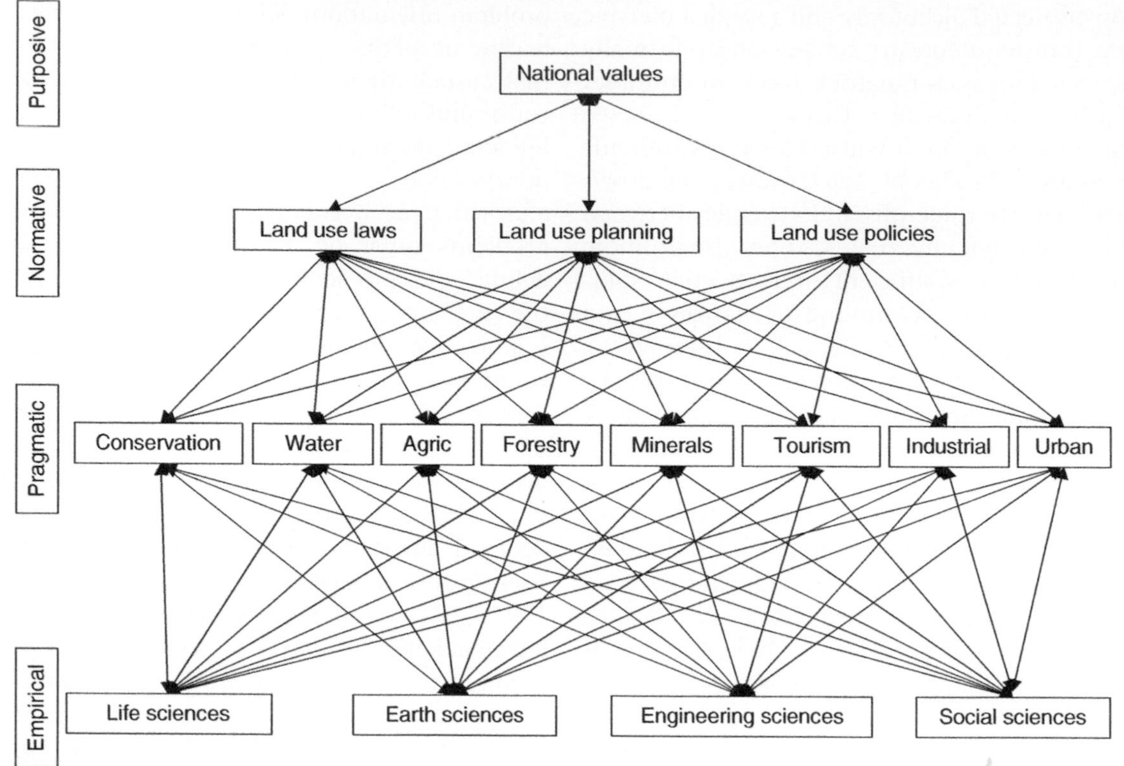

Figure 20.2 Effective conservation planning is a transdisciplinary process which can be conceptualised as a subactivity of land use planning, and which is informed by a diverse range of sciences. Adapted from Reyers et al. (2010).

et al. 2002; Knight et al. 2006a), and not simply a static, technical and prescriptive spatial prioritisation exercise. It forges beyond the simple identification of new protected areas, to become an integral component of broader land use planning (e.g. Pierce et al. 2005; Knight et al. 2006b; Reyers et al. 2007) (Figure 20.2).

A focus on effectiveness

Many applied disciplines struggle to comprehensively achieve their goals (e.g. Pfeffer & Sutton 1999), and conservation planning is no exception. The combination of funding shortfalls, rapid environmental change, and the human and institutional challenges of implementing conservation initiatives effectively presents two major lessons. First, conservation planning initiatives are personally and professionally challenging (e.g. Soulé & Terborgh 1999; Redford & Taber 2000; Cowling & Pressey 2003). For example, Knight (2006) and Beier (2008) discuss the personal challenges of grappling with learning to implement conservation plans effectively, and the emotional toll of failure. Second, this complexity emphasises the urgent need to implement scientifically defensible and cost-efficient methods for decision making, which is clearly demonstrated when conservation plans are challenged by those wishing to develop and destroy important natural areas (Noss et al. 1997), or through the improved cost-efficiency (Wilson et al. 2007) or effectiveness (Knight et al. 2010) in achieving conservation goals provided by spatial prioritisations.

Many, if not most, conservation planning initiatives that reach the implementation stage are only partially effective (Whitten et al. 2001;

Knight et al. 2008), probably due to a diversity of complex factors such as unexpected changes in planning contexts, funding cuts, poor understanding of the links between social and ecological systems, and competition with other land uses. For example, funding for conservation planning initiatives can be terminated through ever-changing political agendas (e.g. Knight 2006) or mainstreaming of maps of important areas for conservation into decision-making institutions can falter despite years of investment of time and resources (Knight et al. 2011a). Society's current collective understanding of how to develop and implement spatial prioritisation techniques far exceeds its ability to consistently deliver effective conservation action (Knight et al. 2006a). This divide between knowing how to define and design a solution to a conservation problem, and effectively delivering that solution (i.e. the 'knowing–doing gap'; Pfeffer & Sutton 1999) forms part of the human condition, and must be consciously acknowledged to improve the future effectiveness of conservation planning initiatives (Knight et al. 2008). Effective conservation planning (as a scientific discipline) maintains a pragmatic, problem-focused approach, in which research plays a fundamentally important role in advancing our understanding, so as to ensure that it most effectively contributes towards achieving conservation goals, and avoids the 'research–implementation gap' (Knight et al. 2008).

Most spatial prioritisation studies in the peer-reviewed literature were not undertaken with an intention to translate their results into conservation action (Knight et al. 2008). Until relatively recently, conservation organisations did not apply these techniques or developed their own approaches and techniques rather than employing those presented in the peer-reviewed literature (Prendergast et al. 1999; Hopkinson et al. 2000). The recent conceptual shift by conservation planners from simply conducting spatial prioritisations to planning for the implementation of conservation action by complementing spatial prioritisations with processes for strategy development and stakeholder collaboration has significantly improved the effectiveness of conservation planning initiatives (Knight et al. 2006b; Rouget et al. 2006; CMP 2007; Gallo et al. 2009; Morrison et al. 2009; Game et al. 2011).

The effectiveness of a conservation planning initiative can be measured as the extent to which stated objectives are achieved (Rodrigues et al. 1999). To be effective, conservation decision making must:

- be based on accurate, precise, spatially explicit and collaborative identification of conservation problems, using, for example, the analytic framework of Clark (2002)
- design suites of complementary instruments (e.g. protected areas, legislation), incentives (e.g. tax rebates, awards) and institutions (i.e. social learning forums) as solutions for protecting natural features, that are integrated into a single landscape management model, such as the Megaconservancy Networks of Rouget et al. (2006)
- deliver cost-efficient solutions across both space and time to ensure a defensible return on investment from conservation activities, for example, mixing suites of private land and formally protected areas for the most effective conservation outcome (Pence et al. 2003)
- progressively transform individual, institutional and societal values and behaviours towards achieving a collaboratively envisaged sustainable future, using techniques such as social marketing (Wilhelm-Rechmann & Cowling 2011).

Accordingly, this chapter presents:

- a suite of characteristics defining, and principles for, effective conservation planning, including both technical principles for spatial prioritisation, and those that ensure effective implementation through collaboration
- an operational model for conservation planning, which provides an explicit process for implementing conservation action

- an approach for explicitly defining conservation planning problems, inclusive of problem orientation (i.e. social assessment) and problem formulation (i.e. for spatial conservation prioritisations)
- future challenges and research directions.

Methodological details for conducting spatial conservation prioritisations are addressed at length elsewhere (e.g. Groves 2003; Sarkar et al. 2006; Margules & Sarkar 2007; Moilanen et al. 2009c; Wilson et al. 2009b) and so are not presented here. We do not address conservation management, other than to state that an unwavering focus on achieving adaptive and collaborative conservation management is an essential foundation for ensuring the effectiveness of conservation planning initiatives (see Margolius & Salafsky 1998; McShane & Wells 2004; Knight et al. 2006a; Smith et al. 2009).

Characteristics and principles of effective conservation planning

Effective conservation planning exhibits specific characteristics. First and foremost, it is a social process focused upon providing pragmatic solutions to real-world problems where societal goals for managing nature are operationalised as effective and cost-efficient decision-making processes (Clark 2002; Knight et al. 2006a). Second, it is a transdisciplinary activity (*sensu* Max-Neef 2005) that collectively and collaboratively defines problems by genuinely valuing and engaging the multiple groups of stakeholders and their types of knowledge (e.g. local, scientific, traditional). It forms a subset of broader land use planning activities (Knight et al. 2006a,b; Reyers et al. 2010; see Figure 20.2). Third, it ensures that the limited resources committed to conservation activities are used cost-efficiently (Wilson et al. 2007). Fourth, it adopts an evidence-based approach, whereby scientific evaluation of the effectiveness of the different approaches and techniques employed throughout a conservation planning process informs decision making (Pullin & Knight 2003). Finally, it is explicitly interactive and adaptive, regarding both individual practitioners behaviour and the operations of organisations, so that conservation methodologies and practices evolve in response to changing knowledge and contexts (e.g. climate change, economic fluctuations, changes in government). Institutionalised processes of social learning, such as communities of practice or learning groups, provide an essential foundation for effective conservation planning initiatives (Daniels & Walker 2001; Salafsky et al. 2002; Knight et al. 2006a).

A suite of principles for guiding spatial prioritisations has been widely endorsed (see Table 20.1; see Pressey et al. 1993; Margules & Pressey 2000; Margules & Sarkar 2007; Wilson et al. 2009a). However, there are few well-established principles of planning practice beyond spatial prioritisation (Knight et al. 2006b). Table 20.2 presents a suite of principles for ensuring the effectiveness of conservation planning initiatives synthesised from the literature and our personal experience. They aim to ensure effective collaboration between diverse groups of stakeholders, culminating in a mutually beneficial social learning process that deepens understanding of problems, opportunities and feasible solutions, whilst minimising conflicts. The Open Standards for the Practice of Conservation provide an example of a widely agreed attempt to synthesise pragmatic principles for conservation planning (CMP 2007).

An operational model for conservation planning

Effective conservation planning is fundamentally a practice, rather than a science, that by necessity integrates a diverse array of disciplines (e.g. landscape ecology, operations research, management science, social marketing), knowledge (e.g. experiential, traditional, scientific), approaches (e.g. expert, systematic), activities (e.g. stakeholder

Table 20.2 Principles supporting an effective collaborative, transdisciplinary approach to conservation planning

Principle	Description
Accountability	The condition whereby stakeholders collectively agree to justify their decisions and actions regarding explicitly stated and collaboratively agreed responsibilities, more generally for upholding the principles outlined in this table, and specifically, the equitable, timely, cost-efficient and effective implementation of conservation action. Mechanisms such as memoranda of understanding, reporting schedules and initiative evaluations facilitate accountability.
Adaptability	The willingness and capacity to adjust to changing conditions whilst learning to implement actions to better achieve conservation planning goals in increasingly effective ways.
Collaboration	The process of pooling appreciations and tangible resources (e.g. funding, labour) by two or more stakeholders to find solutions to a set of problems that neither can solve individually. Aims to empower local people; ensure relevance of solutions; improve collective decision making through consilience; and ultimately improve the effectiveness of conservation planning initiatives. Not to be confused with simply 'informing' or 'involving' stakeholders in conservation planning initiatives, which do not effectively empower stakeholders.
Defensibility	The ability to argue the case for decisions by implementing evidence-based approaches to conservation that demonstrate effectiveness, optimality, cost-efficiency and equity. Assists in avoiding criticism and litigation.
Equity	A condition whereby the interests and inputs of interested and affected stakeholders are fairly and impartially included in a conservation planning initiative. An essential prerequisite for ensuring effective collaboration and consilience (the fusion of different knowledge traditions).
Feasibility	The state or degree exhibited by a proposed or implemented action of being easily or conveniently completed. Is dependent upon the availability, and knowledge of, individuals' and organisations' willingness, capacity, time and resources.
Pragmatism	The condition of being problem and action focused, and so approaching decision making in a realistic way that is founded on practical rather than theoretical considerations. Stands in contrast to positivism (i.e. the philosophy behind modern experimental science).
Resilience	A state achieved by a conservation planning initiative that promotes stability through fostering conditions that encourage adaptability to disturbances and change, such as those of political agendas, funding and staff turnover.
Social learning	A process of iterative reflection which aims to improve useful knowledge that supports collective action through adaptive management and occurs through partnerships where we share our experiences, ideas and environments with others.
Transparency	The degree to which decision making can be openly witnessed and is documented.

collaboration, spatial prioritisation, strategy implementation, protected area management) and stakeholders (e.g. land managers, government officials, researchers) across a range of temporal, geographical, ecological, economic and social scales. Effective conservation planning initiatives are inevitably highly complex activities.

An operational model provides a simplified, generic but detailed functional conceptualisation of a conservation planning process (Knight et al. 2006a; see Figure 20.1). A generic approach is useful because processes for delivering effective conservation planning solutions are rarely unique, although solutions are typically case specific (Murphy & Noon 1992; Clark 2002). Operational models should broadly outline the tools and techniques for conservation planning and avoid attempting to provide universally applicable solutions, as panaceas for natural resource management problems are a myth (Ostrom et al. 2007). For example, collaboration with stakeholders is widely cited as an essential ingredient for effective conservation planning, but the way in which a planning initiative collaborates with a poorly educated, poverty-stricken African community in the tropical rainforests of Rwanda will be different from how collaboration functions with a well-educated, relatively wealthy suburban community in the United Kingdom. Language, culture, religion, education and financial resources (to name but a few factors driving effective implementation) all differ significantly. Operational models should be focused on implementing effective conservation action, and, when applied, be explicitly utilising within the unique characteristics of a planning region's social-ecological system (*sensu* Berkes & Folke 1998).

An operational model is distinctly different from, but should be complemented with, a conceptual framework (Knight et al. 2006a). The latter provides context and integrates key concepts so as to order knowledge and reveal patterns from which models and theories for delivery of feasible conservation solutions can be developed and improved (Rapoport 1985). For example, the operational model in Figure 20.1 outlines a sequence of 'real-world' activities essential for the effective practice of conservation planning where the goal is, say, to ensure the persistence of subtropical thicket in the Maputaland-Pondoland-Albany hotspot (see Pierce et al. 2005; Rouget et al. 2006; Knight et al. 2011a). Having planners apply a body of theory to such problems ensures that the functioning of the social-ecological systems in which the operational model is implemented is well understood. Effective conservation professionals shift consciously and routinely between implementing an operational model and critiquing their conceptual framework to ensure that practice informs theory and vice versa (Lawton 1996; Hobbs & Harris 2001; Sayer & Campbell 2004). We suggest that a conceptual framework comprise a nested suite of conceptual models that include:

- the regional-scale social-ecological system in the context of national and global processes
- a landscape management model for enacting conservation action (e.g. conservation corridors, biosphere reserves)
- the conservation planning process (i.e. the social and decision processes manifesting as the operational model presented here; Clark 2002)
- the role and scope of the conservation planner.

An operational model should be explicitly outlined prior to undertaking a conservation planning initiative (Margolius & Salafsky 1998; Noss 2003; Knight et al. 2006a). Numerous operational models have been developed, including Brunckhorst (2000), Margules & Pressey (2000), Cowling & Pressey (2003), Groves (2003), Knight et al. (2006a), Margules & Sarkar (2007) and Pressey & Bottrill (2008). Early operational models (e.g. Margules & Pressey 2000) focused primarily on spatial prioritisation, whilst later examples have increasingly emphasised collaborative processes (e.g. Pressey & Bottrill 2008). Significant similarities, and differences, exist between them regarding philosophy (e.g. top-down versus bottom-up), activities (e.g. the relative importance of spatial prioritisation versus conservation planning), and degree of stakeholder collaboration (e.g. true collaboration versus simple engagement). The implementation of an operational model should be guided by a suite of principles (Noss 2003) developed and sanctioned by stakeholders (Smith et al. 2009).

Problem definition for conservation planning: orientation and formulation

Conservation planning is a problem-solving activity in which individual planning contexts (i.e. social-ecological systems) differ, meaning that solutions are typically context specific. Conservation planning problems should be explicitly defined at two operational scales (see Figure 20.1). First, the *problem orientation* of a planning region should be assessed to elucidate the broad context in which conservation planning will occur (Brunckhorst 2000; Clark 2002; Cowling & Pressey 2003; Knight et al. 2006a; Cowling & Wilhelm-Rechmann 2007; Cowling et al. 2008; Pressey & Bottrill 2008). This should include developing a sound understanding of the social, cultural, economic, environmental and political context of a planning region, which subsequently underpins the framing and development of conservation planning activities such as mainstreaming, enabling and implementation (see Figure 20.1). Second, and more specifically, a *problem formulation* must be explicitly stated and operationalised through a spatial prioritisation that schedules where and when conservation resources should be allocated to assist in solving the conservation planning problem (Moilanen et al. 2009a; Ferrier & Wintle 2009; Wilson et al. 2009b).

Effective problem formulations are therefore defined within the context of the broader problem orientation (e.g. address vulnerability from the threats identified through the problem orientation). This promotes the design and implementation of feasible conservation actions matched to the complex suite of factors that define the challenges of managing real-world social-ecological systems (Knight & Cowling 2007; Knight et al. 2010; Raymond & Brown 2011). For example, in the subtropical thicket biome within the Maputaland-Pondoland-Albany hotspot, problem orientation revealed that elephants were a keystone species, local land managers were willing to implement covenants on their farms, and decision makers in local municipalities were unfamiliar with geographic information systems (GIS) and lacked sophisticated understanding of environmental legislation. As a result, the problem formulation guided the spatial prioritisation to design corridors that facilitated elephant movements and representatively sampled vegetation types (Rouget et al. 2006), focus on private land linking existing protected areas, not land purchase for formal protected area expansion, and deliver hard-copy maps, not GIS data, to local decision makers who were then trained in their use.

Problem orientation

Problem orientation, variously referred to as scoping (Brunckhorst 2000; Pressey & Bottrill 2008) or social assessment (Cowling & Wilhelm-Rechmann 2007), provides a comprehensive understanding of the social, cultural, economic, political and environmental context of a planning region (Clark 2002; Cowling & Wilhelm-Rechmann 2007). It assists in determining effective ways in which conservation planning problems are framed and the most appropriate decision-making process to arrive at a solution (Clark 2002). For example, government conservation agencies will have a specific decision process for acquiring new additions to formally protected areas that is distinctly different from the appropriate process for signing covenants with rural land managers. Whilst biologically focused spatial prioritisations provide a defensible approach for identifying cost-efficient arrangements of candidate protected areas (the 'where' problem), problem orientation also provides prerequisite information required for implementing effective conservation action (the 'how' problem) (Scott & Csuti 1997). Problem orientations should not be limited to demographic and land use data but should rather be specifically assessed for specific individual planning regions, gathering spatially explicit information where possible. Problem orientation avoids a conservation planner imposing his or her values upon society, instead presenting recommendations (not prescriptions) that reflect society's values (Theobald et al. 2000). Collaborative development of conceptual (or logic) models from the practice of evaluation can provide useful tools that elucidate the elements and processes comprising social-ecological systems (Margolius & Salafsky 1998).

Problem orientation should not be limited simply to understanding conservation management challenges, but should provide detailed insights into the links between human society and the natural environment. Understanding the socio-economic drivers of land use and development trends and projections of future land use pressures, and how to mainstream conservation into production activities, is critically important. Pierce et al. (2005) highlight the importance of understanding the knowledge and capacity of decision makers required to mainstream maps of important conservation areas into the land use planning process in South Africa. Fundamentally, such understanding begins with developing detailed knowledge of the key stakeholders and their power relations, inclusive of interested and affected individuals, groups and organisations. The principal drivers of their behaviour should be identified, specifically their differing perspectives, cultural and environmental values, how these values are expressed, how various situations influence their actions, and the ways in which outcomes and long-term effects of their actions might be quantified (Clark 2002).

Stakeholder data should be complemented with information on the conservation instruments, incentives and institutions that might be feasibly and effectively employed to secure conservation goals (e.g. covenants, tax rebates, social learning groups), and the decision-making processes that provide opportunities for operationalising conservation action (e.g. traditional indigenous practices, policy processes). Problem orientation assessment should also reveal opportunities for linking conservation actions synergistically with initiatives in other sectors (e.g. agriculture, industry), thereby promoting the mainstreaming of conservation goals (Rietbergen-McCracken & Narayan 1998; Cowling & Wilhelm-Rechmann 2007). Assessment of institutional and organisational functioning, capacity, stability and effectiveness to assist in identifying feasible implementation activities and enabling (i.e. capacity-building) requirements should also be undertaken. In short, problem orientation should ultimately offer an appropriate approach for collaborative decision making, and direction for formulating a preliminary landscape management model (Knight et al. 2006a; Rouget et al. 2006) that comprises an optimal mix of instruments, incentives and institutions (Young et al. 1996).

The megaconservancy networks of Rouget et al. (2006) provide an example of landscape management model that comprises:

- a negotiated agreement (contractual or voluntary) between the state and a private land manager
- a financial incentive for the land manager from national government (e.g. a tax rebate)
- a financial incentive from local government (e.g. a rates rebate)
- supportive legislation through the National Environmental Management (Biodiversity and Protected Areas) Acts
- networking and training opportunities for participating land managers.

Other examples include biosphere reserves and conservation corridors. Landscape management models should be explicitly defined prior to conducting a spatial prioritisation (Knight et al. 2006a; Ferrier & Wintle 2009), as they inform problem formulation, determine the technical requirements and appropriate analytical technique, and provide stakeholders with a clear concept of how implementation will be approached.

Problem formulation

Spatial prioritisation analyses are used to strategically allocate resources for conservation actions that are inevitably scarce and often inadequate for ensuring the long-term effectiveness of these activities. Accordingly, spatial prioritisations aim to assist conservation planners to make decisions about 'what' and 'where' conservation should be enacted (Knight et al. 2006a; Wilson et al. 2007), so as to ensure that these resources are utilising as cost-efficiently and effectively as possible (Wilson et al. 2007). Ultimately, spatial prioritisation analyses should not be considered definitive decision-making tools, but rather a source of information that decision makers can consider against a range of other important information (Ferrier & Wintle 2009). The spatial prioritisation software used by conservation planners to identify networks of complementary areas important for achieving conservation goals is driven by a *problem formulation*.

Clear problem formulation articulates stakeholders' goals, and should be tailored to the specific characteristics of a planning region, specifically the opportunities and constraints upon implementing effective conservation action and, pragmatically, the availability of data (Pressey 2004) and other resources. It strikes a balance between the goal of the initiative and what is feasible and acceptable in order to achieve that outcome (i.e. the constraints), and should therefore be defined from a mix of environmental, social, political and economic input. The sociopolitical context provides information on the values society places upon natural features and hence the choice of conservation actions, whilst the ecological context provides information on the natural features of concern and targets for their conservation management (Rondinini & Chiozza 2010). Economic factors, such as the financial or resource requirements to achieve goals, determine the types, timing and extent of proposed conservation actions (Wilson et al. 2009b). A typical example of a problem formulation would be identifying a formal protected area network that comprehensively samples all ecosystems and IUCN Red List species in the most cost-efficient, yet connected, spatial arrangement possible.

Problem formulation is essentially a mathematical problem (Moilanen et al. 2009b) requiring numerous decisions on a range of technical matters. Perhaps the most important of these decisions is identifying the appropriate type of spatial prioritisation analysis (Ferrier & Wintle 2009), as this directly influences the type of conservation planning products developed (Figure 20.3). Ultimately, implementers' needs must be met (Pierce et al. 2005; Knight et al. 2006b). There has been considerable research and development of decision support software and spatial prioritisation techniques that reflect different formulations of the spatial conservation prioritisation problem (see Sarkar et al. 2006; Moilanen et al. 2009c). Three general types of analysis are prominent (Ferrier & Wintle 2009).

Optimal plan development provides one or more spatial solutions for a planning region that are comprehensive and mathematically optimal (or near-optimal) in terms of cost-efficiency. A solution may consist, for example, of a recommended configuration of protected areas where candidate areas are 'selected' or 'not selected' for eventual inclusion in a protected area network. Such solutions are useful for decision makers requiring definitive solutions regarding where conservation should, or should not, be enacted (e.g. those transferring public lands from production to conservation tenures) (Ferrier et al. 2000) or those assessing development applications (Pierce et al. 2005), or may be used as a starting point for subsequent refinement by decision makers (Fernandes et al. 2005). Such analyses have traditionally often employed biological (e.g. species locality) or ecological (e.g. habitat or ecosystem) data alone, although in regions of high spatial turnover in environmental features, such as biodiversity 'hotspots' (Myers et al. 2000), human and social data alone may provide more cost-effective solutions (Cowling et al. 2010).

In contrast, *priority mapping* provides no definitive solutions but presents a range of available choices across a scale of relative importance (or priority). Maps of irreplaceability (Ferrier et al. 2000) provide one such example, where ranked choices are illustrated using a continuous colour scale representing the degree to which individual areas are necessary (or conversely, replaceable) for achieving conservation targets. Such solutions are most useful where implementing organisations make decisions in a highly incremental or opportunistic manner, for example, where funding or opportunities for land acquisition are sporadically available, or land manager willingness to sell (e.g. Knight et al. 2011) or to rezone traditional fishing areas (e.g. Game et al. 2011) is low or variable. The irreplaceability of individual areas

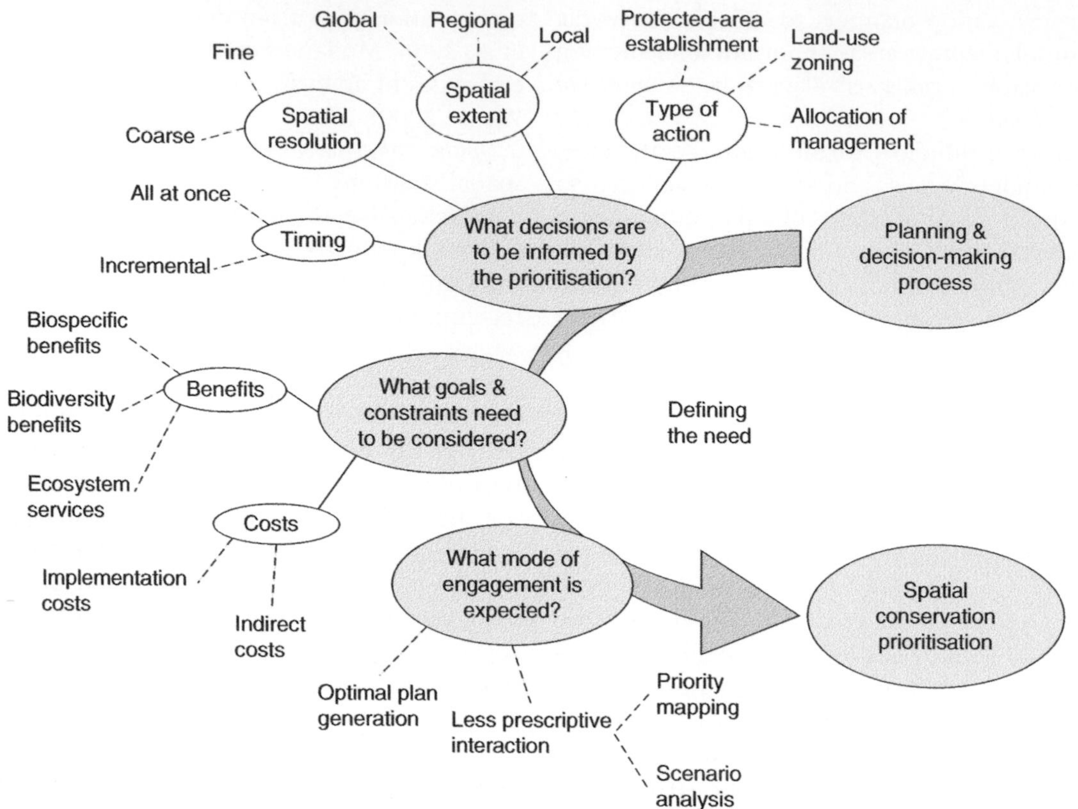

Figure 20.3 Issues requiring consideration in defining the problem formulation for a spatial conservation prioritisation to ensure it delivers outputs that meet the needs of a specific planning process (from Ferrier & Wintle 2009).

changes as decisions are made to conserve areas, so a new irreplaceability map must be generated between each effective implementation activity (e.g. land purchase, covenant signing). Priority mapping can facilitate an 'informed opportunism' approach to implementation (e.g. Noss et al. 2002; Game et al. 2011) where decisions are made incrementally, with emerging opportunities locked in before conducting repeated irreplaceability analyses to derive new solutions.

Finally, in situations where decision makers require feedback on the relative effectiveness of different future contexts, *scenario evaluation* may be more appropriate. This requires presenting decision makers with a small number of alternative futures, for example, presenting three alternative protected area configurations to address alternative global climate change models. These might include alternatives developed using expert judgement alone and/or a negotiation process. This allows the effectiveness of different solutions to be compared by decision makers. The application of the InVEST modelling tool to develop a small number of spatially explicit, stakeholder-defined scenarios that quantify the value of ecosystem services, nature conservation and commodity production in the Willamette Basin, Oregon, United States is a strong example of the utility of scenario evaluation (Nelson et al. 2009).

These three analyses are not mutually exclusive, and any given planning process may require two, or even all three, applied in combination. For example, in providing local government land use planners with information useful for assessing development applications, Rouget et al. (2006) employed a least-cost path analysis to identify a near-optimal plan composed of land parcels that achieved quantitative conservation targets that was then complemented with priority mapping of land parcels outside the near-optimal set of land parcels. This provided the planners with information on 'no-go' areas where development should be restricted to no or low impact activities, in addition to areas of low conservation importance where proposed developments could be negotiated.

Further to deciding upon the appropriate analysis, a range of subsequent decisions must be made regarding, but not limited to, the data with which to conduct the spatial prioritisation (Cowling et al. 2004; Pressey 2004), the surrogates to use for natural features (Rodrigues & Brooks 2007), the targets to serve as quantified goals (Desmet & Cowling 2004), and ways to calculate costs (Naidoo et al. 2006), predict threats (Wilson et al. 2005) and measure the likely effectiveness of different actions. These decisions may be best made by 'expert' conservation planners who understand the technical limitations of spatial prioritisation software. However, it is imperative that these decisions involve staff from implementing organisations, to ensure their understanding and 'buy-in' (Smith et al. 2009), and that the eventual conservation planning products can be mainstreamed into decision-making processes (Knight et al. 2006b).

Quantitative targets should also be defined as an explicit statement of goals. Target definition requires balancing ecological considerations (e.g. long-term population viability; Burgman et al. 2001) and social values (e.g. areas of biocultural importance; Maffi & Woodley 2010) against the pragmatic constraints of data availability (e.g. the feasibility of setting population-based targets for species for which spatially explicit population data are available; Jackson et al. 2004). For example, legislation may define appropriate targets by stipulating a legal requirement for habitat conservation of specific species (e.g. the United States' Endangered Species Act or the European Union's Habitats Directive), or through explicitly stated quantitative targets for individual ecosystems (such as with South Africa's National Environmental Management (Biodiversity) Act of 2004). Legislation may also specify constraints to planning (e.g. land tenure rights or land use zoning, such as protected areas or forestry concessions).

Decisions must also be made as to whether to account for the spatial configuration of existing protected areas (McDonnell et al. 2002), the connectedness of valued natural or social features (Hanski & Ovaskainen 2000), and/or the relative location and configuration of different zones for conservation or extractive use (Watts et al. 2009; Wilson et al. 2010). It might be important that conservation investments are scheduled through time if, for example, the objective is to minimise the loss of ecosystems, but in some cases a static evaluation of conservation priorities will be more appropriate (Costello & Polasky 2004).

Raw outputs from spatial prioritisation analyses may not be user-friendly for decision makers, and may require interpretation and refinement (Pierce et al. 2005; Ferrier & Wintle 2009). For example, Pierce et al. (2005) spent the better part of a year refining and translating spatial prioritisation outputs of Rouget et al. (2006) into conservation planning products for local government land use planners in South Africa. Conservation planning products will vary according to the agreed landscape management model, the appropriate decision-making process for implementing conservation action, and the capacity of practitioners. Examples include briefing notes for politicians, lists of priority areas, hard copy maps, digital maps, interpretive handbooks, training courses for staff to use maps and handbooks (Pierce et al. 2005), and institutionalised decision support systems (e.g. Theobald et al. 2000). Ensuring that conservation planning products can be effectively translated into conservation action requires that they be complemented with processes for stakeholder collaboration for developing a common vision and support for a specific conservation planning initiative and implementation strategy development specifying 'how' conservation should be enacted (Knight et al. 2006a, 2011a).

Future directions for improving the effectiveness of conservation planning

The vast majority of conservation planning research has focused upon the technical aspects of spatial prioritisation (Knight et al. 2006a), and not on the normative, typically human and social dimensions of conservation planning problems and processes (Knight et al. 2008). Accordingly, our understanding of the technical limitations and applications of spatial prioritisation techniques far exceeds our knowledge of how to translate outputs from these analyses into effective conservation action (Knight et al. 2006a). The most significant challenges to improving the effectiveness of conservation planning initiatives are, therefore, not technical but rather operational.

Research focused upon integrating human and social data into spatial prioritisations will transform these techniques from simply assessing biological priority to presenting hypotheses for the likelihood of implementing feasible and effective conservation action. Such approaches will better link spatial prioritisation outputs to conservation action by incorporating data on the behaviour, capacity and willingness of stakeholders to effectively engage conservation planning initiatives (Knight et al. 2010, 2011b; Raymond & Brown 2011).

Experience from South Africa indicates that spatial prioritisation outputs need to be translated into conservation planning products, such as institutionalised decision support systems, maps tailored to user needs, explanatory handbooks and training courses (Pierce et al. 2005; Reyers et al. 2007). Research into the types of products required by implementers, and how best to collaboratively develop these with stakeholders, is required (Theobald et al. 2000; Pierce et al. 2005; Ferrier & Wintle 2009). Ensuring that these conservation planning products are mainstreamed effectively into implementing organisations is a highly challenging task, and one insufficiently researched. Evidence-based approaches for developing mainstreaming techniques for the framing and social marketing of conservation planning initiatives will become increasingly important (Wilhelm-Rechmann & Cowling 2011).

The lack of organisational uptake of conservation planning products (e.g. Knight 2006; Beier 2007; Knight et al. 2011a; Wilhelm-Rechmann & Cowling 2011) points towards the need for a deeper understanding of the institutional contexts hindering the achievement of conservation goals. Current understanding of how to design and implement top-down, multi-organisational collaborative governance structures is poor, but is a fundamental requirement, for example, for aligning the goals of the different organisations responsible for managing terrestrial, freshwater and marine ecosystems across multiple geographical and institutional scales (Stoms et al. 2005). This will require research into the institutional constraints that hinder co-ordinated action across organisations, mechanisms to address scale mismatches between administrative jurisdictions and ecological systems (Briggs 2001; Cumming et al. 2006), and also the requirements for establishing information networks that promote collaborative planning (Bodin & Crona 2009). In this regard, conservation planners can learn much from researchers in natural resource management (e.g. Ostrom 1990; Berkes & Folke 1998).

The need to address the complexity and long-term timeframes of most conservation planning initiatives ensures that social learning and adaptive management are fundamental goals of these processes (Salafsky et al. 2002; Knight et al. 2006a; Grantham et al. 2010). The importance of establishing effective social learning institutions (e.g. Knight et al. 2006a) is both underappreciated and poorly understood by conservation planners generally. It is therefore critically important that research be conducted into the impacts of, and learning opportunities provided by, failure; the adaptive capacity of individuals and organisations (e.g. their ability and willingness to change behaviours); the factors perversely confounding the uptake of recommendations from implementation strategies (e.g. Knight et al. 2011a); and approaches for institutionalising adaptive management. The development and integration of evaluation and monitoring techniques into conservation planning operational models will be essential for achieving this goal, and also requires further research.

Most of the stages represented in conservation planning operational models (see Figure 20.1) include activities that are mainstays of other disciplines. If conservation planning is to become truly transdisciplinary and implement effective conservation actions in the most cost-efficient ways, it will be essential that conservation planners build strong, long-term collaborations with practitioners from other disciplines, notably protected area managers, community development specialists, resource economists, town and regional planners, sociologists and anthropologists, and perhaps most importantly, private land managers, so as to integrate (and avoid duplicating) the best existing knowledge into conservation planning operational models.

Acknowledgements

ATK and KW acknowledge the support of the ARC Centre of Excellence in Environmental Decisions at the University of Queensland, Australia. ATK thanks Richard Cowling, Bruce Campbell, Susan Clark, Kent Redford, Nick Salafsky, Gregg Walker and Matt Keene for insights that helped to formulate these ideas, and Stellenbosch University for funding. KW thanks the Australian Research Council for funding. NS thanks the Danish National Research Foundation for supporting his research through the Center for Macroecology, Evolution and Climate, and the Danish Council for Independent Research – Social Science (grant no. 75-07-0240).

References

Ando, A., Camm, J.D., Polasky, S. & Solow, A.R. (1998) Species distribution, land values and efficient conservation. *Science*, **279**, 2126–2128.

Araújo, M.B., Cabeza, M., Thuiller, W., Hannah, L & Williams, P.H. (2004) Would climate change drive species out of reserves? An assessment of existing reserve-selection methods. *Global Change Biology*, **10**, 1618–1626.

Balmford, A., Bruner, A., Cooper, P., *et al.* (2002) Economic reasons for conserving wild nature. *Science*, **297**, 950–953.

Beier, P. (2007) Learning like a mountain. *Wildlife Professional*, **Winter**, 26–29.

Berkes, F. & Folke, C. (eds.) (1998) *Linking Social and Ecological Systems: Management Practices and Social Mechanisms for Building Resilience*. Cambridge University Press, Cambridge.

Bodin, O. & Crona, B.I. (2009) The role of social networks in natural resource governance: what relational patterns make a difference? *Global Environmental Change – Human and Policy Dimensions*, **19**, 366–374.

Briggs, S.V. (2001) Linking ecological scales and institutional frameworks for landscape rehabilitation. *Ecological Management and Restoration*, **2**, 28–35.

Brunckhorst, D.J. (2000) *Bioregional Planning: Resource Management Beyond the New Millennium*. Harwood Academic Publishers, Amsterdam.

Burgman, M.A., Possingham, H.P., Lynch, A.J., *et al.* (2001) A method for setting the size of plant conservation target areas. *Conservation Biology*, **15**, 603–616.

Ciarleglio, M., Barnes, J.W. & Sarkar, S. (2010) ConsNet – a tabu search approach to the spatially coherent conservation area network design problem. *Journal of Heuristics*, **16**, 537–557.

Clark, T.W. (2002) *The Policy Process*. Yale University Press. New Haven, CT.

CMP (2007) *Open Standards for the Practice of Conservation*, Version 2.0: www.conservationmeasures.org

Coad, L., Burgess N.D., Loucks, C., *et al.* (2010) Reply to Jenkins and Joppa – expansion of the global terrestrial protected area system. *Biological Conservation*, **143**, 5–6.

Costello, C. & Polasky, S. (2004) Dynamic reserve site selection. *Resource and Energy Economics*, **26**, 157–174.

Cowling, R.M. & Pressey, R.L. (2003) Introduction to systematic conservation planning in the Cape Floristic Region. *Biological Conservation*, **112**, 1–14.

Cowling, R.M. & Wilhelm-Rechmann, A. (2007) Social assessment as a key to conservation success. *Oryx*, **41**, 135.

Cowling, R.M., Pressey, R.L., Lombard, A.T., Desmet, P.G. & Ellis, A.G. (1999) From representation to persistence: requirements for a sustainable system of conservation areas in the species rich Mediterranean climate desert of southern Africa. *Diversity and Distributions*, **5**, 51–71.

Cowling, R.M., Pressey, R.L., Sims-Castley, R., *et al.* (2003) The expert or the algorithm? – comparison of priority conservation areas in the Cape Floristic Region identified by park managers and reserve selection software. *Biological Conservation*, **112**, 147–167.

Cowling, R.M., Knight, A.T., Faith, D.P., *et al.* (2004) Nature conservation requires more than a passion for species. *Conservation Biology*, **18**, 1674–1677.

Cowling, R.M., Egoh, B., Knight, A.T., *et al.* (2008) An operational model for mainstreaming ecosystem services for implementation. *Proceedings of the National Academy of Sciences USA*, **105**, 9483–9488.

Cowling, R.M., Knight, A.T., Privett, S.D. & Sharma, G.P. (2010) Invest in opportunity, not inventory in hotspots. *Conservation Biology*, **24**, 633–635.

Cumming, G.S., Cumming, D.H. & Redman, C.L. (2006) Scale mismatches in social-ecological systems: Causes, consequences, and solutions. *Ecology and Society*, **11**(1), 14.

Daniels, S.E. & Walker, G.B. (2001). *Working Through Environmental Conflicts: The Collaborative Learning Approach.* Praeger Publishers, Westport, CO.

Desmet, P.G. & Cowling, R.M. (2004) Using the species-area relationship to set baseline targets for conservation. *Ecology and Society*, **9**(2), 11.

Fernandes, L., Day, J., Lewis, A., *et al.* (2005) Establishing representative no-take areas in the Great Barrier Reef: large-scale implementation of theory on marine protected areas. *Conservation Biology*, **19**, 1733–1744.

Ferrier, S. & Wintle, B.A. (2009) Quantitative approaches to spatial conservation prioritisation: matching the solution to the need. In: *Spatial Conservation Prioritisation: Quantitative Methods and Computational Tools* (eds A. Moilanen, K.A. Wilson &, H.P. Possingham), pp.1–15. Oxford University Press, Oxford.

Ferrier, S., Pressey, R.L. & Barrett, T.W. (2000) A new predictor of the irreplaceability of areas for achieving a conservation goal, its application to real-world planning, and a research agenda for further refinement. *Biological Conservation*, **93**, 303–325.

Ferrier, S., Powell G.V., Richardson, K.S., *et al.* (2004) Mapping more of biodiversity for global conservation assessment. *BioScience*, **54**, 1101–1109.

Gallo, L.A., Marchelli, P., Chauchard, L. & Gonzalez Peñalba, M. (2009) Knowing and doing: research leading to action in the conservation of forest genetic diversity of Patagonian temperate forests. *Conservation Biology*, **23**, 895–898.

Game, E.T., Lipsett-Moore, G., Hamilton, R., *et al.* (2011) Informed opportunism for conservation planning in the Solomon Islands. *Conservation Letters*, **4**, 38–46.

Gelderblom C.M., van Wilgen, B.W., Nel, J.L., Sandwith, T., Botha, M.A. & Hauck, M. (2003) Turning strategy into action: implementing a conservation action plan in the Cape Floristic Region. *Biological Conservation*, **112**, 291–297.

Grantham, H.S., Bode, M., McDonald-Madden, E., Game, E.T., Knight, A.T. & Possingham, H. (2010) Effective conservation planning requires learning and adaptation. *Frontiers in Ecology and the Environment*, **8**, 431–437.

Groves, C.R. (2003) *Drafting a Conservation Blueprint: A Practitioners Guide to Planning for Biodiversity*. Island Press, Washington, D.C.

Hanski, I. & Ovaskainen, O. (2000) The metapopulation capacity of a fragmented landscape. *Nature*, **404**, 755–758.

Hobbs, R.J. & Harris, J.A. (2001) Restoration ecology: repairing the earth's ecosystems in the new millennium. *Restoration Ecology*, **9**, 239–246.

Holling, C.S. (1978) *Adaptive Environmental Assessment and Management*. John Wiley, New York.

Hopkinson, P., Evans, J. & Gregory, R.D. (2000) National-scale conservation assessments at an appropriate resolution. *Diversity and Distributions*, **6**, 195–204.

Jackson, S.F., Kershaw, M. & Gaston, K.J. (2004) Size matters: the value of small populations for wintering waterbirds. *Animal Conservation*, **7**, 229–239.

Justus, J. & Sarkar, S. (2002) The principle of complementarity in the design of reserve networks to conserve biodiversity: a preliminary history. *Journal of Biosciences*, **27**(4) Suppl. 2, 421–435.

Kirkpatrick, J.B. (1983) An iterative method for establishing priorities for the selection of nature reserves: an example from Tasmania. *Biological Conservation*, **25**, 127–134.

Kirkpatrick, J.B., Brown, M.J. & Moscal, A. (1980) *Threatened Plants of the Tasmanian Central East Coast*. Tasmanian Conservation Trust, Hobart, Australia.

Klein, C.J., Wilson, K., Watts, M., *et al.* (2009) Incorporating ecological and evolutionary processes into large-scale conservation planning. *Ecological Applications*, **19**, 206–217.

Knight, A.T. (2006) Failing but learning: writing the wrongs after Redford and Taber. *Conservation Biology*, **20**, 1312–1314.

Knight, A.T. & Cowling, R.M. (2007) Embracing opportunism in the selection of priority conservation areas. *Conservation Biology*, **21**, 1124–1126.

Knight, A.T., Cowling, R.M. & Campbell, B.M. (2006a) An operational model for implementing conservation action. *Conservation Biology*, **20**, 408–419.

Knight, A.T., Driver, A., Cowling, R.M., *et al.* (2006b) Designing systematic conservation assessments that promote effective implementation: best practice from South Africa. *Conservation Biology*, **20**, 739–750.

Knight, A.T., Cowling, R.M., Rouget, M., Balmford, A., Lombard, A.T. & Campbell, B.M. (2008) Knowing but not doing: selecting priority conservation areas and the research-implementation gap. *Conservation Biology*, **22**, 610–617.

Knight, A.T., Cowling, R.M., Difford, M. & Campbell, B.M. (2010) Mapping human and social dimensions of conservation opportunity for the scheduling of conservation action on private land. *Conservation Biology*, **24**, 1348–1358.

Knight , A.T., Cowling, R.M., Boshoff, A.F., Wilson, S.L. & Pierce, S.M. (2011a) Walking in STEP: lessons for linking spatial prioritisations to implementation strategies. *Biological Conservation*, **144**, 202–211.

Knight, A.T., Grantham, H.S., Smith, R.J., McGregor, G.K., Possingham, H.P & Cowling, R.M. (2011b). Land manager willingness-to-sell defines conservation opportunity for protected area expansion. *Biological Conservation*, **144**, 2623–2630.

Lawton, J.H. (1996) Corncrake pine and prediction in ecology. *Oikos*, **76**, 3–4.

Maffi, L. & Woodley, E. (eds) (2010) *Biocultural Diversity Conservation. A Global Sourcebook.* Earthscan, London.

Margoluis, R. & Salafsky, N. (1998) *Measures of Success: Designing, Managing and Monitoring Conservation and Development Projects.* Island Press, Washington, D.C.

Margules, C. & Pressey, R.L. (2000) Systematic conservation planning. *Nature*, **405**, 243–253.

Margules, C. & Sarkar, S. (2007) *Systematic Conservation Planning*. Cambridge University Press, Cambridge.

Margules, C., Nicholls, A.O. & Pressey, R.L. (1988) Selecting networks of reserves to maximise biological diversity. *Biological Conservation*, **43**, 63–76.

Max-Neef, M.A. (2005) Foundations of transdisciplinarity. *Ecological Economics*, **53**, 5–16.

McDonnell, M.D., Possingham, H.P., Ball, I.R. & Cousins, E.A. (2002) Mathematical models for spatially cohesive reserve design. *Environmental Modeling and Assessment*, **7**, 107–114.

McShane, T.O. & Wells, M.P. (eds) (2004) *Getting Biodiversity Projects to Work: Towards More Effective Conservation and Development.* Columbia University Press, New York.

Millennium Ecosystem Assessment (2005) *Ecosystems and Human Well-being: Synthesis.* Island Press, Washington, D.C.

Moilanen, A., Kujala, H. & Leathwick, J.R. (2009a) The zonation framework and software for conservation prioritization. In: *Spatial Conservation Prioritisation: Quantitative Methods and Computational Tools* (eds A. Moilanen, K.A. Wilson &, H.P. Possingham), pp.196–210. Oxford University Press, Oxford.

Moilanen, A., Possingham, H.P. & Polasky, S. (2009b) A mathematical classification of conservation prioritisation problems. In: *Spatial Conservation Prioritisation: Quantitative Methods and Computational Tools* (eds A. Moilanen, K.A. Wilson &, H.P. Possingham), pp.28–42. Oxford University Press, Oxford.

Moilanen, A., Wilson, K.A. & Possingham, H.P. (2009c) *Spatial Conservation Prioritisation: Quantitative Methods and Computational Tools.* Oxford University Press, Oxford.

Morrison, J., Loucks, C., Long, B. & Wikramanayake, E. (2009) Landscape-scale spatial planning at WWF: a variety of approaches. *Oryx*, **34**, 499–507.

Murphy, D.D. & Noon, B.R. (1992) Integrating scientific methods and habitat conservation planning: reserve design for Northern Spotted Owls. *Ecological Applications*, **2**(1), 3–17.

Myers, N., Mittermeier, R.A., Mittermeier, C.G., da Fonseca, G.A. & Kent, J. (2000) Biodiversity hotspots for conservation priorities. *Nature*, **403**, 853–858.

Naidoo, R., Balmford, A., Ferraro, P.J., Polasky, S., Ricketts, T.H. & Rouget, M. (2006) Integrating economic costs into conservation planning. *Trends in Ecology and Evolution*, **21**, 681–687.

Nelson, E., Mendoza, G., Regetz, J., *et al.* (2009) Modeling multiple ecosystem services, biodiversity conservation, commodity production, and tradeoffs at landscape scales. *Frontiers in Ecology and the Environment*, **7**, 4–11.

Noss, R.F. (2003) A checklist for wildlands network designs. *Conservation Biology*, **17**(5), 1270–1275.

Noss, R.F., O'Connell, M.A. & Murphy, D.D. (1997) *The Science of Conservation Planning: Habitat*

Conservation Under the Endangered Species Act. Island Press, Washington, D.C.

Noss, R.F., Carroll, C., Vance-Borland, K. & Wuerthner, G. (2002) A multicriteria assessment of the irreplaceability and vulnerability of sites in the Greater Yellowstone ecosystem. *Conservation Biology*, **16**(4), 895–908.

Ostrom, E. (1990) *Governing the Commons: The Evolution of Institutions for Collective Action*. Cambridge University Press, Cambridge, NY.

Ostrom, E., Janssen, M.A. & Anderies, J.M. (2007) Going beyond panaceas. *Proceedings of the National Academy of Sciences USA*, **104**, 15176–15178.

Pence, G.Q., Botha, M.A. & Turpie, J.K. (2003) Evaluating combinations of on- and off-reserve conservation strategies for the Agulhas Plain, South Africa: a financial perspective. *Biological Conservation*, **112**, 253–274.

Pfeffer, J. & Sutton, R.I. (1999) Knowing "What" to do is not enough: turning knowledge into action. *California Management Review*, **42**, 83–107.

Pierce, S.M., Cowling, R.M., Knight, A.T., Lombard, A.T., Rouget, M. & Wolf, T. (2005) Systematic conservation planning products for land-use planning: interpretation for implementation. *Biological Conservation*, **125**, 441–458.

Prendergast, J.R., Quinn, R.M. & Lawton, J.H. (1999) The gaps between theory and practice in selecting nature reserves. *Conservation Biology*, **13**, 484–492.

Pressey, R.L. (1994) Ad hoc reservations – forward or backward steps in developing representative reserve systems? *Conservation Biology*, **8**, 662–668.

Pressey, R.L. (2002) The first reserve selection algorithm – a retrospective on Jamie Kirkpatrick's 1983 paper. *Progress in Physical Geography*, **26**, 434–441.

Pressey, R.L. (2004) Conservation planning and biodiversity: assembling the best data for the job. *Conservation Biology*, **18**, 1677–1681.

Pressey, R.L. & Bottrill, M.C. (2008) Opportunism, threats and the evolution of systematic conservation planning. *Conservation Biology*, **22**, 1340–1345.

Pressey, R.L. & Nicholls, A.O. (1989) Application of a numerical algorithm to the selection of reserves in semi-arid New South Wales. *Biological Conservation*, **50**, 263–278.

Pressey, R.L. & Taffs, K.H. (2001) Scheduling conservation action in production landscapes: priority areas in western New South Wales defined by irreplaceability and vulnerability to vegetation loss. *Biological Conservation*, **100**, 355–376.

Pressey, R.L. & Tully, S.L. (1994) The cost of ad hoc reservation – a case-study in western New South Wales. *Australian Journal of Ecology*, **19**, 375–384.

Pressey, R.L., Humphries, C.J., Margules, C.R., *et al.* (1993) Beyond opportunism – key principles for systematic reserve selection. *Trends in Ecology and Evolution*, **8**, 124–128.

Pressey, R.L., Cabeza, M., Watts, M., Cowling, R.M. & Wilson, K.A. (2007) Conservation planning in a changing world. *Trends in Ecology and Evolution*, **22**, 583–592.

Pressey, R.L., Watts, M.E., Barrett, T.W. & Ridges, M.J. (2009) The C-Plan conservation planning system: origins, applications, and possible futures. In: *Spatial Conservation Prioritisation: Quantitative Methods and Computational Tools* (eds A. Moilanen, K.A. Wilson & H.P. Possingham), pp.211–234. Oxford University Press, Oxford.

Pullin, A.S. & Knight, T.M. (2003) Support for decision making in conservation practice: an evidence-based approach. *Journal for Nature Conservation*, **11**, 83–90.

Rapoport, A. (1985) Thinking about home environments: a conceptual framework. In: *Home Environments* (eds I. Altman & C.M. Werner), pp.255–261. Plenum Press, New York.

Raymond, C.M. & Brown, G. (2011) Assessing conservation opportunity on private land: socio-economic, behavioral, and spatial dimensions. *Journal of Environmental Management*, **92**, 2513–2523.

Redford, K.H. & Taber, A. (2000) Writing the wrongs: developing a safe-fail culture in conservation. *Conservation Biology*, **14**, 1567–1568.

Reyers, B., Rouget, M., Jonas, Z., *et al.* (2007) Developing products for conservation decision-making: lessons from a spatial biodiversity assessment for South Africa. *Diversity and Distributions*, **13**, 608–619.

Reyers, B., Roux, D., Cowling, R.M., Ginsburg, A.E., Nel, J.L. & O'Farrell, P. (2010) Conservation planning as a transdisciplinary process. *Conservation Biology*, **24**, 957–965.

Rietbergen-McCracken, J. & Narayan, D. (1998) *Participation and Social Assessment: Tools and Techniques*. World Bank, Washington, D.C.

Rodrigues, A.S. & Brooks, T.M. (2007) Shortcuts for biodiversity conservation planning: the effectiveness of surrogates. *Annual Review of Ecology, Evolution, and Systematics*, **38**, 713–737.

Rodrigues, A.S., Tratt, R., Wheeler, B.D. & Gaston, K.J. (1999) The performance of existing networks of conservation areas in representing biodiversity. *Proceedings of the Royal Society B: Biological Sciences*, **266**, 1453–1459.

Rodrigues, A.S., Cerdeira, J.O. & Gaston, K.J. (2000) Flexibility, efficiency, and accountability: adapting reserve selection algorithms to more complex conservation problems. *Ecography*, **23**, 565–574.

Rodrigues, A.S., Andelman, S.J., Bakarr, M.I., *et al.* (2004) Effectiveness of the global protected area network in representing species diversity. *Nature*, **428**, 640–643.

Rondinini, C. & Chiozza, F. (2010) Quantitative methods for defining percentage area targets for habitat types in conservation planning. *Biological Conservation*, **143**, 1646–1653.

Rouget, M., Cowling, R.M., Pressey, R.L. & Richardson, D.M. (2003) Identifying spatial components of ecological and evolutionary processes for regional conservation planning in the Cape Floristic Region, South Africa. *Diversity and Distributions*, **9**, 191–210.

Rouget, M., Cowling, R.M., Lombard, A.T., Knight, A.T. & Kerley, G.I. (2006) Designing large-scale conservation corridors for pattern and process. *Conservation Biology*, **20**, 549–561.

Salafsky, N., Margoluis, R., Redford, K.H. & Robinson J.G. (2002) Improving the practice of conservation: a conceptual framework and research agenda for conservation science. *Conservation Biology*, **16**, 1469–1479.

Sarkar, S., Pressey, R.L., Faith, D.P., *et al.* (2006) Biodiversity conservation planning tools: present status and challenges for the future. *Annual Review of Environment and Resources*, **31**, 123–159.

Sayer, J.A. & Campbell, B.M. (2004) *The Science of Sustainable Development: Local Livelihoods and the Global Environment*. Cambridge University Press, Cambridge.

Scott, J.M. & Csuti, B.A. (1997) Noah worked two jobs. *Conservation Biology*, **11**, 1255–1257.

Scott, J.M., Davis, F., Csuti, B., *et al.* (1993) Gap analysis – a geographic approach to protection of biological diversity. *Wildlife Monographs*, **123**, 1–41.

Smith, R.J., Verissimo, D., Leader-Williams, N., Knight, A.T. & Cowling, R.M. (2009) Let the locals lead. *Nature*, **462**, 280–281.

Soulé, M.E. & Terborgh, J.W. (eds) (1999) *Continental Conservation: Scientific Foundations of Regional Reserve Networks*. Island Press, Washington D.C.

Stoms, D.M., Davis, F.W., Andelman, S.J., *et al.* (2005) Integrated coastal reserve planning: making the land–sea connection. *Frontiers in Ecology and the Environment*, **3**, 429–436.

Theobald, D.M., Hobbs, N.T., Bearly, T., Zack, J.A., Shenk, T. & Riebsame, W.E. (2000) Incorporating biological information in local land-use decision-making: designing a system for conservation planning. *Landscape Ecology*, **15**, 35–45.

Vie, J.C., Hilton-Taylor, C. & Stuart, S.N. (2009) *Wildlife in a Changing World: An Analysis of the 2008 IUCN Red List of Threatened Species*. IUCN, Gland, Switzerland.

Watts, M.E., Ball, I.R., Stewart, R.S., *et al.* (2009) Marxan with Zones: software for optimal conservation-based land- and sea-use zoning. *Environmental Modelling and Software*, **24**, 1513–1521.

Whitten, T., Holmes, D. & MacKinnon, K. (2001) Conservation biology: a displacement behavior for academia? *Conservation Biology*, **15**, 1–3.

Wilhelm-Rechmann, A. & Cowling, R.M. (2011) Social marketing as a tool for implementation in complex social-ecological systems. In: *Exploring Sustainability Science – A Southern African Perspective* (eds M. Burns & A. Weaver), pp.179–204. SUN Press, Stellenbosch, South Africa.

Wilson, K.A., Pressey, R.L., Newton, A., Burgman, M., Possingham, H. & Weston, C. (2005) Measuring and incorporating vulnerability into conservation planning. *Environmental Management*, **35**, 527–543.

Wilson, K.A., Underwood, E.C., Morrison, S.A., *et al.* (2007) Conserving biodiversity efficiently: what to do, where and when. *PLoS Biology*, **5**, e223.

Wilson, K.A., Cabeza, M. & Klein, C.J. (2009a) Fundamental concepts of spatial conservation prioritisation. In: *Spatial Conservation Prioritisation: Quantitative Methods and Computational Tools* (eds A. Moilanen, K.A. Wilson & H.P. Possingham), pp.16–27. Oxford University Press, Oxford.

Wilson, K.A., Carwardine, J. & Possingham, H.P. (2009b) Setting conservation priorities. *Annals of the New York Academy of Sciences*, **1162**, 237–264.

Wilson, K.A., Meijaard, E., Drummond, S., *et al.* (2010) Conserving biodiversity in production landscapes. *Ecological Applications*, **20**, 1721–1732.

Wondolleck, J.M. & Yaffee, S.L. (2000) *Making Collaboration Work: Lessons from Innovation in Natural Resource Management*. Island Press, Washington, D.C.

Young, M.D., Gunningham, N., Elix, J., *et al.* (1996) *Reimbursing The Future: An Evaluation of Motivational Voluntary, Price-Based, Property-Right, and Regulatory Incentives for the Conservation of Biodiversity, Parts 1 and 2*. Biodiversity Series, Paper No. 9. Department of the Environment, Sport and Territories, Canberra.

21

Biological corridors and connectivity

Samuel A. Cushman[1], Brad McRae[2], Frank Adriaensen[3], Paul Beier[4], Mark Shirley[5] and Kathy Zeller[6]

[1]USDA Forest Service, Rocky Mountain Research Station, Flagstaff, AZ, USA
[2]The Nature Conservancy, North America Region1917, Seattle, USA
[3]Department of Biology, University of Antwerp, Antwerp, Belgium
[4]School of Forestry, Northern Arizona University, Flagstaff, AZ, USA
[5]School of Biology, Newcastle University, Newcastle upon Tyne, UK
[6]Panthera, 8 West 40th Street, 18th Floor, NY, USA

hring utan ymbbearh, þæt heo þone fyrdhom ðurhfon ne mihte, locene leoðosyrcan laþan fingrum.

On his shoulder lay braided breast-mail, barring death, withstanding entrance of edge or blade.

Beowulf (Old English epic poem, c. 10th Century)

Introduction

The ability of individual animals to move across complex landscapes is critical for maintaining regional populations in the short term (Fahrig 2003; Cushman 2006), and for species to shift their geographic range in response to climate change (Heller & Zavaleta 2009). As organisms move through spatially complex landscapes, they respond to multiple biotic and abiotic factors to maximize access to resources and mates while minimizing fitness costs such as mortality risks. Habitat fragmentation decreases dispersal success (Gibbs 1998), increases mortality (Fahrig et al. 1995) and reduces genetic diversity (Reh & Seitz 1990; Wilson & Provan 2003). Local populations may decline if immigration is prevented (Brown & Kodric-Brown 1977; Harrison 1991) and may prevent recolonization following local extinction (Semlitsch & Bodie 1998).

The goal of this chapter is to describe the state of the art in quantitative corridor and connectivity modelling. We will review several critical issues in modelling, and provide expert guidance and examples to help practitioners implement effective programmes to preserve, enhance or create connectivity among wildlife populations. We first

Key Topics in Conservation Biology 2, First Edition. Edited by David W. Macdonald and Katherine J. Willis.

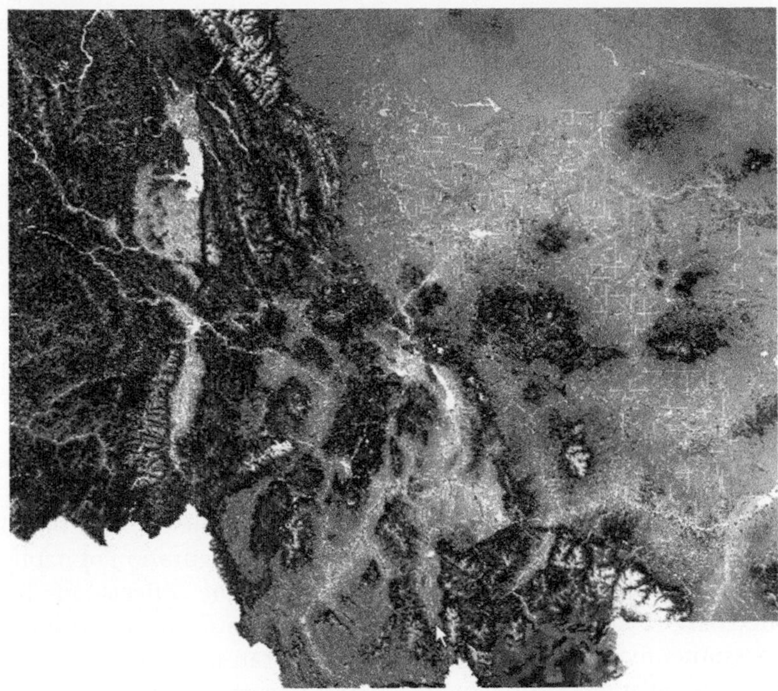

Figure 21.1 Example landscape resistance map for American black bear in an area of the US northern Rocky Mountains encompassing Montana and northern Idaho. Dark areas are low resistance for movement, while light areas are high resistance for movement. The resistance map was developed by Cushman et al. (2006) and validated with independent data by Cushman & Lewis (2010) and in multiple independent study areas by Short Bull et al. (2011).

review the fundamental task of estimating landscape resistance, comparing expert opinion and empirical methods. Next, we describe current methods of predicting connectivity from resistance surfaces. Then we discuss how to develop linkage designs that can maintain connectivity for multiple species, and under changing climate. We conclude with discussion of how effectively to validate connectivity model predictions.

Estimating landscape resistance

Most current methods of predicting population connectivity and mapping areas significant in facilitating animal movements begin with landscape resistance maps (Figure 21.1). Landscape resistance maps depict the cost of movement through any location in the landscape (pixel cell in a raster map) as a function of landscape features of that cell (e.g. high resistance might be assigned to a road or a body of water). In its most basic sense, landscape resistance reflects the local movement cost incurred by an animal. More formally, the resistance reflects the step-wise cost of moving through each cell for least-cost analyses (Singleton et al. 2002) or the relative probability of moving into the cell for circuit theory-based analyses (McRae et al. 2008).

Expert versus empirical estimation

Most published studies using landscape resistance maps have estimated resistance of landscape features to movement based on expert opinion alone (e.g. Compton et al. 2007). However, non-human species perceive landscapes in ways that may not correspond to human assumptions concerning connectivity and habitat quality (With et al. 1997). Using unvalidated expert opinion to develop resistance maps has been a major weakness of most past landscape resistance modelling efforts (Seoane et al. 2005).

Methods for empirically estimating resistance

Habitat quality as surrogate for landscape resistance

Habitat quality can be predicted based on patterns of occupancy in relation to ecological conditions, such as through resource selection functions (e.g. Guerry & Hunter 2002; Weyrauch & Grubb 2004). The simplest way to estimate relationships empirically between population connectivity and environmental conditions is to assume that habitat quality directly equates to population connectivity. Predictions of habitat quality based on patterns of occurrence studies are limited because they do not directly measure biological responses such as mortality, movement and productivity (Cushman 2006). Patterns of species occurrence do not necessarily reflect patterns of fitness with respect to environmental gradients and landscape patterns (Van Horne 1983). More importantly, in the context of connectivity modelling, suitability for occupancy and suitability for dispersal may not be driven by the same factors at the same scales (e.g. Shirk et al. 2010; Wasserman et al. 2010). Habitat selection reflects the behaviour of individual organisms to maximize fitness within home ranges, while population connectivity is driven by dispersal, migration and mating events. These are functionally and biologically different processes. Few studies have formally evaluated the performance of habitat suitability models as surrogates for landscape resistance, but those that have generally have found them to perform poorly (e.g. Shirk et al. 2010; Wasserman et al. 2010). This highlights the importance of not assuming that habitat relationships optimally reflect the landscape features governing population connectivity.

Mark-recapture and experimental movement studies

By quantifying movement rates, distances travelled and routes of animals through complex environments, researchers can quantitatively describe species-specific responses to environmental conditions and landscape structure. For example, a study by Gamble et al. (2007) quantified dispersal in relation to topography and vegetation for several pond-breeding amphibians, demonstrating the value of mark-recapture approaches to evaluating population connectivity. In addition, these methods are well suited for incorporation in manipulative field experiments in which the area and configuration of habitat are controlled to isolate the effects of habitat loss and fragmentation on organism movement and survival rates. For example, Haddad & Baum (1999) used a large-scale experiment to find that three habitat-restricted butterfly species reached higher densities in patches connected by corridors than in similar, isolated patches.

These kinds of studies provide the most reliable inferences about relationships between survival rates, movement and ecological conditions (McGarigal & Cushman 2002). Unfortunately, large-scale manipulative field experiments and mark-recapture meta-population studies are expensive, take several years, and generally suffer from small sample sizes. Another potential limitation is that these studies focus on short-term, fine-scale movement path selection of individual animals, which may not scale up to population-level effects on migration and gene flow.

Telemetry

Advances in wildlife telemetry technology have enabled collection of very accurate and frequent location data for individual animals. Landscape resistance modelling based on telemetry is a powerful technique to address the factors that affect organism movement directly on scales of space and time greater than are possible with mark-recapture and experimental movement studies (e.g. Osborn & Parker 2003; Cushman et al. 2005,

2010a). GPS telemetry data enable direct assessment of the influences of landscape features on movement path selection. For example, Cushman et al. (2010a) modelled the influence of landscape features on elephant movement path selection using telemetry data, showing that elephants (*Loxodonta africana*) select movement paths near water, avoid human settlements and do not cross wildlife cordon fences. Similarly, Cushman & Lewis (2010) used satellite telemetry data to show that American black bears (*Ursus americanus*) choose movement paths that avoid roads and human residences and concentrate activity in forested areas at middle elevations. Directly associating movement paths with landscape features enables the development of species-specific landscape resistance models that are more reliable than those produced by expert opinion.

Landscape genetics

Gene flow among populations is necessary to support the long-term viability of populations, as it maintains local genetic variation and spreads potentially advantageous genes. Thus it is important to infer the functional connectivity among populations and across landscapes (van Dyck & Baguette 2005). The ultimate validation of any method of estimating functional connectivity lies in how well it explains gene flow (Cushman et al. 2006; Shirk et al. 2010; Wasserman et al. 2010; Short Bull et al. 2011). Genetic methods can directly measure dispersal and immigration (Waples 1998; Landguth et al. 2010). Logistical and financial costs associated with tracking individual animals are obviated and because genetic data integrate time and space, slow rates of dispersal through complex landscapes are measurable. Landscape genetic analyses enable direct association of movement cost across resistance surfaces with genetic differentiation, which enables empirical derivation and validation of connectivity maps. For example, Wasserman et al. (2010) used non-invasive monitoring to collect genetic data from several hundred individual American marten (*Martes americana*) across a 4000 square kilometer study area, and were able to use multivariate landscape genetic modelling to identify the landscape features that affect gene flow.

Combining multiple methods to produce robust estimates of resistance

Every method of estimation has its own limitations, so it is valuable to use multiple methods and independent data sets to estimate resistance. The strongest inferences are derived from multiple analyses of different kinds of data that produce a consistent result (Cushman & Lewis 2010). For example, landscape genetics and GPS telemetry are two complementary analyses that can be combined to produce robust estimates of landscape resistance. Using movement data to predict landscape resistance, and comparing that to landscape resistance predicted from landscape genetic analyses of the same species in the same study area, is a useful way to verify the robustness of landscape connectivity hypotheses (e.g. Cushman & Lewis 2010). In addition, such analyses would illuminate the multi-scale drivers of population connectivity, since mating and dispersal movement behaviours are the mechanisms through which gene flow operates in animal populations. At the present time, only a few research programmes have estimated resistance from a combination of approaches (Coulon et al. 2008; Cushman & Lewis 2010; Shanahan et al. 2011; see Box 21.1).

From landscape resistance to population connectivity

While resistance is point specific, connectivity is route specific (Cushman et al. 2008). Therefore, while resistance models can provide the foundation for applied analyses of population connectivity, they do not, in themselves, provide sufficient information to evaluate the existence, strength and location of barriers and movement corridors. Connectivity must be evaluated with respect to the paths, costs and success of moving across a landscape. The resistance model is the

Box 21.1 Combining landscape genetics and telemetry to estimate landscape resistance for American black bear

Cushman et al. (2006) used causal modelling with landscape genetics data to evaluate support for 110 alternative hypotheses describing the effects of landscape variables on population connectivity in an American black bear (*Ursus americanus*) population in northern Idaho, USA. Their analysis rejected hypotheses of isolation by distance and isolation by a geographical barrier, and affirmed a landscape resistance model which predicts that rates of gene flow are related to elevation, forest cover, roads and human development. Cushman & Lewis (2010) used conditional logistic regression to predict landscape resistance based on black bear GPS telemetry data in the same landscape. They used a path-level spatial randomization method to assess the effects of multiple landscape features on movement path selection (e.g. Cushman et al 2010a). The path-level randomization approach provides a robust means to compare the landscape features an animal encounters in its actual path with those that would be encountered in a large sample of available paths of identical length and topology. They found that consistent landscape factors influence genetic differentiation and movement path selection, with strong similarities between the predicted landscape resistance surfaces. Genetic differentiation among individual American black bears is driven by spring movement (mating and dispersal) in relation to residential development, roads, elevation and forest cover. The real value of this study is that it used two independent data sets and different kinds of analyses to validate the results, and it quantified the scale and strength of bear behavioural response to several landscape features. For example, it showed that gene flow is maximum at middle elevations due to impassable snow pack at high elevations in the dispersal season, and concentrated human populations in low-elevation valleys, and that bears strongly avoid roads and human resistances (e.g. near-total avoidance within a 200 m radius buffer around human structures).

foundation for these analyses, but it is explicit consideration of movement paths across the resistance surface that provides the key information for conservation and management.

Identifying corridors using least-cost modelling

In recent years least-cost (LC) modelling (part of graph theory, see below) has become the dominant modelling tool to evaluate functional landscape connectivity, especially in applied studies. This is mainly because:

- it produces an unambiguous corridor or path as an output, whereas most other approaches do not
- it is available in most commercial GIS packages as well as open source software
- LC models generate visually attractive and easy to communicate representations of connectivity (maps) and quantitative metrics of effective distance (cost values) in the same units (meters) as Euclidean distance (Adriaensen et al. 2003; Fagan & Calabrese 2006).

These attributes make LC modelling very well suited for quantitative landscape analyses and for evaluating effects of future scenarios on connectivity.

In LC models the only inputs are the map of sources and targets and the map of resistance values (R; Figure 21.2a). The cost layer is the first and central level of output of a LC analysis (see Figure 21.2a), and provides the functional cost distance values from the designated source to all locations in the geographical extent of the analysis. The least-cost path (Figure 21.2b) is the series of cells in the landscape which results in the minimum cumulative cost value (LC path value) to move from a source cell/cells to the target cell/cells under investigation. The LC path indicates the location of the cheapest route, but gives no information on how cost values are distributed over the landscape. For example, it does not indicate other zones in the landscape resulting in comparable costs (Figure 21.2c) or how wide the LC path zone is (Adriaensen et al. 2003; Pinto & Keitt 2009).

There are several methods available to produce biologically informative measures of landscape connectivity from such cost surfaces. One of these is the combination of several cost

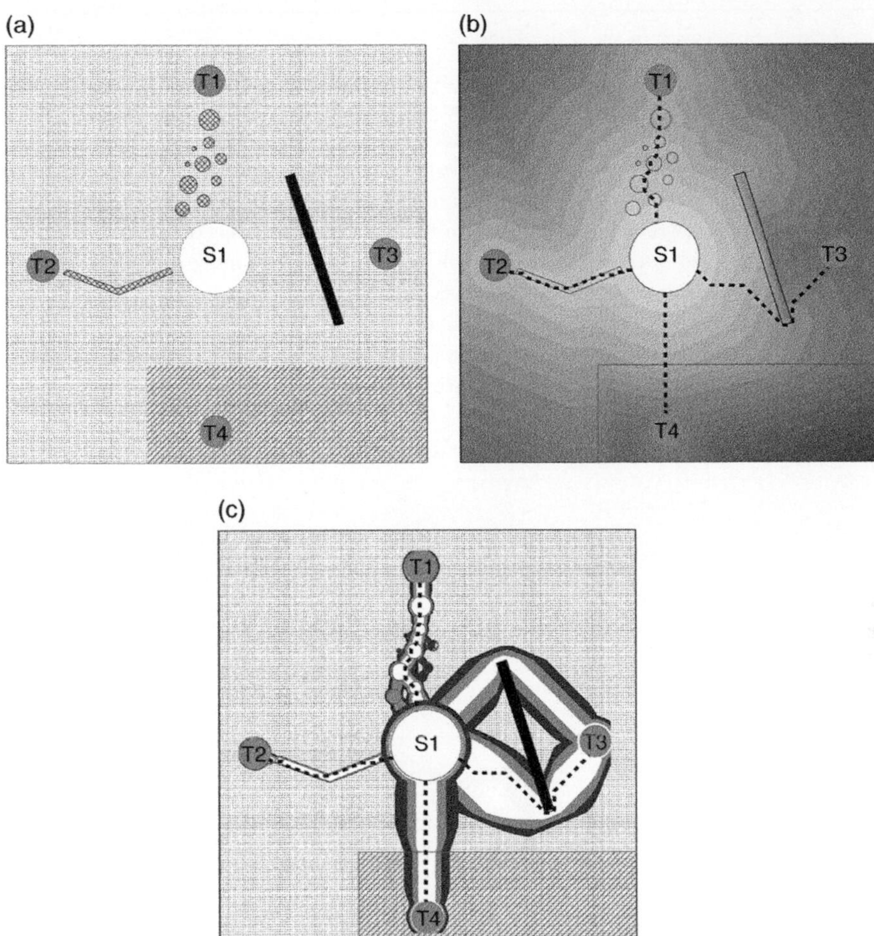

Figure 21.2 Input (a) to output (b-c) in LC modelling using a simple virtual landscape to show the basics of this modelling approach. (a) Landcover map showing a hypothetical landscape (source area S1: forest; target areas T1–4: woodlots; grassland (dotted); intensive cropland (hatched); hedgerow and bushes (cross-hatched); open water (solid black). (b) Cost layer from source area S1 and LC path to 4 target areas (dashed lines). (c) Corridor map showing 1–10% corridor buffers derived from bidirectional cost layers from source area S1. Result of 4 corridor analyses superimposed (targets=T1–4): white, increase in minimum cumulative cost less than 1% of LCP; light grey 1–5%; dark grey 5–10%.

layers into one 'corridor' layer (other names: bidirectional cost layer, conditional minimum transit cost [CMTC, Pinto & Keitt 2009]; see Figure 21.2c), in which the value of each cell is the overall cost to reach the target cell T from source cell S, but with the constraint to go through the cell under investigation. The LC path is a special case of this (with all cells having a value equal to the LC path value and thus the minimum present in the corridor layer). The LC path will always be the path of minimal corridor values but elsewhere in the landscape, there could be other zones with nearly equal cost values (see Figure 21.2c). Corridor maps give a more realistic view of the functionally cheapest routes in the landscape from the designated source to the destination (Adriaensen et al. 2003) (Box 21.2). For example, the width of corridors can be determined by taking percent slices of the landscape representing the lowest cumulative resistance (e.g. Singleton et al. 2002; Spencer et al. 2010) or by limiting

Box 21.2 Landscape connectivity in the Taita Hills

The Restoration and Increase of Connectivity among Fragmented Forest Patches in the Taita Hills, South-east Kenya project (CEPF project 1095347968; Adriaensen et al. 2007) included a detailed analysis of functional landscape connectivity in the area. In this project, evidence of the distribution and population status of bird species in the remaining small cloud forest patches on the hill tops (black patches) was successfully combined with output of LC models to support and prioritize habitat restoration actions in plantations with exotic trees (white patches). Forest restoration is now being implemented in a set of five pilot projects.

Least-cost models were used to model the location of exotic tree plantations in relation to modelled connectivity corridors for forest interior birds (dark grey zones), in order to evaluate their potential roles as stepping stones to promote recolonization after rehabilitation of the plantations. In the map shown, corridors between all pairs of remaining forest plots were superimposed (resistance set R1S5 for eco-type 'sensitive interior forest bird', including the critically endangered Taita thrush *Turdus helleri*).

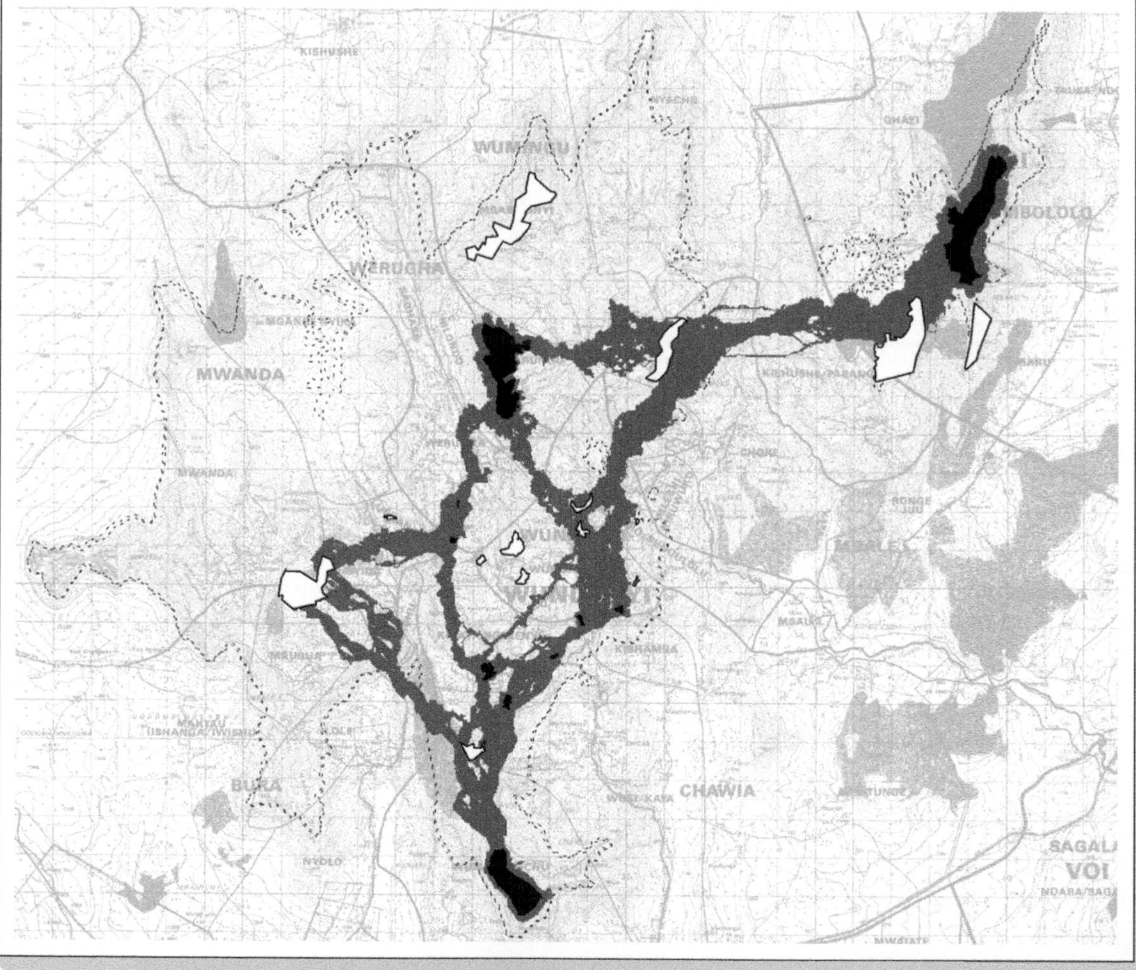

corridors to a maximum cost-weighted cut-off distance above that of the LC path (WHCWG 2010). Regional connectivity assessments can require mapping corridors between hundreds to thousands of core area pairs (e.g. Spencer et al. 2010; WHCWG 2010). The development of GIS tools to automate corridor mapping, including decisions of which pairs

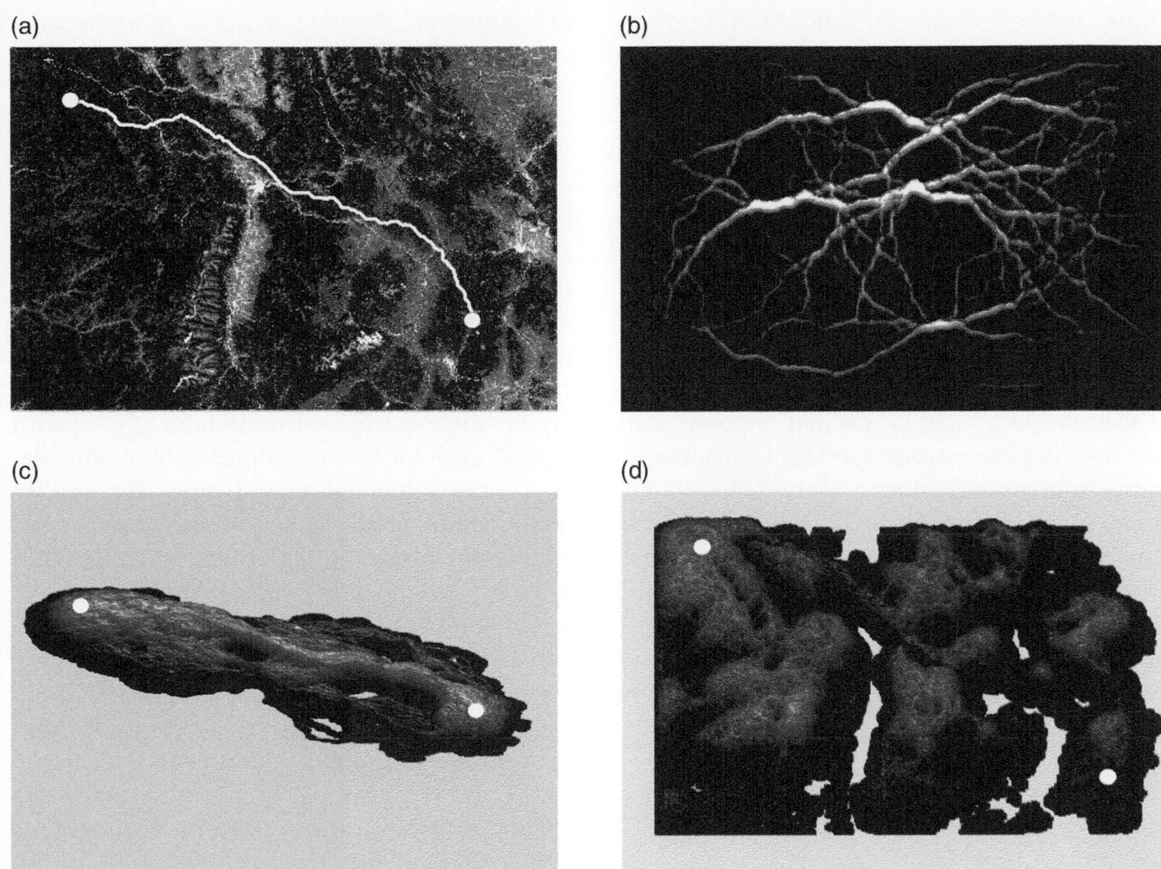

Figure 21.3 Comparison of four connectivity modelling methods applied to a single study area and resistance map. The study area is northern Idaho, USA. The resistance map is shown in panel (a) as a colour scale from blue (low resistance) to red (high resistance), and reflects landscape resistance to black bear gene flow (Cushman et al. 2006). Panel (a) shows a single least-cost path (white line) between two point locations (white dots). Panel (b) shows a factorial least-cost path analysis between several hundred source points. Panel (c) shows the least-cost corridor between the same two source points as in (a). Panel (d) shows the cumulative resistant kernel model of synoptic landscape connectivity.

of core areas to connect (e.g. McRae & Kavanagh 2011), makes this easy.

Factorial least-cost paths

One limitation of traditional LC path and LC corridor analyses is that they are limited to prediction of connectivity between single sources and single destinations (Figure 21.3a). While this may be ideal in the case where one is interested in the lowest cost routes between two focal conservation areas, there are many situations where a more synoptic analysis of connectivity is valuable. For example, it may be that there is a need to calculate corridor connectivity between thousands of sources and a single destination (e.g. Cushman et al. 2010a) or between hundreds of sources and hundreds of destinations distributed across a complex landscape (e.g. Cushman et al. 2008, Cushman et al. 2011; Figure 21.3b). For example, Cushman et al. (2008) used factorial least cost path analysis to predict the most important movement routes for bears between Yellowstone National Park and the Canadian border in the United States Northern Rocky Mountains, showing that there are few major connections and locating several dozen potential barriers. This, in turn, focuses attention on where restoration and mitigation efforts would be most effective. A factorial implementation of least cost paths (e.g. UNICOR; Landguth et al. 2011) permits integration of a vast number of least cost paths to show synoptic connectivity across large and complex landscapes (Figure 21.3b). For example (Cushman et al. 2011) mapped regional corridor networks for several species of conservation concern across a vast area of the United States great plains using UNICOR (Landguth et al. 2011). The analysis identified which species have the most fragmented populations and mapped the most important corridor linkages among population core areas, focusing conservation efforts on the most important locations.

Other ways to analyse connectivity

Ecologists often use the term *graph theory* to refer to a family of analyses in which patches are reduced to nodes at patch centroids, with centroids connected by lines or 'edges' (e.g. Bunn et al. 2000; Urban & Keitt 2001; Minor & Urban 2007). Such graphs underlie many methods in connectivity analysis, including LC corridor modelling. Advances in computing and algorithms borrowed from other disciplines have allowed applications of graph algorithms to continuous landscapes instead of simple networks. Rayfield et al. (2011) review graph-based connectivity measures and provide a framework for classifying them as applications to connectivity conservation.

Circuit theory

Connectivity analyses based on electrical circuit theory use networks of electrical nodes connected by resistors as models for networks of populations, habitat patches or locations on a landscape connected by movement. Because connectivity increases with multiple pathways in electrical networks, distance metrics based on electrical connectivity are applicable to processes (e.g. gene flow; McRae 2006) that respond positively to increasing numbers of pathways. Additionally, previous work has shown that current, voltage and resistance in electrical circuits all have mathematical relationships with random walks (Doyle & Snell 1984; Chandra et al. 1997). Random walks can predict the expected routes that an animal with a preference for low-resistance habitat will take as it moves through a landscape. The precise relationships between circuit theory and random walks mean that circuits can be related to movement ecology and population genetics via random walk and coalescent theories, providing concrete interpretations of connectivity measures (McRae 2006; McRae et al. 2008).

Circuit and LC models represent two extremes in assumptions about movement and connectivity. Least-cost corridors calculate the routes expected to be taken by animals with perfect or near-perfect knowledge of the landscape, whereas current maps generated from circuit models predict movement routes taken by random walkers, with all possible paths contributing to connectivity. Neither will entirely correctly predict movement behaviour of real animals (Spear et al. 2010, and see below) but there are benefits to both models, as we show in the example in Figure 21.4. Least-cost analyses can show what routes/zones would permit the most efficient movement, which can be important for conservation planning; if a large portion of a landscape is likely to be developed, identifying those areas which, if conserved, provide the easiest movement routes will be important. Circuit theory has the advantage of identifying and quantifying 'pinch points' (see Figure 21.4), i.e. constrictions in corridors that, if lost, could sever connectivity entirely. Such areas can be prioritized for early conservation action because options are limited. Circuit algorithms also integrate across all movement pathways to

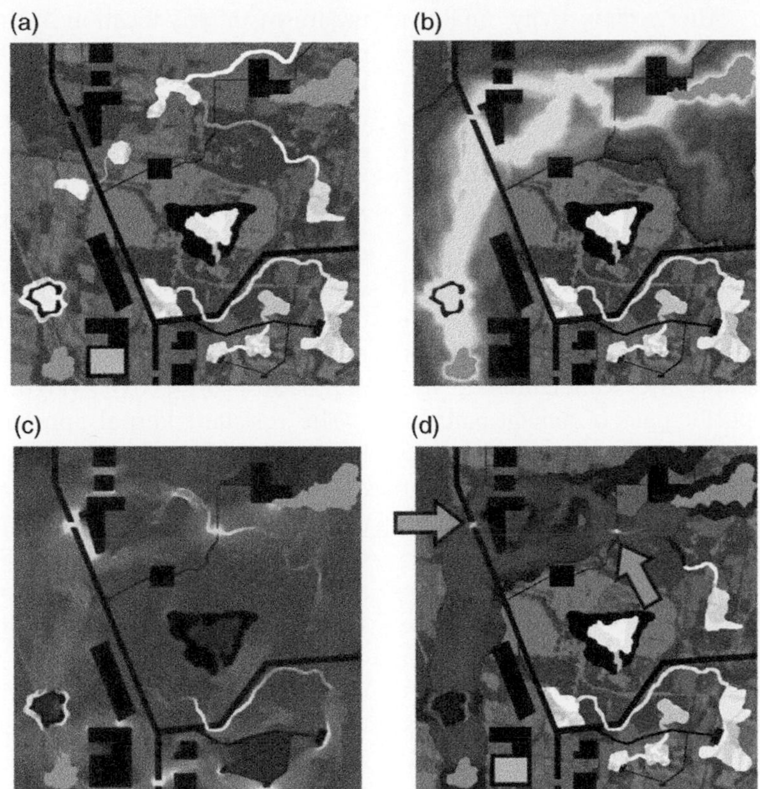

Figure 21.4 Example of how circuit theory can be used to identify and prioritize important areas for connectivity conservation. (a) Simple landscape, with two patches to be connected (green) separated by a matrix with varying resistance to dispersal (low resistance in white, higher resistance in darker shades, and complete barriers in black). (b) Least-cost corridor between the patches (lowest resistance routes in yellow, highest in blue). (c) Current flow between the same two habitat patches derived using Circuitscape (McRae & Shah 2009), with highest current densities shown in yellow (from McRae et al. 2008). Circuit analyses complement least-cost path results by identifying important alternative pathways and 'pinch points', where loss of a small area could disproportionately compromise connectivity. (d) A promising application is restricting circuit analyses to least-cost corridor slices to take advantage of the strengths of both approaches (from McRae & Kavanagh 2011). This hybrid approach shows both the most efficient movement pathways and critical 'pinch points' within them, which glow yellow. These could be prioritized over areas that contribute little to connectivity, such as the corridor at the top right of the map that has been coloured dark blue because it does not provide connectivity between the patches.

provide measures of redundancy, i.e. availability of alternative pathways for movement (see Figure 21.4). New applications allow identification of barriers that have a strong effect on connectivity, which can be useful for highlighting opportunities to restore connectivity, e.g. through re-establishment of natural vegetation or installation of highway crossing structures (McRae, unpublished data).

Centrality analyses

A promising graph-theoretic approach to connectivity modelling is centrality analysis, which ranks the importance of habitat patches or corridors in providing movement across an entire network, i.e. as 'gatekeepers' of flow across a landscape (Carroll et al. 2011). Centrality analyses can be based in LC path, circuit theory or other connectivity analysis methods. The difference is that, instead of mapping corridors or current flow between single pairs of core areas, they add up results from connectivity analyses between all pairs of nodes (sites or cells) on a landscape. Centrality analyses can be applied to raster GIS data or networks to identify core areas, linkages or grid cells that are particularly important for overall connectivity. Because centrality metrics can incorporate connectivity between all pairs of nodes on a landscape, they can eliminate the need to identify specific pairs of habitat patches to connect. For example, betweenness centrality (Freeman et al. 1991) identifies the shortest paths connecting all pairs of nodes in a network, and sums the number of such shortest paths involving each intervening node. This procedure identifies areas lying on a large proportion of the shortest paths in a network, the loss of which can disproportionately disrupt connectivity across the network as a whole. The Connectivity Analysis Toolkit (Carroll 2010) specializes in centrality analysis, and supports metrics based on betweenness, current flow (Newman 2005), maximum flow (Freeman et al. 1991) and minimum-cost flow (Ahuja et al. 1993). It also allows time-series analyses of connectivity across landscapes where habitats shift through time (Phillips et al. 2008).

Resistant kernels

The resistant kernel approach to connectivity modelling is based on least-cost dispersal from some defined set of sources. The model calculates the expected density of dispersing individuals in each pixel around the source, given the dispersal ability of the species, the nature of the dispersal function and the resistance of the landscape (Compton et al. 2007; Cushman et al. 2010b). Once the expected density around each source cell is calculated, the kernels surrounding all sources are summed to give the total expected density at each pixel (see Figure 21.3d). The results of the model are surfaces of expected density of dispersing organisms at any location in the landscape. For example, Cushman et al. (2010b) used resistant kernel modelling to evaluate the interactive effects of roads and human land use change on population connectivity for a large number of pond-breeding species in Massachusetts (USA). The resistant kernel approach quantified expected density of dispersers in the upland environment as functions of breeding population size, dispersal ability and quantified the relative impacts of roads and land use on population connectivity (Figure 21.5).

The resistant kernel approach to modelling landscape connectivity has a number of advantages as a robust approach to assessing current population connectivity (Compton et al. 2007; Cushman et al. 2010b, 2011). First, unlike most approaches to mapping corridors , it is spatially synoptic and provides prediction and mapping of expected migration rates for every pixel in the whole study area, rather than only for a few selected 'linkage zones' (e.g. Compton et al. 2007). Second, scale dependency of dispersal ability can be directly included to assess how species of different vagilities will be affected by landscape change and fragmentation under a range of scenarios (e.g. Cushman et al. 2010b). Third, it is computationally efficient, enabling simulation and mapping at a fine spatial scale across large geographical extents (e.g. Cushman et al. 2010b, 2011).

Individual-based movement models

Individual-based (IB) models explicitly simulate the processes acting on the individual to predict movement. IB models predict movement paths of simulated dispersers based on parameters such as energetic cost of movement in different patch types, turning angles within patches and at patch transitions, movement speeds, duration of movement events, mortality risks in different patch types, and likelihoods of movements between patch types. Thus, IB models usually incorporate much more detail and thus greater realism than other

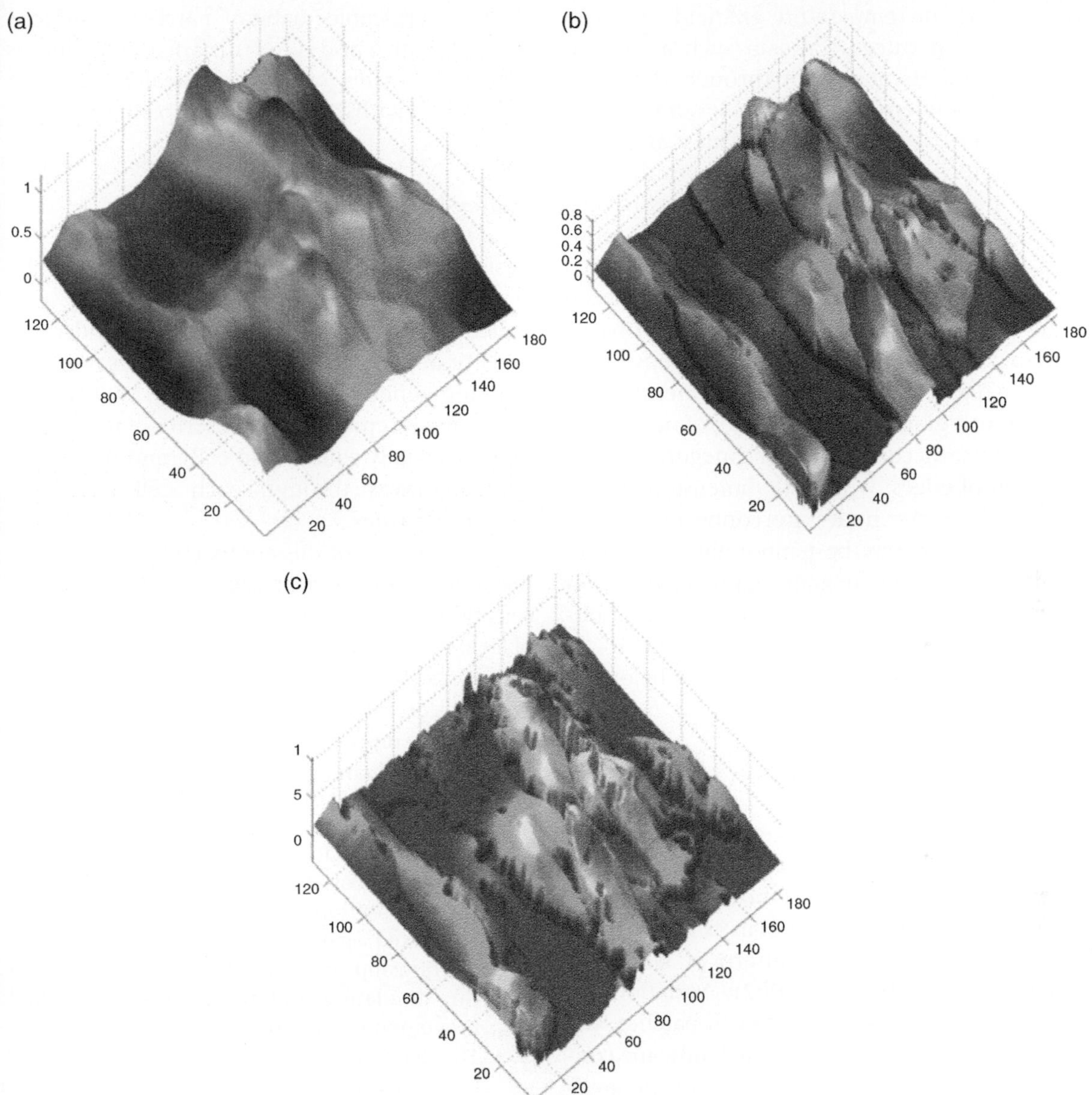

Figure 21.5 Example of resistant kernel results from Cushman et al. (2010b) showing predicted density of dispersing individuals in upland habitat under three hypotheses: (a) connectivity is unaffected by land use and roads and only a function of distance, (b) connectivity is reduced by roads but not by differences in land cover and land use, (c) connectivity is affected by roads and land use/land cover.

connectivity models, such as demographic and dispersal data, in addition to landscape characteristics.

There are three broad categories of models that simulate individual movement (raster based, vector based and network based), which differ according to whether the landscape is represented as fields, features or graphs. Conceived as fields, a landscape is a continuous surface defined by one or more variables (layers) that can be measured at any point within the field. Fields usually model continuous data such as elevation, or temperature gradients, but can also represent categorical data such as habitat classification. If movement through the landscape is dependent on the variables of the field, then raster-based movement rules are most appropriate. Features are discrete entities that occupy positions in space, such as lines (rivers, roads, hedgerows) and polygons (lakes, woodland). The interiors of polygons are considered to be homogeneous. Movement between features is usually simulated using vector-based models. Finally, graphs represent the positional relationship between discrete elements in a landscape; a graph consists of a set of nodes that may represent continuous or categorical data, and a set of edges, which are dimensionless but describe how the nodes are connected to one another. Edges may be temporally referenced, indicating changes in graph connectivity over time. Network-based models are used to simulate IB movement in graphs. Examples of all three of these categories are discussed below.

Movement rules and models

Regardless of whether movement models are raster, vector or network based, they encode a series of rules that predict how the dispersal behaviour of individual animals is expected to interact with the spatial pattern of landscape structure (King & With 2002). Variations in patch quality, boundaries between patches, the nature of the mosaic, and overall landscape connectivity all affect the permeability of the landscape to dispersing individuals (Wiens 1997). The limited empirical information on the behavioural responses of animals to landscape structure (Turner et al. 1995; Lima & Zollner 1996) means that model parameters are usually based on observed habitat preference, dispersal rates in different patches, and how the energetic costs of crossing a landscape affect distance moved as well as direction taken. For example, the rules employed by Boone & Hunter (1996) simulated IB searching behaviour in grizzly bears by encoding permeability into the cells of habitat patches. Highly permeable habitat patches produced straight paths and long distance movements whereas patches of low permeability caused convoluted paths and short displacement.

Raster-based models

Raster- or grid-based representations of the landscape permit the greatest flexibility with which movement interacts with the landscape, and are appropriate where the dispersal matrix is heterogeneous (Wiegand et al. 1999). The landscape is represented as a series of tessellated shapes, usually square grid cells, and the model animal moves through each cell based on movement rules.

An advantage of this approach to modelling is the inclusion of a clear relationship between a cell and its neighbours, facilitating the description of local interactions by state transition rules. Each cell stores its own state variables that influence the decisions made by individuals through the landscape it represents. However, there are three principal disadvantages to raster-based models.

- The resolution of the grid is limited by memory capacity and simulation speed, and raster-based models have a tendency to be computationally demanding.
- The fixed spatial structure implies a fixed relationship between the spatial scale in the simulation and the scale of individual movements of the organism investigated.
- The geometry chosen to represent landscape in raster-based models (i.e. square grid, hexagonal grid, Dirichlet tessellation, etc.) can substantially affect the simulated behaviour of the individual dispersers even if the rules for movement and settlement are the same between different geometries (Holland et al. 2007).

Vector-based models

Vector-based models simulate organisms dispersing through continuous or homogeneous landscapes. If the motivations for these movements are random or quasi-random search patterns, they can be simulated using correlated random walk algorithms (Kareiva & Shigesada 1983). Alternatively, if individual movements are targeted searches for resources with a particular spatial or temporal distribution, movement decisions will be informed by the underlying landscape structure. Finally, if motivation for movement is prompted by the desire to avoid or join conspecifics, it will result in density-dependent movement rules. Where motivations for movement are known and appropriate, IB models benefit from vector-based dispersal simulations, which are less computationally demanding than the raster-based alternative.

Network-based models

Network-based models differ from the other types in that they do not include a continuous representation of the landscape. Rather, connectivity between locations is represented by an edge between nodes. Network-based models usually specify an *a priori* representation of patch size, patch adjacency and other criteria (e.g. Lookingbill et al. 2010). Edges are formed when movement is possible between nodes. Dispersal corridors can be represented as nodes as well as edges in network visualizations of a landscape used as analytic connectivity models (McRae et al. 2008). They calculate walks through the network that minimize total weight, suggesting optimal pathways for dispersal. In IB models, network-based landscapes are utilized probabilistically (Lookingbill et al. 2010; Morzillo et al. 2011), and may result in biologically plausible but analytically suboptimal solutions. Graph-theoretic approaches to network analysis can be applied to the utilized networks of IB models to identify the nodes and edges that maintain cohesion of the network. For example, Gurnell et al. (2006) identified routes of entry for invasive grey squirrels into potential conservation areas for the endangered red squirrel in northern England through network analysis.

Corridors based on shifting climate envelopes

This approach produces 'temporal corridors' that track how a species' climatic envelope (suitable temperature and moisture regimes) might move across a landscape under climate change scenarios. Like some types of individual-based models, this approach avoids the concept of resistance that is central to most previous approaches. The heart of this approach is either a dispersal chain model (Williams et al. 2005) or a network flow model (Phillips et al. 2008), either of which identifies cells with suitable climate envelopes that are spatially contiguous for long enough to allow the species to establish new populations in cells as they become suitable. Although dispersal chain and network flow models are conceptually sound, they depend completely on the outputs of three other models, namely models of future emissions of greenhouse gasses, models of future climate resulting from how the atmosphere and oceans respond to these emissions, and climate envelope models for the focal species. Unfortunately, each of these latter three models is plagued with massive uncertainty (summarized in Beier & Brost 2010). In the future, ensemble modelling (building many alternative corridors based on various combinations of emission scenarios, circulation models and climate envelope models) might identify corridors robust across the range of assumptions in the ensemble.

Beyond single species

From optimal corridors for single species to linkage designs for multiple species

Up to this point, we have described methods of mapping an optimal corridor, or areas important for connectivity, for a single species. Beier et al. (2008) proposed the term *linkages* to denote lands intended to support movement of

Box 21.3 Example of optimizing multispecies linkage

A hypothetical linkage design including optimal corridors for eight focal species, expanded to include patches of modelled breeding habitat for an additional five focal species for which corridor models were not appropriate, and a narrow riparian strand for fishes. Each strand needs to be >1 km wide in order to create large interior spaces free of edge effects, support meta-populations of species needing multiple generations to achieve gene flow through the corridor, and support ecological processes more complex than animal movement.

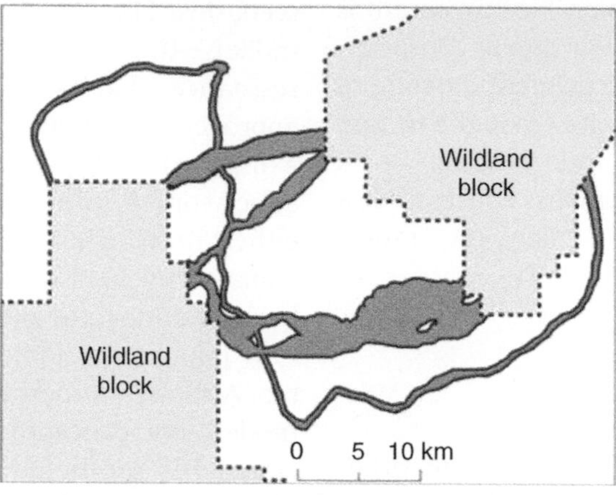

multiple focal species and ecosystem processes. To design linkages, conservation planners can select a suite of representative focal species suitable to serve as a collective umbrella for the entire biota. For instance, each of 27 linkage plans in California and Arizona (Beier et al. 2006, 2007) was designed to meet the needs of 10–30 focal mammals, reptiles, fishes, amphibians, plants and invertebrates. Focal species included species requiring dispersal for meta-population persistence, species with short or habitat-restricted dispersal movements, species tied to an important ecological process (e.g. predation, pollination, fire regime), and species reluctant to traverse barriers in the planning area. Although large carnivores are appropriate focal species and flagships (Servheen et al. 2001; Singleton et al. 2002), most of them are highly mobile habitat generalists and thus inadequate umbrellas for other species (Beier et al. 2009; Minor & Lookingbill 2010).

A simple unweighted union of single-species corridors is an obvious way to produce a linkage design to promote the goal of 'no species left behind' (Beier et al. 2006, 2007; Adriaensen et al. 2007; Cushman et al. 2011) (Box 21.3). But corridor models are not appropriate for some focal species, such as many flying animals, that do not move across the landscape in pixel-to-pixel fashion. To support movement of these species, Beier et al. (2008) recommend draping maps of known or modelled breeding habitat over the union of corridors, and enlarging the union to include patches that would decrease the inter-patch distances that dispersers would need to cross. The linkage design should be further expanded to include major riverine connections, which provide natural corridors for aquatic and some upland organisms, and promote other ecological processes and flows such as movement of sediment, water and nutrients.

Coarse-filter linkage designs for climate change

Climate change poses a challenge to all types of conservation planning, including linkage design. As climate changes, existing land covers in some planning areas will not merely shift but will disappear as plant associations reassemble (Hunter et al. 1988; Lovejoy & Hannah 2005). Linkage designs should be robust to such changes, and should allow species to shift their ranges into and out of the planning area. To address this, one could attempt to model corridors for the shifting climate envelopes of all species (above). A simpler alternative is to design linkages with a coarse-filter approach based on the abiotic drivers of land cover and species distributions (Hunter et al. 1988; Anderson & Ferree 2010). This idea is grounded in the foundational ecological concept (Jenny 1941; Amundson & Jenny 1997) that biodiversity at any point in time is determined by the interaction of the recent species pool with climate, soils and topography.

Beier & Brost (2010) and Brost & Beier (2012) developed multivariate procedures to identify *land facets*, defined as recurring landscape units with uniform topographic and soil attributes, from readily available digital maps of elevation and soils. They used multivariate dissimilarity as a measure of pixel resistance for each land facet type. Finally, they used least-cost modelling to design land facet corridors, and joined these corridors into a linkage design. Other coarse-filter approaches are feasible. For instance, Rouget et al. (2006) suggest that species will shift their ranges by sequentially colonizing areas that lie along the most gentle and monotonic temperature gradients. Assuming these gradients in temperature are conserved in a changing climate, it may be possible to identify corridors along today's most gentle and monotonic temperature gradients, without the need for uncertain models of future climate.

Linkage designs should be produced by a combination of coarse-filter and focal species approaches. In each of three landscapes, Beier & Brost (in preparation) developed two linkages designs – one based on land facets and the other on focal species. The land facet linkage designs included optimal corridors for 25 of 28 focal species, whereas the focal species designs encompassed optimal corridors for 21 of 32 land facets. Neither approach on its own was likely to meet all conservation goals.

Validation of predicted corridors

Corridors resulting from models have sometimes been criticized because they lack supporting movement data (Simberloff et al. 1992; Rosenberg et al. 1997) and because they may contain errors in model parameters or incorrect assumptions (Spear et al. 2010). Therefore, additional vetting of modelled corridors in the field is strongly recommended.

Many field studies have evaluated the efficacy of existing corridors, such as corridors that follow linear features like fencerows or rivers (Hill 1995; Castellón & Sieving 2006), or that were constructed as part of experimental landscapes (Berggren et al. 2002; Haddad et al. 2003). There have also been tests of species' response to conservation action in established corridors (Duke at al. 2001; Shepherd & Whittington 2006). But field testing of modelled corridors, like the ones described in this chapter, have been scarce.

Modelled corridors may cover large spatial extents and span multiple land ownerships and management types, or even national borders, making the collection of field data logistically complex and resource intensive. If corridors are modelled for dispersal movement, capturing infrequent dispersal events is akin to finding a needle in a haystack, so collecting sufficient data to reliably test predicted corridors can be difficult. Finally, modelled corridors can only be truly validated if movement through the corridor is documented along with the outcome for which the corridor was intended, whether that be by successful migration to summer or winter ranges, successful recolonization of habitat patches, safe passage across a road, demographic rescue, or successful breeding and gene flow.

Even if all aspects of linkage cannot be validated, a partial field study will add confidence and transparency to a corridor project. For example, Clevenger et al. (2002) developed two habitat models for black bears, one based on expert opinion and the other based on data from the literature. They identified road crossing zones from these models, and using data on crossings by real bears, they tested if the predicted linkages were used more than would be expected by chance. They found that the linkage models based on data from the literature outperformed the expert opinion models. The authors indicated that the expert opinion models may not have performed as well due to an overestimation of the importance of riparian habitat.

As an additional example of empirical field validation of corridors, Quinby (2006) used existing data from the annual breeding bird survey to test the utility of a proposed corridor. More bird species were found inside the corridor than outside it, confirming its validity. Chardon et al. (2003) used presence/absence data on the speckled wood butterfly from two different landscapes to compare the explanatory power of Euclidean distance and effective-distance connectivity models. They found that cost-distance was better able to predict connectivity than Euclidean distance. Zeller et al. (2011) used interviews with local residents to collect detection/non-detection data on jaguars and seven prey species in a grid-based design. The data were analysed by a site-occupancy model to determine probability of habitat use inside and outside the modelled corridor. It was found that probabilities of habitat use were mostly higher outside the modelled corridor, a conclusion which prompted a redesign of the final corridor.

The fact that there have been few studies to validate corridor models calls for more attention to this topic. Corridor validation techniques not only need to be improved upon, they need to be accessible to researchers and land managers working at different scales and on various species. Bridging the gap between corridor identification and corridor implementation will increasingly depend upon these validation studies, since land managers do not want to be left to implement a corridor of questionable efficacy, or be blamed for creating a sub-par corridor while more appropriate lands are unprotected from development and fragmentation (Hess & Fischer 2001; Morrison & Boyce 2008).

Conclusions

Population connectivity is critical for maintaining viable regional populations in the short term and to enable species to shift their geographic range in response to future climate change and other pressures such as land use change. In this chapter, we described the state of the art in quantitative corridor and connectivity modelling approaches. The first step in most quantitative connectivity analyses is to estimate and map landscape resistance. Traditional expert opinion is less useful for developing landscape resistance maps now that new and effective approaches using empirical data provide a much more reliable and robust means to map landscape resistance. There are a number of ways to predict or describe connectivity from resistance surfaces. Least-cost paths, least-cost corridors, circuit theory, centrality analyses, and resistant kernels are all powerful approaches suitable for different objectives. The efficient application of corridor analyses to future applied conservation problems must develop corridor designs to maintain connectivity for multiple species, and under changing climate. Finally, empirical validation of predicted corridors and linkages is essential to demonstrate their functionality and guide improvement of future corridor designs.

References

Adriaensen, F., Chardon, J.P., de Blust, G., *et al.* (2003) The application of 'least-cost' modelling as a functional landscape model. *Landscape and Urban Planning*, **64**, 233–247.

Adriaensen, F., Githiru, M., Mwang'ombe, J., Matthysen, E. & Lens, L. (2007) *Restoration and*

Increase of Connectivity among Fragmented Forest Patches in the Taita Hills, Southeast Kenya. Part II technical report, CEPF project 1095347968, University of Gent, Gent, Belgium.

Ahuja, R.K., Magnanti, T.L. & Orlin, J.B. (1993) *Network Flows: Theory, Algorithms, and Applications*. Prentice Hall, Englewood Cliffs, NJ.

Amundson, R. & Jenny, H. (1997) On a state factor model of ecosystems. *BioScience*, **47**, 536–543.

Anderson M.G. & Ferree, C.E. (2010) Conserving the stage: climate change and the geophysical underpinnings of species diversity. *PLoS ONE*, **5**(7), e11554.

Beier, P. & Brost, B. (2010) Use of land facets to plan for climate change: conserving the arenas, not the actors. *Conservation Biology*, **24**, 701–710.

Beier, P., Penrod, K., Luke, C., Spencer, W. & Cabanero, C. (2006) South Coast missing linkages: restoring connectivity to wildlands in the largest metropolitan area in the USA. In: *Connectivity Conservation* (eds K.R. Crooks & M.A. Sanjayan), pp.555–586. Cambridge University Press, Cambridge.

Beier, P., Majka, D. & Bayless, T. (2007) *Linkage Designs for Arizona's Missing Linkages*. Arizona Game and Fish Department, Phoenix. www.corridordesign.org/arizona/

Beier, P., Majka, D.R. & Spencer, W.D. (2008) Forks in the road: choices in procedures for designing wildland linkages. *Conservation Biology*, **22**, 836–851.

Beier, P., Majka, D.R. & Newell, S.L. (2009) Uncertainty analysis of least-cost modeling for designing wildlife linkages. *Ecological Applications*, **19**, 2067–2077.

Berggren, A., Birath, B. & Kindvall, O. (2002) Effect of corridors and habitat edges on dispersal behavior, movement rates, and movement angles in Roesel's bush-cricket (*Metrioptera roeseli*). *Conservation Biology*, **16**, 1562–1569.

Boone, R.B. & Hunter, M.L. (1996) Using diffusion models to simulate the effects of land use on grizzly bear dispersal in the Rocky Mountains. *Landscape Ecology*, **11**, 51–64.

Brost, B. & Beier, P. (2012) Use of land facets to design linkages for climate change. *Ecological Applications*, **22**(1), 87–103.

Brost, B.M. & Beier, P. (in press) Comparing linkage designs based on land facets to linkage designs based on focal species. *Journal of Applied Ecology*.

Brown, J.H. & Kodric-Brown, A. (1977) Turnover rates in insular biogeography: effect of immigration on extinction. *Ecology*, **58**, 445–449.

Bunn, A.G., Urban, D.L. & Keitt, T. (2000) Landscape connectivity: a conservation application of graph theory. *Journal of Environmental Management*, **59**(4), 265–278.

Carroll, C. (2010) *Connectivity Analysis Toolkit (CAT) Manual*. Klamath Center for Conservation Research. www.connectivitytools.org

Carroll, C., McRae, B.H. & Brookes, A. (2011) Use of linkage mapping and centrality analysis across habitat gradients to conserve connectivity of Gray Wolf populations in Western North America. *Conservation Biology*, **26**(1), 78–87.

Castellón, T.D. & Sieving, K.E. (2006) An experimental test of matrix permeability and corridor use by an endemic understory bird. *Conservation Biology*, **20**: 135–145.

Chandra, A.K., Raghavan, P., Ruzzo, W.L., Smolensky, R. & Tiwari, P. (1997) The electrical resistance of a graph captures its commute and cover times. *Computational Complexity*, **6**(4), 312–340.

Chardon, J.P., Adriaensen, F. & Matthysen, E. (2003) Incorporating landscape elements into a connectivity measure: a case study for the Speckled Wood Butterfly (*Pararge aegeria* L.). *Landscape Ecology*, **18**, 561–573.

Clevenger, A.P., Wierzchowski, J., Chruszcz, B. & Gunson, K. (2002) GIS-generated, expert-based models for identifying wildlife habitat linkages and planning mitigation passages. *Conservation Biology*, **16**, 503–514.

Compton, B., McGarigal, K., Cushman, S.A. & Gamble, L. (2007) A resistant kernel model of connectivity for vernal pool breeding amphibians. *Conservation Biology*, **21**, 788–799.

Coulon, A., Morellet, N., Goulard, M., Cargnelutti, B., Angibault, J.-M. & Hewston, A.J. (2008) Inferring the effects of landscape structure on roe deer (Capreolus capreolus) movements using a step selection function. *Landscape Ecology*, **23**, 603–614.

Cushman, S.A. (2006) Effects of habitat loss and fragmentation on amphibians: a review and prospectus. *Biological Conservation*, **128**, 231–240.

Cushman, S.A. & Lewis, J. (2010) Movement behavior explains genetic differentiation in American black bear. *Landscape Ecology*, **25**, 1613–1625.

Cushman, S.A., Chase, M. & Griffin, C. (2005) Elephants in space and time. *Oikos*, **109**, 331–341.

Cushman, S.A., Schwartz, M.K., Hayden, J. & McKelvey, K. (2006) Gene flow in complex landscapes: confronting models with data. *American Naturalist*, **168**, 486–499.

Cushman, S.A., McKelvey, K. & Schwartz, M.K. (2008) Using empirically derived source-destination models to map regional conservation corridors. *Conservation Biology*, **23**, 368–376.

Cushman, S.A., Chase, M.J. & Griffin, C. (2010a) Mapping landscape resistance to identify corridors and barriers for elephant movement in southern Africa. In: *Spatial Complexity, Informatics and Wildlife Conservation* (eds S.A. Cushman & F. Huettman), pp.349–368. Springer, Tokyo.

Cushman, S.A., Compton, B.W. & McGarigal, K. (2010b) Habitat fragmentation effects depend on complex interactions between population size and dispersal ability: Modeling influences of roads, agriculture and residential development across a range of lifehistory characteristics. In: *Spatial Complexity, Informatics and Wildlife Conservation* (eds S.A. Cushman & F. Huettman), pp.369–387. Springer, Tokyo.

Cushman, S.A., Landguth, E.L. & Flather, C.H. (2011) *Climate Change and Connectivity: Assessing Landscape and Species Vulnerability*. Final Report to USFWS Great Plains Landscape Conservation Co-operative.

Doyle, P.G. & Snell, J.L. (1984) *Random Walks and Electric Networks*. Mathematical Association of America, Washington, D.C.

Duke, D.L., Hebblewhite, M., Paquet, P.C., Callaghan, C. & Percy, M. (2001) Restoration of a large carnivore corridor in Banff National Park, Alberta. In: *Large Mammal Restoration: Ecological and Sociological Challenges in the 21st Century* (eds D.S. Maehr, R.F. Noss & J.F. Larkin), pp.261–275. Island Press, Washington, D.C.

Fagan, W.F. & J.M. Calabrese. (2006) Quantifying connectivity: balancing metric performance with data requirements. In: *Connectivity Conservation* (eds K.R. Crooks & M.A. Sanjayan), pp.297–317. Cambridge University Press, Cambridge.

Fahrig, L. (2003) Effects of habitat fragmentation on biodiversity. *Annual Review of Ecology, Evolution and Systematics*, **34**, 487–515.

Fahrig, L., Pedlar, J.H., Pope, S.E., Taylor, P.D. & Wegner, J.F. (1995) Effect of road traffic on amphibian density. *Biological Conservation*, **73**, 177–182.

Freeman, L.C., Borgatti S.P. & White, D.R. (1991) Centrality in valued graphs: a measure of betweenness based on network flow. *Social Networks*, **13**, 141–154.

Gamble, L.R., McGarigal, K. & Compton, BW. (2007) Fidelity and dispersal in the pond-breeding amphibian, *Ambystoma opacum*: implications for spatio-temporal population dynamics and conservation. *Biological Conservation*, **139**, 247–257.

Gibbs, J.P. (1998) Amphibian movements in response to forest edges, roads, and streambeds in southern New England. *Journal of Wildlife Management*, **62**, 584–589.

Guerry, A.D. & Hunter, M.L. Jr (2002) Amphibian distributions in a landscape of forests and agriculture: an examination of landscape composition and configuration. *Conservation Biology*, **16**, 745–754.

Gurnell, J., Rushton, S.P., Lurz, P.W., *et al.* (2006) Squirrel poxvirus: landscape scale strategies for managing disease threat. *Biological Conservation*, **131**, 287–295.

Haddad, N.M & Baum, K.A. (1999) An experimental test of corridor effects on butterfly densities. *Ecological Applications*, **9**, 623–633.

Haddad, N.M., Bowne, D.R., Cunningham, A., *et al.* (2003) Corridor use by diverse taxa. *Ecology*, **84**, 609–615.

Harrison, S. (1991) Local extinction in a meta population context: an empirical evaluation. In: *Metapopulation Dynamics: Empirical and Theoretical Investigations*. (eds M.E. Gilpin & I. Hanski), pp. 73–88. Academic Press, London.

Heller, N.E. & Zavaleta, E.A. (2009) Biodiversity management in the face of climate change: a review of 22 years of recommendations. *Biological Conservation*, **142**, 14–32.

Hess, G.R. & Fischer, R.A. (2001) Communicating clearly about conservation corridors. *Landscape and Urban Planning*, **55**, 195–208.

Hill, C.J. (1995) Linear strips of rainforest vegetation as potential dispersal corridors for rainforest insects. *Conservation Biology*, **9**, 1559–1566.

Holland, E.P., Aegerter, J.N., Dytham, C. & Smith, G.C. (2007) Landscape as a model: the importance of geometry. *PLoS Computational Biology*, **3**, 1979–1992.

Hunter, M.L. Jr, Jacobson, G.L. Jr & Webb, T. III (1988) Paleoecology and the coarse-filter approach to maintaining biological diversity. *Conservation Biology*, **2**, 375–385.

Jenny, H. (1941) *Factors of Soil Formation: A System of Quantitative Pedology*. McGraw-Hill, New York.

Kareiva, P.M. & Shigesada, N. (1983) Analyzing insect movement as a correlated random walk. *Oecologia*, **56**, 234–238.

King, A.W. & With, K.A. (2002) Dispersal success on spatially structured landscapes: when do spatial pattern and dispersal behavior really matter? *Ecological Modelling*, **147**, 23–39.

Landguth, E.L., Cushman, S.A., Murphy, M.A. & Luikart, G. (2010) Relationships between migration rates and landscape resistance assessed using individual-based simulations. *Molecular Ecology Resources*, **10**, 854–862.

Landguth, E.L., Hand, B.K., Glassy, J., Cushmann, S.A. & Sawaya, M. (2011) UNICOR: a species connectivity and corridor network simulator. *Ecography*, **35**, 9–14.

Lima, S.L. & Zollner, P.A. (1996) Towards a behavioral ecology of ecological landscapes. *Trends in Ecology and Evolution*, **11**, 131–135.

Lookingbill, T.R., Gardner, R.H., Ferrari, J.R. & Keller, C.E. (2010) Combining a dispersal model with network theory to assess habitat connectivity. *Ecological Applications*, **20**, 427–441.

Lovejoy, T.E. & Hannah, L. (eds) (2005) *Climate Change and Biodiversity*. Yale University Press, New Haven, CT.

McGarigal, K. & Cushman, S.A. (2002) Comparative evaluation of experimental approaches to the study of habitat fragmentation effects. *Ecological Applications*, **12**(2), 335–345.

McRae, B.H. (2006) Isolation by resistance. *Evolution*, **60**, 1551–1561.

McRae, B.H. & Kavanagh, D.M. (2011) *Linkage Mapper Connectivity Analysis Software*. The Nature Conservancy, Seattle, WA. www.waconnected.org/habitat-connectivity-mapping-tools.php

McRae, B.H. & Shah, V.B. (2009) *Circuitscape User Guide*. University of California, Santa Barbara, CA. www.circuitscape.org

McRae, B.H., Dickson, B.G., Keitt, T.H. & Shah, V.B. (2008) Using circuit theory to model connectivity in ecology and conservation. *Ecology*, **10**, 2712–2724.

Minor, E.S. & Lookingbill, T.R. (2010) Network analysis of protected-area connectivity for mammals in the United States. *Conservation Biology*, **24**(6), 1549–1558.

Minor, E.S. & Urban, D.L. (2007) Graph theory as a proxy for spatially explicit population models in conservation planning. *Ecological Applications*, **17**, 1771–1782.

Morrison, S.A. & Boyce, W.M. (2009) Conserving connectivity: some lessons from Mountain Lions in Southern California. *Conservation Biology*, **23**(2), 275–285.

Morzillo, A.T., Ferrari, J.R. & Liu, J.G. (2011) An integration of habitat evaluation, individual based modeling, and graph theory for a potential black bear population recovery in southeastern Texas, USA. *Landscape Ecology*, **26**, 69–81.

Newman, M.E. (2005) A measure of betweenness centrality based on random walks. *Social Networks*, **27**, 39–54.

Osborn, F.V. & Parker, G.E. (2003) Linking two elephant refuges with a corridor in the communal lands of Zimbabwe. *African Journal of Ecology*, **41**, 68–74.

Phillips, S.J., Williams, P., Midgley, G. & Archer, A. (2008) Optimizing dispersal corridors for the Cape Proteaceae using network flow. *Ecological Applications*, **18**, 1200–1211.

Pinto N. & Keitt, T.H. (2009) Beyond the least-cost path: evaluating corridor redundancy using a graph theoretic approach. *Landscape Ecology*, **24**, 253–266.

Quinby, P.A. (2006) Evaluating regional wildlife corridor mapping: a cast study of breeding birds in Northern New York State. *Adirondack Journal of Environmental Studies*, **13**, 27–33.

Rayfield, B., Fortin, M.J. & Fall, A. (2011) Connectivity for conservation: a framework to classify network measures. *Ecology*, **92**, 847–858.

Reh, W. & Seitz, A. (1990) The influence of land use on the genetic structure of populations of the common frog (Rana temporaria). *Biological Conservation*, **54**, 239–249.

Rosenberg, D.K., Noon, B.R. & Meslow, E.C. (1997) Biological corridors: form, function, and efficacy. *BioScience*, **47**, 677–687.

Rouget, M., Cowling, R.M., Lombard, A.T., Knight, A.T. & Kerley, G.I. (2006) Designing large-scale conservation corridors for pattern and process. *Conservation Biology*, **20**, 549–561.

Semlitsch, R.D., Bodie, J.R. (2003) Biological criteria for buffer zones around wetlands and riparian habitats for amphibians and reptiles. *Conservation Biology*, **17**, 1219–1228.

Seoane, J., Bustamante, J. & Diaz-Delgado, R. (2005) Effect of expert opinion on the predictive ability of environmental models of bird distribution. *Conservation Biology*, **19**, 512–522.

Servheen, C., Walker, J.S. & Sandstrom, P. (2001) Identification and management of linkage zones for grizzly bears between the large blocks of public land in the northern Rocky Mountains. In: *Proceedings of the 2001 International Conference on Ecology and Transportation* (eds C.L. Irwin, P. Garrett & K.P. McDermott), pp. 161–179. Center for Transportation and the Environment, North Carolina State University, Raleigh, NC.

Shanahan, D.F., Possingham, H.P. & Riginos, C. (2011) Models based on individual-level movement predict spatial patterns of genetic relatedness for two Australian forest birds. *Landscape Ecology*, **26**, 137–148.

Shepherd, B. & Whittington, J. (2006) Response of wolves to corridor restoration and human use management. *Ecology and Society*, **11**. www.ecologyandsociety.org/vol11/iss2/art1

Shirk, A., Wallin, D.O., Cushman, S.A., Rice, R.C. & Warheit, C. (2010) Inferring landscape effects on gene flow: a new multi-scale model selection framework. *Molecular Ecology*, **19**, 3603–3619.

Short Bull, R.A., Cushman, S.A., Mace, R., *et al.* (2011) Why replication is important in landscape genetics: American black bear in the Rocky Mountains. *Molecular Ecology*, **20**(6), 1092–1107.

Simberloff, D., Farr, J.A., Cox, J. & Mehlman, D.W. (1992) Movement corridors: conservation bargains or poor investments? *Conservation Biology*, **6**, 493–504.

Singleton, P.H., Gaines, W. & Lehmkuhl, J.F. (2002) *Landscape Permeability for Large Carnivores in Washington: A Geographic Information System Weighted-Distance and Least-Cost Corridor Assessment*. USDA Forest Service Research Paper, PNW-RP 549. Pacific Northwest Field Station, OR.

Spear, S.F., Balkenhol, N., Fortin, M.J., McRae, B.H. & Scribner, K. (2010) Use of resistance surfaces for landscape genetic studies: considerations for parameterization and analysis. *Molecular Ecology*, **19**, 3576–3591.

Spencer, W.D., Beier, P., Penrod, K. (2010) *California Essential Habitat Connectivity Project: A Strategy for Conserving a Connected California*. Report prepared for California Department of Transportation and California Department of Fish and Game. www.dfg.ca.gov/habcon/connectivity

Turner, M.G., Arthaud, G.J., Engstrom, R.T., *et al.* (1995) Usefulness of spatially explicit population models in land management. *Ecological Applications*, **5**, 12–16.

Urban, D. & Keitt, T. (2001) Landscape connectivity: a graph-theoretic perspective. *Ecology*, **82**, 1205–1218.

Van Dyck, H. & Baguette, M. (2005) Dispersal in fragmented landscapes: routine or special movements? *Basic and Applied Ecology*, **6**, 535–545.

Van Horne, B. (1983) Density as a misleading indicator of habitat quality. *Journal of Wildlife Management*, **47**, 893–901.

Waples, R.S. (1998) Separating the wheat from the chaff: patterns of genetic differentiation in high gene flow species. *Journal of Heredity*, **89**, 438–450.

Wasserman, T.N., Cushman, S.A., Schwartz, M.K. & Wallin, D.O. (2010) Spatial scaling and multimodel inference in landscape genetics: *Martes americana* in northern Idaho. *Landscape Ecology*, **25**, 1601–1612.

Weyrauch, S.L. & Grubb, T.C. (2004) Patch and landscape characteristics associated with the distribution of woodland amphibians in an agricultural fragmented landscape: an information-theoretic approach. *Biological Conservation*, **115**(3), 443–450.

WHCWG (2010) *Washington Connected Landscapes Project: Statewide Analysis*. Washington Wildlife Habitat Connectivity Working Group, Washington Departments of Fish and Wildlife, and Transportation, Olympia, WA. www.waconnected.org

Wiegand, T., Moloney, K.A., Naves, J. & Knauer, F. (1999) Finding the missing link between landscape structure and population dynamics: a spatially explicit perspective. *American Naturalist*, **154**, 605–627.

Wiens, J.A. (1997) Metapopulation dynamics and landscape ecology. In: *Metapopulation Dynamics, Ecology, Genetics, and Evolution* (eds I. Hanski & M. Gilpin), pp.43–62. Academic Press, London.

Williams, P., Hannah, L., Andelman, S., *et al.* (2005) Planning for climate change: identifying minimum-dispersal corridors for the Cape Proteaceae. *Conservation Biology*, **19**, 1063–1074.

Wilson, P.J. & Provan, J. (2003) Effect of habitat fragmentation on levels and patterns of genetic diversity in natural populations of peat moss Polytrichum commone. *Proceedings of the Royal Society Series B: Biological Sciences*, **270**, 881–886.

With, K.A., Gardner, R.H. & Turner, M.G. (1997) Landscape connectivity and population distributions in heterogeneous environments. *Oikos*, **78**, 151–169.

Zeller, K.A., Nijhawan, S., Salom-Pérez, R., Potosme, S.H. & Hines, J.E. (2011) Integrating occupancy modeling and interview data for corridor identification: a case study for jaguars in Nicaragua. *Biological Conservation*, **144**, 892–901.

22

Righting past wrongs and ensuring the future: challenges and opportunities for effective reintroductions amidst a biodiversity crisis

Axel Moehrenschlager[1], Debra M. Shier[2], Tom P. Moorhouse[3] and Mark R. Stanley Price[4]

[1]Centre for Conservation Research, Calgary Zoological Society, Calgary, Alberta, Canada
[2]Applied Animal Ecology Division, San Diego Zoo Institute for Conservation Research, Escondido, CA, USA
[3]Wildlife Conservation Research Unit, Department of Zoology, Recanati-Kaplan Centre, University of Oxford, Oxford, UK
[4]Wildlife Conservation Research Unit, Department of Zoology, Recanati-Kaplan Centre, University of Oxford, Oxford, UK *and* Al Ain Zoo and Aquarium, Abu Dhabi

"The future is not some place we are going, but one we are creating. The paths are not to be found, but made. And the activity of making them changes both the maker and the destination."

John H. Schaar

Emerging challenges and opportunities for reintroductions

Compared with 30 years ago, reintroductions and the science behind them have come of age. Early reintroductions, such as of bison onto the Great Plains of North America in 1907 (Kleiman 1989), released large, charismatic species with the common attitude that 'you open the gate, out it goes and all is fine'. We have learnt much since then; now reintroductions are better justified, planned, executed, monitored and, perhaps as a consequence, more frequent.

A major factor in this evolution was the publication of IUCN's Reintroduction Guidelines (IUCN 1998). Brief and translated into seven major languages, they set out the principles of responsible reintroduction design. Their impact has been greatly enhanced by the development of detailed

Key Topics in Conservation Biology 2, First Edition. Edited by David W. Macdonald and Katherine J. Willis.

and highly prescriptive guidelines for taxa as diverse as rhinoceroses, primates and crocodiles.

Known reintroduction projects increased from 126 between 1900 and 1992, to 218 by 1998, to 489 by 2005, and the number of peer-reviewed publications describing them increased more than 10-fold from the early 1990s to 2005 (Seddon et al. 2007). A recent review found that 653 reintroduction programmes for 629 plant species have occurred in 39 countries, and 77% of such reintroductions have occurred in Europe and North America (Godefroid & Vanderborght 2011). Despite a global volunteer body of 200 expert practitioners, the IUCN SSC Reintroduction Specialist Group now increasingly struggles to track all reintroduction programmes. The diversity of taxa involved has also increased. For example, two recent compilations of 134 case histories (Soorae 2008, 2010) now include 13 invertebrate and 13 fish examples, although these groups have traditionally been under-represented, compared both with other taxa and with their diversity in nature (Seddon et al. 2005).

The near-exponential growth in the frequency of reintroductions surely indicates that reintroductions are now a highly effective tool to combat the increasing loss of global biodiversity. But are they? It is less easy to state whether the success rate of reintroductions has increased in parallel with the number of attempts. While the 134 case history accounts rate 96% of reintroductions as 'highly successful', 'successful' or 'partially successful', these represent a biased sample of reintroductions and utilize subjective success measures. In fact, reintroductions often fail and, of 116 reintroductions evaluated by Fischer & Lindenmayer (2000), 26% were successful, 27% failures and 47% had unknown success. Reintroductions have increased in frequency not only because they can be effective but also because they are popular. For the public, reintroductions can represent an opportunity to make a positive impact within a world where negative stories regarding environmental degradation permeate the headlines. For example, the reintroduction of black-footed ferrets to Canada in 2009, after the species' national extirpation in the late 1930s, was marked by ministerial announcements, aboriginal ceremonies and wide-ranging press including international television documentaries.

The preponderance of emerging reintroduction programmes is at least equalled by the number and diversity of challenges that threaten biodiversity today. Lessons learned from classic reintroduction programmes, where extirpated species are returned to historic ranges, are needed not only to guide the reintroductions of the future but also to inform the complex and often controversial conservation translocation alternatives. Conservation translocation can be defined as the human-mediated movement of a species from one part of its range to another geographical location for purposes including:

- assisted colonization, where species such as conifers in British Columbia, Canada (Marris 2009), are moved beyond historic ranges to protect them from human-induced threats.
- ecological replacement, where species such as the giant tortoise (*Aldabrachelys gigantean*), which was captive bred for releases onto Ile aux Aigrettes and Round Island in the Indian Ocean since 2000 (Hansen et al. 2010), are introduced outside historic ranges to fill the ecological niche of extinct species.
- community construction where, in the future, entirely novel ecosystems may be assembled to retain biodiversity and ecological functions (Seddon 2010; see Chapter 23).

These challenges have prompted the current revision of the global IUCN Reintroduction Guidelines to be published in 2012–13, which will aim to prepare the practitioner for the myriad issues and opportunities that certainly lie ahead.

At this time of need, it may be tempting to run before learning to walk. This is risky in many ways. The irresponsible movement of organisms could lead to the suffering of individual animals, reduced viability of extant populations including of plant species, no overall conservation gain in countering severe extinction risks, and threats to recipient ecosystems. Therefore, we should now pause, reflect and ask questions which are both fundamental and difficult. What are the risks? When is the right or the wrong time to engage in reintroductions? How can our

Table 22.1 Summary of risks and benefits from reintroduction programmes

Level at which risk/benefit operates	Risks	Benefits
Individual	• Stress of animals housed under captive conditions prior to release • Stress, suffering, starvation or mortality of animals postrelease • Negative public opinion due to perceived welfare concerns	• Release from captivity of individuals
Population	• Loss of reintroduced population from existing factors (e.g. temporally unpredictable fluctuations in prey/forage availability) • Loss of population due to a failure of management, in particular a return of the original cause of the species' decline • Onset of disease in the reintroduced population, either due to exposure to sympatric species or subclinical infections becoming pathogenic in response to stress from the reintroduction process	• Establishment of viable population of species of conservation concern • Improved attitude towards species conservation activities from successful reintroduction
Ecosystem	• Spread of disease from reintroduced population to susceptible sympatric populations • Unpredictable community effects such as guild shift altering predator prey dynamics/ reintroduced herbivore or plant species altering vegetation composition • Conflict with local stakeholders (e.g. due to livestock depredation or exclusion from previously available land)	• Re-establishment of keystone species with net benefit to ecosystem as a whole (e.g. top-down control of herbivores by large carnivores or re-establishment of prey base) • Legal protection of wide area with conservation benefits for a large number of species due to re-establishment of a charismatic and/or keystone protected species

techniques be refined? How can we defensibly classify 'success' in reintroductions? Are reintroductions on the verge of a disciplinary shift within the conservation toolbox?

What are the risks?

Risks associated with reintroductions operate at a number of levels (Table 22.1). Animal reintroductions balance species conservation benefits against short-term decreases in individual welfare amongst members of the release cohort, because it is highly unlikely that all released individuals will survive (Griffith et al. 1989; Fischer & Lindenmayer 2000), and reintroduced animals often have a greater likelihood of stress (Teixeira et al. 2007), suffering or mortality both before and after release.

Mortality following release in reintroductions is typically high, and thus release groups often fail to establish and grow into viable populations (Griffith et al. 1989; Beck et al. 1994; Wolf et al. 1998; Godefroid et al. 2011). For example, known mortality rates of captive-bred grey partridges (*Perdixperdix*) were 38% after just 13 days and 42–56% after only 8 weeks (Rantanen et al. 2010a). These captive-bred birds exhibited poor predator vigilance and birds suffering high postrelease mortality rates had selected habitats on field margins where predator abundance was likely higher than in crop fields (Rantanen et al. 2010b). Direct translocation or head-starting of northern water snakes (*Nerodeasipedonsipedon*) yielded annual survival rates of 19.6% and 16.0% respectively, which were about three times lower than those of resident snakes (Roe et al. 2010). European mink (*Mustelalutreola*) that were released onto an Estonian island following American mink (*Neovisonvison*) control had mortality rates of 25% after 16 days, and 50% after 38 days, primarily because these captive-bred animals undertook extensive movements, often through unsuitable habitat (Maran et al. 2009).

Similarly, all animals intended for reintroduction require at least some period in captive conditions, even if this simply consists of confinement during transport, which may not be representative of the species' preferred social structure (Morgan & Tromborg 2007) and may have welfare implications (Olsson & Westlund 2007) or detrimental consequences for the physiological condition of the individuals (e.g. Gelling et al. 2010). For example, group size in water voles (*Arvicola amphibius*), housed in laboratory cages prior to a UK reintroduction, has been shown to correlate negatively with immunocompetence, such that individuals in larger groups experienced a larger degree of immunosuppression than did individuals housed in smaller groups or individually (Gelling et al. 2010). Water voles in the wild typically hold individual ranges during the breeding season (Moorhouse et al. 2008), and Gelling et al.'s (2010) study highlights the need to consider life-history strategies when choosing housing systems.

Welfare considerations post release may lead critics and/or the public to call for the reintroduction to be stopped, which happened during the early stages of the reintroduction of the Colorado lynx. Although the project was ultimately heralded a 'success' (Program 2010), five newly released lynx starved to death during the first year, leading to a public outcry and uncertainty over the future of not only that reintroduction but similar plans in other states (Kloor 1999).

Risks to individuals are compounded in situations where the likelihood of extinction of the target population is predicted to be high prior to the reintroduction, whether due to factors intrinsic to the species (such as slow rates of population increase, which will extend the time when it is at risk of stochastic extinction; Griffith et al. 1989) or extrinsic factors such as unpredictable fluctuations in populations of prey species.

Even 'successful' reintroductions may carry risks because the introduction of one or more species has the potential to affect resident species in the destination ecosystem. For example, the release cohort, and the subsequently established population, may represent a disease or a hybridization risk to extant, sympatric populations, especially if the animals are translocated from distant geographic areas. Minimizing disease transfer should be a reasonable goal of species recovery programmes but disease screening cannot be exhaustive (Mathews et al. 2006) and it is not always clear which diseases should be screened for or what are the risks of their spreading.

Even from this relatively short discussion, it is clear that the risks involved in reintroduction, for both individuals and ecosystems, can be substantial and so nearly all aspects of reintroductions require careful consideration, including whether, and when, to reintroduce in the first place.

When is the right time to start a reintroduction?

The guidelines from the IUCN/SSC Reintroduction Specialist Group provide a valuable resource describing the preconditions for, and stages of, any reintroduction project (IUCN 1998). However, no guidelines exist that allow policy makers or institutions to assess whether or not engaging in a reintroduction is desirable in the first place. It is unclear, given an initial perceived need for reintroduction, whether a reintroduction should be performed simply because conditions appear to be favourable. This uncertainty arises because, as described above, reintroductions are inherently risky, with no guarantee of success, and so the decision to engage is necessarily context specific and dependent upon risk tolerance. Moreover, as reintroductions are likely to be driven by passionate champions, they may not necessarily deliver conservation where and when most needed on objective criteria.

In practice, all reintroductions represent a mixture of risks and benefits, both of which will pertain to the species or taxa in question. For example, the reintroduction of grey wolves (*Canis lupus*) to Yellowstone National Park in 1995 has been credited with top-down benefits such as restoring ecosystem processes through the regulation of elk browsing (Mao et al. 2005) and buffering populations of carrion eaters against the effects of climate change (Wilmers & Getz 2005). However, wolf depredation on livestock, although a small economic cost to the local livestock industry (<0.01% of the annual gross income; Muhly & Musiani 2009), has created conflict between livestock producers and conservation organizations which necessitated the establishment of compensation schemes (Naughton-Treves et al. 2003). Similarly reintroductions and population expansion of prairie dogs (*Cynomys spp.*) in North America can benefit a wide range of species. Black-tailed prairie dog (*C. ludovicianus*) declines may contribute to an overall decline in grassland vertebrate communities (Ceballos et al. 2010) and the reintroduction success of critically endangered black-footed ferrets (*Mustela nigripes*) is principally dictated by prairie dog densities (Jachowski et al. 2011). However, prairie dogs are extremely susceptible to sylvatic plague (*Yersinia pestis*), which they can pass both to other wildlife, including black-footed ferrets, and, more rarely, to humans (Gage et al. 1992; Rocke et al. 2008).

Recent debate has also arisen regarding the value and efficacy of plant reintroductions. Analysis of 249 plant species reintroductions revealed only 52% individual plant survival, 19% flowering rates and 16% fruiting rates (Godefroid et al. 2011). Moreover, success rates after reintroductions were found to decline over 4 years. Many unsuccessful attempts were found to be unpublished, which further led Godefroid et al. (2011) to question the true success of plant reintroductions. Other data, however, demonstrate that 92% of plant reintroductions yielded surviving populations, that 33% of reintroductions yielded a next generation, and in 16% of cases the next generation had reproductive individuals (Guerrant 2012). These data, along with the persistence of some reintroduced populations for over 24 years, compelled Albrecht et al. (2011) to caution against completely dismissing plant reintroductions. Nevertheless, Godefroid (2011) believes that the cited survival rates are too low and, like Dalrymple et al. (2011), still maintains that evaluations based on published plant reintroductions are overly optimistic.

As seen from the above examples, the decision on whether or not to reintroduce a species inevitably requires an assessment of the trade-offs between the likely risks and benefits. Table 22.1 provides a summary of these, operating at a variety of levels. Managing the cost–benefit trade-off of a putative reintroduction is not straightforward, and requires thorough knowledge of the ecology and, if applicable, behaviour of the species in question as well as a thorough *a priori* assessment of the potential risks and likelihood of success. Modelling tools such as population viability analysis (PVA) are increasingly employed to make management decisions in reintroductions (e.g. Zeoli et al. 2008; Moorhouse et al. 2009; Schaub et al. 2009). However, any such *a priori* analyses can only determine a certain probability of subsistence for the target species of potential reintroductions; they do not incorporate risks of ecological impacts (either beneficial or harmful) of a successful reintroduction to other species/human enterprises, and they ultimately still leave to human judgement the question of what risk of extinction is deemed acceptable.

There is a continuum of desirability for putative reintroduction projects under different circumstances. Although the risk of inaction may appear overwhelmingly large as it can, in the most extreme, lead to the extinction of the species, each project must be assessed individually, concentrating on the motivation for reintroduction and the likelihood of a positive outcome.

In Box 22.1 we present some broad rules to aid decision making, listing motivations for reintroduction and dividing them into those

Box 22.1 Examples of inappropriate and desirable motives for reintroductions

Reasons that constitute, in isolation or as the principal motivator, an inappropriate case for reintroduction

Dumping: An existing captive breeding organization has produced or acquired a large surplus of individuals or wants to justify housing a captive collection on conservation grounds.

Disguised Rehabilitation: Previously injured or confiscated animals have been rehabilitated and (unjustifiably) claiming a reintroduction need would allow for the return of individuals to the wild.

Shirking Responsibility: Plants or animals that are protected by law or regulations that obstruct land use development are translocated or moved into captivity, and are later reintroduced into degraded habitat instead of employing less invasive mitigation measures or ceasing development.

Saving Face: A jurisdiction (e.g. a country, province, state, district or park) is embarrassed publicly about the loss of a species and initiates reintroductions too quickly without adequately considering IUCN guidelines.

PR Ploy: A reintroduction would be an excellent public relations opportunity to raise the profile of an organization, area or cause.

Money Grab: A reintroduction would attract funding, either for that species in a wider context, for a particular ecosystem, or for a broader conservation programme.

Academic Greed: A reintroduction would allow for excellent experimentation opportunities that could yield publications to bolster academic credentials of particular individuals or organizations.

Situations where the case for reintroduction is highest relative to less desirable alternatives

Species versus Subspecies: The reintroduction candidate is an imperilled species instead of subspecies or regional adaptation. (Bearing in mind, however, that taxonomic classification is dynamic, and threat designation on national/ international levels can be delayed due to slow (re) assessments of status, and that political stalling of controversial species can also affect listing. Where possible true biological status – for example whether the candidate comprises an evolutionarily significant unit – should be taken into account.)

Global versus Local Status: Successful reintroduction in the target area would improve the global IUCN status for the species, instead of merely restoring a geographic isolate.

Global versus Local Lessons: The reintroduction will test theories or techniques that could convincingly yield lessons for reintroductions of similar or other taxa elsewhere.

Low versus High Animal Welfare Effects: The reintroduction would cause minimal stress, suffering or mortality among animals in source, captive or released populations.

Ecosystem versus Species Function: The reintroduction candidate could provide a convincing ecosystem function above and beyond the expected benefits from the return of a single species.

Designated versus General Conservation Resources: The reintroduction attracts unique financial or logistical resources, and does not detract from less 'glamorous' conservation methods that might be more urgent/effective such as habitat protection for other imperilled species in the ecosystem.

Umbrella versus Species Protection: Formalized protection of the reintroduction candidate could bring umbrella protection to ecosystem (for example through creation of a protected area or restriction of activities that are harmful to the ecosystem).

that we consider indefensible and those that we consider relatively desirable. Clearly, any adequately motivated reintroduction must still consider and satisfy the pre-project activities outlined in the IUCN Reintroduction Specialist Group guidelines before proceeding.

Strategies to improve reintroduction techniques

The exponential increase in the number of reintroductions over the last 20 years has developed in response to changes in land use practices that have resulted in degradation, fragmentation or complete habitat loss (Armstrong & Seddon 2007; Seddon et al. 2007). As climate change and other human activities continue to alter landscapes, often leaving plants and animals in habitat fragments too small or degraded to sustain them, conservationists will resort to reintroduction more frequently. Yet, reintroduction science is not yet adequately developed to predict and mitigate the consequences of disruption of the organism–native ecosystem relationship which is inevitably entailed in relocation.

The typically high level of mortality following release described above is not only a conservation concern; it raises an ethical dilemma concerning the welfare of animals. Should one relocate animals for conservation in cases where we know most may not survive the experience (Cayford & Percival 1992)? Because conservation biologists are focused on preserving ecosystems, species and/or populations, reintroduction practitioners may accept low initial survival as a price worth paying if it might ultimately lead to restoration success (Harrington et al. submitted). Others may not be as forgiving. Animal welfare researchers give the fates of individuals higher priority than the fates of populations or species, and would likely argue against this conservation strategy when high levels of mortality occur. Though traditionally there has been relatively little interaction between conservation biologists and animal welfare researchers, interdisciplinary dialogue and research have been increasing and proponents of both sides would surely favour improving reintroduction techniques to enhance postrelease survival of individuals and the establishment of populations (Harrington et al. submitted).

How can we improve reintroduction techniques to enhance release success? Reintroduction is a management action, and the incorporation of rigorous scientific methods into reintroduction planning is relatively novel. Unfortunately, a majority of reintroductions today are conducted without controlled experimental designs or postrelease monitoring, and therefore lack the empirical data needed to evaluate and improve the technique. For decades, scientists have recognized the need for better postrelease monitoring and result reporting following reintroduction (Griffith et al. 1989; Sarrazin & Barbault 1996; Fischer & Lindenmayer 2000), and recently others have suggested several possible directions for improving reintroduction outcomes, including hypothesis testing through replicated controlled experiments, meta-replication (several separate studies testing the same process) and adaptive management (Seddon et al. 2007; Swaisgood 2010). Structured decision making through active adaptive management may be the key.

Though adaptive management is widely promoted in conservation biology, in reality it is rarely implemented (Sutherland 2006). Adaptive management is an iterative process that incorporates scientific methodologies in the design, planning, implementation and evaluation of management strategies (Schreiber et al. 2004). Adaptive management begins with specific objectives determined through stakeholder (managers, scientists, land owners, regulatory agency representatives) co-operation. Modelling can then be conducted with existing knowledge to promote consensus and identify informational gaps. Once it is determined where greater certainty would lead to better management, specific hypotheses can be articulated and evaluated through controlled experimental design and monitoring (Schreiber et al. 2004; Nichols & Williams 2006). This hypothetico-deductive process allows for 'directed' as opposed to 'trial and error' learning (Walters 1997). Management strategies are subsequently refined following evaluation of results, incorporating 'lessons learned'. By repeating this cycle and increasing the body of knowledge about the system in question, reintroduction practitioners will be able to refine their methods and increase survival following release.

An example of a successful adaptive management programme is the recovery of the North Island kokako *Callaeascinerea wilsoni* in New Zealand (Innes et al. 1999). Researchers used controlled experiments to test the hypothesis that introduced predatory and/or browsing mammals were causing the kokako decline. They found that reduction of introduced predators, ship rats (*Rattus rattus*) and brushtailed possums (*Trichosurus vulpecula*), resulted in immediate increases of kokako fitness (Innes et al. 1999). These results were incorporated into management of the species and facilitated recovery (Brown et al. 2004).

While each step in adaptive management is important, articulating and evaluating specific hypotheses through controlled experimental design and mid- to long-term monitoring are what ultimately allow for learning and refining techniques (Nichols & Williams 2006). To date, only a small fraction of reintroductions (12% of published reports as of 2007) incorporate rigorous experimental tests of explicit hypotheses (Seddon et al. 2007). Some ecologists contend that it is challenging, perhaps impossible, to perform replicated controlled experiments that address many large-scale ecological questions (Bennett & Adams 2004) and even more difficult when working with a species that is listed as imperilled globally or nationally (Seddon et al. 2007). Indeed, even modest questions can be difficult to address with highly endangered species. Captive breeding for the reintroduction of the critically endangered Vancouver Island marmot, which declined to fewer than 100 individuals in wild and captive populations combined, benefited from critical behavioural research (Casimir et al. 2007), but concerns regarding the marking of captive males first delayed necessary behavioural observations for 2 years because potentially harmful effects on any individuals might compromise the global population.

Not surprisingly, most recent reviews that propose strategies for improving reintroduction methodology focus on questions at the population, meta-population and ecosystem levels to facilitate species persistence (e.g. What habitat conditions are needed for persistence of the reintroduced population? How will genetic make-up affect persistence of the reintroduced population?) (Armstrong & Seddon 2007). Yet, reintroduction failures are most apparent during the establishment phase – the first days to weeks following release (Beck et al. 1994; Armstrong & Seddon 2007) – and without successful establishment one cannot examine persistence. For animal reintroductions, during the establishment phase poor release performance is primarily due to the disruption of the relationship between the individual translocated animals and their environment, rather than large-scale ecological processes. Clearly, a better understanding of animal behaviour at the individual level could improve the success of reintroduction releases, but the standard conservation approach of focusing primarily on population-level outcomes may be hindering the progress of reintroduction science.

Behavioural ecology is the interface between an animal and its environment, including other animals, and high mortality following release has long been blamed on the suboptimal behavioural responses of the individuals reintroduced (Kleiman 1989). In particular, predation is a major cause of mortality during establishment, especially among captive-bred releases (Short & Smith 1994; Fischer & Lindenmayer 2000). Newly released animals find themselves in completely unfamiliar and dangerous surroundings without the benefit of escape refuges or experience with local foraging patches, competitors or predators.

It is easy to appreciate how understanding the target species' behavioural ecology can be important for improving reintroduction success of captive-reared animals. Captive-reared animals are thought to possess ineffective survival skills because while in captivity they are not subject to the threats and harsh conditions that exist in nature (Beck et al. 1994; McPhee 2003a; Mathews et al. 2005). Not only are specific survival skills absent in the behavioural repertoire of many captive-reared animals, but skills of wild-caught animals can erode while they are in captivity. Furthermore, under captive conditions there are additional general effects such as stress, impaired cognitive function and ineffective social behaviour that can influence learning and/or performing these critical skills (Teixeira et al. 2007; Zidon et al. 2009; Swaisgood 2010).

Translocations of animals from extant wild populations to release sites have proven to be more successful than captive releases (Fischer & Lindenmayer 2000; Jule et al. 2008), but suffer from similar, though less severe, behavioural issues. Behavioural deficiencies are evident across taxa in foraging, locomotion, site fidelity, social interactions, mating, nesting and antipredator skills (Miller et al. 1999; Rabin 2003). Captive-born golden lion tamarins (*Leontopithecus rosaliarosalia*) show deficient foraging and locomotory skills up to 2 years after release to the wild compared to same-aged wild-born offspring (Stoinski et al. 2003; Stoinski & Beck 2004). Wild Coho salmon (*Oncorhynchus kisutch*) are more aggressive and outcompete hatchery-reared males for females during the breeding season (Fleming & Fross 1993), and development of effective antipredator behaviour in kangaroo rats requires predator experience (Yoerg & Shier 1997). In relation to survival skills, the social context and timing of antipredator skill development may play an important role in determining release timing for translocated juveniles (Shier 2006). For example, juvenile black-tailed prairie dogs are highly social and learn antipredator behaviour from experienced social group members (Hoogland 1995). Prairie dog translocations conducted late in the summer yielded higher juvenile survival compared to releases early in the summer (Shier 2006). Translocation later in the summer may allow juveniles more time to develop and hone their survival skills through interactions with experienced kin.

Almost every aspect of a species' behavioural ecology offers opportunities for improving release techniques (e.g. territoriality, foraging, temperament, social behaviour, communication) but understanding the target species' dispersal biology may prove to be a close second to antipredator behaviour in degree of importance (Stamps & Swaisgood 2007; Swaisgood 2010). Immediate rejection of the release site followed by long-distance movements ('dispersal') away from the site often plague reintroductions (Griffith et al. 1989; Kleiman 1989; Miller et al. 1999). For example, 60% of hard-released dormice (*Muscardinusavellanarius L.*) dispersed immediately following release (Bright & Morris 1994) and 77% of reintroduced cage-reared black-footed ferrets moved further than 7 km within a 12-hour period (Biggins et al. 1999).

Excessive postrelease movements increase the cumulative risks associated with exposure to predators and aggressive unfamiliar conspecifics (Moehrenschlager & Macdonald 2003; Linklater & Swaisgood 2008) and also may divert time and energy away from the establishment of translocated individuals in a novel environment, including finding or creating shelter, such as dens or burrows (Moehrenschlager & Macdonald 2003; Shier 2006). Occasional long-distance movements, such as the 191 km movement of one swift fox that traversed among reintroduced populations (Ausband & Moehrenschlager 2009), may occur and aid population connectivity, but are not necessarily favourable during reintroduction release phases. In swift fox (*Vulpesvelox*) reintroductions in Canada, increased travel distance before settlement reduced postrelease fitness in terms of survival and reproductive success. For the foxes in this study, increased travel distances delayed settlement and mating, and probably increased predation risk as they were unfamiliar with their release environment, including escape dens which are crucial for the foxes to evade coyotes and golden eagles

(Moehrenschlager et al. 2007). Therefore, understanding how animals make settlement decisions and choose habitat may be critical for establishing released animals quickly and safely in a new environment.

A few case studies provide further, compelling evidence that understanding the target species' behavioural ecology, and implementing an adaptive management approach utilizing controlled experimental designs, can both profoundly improve reintroduction outcomes. In the black-footed ferret programme, researchers used experiments in captivity and during experimental releases into the species' historic range in Wyoming, USA to test several variables including captive enrichment and pen size, experience with live prey and antipredator training (Biggins et al. 1999; Vargas & Anderson 1998, 1999). These experiments showed, among other things, that ferrets require exposure to live prey during ontogeny to learn effective foraging skills (i.e. make effective kills) and that ferrets housed in outdoor pens compared to indoor cages had higher release success. Combined with postrelease monitoring, researchers were able to refine their rearing and release techniques to devise a more effective reintroduction programme. From small-scale captive-release studies and large-scale controlled experimental translocations, researchers learned that for black-tailed prairie dogs:

- antipredator training is effective (Shier & Owings 2006).
- experienced prairie dog demonstrators enhance training to the extent that captive-reared and socially trained juveniles perform as well as wild-reared juveniles following release (Shier & Owings 2007).
- prairie dogs are five times more likely to survive and achieve higher reproductive success following a translocation if they are moved in family groups (Shier 2006).
- releases are most effective in late summer when juveniles are older and more predator aware, and adult females have recovered from the energetic demands of reproduction (Shier 2006).

Conservation biologists, and therefore reintroduction practitioners, have, in general, been reluctant to embrace disciplines such as behavioural ecology because they appear to require developing protocols species by species. However, general reintroduction methods may be drawn up for groups of species with similar behavioural ecology. For example, for any species in which social interactions influence fitness (e.g. species that exhibit kin selection, reciprocity, communal nesting, coalition formation), maintaining social groups during reintroduction may improve postrelease performance (Shier 2006; Shier & Swaisgood 2012). Perhaps surprisingly, maintaining social relationships may be just as important for some territorial species as for highly social ones. In solitary and territorial Stephens' kangaroo rats (*Dipodomys stephensi*), individuals translocated with familiar neighbours survived at higher rates and had 24 times more offspring compared to kangaroo rats translocated without neighbours (Shier & Swaisgood 2012). These results may be explained in part by postrelease behaviour. Immediately following release, kangaroo rats translocated without known neighbours fought more and spent less time foraging compared to kangaroo rats translocated with known neighbours. Like unfamiliar kangaroo rats, territorial black rhinos translocated without attention to familiarity among founders exhibit high levels of intraspecific aggression following release, leading to serious injury and death (Linklater & Swaisgood 2008). Thus, choosing the right number and composition of animals and providing sufficient space at the release site can reduce these problems.

While the global IUCN Reintroduction Guidelines outline the key conditions necessary for responsible reintroductions, they do not aim to address release- or taxon-specific issues. Recent evaluations of plant reintroductions are of interest in stirring novel reintroduction considerations and experimental opportunities. For example, meta-data analysis of 301 attempted reintroductions of 128 taxa found that sourcing founders from wild instead of captive populations, removing original causes of species decline prior to propagule introduction, and reintroducing within the historic range of a species did not increase reintroduction success (Dalrymple et al. 2011). We believe that the following general questions concerning the release stage of plants or animals within a broader reintroduction plan are not only necessary to address, but also present fruitful opportunities for modelling or experimentation that can advance reintroductions in general (Harrington et al. submitted).

- What are the indicators of reintroduction success, what techniques will be used to monitor the population, and with what regularity, over the short, mid and long term?
- How many individuals need to be released initially, and how many supplementary releases might be required/justifiable?
- Will founder plants or animals come from genetically diverse captive or wild populations?
- What sex ratio should be specified for released animals?
- Will seeds or seedlings be used for released plants?
- What life stage(s) and social units should animal release cohorts comprise?
- What density (i.e. how many per unit area) and spacing should be allowed for the released cohort?
- Have any animal behaviours required for survival or reproduction been lost during captivity/captive breeding, and, if so, can they be taught prior to release?
- What physical structures need to be in place to permit individuals to make the transition from captivity to familiarity with the new habitat (e.g. temporary shelters or holding pens; see Moorhouse et al. 2009 for an example)?
- Will supplementary feeding be required for animals or removal of competitors for plants and for what period?
- Should strategies such as animal recapture, predator exclusion or predator control be employed to address potential welfare challenges amongst the release cohort or their immediate descendants?

In summary, to improve reintroduction methodologies and advance reintroduction science, we advocate the use of structured decision making through controlled experiments within the general framework of active adaptive management. Incorporating the study of behavioural ecology is likely the key to manipulating the pre- and postrelease environment to enhance release success, and can provide general principles to direct reintroduction research and supplement currently established guidelines (IUCN 1998).

A novel approach to assessing 'programme success' in reintroductions

Adequate planning, modelling, experimentation and adaptive management are key to optimizing reintroduction techniques, but ultimately all reintroduction practitioners should be asked the question: 'So, has your reintroduction programme been successful?'. Despite increasing frequency of reintroductions, no standardized categorization of 'success' has been developed. Ultimately, reintroductions of plants or animals should aim to establish viable populations (IUCN 1998), but qualifications of success can range from the contribution of a single individual to a population (Fischer & Lindenmayer 2000), through breeding by the first wild-born generation, to the establishment of a self-sustaining population (Seddon 1999). Generally, evaluation of success has relied upon survey responses or opinions of species experts involved in the programmes (Wolf et al. 1996; Breitenmoser et al. 2001; Soorae 2008, 2010) or upon a subjective interpretation of population sustainability by reviewers who are constrained by what information is available within published papers (Fischer & Lindenmayer 2000; Godefroid et al. 2011). The dynamic nature of animal and plant populations also means that unsuccessful reintroductions may be considered successful at some previous or subsequent point in time (Seddon 1999; Jule et al. 2008).

The idea that, like beauty, reintroduction success may be primarily 'in the eye of the beholder' may be unacceptable to critics – who challenge the role of reintroductions on animal welfare, ethical or financial grounds – as it might be to reintroduction supporters such as funders, governments or the general public. Subjective assessments based on inconsistent criteria would not be deemed acceptable for global species status assessments. Given that the IUCN Red List is the most authoritative and objective system for classifying extinction risk (Butchart et al. 2005), an application of its criteria to reintroductions would be desirable. Red List risk assessments use five criteria: high decline rate; small range area and decline; small population size and decline; very small population size; and unfavourable quantitative analysis, which generally pertains to population viability analyses (Mace et al. 2008). Crucial geographic components include the 'extent of occurrence' and 'area of occupancy'. Extent of occurrence is defined as the area contained within the shortest continuous boundary that can be drawn to encompass all the known, inferred or projected locations of a species. Area of occupancy is the area within the extent of occurrence where, given the patchy distribution of populations, the species is actually found (Mace et al. 2008). Some reintroduction evaluations incorporate geographic measures (e.g. Carroll et al. 2003), but many erroneously concentrate only on population abundance and trend.

The criteria for estimating sustainability of extant species should also be applicable to reintroduced populations. Indeed, for the Mauritius kestrel (*Falco punctatus*) and black-footed ferret in North America, which both became extinct in the wild, assessments of global species status and reintroduction status would be identical since all wild populations were established from reintroductions.

Red List criteria were primarily designed to categorize species extinction risk based on past or projected population demography, but their primary function has not been to monitor the growth or decline of populations. However, Red List Indices (RLIs) track the relative rate that sets of species change threat status over time (Butchart et al. 2005) and thereby can be used to evaluate progress towards biodiversity protection targets (Mace et al. 2008). Similarly, on a single species level, changes in Red List-based threat status assessments of a reintroduced population over time can indicate if a population is moving towards sustainability.

Red List criteria were designed to evaluate species globally, but most reintroductions operate on a regional or local scale. As an illustration, the global population of water voles is listed by the IUCN as 'least concern' because despite 'ongoing declines in some range states (such as Britain, Italy and The Netherlands), the overall population trend is believed to be stable at the global level' (Batsaikhan et al. 2008). However within the UK, which has witnessed a substantial decline in national water vole population size (Strachan et al. 2011), the species is afforded legal protection and has been the subject of substantial public concern and ongoing conservation actions which have included reintroductions (e.g. Moorhouse et al. 2009). This is a good example of when a species can be of great concern at a local level but escape the attention of the globally focused IUCN Red List. However, the IUCN's Regional Guidelines for Species Assessment (IUCN 2003) were developed to conduct species assessments on sub-global levels and could consequently be used to evaluate the status of reintroduced populations on the scale of a continent, country, state or province (IUCN 2003).

IUCN Red List criteria can be applied regionally within any geographically defined area as long as the regional population is isolated from conspecifics outside that region (IUCN 2003). The spatial scale of reintroductions depends upon the ecology of plants or animals, their range requirements, the extent of available habitat, and the ability to disperse or migrate. Accordingly, initial habitat models for potential beaver reintroductions into the UK concentrated on an area of less than 2500 km^2 in Norfolk (South et al. 2001), but simulated the potential movement of reintroduced individuals. Practitioners should choose the most biologically meaningful spatial scale that encompasses the potential expanse of a small but growing reintroduced population. Within that region, evaluators should apply IUCN Red List criteria to assess and to assign a Red List-equivalent risk status, at various time intervals following the reintroduction.

We suggest that regional reintroduction success be classified in terms of the degree to which the reintroduced population improves in threat category from the 'regionally extinct' status. A positive change in the threat status of reintroduced populations should indicate a measure of 'success' but the number of 'steps' through the threat categories should qualify its magnitude (Table 22.2). For example, a reintroduction that results in a population with characteristics that qualify for 'endangered' Red List status would be considered as having 'good success' whereas a reintroduction that results in a regional classification of 'least concern' would be scored as having 'excellent success'.

Progressive steps towards recovery need to be based on a complete assessment of reintroduced species status using all Red List criteria, but selected recovery parameters are given for key categories in Table 22.2. Like global species assessments, evaluations can be applied at any point of the reintroduction, and can be repeated over time. While the effectiveness of release techniques should be evaluated frequently, we propose that the outcome of reintroduction programmes should generally be evaluated after a population has been monitored for at least 5 years, even if releases have not been conducted for the entire period. This time period is consistent with guidelines for evaluating the downlisting of extirpated species to 'critically endangered' (IUCN Standards and Petitions Subcommittee 2010), and allows for an adequate determination of population distribution, abundance and trend.

The key difference between this and other success classification schemes is that this method is objective and defensible, as it uses globally accepted classification criteria. An assessment of relevant demographic or spatial thresholds (see Table 22.2) can also guide regional recovery planning; for example, the Canadian swift fox recovery team developed the following reintroduction goal: 'By 2026, restore a self-sustaining swift fox population of 1000 or more mature, reproducing foxes that does not experience greater than 30% population reduction in any 10-year period' (Pruss et al. 2008).

Species that appear to have stable populations regionally could still experience precipitous declines globally, and globally stable species could be compromised regionally (Mace et al. 2008). Regional reintroductions can profoundly help the global status of species, but often the relative scale of reintroductions may be too small to improve global status convincingly. For example, cheetah reintroductions into 37 reserves have impressively increased animal abundance by 258 individuals in South Africa (Lindsey et al. 2011), but the species' Red List 'vulnerable' status remains unaffected. Reintroductions outside extant species ranges could primarily impact global status by increasing the extent of occurrence. If reintroductions occur more centrally, they could increase the area of occupancy. Reintroductions can also increase the number of subpopulations and thereby projected population viability. For example, the application of optimal breeding approaches (Smith et al. 2011) to support a fourth whooping crane release site in Louisiana may eventually increase the number of subpopulations and extent of occurrence and move the species towards a Red List status of 'vulnerable' instead of 'endangered'.

In addition to evaluating reintroduction success regionally, the global benefit of reintroductions should also be determined. To do so, species should be assessed using all Red List global criteria on two levels: the first should classify the risk status of the species excluding populations that have been created through reintroductions, and the second should include reintroduced populations. The difference in species status between these assessments can qualify the magnitude by which reintroductions have contributed to the global recovery of the species (see Table 22.2). For example, a comparison that finds wild populations categorized as

Table 22.2 Framework for evaluating success in reintroductions on regional and global scales. Select IUCN thresholds have been adapted and modified from Butchart et al. (2005) for application to regional reintroduction evaluation

Evaluation of success	Regional status of target species	Criterion B1: small range (extent of occurrence)	Criterion B2: small range (area of occupancy)	Criterion D1: very small population (mature individuals)	Criterion E: quantitative analysis (estimated extinction risk)	Global status of target species
	Reintroduction of the target species has resulted in a population with abundance, trend, extent of occurrence, area of occupancy and/or population viability that an application of global and regional IUCN Red List criteria at the scale of the defined region would …	Area given below should be exceeded. Also fewer than 2 of the following should be present: (a) severe fragmentation/few localities; (b) continuing decline; (c) extreme fluctuation	Area given below should be exceeded. Also fewer than 2 of the following should be present: (a) severe fragmentation/few localities; (b) continuing decline; (c) extreme fluctuation			In addition to remnant wild population(s), reintroduction(s) has/have increased the abundance, population trend, extent of occurrence, area of occupancy and/or population viability of the global population sufficiently that an application of IUCN Red List criteria would …
Excellent	… warrant downlisting of the species from 'regionally extinct' to 'near threatened' or 'least concern'	$>20000\,km^2$	$>2000\,km^2$	>1000	<10% in 100 years	… warrant downlisting of the species across 4 threat categories (i.e. from 'extinct in the wild' to 'near threatened' or 'least concern')

Very good	… warrant downlisting of the species from 'regionally extinct' to 'vulnerable'	> 5000 km^2	> 500 km^2	> 250	< 20% in 20 years/5 generations	… warrant downlisting of the species across 3 threat categories globally (e.g. 'critically endangered' to 'near threatened'/'least concern')
Good	… warrant downlisting of the species from 'regionally extinct' to 'endangered'	>100 km^2	>10 km^2	>50	<50% in 10 years/3 generations	… warrant downlisting of the species across 2 threat categories globally (e.g. 'extinct in the wild' to 'endangered')
Fair	… warrant downlisting of the species from 'regionally extinct' to 'critically endangered'	>5 years of a documented reintroduced population, or time to the viable production of offspring, whichever is longer				… warrant downlisting of the species across 1 threat category globally (e.g. 'vulnerable' to 'near threatened'/'least concern')

'critically endangered' without reintroduced populations but 'vulnerable' if reintroduced populations are included could be classified as having 'good' success in contributing to the global recovery of the species (see Table 22.2).

We consider reintroductions that would cause any change in threat status on a regional level as having a degree of success, regardless of contributions on a global level. However, reintroductions should be evaluated in terms of their contribution to the threat status of species both regionally and globally to fully encompass the potential benefits that reintroductions can bring to species recovery.

Emerging needs: function, form and focus

Up to this point, we have tackled questions regarding risks, the initiation of reintroductions, the refinement of reintroduction techniques and evaluations of reintroduction programme success. We have examined key components that comprise the status quo of reintroduction science and proposed crucial advancements where appropriate. In our collective experience, however, we recognize that the backdrop of reintroduction knowledge, need and practice is increasingly dynamic and shifting beyond traditional considerations. For example, increased emphasis is being placed on the meta-population context surrounding reintroductions, i.e. how heavily source populations should be harvested, how individuals are optimally allocated among release sites, and whether translocations should occur among isolated populations (Armstrong & Seddon 2007). The ecological function, i.e. the behavioural or physiological effects on other organisms, of reintroduced species should be investigated as the ecosystem-level effects of reintroduced species may differ from historical ones in altered ecosystems (see Chapter 2). Our knowledge must evolve, and for the remainder of this chapter we examine the increasing challenges and possible responses of the future, particularly within the context of emerging infectious diseases, increasing habitat loss and climate change.

While the early emphasis for reintroductions was on restoring a key or flagship species, with usually considerable public emotion or sentiment towards non-carnivore species returning to their homeland, there is now increasing interest in reintroductions that restore ecological roles and relationships. The first wolves (*Canis lupus*) returned to Yellowstone National Park in 1995 and 1996, some 4 years after the US Congress had mandated an intensive environmental review of the prospect. During the intervening years, public outreach and consultation comprised 130 public meetings, almost 750,000 pamphlets distributed and 180,000 public comments received for analysis. Wolf release was then feasible only with a dispensation as an experimental population which allowed greater management flexibility in that ranchers could shoot wolves seen attacking livestock (Smith & Bangs 2009). Within a few years of the wolf population establishing, diverse ecological impacts were evident, including significant willow regrowth along river banks, because the numbers of the favourite prey, elk (*Cervus elaphus*), were reduced but also because elk were avoiding the high-risk areas of willow when wolves were present (Beyer et al. 2007). Further consequent effects on biodiversity demonstrate the effects of a reintroduced top carnivore in restoring ecosystem processes (Smith & Bangs 2009).

Despite the increasing prevalence of reintroductions and the opportunities they undoubtedly represent, there are some reasons for caution. The first is the assumption that a returning species will find its earlier niche vacant and waiting for it. We know that vacant niche space can be nibbled away by other members of the community, so a returning species may have to compete for its previous niche or, indeed, as with the Arabian oryx (*Oryx leucoryx*) in Oman, establish itself with a subtly different ecology, distribution or habits from its ancestors (Stanley Price 1989). The second is that the performance of a returning species may be further complicated if the released individuals have been in captivity for several generations. A study on Oldfield mice (*Peromyscus polionotus subgriseus*) showed that the more generations a population had been in captivity, the less likely an individual was to take cover in the presence of a predator, and the variability in predator response behaviours also increased similarly (McPhee 2003b). The butterfly *Pieris brassicae*, bred in captivity for 100–150 generations, showed morphological adaptations to captivity in the form of increased investment in reproduction and reduced capacity for flight, compared to wild-caught individuals (Lewis & Thomas 2001). In such cases, one must be wary of releasing into the wild individuals that are to some, but unknown, extent mere facsimiles of the aboriginal populations.

Recent reintroductions are often portrayed as efforts to reverse past ecological abuses, and the availability of animals to return to the wild is often a testament to the foresight of previous conservationists and the ability of institutions, often zoos, in building up populations to provide animals. But new efforts will be needed as we begin to appreciate the scale of overall declining biodiversity from the major impacts of climate change, habitat fragmentation and absolute loss, pollution and disease, and their interactions. Emergent diseases, as a major threat to biodiversity, are enigmatic in their suddenness, severity and obscure causation (McCallum 2008). Evidence for this was the sudden appearance and rapid spread of the infectious cancer causing facial tumour disease in the Tasmanian devil (*Sarcophilus harrisii*). Where the disease occurs, numbers have declined by 90% with an almost complete absence of animals older than 2 years, the age at which they start to breed; at current rates, extinction within 5–10 years is forecast (McCallum 2008).

Chytridiomycosis in amphibians is evidence of the interaction between major threats of climate change and fungal disease (Pounds et al. 2006). It has stimulated unprecedented actions and collaborations, especially in the captive-breeding world, to take individuals of critically exposed and endangered amphibians into captivity. Here they must be kept under conditions of very high biosecurity to counter the threat of the water-borne fungus entering the system, and of potentially releasing the fungus from infected animals into the outside water environment, but captive animals can be cured of the fungus. The ultimate aim is to return these species to the wild. However, the time horizon for this cannot be estimated while the fungus cannot be eliminated from the wild; there is no means of treating the symptoms of the disease out of protected environments. Near-certainty on the role of climate warming, which is also leading to near-optimum conditions for the growth of the chytrid fungus, has resulted in the extinction of 67% of the 110 species of *Atelopus* in the American tropics (Pounds et al. 2006); these causes cannot be reversed in anything but the long term.

Such crises force conservationists to make longer term commitments to the intensive management of endangered species, and we must acknowledge that in the future more species will require some level of active conservation support in the supposed 'wild'. The breeding systems of birds are particularly amenable to interventionist management; in both the endangered Mauritius kestrel (*Falco punctatus*) and the echo parakeet (*Psittacula eques*) in Mauritius, the survival prospects of undernourished chicks can be improved either by supportive feeding in the nest or by removal of such chicks for hand-rearing and subsequent return to the wild through fostering or cross-fostering (Jones 2004). Such activities blur the formerly rigid distinction between an animal being either in or out of the wild, because technologies and techniques previously only found in zoos are now applied in the field.

On the verge of a disciplinary shift? Beyond single species and beyond the historic range

The precondition for a reintroduction, that suitable habitat remains within historic range (IUCN 1998), presumes that conditions have remained more or less constant during the period of the species' absence from the wild. Increasingly, this is an invalid presumption as the scale and impacts of climate change are seen and felt. Under the relatively moderate B1 future climate scenario (Intergovernmental Panel on Climate Change 2007), 4–20% of global land area is predicted to experience novel climates by 2100 AD, with the tropics and subtropics most liable to be affected, where biodiversity levels are greatest. Higher latitude biomes will be relatively unaffected (Roberts & Hamann 2012). Hence, we must expect novel climates to be accompanied by no-analogue communities and ecological surprises (Williams & Jackson 2007).

Apart from making it more challenging to meet this reintroduction precondition, the prospect of wide-ranging climate change means that many species that are naturally rare, specialized and of limited dispersal ability or mobility may be at risk of extinction. It is a short logical step then to propose that humans should deliberately move the most vulnerable species from their present ranges into areas that may be more suitable in future. While generally contravening the reintroduction principle that species should not be moved outside their inferred historical range, such moves have been considered valid 'conservation introductions' under very specific conditions (IUCN 1998). Current debate labels such moves as 'assisted colonization', 'assisted migration' or 'managed relocation' (e.g. McLachlan et al. 2007; Loss et al. 2011).

At the species level, choices would have to be made as to which species are most eligible to be moved. This will be reflected in a combination of factors, including their exposure to the effects of climate change, their sensitivity to it and the extent to which they might adapt (Williams et al. 2007). Beyond this, we will be in the realms of ignorance and uncertainty, with regard to the behaviour of the species both in its present, source range and in some alternative destination.

On the basis that any reintroduction is a long-term and expensive operation, with success far from guaranteed, we may assume that an assisted colonization in the face of ongoing climate change will be no different. Thus, we need to consider the possible fates for a species facing climate change in its present range.

- It dies out because it cannot adapt to new conditions.
- It adapts successfully.
- It moves itself because new areas of more suitable conditions are within its ranging or maximum dispersal distance.
- It is deliberately moved to an alternative site.
- Management support is provided to the species to enable it to persist in its present range.

The likelihood of the third response above depends on the efficacy of long-range dispersal. Rare long-range dispersal events in plants are more important than often supposed, and are necessary to explain the observed rates of tree migration in postglaciation periods (Higgins & Richardson 1999). On the other hand, for survival purposes, under conditions of changing climate, a plant species will depend on its dispersal kernel, which integrates the effects of seed dispersal, seedling germination and survivorship (Higgins et al. 2003). Forecasting migration rates for plants is generally fraught with multiple uncertainties (Higgins et al. 2003).

Genetic adaptation to climate change can be demonstrated, usually by looking at changes in species across space and sometimes also across time (Hoffmann & Sgró 2011). But rarely are claimed evolutionary responses supported by genetic evidence (Gienapp et al. 2008), for phenotypic responses may be greater and swifter. In a rare example, when Canadian red squirrels showed advanced parturition dates under conditions of warmer winters and more abundant food, phenotypic plasticity in the squirrels was responsible for an 87% observed change in parturition dates, with an evolutionary response accounting for only the remaining 13% (Realé et al. 2003).

The adaptability of a species will also depend on its genetic variability across its total range, which in some will offer considerable scope for natural selection. The fossil record indicates the extent of some species' persistence in the face of more extreme and faster climate change than we expect for the 21st century, with conclusions that extinctions were fewer than might have been expected (Willis et al. 2010). Conservation genetics may contribute to an integrated strategy of conservation in the face of climate change (Loss et al. 2011). Further, gardeners have been moving plants outside their native ranges for years; many European flowering plants thrive in gardens up to 1000 km north of their native ranges, albeit with varying degrees of management support (Van der Keken et al. 2008).

Novel ecosystems, due to species 'occurring in combinations and relative abundances that have not occurred previously in a given biome, and which are caused by human action, environmental change and the impacts of deliberate and inadvertent introduction of species' (Hobbs et al. 2006), are distinct from the conventional spectrum of ecosystem classification from pristine to degraded (Lindenmayer et al. 2008). Species' absolute and relative abundances can change over relatively few years of changing vegetation cover and pattern, raising the prospect of unprecedented ecological relationships (Lindenmayer et al. 2008). We should expect novel ecosystems to increase rapidly due to the effects of climate change, and these may be concentrated in regions of high ecological complexity and diversity (Williams & Jackson 2007). We will be faced with practical issues of land management for specific objectives, and ethical ones relating to the relative ecological values of native versus exotic species (Lindenmayer et al. 2008). These pressures may require us to address future challenges with communities designed for specific purposes (MacMahon & Holl 2001).

There are other solutions to climate-threatened species at the landscape level. In response to overall reductions in habitats for biodiversity, whether through absolute loss or fragmentation, the case for designing landscapes for better connectivity is strong, the argument being that species can then disperse and small populations can become incorporated into larger ones. Hence, the case for expanding and linking protected areas (Hannah et al. 2007), developing corridors (Haddad et al. 2003) and developing matrix landscapes (Fahrig 2001). What remains unclear now is the extent to which wild species use such human-designed landscapes in the ways that human managers desire and expect.

However, if the decision is taken to move a species deliberately, a complex set of concerns come into play.

- Can we predict accurately enough the future climates to be expected at a site scale that is meaningful for the translocated species?
- Even if we are comfortable with the point above, are we sure the climate at the destination site has stabilized, or is it still changing so that the translocated species may have to be moved again after some years?
- Why are we moving the species? Is it to save it from extinction? If so, then how many individuals do we want at the new site, and can we assess or guarantee that we can reach that number? To what extent will conservation support and management be necessary or feasible in the interest of 'saving the species' (Redford et al. 2011)?
- If the translocated species is intended to fulfil some desired ecological function, is our level of ecological knowledge adequate to ensure it? What unintended ecological effects would be acceptable or not tolerated?
- What are the chances of the species not establishing and dying out at the new site?
- Can we predict whether the species is a latent invasive pest? While some general principles about the performance of invasives are known, permitting some clues as to which releases are most likely to lead to serious problems (Mueller & Hellmann 2008; Ricciardi & Simberloff 2009), it is also very hard to predict which species might become invasive in the long term (Crooks 2005). Experiences from invasive biology and biological control are salutary, with the human redistribution of species 'an extended tragedy of errors' (Macdonald et al. 2007).
- If the translocation yields undesirable and unacceptable consequences, is it feasible to reverse or to control the released population? Once established, the costs of removing introduced populations are very high: for example, in 2002–2003 alone, the New Zealand government spent NZ$80 million controlling five pest mammal species (Parkes & Murphy 2003).

Combining issues at both source and destination, these are a formidable set of problems, for which a linear model for decision making (Hoegh-Guldberg et al. 2008) will be inadequate. Hoegh-Guldberg et al. (2008) provide a first general model for assisted colonization, but what is needed is a systematic approach to the uncertainty and risk which are present in every reintroduction and may be critical in determining the outcome. Active adaptive management offers a means to acknowledge and handle our imperfect understanding of natural systems; based in decision theory, it can improve the prospects of short-term success while increasing learning for the longer term when designing a translocation (Rout et al. 2009).

Within the space of 50 years, reintroductions have progressed from being a largely unscientific but well-meaning way of redressing extirpations to a responsible tool for science-based conservation. The lessons learned and the framework of principles behind them are now of great potential significance as we strive to maintain the levels of biodiversity specified in the Aichi Targets for Biodiversity Conservation (CBD 2010) in the face of climate change and the progressive loss of wildlife habitats. Given our state of knowledge and appreciation of risks, we should proceed to move species only with prudence and pragmatism, experimenting wisely and learning more.

One feature in particular has emerged as crucial time and again in this chapter: the need for increased rigour in all aspects of reintroduction ecology. Rigour is required in assessing whether a reintroduction is desirable in the first place, in the creation of criteria for ascribing 'success' and in creating the framework by which we are collectively able to learn from unsuccessful reintroductions and so improve the discipline. In this chapter, we have outlined some of the ways in which we think rigour could be implemented to improve the success rates – and their definition – of reintroductions. In this way, reintroductions may finally fulfil their potential as an integral part of the armoury available to conservation biologists attempting to meet the future needs of the global community.

"When we try to pick out anything by itself we find that it is bound fast by a thousand invisible cords that cannot be broken, to everything in the universe."

John Muir

References

Albrecht, M.A., Guerrant, E.O. Jr, Maschinski, J. & Kennedy, K.L. (2011) A long-term view of rare plant reintroduction. *Biological Conservation*, **144**, 2557–2558.

Armstrong, D.P. & Seddon, P.J. (2007) Directions in reintroduction biology. *Trends in Ecology and Evolution*, **23**, 20–25.

Ausband, D. & Moehrenschlager, A. (2009) Long-range juvenile dispersal and its implications for the conservation of reintroduced swift fox populations in the USA and Canada. *Oryx*, **43**(1), 73–77.

Batsaikhan, N., Henttonen, H., Meinig, H., *et al.* (2008) Arvicolaamphibius. In: *IUCN Red List of Threatened Species*, Version 2011.1. www.iucnredlist.org

Beck, B.B., Rapaport, L.G. & Wilson, A.C. (1994) Reintroduction of captive-born animals. In: *Creative Conservation* (ed. A. Feistner), pp.265–286. Chapman and Hall, London.

Bennett, L.T. & Adams, M.A. (2004) Assessment of ecological effects due to forest harvesting: approaches and statistical issues. *Journal of Applied Ecology*, **41**, 585–598.

Beyer, H.L., Merrill, E.H., Varley, N. & Boyce, M.S. (2007). Willow on Yellowstone's northern range: evidence for a trophic cascade? *Ecological Applications*, **17**(6), 1563–1571.

Biggins, D., Vargas, A., Godbey, J.L. & Anderson, S.H. (1999) Influences on pre-release experience on reintroduced black-footed ferrets (*Mustelanigripes*). *Biological Conservation*, **89**, 121–129.

Breitenmoser, U., Breitenmoser-Wursten, C., Carbyn, L. N. & Funk, S. M. (2001) Assessment of carnivore reintroductions. *Carnivore Conservation*, **5**, 241–281.

Bright, P.W. & Morris, P.A. (1994) Animal translocation for conservation: performance of dormice in relation to release. *Journal of Applied Ecology*, **31**, 699–708.

Brown, K.P., Empson, R., Gorman, N. & Moorcroft, G. (2004) *North Island Kokako (Callaeas cinerea wilsoni)*

Translocations and Establishment on Kapiti Island, New Zealand. New Zealand Department of Conservation, Auckland, pp.1–24.

Butchart, S.H.M., Stattersfield, A.J., Baillie, J., *et al.* (2005) Using Red List Indices to measure progress towards the 2010 target and beyond. *Philosophical Transactions of the Royal Society B:Biological Sciences*, **360**, 255–268.

Carroll, C., Phillips, M.K., Schumaker, N. & Smith, D.W. (2003) Impacts of landscape change on wolf restoration success: planning a reintroduction program based on static and dynamic spatial models. *Conservation Biology*, **17**(2), 536–548.

Casimir, D.L., Moehrenschlager, A. & Barclay, R.M.R. (2007) Factors influencing reproduction in captive Vancouver Island marmots: implications for captive breeding and reintroduction programs. *Journal of Mammalogy*, **88**, 1412–1419.

Cayford, J. & Percival, S. (1992) Born captive, die free. *New Scientist*, **1807**, 29–33.

CBD (2010) *Strategic Plan for Biodiversity 2010–2020 and the Aichi Targets*. www.cbd.int/sp2020

Ceballos, G., Davidson, A., List, R., *et al.* (2010) Rapid decline of a grassland system and its ecological and conservation implications. *PLoS ONE*, **5**(1), e8562.

Crooks, J.A. (2005) Lag times and exotic species: the ecology and management of biological invasions in slow-motion. *Ecoscience*, **12**, 316–329.

Dalrymple, S.E., Stewart, G.B. & Pullin, A.S. (2011) *Are Re-introductions an Effective Way of Mitigating Against Plant Extinctions?* CEE review 07-008 (SR32). Collaboration for Environmental Evidence. www.environmentalevidence.org/SR32.html

Fahrig, L. (2001) How much habitat is enough? *Biological Conservation*, **100**, 65–74.

Fischer, J. & Lindenmayer, D.B. (2000) An assessment of the published results of animal relocations. *Biological Conservation*, **96**, 1–11.

Fleming, I.A. & Fross, M.R. (1993) Breeding success of hatchery and wild coho salmon (Oncorhynchuskisutch) in competition. *Ecological Applications*, **3**, 230–245.

Gage, K.L., Lance, S.E., Dennis, D.T. & Montenieri, J.A. (1992) Human plague in the United States: a review of cases from 1988–1992 with comments on the likelihood of increased plague activity. *Border Epidemiological Bulletin*, **19**, 1–10.

Gelling, M., Montes, I., Moorhouse, T.P. & Macdonald, D.W. (2010) Captive housing during water vole (Arvicola terrestris) reintroduction: does short-term social stress impact on animal welfare? *PLoS ONE*, **5**(3), e9791.

Gienapp, P., Teplitsky, C., Alho, J.S., Mills, J.A. & Merila, J. (2008) Climate change and evolution: disentangling environmental and genetic responses. *Molecular Ecology*, **17**, 167–178.

Godefroid, S. (2011) Response to Albrecht *et al.* *Biological Conservation*, **144**, 2559.

Godefroid, S. & Vanderborght, T. (2011) Plant reintroductions: the need for a global database. *BioScience*, **20**, 3683–3688.

Godefroid, S., Piazza, C., Rossi, G., *et al.* (2011) How successful are plant species reintroductions? *Biological Conservation*, **144**, 672–682.

Griffith, B., Scott, J.M., Carpenter, J.W. & Reed, C. (1989) Translocation as a species conservation tool – status and strategy. *Science*, **245**, 477–480.

Guerrant, E.O. Jr. (2012) Characterizing two decades of rare plant reintroductions. In: *Plant Reintroduction in a Changing Climate: Promises and Perils*. (eds J. Maschinski & K.E. Haskins). Island Press, Washington, D.C.

Haddad, N.M., Bowne, D.R., Cunningham, A., *et al.* (2003) Corridor use by diverse taxa. *Ecology*, **84**(3), 609–615.

Hannah, L., Midgley, G., Andelman, S., Araújo M, *et al.* (2007) Protected area needs in a changing climate. *Frontiers in Ecology and Environment*, **5**(3), 131–138.

Hansen, D.M., Donlan, J., Griffiths, C.J. & Campbell, K.J. (2010) Ecological history and latent conservation potential: large and giant tortoises as a model for taxon substitutions. *Ecography*, **33**, 272–284.

Harrington, L.A., Moehrenschlager, A., Gelling, M., Hughes, J., Atkinson, R. & Macdonald, D.W. (Submitted). Welfare and ethics in animal reintroductions.

Higgins, S.I. & Richardson, D.M. (1999) Predicting plant migration rates in a changing world: the role of long-distance dispersal. *American Naturalist*, **153**(5), 464–475.

Higgins, S.I., Clark, J.S., Nathan, R., *et al.* (2003) Forecasting plant migration rates: managing uncertainty for risk assessment. *Journal of Ecology*, **91**(3), 341–347.

Hobbs, R.J., Arico, S., Aronson, J., *et al.* (2006) Novel ecosystems: theoretical and management aspects of the new ecological world order. *Global Ecology and Biogeography*, **15**, 1–7.

Hoegh-Guldberg, O., Hughes, L., McIntyre, S., *et al.* (2008) Assisted colonization and rapid climate change. *Science*, **321**, 345–346.

Hoffmann, A.A. & Sgró, C.M. (2011) Climate change and evolutionary adaptation. *Nature*, **470**, 479–485.

Hoogland, J.L. (1995) *The Black-Tailed Prairie Dog: Social Life of a Burrowing Mammal*. University of Chicago Press, Chicago, IL.

Innes, J.G., Hay, R., Flux, I., Bradfield, P., Speed, H. & Janesn, P. (1999) Successful recovery of North Island kokako *Callaeascinereawilsoni* populations. *Biological Conservation*, **87**, 201–214.

Intergovernmental Panel on Climate Change (2007) *Climate Change 2007: The Physical Science Basis. Summary for Policymakers*. Intergovernmental Panel on Climate Change. Geneva, Switzerland.

IUCN (1998) *IUCN Guidelines for Re-Introductions*. IUCN/SSC Re-introduction Specialist Group, Gland, Switzerland.

IUCN (2003) *Guidelines for Application of IUCN Red List Criteria at Regional Levels: Version 3.0*. IUCN Species Survival Commission, Gland, Switzerland.

IUCN Standards and Petitions Subcommittee (2010) *Guidelines for Using the IUCN Red List Categories and Criteria. Version 8.1.* http://intranet.iucn.org/web files/doc/SSC/RedList/ RedListGuidelines.pdf

Jachowski, D.S., Gitzen, R.A., Grenier, M.B., Holmes, B. & Millspaugh, J.J. (2011) The importance of thinking big: large-scale prey conservation drives black-footed ferret reintroduction success. *Biological Conservation*, **14**, 1560–1566.

Jones, C. (2004) Conservation management of endangered birds. In: *Bird Ecology and Conservation: A Handbook of Techniques* (eds W.J. Sutherland, I. Newton & R.E. Green). Oxford University Press, Oxford.

Jule, K.R., Leaver, L.A. & Lea, S.E. (2008) The effects of captive experience on reintroduction survival in carnivores: a review and analysis. *Biological Conservation*, **141**, 355–363.

Kleiman, D.G. (1989) Reintroduction of captive mammals for conservation. *BioScience*, **39**, 152–161.

Kloor, K. (1999) Lynx and biologists try to recover after disastrous start. *Science*, **285**, 320–321.

Lewis, O.T. & Thomas, C. (2001) Adaptations to captivity in the butterfly *Pierisbrassicae* (L) and the implications for *ex-situ* conservation. *Journal of Insect Conservation*, **5**, 55–63.

Lindenmayer, D.B., Fischer, J., Felton, A., *et al.* (2008) Novel ecosystems resulting from landscape transformation create dilemmas for modern conservation practice. *Conservation Letters*, **1**, 129–135.

Lindsey, P., Tambling, C.J., Brummer, R., *et al.* (2011) Minimum prey and area requirement of the vulnerable cheetah *Acinonyx jubatus*: implications for reintroduction and management of the species in South Africa. *Oryx*, **45**(4), 587–599.

Linklater, W.L. & Swaisgood, R. (2008) Reserve size, conspecific density, and translocation success for black rhinoceros. *Journal of Wildlife Management*, **72**, 1059–1068.

Loss, S.R., Terwilliger, L.A. & Peterson, A.C. (2011) Assisted colonization: integrating conservation strategies in the face of climate change. *Biological Conservation*, **144**(1), 92–100.

Macdonald, D.W., King, C.M. & Strachan, R. (2007) *Introduced Species and the Line Between Biodiversity Conservation and Naturalistic Eugenics*. Blackwell Publishing, Oxford.

Mace, G.M., Collar, N.J., Gaston, K.J., *et al.* (2008) Quantification of extinction risk: IUCN's system for classifying threatened species. *Conservation Biology*, **22**(6), 1424–1442.

MacMahon, J.A. & Holl, K.D. (2001) Ecological restoration – a key to conservation biology's future. In: *Conservation Biology: Research Priorities for the Next Decade* (eds M. Soulé & G. Orians), pp.245–269. Island Press, Washington, D.C.

Mao, J.S., Boyce, M.S., Smith, D.W., *et al.* (2005) Habitat selection by elk before and after wolf reintroduction in Yellowstone National Park. *Journal of Wildlife Management*, **69**, 1691–1707.

Maran, T., Põdra, M., Põlma, M. & Macdonald, D.W. (2009) The survival of captive-born animals in restoration programmes – case study of the endangered European mink *Mustelalutreola*. *Biological Conservation*, **142**, 1685–1692.

Marris, E. (2009) Planting the forest of the future. *Nature*, **459**, 906–908.

Mathews, F., Orros, M., McLaren, G., Gelling, M. & Foster, R. (2005) Keeping fit on the ark: assessing the suitability of captive-bred animals for release. *Biological Conservation*, **121**, 569–577.

Mathews, F., Moro, D., Strachan, R., Gelling, M. & Buller, N. (2006) Health surveillance in wildlife reintroductions. *Biological Conservation*, **131**, 338–347.

McCallum, H. (2008) Tasmanian devil facial tumour disease: lessons for conservation biology. *Trends in Ecology ahd Evolution*, **23**(11), 631–637.

McLachlan, J.S., Hellmann, J.J. & Schwartz, M.W. (2007) A framework for debate of assisted migration in an era of climate change. *Conservation Biology*, **21**(2), 297–302.

McPhee, M.E. (2003a) Generations in captivity increases behavioral variance: considerations for captive breeding and reintroduction programs. *Biological Conservation*, **115**, 71–77.

McPhee, M.E. (2003b) Effects of captivity on response to a novel environment in the Oldfield mouse (Peromyscuspolionotussubgriseus). *International Journal of Comparative Psychology*, **16**(2), 85–94.

Miller, B., Ralls, K., Reading, R., Scott, J. & Estes, J. (1999) Biological and technical considerations of carnivore translocation: a review. *Animal Conservation*, **2**, 59–68.

Moehrenschlager, A. & Macdonald, D.W. (2003) Movement and survival parameters of translocated and resident swift foxes *Vulpesvelox*. *Animal Conservation*, **6**, 199–206.

Moehrenschlager, A., List, R. & Macdonald, D.W. (2007) Escaping interspecific killing: Mexican kit foxes survive while coyotes and golden eagles kill Canadian swift foxes. *Journal of Mammalogy*, **88**, 1029–1039.

Moorhouse, T.P., Gelling, M. & Macdonald, D.W. (2008) Effects of forage availability on growth and maturation rates in water voles. *Journal of Animal Ecology*, **77**, 1288–1295.

Moorhouse, T.P., Gelling, M. & Macdonald, D.W. (2009) Effects of habitat quality upon reintroduction success in water voles: evidence from a replicated experiment. *Biological Conservation*, **142**, 53–60.

Morgan, K.N. & Tromborg, C.T. (2007) Sources of stress in captivity. *Applied Animal Behaviour Science*, **102**, 262–302.

Mueller, J.M. & Hellmann, J.J. (2008) An assessment of invasion risk from assisted migration. *Conservation Biology*, **22**, 562–567.

Muhly, T.B. & Musiani, M. (2009) Livestock depredation by wolves and the ranching economy in the Northwestern US. *Ecological Economics*, **68**, 2439–2450.

Naughton-Treves, L., Grossberg, R. & Treves, A. (2003) Paying for tolerance: rural citizens' attitudes toward wolf depredation and compensation. *Conservation Biology*, **17**, 1500–1511.

Nichols, J.D. & Williams, B.K. (2006) Monitoring for conservation. *Trends in Ecology and Evolution*, **21**, 668–673.

Olsson, I.A. & Westlund, K. (2007) More than numbers matter: the effect of social factors on behaviour and welfare of laboratory rodents and non-human primates. *Applied Animal Behaviour Science*, **103**, 229–254.

Parkes, J. & Murphy, E. (2003) Management of introduced mammals in New Zealand. *New Zealand Journal of Zoology*, **30**(4), 335–359.

Pounds, J.A., Bustamante, M.R., Coloma, L.A., *et al.* (2006) Widespread amphibian extinctions from epidemic disease driven by global warming. *Nature*, **439**(7073), 161–167.

Program, T.C. (2010) Success of the Colorado Division of Wildlife's lynx reintroduction program. http://wildlife.state.co.us/SiteCollectionDocuments/DOW/Research/Mammals/ColoradoLynxReintroductionAssessment_090710.pdf

Pruss, S., Fargey, P. & Moehrenschlager, A., Canadian Swift Fox Recovery Team. (2008) *National Swift Fox Recovery Strategy*. Species at Risk Act Recovery Strategy Series. Parks Canada Agency.

Rabin, L.A. (2003) Maintaining behavioural diversity in captivity for conservation: natural behaviour management. *Animal Welfare*, **12**, 85–94.

Rantanen, E.M., Buner, F., Riordan, P., Sotherton, N.W. & Macdonald, D.W. (2010a) Vigilance, time budgets and predation risk in reintroduced captive-bred grey partridges *Perdixperdix*. *Applied Animal Behaviour Science*, **127**, 43–50.

Rantanen, E.M., Buner, F., Riordan, P., Sotherton, N.W. & Macdonald, D.W. (2010b) Habitat preferences and survival in wildlife reintroductions: an ecological trap in reintroduced grey partridges. *Journal of Applied Ecology*, **47**, 1357–1364.

Realé, D., McAdam, A.G., Boutin, S. & Berteaux, D. (2003) Genetic and plastic responses of a northern mammal to climate change. *Proceedings of the Royal Society B: Biological Sciences*, **270**, 591–596.

Redford, K.H., Amato, G., Baillie, J., *et al.* (2011) What does it mean to successfully conserve a (vertebrate) species? *BioScience*, **61**(1), 38–48.

Ricciardi, A. & Simberloff, D. (2009) Assisted colonization is not a viable conservation strategy. *Trends in Ecology and Evolution*, **24**(5), 248–253.

Roberts, D.R. & Hamann, A. (2012) Predicting potential climate change impacts with bioclimate envelope models: a palaeoecological perspective. *Global Ecology and Biogeography*, **21**, 121–133.

Rocke, T.E., Smith, S.R., Stinchcomb, D.T. & Osorio, J.E. (2008) Immunization of black-tailed prairie dog against plague through consumption of vaccine-laden baits. *Journal of Wildlife Diseases*, **44**, 930–937.

Roe, J.H., Frank, M.R., Gibson, S.E., Attum, O. & Kingsbury, B.A. (2010) No place like home: an experimental comparison of reintroduction strategies using snakes. *Journal of Applied Ecology*, **47**, 1253–1261.

Rout, T.M., Hauser, C.E. & Possingham, H.P. (2009) Optimal adaptive management for the translocation of a threatened species. *Ecological Applications*, **19**(2), 515–526.

Sarrazin, F. & Barbault, R. (1996) Reintroduction: challenges and lessons for basic ecology. *Trends in Ecology and Evolution*, **11**, 474–478.

Schaub, M., Zink, R., Beissmann, H., Sarrazin, F. & Arlettaz, R. (2009) When to end releases in reintroduction programmes: demographic rates and population viability analysis of bearded vultures in the Alps. *Journal of Applied Ecology*, **46**, 92–100.

Schreiber, S.G., Bearlin, A.R., Nicol, S.J. & Todd, C.R. (2004) Adaptive management: a synthesis of current understanding and effective application. *Ecological Management and Restoration*, **5**, 177–182.

Seddon, P.J. (1999) Persistence without intervention: assessing success in wildlife reintroductions. *Trends in Ecology and Evolution*, **14**, 503.

Seddon, P.J., Soorae, P.S. & Launay, F. (2005) Taxonomic bias in reintroduction projects. *Animal Conservation*, **8**, 51–58.

Seddon, P.J., Armstrong, D.P. & Maloney, R.F. (2007) Developing the science of reintroduction biology. *Conservation Biology*, **21**, 303–312.

Seddon, P.J. (2010) From reintroduction to assisted colonization: moving along the conservation translocation spectrum. *Restoration Ecology*, **18**, 796–802.

Shier, D.M. (2006) Effect of family support on the success of translocated black-tailed prairie dogs. *Conservation Biology*, **20**, 1780–1790.

Shier, D.M. & Owings, D.H. (2006) Effects of predator training on behavior and post-release survival of captive prairie dogs (*Cynomys ludovicianus*). *Biological Conservation*, **132**, 126–135.

Shier, D.M. & Owings, D.H. (2007) Effects of social learning on predator training and post-release survival in juvenile black-tailed prairie dogs (*Cynomys ludovicianus*). *Animal Behaviour*, **73**, 567–577.

Shier, D.M. & Swaisgood, R.R. (2012) Fitness costs of neighborhood disruption in translocations of a solitary mammal. *Conservation Biology*, **26**(1), 116–123.

Short, J. & Smith, A. (1994) Mammal decline and recovery in Australia. *Journal of Mammalogy*, **75**, 288–297.

Smith, D.H., Converse, S.J., Gibson, K.W., *et al.* (2011) Reducing hatching failure in the whooping crane conservation breeding program. *Journal of Wildlife Management*, **75**(3), 501–508.

Smith, D.W. & Bangs, E.E. (2009) Reintroduction of wolves to Yellowstone National Park: history, values and ecosystem restoration. In: *Reintroduction of Top-Order Predators* (eds M.W. Hayward & M.J. Somers), pp.92–105. Wiley-Blackwell, Oxford.

Soorae, P.S. (ed.) (2008) *Global Re-Introduction Perspectives: Re-Introduction Case-Studies from Around the Globe*. IUCN/SSC Re-introduction Specialist Group, Abu Dhabi, UAE.

Soorae, P.S. (ed.) (2010) *Global Re-Introduction Perspectives: Additional Case-Studies from Around the Globe*. IUCN/SSC Re-introduction Specialist Group, Abu Dhabi, UAE.

South, A.B., Rushton, S.P., Macdonald, D.W. & Fuller, R. (2001) Reintroduction of the European beaver (*Castor fiber*) to Norfolk, U.K.: a preliminary modelling analysis. *Journal of Zoology*, **254**, 473–479

Stamps, J.A. & Swaisgood, R.R. (2007) Someplace like home: experience, habitat selection and conservation biology. *Applied Animal Behaviour Science*, **102**, 392–409.

Stanley Price, M.R. (1989) *Animal Reintroductions: The Arabian Oryx in Oman*. Cambridge Studies in Applied Ecology and Resource Management. Cambridge University Press, Cambridge.

Stoinski, T.S. & Beck, B.B. (2004) Changes in locomotor and foraging skills in captive-born, reintroduced golden lion tamarins (Leontopithecusrosaliarosalia). *American Journal of Primatology*, **62**, 1–13.

Stoinski, T.S., Beck, B.B., Bloomsmith, M.A. & Maple, T.L. (2003) A behavioral comparison of captive-born, reintroduced golden lion tamarins and their wild-born offspring. *Behaviour*, **140**, 137–160.

Strachan, R., Moorhouse, T.P. & Gelling, M.G. (2011) *Water Vole Conservation Handbook*, 3rd edn. Wildlife Conservation Research Unit, University of Oxford, Oxford.

Sutherland, W.J. (2006) Predicting the ecological consequences of environmental change: a review of methods. *Journal of Applied Ecology*, **43**, 599–616.

Swaisgood, R.R. (2010) The conservation-welfare nexus in reintroduction programs: a role for sensory ecology. *Animal Welfare*, **19**, 125–137.

Teixeira, C.P., de Azevedo, C.S., Mendl, M., Cipreste, C.F. & Young, R.J. (2007) Revisiting translocation and reintroduction programmes: the importance of considering stress. *Animal Behaviour*, **73**, 1–13.

Van der Keken, S., Hermy, M., Vellend, M., Knapen, A. & Verhuyen, C. (2008) Garden plants get a head start on climate change. *Frontiers in Ecology and Environment*, **6**(4), 212–216.

Vargas, A. & Anderson, S.H. (1998) Ontogeny of black-footed ferret predatory behavior towards prairie dogs. *Canadian Journal of Zoology*, **76**, 1696–1704.

Vargas, A. & Anderson, S.H. (1999) Effects of experience and cage enrichment on predatory skills of black-footed ferrets (*Mustela nigripes*). *Journal of Mammalogy*, **80**, 263–269.

Walters, C. (1997) Challenges in adaptive management of riparian and coastal ecosystems. *Conservation Ecology*, **1**(2), 1.

Williams, J.W. & Jackson, S.T. (2007) Novel climates, no-analog communities, and ecological surprises. *Frontiers in Ecology and Environment*, **5**(9), 475–482.

Williams, J.W., Jackson, S.T. & Kutzbach, J.E. (2007) Projected distributions of novel and disappearing climates by 2100 AD. *Proceedings of National Academy of Sciences USA*, **104**(14), 5738–5742.

Willis, K.J., Bennett, K.D., Bhagwat, S.A. & Birks, H.J.B. (2010) 4° C and beyond: what did this mean for biodiversity in the past? *Systematics and Biodiversity*, **8**(1), 3–9.

Wilmers, C.C. & Getz, W.M. (2005) Gray wolves as climate change buffers in Yellowstone. *PLoS Biology*, **3**, 571–576.

Wolf, C.M., Griffith, B., Reed, C. & Temple, S.A. (1996) Avian and mammalian translocations: update and reanalysis of 1987 survey data. *Conservation Biology*, **10**, 1142–1154.

Wolf, C.M., Garland, T. Jr & Griffith, B.J. (1998) Predictors of avian and mammalian translocation success: reanalysis and phylogenetically independent contrasts. *Biological Conservation*, **86**, 243–255.

Yoerg, S.I. & Shier, D.M. (1997) Maternal presence and rearing condition affect responses to a live predator in kangaroo rats, *Dipodomys heermanni arenae*. *Journal of Comparative Psychology*, **111**, 362–369.

Zeoli, L.F., Sayler, R.D. & Wielgus, R. (2008) Population viability analysis for captive breeding and reintroduction of the endangered Columbia basin pygmy rabbit. *Animal Conservation*, **11**, 504–512.

Zidon, R., Saltz, D., Shore, L.S. & Motro, U. (2009) Behavioral changes, stress, and survival following reintroduction of Persian fallow deer from two breeding facilities. *Conservation Biology*, **23**, 1026–1035.

23

Rewilding

Chris Sandom[1], C. Josh Donlan[2], Jens-Christian Svenning[3] and Dennis Hansen[4]

[1]Ecoinformatics & Biodiversity Group, Department of Bioscience, Aarhus University, Ny Munkegade 114, Aarhus, Denmark
[2]Advanced Conservation Strategies, Midway, UT, USA *and* Cornell University, Department of Ecology & Evolutionary Biology, Ithaca, NY, USA
[3]Ecoinformatics & Biodiversity Group, Department of Bioscience, Aarhus University, Ny Munkegade 114, Aarhus, Denmark
[4]Institute of Evolutionary Biology and Environmental Studies, University of Zurich, Winterthurerstrasse 190, Zurich, Switzerland

'A thing is right when it tends to preserve the integrity, stability, and beauty of the biotic community. It is wrong when it tends otherwise.'

(Aldo Leopold, 1887–1948, from his *Sand County Almanac*, 1949)

Introduction: in need of the wild

The natural world provides humanity with critical goods and services on which people depend for their livelihoods and well-being. Far more than any other species in the history of life on earth, humans are altering the global environment by eliminating species and drastically changing ecosystem function and services (Millennium Ecosystem Assessment 2005; Barnosky et al. 2011). Earth is now nowhere pristine. Human economics, politics, demographics and chemicals pervade every ecosystem; even the largest protected areas require management to prevent the loss of biodiversity (Newmark 1995; Berger 2003). Conservation targets, such as the aim to reduce significantly the loss of biodiversity by 2010, set by the Convention on Biological Diversity (CBD) in 2002, are not being achieved (Butchart et al. 2010). This continued degradation of the natural world is threatening the ecosystem services on which humanity depends (Millennium Ecosystem Assessment 2005). Furthermore, conservation has been characterized as a 'doom and gloom' discipline (Myers 2003) for acquiescing to a default goal of exposing and merely slowing the rate of biodiversity loss, minimizing excitement for conservation and even actively discouraging it (Redford & Sanjayan 2003).

In light of these difficulties there is a growing consensus that biodiversity conservation must move away from merely managing loss and toward active restoration (Dobson et al. 1997;

Key Topics in Conservation Biology 2, First Edition. Edited by David W. Macdonald and Katherine J. Willis.

Young 2000; Choi 2007; CBD 2010). To tackle this change in focus, two complementary and proactive approaches have emerged over the past decade: rewilding and paying for ecosystem services (Costanza et al. 1997; Soulé & Noss 1998; Donlan et al. 2005; Naidoo et al. 2008; Benayas et al. 2009). The latter strives to connect ecosystem function with human welfare, while the former seeks to restore ecosystem functions and services by reintroducing the extirpated species that drive these processes. Together these two approaches can help align human needs and conservation priorities; harmony between these two areas of concern will be essential for the success of any serious environmental initiative (Macdonald et al. 2007).

What is rewilding? Origins and purpose

In the late 1980s, the eminent conservation biologist Michael Soulé collaborated with the wilderness activist Dave Forman to create the Wildlands Project,[1] giving rise to the concept of rewilding (Soulé & Noss 1998). Rewilding was defined as the scientific argument for continental-scale conservation (Soulé & Terborgh 1999), based on the regulatory roles of keystone species and especially large predators. There are three pillars of rewilding.

- Large, protected core reserves
- Connectivity
- Keystone species

In relation to the third pillar, measuring the relative importance of species for ecosystem functioning has progressed tremendously since the concept of keystone species was first proposed by Paine (1969). He suggested this term for species which in proportion to their biomass have a disproportionately large impact in an ecosystem; however, today there is an equally important focus on species as ecosystem engineers (Hastings et al. 2007), and topological keystone species that structure flows of energy or mutualistic interactions in ecosystems (Jordán 2009). Therefore, we here choose to rename the third pillar 'species reintroduction to restore ecosystem functioning'.

Over the past decade, a strong scientific justification has emerged for the restoration of ecosystem function through the reintroduction of keystone species to regionally connected networks of large protected areas, embodying the pillars of rewilding (Soulé & Terborgh 1999; Terborgh & Estes 2010; Estes et al. 2011). The socio-economic benefits from functioning ecosystems and the services they provide humanity are considerable (Sukhdev et al. 2010). To meet biodiversity conservation targets and the global demand for ecosystem services, a globally distributed and connected network of functioning ecosystems, preserved in landscape-scale protected areas, is required (Soulé & Terborgh 1999). To achieve this goal, ecological restoration is required to restore and connect functioning ecosystems. Benayas et al. (2009) analysed 89 restoration projects and estimated that restoration typically increased the provision of biodiversity by 44% and ecosystem services by 25%. While this return must be improved (Bullock et al. 2011), ecological restoration programmes promoting the three key principles of rewilding will help conserve and restore biodiversity, as well as improving the critical ecosystem services on which humanity depends.

Rewilding: science and situation

Rewilding falls within the general framework of restoration ecology, but differs from a traditional view of habitat restoration and species reintroduction (see Chapter 22). Habitat restoration programmes typically seek to improve abiotic and biotic conditions for the benefit of specific threatened species (Miller & Hobbs 2007), while the goal in most species reintroduction programmes is to re-establish viable populations of target species (Soorae 2008, 2010). Both actions aim to aid the conservation of specific threatened species and so conserve biodiversity. In contrast, rewilding explicitly seeks to restore missing or

[1] Now the Wildlands Network, www.twp.org

dysfunctional ecological processes and ecosystem function via a process of species reintroduction.

Achieving rewilding goals requires a comprehensive understanding of ecosystem architecture and ecological processes. Much of this knowledge can be gained from existing communities. However, as humans and their impacts now pervade nearly all ecosystems, with particularly devastating effects on the vertebrate megafauna, vital knowledge must also be extracted by reconstructing past environments. Starting in the late 1970s, scientists began unravelling the ecological implications of species extirpation (Estes et al. 2011). However, determining the drivers behind species range contractions, or uncovering the causes of extinctions, is often complex and becomes increasingly so as the temporal scale increases. Ever since Martin (1967) implicated humans in the Pleistocene megafauna extinctions, potential rewilding baselines have stretched back over 50,000 to >100,000 years (Flannery 2002; Svenning 2002; Donlan et al. 2006).

Controversy has raged over:

- why the mammalian megafauna (body mass >44 kg) were lost primarily?
- exactly when the Pleistocene extinctions occurred?
- why the period and magnitude of extinctions varied between the continents, with the Americas and Australia suffering the worst effects?
- to what extent humans or environmental change were the major drivers (Koch & Barnosky 2006)?

The fourth point of controversy is the one of greatest interest to us, if we want to argue that rewilding is essentially about mitigating anthropogenic ecosystem impacts – and we firmly believe that the available evidence points towards a major human role in the majority of Pleistocene megafauna extinctions. The key problem with the alternative explanations involving environmental changes is that they require the late Pleistocene changes to be unique in how they alone, and not a single one of the many earlier equally dramatic changes over the last 1.8 million years, resulted in similarly widespread megafaunal extinctions (Barnosky et al. 2004; Koch & Barnosky 2006). In contrast, the anthropogenic arguments must explain how the arrival of a new bipedal predator, with only Stone Age technology, could drive so many megafauna over the extinction precipice (Grayson & Meltzer 2003). However, we note that in any case it is clear that the current megafauna-poor situation is highly unusual on the time-scale that current mammal species diversity evolved (middle to late Cenozoic) (e.g. Agustí & Antón 2002).

The body of evidence suggesting a strong anthropogenic effect in many of the continental and oceanic island megafauna extinctions is growing (Burney & Flannery 2005; Gillespie 2008). Even in cases where environmental changes caused dramatic range shifts or contractions, newly appeared humans were likely the most important difference between this and earlier periods of environmental change, and thus a major driver of the megafaunal demise (Koch & Barnosky 2006). For example, it is probably not a coincidence that in the higher altitudes of New Guinea, one of the places early *Homo* species never reached, the megafauna survived until after modern humans arrived (Fairbairn et al. 2006; Corlett 2010), while many regional, climatically similar Indomalaysian islands were defaunated much earlier in the presence of early *Homo* species. With this mounting evidence, it is difficult to ignore the cascading effects these anthropogenic megafaunal extinctions are likely to have had on ecosystem function up to today (Johnson 2009). For instance, the ecology of many large-seeded trees in the Americas, such as jicaro (*Crescentia alata*), is now viewed as anachronistic due to the 10,000-year absence of the seed-dispersing megafauna, such as the gomphotheres (*Cuvieronius* and *Stegomastodon* spp.) (Janzen & Martin 1982). It is furthermore beyond any doubt that humans have directly decimated megafaunas in many parts of the world during the Holocene, including in historic times (e.g. Vuure 2005; Elvin 2006; Crowley 2010; Bar-Oz et al. 2011).

This extended period of anthropogenic influence and the remaining uncertainties regarding the megafaunal demise (Grayson 2007) have led to divergent visions for rewilding. Donlan et al. (2006) stoked fierce debate when they introduced the concept of Pleistocene rewilding, the reintroduction of processes and species lost within the Pleistocene. Critics of this approach question the arguments for reintroducing species and recreating landscapes that have been absent for over 10,000 years (Rubenstein et al. 2006; Caro & Sherman 2009; Oliveira-Santos & Fernandez 2010; Richmond et al. 2010). These questions are indeed valid for any temporal period; why should any lost landscapes or ecosystems be recreated?

However, rewilding is fundamentally a future-oriented proposal that seeks to learn from the past rather than recreate it (Dobson et al. 1997; Choi 2007; Hobbs & Cramer 2008; Macdonald 2009; Hansen 2010). For example, red deer (*Cervus elephus*) in the hills of the Scottish Highlands and mustang horses (*Equus caballus*) on the plains of North America, reintroduced to North America after a 10,000-year absence, pose a conservation threat through overgrazing (Russell 2004; Hobbs 2009). Analysis of the respective ecosystem structures and their environmental history clearly highlights the absence of apex consumers and recommends the reintroduction of the wolf (*Canis lupus*) to the Scottish Highlands and the lion (*Panthera leo*) to North America to restore top-down forcing of the grazing communities (the extinct North American lion is a subspecies belonging to an extinct Holarctic subspecies of the extant African-South Asian lion; Burger et al. 2004; Donlan et al. 2006; Nilsen et al. 2007; Barnett et al. 2009; Estes et al. 2011; Sandom et al. 2011). These actions would reduce the need for perpetual anthropogenic intervention, resulting in a naturally regulated, ecologically functioning and wilder landscape.

As these examples suggest, the temporal distance to which ecologists and rewilding practitioners gaze into the past to learn these lessons should be determined on a case-by-case basis, depending on the conservation problem being tackled, the period in which the relevant functional species were lost, and whether humans are implicated in their demise.

Developing the rewilding manual: putting rewilding into practice?

Restoring function

Four initial steps are required to instigate a rewilding project:

- identification of the issue of conservation concern (e.g. overgrazing by large ungulates)
- identification of the missing ecological processes (e.g. predation)
- identification of the functional characteristics required to restore the missing processes (e.g. large apex consumers)
- selection and reintroduction of the most suitable species to restore the missing or dysfunctional processes.

A knowledge of complete species assemblages, functional community structures, and response to changing conditions throughout the late Quaternary (130,000 years ago to the present day) would be the gold standard as a baseline from which to determine which extinct taxa to reintroduce or replace. However, with inevitably imperfect empirical knowledge, theoretical ecology can be useful. For example, network theory can be used to map and analyse ecosystem architecture and function (Bascompte 2009), facilitating comparisons between dysfunctional ecosystems with intact extant and reconstructed past reference ecosystems (Egan & Howell 2001). Such comparisons can help identify missing interactions (ecological processes) and the species that facilitate them. Furthermore, analysis of ecological networks can help identify the most effective keystone species to reintroduce (Ebenman & Jonsson 2005). For instance, analysis of food webs indicates they are relatively robust to random extinction events, but highly susceptible to the loss of well-connected species (Dunne et al. 2002; Allesina & Bodini 2004; Bascompte 2009; Terborgh & Estes 2010; Estes et al. 2011).

To minimize the risks associated with rewilding, species recently extirpated from an ecosystem would ideally be reintroduced. In some cases translocations of species outside their current biogeographical range may be required as taxon substitutes to ecologically replace a globally extinct species, e.g. Asian camel (*Camelus bactrianus*) as replacements for their extinct North American counterpart (*Camelops hesternus*) or Asian elephant (*Elephas maximus*) as replacements for the extinct proboscideans in Europe (straight-tusked elephant, *Elephas antiquus*) or the Americas (mammoths, mastodons, gomphotheres) (Martin 2005; Zimov 2005; Donlan et al. 2006; Svenning 2007; Griffiths & Harris 2010; Griffiths et al. 2010; Hansen et al. 2010; Griffiths et al. 2011). However, the threat posed by invasive alien species demonstrates the risks posed by species introductions (Long 2003; Macdonald & Burnham 2010) and so concern has been raised about the use of taxon substitutes (Rubenstein et al. 2006). Alien species become invasive when they interact strongly with the native ecosystem, altering function and process and threatening native biodiversity (Shine et al. 2000). Efforts to reduce the economic impact of pest species through biological control have often ended in disaster as a result of unexpected interactions between the introduced species and the environment. In contrast, a well-selected taxon substitute will interact strongly with the native environment but support and maintain biodiversity (Hansen et al. 2010). We note that megafaunal taxon substitutes will be easier to control, if needed, than many plants, small animals and micro-organisms.

Species reintroduction and taxon substitution present risks; however, failing to restore keystone interactions also risks a wave of secondary extinctions (Terborgh & Winter 1980; Soulé et al. 1988). Ripple & Beschta (2006) observed that where cougar (*Puma concolor*) density was reduced by disturbance from tourism in Zion Canyon, Zion National Park, USA, the deer population increased by 750%. The increased browsing pressure prevented woodland regeneration and significantly reduced the abundance of hydrophytic plants, wildflowers, amphibians, lizards and butterflies compared to the local North Creek that maintained stable cougar densities. The effects of species extirpation can also be long-lasting (Borrvall & Ebenman 2006). For instance, patchy distribution of various large-seeded American trees may be the ill effects of a 10,000-year absence of megafaunal seed dispersal agents (Janzen & Martin 1982).

Assisted colonization (assisted migration) has been proposed to translocate populations of plants and animals particularly threatened by climate change (Hoegh-Guldberg et al. 2008). As such, assisted colonization is a 'kindred spirit' of rewilding (Hansen 2010), and there could even be cases where rewilding could be combined with assisted colonization, using 'climatic refugee' species in one region as substitutes for extinct species elsewhere. The fear that taxon substitution may create new invasive species is also a concern applied to the concept of assisted colonization (Ricciardi & Simberloff 2009) (see Chapter 22) and in either case risks need to be carefully considered (cf. Morueta-Holme et al. 2010). Secondary extinctions caused by ecosystem dysfunction or poorly considered species reintroductions or taxon substitutions can perhaps be best avoided through cautious action. Such action would include rigorous species selection protocols and geographically limited experimental translocations, followed by, if successful, carefully monitored wider translocations.

Some translocations present greater risks than others. As such, greater degrees of caution need to be applied to the introduction of a taxon substitute replacing a species lost in the Pleistocene compared to the reintroduction of a recently extirpated species to a relatively intact ecosystem. Species selection must be approached pragmatically and on a case-by-case basis, with a full consideration of local ecological and human requirements. However, ultimately species selection should be driven by identifying which species can effectively and efficiently restore missing ecosystem processes and function, regardless of whether the conservation action is a species reintroduction or taxon substitution.[2]

Rewilding: scenario planning and ecological experiments

One method of progressing rewilding to a mainstream management option is to test *a priori* hypotheses with quantifiable outcomes within rewilding projects. Scenario planning and the 'three horizons' analysis (Curry & Hodgson 2008), useful tools in conservation (Kass et al. 2011), allow long-term rewilding projects to be considered in three phases:

- *first horizon*: the current, functionally deficient ecosystem in need of restoration
- *third horizon*: a projected future scenario where the ecosystem is restored to a functional and self-sustaining state
- *second horizon*: a transition state between the first and third horizons.

Here we explore the proposed restoration of the Caledonian pine forest in the Scottish Highlands as an example.

FIRST HORIZON

Six thousand years ago, the Caledonian pine forest covered 50% of the Scottish Highlands. Initially through natural climate change, and more recently through centuries of anthropogenic exploitation, semi-natural woodland cover has been reduced to just 1.7% of Scotland (Forestry Commission 2001; Smout et al. 2008; Hobbs 2009). The mammal community has also been heavily reduced with the loss of the entire guild of large carnivores – wolf (*Canis lupus*), bear (*Ursus arctos*) and lynx (*Lynx lynx*) – as well as many of the large herbivores including wild boar (*Sus scrofa*), aurochs (*Bos primigenius*) and European elk (moose) (*Alces alces*) (Yalden & Barrett 1999). The population of the only remaining large mammal, the red deer, has expanded to carrying capacity in response to the loss of the large carnivore guild, reduced competition from an impoverished large herbivore guild and favourable management practices employed by land owners (Clutton-Brock et al. 2004). Under high browsing and grazing pressure, forest regeneration is heavily restricted (Palmer & Truscott 2003).

THIRD HORIZON

Instead of a status quo scenario where native pine forest habitat will continue to decline (Hobbs 2009), we envision an alternative state achieved within the next 100 years, with a restored, ecologically functioning and self-sustaining forest at the landscape scale with active disturbance, regeneration and dispersal processes.

SECOND HORIZON

To reinvigorate natural woodland regeneration, at least three processes must be restored:

- seed production
- the disturbance regime
- top-down control of the large herbivore guild (Sandom et al. 2011; Sandom et al. 2012; Sandom et al. forthcoming a).

Three species have the desired functional characteristics to restore these processes, and should thus be the main targets for reintroduction: Scots pine (*Pinus sylvestris*), wild boar and wolves. Scots pine, the dominant tree species of this ecosystem, is needed to re-establish a wider distribution of a viable seed source; the wild boar is an ecosystem engineer to reinvigorate the patch-scale disturbance regime; and the wolf, the keystone top predator, would restore top-down forcing of the red deer population.

[2] Here, when we use the word 'reintroduction' it refers specifically to ecosystem functioning rather than taxa.

To determine whether the reintroduction of these species would be sufficient to reinvigorate woodland regeneration and restore the wider ecosystem at the landscape scale by promoting rapid woodland expansion, it is helpful to begin by experimenting with computational modelling and fenced areas to minimize the threat of unforeseen and detrimental interactions (Boyce et al. 2007; Manning et al. 2009). These tools can also provide manageable 'stepping stone' goals that make rewilding projects more viable (Manning et al. 2006). Fenced deer exclusion areas have illustrated that woodland regeneration is viable under current environmental conditions in the absence of any browsing pressure (Gong et al. 1991). However, the lack of large ungulates reduces germination niche availability, limiting long-term regeneration in fenced exclusion areas (Warren 2009). Sandom et al (forthcoming a,c) quantified the rate, distribution and impact of rooting behaviour by wild boar in a fenced Scottish Highland landscape on the Alladale Wilderness Reserve (see Figure 23.2). Rooted areas, typically created beneath a woodland canopy in bracken-dominated vegetation, recorded 3.6 times greater seedling regeneration than unrooted areas, reduced bracken frond density by 69% and significantly increased forb species richness. However, wild boar also pose a threat to sapling and mature trees through uprooting and bark stripping. Rooting rate was quantified to typically fall between 22 and 75 m^2/boar/week (Sandom et al. forthcoming b). At densities unlikely to exceed 4/km^2 (Howells & Edwards-Jones 1997), rooted area will accumulate between 4500 and 15,600 m^2/km^2/year. Concentrated beneath a woodland canopy, wild boar will increase patch scale heterogeneity in vegetation structure, but only promote a slow rate of woodland expansion.

Nilsen et al. (2007) and Sandom et al. (2011) modelled the potential impact of a wolf reintroduction on the Scottish deer herd. Nilsen et al. (2007) concluded that a wolf reintroduction may result in a 50% reduction of the deer herd. Sandom et al. (2011) explored the predator–prey interaction within a landscape-scale fenced reserve scenario, the fence proposed as a means to mitigate human–wildlife conflict. The limited space availability of such a scenario highlighted the importance of factors affecting maximum wolf population density. If, for instance, social interactions and intraspecific competition limit maximum wolf density to one pack per 200 km^2, which is the average pack territory size at high deer densities (Fuller et al. 2003), predator–prey dynamics are likely to be weak, with limited effect on deer density. Yet, where wolf density is comparable with or greater than the highest wolf densities recorded, e.g. 92/1000 km^2 on Isle Royale in Lake Superior in North America (Fuller et al. 2003), wolves have the potential to control a deer population at densities that would allow woodland regeneration.

These data indicate that the simple reintroduction of the three target species (Scots pine, wild boar and wolf) is unlikely to result in a rapid transition between the first and third horizons at the landscape or regional scale within the 100-year time period specified. While these species may have favourable impacts on restoring the key ecological processes, the scale at which they operate is too small to achieve rapid forest expansion. The second horizon is likely to require extensive tree planting to re-establish and widely distribute seed source and wild boar would need to be initially stocked at artificially high densities in proposed woodland expansion zones to create sufficient bare ground patches to aid planting and natural regeneration. The total extent of the impact of wolves on deer density and behaviour remains to be seen. However, all of these species are likely to have a positive effect in meeting long-term rewilding objectives of a self-sustaining functional forest and this encourages exploration of the next step: the creation of a regional-scale fenced rewilding project. Sandom et al. (2011) suggest the creation of a 600 km^2 or larger enclosed reserve, including populations of wild boar and wolves, that would represent an exciting rewilding opportunity (Macdonald et al. 2000; Manning et al. 2009; Sandom et al. 2011).

Management requirements: questions of scale

Rewilding seeks ultimately to remove the need for anthropogenic management in conservation areas. However, even within large national parks management is required to prevent species extinction (Newmark 1995; Berger 2003), and as the Scottish example demonstrates, the reintroduction of processes may not be sufficient to restore heavily degraded ecosystems quickly and easily. As a result, rewilding should not only aim to restore vast wildernesses, but also be seen as an appropriate management action in much smaller conservation or restoration areas, restoring dynamics, often with the specific aim of preventing secondary extinctions (Griffiths & Harris 2010; Hansen 2010). Along a spatial gradient, the necessary management levels are likely to decrease from relatively high levels in small-scale projects to comparatively low levels for large-scale projects. The exact shape of these relationships will differ depending on the specific needs of the project. Figure 23.1a illustrates several such patterns. For example, for supplemental feeding of herbivorous megafauna, we could expect management levels to decrease rapidly with increasing area, Wallis de Vries (1995) highlighted that feeding is often used to maintain high ungulate densities on reserves much smaller than 10,000 ha, the area thought needed to support such communities (curve a), or in a more linear fashion (line b), for instance when controlling a population of introduced top predators. With regard to dynamics that rely on maintaining populations above certain minima, for example maintaining genetic diversity, we can expect relationships close to curve (c). The wolves confined to the 544 km^2 Isle Royale have illustrated that a wolf population is viable at this scale, but inbreeding has led to genetic deterioration (Räikkönen et al. 2009); unless the area is of a size with a carrying capacity above these minima, comparatively high management levels are required. With two management factors, e.g. maintaining genetic diversity *and* providing supplementary feeding, the relationship may be a step-wise one (curve d) where both are required in smaller areas but only maintaining genetic diversity is necessary in larger areas.

For temporal patterns, management trajectories are likely to vary quite dramatically, as highlighted in Figure 23.1b. In an ideal case (a), a relatively high initial management level (introducing the substitutes) rapidly decreases and remains low thereafter. A variation of this trajectory (b) could be induced by increasing management again for shorter or longer periods of time to prevent undesirable trajectories or phase shifts. As for spatial patterns, many temporal management trajectories are likely to be stepwise, with high initial levels that subsist until a threshold has been reached (c) (e.g. supplementary introductions until population establishment). Especially for smaller areas, some management is likely to be seasonally cyclic (d), e.g. off-site husbandry during winter or dry seasons.

The prospect of restoring the Barbary lion offers a good example of a possible rewilding project proposing a pragmatic approach (Macdonald et al. 2010). The recent historical proximity of lions to Europe is caught in the observation by Ormsby (1864) that '... a man who has dined on Monday in London can, if he likes, by making the best use of express trains and quick steamers, put himself in a position to be dined on by a lion in Africa on the following Friday evening ...'. The lion Ormsby had in mind was the North African Barbary lion or Atlas lion, renowned for the males' dark and copious manes. However, recent molecular work revealed that Barbary lions were phylogenetically distinct, and probably no pure specimens survive in captivity (Barnett et al. 2006). If they were to be reintroduced, the best option as a step towards restoring a functioning predator community in the Atlas Mountains might be to introduce their closest living relatives, the Asiatic lion (*P. l. persica*) and closely related West African lion to North Africa (Barnett et al. 2007; Bertola et al. 2011). Indeed, between the Middle

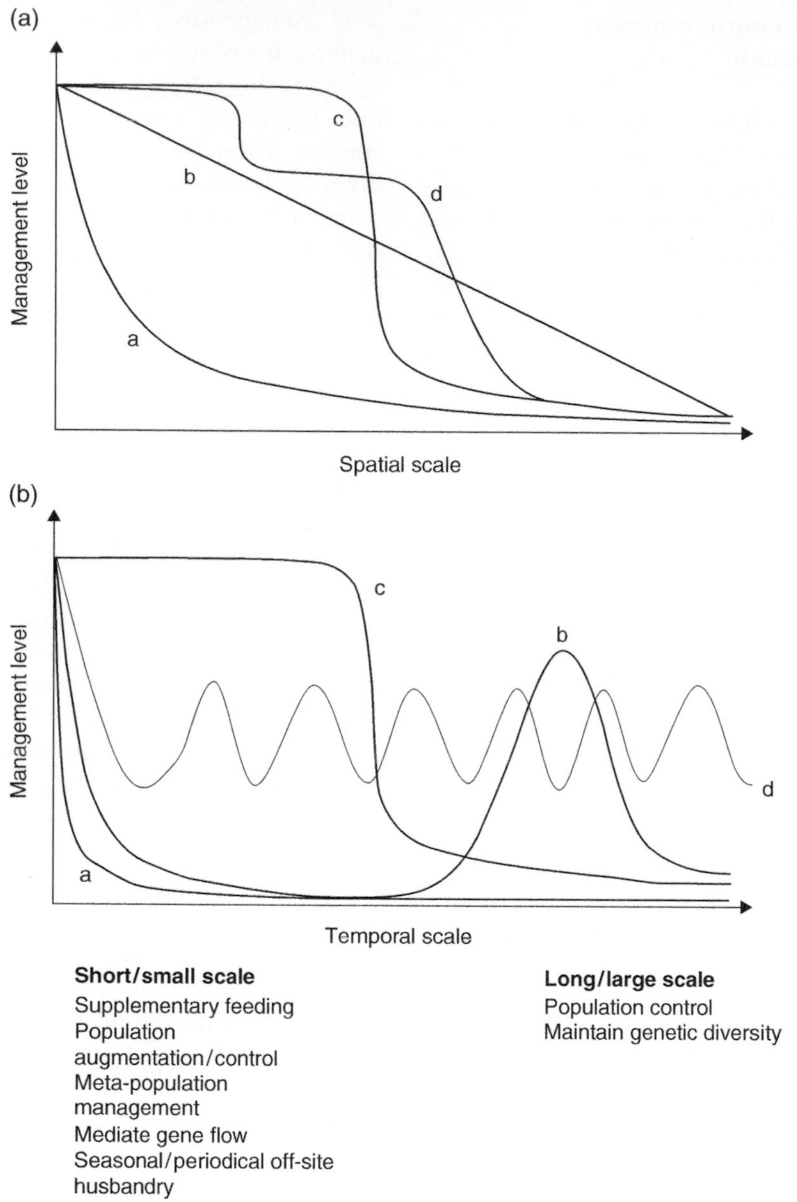

Figure 23.1 Hypothetical trajectories of required levels of human management in rewilding projects vary with scale, with spatial scale varying between projects, and temporal scale within a project. The specific management trajectories (denoted by small letters) are discussed in the main text.

and High Atlas lies a rocky mountainous area where green oaks dominate a landscape in which the endangered Barbary leopard, (*P. p. panthera*) may still survive alongside a prey base that could be restored. There has been talk that a fenced enclave, perhaps as large as 100 km², could hold a managed group of lions (loosely similar to the fenced reserves for African wild dogs that Davies-Mostert *et al.* (2009) describe). Eventually, perhaps scientific planning and changing attitudes would allow the fence to be breached.

Where and when is rewilding appropriate?

Rewilding to date: how successful?

In practice, rewilding provides the opportunity to meet conservation objectives and test theoretical ecology. However, because rewilding often requires large areas and species reintroduction (particularly of the much persecuted and still feared megafauna), putting rewilding into practice presents numerous challenges. Figure 23.2 illustrates a limited selection of projects that contain one or more of the three pillars of rewilding. Covering all five inhabited continents as well as numerous oceanic islands, they offer interesting examples of how rewilding can be applied to tackle problems in a variety of novel ways that we describe below.

Resurrection of ecosystem functioning

Dysfunctional island ecosystems around the world are at the forefront of implementing the most controversial aspect of rewilding – taxon substitution. The comparatively simple island ecosystems, where the few, large megavertebrates weighed hundreds rather than thousands of kilograms (Hansen & Galetti 2009), and where many of these animals went extinct only a few hundred years ago, make islands prime laboratories for rewilding projects (Hansen 2010).

The three Mascarene Islands in the Western Indian Ocean are an excellent example. Here, replacing recently extinct endemic giant tortoises has resurrected extinct seed dispersal interactions (Hansen et al. 2008; Griffiths et al. 2011), and is reinstating a herbivory regime that is likely to benefit native plants, while controlling invasive alien plants (Griffiths et al. 2010).

At a larger scale, in Oostvaarderplassen, a 6000-hectare fenced nature reserve a few hours outside Amsterdam, scientists have introduced Heck cattle, red deer and Konik horses in an attempt to restore the guild of large herbivores that were once present throughout Europe. The presence of a restored grazing guild at high densities has limited the extent of woodland regeneration and has challenged the long-held concept that a 'wild Europe' would be dominated by a closed forest (Olff et al. 1999; Bakker et al. 2004; Vera 2009). Although the validity of the heavily grazed, half-open forest European landscape, proposed by Vera (2000), is still debated (Svenning 2002; Hodder et al. 2005), Oostvaarderplassen has clearly demonstrated impressive, positive biodiversity outcomes at the local to landscape scale, including the first breeding pair of white-tailed eagles in The Netherlands since the Middle Ages (Curry 2010).

At even larger scales, the Rewilding Europe Initiative is seeking to restore missing species and function to 10 100,000-hectare core areas by 2020. In four of the five projects already initiated (Western Iberia, Eastern Carpathians, Southern Carpathians and Velebit), rural land abandonment has been identified as a cause of a reduced grazing regime that threatens biodiversity conservation objectives (Rewilding Europe Initiative 2011). A proposed solution is the reintroduction of wild grazers in the form of horses, bison and bovid substitutes for the extinct aurochs ('rewilded' domesticated conspecifics). Predator reintroduction may swiftly follow, especially if these proposed reserves can be effectively connected, with particular attention paid towards the threatened Iberian (*Lynx pardinus*) and Eurasian lynx (*Lynx lynx martinoi*) (Macdonald et al. 2010).

In Siberia, scientists are trying to restore the mammoth steppe – once one of the world's most extensive ecosystems – by introducing Yakutian horses, musk ox (*Ovibos moschatus*), bison (*Bison bison*) and other large herbivores (Stone 1998; Zimov 2005). They hope eventually to introduce the endangered Siberian tiger (*Panthera tigris altaica*) and thus the important process of predation. This Pleistocene Park also has important implications for climate change if its concept is broadly implemented in the region: frozen Siberian soils lock up over 500 gigatons of organic carbon (over twice as much

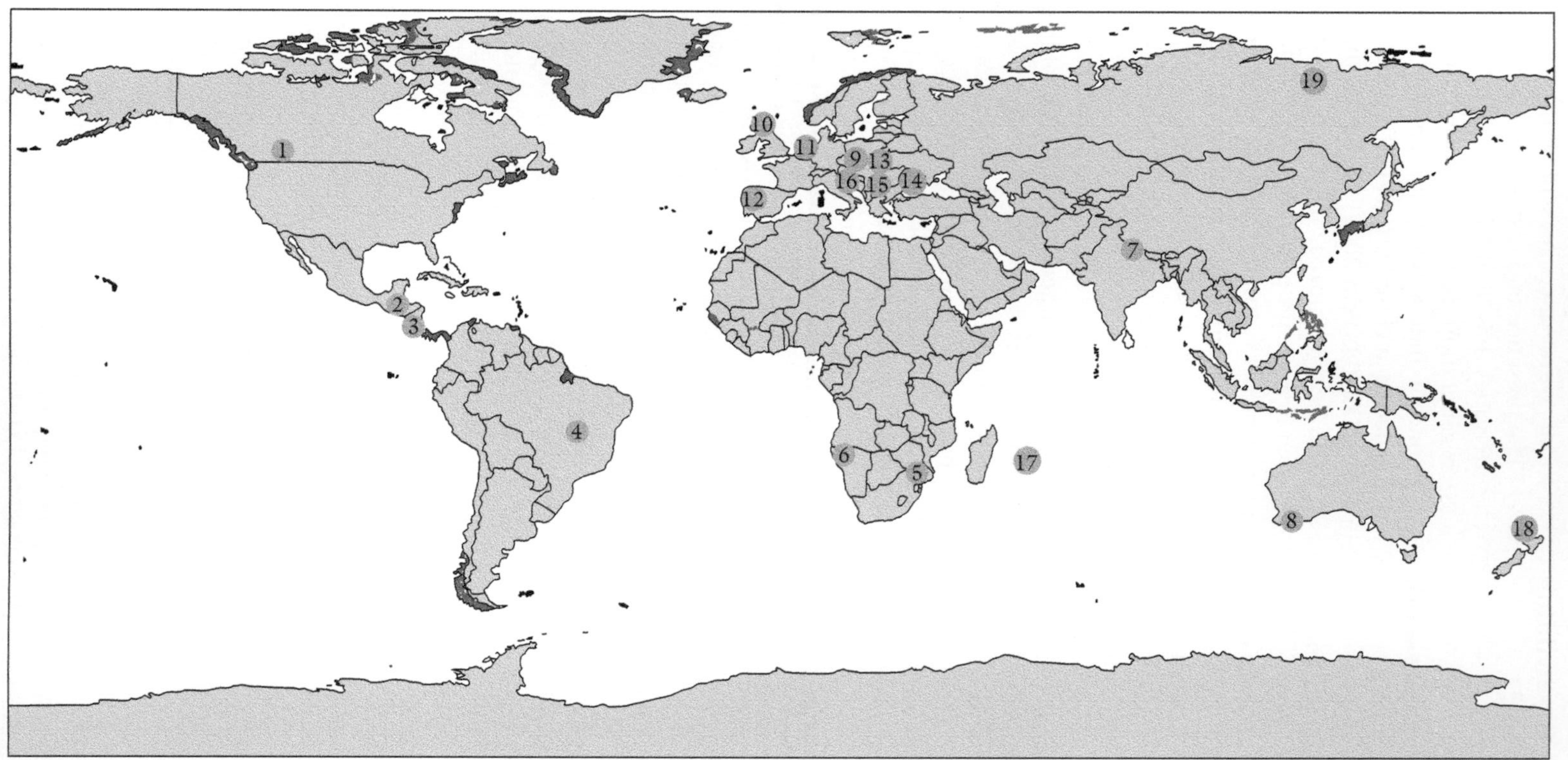

Figure 23.2 Rewilding projects in practice. 1. Yellowstone to Yukon. 2. Paseo Pantera. 3. Area de Conservacion Guanacaste. 4. Cerrado-Pantanal. 5. Great Limpopo Transfrontier Conservation Area. 6. Iona-Skeleton Coast TFCA. 7. Terai Arc Landscape. 8. Gondwana Link. 9. European Green Belt. 10. Alladale Wilderness Reserve. 11. Oostvaarderplassen. 12. Western Iberia. 13. Eastern Carpathians. 14. Danube Delta. 15. Southern Carpathians. 16. Velebit. 17. Mascarene Islands. 18. New Zealand. 19. Pleistocene Park.

as the world's rainforests). As the permafrost melts, microbial activity will release these carbon stores into the atmosphere, exacerbating climate change. Restoring the ancient grassland ecosystem could prevent permafrost thawing by increasing soil stability with a root system and increasing albedo, helping to combat climate change (Zimov 2005; Nicholls 2006).

Large core areas and connectivity

Large core areas and well-connected ecological networks are required for rewilding and the provision of ecosystem services (see Chapter 21). Fraser (2009) offers an excellent review of numerous projects with such ambitions. Establishing vast conservation areas, particularly those crossing national boundaries, presents considerable challenges. Kruger National Park in South Africa covers 2 million hectares and is one of the largest national parks. Yet even at this prodigious scale, it may be insufficient to support natural dynamics of a confined elephant population (van Aarde et al. 1999). Restricted dispersal and the provision of artificial water sources are thought to contribute to rapidly increasing elephant population densities (Slotow et al. 2005; van Aarde & Jackson 2007). At high densities, these ecosystem engineers can exert a potentially harmful disturbance to vegetation structure on the landscape scale. For instance, Cumming (1997) illustrated that when elephant density exceeded $0.5\,km^2$ savanna woodland is converted to shrub- and grassland with an accompanying loss of biodiversity. This megamammal clearly requires conservation management at the regional and continental scales to ensure a wide spatial and temporal distribution of its disturbance. Recent plans to create huge transfrontier conservation areas (TFCAs), which merge or link great conservation areas across national borders, hold great promise which so far has been hard to turn into reality.

A good example of this comes from south-eastern Africa. The proximity of Kruger National Park to parks in Zimbabwe and Mozambique prompted a proposal to remove the fences between them to create a transfrontier park covering 3.5 million hectares, with the ultimate ambition to create the Great Limpopo Transfrontier Conservation Area, covering an impressive 10 million hectares (Figure 23.3a) (Wolmer 2003). However, despite the considerable excitement this proposed super-park generated when initiated in 2001, little activity has taken place on the ground in the decade following its proposal. The challenges faced involve a combination of political mistrust, social resentment over the limited benefits passed on to local communities, and anger at the uneven distribution of economic benefits between countries, as well as the threat posed by wandering megafauna in regions no longer accustomed to coping with them (Fraser 2009).

Nevertheless, the Great Limpopo TFCA remains a flagship of the transfrontier park concept, and its advocates are certainly not alone in struggling with the challenges that these bold plans pose. Supporters of the the Paseo Pantera in Central America had equally grand ambitions, attempting to connect 600 protected areas covering eight countries. The original conservation objectives were combined with efforts to promote rural community development, but an uneven balance and insufficient alignment between these equally important objectives meant that despite the deployment of considerable resources, few of the original goals were reached (Kaiser 2001; Fraser 2009).

Other TFCA programmes have achieved more significant progress by successfully aligning conservation and human welfare objectives. The Area de Conservacion Guanacaste in Central America, led by the eminent conservation biologist Daniel Janzen, had as its primary objective to restore dry tropical forest and connect it to areas of surrounding rainforest to restore a functioning ecosystem. Presented with these challenges, Janzen used novel and often

(a)

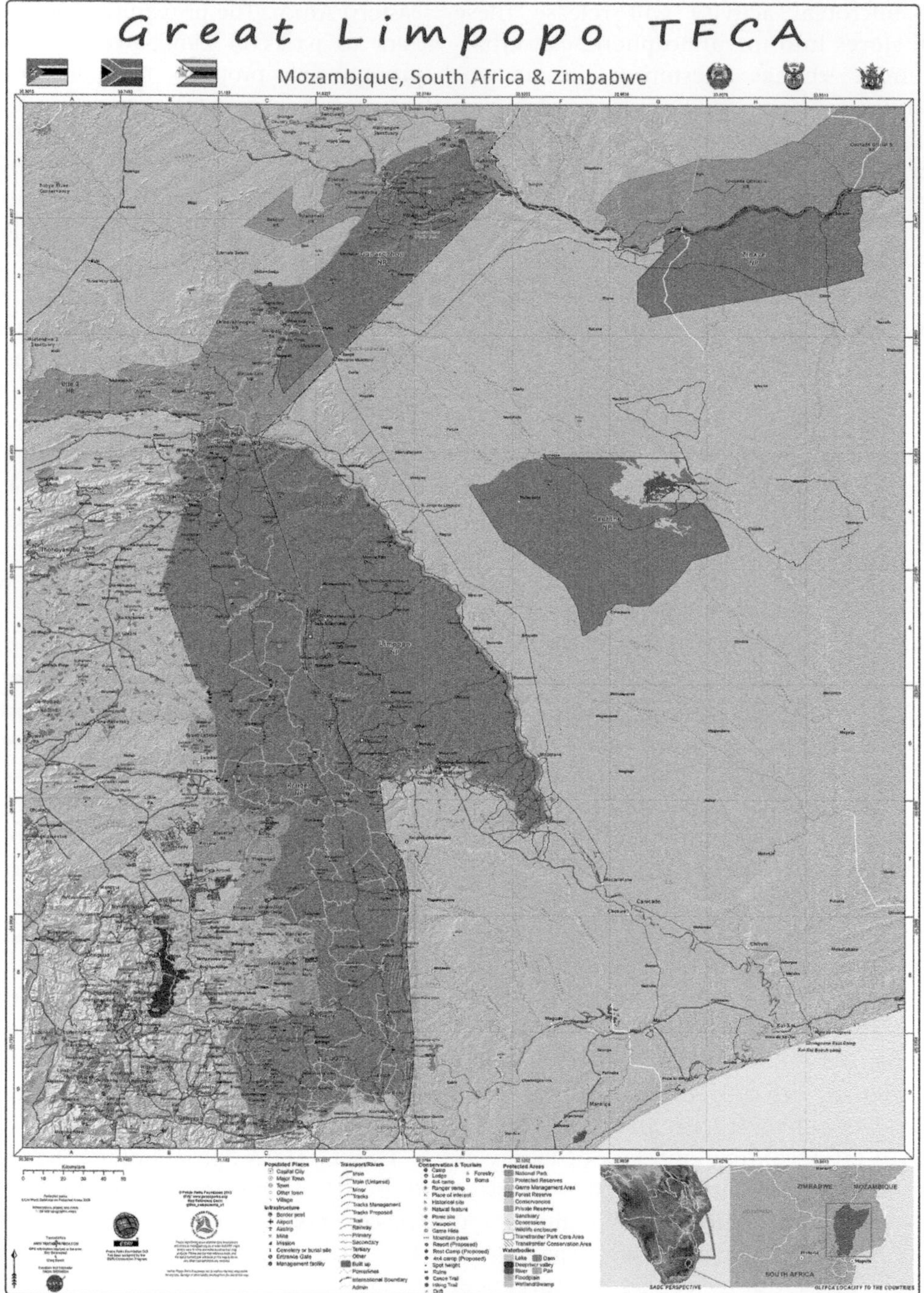

(b)

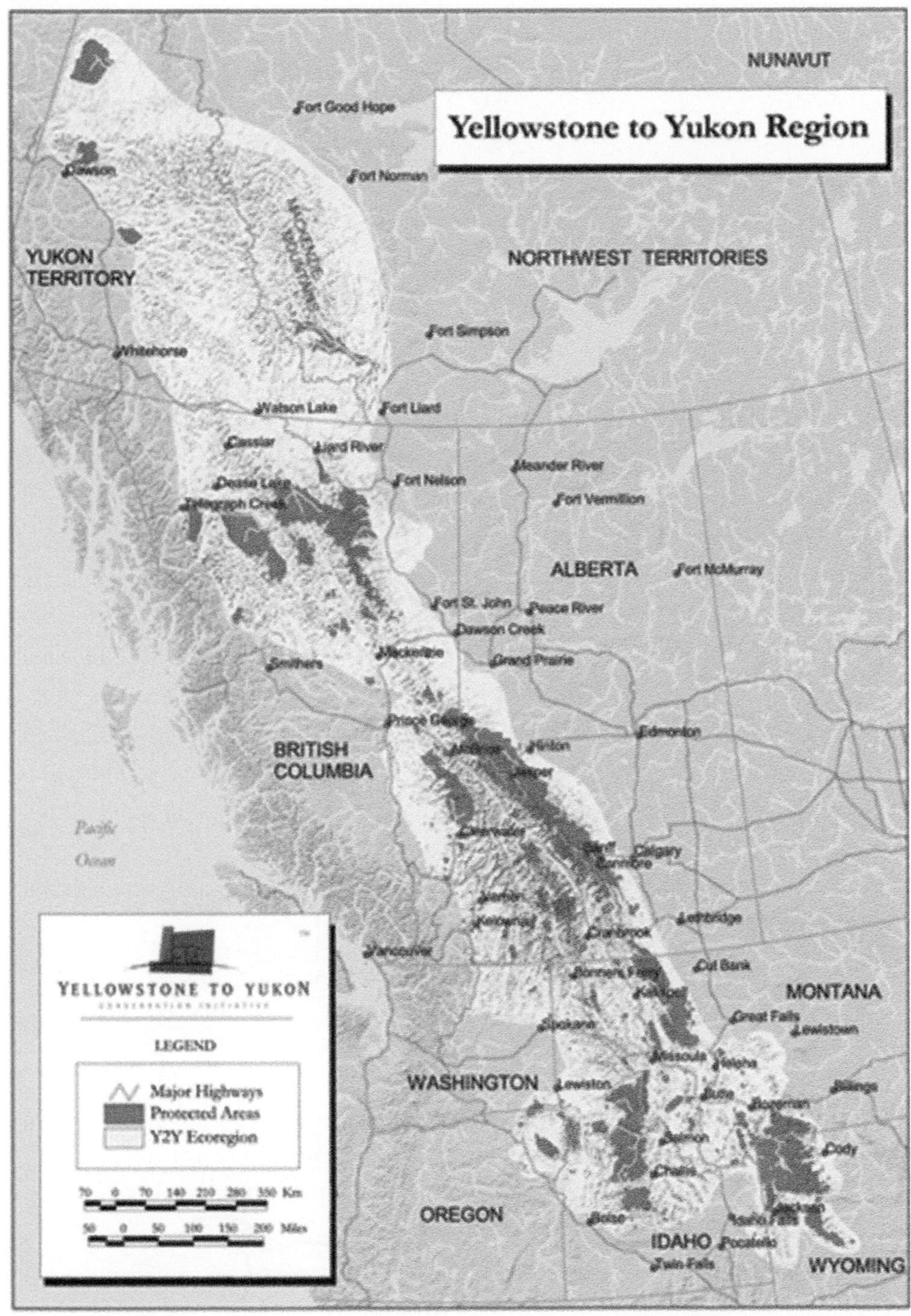
Yellowstone to Yukon Region
NUNAVUT
Fort Good Hope
Dawson
Fort Norman
YUKON TERRITORY
NORTHWEST TERRITORIES
Fort Simpson
Whitehorse
Watson Lake
Fort Liard
Cassiar
Liard River
Fort Nelson
Meander River
Fort Vermillion
Dease Lake
Telegraph Creek
ALBERTA
Fort McMurray
Fort St. John
Peace River
Dawson Creek
Mackenzie
Smithers
Grand Prairie
Prince George
Hinton
Edmonton
BRITISH COLUMBIA
Jasper
Pacific Ocean
Banff
Canmore
Calgary
Lethbridge
Cranbrook
Vancouver
Cut Bank
Bonners Ferry
Kalispell
MONTANA
Great Falls
Lewistown
Spokane
Missoula
Helena
WASHINGTON
Lewiston
Butte
Bozeman
Billings
Salmon
Cody
Challis
OREGON
Boise
Jackson
Idaho Falls
IDAHO
Pocatello
Twin Falls
WYOMING
YELLOWSTONE TO YUKON
LEGEND
Major Highways
Protected Areas
Y2Y Ecoregion
70 0 70 140 210 280 350 Km
50 0 50 100 150 200 Miles

Figure 23.3 (a) The Great Limpopo Transfrontier Conservation Area. (Peace Parks Foundation 2011); (b) Yellowstone to Yukon (Yellowstone to Yukon 2011); (c) European Green Belt (European Green Belt 2011).

controversial approaches. The overly frequent fire regime was tackled by restoring domestic cattle to graze and reduce fuel availability within the landscape; cattle removal had been the conservation management plan until that point. Restoring forest to cleared regions was achieved by using a non-native plantation tree species to foster rainforest species in the shade beneath. Once the trees had begun to grow, cattle were then removed once more. Social issues were addressed by providing jobs to local people paid for by a large endowment generated through funding raised internationally, which will aid running the reserve and the jobs it supports long into the future (Janzen 2000; Fraser 2009).

Other successful projects include the Iona-Skeleton Coast TFCA. The Namibian government has written conservation into its constitution and supports the establishment of conservation areas within local communities with considerable success, providing lasting conservation and public benefits (Constitution of the Republic of Namibia 1990; Fraser 2009). The Terai Arc Landscape in Nepal has sought to meet rewilding objectives by encouraging communities to engage in habitat restoration to build corridors between protected areas. By supporting ecotourism, biogas facilities and sustainable livelihood initiatives, Nepal is able to use rewilding to bring about positive social change (Baral & Heinen 2007; Fraser 2009).

Numerous other projects are also promoting the conservation and community benefits of core areas and connectivity. The Yellowstone to Yukon Project was conceived in 1997 to help connect the national parks in north west North America and could create a conservation area covering 130 million hectares (Figure 23.3b). Yellowstone National Park is already famous for its wolf reintroduction that has been at the forefront of the rewilding debate (Smith et al. 2003; White & Garrott 2005; Beyer et al. 2007; Ripple & Beschta 2007). In South America, the Cerrado-Pantanal ecological corridors project is seeking to connect the last remnants of the savanna of the Cerrado with the wetlands of the Pantanal to conserve what has been described as the Brazilian Serengeti. Local community involvement and a dedication to collecting the ecological data needed to design an efficient network have led to the creation of 400 km-long corridors and the establishment of a 2000 km^2 biodiversity protection area (Klink & Machado 2005; Fraser 2009), perhaps the first step in the visionary rewilding proposed by Galetti (2004) to replace extinct Pleistocene megafauna with extant species? The European Green Belt is a continental-scale connectivity project that aims to capitalize on the green and wild land that was once the 'iron curtain' dividing east and west. This corridor could connect as many as 3272 protected areas in 23 countries and could form the backbone of a European network of rewilded areas (Figure 23.3c Terry et al. 2006). The Gondwana Link in Australia is attempting to restore native flora and fauna and reinstate critical processes that prevent the drying out of the landscape (Jonson 2010), essential steps for both conservation and human needs.

Conclusion

Rewilding is a future-oriented restoration proposal that seeks to learn from the past to meet the conservation challenges of today, and secure the ecosystem resilience of tomorrow. It is an ambitious and proactive approach to inspire conservation practitioners to engage with the realities of biodiversity conservation in the 21st century (Macdonald et al. 2000). Specifically and uniquely, rewilding seeks to achieve this through the reintroduction of extirpated species and the replacement of globally extinct species to resurrect ecological processes – in particular top-down trophic effects – and thereby restore ecosystem function. Ultimately it challenges the traditional view of the native species by exploring historic species distributions when they were less affected by anthropogenic environmental change and uses these lessons to help restore ecosystem function. Rewilding thus also directly links with the need for other controversial, but potentially essential, conservation measures, such as assisted migration of species threatened by environmental change to areas outside their natural biogeographical ranges.

Gazing into the past to better understand ecosystem function reminds us that missing ecological and evolutionary functions that were present 100 or 30,000 years ago still affect today's ecosystems. In this context, the implications of today's management actions must be considered equally far into the future, beyond immediate conservation or restoration needs. Taking a long-term and positive view, we argue that rewilding is a key element to restricting the current, early part of the Anthropocene Era, with its negative biodiversity trends, to as short and ultimately reversible an epoch as possible.

Pragmatism must be applied to the application of rewilding principles. As with any reintroduction programme, potential interactions between the restored ecosystem component and human presence within the landscape must be carefully considered prior to any action on the ground. For instance, where human–wildlife conflict is likely, as with the often controversial measure of restoring apex consumers (e.g. Quammen 2004), the use of mitigating strategies such as fenced reserves may be necessary to align conservation goals and human needs. Given these potential conflicts, it is essential to be pragmatic by applying appropriate resources to projects that present little risk and great reward while not hiding from the challenges posed by rewilding the human landscape that must be achieved for a global provision of ecosystem function and services. While challenges and risks are present, there are clear scientific, economic and social justifications for considering such bold conservation actions.

Rewilding seeks to inspire a generation to set something right and to redress the major wounds of past and present abusive land uses for a brighter and more sustainable future.

References

Agustí, J. & Antón, M. (2002) *Mammoths, Sabertooths, and Hominids: 65 Million Years of Mammalian Evolution in Europe*. Columbia University Press, New York.

Allesina, S. & Bodini, A. (2004) Who dominates whom in the ecosystem? Energy flow bottlenecks and cascading extinctions. *Journal of Theoretical Biology*, **230**, 351–358.

Bakker, E.S., Olff, H., Vandenberghe, C., *et al.* (2004) Ecological anachronisms in the recruitment of temperate light-demanding tree species in wooded pastures. *Journal of Applied Ecology*, **41**, 571–582.

Baral, N. & Heinen, J.T. (2007) Decentralization and people's participation in conservation: a comparative study from the Western Terai of Nepal. *International Journal of Sustainable Development and World Ecology*, **14**, 520–531.

Barnett, R., Yamaguchi, N., Barnes, I. & Cooper, A. (2006) Lost populations and preserving genetic diversity in the lion Panthera leo: implications for its ex situ conservation. *Conservation Genetics*, **7**, 507–514.

Barnett, R., Yamaguchi, N., Shapiro, B. & Nijman, V. (2007) Using ancient DNA techniques to identify the origin of unprovenanced museum specimens, as illustrated by the identification of a 19th century lion from Amsterdam. *Contributions to Zoology*, **76**, 87–94.

Barnett, R., Shapiro, B., Barnes, I.A., *et al.* (2009) Phylogeography of lions (Panthera leo ssp.) reveals three distinct taxa and a late Pleistocene reduction in genetic diversity. *Molecular Ecology*, **18**, 1668–1677.

Barnosky, A.D., Koch, P.L., Feranec, R.S., Wing, S.L. & Shabel, A.B. (2004) Assessing the causes of Late Pleistocene extinctions on the continents. *Science*, **306**, 70–75.

Barnosky, A.D., Matzke, N., Tomiya, S., *et al.* 2011. Has the Earth's sixth mass extinction already arrived? *Nature*, **471**, 51–57.

Bar-Oz, G., Zeder, M. & Hole, F. (2011) Role of mass-kill hunting strategies in the extirpation of Persian gazelle (Gazella subgutturosa) in the northern Levant. *Proceedings of the National Academy of Sciences USA*, **108**, 7345–7350.

Bascompte, J. (2009) Disentangling the Web of Life. *Science*, **325**, 416–419.

Benayas, J.M., Newton, A.C., Diaz, A. & Bullock, J.M. (2009) Enhancement of biodiversity and ecosystem services by ecological restoration: a meta-analysis. *Science*, **325**, 1121–1124.

Berger, J. (2003) Is it acceptable to let a species go extinct in a national park? *Conservation Biology*, **17**, 1451–1454.

Bertola, L.D., van Hooft, W.F., Vrieling, K., *et al.* (2011) Genetic diversity, evolutionary history and implications for conservation of the lion (Panthera leo) in West and Central Africa. *Journal of Biogeography*, **38**, 1356–1367.

Beyer, H.L., Merrill, E.H., Varley, N. & Boyce, M.S. (2007) Willow on Yellowstone's northern range: evidence for a trophic cascade? *Ecological Applications*, **17**, 1563–1571.

Borrvall, C. & Ebenman, B. (2006) Early onset of secondary extinctions in ecological communities following the loss of top predators. *Ecology Letters*, **9**, 435–442.

Boyce, M., Rushton, S. & Lynam, T. (2007) Does modelling have a role in conservation? In: *Key Topics in Conservation Biology* (eds D.W. Macdonald & K. Service), pp. 134–144. Wiley-Blackwel, Oxford.

Bullock, J.M., Aronson, J., Newton, A.C., Pywell, R.F. & Rey-Benayas, J.M. (2011) Restoration of ecosystem services and biodiversity: conflicts and opportunities. *Trends in Ecology and Evolution*, **26**(10), 541–549.

Burger, J., Rosendahl, W., Loreille, O., *et al.* (2004) Molecular phylogeny of the extinct cave lion Panthera leo spelaea. *Molecular Phylogenetics and Evolution*, **30**, 841–849.

Burney, D.A. & Flannery, T.F. (2005) Fifty millennia of catastrophic extinctions after human contact. *Trends in Ecology and Evolution*, **20**, 395–401.

Butchart, S.H., Walpole, M., Collen, B., *et al.* (2010) Global biodiversity: indicators of recent declines. *Science*, **328**, 1164–1168.

Caro, T. & Sherman, P. (2009) Rewilding can cause rather than solve ecological problems. *Nature*, **462**, 985.

CBD (2010) *Revised and Updated Strategic Plan: Technical Rationale and Suggested Milestones and Indicators*. CBD, Montreal.

Choi, Y.D. (2007) Restoration ecology to the future: a call for new paradigm. *Restoration Ecology*, **15**, 351–353.

Clutton-Brock, T.H., Coulson, T. & Milner, J.M. (2004) Red deer stocks in the Highlands of Scotland. *Nature*, **429**, 261–262.

Constitution of the Republic of Namibia (1990) *Chapter 11: Principles of State Policy, Article 95, Promotion of the Welfare of the People*.

Corlett, R.T. (2010) Megafaunal extinctions and their consequences in the tropical Indo-Pacific. In: *Terra Australis 32: Altered Ecologies: Fire, Climate and Human Influence on Terrestrial Landscapes* (eds S.G. Haberle, J. Stevenson & M. Prebble), pp. 131–177. ANU E-Press, Canberra.

Costanza, R., d'Arge, R., de Groot, R., *et al.* (1997) The value of the world's ecosystem services and natural capital. *Nature*, **387**, 253–260.

Crowley, B.E. (2010) A refined chronology of prehistoric Madagascar and the demise of the megafauna. *Quaternary Science Reviews*, **29**, 2591–2603.

Cumming, D.H., Fenton, M.B., Rautenbach, I.L., *et al.* (1997) *Elephants, Woodlands and Biodiversity in Southern Africa*. Open Journals Publishing, Tygervalley, South Africa.

Curry, A. (2010) *Where the Wild Things Are*: http://daughternumberthree.blogspot.com/2010/2002/discover-magazine-march-2010.html

Curry, A. & Hodgson, A. (2008) Seeing in multiple horizons: connecting futures to strategy. *Journal of Futures Studies*, **13**, 1–20.

Davies-Mostert, H., Mills, M. & Macdonald, D. (2009) A critical assessment of South Africa's managed metapopulation recovery strategy for African wild dogs and its value as a template for large carnivore conservation elsewhere. In: *Reintroduction of Top-Order Predators* (eds M. Hayward & M. Somers), p.10. Wiley-Blackwell, Oxford.

Dobson, A.P., Bradshaw, A.D. & Baker, A.J. (1997) Hopes for the future: restoration ecology and conservation biology. *Science*, **277**, 515–522.

Donlan, J., Berger, J., Bock, C.E., *et al.* (2005) Re-wilding North America. *Nature*, **436**, 913–914.

Donlan, C. J., Berger, J., Bock, C.E., *et al.* (2006) Pleistocene rewilding: an optimistic agenda for twenty-first century conservation. *American Naturalist*, **168**, 660–681.

Dunne, J.A., Williams, R.J. & Martinez, N.D. (2002) Network structure and biodiversity loss in food webs: robustness increases with connectance. *Ecology Letters*, **5**, 558–567.

Ebenman, B. & Jonsson, T. (2005) Using community viability analysis to identify fragile systems and keystone species. *Trends in Ecology and Evolution*, **20**, 568–575.

Egan, D. & Howell, E.A. (2001) *The Historical Ecology Handbook: A Restorationist's Guide to Reference Ecosystems*. Island Press, Washington, D.C.

Elvin, M. (2006) *The Retreat of the Elephants: An Environmental History of China*. Yale University Press, New Haven, CT.

Estes, J.A., Terborgh, J., Brashares, J.S., *et al.* (2011) Trophic downgrading of planet Earth. *Science*, **333**, 301–306.

European Green Belt (2011) www.europeangreenbelt.org/005.database_gallery.maps.html

Fairbairn, A.S., Hope, G.S. & Summerhayes, G.R. (2006) Pleistocene occupation of New Guinea's highland and subalpine environments. *World Archaeology*, **38**, 371–386.

Flannery, T. (2002) *The Future Eaters: An Ecological History of the Australasian Lands and People*. Grove Press, New York.

Forestry Commission (2001) *National Inventory of Woodlands and Trees: Scotland – Highland Region*. Forestry Commission, Edinburgh.

Fraser, C. (2009) *Rewilding the World: Dispatches from the Conservation Revolution*. Henry Holt and Company, New York.

Fuller, T.K., Mech, L.D. & Cochrane, J.F. (2003) Wolf Population Dynamics. In: *Wolves: Behavior, Ecology, and Conservation* (eds L.D. Mech & L. Boitani). University of Chicago Press, Chicago, IL.

Galetti, M. (2004) Parks of the Pleistocene: recreating the Cerrado and the Pantanal with megafauna. *Natureza and Conservação*, **2**, 93–100.

Gillespie, R. (2008) Updating Martin's global extinction model. *Quaternary Science Reviews*, **27**, 2522–2529.

Gong, Y.L., Swaine, M.D. & Miller, H.G. (1991) Effects of fencing and ground preparation on natural regeneration of native pinewood over 12 years in Glen Tanar, Aberdeenshire. *Forestry*, **64**, 157–168.

Grayson, D.K. (2007) Deciphering North American Pleistocene extinctions. *Journal of Anthropological Research*, **63**, 185–213.

Grayson, D.K. & Meltzer, D.J. (2003) A requiem for North American overkill. *Journal of Archaeological Science*, **30**, 585–593.

Griffiths, C.J., Hansen, D.M., Jones, C.G., Zuël, N. & Harris, S. (2011) Resurrecting extinct interactions with extant substitutes. *Current Biology*, **21**, 762–765.

Griffiths, C.J. & Harris, S. (2010) Prevention of secondary extinctions through taxon substitution. *Conservation Biology*, **24**, 645–646.

Griffiths, C.J., Jones, C.G., Hansen, D.M., *et al.* (2010) The use of extant non-indigenous tortoises as a restoration tool to replace extinct ecosystem engineers. *Restoration Ecology*, **18**, 1–7.

Hansen, D.M. (2010) On the use of taxon substitutes in rewilding projects on islands. In: *Islands and Evolution*. (eds V. Pérez-Mellado & C. Ramon), pp.111–146. Institut Menorquí d'Estudis, Menorca.

Hansen, D.M. & Galetti, M. (2009) The forgotten megafauna. *Science*, **324**, 42–43.

Hansen, D.M., Kaiser, C.N. & Muller, C.B. (2008) Seed dispersal and establishment of endangered plants on Oceanic islands: the Janzen–Connell model, and the use of ecological analogues. *PLoS ONE*, **3**(5), e2111.

Hansen, D.M., Donlan, C.J., Griffiths, C.J. & Campbell, K.J. (2010) Ecological history and latent conservation potential: large and giant tortoises as a model for taxon substitutions. *Ecography*, **33**, 272–284.

Hastings, A., Byers J.E., Crooks, J.A., *et al.* (2007) Ecosystem engineering in space and time. *Ecology Letters*, **10**, 153–164.

Hobbs, R. (2009) Woodland restoration in Scotland: ecology, history, culture, economics, politics and change. *Journal of Environmental Management*, **90**, 2857–2865.

Hobbs, R.J. & Cramer, V.A. (2008) Restoration ecology: interventionist approaches for restoring and maintaining ecosystem function in the face of rapid environmental change. *Annual Review of Environment and Resources*, **33**, 39–61.

Hodder, K., Bullock, J., Buckland, P. & Kirby, K. (2005) *Large Herbivores in the Wildwood and Modern Naturalistic Grazing Systems*. English Nature, Peterborough.

Hoegh-Guldberg, O., Hughes, L., McIntyre, S., *et al.* (2008) Assisted colonization and rapid climate change. *Science*, **321**, 345–346.

Howells, O. & Edwards-Jones, G. (1997) A feasibility study of reintroducing wild boar Sus scrofa to Scotland: are existing woodlands large enough to support minimum viable populations. *Biological Conservation*, **81**, 77–89.

Janzen, D.H. (2000) Costa Rica's Area de Conservación Guanacaste: a long march to survival through non-damaging biodevelopment. *Biodiversity*, **1**, 7–20.

Janzen, D.H. & Martin, P.S. (1982) Neotropical anachronisms – the fruits the gomphotheres ate. *Science*, **215**, 19–27.

Johnson, C.N. (2009) Ecological consequences of Late Quaternary extinctions of megafauna.

Proceedings of the Royal Society B: Biological Sciences, **276**, 2509–2519.

Jonson, J. (2010) Ecological restoration of cleared agricultural land in Gondwana Link: lifting the bar at 'Peniup'. *Ecological Management and Restoration*, **11**, 16–26.

Jordán, F. (2009) Keystone species and food webs. *Philosophical Transactions of the Royal Society B: Biological Sciences*, **364**, 1733–1741.

Kaiser, J. (2001) Bold corridor project confronts political reality. *Science*, **293**, 2196–2199.

Kass, G., Shaw, R., Tew, T. & Macdonald, D.W. (2011) Securing the future of the natural environment: using scenarios to anticipate challenges to biodiversity, landscapes and public engagement with nature. *Journal of Applied Ecology*, **48**(6), 1518–1526.

Klink, C.A. & Machado, R.B. (2005) Conservation of the Brazilian Cerrado Conservación del Cerrado Brasileño. *Conservation Biology*, **19**, 707–713.

Koch, P.L. & Barnosky, A.D. (2006) Late quaternary extinctions: state of the debate. *Annual Review of Ecology Evolution and Systematics*, **37**, 215–250.

Long, J.L. (2003) *Introduced Mammals of the World: Their History, Distribution and Influence*. CABI Publishing, Wallingford.

Macdonald, D.W. (2009) Lessons learnt and plans laid: seven awkward questions for the future of reintroductions. In: *Reintroduction of Top-Order Predators* (eds M. Hayward & M. Somers). Wiley-Blackwell, Hoboken, NJ.

Macdonald, D.W. & Burnham, D. (2010) *The State of Britain's Mammals: A Focus on Invasive Species*. People's Trust for Endangered Species, London.

Macdonald, D.W., Mace, G. & Rushton, S. (2000) British mammals: is there a radical future? In: *Priorities for the Conservation of Mammalian Diversity: Has the Panda had its Day?* (eds A. Entwistle & N. Dunstone). Cambridge University Press, Cambridge.

Macdonald, D.W., Collins, N.M. & Wrangham, R. (2007) Principles, practice and priorities: the quest for 'alignment'. In: *Key Topics in Conservation Biology* (eds D.W. Macdonald & K. Service). Wiley-Blackwell, Oxford.

Macdonald, D.W., Loveridge, A.J. & Rabinowitz, A. (2010) Felid futures: crossing disciplines, borders, and generations. In: *The Biology and Conservation of Wild Felids* (eds D.W. Macdonald & A.J. Loveridge), pp. 599–649. Oxford University Press, Oxford.

Manning, A.D., Lindenmayer, D.B. & Fischer, J. (2006) Stretch goals and backcasting: approaches for overcoming barriers to large-scale ecological restoration. *Restoration Ecology*, **14**, 487–492.

Manning, A.D., Gordon, I.J. & Ripple, W.J. (2009) Restoring landscapes of fear with wolves in the Scottish Highlands. *Biological Conservation*, **142**, 2314–2321.

Martin, P.S. (1967) Pleistocene overkill. In: *Pleistocene Extinctions: The Search for a Cause* (eds P.S. Martin and H.E. Wright). Yale University Press, New Haven, CT.

Martin, P.S. & Greene, H.W. (2005) *Twilight of the Mammoths – Ice Age Extinctions and the Rewilding of America*. California University Press, Berkeley, CA.

Millennium Ecosystem Assessment (2005) *Ecosystems and Human Well-being: Biodiversity Synthesis*: www.millenniumassessment.org/en/Synthesis.html

Miller, J.R. & Hobbs, R.J. (2007) Habitat restoration – do we know what we're doing? *Restoration Ecology*, **15**, 382–390.

Morueta-Holme, N., Fløjgaard, C. & Svenning, J.C. (2010) Climate change risks and conservation implications for a threatened small-range mammal species. *PLoS ONE*, **5**, e10360.

Myers, N. (2003) Conservation of biodiversity: how are we doing? *Environmentalist*, **23**, 9–15.

Naidoo, R., Balmford, A., Costanza, R., *et al.* (2008) Global mapping of ecosystem services and conservation priorities. *Proceedings of the National Academy of Sciences USA*, **105**, 9495–9500.

Newmark, W.D. (1995) Extinction of mammal populations in Western North-American national-parks. *Conservation Biology*, **9**, 512–526.

Nicholls, H. (2006) Restoring nature's backbone. *PLoS Biology*, **4**, e202.

Nilsen, E.B., Milner-Gulland, E.J., Schofield, L., Mysterud, A., Stenseth, N.C. & Coulson, T. (2007) Wolf reintroduction to Scotland: public attitudes and consequences for red deer management. *Proceedings of the Royal Society B: Biological Sciences*, **274**, 995–1003.

Olff, H., Vera F.W., Bokdam J., *et al.* (1999) Shifting mosaics in grazed woodlands driven by the alternation of plant facilitation and competition. *Plant Biology*, **1**, 127–137.

Oliveira-Santos, L.G. & Fernandez, F.A. (2010) Pleistocene rewilding, Frankenstein ecosystems, and an alternative conservation agenda. *Conservation Biology*, **24**, 4–5.

Paine, R.T. (1969) A note on trophic complexity and community stability. *American Naturalist*, **103**, 91–93.

Palmer, S. C. & Truscott, A.M. (2003) Browsing by deer on naturally regenerating Scots pine (*Pines sylvestris L.*) and its effects on sapling growth. *Forest Ecology and Management*, **182**, 31–47.

Peace Parks Foundation (2011) *Great Limpopo TFCA*: www.peaceparks.org/

Quammen, D. (2004) *Monsters of God: The Man-Eating Predator in the Jungles of History and the Mind*. Hutchinson, London.

Räikkönen, J., Vucetich, J.A., Peterson, R.O. & Nelson, M.P. (2009) Congenital bone deformities and the inbred wolves (*Canis lupus*) of Isle Royale. *Biological Conservation*, **142**, 1025–1031.

Redford, K. & Sanjayan, M.A. (2003) Retiring Cassandra. *Conservation Biology*, **17**, 1473–1474.

Rewilding Europe Initiative (2011) *Rewilding Europe Initiative*. http://rewildingeurope.com/about-us/wild-europe-initiative/

Ricciardi, A. & Simberloff, D. (2009) Assisted colonization is not a viable conservation strategy. *Trends in Ecology and Evolution*, **24**, 248–253.

Richmond, O.M., McEntee, J.P., Hijmans, R.J. & Brashares, J.S. (2010) Is the climate right for Pleistocene rewilding? Using species distribution models to extrapolate climatic suitability for mammals across continents. *PLoS ONE*, **5**(9), e12899.

Ripple, W.J. & Beschta, R.L. (2006) Linking a cougar decline, trophic cascade, and catastrophic regime shift in Zion National Park. *Biological Conservation*, **133**, 397–408.

Ripple, WJ. & Beschta, R.L. (2007) Restoring Yellowstone's aspen with wolves. *Biological Conservation*, **138**, 514–519.

Rubenstein, D.R., Rubenstein, D.I., Sherman, P.W. & Gavin, T.A. (2006) Pleistocene park: does re-wilding North America represent sound conservation for the 21st century? *Biological Conservation*, **132**, 232–238.

Russell, M.L. (2004) Wild horses: legends or burdens on our rangelands? *Rangelands*, **26**, 40–42.

Sandom, C.J., Bull, J., Canney, S. & Macdonald, D.W. (2011) Exploring the value of wolves (*Canis lupus*) in landscape-scale fenced reserves for ecological restoration in the Scottish Highlands. In: *Fencing for Conservation: Restriction of Evolutionary Potential or a Riposte to Threatening Processes?* (eds M. Somers & M. Hayward). Springer, New York.

Sandom, C.J., Hughes, J. & Macdonald, D.W. (2012) Rooting for Rewilding: Quantifying Wild Boar's Sus scrofa Rooting Rate in the Scottish Highlands. Restoration Ecology.

Sandom, C.J., Hughes, J. & Macdonald, D.W. (forthcoming a) *Wild Boar Habitat Preference and Foraging Strategy in a Scottish Highland Rewilding Project*. WildCRU, Oxford.

Shine, C., Williams, N. & Gündling, L. (2000) *A Guide to Designing Legal and Institutional Frameworks on Alien Invasive Species*. IUCN, Gland, Switzerland.

Slotow, R., Garai, M., Reilly, B., Page, B. & Carr, R. (2005) Population dynamics of elephants re-introduced to small fenced reserves in South Africa: research article. *South African Journal of Wildlife Research*, **35**, 23–32.

Smith, D.W., Peterson, R.O. & Houston, D.B. (2003) Yellowstone after wolves. *BioScience*, **53**, 330–340.

Smout, T.C., MacDonald, A.R. & Watson, F. (2008) *A History of the Native Woodlands of Scotland, 1500–1920*. Edinburgh University Press, Edinburgh.

Soorae, P.S. (ed.) (2008) *Global Re-introduction Perspectives: Re-Introduction Case-Studies from Around the Globe*. IUCN/SSC Re-introduction Specialist Group, Abu Dhabi, UAE.

Soorae, P.S. (ed.) (2010) *Global Re-introduction Perspectives: Additional Case-Studies from Around the Globe*. IUCN/SSC Re-introduction Specialist Group, Abu Dhabi, UAE.

Soulé, M.E. & Noss, R.F. (1998) Rewilding and biodiversity: complementary goals for continental conservation. *Wild Earth*, **Fall**, 22.

Soulé, M.E. & Terborgh, J. (eds) (1999) *Continental Conservation: Scientific Foundations of Regional Reserve Networks*. Island Press, Washington, D.C.

Soulé, M.E., Bolger D.T., Allison, C.A., Wright, J., Sorice, M. & Hill, S. (1988) Reconstructed dynamics of rapid extinctions of chaparral-requiring birds in urban habitat islands. *Conservation Biology*, **2**, 75–92.

Stone, R. (1998) A bold plan to re-create a long-lost siberian ecosystem. *Science*, **282**, 31–34.

Sukhdev, P., Wittmer, H., Schroter-Schlaack, C., *et al.* (2010) *The Economics of Ecosystems and Biodiversity. Mainstreaming the Economics of Nature: A Synthesis of the Approach, Conclusions and Recommendations of TEEB*. TEEB, Bonn.

Svenning, J.C. (2002). A review of natural vegetation openness in north-western Europe. *Biological Conservation*, **104**, 133–148.

Svenning, J.C. (2007) Plesitocene re-wilding' merits serious consideration also outside North America. *IBS Newsletter*, **5**, 3–9.

Terborgh, J. & Estes, J.A. (eds) (2010) *Trophic Cascades: Predators, Prey, and Changing Dynamics of Nature*. Island Press, Washington, D.C.

Terborgh, J. & Winter, B. (1980) Some causes of extinction. In: *Conservation Biology: An Evolutionary-Ecological Perspective*. (eds M.E. Soulé & B.A. Wilcox). Sinauer Associates Inc, Sunderland, MA.

Terry, A., Ullrich, K. & Riecken, U. (2006) *The Green Belt of Europe From Vision to Reality*. IUCN, Gland, Switzerland.

Van Aarde, R.J. & Jackson, T.P. (2007) Megaparks for metapopulations: addressing the causes of locally high elephant numbers in southern Africa. *Biological Conservation*, **134**, 289–297.

Van Aarde, R., Whyte, I. & Pimm, S. (1999) Culling and the dynamics of the Kruger National Park African elephant population. *Animal Conservation*, **2**, 287–294.

Vera, F. (2000) *Grazing Ecology and Forest History*. CABI, Wallingford.

Vera, F. (2009) Large-scale nature development – the Oostvaardersplassen. *British Wildlife*, **20**, 28–36.

Van Vuure, C. (2005) *Retracing the Aurochs: History, Morphology, and Ecology of an Extinct Wild Ox*. Pensoft Publishers, Bulgaria.

Wallis de Vries, M.F. (1995) Large herbivores and the design of large-scale nature reserves in Western Europe. *Conservation Biology*, **9**, 25–33.

Warren, C.R. (2009) *Managing Scotland's Environment*. Edinburgh University Press, Edinburgh.

White, P.J. & Garrott, R.A. (2005) Yellowstone's ungulates after wolves – expectations, realizations, and predictions. *Biological Conservation*, **125**, 141–152.

Wolmer, W. (2003) Transboundary conservation: the politics of ecological integrity in the Great Limpopo Transfrontier Park*. *Journal of Southern African Studies*, **29**, 261–278.

Yalden, D.W. & Barrett, P. (1999) *The History of British Mammals*. T & A.D. Poyser, London.

Yellowstone to Yukon (2011) *Yellowstone to Yukon Conservation Initiative*. www.y2y.net/home.aspx

Young, T.P. (2000) Restoration ecology and conservation biology. *Biological Conservation*, **92**, 73–83.

Zimov, S.A. (2005) Pleistocene park: return of the mammoth's ecosystem. *Science*, **308**, 796–798.

24

Disease control

Peter D. Walsh

Department of Archaeology and Anthropology, University of Cambridge, Cambridge, UK

...the only thing we have to fear is...fear itself — nameless, unreasoning, unjustified terror which paralyzes needed efforts to convert retreat into advance.

Franklin Delano Roosevelt

Introduction

Infectious disease is a growing threat to wildlife, particularly to species already endangered by habitat loss or overhunting. The global explosion of human abundance and resource use, as well as the globalization of movement by humans and other invasive species, have fundamentally altered wildlife disease dynamics around the world, increasing rates at which pathogens endemic in other hosts are transmitted to endangered species ('spillover'), introducing new hosts and pathogens, and reducing and fragmenting wildlife populations to the point where disease can more easily deliver a knock-out punch (Smith et al. 2006, 2009).

The resulting disease threat is taxonomically diverse in terms of both the pathogens involved and the wildlife hosts affected. Viruses are major players: canine distemper, rabies, herpes and Ebola viruses cause virulent infections that have been implicated in die-offs of endangered species such as black footed ferrets (Carpenter et al. 1976), Ethiopian wolves (Haydon et al. 2006), African wild dogs (van de Bildt et al. 2002), sea turtles (Chaloupka et al. 2009) and African apes (Walsh et al. 2003). Bacteria such as *Bacillus* (anthrax), *Yersinia* (plague) and *Pneumoniae* have also caused high mortality outbreaks in endangered chimpanzees (Leendertz et al. 2004), black-footed ferrets (Rocke et al. 2008) and bog turtles (Brenner et al. 2002). Chytrid fungi are thought to be a prime mover in global amphibian decline (Pounds et al. 2006). Chronically infectious blood parasites such as malaria and *Trichomonas*, gut parasites such as *Strongyloides*, *Campylobacter* and *Escherichia*, and viruses such as feline leukaemia and simian immunodeficiency tend not to cause synchronous die-offs that grab public attention, but their long-term impact on the survival and reproduction of endangered species such as the Madagascar pink pigeon (Swinnerton et al. 2005), Iberian lynx (Lopez et al. 2009) and chimpanzees (Kaur et al. 2011) may be equally severe. And, in a truly bizarre case, the transfer of facial tumour disease-infected cells during aggressive encounters is rapidly exterminating Tasmanian devils (Hamede et al. 2009).

Key Topics in Conservation Biology 2, First Edition. Edited by David W. Macdonald and Katherine J. Willis.

The options for controlling these disease threats are many, ranging from maximally invasive approaches such as culling, translocation and quarantine to less invasive medical interventions like birth control and treatment, to environmental manipulations that reduce rates of contact between hosts. However, in this chapter I focus exclusively on another disease control option: vaccination. Why? One reason is that several recent reviews have expertly and exhaustively covered these options (Laurenson et al. 2005). Another is that since 1966, when Jane Goodall gave polio vaccine-spiked bananas to her study chimpanzees (Goodall 1986), endangered wildlife vaccination has been the subject of fierce controversy. Some opponents object to vaccination because it upsets the 'natural balance' while others have concerns about safety, cost or feasibility that are discussed later in this chapter. The consequence of this opposition is that vaccination has not attained broad acceptance as a tool for endangered species conservation.

Several recent developments undercut some of the arguments made by vaccination opponents. First, the situation has now become so dire for many endangered species that tipping the natural balance a bit no longer seems so terrible compared to the likely outcome of not intervening: extinction. Second, the biotech revolution has greatly increased the safety and availability of vaccines. Consequently, there are a growing number of examples suggesting that endangered species can be vaccinated safely and effectively. Third, new 'smart' vaccination strategies that exploit the structure of disease transmission networks have the potential to substantially reduce vaccination coverage rates and, therefore, the cost of disease control. The use of barrier vaccine to control a 2003–4 rabies epidemic in Ethiopian wolves is an already classic example (Haydon et al. 2006).

Although these new developments in vaccine technology and epidemiology have already reshaped the thinking of many disease control experts, they have yet fully to convince the broader conservation community. Therefore, the first part of this chapter clarifies the issues on vaccine safety. Next, I describe smart vaccination strategies that exploit the structure of social contact networks to improve disease control efficiency and lower cost. Finally, I present a case study on African apes, which are increasingly threatened by disease and have a demography, social structure and enduring charisma that make them compelling 'poster children' for illustrating both the general benefits of vaccination and the particular benefits of smart vaccination.

Vaccine safety

Attitudes about the safety and effectiveness of vaccinating endangered species have been strongly coloured by vaccination efforts in carnivores. In the early 1970s, four individuals taken into captivity from the last known wild population of black-footed ferrets died after being vaccinated with a canine distemper vaccine that had previously been used safely on domestic ferrets (Carpenter et al. 1976). Shortly thereafter, vaccine-induced distemper fatalities occurred in another critically endangered species taken into captivity: the African wild dog (McCormick 1983). There have also been widely reported cases of vaccine failure. For example, African wild dogs have died when vaccination did not provide protection against natural infection during outbreaks of both rabies (Woodroffe 2001) and canine distemper (van de Bildt et al. 2002). To further muddy the waters, claims were made that stress from immobilization, vaccination or radio-collaring had immunosuppressed African wild dogs in the Serengeti, allowing rabies infection to then kill them (Burrows 1992; Burrows et al. 1994). Although careful analyses have thoroughly debunked these claims (Macdonald et al. 1992; de Villiers et al. 1995; Woodroffe 2001), they have contributed greatly to a negative perception of medical intervention amongst many in the broader endangered species conservation community.

So is the early experience in endangered carnivores indicative of wider trends in endangered species vaccination? Have all vaccines proven to be equally unsafe or ineffective? Accurately answering these questions requires a little background on vaccine technology. Until the 1980s there were only two widely used technologies: inactivated vaccines and live attenuated vaccines. Inactivated or 'killed' vaccines are non-infectious agents produced by exposing the target pathogen to chemicals, heat, radiation or antibiotics. Examples of inactivated vaccines used on wildlife include the rabies vaccine used to protect Ethiopian wolves during the 2003–4 outbreak (Knobel et al. 2008) and foot and mouth vaccines given regularly to captive and reintroduced Arabian oryx (Kilgallon et al. 2008), which like California condors and black-footed ferrets were once extinct in the wild. The canine distemper vaccine given to black-footed ferrets and African wild dogs was also a live attenuated vaccine. In addition, 65 critically endangered mountain gorillas in Rwanda and the Democratic Republic of Congo were safely vaccinated with live attenuated vaccine during a fatal measles outbreak (Hasting et al. 1991). Subsequent vaccination of three endangered ungulates, the Arabian oryx, sand gazelle and hirola antelope, with live attenuated rinderpest vaccine (Zafar-ul Islam et al. 2010; Butynski 2000) was part of the recently successful global eradication campaign, joining smallpox vaccination as the only campaigns to eradicate a viral disease entirely (Morens et al. 2011).

Modern genetic engineering methods have now introduced several new types of vaccines (Babiuk 1999). Recombinant vectored vaccines are made by inserting a small amount of material from the target pathogen into a non-virulent vector. The objective is a recombinant that presents the antigen 'signature' of the virulent pathogen to the host immune system but causes only the mild infection typical of the vector. A canary pox-vectored vaccine that had protected domestic cats against feline leukaemia in captive trials was recently given to Iberian lynx during an outbreak that reached an estimated 29% prevalence in the largest remaining subpopulation of this critically endangered species (Lopez et al. 2009). A different canary pox-vectored vaccine against canine distemper also proved immunogenic in endangered Channel Island foxes (Coonan et al. 2005), and a raccoon pox-vectored, orally delivered vaccine conferred protective immunity to plague challenge in captive trials on black-tailed prairie dogs, the main prey of and infection reservoir for critically endangered black-footed ferrets (Mencher et al. 2004).

Subunit and DNA vaccines are two more recent developments that include enough viral information about antigen protein structure for the host immune system subsequently to recognize the pathogen but not the additional machinery necessary for cell infection or replication. In subunit vaccines, this is achieved through insertion of protein coding sequences into the genome of a host cell line. Protein produced by the host cell is then purified and used as a vaccine. For example, a new hepatitis B vaccine uses envelope protein grown in yeast cells. In virus-like particle (VLP) vaccines, often grown in insect cells, viral envelope or capsid proteins self-assemble into particles that are structurally similar to real viral capsids but lack the internal genetic material (Noad & Roy 2003). DNA vaccines include not the antigenic protein itself but only a DNA sequence encoding the protein, often delivered in a bacterial plasmid that inserts itself into the host nucleus which then replicates the gene sequences carried by the plasmid (Donnelly et al. 2005).

An attractive feature of subunit and DNA vaccines is that they evade the pre-existing immunity problems that can occur when previous exposure to a vaccine vector or a cross-reactive pathogen allows the host immune system to neutralize a vaccine before it is effective. The lack of pre-existing immunity problems also means that, unlike many vectored vaccines, a single subunit or DNA vaccine platform could be used as a vehicle for developing vaccines against numerous pathogens.

Although these vaccines have yet to be used on endangered species in the wild, captive trials on several endangered species have shown high immunogenicity, including trials of a hepatitis B vaccine in orang utans (Davis et al. 2000). More importantly, DNA vaccines have been protective in wild challenge of California condors by West Nile virus (Chang et al. 2007) and captive challenge of black-footed ferrets by sylvatic plague (Rocke et al. 2008).

The safety profiles of these different types of vaccines correlate with their degree of infectiousness. Because they lack potential for uncontrolled infection, back-mutation to virulence or infection of non-target species, inactivated and subunit vaccines are usually considered to be very safe. The same benefits apply to DNA vaccines, although there are concerns about a range of negative effects that might result if the bacterial plasmids used to carry pathogen DNA into the host nucleus incorporate themselves into the host nuclear genome (Nichols et al. 1995). On the other hand, live attenuated and vectored vaccines are more often cited as potentially dangerous, particularly because of the possibility for uncontrolled infection. This might occur if an endangered species had some critical immunological difference from the species in which the vaccine was safety tested, as was the case in black-footed ferrets (Wimsatt et al. 2006), or because the immune systems of chronically immune-stressed wild animals are less robust than the pampered captive animals used in vaccine trials. For instance, simian immunodeficiency has recently been shown to be highly virulent in wild chimpanzees even though years of research on captive chimpanzees had suggested very low virulence (Keele et al. 2009). That immunological competence may be at play in African wild dogs is suggested by the fact that all cases of vaccine-induced canine distemper have been in pups, not adults, and that vaccination has caused fatal infection at some captive facilities but not others (Woodroffe & Ginsburg 1997). When baits are used to deliver oral vaccine to wildlife species, there is also the potential to infect non-target species in which the vaccine has not been safety tested. Both the creation of new 'superviruses' through the recombination of vectored vaccines with co-infecting pathogens and back-mutation of replication-defective vaccines to replication competence and virulence are also matters of concern.

Why are live vaccines so commonly used if they entail greater safety risks? One answer is that, by entering cells, live vaccines induce the cell-mediated immunity critical to controlling some pathogens. Another is that replication within the host results in a large concentration of circulating pathogen and, therefore, stimulates immune responses which are more robust and long-lived than those typically induced by non-infectious vaccines. This makes live vaccines particularly attractive for wildlife applications, where it may be difficult to achieve the multiple doses sometimes required for inactivated, subunit or DNA vaccines to reach full potency. A study showing that multiple doses of inactivated vaccine protected South African wild dogs against rabies suggests that one reason vaccination failed to protect wild dogs in Tanzania was that only one dose was applied (Hofmeyr et al. 2004). Results from the Ethiopian wolf vaccination programme showed that animals that received a booster vaccination were more likely to be seropositive to rabies 6–12 months after vaccination than animals that did not, although sample sizes were small (Knobel et al. 2008). Current oral vaccination programmes rely primarily on live vaccines precisely because the amount of vaccine crossing the mucosal barrier of the mouth, throat or gastrointestinal tract is typically too small to produce a robust immune response without the amplifying effect of infection.

As advances in vaccine technology accumulate, prospects for safe and effective endangered species vaccination will only improve. New adjuvants – compounds that boost the immune response to vaccines (Dubensky et al. 2000; Pashine et al. 2005) – promise to improve vaccine immunogenicity, particularly for subunit and DNA vaccines (e.g. Chen et al. 2000; Nguyen et al. 2009; Dubensky et al. 2000). New subunit vaccine formulations using microparticles and nanoparticles (Singh et al. 2007), microemulsions (Jadhav et al. 2006) or cross-linked crystals (St Clair et al. 1999) also offer the potential for timed release of vaccine. This may not only boost immunogenicity but also obviate the need for multiple vaccine doses. Linking of adjuvants directly to vaccine antigens may also ensure that both arrive at target cells in optimal concentrations.

Another promising area is the use of non-invasive faecal, urinary or salivary assays in monitoring the health state of wild species from which it is not feasible to draw blood. Non-invasive assays of faecal antibodies have, thus far, been used only in disease diagnosis of pathogens such as simian immunodeficiency virus (SIV), malaria and human respiratory viruses (Keele et al. 2009; Liu et al. 2010; Koendgen et al. 2010). However, they have excellent potential to assay whether vaccination has induced humoral immunity or whether antibody titres have waned to the point where a booster is necessary. Similar potential for non-invasive assay of vaccine immunogenicity lies in the polymerase chain reaction (PCR) amplification of pathogen DNA or RNA, which are often excreted in faeces, urine or saliva during the postvaccination infection caused by live viruses.

New delivery technologies also promise to further improve vaccine safety. Darting technologies such as the Biobullet, a biodegradable plastic pellet containing lyopholized (freeze-dried) vaccine, have been used extensively on domestic and wild hoofed stock in the United States (Jessup et al. 1992; Aune et al. 2002). The Biobullet reduces the potential for vital organ puncture and can be delivered from 50–75 m.

Recent years have also seen great progress in oral vaccination technology (Cross et al. 2007; Simerska et al. 2009). In the many species that are not easily darted, oral vaccination has the potential to vaccinate enough animals to produce herd immunity, i.e. protecting even unvaccinated animals by breaking the transmission chain. A prime example is the virtual eradication of fox rabies from western Europe using oral, *Vaccinia* virus-vectored vaccine (Brochier et al. 1991). New recombinant technologies are now producing live attenuated vaccines that are even less prone to reversion to virulence and, therefore, safer (Lauring et al. 2010). Research on oral delivery of subunit vaccines that do not pose an infection risk to either healthy or immunosuppressed individuals is also progressing rapidly (e.g. Takamura et al. 2004; Connor et al. 1996; Streatfield 2006). These advances may reduce disease risk in both target and non-target species. Strides are also being made on environmentally stable oral vaccine formulations as well as baits that are readily and selectively taken by target hosts (Ballesteros et al. 2007). Vaccine adjuvants and preservatives such as thimerosal that are known to cause health complications are being phased out in favour of more benign alternatives (Fombonne 2008; Harandi et al. 2009; Chatterjee & O'Keefe 2010).

Finally, future years are likely to see an increasing role for 'transmissible vaccines' (infectious agents that spread between hosts), which offer the potential to vaccinate large populations at low cost. Good examples are the transmissible vaccines under development for the control of myxomatosis and rabbit haemorrhagic disease in rabbits (Angulo & Bárcena 2007). Transmissible vaccines are controversial because of the potential for reversion or recombination to virulence and spillover into non-target hosts (Heinsbroek & Ruitenberg 2010). However, that transmissible vaccines can be used safely under some circumstances is evident in that a transmissible vaccine is currently the primary weapon in the global fight against polio. Research is advancing rapidly on methods for making transmissible vaccines safer. One approach being pioneered by HIV researchers, therapeutic interfering particles, engineers a pathogen truncated to reduce virulence and to be transmissible only in the presence of wild-type pathogen (Metzger et al. 2011).

Networks and disease control efficiency

In *in situ* wildlife vaccination programmes, the objective is typically not just to protect each individual vaccinated but to vaccinate enough individuals to prevent long chains of transmission and, thereby, to suppress or eradicate endemic diseases or to prevent the amplification of spillover epizootics. The critical proportion of the population necessary to achieve such 'herd immunity' has historically been estimated using epidemiological models based on the mass action assumption, i.e. that contacts between susceptible and infected individuals are randomly distributed within the population (Anderson & May 1979). However, it has become obvious in recent years that networks of disease transmission in wildlife populations tend not to be random but highly structured (Keeling & Eames 2005; Craft & Caillaud 2011). This is good news because 'smart' vaccination strategies that exploit network structure have the potential to achieve herd immunity at substantially lower coverage than the roughly 70% typically predicted by mass action models.

Types of network heterogeneity

At least three forms of transmission network heterogeneity can be exploited. Perhaps the best recognized is the clustering of contacts, which in wildlife populations is often spatial. In species with exclusive territory defence, for example, the vast majority of disease-transmitting contacts may be between nearest neighbours. The tendency for epizootics to spread in coherent spatial waves in such systems presents the opportunity to greatly improve efficiency. Vaccinating a relatively small number of animals along a movement corridor separating subpopulations can block epizootic spread. A textbook example was the 2004–5 use of barrier vaccination to block the spread of rabies from the valley holding the second largest remaining Ethiopian wolf population to the valley holding the largest. Simulation models suggest that barrier vaccination effectively controlled the outbreak with a 37% coverage rate (Haydon et al. 2006).

The spatial clustering of contacts also brings the issue of vaccination efficiency in many endangered species firmly under the umbrella of percolation theory, which deals with systems in which interactions are well represented as occurring only between nearest neighbour nodes on a lattice. The central insight of percolation theory is that the probability that a chain of interactions will span the lattice does not increase smoothly with the proportion of 'occupied' nodes. Rather, at some critical threshold occupancy, an abrupt phase change takes the system from a set of small, unconnected clusters to one giant, highly connected component. In the disease context, this means that outbreak size does not increase smoothly with connectivity but undergoes an abrupt transition from only a few infected individuals to a population-wide epizootic (Meyers 2007). A nice example is the spread of plague between neighbouring systems of gerbil burrows in Kazakhstan (Davis et al. 2008), where large outbreaks only occur when the occupancy rate for burrow systems passes the critical percolation threshold. Percolation theory has implications not just for species that are strictly territorial but for any species in which individuals or social groups occupy a persistent home range whose width does not approach the scale of the entire population. Although percolation has yet to be applied to endangered species vaccination, it has aroused intense interest in epidemiological modelling circles because achieving herd immunity by pushing population connectivity below the percolation threshold may require substantially less vaccination coverage than that predicted by mass action models (Sander et al. 2002; Kenah & Miller 2011).

An important constraint on the applicability of barrier and percolation vaccination strategies to endangered species is the tendency for otherwise sedentary species to occasionally make long-distance movements during a dispersive life-history phase, as a side-effect of disease, or simply in search of food, water or mates. Such long-distance movements embue networks with a tendency towards more explosive outbreak dynamics than might be predicted from normal sedentary ranging patterns. For example, the tendency for rabies-infected animals to go 'walkabout' (Hampson et al. 2007) is an important factor in evaluating whether vaccinating domestic dogs can prevent rabies spillover into wildlife (Lembo et al. 2010). Similarly, canine distemper outbreaks in Serengeti lions appear to reach large size because of occasional long-distance forays by pride members (Craft et al. 2011) while the spread of rabies in the north eastern United States has been accelerated by infected raccoons 'hitchhiking' on garbage trucks (Smith et al. 2002).

A second important form of network heterogeneity is contact rate variance, often discussed in terms of the shape of the network degree distribution: the frequency distribution for the number of contacts per individual. Networks with long-tailed degree distributions in which a few 'superspreaders' have a disproportionate number of contacts tend to show more explosive disease dynamics than random networks (e.g. Porphyre et al. 2008) because superspreader infection can quickly propagate disease throughout the system (Lloyd-Smith et al. 2005). Superspreading can stem from either host behaviour or host immunological status, which affects both the duration of infection and the infectiousness of shed body fluids. Superspreading has sparked excitement because the targeted vaccination of superspreaders holds the potential for highly efficient disease control (Galvani & May 2005). The biggest challenge in adapting the concept for disease control in endangered species may be identifying *a priori* who the superspreaders are. One possible strategy in this respect is the preferential targeting of easily identifiable age/sex classes which are less immunologically competent, particularly social or disproportionately prone to behaviours such as play fighting or copulation that carry a high disease transmission risk (Bolzoni et al. 2007; Cross et al. 2009). One option that has not been thoroughly explored in the wildlife context is vaccination at activity hot spots such as water holes, mineral licks or fruiting trees (Scoglio et al. 2010).

A third important property is network modularity: the tendency for networks to be partitioned into distinct communities within which interaction is high but between which interaction is low (Newman 2006). Modularity can involve social groups of related individuals, which in species such as humans and elephants can be nested through several hierarchical levels from the nuclear family up through the clan (Wittemyer et al. 2005). Modularity can also be induced by fidelity of unrelated individuals to rare but high-value resources such as water holes or nesting islands. Modularity is important from a disease control perspective in that vaccinating individuals who bridge communities may be a highly efficient means of achieving herd immunity (Salathé & Jones 2010). In wildlife, bridge individuals might be identifiable through the position of their core activity areas on the margin between major activity hot spots such as water holes or by their age, if natal dispersal between communities occurs at a characteristic age. Even if bridge individuals cannot be identified, modularity could be useful if persistent social groups or communities defined by resource hot spots can be identified. In this case, levels of vaccination that might be ineffective at inducing herd immunity when spread across many modules may be highly effective if concentrated within a subset of modules. In fact, in the common case in which interactions between modules are spatially clustered, paying attention to modularity might even enhance the effectiveness of barrier vaccination.

Estimating network structure

There are several methods for collecting the network structure data necessary to implement smart vaccination. The first is direct observation.

One can simply observe the behaviour of animals that either have individually identifiable morphological characteristics or have been marked with individual identifiers (e.g. Stoinski et al. 2003; Sundaresan et al. 2007). For easily transmissible pathogens, it may be adequate to record social group membership. For diseases that require direct physical contact for transmission (e.g. sexual or blood-borne), it may also be necessary to record potentially disease-transmitting social contacts. This may work well for species that live in open habitat, have small home ranges, are diurnally active and/or interact frequently but not so well for species that do not meet one or more of these criteria. For example, 1294 hours of observations on Serengeti lions yielded only 36 interactions between social groups (Craft et al. 2011).

Another well-developed method that has recently been adapted to estimating disease network structure is mark-capture-recapture analysis. For example, mark-capture-recapture analyses have been used to show that long-distance movements tend to mix chronic wasting disease over a large area (Lowe & Allendorf 2010) and that superspreading plays an important role in the transmission of tuberculosis amongst brush-tailed possums (*Trichosurus vulpecula*) (Porphyre et al. 2008).

Radiotelemetry is a promising alternative for estimating contact network structure in difficult-to-observe animals. GPS telemetry devices that record animal positions on a regular schedule have now been miniaturized to the point that they can be attached to a wide range of species. These positional data can be used to estimate rates of long-distance dispersal or movement between social groups. Movement data from radio-collared African buffalo (*Syncerus caffer*) suggest that the effect of drought on rates of movement between herds may have a critical effect on the transmission of tuberculosis (Cross et al. 2004, 2005). Proximity sensors can also record cases in which tagged individuals approach each within some specified distance (Salathé et al. 2010). For example, proximity sensor data have been used to show that European rabbit populations are highly modular (Marsh et al. 2011) and that superspreading plays a major role in the transmission of tuberculosis between badgers and cattle (Böhm et al. 2009) but not in the spread of facial tumour disease amongst Tasmanian devils (Hamede et al. 2009). The downsides of telemetry include its expense (e.g. GPS tags cost $1000–$3000) and that immobilizing animals for tag attachment can be logistically difficult and dangerous to the animal, although advances in veterinary methods have greatly attenuated immobilization risk (West et al. 2007). Studies on African wild dogs do not support previous claims that immobilization for telemetry collar attachment causes chronic stress, immunosuppression or consequent death from canine distemper virus (de Villiers et al. 1995).

A final option is the genetic reconstruction of contact networks. New non-invasive methods now make it possible to amplify both host DNA from hair and faeces and pathogen DNA or RNA from host faeces. And the laboratory equipment necessary for these analyses is now widely accessible and the per sample cost reasonable (at most, in the low tens of dollars) and continually falling. Host DNA can be used to estimate rates of long-distance movement either directly by parentage analyses or indirectly by estimating rates at which pair-wise genetic distances increase with geographic distance between sampling sites or through analyses of spatial structure in allele prevalence (Broquet & Petit 2009). Analyses of the rate of pair-wise DNA sequence divergence in space found only weak evidence of spatial structuring of movements in deer threatened by chronic wasting syndrome (Cullingham et al. 2010). New phylogenetic reconstruction methods can also provide several kinds of useful information about contact network structure (Archie et al. 2009). The phylogenetic affinities of pathogens infecting an endangered species provide information about the source of pathogen spillover. The clustering of both intestinal bacteria (Goldberg et al. 2007) and respiratory viruses (Koendgen et al. 2008) sampled from wild habituated chimpanzees firmly inside human virus clades pointed the finger strongly at virus spillover from humans.

Finally, an emerging use of molecular data is the application of phylogenetic methods to reconstruct fine-scaled structure of pathogen transmission networks. Coalescent methods have been used to identify clusters of HIV infection derived from a single superspreading event or associated with particularly geographic areas or socio-economic strata (Lewis et al. 2008; Bon et al. 2010). This approach may be particularly useful for designing endangered species vaccination programmes if it can be used to identify age/sex classes that act as superspreaders or landscape features that define different population modules (Real & Biek 2007) or which act as hot spots for transmission.

Disease control in African apes

The importance of endangered species vaccination was highlighted in 2002, with reports of large gorilla and chimpanzee die-offs close to and concurrent with human Ebola virus outbreaks along the border between Gabon and Republic of Congo (Walsh et al. 2003). At about that time, results on the first vaccines to successfully protect captive macaques from Ebola virus challenge were published (Sullivan et al. 2003), leading to the suggestion that these vaccines should also be used to protect wild gorillas and chimpanzees (Walsh et al. 2003). The avalanche of vitriol that ensued made it clear why Goodall's 1966 polio vaccination effort (Goodall 1986) and the Mountain Gorilla Project's 1988 measles vaccination project (Hasting et al. 1991) had not been followed by further programmes. A range of objections were voiced, including that Ebola impact had been exaggerated, that it would upset the 'natural balance' and that it would prevent the evolution of Ebola resistance. Safety concerns were also expressed, including anxiety about the negative health effects of vaccines and the potential for vaccine spillover into non-target species. Ape-specific concerns were also expressed, such as the danger of darting adult male 'silverback' gorillas and the catastrophic consequences for gorilla tourism if vaccination made habituated gorillas afraid of people. Some erroneously assumed that vaccination required immobilization (it does not). There were also doubts about the logistical feasibility of vaccination, both because of the difficulty of approaching gorillas and chimpanzees in the wild and because of the prevailing view that Ebola outbreaks emerge unpredictably in space and time. Finally, because wild apes are so hard to immobilize, it was assumed that it was impossible to document rigorously whether vaccination was successful in evoking immunity.

The 'not natural' argument is particularly unconvincing. Central Africa is awash with negative human impacts, from uncontrolled bushmeat hunting and deforestation to the unprecedented levels of disease spillover caused by exploding human populations. The natural balance is already upset. What really matters now is whether apes will survive without intervention. In order to evaluate critically the other objections in the light of the best science, a loosely structured volunteer organization named VaccinApe was established.

The first question was whether the population impact of Ebola was sufficiently great to warrant vaccination. Data showing Ebola mortality rates of 95% in about 600 individually known and monitored gorillas from two different populations. Larger scale nest survey data suggested that Ebola had killed thousands of gorillas and chimpanzees in surrounding areas (Caillaud et al. 2006; Bermejo et al. 2006). These data ultimately led to the 2007 upgrading of western gorillas to Critically Endangered on the IUCN's Red List of threatened species (Walsh et al. 2007). The patchy, 'all or none' pattern of mortality revealed by the surveys also called into question whether resistance to Ebola would evolve, given that the vast majority of survivors lived in non-outbreak areas and had likely never been exposed to the virulent Zaire strain of Ebola that caused the outbreaks.

Data on the spatial structure of both outbreaks and the Ebola virus genome also did not support the claim that emergence was unpredictable. Rather, Ebola Zaire appeared to have spread in a highly coherent wave across the region (Walsh et al. 2005; Biek et al. 2006) in a reservoir host that appears to be fruit bats (Leroy et al. 2005). Patterns of mortality observed in contiguous gorilla social groups at the Lossi outbreak site also implied that most gorillas are infected in secondary waves of gorilla infection rather than through 'massive spillover' from bats (Bermejo et al. 2006). Substantial transmission between gorilla social groups is also suggested by survey data showing density-dependent ape mortality (Walsh et al. 2009). These results implied that the efficiency of vaccination efforts might be greatly increased by targeting gorilla vaccination just ahead of spreading waves of infection.

In parallel to these data analyses, a long series of expert workshops and consultations explored which vaccine to use and how to deliver it. Two developments during this process had particular influence on deliberations. Because of the potential to vaccinate large numbers of wild apes, initial focus was on oral vaccination. Sugary baits for the oral delivery of vaccine were successfully tested on zoo chimpanzees and gorillas. However, field trials made it clear that finding baits that wild gorillas would readily eat was a difficult task. Expert advice also made it clear that the cost and technical challenges associated with formulating an environmentally stable oral vaccine, conducting safety trials in both apes and non-target species, and manufacturing large numbers of baits for broadcast would be substantial. Priorities also shifted when molecular and demographic analyses made it clear that respiratory disease spillover from humans was responsible for about half of deaths in chimpanzees and gorillas habituated to human approach for research and tourism (Koendgen et al. 2008; Williams et al. 2008).

These developments lead a shift to a more incremental strategy. Instead of ambitiously beginning with oral vaccination using a live vaccine, the team opted to demonstrate proof of concept by using a less controversial vaccine delivery method, darting, and a non-infectious vaccine, a VLP vaccine being developed for human use by Integrated Biotherapeutics with the support of the US National Institutes of Health. In early 2011, captive trials (without Ebola challenge) on six chimpanzees showed that the vaccine was safe and immunogenic. Despite decades of using chimpanzees as a model system for testing human vaccines, this was the first trial in which captive chimpanzees were used to test a vaccine destined for use on wild apes.

Because the idea of Ebola vaccination seemed to be alarming to managers, initial field vaccination efforts were targeted at a human respiratory virus. Measles was chosen because it had previously caused deaths in habituated gorillas and because measles vaccine has an exceptionally good safety record, both in thousands of chimpanzees and gorillas held in US captive facilities and in hundreds of millions of human children. Great pains were taken to design the study so that darting was as safe as possible for both gorillas and the darting team, for instance, using a blowpipe to reduce the tissue and psychological trauma caused by dart impact, following strict protocols on when and who to dart, and developing a detailed emergency contingency plan. Faecal samples were used to measure the effect of darting on gorilla immune response to vaccination and stress, and behavioral data to assess the effect on approachability by humans. In April and May 2011, the team vaccinated without incident 18 western gorillas in two habituated groups in Dzanga-Sangha National Park, Central African Republic. This was the first medical intervention of any kind in wild western gorillas, by far the more abundant of the two currently recognized gorilla species.

The near-term goal of VaccinApe is to use dart vaccination to immunize gorillas and chimpanzees at all habituation sites against both Ebola and high-risk respiratory pathogens for which safe and effective vaccines are now available. The longer term goal is to develop oral vaccines and delivery methods that can allow the vaccination of the much larger population of unhabituated apes.

Conclusion

Do the safety and effectiveness problems experienced during early attempts at carnivore vaccination reflect the inherent unsuitability of vaccination as a tool for endangered species conservation? No; severe problems have been the exception rather than the rule. Vaccines against a wide range of pathogens have now been safe and effective in many endangered species, ranging from California condors to Arabian oryx to orang utans. Advances in vaccine technology, vaccine delivery methods and smart vaccination strategies also promise to make vaccination an ever safer and more cost-effective conservation tool. To be sure, each case must be approached systematically. And there are still lingering challenges such as the ethical considerations that have limited many endangered species trials to evaluating vaccine safety and immunogenicity rather than protection against pathogen challenge. But these same limitations have not prevented the licensing and human use of a wide range of vaccines that are not ethical to test in humans. The damage done by the disease itself has been judged to outweigh the risks posed by the vaccine. This same logic should hold in endangered wildlife applications because vaccination could be a potent tool in the fight to prevent the extinction of many endangered species.

References

Anderson, R.M. & May, R.M. (1979) Population biology of infectious diseases: Part I. *Nature*, **280**, 361–367.

Angulo, E. & Bárcena, J. (2007) Towards a unique and transmissible vaccine against myxomatosis and rabbit haemorrhagic disease for rabbit populations. *Wildlife Research*, **34**, 567–577.

Archie, E.A., Luikart, G. & Ezenwa, V.O. (2009) Infecting epidemiology with genetics: a new frontier in disease ecology. *Trends in Ecology and Evolution*, **24**, 21–30.

Aune, K., Kreeger, T. & Roffe, T. (2002) Overview of delivery systems for the administration of vaccines to elk and bison of the Greater Yellowstone Area. In: *Brucellosis in Elk and Bison in the Greater Yellowstone Area* (ed. T. Kreeger), pp.66–79. Wyoming Game and Fish Department, Cheyenne, WY.

Babiuk, L.A. (1999) Broadening the approaches to developing more effective vaccines. *Vaccine*, **17**, 1587–1595.

Ballesteros, C., Pérez de la Lastra, J.M. & de la Fuente, J. (2007) Recent developments in oral bait vaccines for wildlife. *Recent Patents on Drug Delivery and Formulation*, **1**, 230–235.

Bermejo, M., Rodriguez-Teijeiro, J.D., Illera, G., Barroso, A., Vila, C. & Walsh, P.D. (2006) Ebola outbreak killed 5000 gorillas. *Science*, **314**, 1564.

Biek, R., Walsh, P.D., Leroy, E.M. & Real, L.A. (2006) Recent common ancestry of Ebola Zaire virus found in a bat reservoir. *PLoS Pathogens*, **2**(10), e90.

Böhm, M., Hutchings, M.R. & White, P.C. (2009) Contact networks in a wildlife–livestock host community: identifying high-risk individuals in the transmission of bovine TB among badgers and cattle. *PLoS ONE*, **4**(4), e5016.

Bolzoni, L., Real, L. & de Leo, G. (2007) Transmission heterogeneity and control strategies for infectious disease emergence. *PLoS ONE*, **2**(8), e747.

Bon, I., Ciccozzi, M., Zehender, G., *et al.* (2010) HIV-1 subtype C transmission network: the phylogenetic reconstruction strongly supports the epidemiological data. *Journal of Clinical Virology*, **48**, 212–214.

Brenner, D., Lewbart, G., Stebbins, M. & Herman, D.W. (2002) Health survey of wild and captive bog turtles (*Clemmys muhlenbergii*) in North Carolina and Virginia. *Journal of Zoo and Wildlife Medicine*, **33**, 311–316.

Brochier, B., Kieny, M.P., Costy F, *et al.* 1991. Large-scale eradication of rabies using recombinant vaccinia rabies vaccine. *Nature*, **354**, 520–522.

Broquet, T. & Petit, E.J. (2009) Molecular estimation of dispersal for ecology and population genetics. *Annual Review of Ecology and Evolutionary Systematics*, **40**, 193–216.

Burrows, R. (1992) Rabies in wild dogs. *Nature*, **359**, 277.

Burrows, R., Hofer, H. & East, I.M. (1994) Demography, extinction and intervention in a small population: the case of the Serengeti wild dogs. *Proceedings of the Royal Society B: Biological Sciences*, **256**, 281–292.

Butynski, T.M. (2000) *Independent Evaluation of Hirola Antelope (Beatragus hunteri) Conservation Status and Conservation Action in Kenya*. Kenya Wildlife Service and Hirola Management Committee, Nairobi: www.cf.tfcg.org/pubs/Hirola%20Evaluation%20Report.pdf

Caillaud, D., Levrero, F., Cristescu, R., *et al.* (2006). Gorilla susceptibility to Ebola virus: the cost of sociality. *Current Biology*, **16**, R489–R491.

Carpenter, J.W., Appel, M.J., Erickson, R.C. & Novilla, M.N. (1976) Fatal vaccine-induced canine distemper virus infection in black-footed ferrets. *Journal of the American Veterinary Medical Association*, **169**, 961–4.

Chaloupka, M., Balazs,, G.H. & Work, T.M. (2009) Rise and fall over 26 years of a marine epizootic in Hawaiian green sea turtles. *Journal of Wildlife Diseases*, **45**, 1138–1142.

Chang, G.J., Davis, B.S., Stringfield, C. & Lutz, C. (2007) Prospective immunization of the endangered California condors (*Gymnogyps californianus*) protects this species from lethal West Nile virus infection. *Vaccine*, **25**, 2325–2330.

Chatterjee, A. & O'Keefe, C. (2010) Current controversies in the USA regarding vaccine safety. *Expert Review of Vaccines*, **9**, 497–502.

Chen, C.H., Wang, T.L., Hung, C.F., *et al.* (2000) Enhancement of DNA vaccine potency by linkage of antigen gene to an HSP70 gene. *Cancer Research*, **60**, 1035–1042.

Connor, M.E., Zarley, C.D., Hu, B., *et al.* (1996) Virus-like particles as a rotavirus subunit vaccine. *Journal of Infectious Diseases*, **174**, S88–92.

Coonan, T.J., Rutz, K., Garcelon, D.K., Latta, B.C., Gray, M.M. & Ashehoug, E.T. (2005) Progress in island fox recovery efforts on the northern Channel Islands. In: *Proceedings of the Sixth California Islands Symposium* (eds D. Garcelon & C. Schwemm), pp.263–273. National Park Service Technical Publication CHIS–05–01, Institute for Wildlife Studies, Arcata.

Craft, M.E. & Caillaud, D. (2011) Network models: an underutilized tool in wildlife epidemiology? *Interdisciplinary Perspectives on Infectious Diseases* (epub ahead of print) doi:10.1155/2011/676949.

Craft, M.E., Volz, E., Packer, C. & Meyers, L.A. (2011) Disease transmission in territorial populations: the small-world network of Serengeti lions. *Journal of the Royal Society Interface*, **59**, 776–78.

Cross, M.L., Buddle, B.M. & Aldwell, F.E. (2007) The potential of oral vaccines for disease control. *Veterinary Journal*, **174**, 472–480.

Cross, P.C., Lloyd-Smith, J.O., Bowers, J.A., Hay, C.T., Hofmeyr, M. & Getz, W.M. (2004) Integrating association data and disease dynamics in a social ungulate: bovine tuberculosis in African buffalo in the Kruger National Park. *Annales Zoologici Fennici*, **41**, 879–892.

Cross, P.C., Lloyd-Smith, J.O. & Getz, W.M. (2005) Disentangling association patterns in fission-fusion societies using African buffalo as an example. *Animal Behaviour*, **69**, 499–506.

Cross, P.C., Drewe, J., Patrek, V., *et al.* (2009) Wildlife population structure and parasite transmission: implications for disease management. In: *Management of Disease in Wild Mammals* (eds R.J. Delahay, G.C. Smith & M.R. Hutchings), pp.9–30. Springer, Tokyo.

Cullingham, C.I., Merrill, E.H., Pybus, M.J., Bollinger, T.K., Wilson, G.A. & Coltman, D.W. (2010) Broad and fine-scale genetic analysis of white-tailed deer populations: estimating the relative risk of chronic wasting disease spread. *Evolutionary Applications*, **4**(1), 116–131.

Davis, H.L., Suparto, I., Weeratna, R., *et al.* (2000) CpG DNA overcomes hypo-responsiveness to hepatitis B vaccine in Orangutans. *Vaccine*, **18**, 1920–1924.

Davis, S., Trapman, P., Leirs, H., Begon, M. & Heesterbeek, J.A. (2008) The abundance threshold for plague as a critical percolation phenomenon. *Nature*, **454**, 635–637.

De Villiers, M.S., Meltzer, D.G., van Heerden, J., Mills, M.G., Richardson, P.R. & van Jaarsveld, A.S. (1995) Handling-induced stress and mortalities in African wild dogs (*Lycaon pictus*). *Proceedings of the Royal Society B: Biological Sciences*, **262**, 215–220.

Donnelly, J.J., Wahren, B. & Liu, M.A. (2005) DNA vaccines: progress and challenges. *Journal of Immunology*, **175**, 633–639.

Dubensky, T.W., Liu, M.A. & Ulmer, J.B. (2000) Delivery systems for gene-based vaccines. *Molecular Medicine*, **6**, 723–732.

Fombonne, E. (2008) Thimerosal disappears but autism remains. *Archives of General Psychiatry*, **65**, 15–6.

Galvani, A.P. & May, R.M. (2005) Epidemiology: dimensions of superspreading. *Nature*, **438**, 293–295.

Goldberg, T.L., Gillespie, T.R., Rwego, I.B., Wheeler, E., Estoff, E.L. & Chapman, C.A. (2007) Patterns of gastrointestinal bacterial exchange between chimpanzees and humans involved in research and tourism in western Uganda. *Biological Conservation*, **135**, 511–517.

Goodall, J. (1986) *The Chimpanzees of Gombe: Patterns of Behavior*. Harvard University Press, Cambridge, MA.

Hamede, R.K., Bashford, J., McCallum, H. & Jones, M. (2009) Contact networks in a wild Tasmanian devil (*Sarcophilus harrisii*) population: using social network analysis to reveal seasonal variability in social behaviour and its implications for transmission of devil facial tumour disease. *Ecology Letters*, **12**, 1147–1157.

Hampson, K., Dushoff, J., Bingham, J., *et al.* (2007) Synchronous cycles of domestic dog rabies in sub-Saharan Africa and the impact of control efforts. *Proceedings of the National Academy of Sciences USA*, **104**, 7717–7722.

Harandi, A.M., Davies, G. & Olesen, O.F. (2009) Vaccine adjuvants: scientific challenges and strategic initiatives. *Expert Review of Vaccines*, **8**, 293–298.

Hasting, B.E., Kenny, D., Löwenstine, L.J. & Foster, J.W. (1991) Mountain gorillas and measles: ontogeny of a wildlife vaccination program. *Proceedings of the American Association of Zoo Veterinarians Annual Meeting*, 198–205.

Haydon, D.T., Randall, D.A., Matthews L, *et al.* (2006) Low-coverage vaccination strategies for the conservation of endangered species. *Nature*, **443**, 692–695.

Heinsbroek, E. & Ruitenberg, E.J. (2010) The global introduction of inactivated polio vaccine can circumvent the oral polio vaccine paradox, *Vaccine*, **28**, 3778–3783.

Hofmeyr, M., Hofmeyr, D., Nel, L. & Bingham, J. (2004) A second outbreak of rabies in African wild dogs (*Lycaon pictus*) in Madikwe Game Reserve, bait vaccines for wildlife. *Recent Patents on Drug Delivery and Formulation*, **1**, 230–235.

Jadhav, K.R., Shaikh, I.M., Ambade, K.W. & Kadam, V.J. (2006) Applications of microemulsion based drug delivery system. *Current Drug Delivery*, **3**, 267–273.

Jessup, D., Deforge, J.R. & Sandberg, S. (1992) *Biobullet Vaccination of Captive and Free-Ranging Bighorn Sheep*. Proceedings of the 2nd International Game Ranching Symposium, pp. 429–434.

Kaur, T., Singh, J., Huffman, M.A, *et al.* (2011) Campylobacter troglodytis sp. nov., isolated from feces of human–habituated wild chimpanzees (*Pan troglodytes schweinfurthii*) in Tanzania. *Applied and Environmental Microbiology*, **77**, 2366–2373.

Keele, B.F., Jones, J.H., Terio, K.A., *et al.* (2009) Increased mortality and AIDS-like immunopathology in wild chimpanzees infected with SIVcpz. *Nature*, **460**, 515–519.

Keeling, M.J. & Eames, K.T. (2005) Networks and epidemic models. *Journal of the Royal Society Interface*, **2**, 295–307.

Kenah, E. & Miller, J.C. (2011) Epidemic percolation networks, epidemic outcomes, and interventions. *Interdisciplinary Perspectives on Infectious Diseases* (epub ahead of print) doi:10.1155/2011/543520.

Kilgallon, C.P., O'Donovan, D., Wernery, U. & Alexandersen, S. (2008) Temporal assessment of seroconversion in response to inactivated foot-and-mouth disease vaccine in Arabian oryx (*Oryx leucoryx*). *Veterinary Record*, **163**, 717–720.

Knobel, D.L., Fooks, A.R. & Brookes, S.M., *et al.* (2008) Trapping and vaccination of endangered Ethiopian wolves to control an outbreak of rabies. *Journal of Applied Ecology*, **45**, 109–116.

Koendgen, S., Kuhl, H., N'Goran, P.K., *et al.* (2008) Pandemic human viruses cause decline of endangered great apes. *Current Biology*, **18**, 260–264.

Koendgen, S., Schenk, S., Pauli, G., Boesch, C. & Leendertz, F.H. (2010) Noninvasive monitoring of respiratory viruses in wild chimpanzees. *Ecohealth* (epub ahead of print) doi:10.1007/s10393-010-0340-z.

Laurenson, M.K., Mlengeya, T., Shiferaw, F. & Cleaveland, S. (2005) Approaches to disease control in domestic canids for the conservation of endangered wild carnivores. In: *Proceedings of the Southern and East African Experts Panel on Designing Successful Conservation and Development Interventions at the Wildlife/Livestock Interface* (eds S.A. Osofsky, S. Cleaveland, W.B. Karesh, *et al*). IUCN, Gland, Switzerland.

Lauring, A.S., Jones, J.O. & Andino, R. (2010) Rationalizing the development of live attenuated virus vaccines. *Nature Biotechnology*, **28**, 573–579.

Leendertz, F.H., Ellerbrok, H., Boesch, C., *et al.* (2004) Anthrax kills wild chimpanzees in a tropical rainforest. *Nature*, **430**, 451–452.

Lembo, T., Hampson, K., Kaare, M.T., *et al.* (2010) The feasibility of canine rabies elimination in Africa: dispelling doubts with data. *PLoS Neglected Tropical Diseases*, **4**, e626.

Leroy, E.M., Kumulungui, B., Pourrut, X., *et al.* (2005) Fruit bats as reservoirs of Ebola virus. *Nature*, **438**, 575–576.

Lewis, F., Hughes, G.J., Rambaut, A., Pozniak, A. & Leigh Brown, A.J. (2008) Episodic sexual transmission of HIV revealed by molecular phylodynamics. *PLoS Medicine*, **5**(3), e50.

Liu, W., Li, Y., Learn, G.H., *et al.* (2010) Origin of the human malaria parasite Plasmodium falciparum in gorillas. *Nature*, **467**, 420–425.

Lloyd-Smith, J.O., Schreiber, S.J., Kopp, P.E. & Getz, W.M. (2005) Superspreading and the effect of individual variation on disease emergence. *Nature*, **438**, 355–359.

Lopez, G., Lopez-Parra, M. & Fernandez, L. (2009) Management measures to control a feline leukemia virus outbreak in the endangered Iberian lynx. *Animal Conservation*, **12**, 173–182.

Lowe, W.H. & Allendorf, F.W. (2010) What can genetics tell us about population connectivity? *Molecular Ecology*, **19**, 3038–3051.

Macdonald, D.W., Artois, M., Aubert, M., *et al.* (1992) Cause of wild dog deaths. *Nature*, **360**, 633–634.

Marsh, M.K., McLeod, S.R., Hutchings, M.R. & White, P.C. (2011) Use of proximity loggers and network analysis to quantify social interactions in free-ranging wild rabbit populations. *Wildlife Research*, **38**, 1–12.

McCormick, A.E. (1983) Canine distemper in African cape hunting dogs (*Lycaon pictus*) possibly vaccine induced. *Journal of Zoo Animal Medicine*, **14**, 66–71.

Mencher, J.S., Smith, S.R., Powell, T.D., Stinchcomb, D.T., Osorio, J.E. & Rocke, T.E. (2004) Protection of black-tailed prairie dogs (*Cynomys ludovicianus*) against plague after voluntary consumption of baits containing recombinant raccoon poxvirus vaccine. *Infection and Immunity*, **72**, 5502–5505.

Metzger, V.T., Lloyd-Smith, J.O. & Weinberger, L.S. (2011) Autonomous targeting of infectious superspreaders using engineered transmissible therapies. *PLoS Computational Biology*, **7**(3), e1002015.

Meyers, L.A. (2007) Contact network epidemiology: bond percolation applied to infectious disease prediction and control. *Bulletin of the American Mathematical Society*, **44**, 63–86.

Morens, D.M., Holmes, E.C., Davis, A.S. & Taubenberger, J.K. (2011) Global rinderpest eradication: lessons learned and why humans should celebrate too. *Journal of Infectious Disease*, **204**(4), 502–505.

Newman, M.E. (2006) Modularity and community structure in networks. *Proceedings of the National Academy of Sciences USA*, **103**, 8577–8582.

Nguyen, D.N., Green, J.J., Chan, J.M., Langer, R. & Anderson, D.G. (2009) Polymeric materials for gene delivery and DNA vaccination. *Advanced Materials*, **21**, 847–867.

Nichols, W.W., Ledwith, B.J., Manam, S.V. & Troilo, P.J. (1995) Potential DNA vaccine integration into host cell genome. *Annals of the New York Academy of Sciences*, **772**, 30–39.

Noad, R. & Roy, P. (2003) Virus-like particles as immunogens. *Trends in Microbiology*, **11**, 438–444.

Pashine, A., Valiante, N.M. & Ulmer, J.B. (2005) Targeting the innate immune response with improved vaccine adjuvants. *Nature Medicine*, **11**, S63–S68.

Porphyre, T., Stevenson, M., Jackson, R. & McKenzie, J. (2008) Influence of contact heterogeneity on TB reproduction ratio R-0 in a free-living brushtail possum *Trichosurus vulpecula* population. *Veterinary Research*, **39**, 31–42.

Pounds, J.A., Bustamente, M.R., Coloma, L.A., *et al.* (2006) Widespread amphibian extinctions from epidemic disease driven by global warming. *Nature*, **439**, 161–167.

Real, L.A. & Biek, R. (2007) Spatial dynamics and genetics of infectious diseases on heterogeneous landscapes. *Journal of the Royal Society Interface*, **4**, 935–948.

Rocke, T.E., Smith, S., Marinari, P., Kreeger, J., Enama, J.T. & Powell, B.S. (2008) Vaccination with F1-V fusion protein protects black-footed ferrets (*mustela nigripes*) against plague upon oral challenge with Yersinia pestis. *Journal of Wildlife Diseases*, **44**, 1–7.

Salathé, M. & Jones, J.H. (2010) Dynamics and control of diseases in networks with community structure. *PLoS Computational Biology*, **6**, e1000736.

Salathé, M., Kazandjieva, M., Lee, J.W., Levis, P., Feldman, M.W. & Jones, J.H. (2010) A high resolution human contact network for infectious disease transmission. *Proceedings of the National Academy of Sciences USA*, **107**, 22020–22025.

Sander, L.M., Warren, C.P., Sokolov, I.M., Simon, C. & Koopman, J. (2002) Percolation on heterogeneous networks as a model for epidemics. *Mathematical Biosciences*, **180**, 293–305.

Scoglio, C., Schumm, W., Schumm, P., *et al.* (2010) Efficient mitigation strategies for epidemics in rural regions. *PLoS ONE*, **5**, e11569.

Simerska, P., Moyle, P.M., Olive, C. & Toth, I. (2009) Oral vaccine delivery – new strategies and technologies. *Current Drug Delivery*, **6**, 347–358.

Singh, M., Chakrapani, A. & O'Hagan, D. (2007) Nanoparticles and microparticles as vaccine-delivery systems. *Expert Review of Vaccines*, **6**, 797–808.

Smith, D.L., Lucey, B., Waller, L.A., Childs, J.E. & Real, L.A. (2002) Predicting the spatial dynamics of rabies epidemics on heterogeneous landscapes. *Proceedings of the National Academy of Sciences USA*, **99**, 3668–3672.

Smith, K.F., Sax, D.F. & Lafferty, K.D. (2006) Evidence for the role of infectious disease in species extinction and endangerment. *Conservation Biology*, **5**, 1349–1357.

Smith, K.F., Acevedo-Whitehouse, K. & Pedersen, A.B. (2009) The role of infectious diseases in biological conservation. *Animal Conservation*, **12**, 1–12.

St Clair, N., Shenoy, B., Jacob, L.D. & Margolin, A.L. (1999) Cross-linked protein crystals for vaccine delivery. *Proceedings of the National Academy of Sciences USA*, **96**, 9469–9474.

Stoinski, T.S., Hoff, M.P. & Maple, T.L. (2003) Proximity patterns of female western lowland gorillas (*Gorilla gorilla gorilla*) during the six months after parturition. *American Journal of Primatology*, **61**, 61–72.

Streatfield, S.J. (2006) Mucosal immunization using recombinant plant-based oral vaccines. *Methods*, **38**, 150–157.

Sullivan, N.J., Geisbert, T.W., Geisbert, J.B., *et al.* (2003) Accelerated vaccination for Ebola virus haemorrhagic fever in non-human primates. *Nature*, **424**, 681–684.

Sundaresan, S.R., Fischhoff, I.R., Dushoff, J., Rubenstein, D.I. (2007) Network metrics reveal differences in social organization between two fission-fusion species, Grevy's zebra and onager. *Oecologia*, **151**, 140–149.

Swinnerton, K.J., Greenwood, A.G., Chapman, R.E. & Jones, C.G. (2005) The incidence of the parasitic disease trichomoniasis and its treatment in reintroduced and wild Pink pigeons Columba mayeri. *Ibis*, **147**, 772–782.

Takamura, S., Niikura, M., Li, T.C., *et al.* (2004) DNA vaccine-encapsulated virus-like particles derived from an orally transmissible virus stimulate mucosal and systemic immune responses by oral administration. *Gene Therapy*, **11**, 628–635.

Van de Bildt, M.W., Kuiken, T., Visee, A.M., Lema, S., Fitzjohn, T.R. & Osterhaus, A.D. (2002) Distemper outbreak and its effect on African wild dog conservation. *Emerging Infectious Diseases*, **8**, 212–213.

Walsh, P.D., Abernethy, K.A., Bermejo, M., *et al.* (2003) Catastrophic ape decline in western equatorial Africa. *Nature*, **422**, 611–614.

Walsh, P.D., Biek, R. & Real, L.A. (2005) Wave-like spread of Ebola Zaire. *PLoS Biology*, **3**, e371.

Walsh, P.D., Tutin, C.E., Oates, J.F., *et al.* (2007) Gorilla gorilla. In: *IUCN Red List of Threatened Species*. www.iucnredlist.org

Walsh, P.D., Rodríguez-Teijeiro, J.D. & Bermejo, M. (2009) Disease avoidance and the evolution of primate social connectivity: Ebola, bats, gorillas, and chimpanzees. In: *Primate Parasite Ecology: The Dynamics And Study Of Host–Parasite Relationships* (eds M.A. Huffman & C.A. Chapman), pp. 183–198. Cambridge University Press, Cambridge.

West, G., Heard, D. & Caulkett, N. (2007) *Zoo Animal & Wildlife Immobilization and Anesthesia*. Blackwell Publishing, Ames, IA.

Williams, J.M., Lonsdorf, E.V., Wilson, M.L., Schumacher-Stankey, J, Goodall, J. & Pusey, A.E. (2008) Causes of death in the Kasekela chimpanzees of Gombe National Park, Tanzania. *American Journal of Primatology*, **70**, 766–777.

Wimsatt, J., Biggins, D.E., Williams, E.S. & Becerra, V.M. (2006) The quest for a safe and effective canine distemper virus vaccine for black-footed ferrets. In: *Recovery of the Black-Footed Ferret: Progress and Continuing Challenges*. (eds J.E. Roelle, B.J. Miller, J.L. Godbey & D.E. Biggins). US Geological Survey, Herndon, VA.

Wittemyer, G., Douglas-Hamilton, I. & Getz, W.M. (2005) The socioecology of elephants: analysis of the processes creating multitiered social structures. *Animal Behaviour*, **69**, 1357–1371.

Woodroffe, R. (2001) Assessing the risks of intervention: immobilization, radio-collaring and vaccination of African wild dogs. *Oryx*, **35**, 234–244.

Woodroffe, R. & Ginsberg, J. (1997) Research and monitoring: information for wild dog conservation. In: *The African Wild Dog – Status Survey and Conservation Action Plan* (eds R. Woodroffe, J.R. Ginsberg & D.W. Macdonald), pp. 58–74. IUCN, Gland, Switzerland.

Zafar-ul Islam, M., Ismail, K. & Boug, A. (2010) Catastrophic die-off of globally threatened Arabian oryx and sand gazelle in the fenced protected area of the arid central Saudi Arabia. *Journal of Threatened Taxa*, **2**, 677–684.

25

Elephants in the room: tough choices for a maturing discipline

David W. Macdonald[1] and Katherine J. Willis[2]

[1]Wildlife Conservation Research Unit, Department of Zoology, Recanati-Kaplan Centre, University of Oxford, Oxford, UK
[2]Biodiversity Institute, Oxford Martin School, Department of Zoology, University of Oxford, UK

'But facts are chiels that winna ding, An' downa be disputed.'[1]
(Robert Burns, *A Dream*, 1786)

Introduction

Wildlife conservation, and the preservation and restoration of biological diversity from genes through to whole landscapes, should be an evidence-based activity. It is built on an interdisciplinary foundation for which natural science is necessary but not sufficient, and which leads from principles to practice along a route involving natural and social science, evaluation, judgement and, inexorably, politics. All these linkages, and the blurred boundary between the scientific and political consequences of conservation decisions, mean that most conservation issues are permeated by profound trade-offs, perplexing dilemmas and sometimes unmentioned truths: metaphorical elephants in the room. Rather than let these Thought Elephants skulk, unspoken, our purpose here is to flush out a sample of them, with the hope that their trumpetings will illustrate how implementing conservation involves choosing actions within a wider framework of ideas and policies, such that selecting one option inevitably involves not selecting another (reminiscent of the painful bind of 'Pareto efficiency' where no one can be made better off without at least one person being made worse off). The greater the extent to which conservation, and wider environmental, choices are recognized as being not only technically complex but also central societal decisions, the tougher they become, and the more obviously entwined with all the major issues affecting the human enterprise. That is why the concluding chapter of the first volume of *Key Topics in Conservation Biology* (Macdonald et al. 2006), and now the vast majority of chapters in this volume, argue for alignment between the different

[1] But facts are fellows that will not be overturned/ And cannot be disputed.

Key Topics in Conservation Biology 2, First Edition. Edited by David W. Macdonald and Katherine J. Willis.

factors, often measured in incommensurable currencies, that must be reconciled in reaching a solution to any problem in wildlife conservation. Continuing that journey, we illustrate answers to questions that often arise *en route* to the 'least worst' compromise for nature: why conserve biodiversity, how much do we need, how to do it and what compromises need to be made?

If a goal is to conserve nature (or at least those parts that humans favour), then how much of it? Variants of this question include how much of particular elements of nature 'we' (that is, society globally, nationally, locally) need and want. Furthermore, how is that conservation to be undertaken? Amongst the many aspects of this question, we consider some distinctions between conservation in protected areas and in unprotected areas. In answering the question of what compromises need to be made, we illustrate our view that each case has to be tackled on its own merits, by reference to two topical and unresolved example dilemmas: regulation to conserve lions and killing badgers to control disease in cattle. These cases not only illustrate the particularity of each issue but also that while alignment remains a helpful goal, almost all conservation choices are tough. Overall, they demand attention to the impact of a growing human population and, more particularly, its consumption, ecological footprint and demography, and so we conclude this essay with a discussion of these issues. We begin, however, by explaining what, exactly biodiversity is and why we need to conserve it.

Introducing the Elephants

Throughout this essay we have penned Thought Elephants in boxes. The purpose is to reveal some of the difficult issues that often remain unspoken and invisible when key topics in conservation biology are discussed – the 'elephants in the room' which increasingly face conservation practitioners. These Thought Elephants are phrased as questions, none of which has an easy answer, and presented as food for thought.

What is biodiversity?

Biodiversity, a contraction of biological diversity, is defined by the Convention on Biological Diversity (CBD) as the variability among living organisms from all sources, including terrestrial, marine and other aquatic ecosystems, and the ecological complexes of which they are part, thus embracing the diversity within species, between species and of ecosystems (CBD 2010). Diversity within a species, or genetic diversity, is biodiversity's lowest common denominator. Environments change and genetic diversity enables life to change with them (Lande & Shannon 1996); its impacts reverberate through population, community and ecosystem levels (Hughes et al. 2008). Reduced genetic diversity diminishes the ability of species to persist in their current state and to undergo evolutionary adaptation in response to changing conditions (Hendry et al. 2010; Sgrò et al. 2011). Loss of genetic diversity, particularly for island populations, may lead to declines in population size or extinction in the wild (Frankham 2003). For example, a study examining the genetic diversity of the heath hen (*Tympanuchus cupido cupido*) from museum examples revealed that 30 years prior to their extinction in the US in 1932, they had extremely low levels of genetic diversity compared to extant populations of prairie chickens, *Tympanuchus* sp. (their closest living relatives) (Johnson & Dunn 2006). Worryingly, this study also demonstrated that current populations of greater prairie chickens are isolated and losing genetic variation due to drift such that within the next 40 years, these populations will reach levels of genetic variation as low as those associated with the extinction of the heath hen.

Species are the units of biodiversity of which most people are most aware. Suites of interdependent species, interacting together and with the environment, form ecosystems. Ecosystems are characterized by their processes (fluxes of energy and matter, including carbon uptake, nutrient cycling and oxygen production), properties (such as resistance to invaders and resilience to environmental changes) and their maintenance, determined by a number of abiotic controls (e.g. climate, geomorphological processes, region, soil or sediment type) as well as biotic elements – the number and (functional) type of organisms, from virus to top vertebrate predator, and their interactions (Hooper et al. 2005; Reiss et al. 2009).

'Biodiversity', then, is a measure that can be applied equally to the differences between the interacting suites of organisms that comprise ecosystems and to the variations within a single species (Chapin et al. 2000). A crude hierarchy of dependence between these scales dictates that if an ecosystem disappears or is degraded then, inevitably, the taxa (e.g. families, genera, species) it supports will be lost or reduced too, as will the genetic diversity between their populations and individuals.

A populist misunderstanding is that more biodiversity is always better. What often matters to conservation is not maximizing biodiversity (after all, adding invasive species is a bad thing) but ensuring its appropriateness to a given environment, not forgetting that a shifting baseline has progressively corroded contemporary perceptions of what is natural (Willis & Birks 2006). Thus tropical forests, with up to 20,000 species of vertebrate (according to the Millenium Ecosystem Assessment), are truly exciting to those who treasure nature but they are inherently no more so than deserts, with fewer than 8000 species in a somewhat larger area. Of course, there are different metrics to value: although the evolutionary, ethical or existence values of the biodiversity in, for example, forests and deserts may be inseparable (or, at least, incommensurable, e.g. phylogenetic eccentricity and beauty are not easily ranked), they are very different in terms of the utilitarian value of the ecosystem services that they provide and therefore their material contribution to human well-being. So, in ethical (as distinct from monetary) terms, the global threat to tropical forests is tragic, but the inherent value of the much less biodiverse or productive natural communities in deserts makes their loss no less tragic (and it is threatened, e.g. of the 14 large vertebrates in the Sahara, 13 have suffered range collapses; Durant et al. 2012) and despair for desert charismatics should not cause us to forget the fungi, cyanobacteria and lichens that form a fragile cryptobiotic crust which assists water retention and thus plant growth, and may take centuries to regrow if damaged (Belnap et al. 2001).

First Elephant

A paradox for conservation is the need to kill some of evolution's greatest successes (viruses, weeds, pests) while nurturing species that may have had their day (e.g. giant pandas), and making a virtue of rarity. Is there a scientific rationale for attaching less value to the Ebola virus than to the gorilla, or to being more motivated to reintroduce the beaver to the UK than the obligate ectoparasitic beaver beetle *Platypsyllus castoris* that was lost with it?

Why conserve biodiversity?

The world's biological diversity is, irrefutably, under increasing threat due to human activities (e.g. Chapin et al. 2000; McKee et al. 2004; Luck 2007; May in press) but why does this matter? May (2011) summarizes the three widely recognized categories of answer: ethical (the responsibility to bequeath to future generations a planet as rich in natural wonders as the one inherited), narrowly utilitarian (species may have as yet undiscovered benefits for biotechnology which may be of use to humans) and broadly utilitarian (ecosytems provide services required for human survival). Of these three (and they generally interact because, in nature, beauty and fascination do not preclude utility), May emphasizes ethics as a primary motivator, and we will touch on this lightly before dwelling at greater length on broadly utilitarian factors.

Second Elephant

The weakness of intergenerational fairness as the primary ethical driver for conservation is that future generations, as arbiters of the performance of their forebears, have no recourse: what retrospective sanctions can they exert on the perpetrators of their misery, considering that those that they might wish to punish for profligacy are no longer alive to feel the pain, or even the embarrassment of their selfishness? What recourse do future generations have, other than to respect or despise us?

The financial contribution of nature to underpinning the human enterprise is increasingly recognized (Kumar 2010). Despite the obvious power of monetary arguments (Pearce et al. 2006), and the optimism that they will empower conservation (see Chapter 4), the potency of ethical values should not be underestimated (notwithstanding their nebulous substance). To quote Thompson & Starzomski (2007): 'Public support for maximizing biodiversity is the driving force for biodiversity maintenance and conservation, and is based on aesthetic, ethical, and spiritual values' (see Chapters 6 and 7). Recognizing these subjective values in ecosystems, landscapes, species and other aspects of biodiversity is a feature of the human condition. However, different cultures, and different sections of any society, may have radically different values and in the case of conservation (like any other societal choice), the imposition of one group's values on another – and associated issues of 'cultural relativism' – raise awkward dilemmas. In short, existence values (for all their subjectivity) may galvanize public opinion, but generally need the force of policy, regulation and policing to resist the power of monetary gain (see below).

Third Elephant

Lots of people, and indeed entire cultures, do not value many of the species that others (notably, many conservationists) value, and so liberal-minded respect for other people's views becomes problematic. Similarly, with cultural relativism (e.g. belief in transmogrification of enemy villagers into villainous carnivores) (see Chapter 7): some beliefs are indisputably wrong, but must they be respected? Is there a point at which it becomes acceptable for advocates of conservation to forego liberal tolerance of alternative values and, instead, enforce their determination that the conservation ethic will prevail over the views of other people?

Earlier rationalizations for conservation commonly emphasized narrow utilitarian potential (e.g. Madagascar's rosy periwinkle (*Catharanthus roseus*) used by western medicine to treat childhood leukaemia) but tomorrow's medicines are increasingly likely to be designed from the molecules up, rather than emerging from high-tech bioprospecting (May 2011). Broad utilitarian arguments recognize the relationship between altered species diversity, ecosystem function and societal costs, resulting in the loss of *ecosystem services* (the processes and conditions of natural ecosystems that support human activity and contribute to human well-being) (Ehrlich & Walker 1998; Chapin et al. 2000; Millennium Ecosystem Assessment 2005; Foley et al. 2005). Such services include those that regulate ecosystems, including, for example, the maintenance of breathable air, soil fertility, climate regulation and natural pest control, and those that provide tangible goods such as food, timber (construction and fuel) and fresh water. Intangible benefits are also viewed as important within a ecosystem service framework; these include the value obtained from aesthetic and cultural activities such as recreational landscapes and sacred sites (Turner & Daily 2008; Turner et al. 2010) (see Chapter 9).

By analogy with capital investment, until the Industrial Revolution (and not withstanding the Pleistocene extinctions and historical excesses of overuse and persecution), humans generally 'lived off the interest', using only the renewable portion of the services provided by the biosphere, leaving the natural capital largely intact. Now, however, humanity's resource use is eroding this natural capital (Foley et al. 2005) with activities that appropriate nearly one-third to one-half of global ecosystem production (net primary productivity) (Vitousek et al. 1986). This prompts the question of whether the human enterprise is degrading the global environment in ways that may ultimately undermine ecosystem services, human welfare and the long-term sustainability of human societies (e.g. Foley et al. 2005; Schade & Pimentel 2010). In tackling this question, the Millennium Ecosystem Assessment, the first global audit of ecosystems, identified 24 categories of ecosystem service (broadly grouped under provisioning, regulating and cultural), of which 15 were either degraded or used unsustainably; four had improved, including crops, livestock and aquaculture (Millennium Ecosystem Assessment 2005). In short, modern land use practices, while increasing the short-term supplies of material goods, may ultimately undermine many ecosystem services, even on regional scales or globally (Foley et al. 2005; Schade & Pimentel, 2010).

But is it possible to place a monetary value on ecosystem services? Historically, the economic value of ecosystem services has been treated largely as an externality and, *de facto*, ignored (Costanza et al. 1997; Kumar 2010). In a first attempt to rectify this, Constanza et al. (1997) valued the services provided by the global biosphere at US$16–54 trillion per annum. An oft-quoted case study (among others in Foley et al. 2005) is New York City's purchase for US$ 1 billion of watersheds in the Catskill Mountains, which provide water purification services, as a cheaper alternative to building a filtration plant for US$6–8 billion plus annual operating costs of US$300 million. Pollination services are another instance where attempts have been made to place a true monetary value on the resource provided. Given that 75% of globally important crops benefit from animal pollination (Lonsdorf et al. 2011), this is an essential ecosystem service and current estimates suggest it is worth €195 billion worldwide (Allsopp et al. 2008). Hand pollination is already necessary in pear orchards in China, and bees are routinely trucked around the US to compensate for the loss of their wild cousins (these costs are an unintended consequence of profligate use of neonicotinoid pesticides). To quote TEEB (Kumar 2010), 'Natural resources are economic assets, whether or not they enter the marketplace ... Failure to incorporate the values of ecosystem services and biodiversity into economic decision making has resulted in the perpetuation of investments and activities that degrade natural capital'.

Whilst it is widely acknowledged that biodiversity has a key role in delivering a number of valuable (but not all) ecosystem services (Kumar 2010) and that declining biodiversity has large-scale economic implications, the relationships are rarely clear (Pfisterer & Schmid 2002; Reiss et al. 2009). Ehrlich & Walker (1998) couch uncertainty of the effects of species loss in terms of the 'rivet popper hypothesis' and 'redundancy hypothesis'. The former famously likens species to rivets in an aeroplane: sooner or later 'rivet popping' (forcing species to extinction) will be catastrophic, although predicting which will be the crucial rivet is difficult. The 'redundancy hypothesis' focuses attention on functional groups that comprise only one or two species (i.e. which have little redundancy) and whose loss might imperil the whole system (e.g. O'Gorman et al. 2011). Given the 'uncertainties and complexities in the relationships between biodiversity and ecosystem services, policy decisions should have a large "insurance" bias toward protection of biodiversity ...' (Ehrlich & Walker 1998). Furthermore, global mapping of key ecosystem service provision indicates strong spatial concordance of global biodiversity priorities and ecosystem service value (Turner et al. 2007). As an aside, it would be prudent to future-proof utilitarian ecosystem services; for example, a century ago nobody would have anticipated worrying about the provision of bio-ethanol, and if hydrogen becomes a future fuel, humans might value methane fermentation services more in the future than currently.

Nobody knows how much biological diversity can be lost whilst still safeguarding the ecosystem services upon which humans depend (May in press). Given the complexity of ecosystems and incomplete knowledge of their mechanics and of which species affect which ecosystem properties, prudence invokes the precautionary principle (Ehrlich & Walker 1998; Chapin et al. 2000; Hooper et al. 2005; Duffy 2009; May in press).

How much biodiversity do we need or want?

The question of how much biodiversity we need is important because progress in conservation must ultimately be gauged against the answer, and it lies at the core of many complex environmental policy decisions (Wilhere 2008). The answer provides a target level of biodiversity (however measured, and at a given scale) below which it would be functionally precarious to dip; a bottom line for the minimum threshold for conservation 'success'.

Fourth Elephant

The transitions from science through judgement to policy are blurred, and not tidily compartmentalized. Do scientists' protests that they will not venture beyond evidence into advocacy sometimes, perhaps, ring hollow? Indeed, while the pursuit of science should be apolitical, why shouldn't scientists be advocates?

To preserve biodiversity is prudent, and not to do so, precarious. Close kin to the 'need' question, however, is the question of how much biodiversity we *want*. This latter question largely concerns policy, and recognizes the complications that arise because the reality of human affairs is that accepting one option involves rejecting others; thereby decisions about biodiversity become, instantly, political and require many trade-offs (Macdonald et al. 2006) (see Chapter 4). Where land is dedicated to biodiversity, the opportunity of some other uses is thwarted, a commonplace which prompted Norton-Griffiths (2007) (see Chapter 4) to ask, provocatively, '... how many wildebeest *do* you need? How many elephants is enough?'. While the amount of biodiversity 'we' want might, for ethical or aesthetic reasons, be 'as much as possible', the amount we need from a broad utilitarian standpoint might be sufficient to preserve the supply of ecosystem services to the human population (a statement instantly raising further questions about what constitutes sufficiency, and about the size of the human population). Norton-Griffiths (2007) mused that 300,000 of the current 1.5 million wildebeest might satisfy the tourist experience of mass migrations.

Remembering utilitarian services, and the prejudices of charisma, one might add balance by asking similar questions about less charismatic organisms; for example, how many nitrogen-fixing bacteria do we need?

Fifth Elephant

Protecting biodiversity by some people will sometimes, and perhaps often, be at a cost, especially an opportunity cost, to others. There will be losers, and there may be a lot of local losers. Often, somebody will have to pay to offset the loss but how big will the bill be, and who should pay it?

Models are available to estimate, for any given species, minimum viable population sizes, which are expressed in terms of the probability of extinction (Soulé 1987; Shaffer 1987). However, these models are complex and recent work indicates that simple species–area relationships/population size models can be extremely poor predictors of the probability of extinction (Prugh et al. 2008; He & Hubbell 2011). Leaving the models aside, it would also be a delusion to consider this a purely objective process; deciding on what probability of extinction is acceptable is a societal choice ultimately based (at least partly subjectively) on ethics (i.e. a quantitative expression of what society ought to do; Prugh et al. 2008; Wilhere 2008), and remembering that the utilitarian would point out that there could be more to mourn in the passing of cyanobacteria than of elephants.

How to conserve biodiversity?

A simple taxonomy of conservation options might consider, first, whether to focus principally on the heartland of a species' remaining range or the fringes of its distribution where it may be particularly imperilled. This choice has surfaced in the context of tigers, where one view had it that focusing effort on the 6% of the tiger's range that encompassed 70% of their numbers offered the best chance of success, whereas spending money on the remaining 94% amounted to little more than distracting frittering (Walston et al. 2010). An alternative view is that some fringe areas are crucial, if frail, outposts of survival and should be defended to the last; Farhadinia (2004) argues for Persian leopards and Cuellar et al. (submitted) for relict populations of guanaco in Bolivia.

Sixth Elephant

The triage principle suggests that we should devote effort to the most viable patient – perhaps the most robust subpopulations of a species or the most safeguarded protected areas. But is it acceptable to abandon the potentially saveable 'patients' at the fringes of a species' range or beyond protected areas?

Protected areas

Protected areas are the bulwark of biodiversity conservation globally. Approximately 13% of the earth's terrestrial surface is under some form of formal protection (Jenkins & Joppa 2009) with a commitment to increase this to 15–20% by 2020 (Stokstad 2010). Marine protected areas cover 1.17% of the world's oceans (4.21 million km^2) and the aim is to increase this to 10% of the world's marine and coastal ecological regions by 2020 (Spalding et al. 2010; Fox et al. 2012) (see Chapter 10). Schemes to identify priority areas for protection can function at scales from global measures that evaluate whole landscapes through to local presence of individual species. 'Hot spots' identify the number of species/endemics/threats in a region (e.g. through land use change; Myers et al. 2000; Brooks et al. 2006; Mittermeier et al. 2008) whereas 'ecoregions' are intended to conserve diversity of geographically distinct assemblages of natural communities (e.g. WWF ecoregions). To date, 34 hot spots have been identified globally and 825 terrestrial ecoregions, 426 freshwater ecoregions and 229 coast and shelf marine ecoregions (www.worldwildlife.org/science/ecoregions/item1847.html). In addition, there are areas recognized for their importance to particular species, for example Important Bird Areas (www.birdlife.org/action/science/sites/) or the habitat that they provide along that species' migratory route. The global registry of migratory species (www.groms.de), for example, currently contains a list of 2880 migratory species (of various taxa) in digital format and digital global migration maps, detailing important landscapes/routes for 545 vertebrate species. Another globally recognized landscape-scale conservation scheme is the UNESCO Man and Biosphere programme: 560 biosphere reserves in over 100 countries protect a combination of biotic and cultural diversity (www.unesco.org). Key biodiversity areas (KBAs) contain a number of overlapping distribution ranges of species with high conservation priorities (http://iucn.org/about/union/secretariat/offices/iucnmed/iucn_med_programme/species/key_biodiversity_areas/).

Despite the complexities of these different approaches, there is an impressive distribution of protected areas at all scales (see www.protectedplanet.net/). A similar landscape-scale approach is being applied to marine biodiversity hot spots (Roberts et al. 2002), and this has led to the creation of marine protected areas although these account for only a paltry 1% or so of the oceans (see Chapter 10).

Protected species

Schemes to determine the global distribution of species and the severity of threats to them have informed the Global Biodiversity Information Facility (GBIF) and IUCN Red List of threatened species. GBIF (http://data.gbif.org) is a data repository and portal for all geo-tagged species data that currently contains records of a staggering 330 million species' occurrences worldwide. The IUCN 2010 Red List of threatened species (www.iucnredlist.org/) is scarcely less impressive, containing assessments of ~58,000 threatened species globally and spatial distribution maps for ~28,000 of them, along with a measure of their status ranging from 'extinct in the wild' through to 'vulnerable' and 'of least concern' (Mace et al. 2006). The EDGE database (www.edgeofexistence.org/) highlights those species that are both endangered (as identified in the IUCN Red List) *and* evolutionarily distinctive globally. Increasingly, web-based tools offer increasing sophistication for mapping conservation priorities (e.g. C-Plan (Pressey et al. 2009); Marxan (Ball et al. 2009); and Zonation (Moilanen 2007) (see Chapters 18 and 20).

Against the achievements of selecting and implementing protected areas (and the stated aim of the CBD (2010) to increase them), challenges for the next decade include the following.

Gathering information on the evolutionary derivation of species/habitats

It has been clear for more than a decade that insufficient is known of the ecological and/or evolutionary processes involved to inform responsible biodiversity conservation, and to make secure predictions about the best interventions (Mace et al. 1998, 2006). This necessitates understanding both the present ecological requirements of the species/habitat and how they evolved, in the context of former distribution/natural variability. What are the natural baseline conditions and how much of the present landscape/distribution of a species reflects earlier human activities?

For example, when a region is declared an IUCN Biosphere Reserve, it is divided into zones, each with its own category of management plan ranging from zone 1 where no human presence is allowed except for scientific research, through to zone 5 where human activities are allowed, including indigenous use of resources in the reserve (e.g. timber and non-timber forest products). For example, in Mexico's Sierre di Manatalan reserve pine forests, their designation as a zone 1 IUCN Biosphere Reserve would result in the complete removal of human presence. However, it has been widely assumed that this is a secondary forest shaped by burning by people. Putting aside the question of how this anthropogenic origin would affect the value attributed to the forest, there was therefore concern that removing human influence would, perversely, threaten the forest's persistence. This prompted research into the processes shaping the forest, revealing that it long predated human influence and was a natural response to the semi-arid climate that depends crucially on naturally occurring fires at 40–60-year intervals (Figueroa-Rangel et al. 2008). Conversely, since indigenous people had lived in the forest for ~150 years without detriment to it, the question arose as to whether the forest's conservation necessitated removing them from the core zone (Willis & Bhagwat 2009).

Determining how anthropogenic climate change will shift the targets for conservation

Computer models can reveal how species' habitats, and thus the envelope encompassing their range margins, are likely to shift (e.g. Carroll et al. 2010). Policy predicated on the 'King Canute' option of striving to stem the tide of change (literally in the case of coastal conservation) may be both futile and betray a distorted understanding of ecological dynamics (Hobbs et al. 2009). Throughout palaeontological history, intervals of rapid climate change have involved rapid turnover of species, range shifts and the formation of novel ecosystems (Williams & Jackson 2007; Willis & Macdonald 2011).

The notion of change management is familiar in businesses and institutions, but less so amongst conservationists facing changing ecosystems, where the priority becomes conserving ecosystem function and evolutionary process during changes in community composition (Mace & Purvis 2008; Seastedt et al. 2008). The task of restoring the trophic structure to a region may require radical interventions, for example through rewilding(see Chapter 23). Priorities will include maintaining a diverse soil biota and conserving regions of high genetic diversity to enable plasticity in response to environmental change (Hellmann & Pineda-Krch 2007; Willis et al. 2010; Sgrò et al. 2011). In Europe, the greatest genetic diversity often occurs where plants and animals were isolated in refugia during the cold phases of the Quaternary (Petit et al. 2008), yet many of these regions are now heavily affected by farming and fall outside protected areas (Taberlet et al. 2011). It is clear that in order to conserve biodiversity in a changing climate, there needs to be a broader spectrum of conservation methods that move beyond the current tendency to try and maintain the status quo, since with current and future climate change this may no longer be possible.

Measuring and identifying success

Too many protected area/species management programmes still lack measurements of success against which to judge value for investment. Crucial indicators are: (a) has long-term sustainability of the targeted area, species or ecosystem service(s) been delivered and, increasingly, (b) what have been the effects on poverty and local livelihoods? In one case study, Nantu National Park in Sulawesi, an investment in conservation of c.£67,300 pa for 12 years was associated with a c.6.5-fold reduction in the rate of forest loss within the park in comparison to that beyond its borders, and it appeared that the effects of forest guards had contributed more to this than had provision of alternative livelihoods (Macdonald et al. 2011).

However, few, if any, structured frameworks and methodologies exist for rapid quantitative assessment of the long-term sustainability or impact of linked conservation-development programmes. Equally, no tools are available to use the quantitative information on 'what has worked before' to plan future projects based on a probability of success. These questions are complex, often interlinked and multiscale and require a combination of natural and social sciences data to chart relationships between inputs and outcome.

The term 'protected area' merits a moment's pause – protected by what means? A ready criticism is that the protection offered to many areas may be ineffectual – the 'paper parks'. However, a polite request to respect protected status may not suffice, and the question arises as to just how far societies are prepared to go to protect wildlife (see Chapter 16). Rithe (2012) reports that despite the best efforts of India's legislature, and whilst under global scrutiny, steel foot-traps and snares nonetheless threaten tigers in almost all India's wildlife reserves (a search, in September 2011, revealed more than 100 spring traps and snares in two famous tiger reserves, Bandipur and Nagarahole) (Figure 25.1). In desperation, the Maharashtra government, in May 2012, adopted a shoot-on-sight policy against poachers. Quite apart from the straightforward ethical test of whether you value tigers sufficiently to shoot the people who would kill them (themselves probably the hapless foot-soldiers of international organized crime), this decision opens another morass of complications (for example, under similar circumstances the forest guards in the Kaziranga National Park in Assam enjoy immunity against prosecution for use of firearms, but this may not apply to guards in others Indian parks).

(a)

(b)

Figure 25.1 (a) Gin-trapped tiger (© K. Riche). (b) Gin-trapped lion (© G. Hemson). Horrors from distant continents (tiger in Madhya Pradesh, India, lioness in Makgadikadi, Botswana) illustrate how different pressures (respectively, Asian traditional medicine and cattle killing) have similarly agonizing, and endangering, outcomes for big cats.

It is notable that each of the foregoing challenges necessitates explicit scrutiny of the role of human use of the land and its resources (particularly species) within the protected area. Indeed, the disquieting ethical test of the value of the lives of tigers versus those of people who would kill them is a microcosmic version of the far more daunting, global ethical test we pose in the form of the final elephant at the end of this essay. If the human dimension plays such an important role for areas and species that are, at least notionally, theoretically protected, then these issues can surely only be larger outside protected areas.

Seventh Elephant

When you call for protection, just how far are you prepared to go? Ultimately, human lives are at stake.

Non-protected areas

Outside protected areas (i.e. ~88% of the earth's terrestrial surface), the question of biodiversity conservation needs a different perspective. Land use change is one of the biggest global threats to biodiversity and the vast majority of this is occurring outside protected areas (indeed, the placement of some parks is precisely where there are fewest alternative economic uses for the land). Factors responsible for land use change include conversion of land for agriculture, extractive industry, biofuels and urbanization. More often than not, this land use change is also legal; concessions are routinely granted to convert land by local/regional governments. In most regions, the only legislative requirement to assess the impact of the conversion on the land is through the Environment Impact Assessment, a process that occurs once the decision has been made where to cite the facility.

Outside protected areas, therefore, the question in the minds of those planning to change land use is, in practice, 'where can we damage?', quickly followed, one hopes, by the question of 'how can we minimize or mitigate that damage?'. From the perspective of conservation and sustainability, the answer prompts debates relating to the size and shape of land parcels outside protected areas and their role in providing an environment that can enable persistence of biodiversity and corridors between protected areas. Biodiversity can persist in the face of extreme land use change (Willis &

Bhagwat 2009). In a study on 972 forest butterflies in Africa (west of the Dahomey gap), for example, Larsen (2008) found that 97% of the species ever recorded in the area were still present between 1990 and 2006 although during the preceding 150 years, the forests had shrunk to 13% of their former distribution. Similarly, a meta-analysis on studies of tropical agricultural landscapes examining species abundance and diversity in fields with crops ranging from banana and benzoin to jungle rubber and coffee (Bhagwat et al. 2008) revealed that richness, for some groups (e.g. bats, lower plants and some birds), was higher on the agricultural matrix. Indeed, in Borneo, some managed forest may support higher densities of orang utans than does pristine forest (returning us to the point that more biodiversity is not necessarily more 'natural'; Husson et al. 2008; Ancrenaz et al. 2010). Indeed, in Bhagwat et al.'s (2008) study, many endemics and rarer plants were missing from the agricultural landscapes and there was a significant difference in community composition (similarity was often around 50% or less).

Eighth Elephant

As the land fills up with people, the number of candidate wildlife corridors shrinks, so there is a race to establish corridors before people settle in them, thereby blocking settlement. In this way, conservation plays development in a game of 'landscape noughts and crosses'. But who is to decide the rules of this game?

This loss of endemics/rare species and change in community composition on converted landscapes has led to a heated debate over the relative merits of land sharing or land sparing. Land-sharing aims to promote dual use of landscapes, i.e. to create 'nature-friendly' landscape management techniques. This might result in less productive output for agricultural purposes but the argument is that it supports both biodiversity and a provisioning service (Perfecto & Vandermeer 2010). In a study comparing bird and dung beetle diversity in Sabah, Malaysian Borneo, for example, comparison of unlogged, once-logged and twice-logged forest revealed that even in twice-logged forest, over 75% of bird and dung species persisted. However, there was far more impact (and species loss) if the logging was done early in the season and the second cut was intensive. The same principles of land sharing underlie attempts, using agri-environment schemes, to find ways of managing non-agricultural productive habitats on European farmland better to integrate conservation alongside food security (Merckx et al. 2009).

In contrast, the land-sparing approach grows from the conclusion that biodiversity cannot be properly conserved in dual-use landscapes and opts instead to set land aside for biodiversity while maximizing the agricultural (or other) productivity of land where that use is prioritized. A recent study by Phalan et al. (2011), for example, comparing crop yields to densities of trees and birds across different agricultural landscapes in south west Ghana and northern India, demonstrated that more species were negatively affected by farming than benefitted from it (of course, a land manager would consider whether the species affected were benign and desirable, or pests). They concluded that land sparing is the best solution for biodiversity, partitioning the landscape between protection and intensive agriculture. This approach, however, is not without criticism: many regions of the world, especially those with high population density/low income, lack the means to protect areas effectively but may have a long record of sustainable land sharing. Poor soils or low rainfall also mean that high intensive yields are not possible in many regions. In reality, use of agrochemicals is likely to be the default method for such regions to obtain high yields which in itself could possibly have an even more negative effect on biodiversity (Fisher et al. 2011).

Interestingly, the EU is scoping the possibility of high-input intensive farmers purchasing offset credits so that, rather than 'greening' their own land, they enable other landowners to do so in more stressed or marginal environments.

Where to conserve biodiversity?

Conservation can operate at scales from species (even populations) to landscapes (see Chapter 2). At a 'big picture' scale, priority areas for biodiversity conservation (or biodiversity 'hot spots') might, together, encompass and maintain populations of all (known or extant) taxa (Margules et al. 2002; Funk & Fa 2010). Two problems thwart this process: first, gaps and biases in the information on which choices can be made, and second the selection of criteria for prioritization. Illustrating the former, national Red Lists have been developed to monitor trends in the status of threatened species as an indicator of patterns of biodiversity loss throughout a species' range (Butchart et al. 2005). In a review of these lists from 109 countries, Zamin et al. (2010) found that their comprehensiveness within a given country was positively correlated with GDP and negatively correlated with total vertebrate richness and threatened vertebrate richness. In short, regions with the greatest and most vulnerable biodiversity receive the least conservation attention. With respect to the latter, the metrics used to define biodiversity hot spots remain controversial (Possingham & Wilson 2005; Funk & Fa 2010). The areas given priority will differ between different taxa, and species-rich areas are often not those containing large densities of endemics, which again may differ from those containing large numbers of threatened species (Funk & Fa 2010). Criteria include the number of endemics or total species present, degree of threat, population viability, ecological and evolutionary processes, and economic costs and benefits of conservation (Lamoreux et al. 2006).

Taken together, gaps in knowledge and incommensurable priorities necessitate simple surrogate metrics of biodiversity (Margules & Pressey 2000). There is hope in the fact that global patterns of terrestrial species richness and of endemism are highly correlated amongst four terrestrial vertebrate classes (Lamoreux et al. 2006). The correlation between richness and endemism is low, but aggregate regions selected for high levels of endemism nevertheless select more species than expected by chance, indicating that global distribution patterns of endemism are a useful surrogate for the conservation of all terrestrial vertebrates (Orme et al. 2005; Lamoreux et al. 2006). By combining different approaches, one can create a global map of priority ecoregions (geographic regions representing natural units of distinct communities and species assemblages), which might be expected to preserve a good representation of the world's biodiversity and threatened species (Funk & Fa 2010).

The goal of maximizing gains from each biodiversity dollar by identifying baskets of priority species that can efficiently be conserved together extends far beyond the formally recognized Conservation International hot spots. For example, Valenzuela-Galván et al. (2008) identified the smallest set of $2° \times 2°$ grid cells that, if protected, would conserve all 47 north American terrestrial carnivores (seven cells were needed to represent all carnivores at least once, 18 to represent each at least three times, and 84 to represent at least 10% of each species regional range), and Macdonald et al. (see Chapter 16) identified, at the level of 1° grid cells, considerable spatial overlap between clusters of threatened primates and felids. More than 80% of the land where at least one threatened species of either primate or felid occurs also contains at least one threatened species of the other taxon (over 60% of these 'feliprime spots' lie outside Conservation International's hot spots) (Figure 25.2).

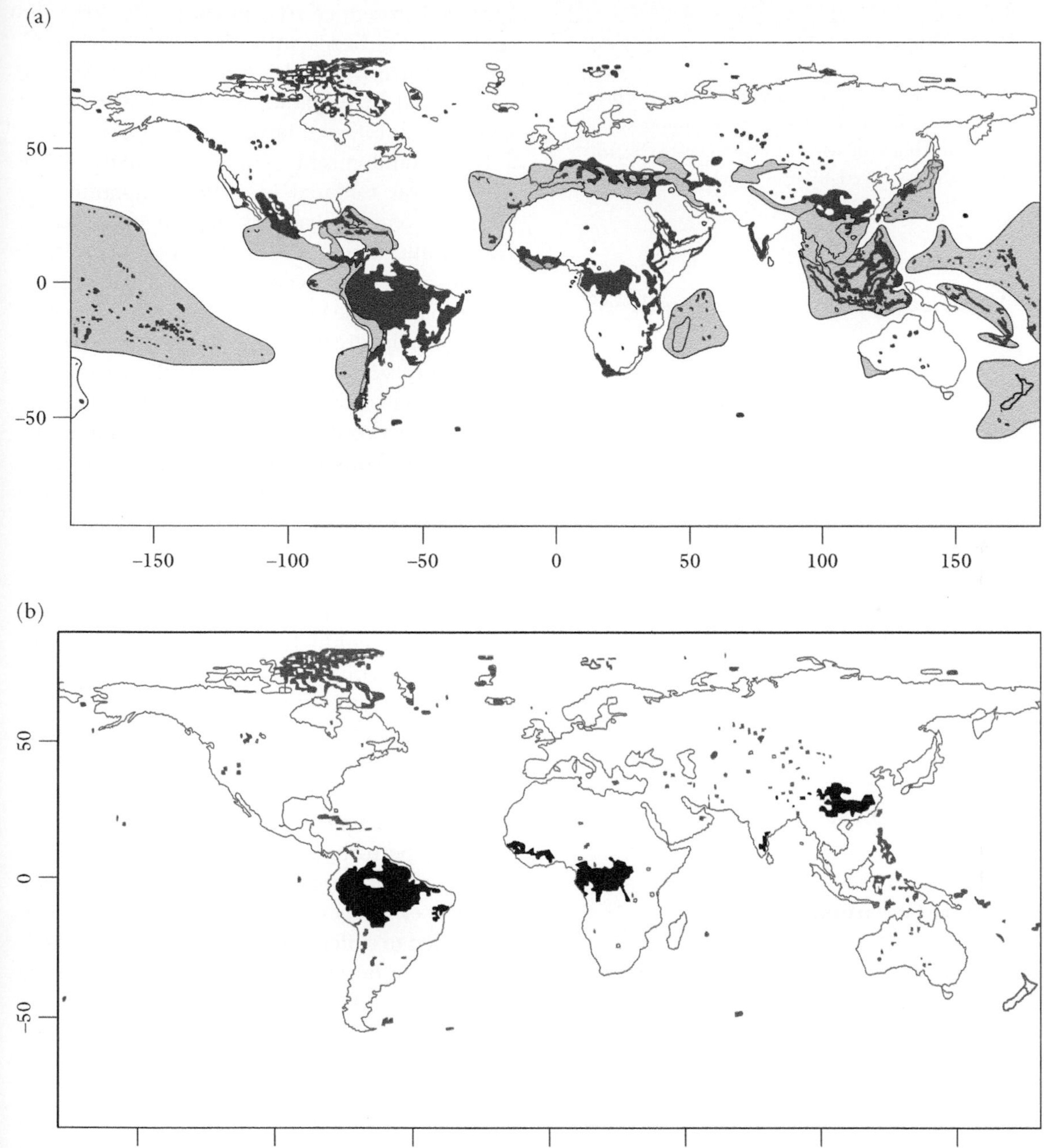

Figure 25.2 (a) Conservation International biodiversity hot spots (shaded in grey) and 'feliprime spots' (where at least one threatened primate and at least one threatened felid occur, shaded in black). (b) Those areas of the felid-primate hot spots which are *not* contained in the Conservation International biodiversity hot spots (reproduced from Macdonald et al. 2012).

Ninth Elephant

Dogmatic proprietorial adherence to different prioritization systems by different organizations may politicize the choice of where to conserve biodiversity. How can the wasted effort of competitive conservation be minimized?

What compromises are required?

Compromise is not necessarily a tawdry outcome. The reality for conservation, and for all other worthwhile elements of the human enterprise, is that the 'least worst' case may be our best option. Two cases, from very different aspects of conservation and different parts of the world, illustrate the inescapable interdisciplinarity, difficulty of alignment and threats of unintended consequences that create a reality where simple caricatures of the conflicting stakeholders are unhelpful, and tough choices are inevitable. We present these examples to expose the complexity of conservation problems, illustrating that biological knowledge is necessary, but not sufficient, to solve them.

Bovine tuberculosis

Bovine tuberculosis (bTB) was first identified in a badger in England in 1971. Badgers (*Meles meles*), which may have contracted bTB (*Mycobacterium bovis*) originally from cattle, are now a maintenance host for it. Controlling bTB in cattle has been a challenge since Robert Koch discovered it in 1882; in those days many people died from milk-borne *M. bovis*. In 1950 a national compulsory TB eradication scheme began. The number of test reactor cows fell from nearly 15,000 in 1961 to 569 in 1982. Continuing higher incidence in south west England implicated badgers, so from 1973 to 1998, cattle test-and-slaughter was complemented with a succession of badger-culling strategies. None appeared to reduce bTB in cattle. By 2010 the percentage of herds suffering bTB infections in the south west was 7% and 25,000 cattle were slaughtered in England, at an average cost of £30,000 per infected farm (Figure 25.3).

The Randomized Badger Culling Trial (RBCT) was initiated in 1998 to quantify the impact of culling badgers on the incidence of TB in cattle (Krebs 1997). This cost nearly £50 million, and its crucial result is a comparison between the cattle herd breakdowns (i.e. detected infections) in areas where badgers were killed proactively versus those where they were not killed. An idea advanced to explain the failure of earlier badger control was the perturbation effect, whereby the consequences of killing some badgers increased the ranging behaviour and susceptibility of the survivors, so that transmission of disease to cattle could be, perversely, worsened (Macdonald et al. 1996; Woodroffe et al. 2006; Riordan et al. 2011). Therefore, it is relevant to compare the change in herd breakdowns within the culling area and in a 2 km perimeter surrounding it. The results (Table 25.1) show that while culling was under way, herd breakdowns decreased in proactive core areas (*relative* to no-cull areas; in *absolute* terms they got worse in both areas), but increased in the perimeter. The perturbation effect had waned by 18 months after culling treatments ended, although, over the entire period, farmers in the perimeter still tended to suffer a worsened breakdown rate – a cost borne by perimeter farmers, offsetting the benefit accrued by core farmers. Importantly, over 9 years, and taking together *both* the core and perimeter of an extrapolated 150 km^2 circular control zone, the estimated net benefit varied from 3% to 21% (Table 25.2). In 2007, the authors of the RBCT concluded: 'These results combined with evaluation of alternative culling methods, suggest that badger culling is unlikely to contribute effectively to the control of cattle TB in Britain' (Bourne et al. 2007). But by 2010, with a longer run of data, society needed to reconsider whether killing badgers was worthwhile, in the context of the annual cost of bTB to the British taxpayer of £90 million.

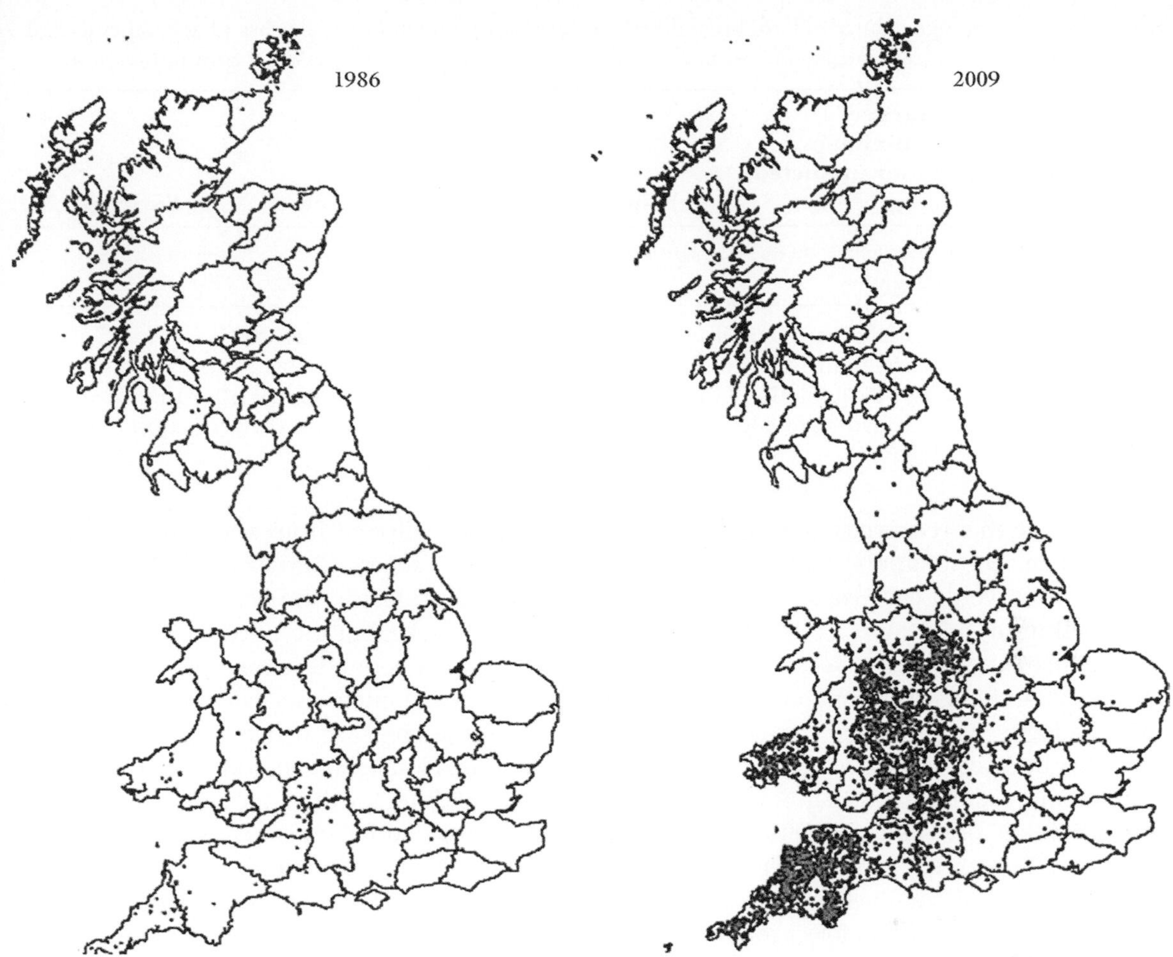

Figure 25.3 Bovine TB cases in England, showing isolated cases in 1986 and more widespread patterns in the south-west in 2009. Maps from DEFRA: http://www.defra.gov.uk/publications/files/pb13601-bovinetb-eradication-programme-110719.pdf).

Table 25.1 Percentage change in herd breakdowns within the culling area and in a 2 km wide perimeter surrounding it (based on Jenkins et al. 2010). Reproduced from Macdonald & Burnham 2011)

	(a) During trial	(b) Postcull years	(c) Entire period
RBCT trial area	−12.4 to −32.7% Est −23.2%	−19.1 to −42.0% Est −31.5%	−20.2 to −33.9% Est −27.4%
2 km perimeter	−0.6 to +56.0% Est +24.5%	−27.4% to +26.0% Est −4.4%	−14.6 to +37.4% Est +8.3%

RBCT, Randomized Badger Culling Trial.
(a) During trial: the 4–7 years of the RBCT starting after the first proactive cull (culling ended in 2005) until 1 year after the last proactive cull.
(b) Postcull years: from 1 year after the last proactive cull to February 2011.
(c) Entire period: from the completion of the initial proactive cull until February 2011.

Table 25.2 Percentage change in herd breakdowns of core and perimeter over 9 years (5 annual culls and then 4 more years) extrapolated from estimates in Table 25.1 (C.A. Donnelly, personal communication)

Extrapolated consequences of badger culling in minimum licenseable cull area for predicted herd breakdowns	(a) Cull years	(b) Postcull years	(c) Entire period
150 km² circular culling area plus 2 km wide perimeter surrounding it	+9 to −17% Est −4%	−8 to −33% Est −21%	−3 to −21% Est −12%

(a) 5 years from the initial proactive cull (assuming annual culling, with 5 such culls).
(b) The following 4 years: from 12 to 60 months after the final proactive cull.
(c) After 9 years: from the completion of the initial proactive cull until 60 months after the 5th annual cull (if the removal area had 'hard edges' and thus no perimeter perturbation, the reduction in bTB incidence might be 20–34%).

As part of this reconsideration, government launched a consultation regarding culls in England that would each cover at least 70% of land within a minimum of 150 km² (larger area reduces the relative size of the perturbed perimeter) and might each involve killing about 1500 badgers over 4 years, the culling being done by licensed groups of farmers or their agents. One estimate was that 5 years' culling with 2.5 years' post-culling might prevent 22.6 confirmed herd breakdowns, saving £610,200 (against the cost of conducting five annual culls estimated as £2.14 million for cage trapping, or £562,500 by farmers shooting; Jenkins et al. 2010). Whether such culling is worthwhile depends not only on whether society judges these gains to merit the costs (financially and in badgers), and over what time-scale. The huge practical task and financial commitment facing the farmers, the likelihood of protest, use of an untested method (shooting), and the international legality (under the Berne Convention) of scaling up this type of control to a sufficient proportion of the 39,000 km² of Britain (c.30% of England) in which bTB is endemic to make significant inroads into the national problem (such scaling up might involve killing about 100,000 badgers over 4 years – there are an estimated 360,000 badgers in the UK) all need to be considered. Is this plan the least worst option, in comparison to, for example, vaccinating cattle (which requires developing a molecular tag to differentiate infected from vaccinated cattle) or vaccinating badgers (Corner et al. 2010)? It's a tough call, considering the livelihoods and human anguish at stake, but the uncertainties, the costs and seemingly poor return on capital, the impacts on badgers and intense societal disquiet combine to make the proposed badger cull unpromising. Each group of stakeholders will view the ocean of possibilities from the dynamic perspectives of different rising and falling crests of waves of opinion. Our point is less about the final outcome (in late 2012 the cull was shelved) – there is no unarguably 'right' answer, and all options involve some losers – but more to illustrate the perplexing judgements that characterize almost all conservation issues. The badger case requires serious attention to issues as diverse as ecology, immunology, agriculture, economics, national and international law, civil rights, ethics and politics.

Tenth Elephant

Are there some problems which are simply intractable for the time being, however sad it may be? For example, it appears that there is no tidy solution now to controlling bTB in badgers, at least not one that society is likely to judge acceptable. The understandable desire to do something should not be at the cost of doing the wrong thing.

Lions, from trophies to cans

Lions are in peril: from a Stone Age distribution across much of Africa and Eurasia (Yamaguchi et al. 2004), their populations have plummeted, reaching perhaps 100,000 animals in the late 19th century, and declining to a population of perhaps 32,000 today, distributed in patchy and often threatened enclaves (Riggio et al. submitted). The key drivers of lion decline are lost habitats through land use change, conflict with people, trophy hunting and medicinal trade in their bodies – all are anthropogenic and all interact, often in unexpected ways.

In general, agriculture is more profitable than wildlife in the lions' range, leading to lost natural habitats and prey and increased conflict due to lions killing stock. The number of lions killed in agricultural conflict is unknown (e.g. Hazzah et al. 2009) but certainly much higher than the 760 lions shot by trophy hunters in Africa in 2011. Lion hunts attract the highest prices of all trophy species (US$24,000–71,000) (although the total income generated by lion hunting is not as important as that generated by elephants (Lindsey et al. 2012). Although considered irredeemably unethical by some, it is possible in principle for sport hunting of lions to be sustainable (e.g. Whitman et al. 2004), but it is clear that in practice it is often not so. Loveridge et al. (2007) detail the case for western Zimbabwe where mistaken estimates of lion numbers, together with financial imperatives to maximize short-term profits, led to unsustainable trophy harvests surrounding Hwange National Park. However, a 5-year moratorium resulted in recovery of lion numbers, readjustment and stabilization of their perturbed social system and subsequently a new hunting quota (reduced from 60 to four males pa; Loveridge et al. 2007; Davidson et al. 2011). In Tanzania trophy hunting may be driving declines in lion numbers (Packer et al. 2011). However, hunting may also contribute to lion conservation by motivating tolerance of lions on land from which they would otherwise be extirpated.

The argument is that there are many areas inhabited by lions that are too inhospitable or otherwise unsuitable for phototourism, where trophy hunting provides the reason for maintaining lion populations. The argument runs that if the trophy market was lost, then land owners and governments might no longer tolerate lions, and be more inclined to look for other (agricultural) land uses that would be inimical to biodiversity more generally. Net returns from livestock in semi-arid rangelands average $10–30/km^2$ pa, whereas the estimate for trophy hunting is $24–164/km^2$ pa (Lindsey et al. 2012), although it is hard to predict how these figures would change as the trophy-hunting industry adapted to the loss of lion hunting. This raises the at first counter-intuitive argument that trophy hunting is what keeps lions alive in considerable parts of their remaining range. This argument is made for much of northern Botswana and South Sudan (where private sector hunting is seen as the only restraint on land purchases by Qatar for food production). Another proposition is that hunting zones form corridors between protected areas (PAs) (e.g. in Cameroon three PAs are linked by hunting zones that would otherwise be speedily converted to cotton production). Of course, sooner or later agricultural technology, such as GMOs, will facilitate a use for these marginal lands that is more profitable even than lion hunting, and so if conservation relies solely on market-based arguments in the long-run it is likely doomed anyway.

Eleventh Elephant

Despite the power of revealing the full financial value of biodiversity, if monetized value becomes the sole arbiter of what wildlife is to be conserved, many of the most treasured wild species will eventually be lost. Can market forces really be entrusted with the fate of irreplaceable species?

This land use argument links, in unexpected ways, to concerns about an emerging threat to lions from increasing demand for their body parts from traditional Asian medicine (TAM). One way to staunch the international legal trade stimulated by this demand would be to uplift lions to Appendix 1 of CITES (remembering that CITES restrictions on legal trade have not controlled illegal trade in tiger bone, rhino horn or ivory). As well as banning medicinal trade, this would, however, make it illegal for lion trophy hunters from the USA to take their trophies home unless they applied to the US Fish and Wildlife Service under the Endangered Species Act (ESA) for an import licence. While a so-called non-detrimental exemption could be issued by the US ESA to allow a hunter to import their CITES Appendix 1 lion trophy, this regulatory bar is set very high (the exporting country must demonstrate national-level monitoring of the harvest for 3 consecutive years against nationwide total population counts). Another complication is that the strictures that apply to importing Appendix 1 species do not apply to Appendix 2 species, which include farmed individuals. A perverse outcome could be, therefore, that it was possible to import trophies of farmed (so-called 'canned') lions but not so-called (some would say, ironically) 'fair-chase' wild lions. Of the 760 lions shot as trophies in 2010, some 40% were farmed (K. Marnewick, personal communication). The outcome, unintended and dreaded by anti-hunters and 'ethical' hunters alike, could therefore be to ruin fair-chase hunting while enhancing the market for farmed lions.

Canned lion hunting lies at the confluence of loathing by almost all interlocutors in the debate. It is despised by the traditional ethic of fair-chase hunting, deplored by anti-hunting organizations, abhorred by welfare organizations, perceived by regulators to be linked to organized crime, and feared by conservationists for, *inter alia*, fostering the demand (for wild lion parts) by TAM. Of course, a counter-argument is that providing the market with farmed lion products could take pressure off wild lions – a principle known as substitutability which appears rarely to work in wildlife trade (see Chapter 5). This is vividly illustrated with respect to the farming of bears for their bile (which merely enhances a luxury high-end market for the wild product; Dutton et al. 2011). Of course, an alternative view emphasizes that the production of canned lions is a legal activity (there are an estimated 3600 lions held in farms in South Africa in some 167 facilities across nine provinces, generating 225 full-time jobs and a revenue of US$12.5 million). This is a prime example of a generic question in conservation: can values (ethics) moderate the free market (which, if unfettered, makes farmed lions irresistible)?

The most thorough analysis yet attempted, by Lindsey et al. (2012), concludes that banning lion hunting could render trophy hunting as a whole unviable in 59,538 km^2 of Africa, thereby posing a risk that those areas would be lost as lion habitat (this representing a minimum of 11.5% of areas where lions are hunted, or 3.6% of the lion's range). Again, it is hard to predict how the trophy-hunting industry might change its commercial behaviour, and thus the financial analysis, if it lost access to wild lions. Lindsey et al. (2012) also cite examples where money from lion hunting supplements the state funding of PAs (e.g. $380–400,000 pa to Niassa National Reserve in Mozambique, $546,000 pa to the Save Valley anti-poaching effort in Zimbabwe). They suggest that, rather than banning lion hunting and risking the loss of associated revenue to conservation along with a motive to protect lion habitat from other uses, a better compromise would be to set stringent quotas (e.g. 0.5 lions/1000 km^2 for relatively high-density populations). For context, Dickman (personal communication) reports >6 lions per 100 km^2 pa killed annually due to conflict with pastoralists in her Ruaha study area.

A critical issue is whether hunting can be regulated and policed effectively; if not, then the quotas are ineffective (parallel arguments are pivotal in other trade issues, such as ivory) (Wasser et al. 2010) (see Chapter 5). In essence, lions have to be made so valuable to professional hunters – and also local communities – that these stakeholders exert irresistible peer pressure on renegades. Another paradox, at the intersection of human–lion conflict with trophy hunting, arises from the fact that protected areas, however expansive, will always have an edge, and generally lions on the edge will conflict with local farmers (Loveridge et al. 2010). Often this will necessitate killing lions and, insofar as this is done legally, the options are to do it for a cost to governments as a form of problem animal control, or to do it at a profit with the private sector hunters: probably the choice will be made by world opinion of hunting.

Twelfth Elephant

Options that at first might be perceived as anathema to conservation may turn out to offer the least worst case – such as trophy hunting providing a motive to protect lions. A balance between purism and pragmatism may increasingly be required to deliver conservation, leading to unaccustomed bedfellows. But how broadly should conservationists forge their alliances, and is there a point beyond which we should not venture?

These case studies reveal two pervasive generalizations, about which society in general, and many conservationists in particular, may be in perilous denial. First, now that it is recognized as a global issue, wildlife conservation must fight its corner alongside the big issues of the day – food and water security, carbon pollution, health, national security, economic growth and development. Sometimes this will bring into painful focus the inescapability of trade-offs between conservation and other societal goods.

Thirteenth Elephant

May more resources for conservation sometimes mean fewer for another worthy cause?

Furthermore, it is often assumed that with the raised awareness of wider environmental issues, biodiversity will be swept to safety on the back of wider environmental protection, for example, forest species being protected as a co-benefit of REDD-like mechanisms. This is by no means a safe assumption (e.g. protecting a habitat does not necessarily protect the threatened species within it ;Collins et al. 2011), so a major question is how is biodiversity conservation to be bundled securely into wider environmental protection and societal development (although there is potential: de Barros et al. (submitted) show that high-priority areas for jaguar conservation in Brazil are positively associated with the carbon stocks that may attract carbon funding)? Second, and in the context of shared gains for conservation and human development (especially poverty alleviation), there is an urgent quest for 'win–win' outcomes, enabling poor people to prosper while wildlife flourishes alongside them. While a decade or more of conservation rhetoric has promoted this obviously desirable synergy, successful examples of biodiversity delivering poverty alleviation are not numerous and the reality may be that on a large scale, win–wins between development and conservation are inherently rare and much of the future of conservation and development will be a quest for better ways of sharing losses.

Fourteenth Elephant

It is possible that the much hoped-for synergy of conservation and poverty alleviation may not always be possible, and that there will be cases where conservation is a block on development. Is that acceptable and, if so, who will pay the opportunity cost?

These two huge issues draw us, finally and inevitably, to the topic of the human environmental footprint. None of the key drivers mentioned above is mutually independent, all are anthropogenic, and all exacerbated by the rapidly increasing human population as pointed out by the famous French naturalist Georges Buffon (Buffon 1785), who presciently illustrated his argument with the lion, as have we 200 years later:

> 'Man's industry augments in proportion to his numbers; but that of the other animals remains always the same. All the destructive species, as that of the lion, seem to be banished to distant regions, or reduced to a small number, not only because mankind have increased, but because they have also become more powerful, and have invented formidable arms which nothing can resist' (Buffon 1785).

The fear, in this context, is that much of conservation is at the scale of a metaphorical sticking-plaster in the face of crippling injury.

The final elephant

The metaphorical elephants that we have flushed from the rooms in conservation's memory palace are, like members of a herd of real elephants, linked by common descent. In this case, the relatedness of Thought Elephants is through the kinship of conflict, direct or indirect, with the human enterprise. Specifically, although each issue has its nuances and technical intricacies, each ultimately abuts against the numbers of people, and their consumption, and consequently the dwindling insufficiency of remaining space and resource for wildlife. We need not rely on Thought Elephants to illustrate this – real ones make the point sufficiently. There is a threshold for human population density of about 16 people per km^2 above which elephant survival is jeopardized (Hoare & du Toit 1999). In these areas, therefore, we could easily ask the question whether increasing the human population by one more person per km^2 would be as attractive an option as ensuring the survival of elephants. This albeit grossly simplified point brings us to the largest elephant in the room, the rampaging tusker in full musth that is human population. Conservation is, by definition, an attempt to ameliorate the consequences of human actions on nature, and ultimately many of these consequences stem from the size of the global human population footprint (Speidel et al. 2007). Increasing human populations, and impact, risks the double-whammy of disaster for both biodiversity and people (Schade & Pimentel 2010).

Returning to the questions of how much biodiversity we need (or want), which transcend science into policy, Ehrlich & Ehrlich (1997) point out that the same can be asked about people, and raise a choice to have fewer humans each with a better quality of life. This is an arena in which commentators find it perilously easy to slip into grotesque simplification, prejudice and insensitivity, each of which we will strive to avoid. Nevertheless, while intellectually exacting, there is no ethical impropriety in pondering possible interactions, and trades-off, between numbers of people, their quality of life and their impacts on nature. Macdonald (in press) suggests that one way of thinking about this could be as an environmental analogue of QALYs – the quality-adjusted life-years used to inform tough decisions in medicine. The WWF (2008) estimates that five worlds' worth of resources are required for the world's population to live at the current standard enjoyed in North America. Five worlds (or the technology to provide their resource equivalent) are currently not available, and so there is a clear trade-off between the quantity of people and their standard of living. Turning the algebra on its head, to maintain the current world population equitably and sustainably might require each person to have an environmental footprint roughly one-fifth that of a contemporary American, or for the world to be populated by a fifth of the current number of people, all at levels of consumption typical of contemporary Americans.

Fifteenth Elephant

Is it the case that wildlife conservation, and perhaps even sufficiency of ecosystem services, is impossible with current trends in human population size, and even less possible if all human populations are to enjoy the levels of consumption currently typical of more developed countries?

Sustainability is a slippery notion, but a widely used definition of sustainable development is '... development that meets the needs of the present without compromising the ability of future generations to meet their own needs'(World Commission on Environment and Development 1987). Accordingly, future generations should have no less of the means to meet their needs than do their predecessors, but how is a generation to judge whether it is bequeathing an adequate productive base for its successor? Tackling this question highlights difficulties with the metrics used to judge the wealth of nations, such as gross domestic product (GDP) and the United Nations' Human Development Index (HDI), both of which routinely underprice the erosion of natural resources caused by economic growth (Dasgupta 2010). Dasgupta (2010) describes this underpricing as 'nature's subsidies': imagine the government of a forest-rich country can earn revenue by granting timber concessions to private firms. If the resultant upland deforestation causes soil erosion and affects water supply downstream, then the timber firm should compensate downstream farmers – an unlikely outcome if the victims are scattered villagers with little influence. By not compensating the downstream stakeholders, the timber firm's operating cost is less than the full social cost of deforestation (i.e. the firm's logging costs *plus* the cost of compensating the damage suffered by all who are adversely affected); in effect, the downstream community is subsidizing the timber exports! Dasgupta (2010) concludes: 'Development policies that ignore our reliance on natural capital are seriously harmful – they do not pass the mildest test for equity among contemporaries, nor among people separated by time and uncertain contingencies'.

Sixteenth Elephant

Considering that people's concern for their descendents beyond their grandchildren may diminish steeply, is it likely that motivation to endure losses, or even to forego gains, for the sake of the quality of life of future generations may wane rapidly across generations to a level not much different to that felt for unknown contemporaries, and perhaps even less, insofar as those contemporaries exist whereas the future is unknown?

Macdonald (in press), deploring the likelihood of famine, plague and warfare as the outcomes of this dilemma, suggests that the only hopeful road is one of phased population reduction over many generations and concomitant curbs on consumption (along with a hope for similarly near-miraculous levels of technological innovation). This would require an intergenerational pact for which there may be no precedent in human history or evolution.

Our herd of Thought Elephants is crowding the room, trumpeting the complexity of issues that link conservation science and policy. If biodiversity, nature, wilderness are to be conserved, whether for utilitarian or ethical reasons, then the evidence so painstakingly gathered by conservation scientists will need to be acted upon by a global society that adopts an ecological, evidence-based variant of President J.F. Kennedy's famous exhortation: Ask not what the world can do for you, but what you can do for the world. Informing the answers will be *the* key topic for the emerging generation of conservationists, and an immense challenge because, as our opening quotation from Robbie Burns made plain, facts are stubborn and, like Thought Elephants, not easily herded.

Acknowledgements

We are rateful for insightful discussions, and comments on earlier drafts from Dawn Burnham, Amy Dickman, Christl Donnelly, Andrew Loveridge, Ewan Macdonald, Gus Mills, Chris Newman, Christopher O'Kane and Nobby Yamaguchi. In particular, we thank Tom Moorhouse for his tireless help as a critical friend, a mahout in the herding of Thought Elephants.

References

Allsopp, M.H., de Lange, W.J. & Veldtman, R. (2008) Valuing insect pollination services with cost of replacement. *PLoS ONE*, **3**, e3128.

Ancrenaz, M., Ambu, L., Sunjoto, I., *et al.* (2010) Recent surveys in the forests of Ulu Segama Malua, Sabah, Malaysia, show that orang-utans (P. p. morio) can be maintained in slightly logged forests. *PLoS ONE*, 5, e11510.

Ball, I.R., Possingham, H.P. & Watts, M. (2009) Marxan and relatives: software for spatial conservation optimization. In: *Spatial Conservation Prioritization: Quantitative Methods and Computational Tools* (eds A. Moilanen, K.A. Wilson & H.P. Possingham). Oxford University Press, Oxford.

Belnap, J., Kaltenecker, J.H., Rosentreter, R., Williams, J., Leonard, S. & Eldridge, D. (2001) *Biological Soil Crusts: Ecology and Management*. BLM technical reference 1730–2. National Applied Resource Science Center, US Bureau of Land Management, Denver, CO.

Bhagwat, S.A., Willis, K.J., Birks, H.J.B. & Whittaker, R.J. (2008) Agroforestry: a refuge for tropical biodiversity? *Trends in Ecology and Evolution*, **23**, 261–267.

Bourne, F.J., Donnelly, C.A., Cox, D.R., *et al.* (2007) *Bovine TB: The Scientific Evidence*. Independent Scientific Group on Cattle TB. http://archive.defra.gov.uk/foodfarm/farmanimal/diseases/atoz/tb/isg/report/final_report.pdf

Brooks, T.M., Mittermeier, R.A., da Fonseca, G.A.B., *et al.* (2006) Global biodiversity conservation priorities. *Science*, **313**, 58–61.

Buffon, G. (1785) *Natural History, General and Particular*, Second English Edition Volume 5 (translated into English by W. Smellie). W. Strahan and T. Cadell, London.

Butchart, S.H.M., Stattersfield, A.J., Baillie, J., *et al.* (2005) Using Red List indices to measure progress towards the 2010 target and beyond. *Philosophical Transactions of the Royal Society B: Biological Sciences*, **360**, 255–268.

Carroll, C., Johnson, D.S., Dunk, J.R. & Zielinski, W.J. (2010) Hierarchical Bayesian spatial models for multispecies conservation planning and monitoring. *Conservation Biology*, **24**, 1538–1548.

CBD. (2010) Global Biodiversity Outlook 3. www.cbd.int/gbo/

Chapin III, F.S., Zavaleta, E.S., Eviner, V. T., *et al.* (2000) Consequences of changing biodiversity. *Nature*, **405**, 234–242.

Collins, M.B., Milner-Gulland, E.J., Macdonald, E.A. & Macdonald, D.W. (2011) Pleiotropy and charisma determine winners and losers in the REDD+game: all biodiversity is not equal. *Tropical Conservation Science*, **4**, 261–266.

Corner, L.A.L., Costello, E., O'Meara, D., *et al.* (2010) Oral vaccination of badgers (Meles meles) with BCG and protective immunity against endobronchial challenge with Mycobacterium bovis. *Vaccine*, **28**, 6265–6272.

Costanza, R., Darge, R., Degroot, R., *et al.* (1997) The value of the world's ecosystem services and natural capital. *Nature*, **387**, 253–260.

Cuéllar, E., Johnson, P. & Macdonald, D.W. (submitted) Diet composition of cattle and guanaco in the relict Chacoan savannas of Bolivia.

Dasgupta, P. (2010) Nature's role in sustaining economic development. *Philosophical Transactions of the Royal Society B:Biological Sciences*, **365**, 5–11.

Davidson, Z., Valeix, M., Loveridge, A.J., Madzikanda, H. & Macdonald, D.W. (2011) Socio-spatial behaviour of an African lion population following perturbation by sport hunting. *Biological Conservation*, **144**, 114–121.

De Barros, A.E., Macdonald, E.A., Matsumoto, M.H., *et al.* (submitted) "Hotspots" to "REDDspots": optimis ing carbon, jaguars and biodiversity conservation.

Duffy, J.E. (2009) Why biodiversity is important to the functioning of real-world ecosystems. *Frontiers in Ecology and Environment*, **7**, 437–444.

Durant, S.M., Pettorelli, N., Bashir, S., *et al.* (2012) Forgotten biodiversity in desert ecosystems. *Science*, **336**, 1379–1380.

Dutton, A.J., Hepburn, C. & Macdonald, D.W. (2011) A stated preference investigation into the Chinese demand for farmed vs. wild bear bile. *PLoS ONE*, **6**, e21243.

Ehrlich, P. & Ehrlich, A.H. (1997) The population explosion: why should we care and what should we do about it? *Environmental Law*, **27**, 1187–1208.

Ehrlich, P. & Walker, B. (1998) Rivets and redundancy. *Bioscience*, **48**, 387–387.

Farhadinia, M.S. (2004) The last stronghold: cheetah in Iran. *Cat News*, **40**, 11–14.

Figueroa-Rangel, B.L., Willis, K.J. & Olvera-Vargas, M. (2008) 4200 years of pine-dominated upland forest dynamics in west-central Mexico: human or natural legacy? *Ecology*, **89**, 1893–1907.

Fisher, B., Polasky, S. & Sterner, T. (2011) Conservation and human welfare: economic analysis of ecosystem services. *Environmental and Resource Economics*, **48**, 151–159.

Foley, J.A., Defries, R., Asner, G.P., *et al.* (2005) Global consequences of land use. *Science*, **309**, 570–574.

Fox, H.E., Soltanoff, C.S., Mascia, M.B., *et al.* (2012) Explaining global patterns and trends in marine protected area (MPA) development. *Marine Policy*, **36**, 1131–1138.

Frankham, R. (2003) Genetics and conservation biology. *Comptes Rendus Biologies*, **326**(Supplement 1), 22–29.

Funk, S.M. & Fa, J.E. (2010) Ecoregion prioritization suggests an armoury not a silver bullet for conservation planning. *Plos ONE*, **5**, e8923.

Hazzah, L., Mulder, M.B. & Frank, L. (2009) Lions and warriors: social factors underlying declining African lion populations and the effect of incentive-based management in Kenya. *Biological Conservation*, **142**, 2428–2437.

He, F. & Hubbell, S.P. (2011) Species-area relationships always overestimate extinction rates from habitat loss. *Nature*, **473**, 368–371.

Hellmann, J.J. & Pineda-Krch, M. (2007) Constraints and reinforcement on adaptation under climate change: selection of genetically correlated traits. *Biological Conservation*, **137**, 599609.

Hendry, A.P., Lohmann, L.G., Conti, E., *et al.* (2010) Evolutionary biology in biodiversity science, conservation, and policy: a call to action. *Evolution*, **64**, 1517–1528.

Hoare, R.E. & du Toit, J.T. (1999) Coexistence between people and elephants in African savannas. *Conservation Biology*, **13**, 633–639.

Hobbs, R.J., Higgs, E. & Harris, J.A. (2009) Novel ecosystems: implications for conservation and restoration. *Trends in Ecology and Evolution*, **24**, 599–605.

Hooper, D.U., Chapin, F.S., Ewel, J.J., et al. (2005) Effects of biodiversity on ecosystem functioning: a consensus of current knowledge. *Ecological Monographs*, **75**, 3–35.

Hughes, A.R., Inouye, B.D., Johnson, M.T.J., Underwood, N. & Vellend, M. (2008) Ecological consequences of genetic diversity. *Ecology Letters*, **11**, 609–623.

Husson, S.J., Wich, S.A., Marshall, A.J., *et al.* (2008) Orangutan distribution, density, abundance and impacts of disturbance. In: *Orangutans: Geographic Variation in Behavioral Ecology and Conservation* (eds S.A. Wich, S. Atmoko, T.M. Setia, *et al.*). Oxford University Press, Oxford.

Jenkins, C.N. & Joppa, L. (2009) Expansion of the global terrestrial protected area system. *Biological Conservation*, **142**, 2166–2174.

Jenkins, H.E., Woodroffe, R. & Donnelly, C.A. (2010) The duration of the effects of repeated widespread badger culling on cattle tuberculosis following the cessation of culling. *Plos ONE*, **5**, e9090.

Johnson, J. & Dunn, P. (2006) Low genetic variation in the heath hen prior to extinction and implications for the conservation of prairie-chicken populations. *Conservation Genetics*, **7**, 37–48.

Krebs, J. (1997) *Bovine Tuberculosis in Cattle and Badgers*. MAFF Publications, London.

Kumar, P.E. (2010) *The Economics of Ecosystems and Biodiversity: Ecological and Economic Foundations*. Earthscan, London.

Lamoreux, J.F., Morrison, J.C., Ricketts, T.H., *et al.* (2006) Global tests of biodiversity concordance and the importance of endemism. *Nature*, **440**, 212–214.

Lande, R. & Shannon, S. (1996) The role of genetic variation in adaptation and population persistence in a changing environment. *Evolution*, **50**, 434–437.

Larsen, T. (2008) Forest butterflies in West Africa have resisted extinction… so far (Lepidoptera: Papilionoidea and Hesperioidea). *Biodiversity and Conservation*, **17**, 2833–2847.

Lindsey, P.A., Balme, G.A., Booth, V.R. & Midlane, N. (2012) The significance of African lions for the financial viability of trophy hunting and the maintenance of wild land. *Plos ONE*, **7**, e29332.

Lonsdorf, E., Ricketts, T.H., Kremen, C., Winfree, R., Greenleaf, S. & Williams, N.M. (2011) Crop pollination services. In: *Natural Capital: Theory and*

Practice of Mapping Ecosystem Services (ed. P. Karieva). Oxford University Press, Oxford.

Loveridge, A.J., Searle, A.W., Murindagomo, F. & Macdonald, D.W. (2007) The impact of sport-hunting on the population dynamics of an African lion population in a protected area. *Biological Conservation*, **134**, 548–558.

Loveridge, A J., Hemson, G., Davidson, Z. & Macdonald, D.W. (2010) African lions on the edge: reserve boundaries as 'attractive sinks'. In: *Biology and Conservation of Wild Felids* (eds D.W. Macdonald & A.J. Loveridge). Oxford University Press, Oxford.

Luck, G.W. (2007) A review of the relationships between human population density and biodiversity. *Biological Reviews of the Cambridge Philosophical Society*, **82**, 607–645.

Macdonald, D.W. (in press) From ethology to biodiversity: case studies of wildlife conservation. *Nova Acta Leopoldina N.F.*, **111**(380).

Macdonald, D.W. & Burnham, D. (2011) *The State of Britain's Mammals 2011*. PTES, London.

Macdonald, D.W., Mitchelmore, F. & Bacon, P.J. (1996) Predicting badger sett numbers: evaluating methods in East Sussex. *Journal of Biogeography*, **23**, 649–655.

Macdonald, D.W., Collins, N.M. & Wrangham, R. (2006) Principles, practice and priorities: the quest for 'alignment'. In: *Key Topics in Conservation Biology* (eds D.W. Macdonald & K. Service). Wiley-Blackwell, Oxford.

Macdonald, D.W., Collins, M., Johnson, P.J., *et al.* (2011) Wildlife conservation and reduced emissions from deforestation in a case study of Nantu National Park, Sulawesi 1. The effectiveness of forest protection – many measures, one goal. *Environmental Science and Policy*, **14**, 697–708.

Macdonald, D.W., Burnham, D., Hinks, A.E. & Wrangham, R. (2012) A problem shared is a problem reduced: seeking efficiency in the conservation of felids and primates. *Folia Primatologica*. DOI:1159/00034 2399

Mace, G.M. & Purvis, A. (2008) Evolutionary biology and practical conservation: bridging a widening gap. *Molecular Ecology*, **17**, 9–19.

Mace, G.M., Balmford, A. & Ginsberg, J.R.E. (1998) *Conservation in a Changing World*. Cambridge University Press, Cambridge.

Mace, G.M., Possingham, H.P. & Leader-Williams, N. (2006) Prioritizing choices in conservation. In: *Key Topics in Conservation Biology* (eds D.W. Macdonald & K. Service). Wiley-Blackwell, Oxford.

Margules, C. & Pressey, R.L. (2000) Systematic conservation planning. *Nature*, **405**, 243–253.

Margules, C., Pressey, R. & Williams, P. (2002) Representing biodiversity: data and procedures for identifying priority areas for conservation. *Journal of Biosciences*, **27**, 309–326.

May, R.M. (2011) Why should we be concerned about loss of biodiversity? *Comptes Rendus Biologies*, **334**, 346–350.

May, R.M. (in press) Why should we be concerned about loss of biodiversity? *Comptes Rendus Biologies*.

McKee, J.K., Sciulli, P.W., Fooce, C.D. & Waite, T.A. (2004) Forecasting global biodiversity threats associated with human population growth. *Biological Conservation*, **115**, 161–164.

Merckx, T., Feber, R.E., Riordan, P., et al. (2009) Optimizing the biodiversity gain from agri-environment schemes. *Agriculture Ecosystems and Environment*, **130**, 177–182.

Millennium Ecosystem Assessment. (2005) *Ecosystems and Human Well-Being: Health Synthesis*. Island Press,Washington, D.C. www.maweb.org

Mittermeier, R.A., Ganzhorn, J.U., Konstant, W.R., *et al.* (2008) Lemur diversity in Madagascar. *International Journal of Primatology*, **29**, 1607–1656.

Moilanen, A. (2007) Landscape zonation, benefit functions and target-based planning: unifying reserve selection strategies. *Biological Conservation*, **134**, 571–579.

Myers, N., Mittermeier, R.A., Mittermeier, C G., da Fonseca, G.A.B. & Kent, J. (2000) Biodiversity hotspots for conservation priorities. *Nature*, **403**, 853–858.

Norton-Griffiths, M. (2007) How many wildebeest do you need? *World Economics*, **8**, 41–64.

O'Gorman, E.J., Yearsley, J.M., Crowe, T.P., Emmerson, M.C., Jacob, U. & Petchey, O.L. (2011) Loss of functionally unique species may gradually undermine ecosystems. *Proceedings of the Royal Society B: Biological Sciences*, **278**, 1886–1893.

Orme, C.D.L., Davies, R.G., Burgess, M., *et al.* (2005) Global hotspots of species richness are not congruent with endemism or threat. *Nature*, **436**, 1016–1019.

Packer, C., Brink, H., Kissui, B.M., Maliti, H., Kushnir, H. & Caro, T. (2011) Effects of trophy hunting on lion and leopard populations in Tanzania. *Conservation Biology*, **25**, 142–153.

Pearce, D., Hecht, S. & Vorhies, F. (2006) What is biodiversity worth? Economics as a problem and a solution. In: *Key Topics in Conservation Biology* (eds

D.W. Macdonald & K. Service). Wiley-Blackwell, Oxford.

Perfecto, I. & Vandermeer, J. (2010) The agroecological matrix as alternative to the land-sparing/agriculture intensification model. *Proceedings of the National Academy of Sciences USA*, **107**, 5786–5791.

Petit, R.J., Hu, F.S. & Dick, C.W. (2008) Forests of the past: a window to future changes. *Science*, **320**, 1450–1452.

Pfisterer, A.B. & Schmid, B. (2002) Diversity-dependent production can decrease the stability of ecosystem functioning. *Nature*, **416**, 84–86.

Phalan, B., Onial, M., Balmford, A. & Green, R.E. (2011) Reconciling food production and biodiversity conservation: land sharing and land sparing compared. *Science*, **333**, 1289–1291.

Possingham, H.P. & Wilson, K.A. (2005) Biodiversity – turning up the heat on hotspots. *Nature*, **436**, 919–920.

Pressey, R.L., Watts, M.E., Barnett, T.W. & Ridges, M.J. (2009) The C-Plan conservation planning system: origins, applications and possible futures. In: *Spatial Conservation Prioritization: Quantitative Methods and Computational Tools* (eds A. Moilanen, K.A. Wilson & H.P. Possingham). Oxford University Press, Oxford.

Prugh, L.R., Hodges, K.E., Sinclair, A.R.E. & Brashares, J.S. (2008) Effect of habitat area and isolation on fragmented animal populations. *Proceedings of the National Academy of Sciences USA*, **105**, 20770–20775.

Reiss, J., Bridle, J.R., Montoya, J.M. & Woodward, G. (2009) Emerging horizons in biodiversity and ecosystem functioning research. *Trends in Ecology and Evolution*, **24**, 505–514.

Riggio, P., Jacobson, A.J., Dollar, L., *et al.* (submitted) The size of savannah Africa: a lion's (Panthera leo) view. *PLoS ONE*.

Riordan, P., Delahay, R.J., Cheeseman, C., Johnson, P.J. & Macdonald, D.W. (2011) Culling-induced changes in badger (Meles meles) behaviour, social organisation and the epidemiology of bovine tuberculosis. *Plos ONE*, 6.

Rithe, K. (2012) They kill tigers, don't they? *Sanctuary Asia*, **June**, 62–67.

Roberts, C.M., McClean, C.J., Veron, J.E.N., *et al.* (2002) Marine biodiversity hotspots and conservation priorities for tropical reefs. *Science*, **295**, 1280–1284.

Schade, C. & Pimentel, D. (2010) Population crash: prospects for famine in the twenty-first century. *Environment, Development and Sustainability*, **12**, 245–262.

Seastedt, T.R., Hobbs, R.J. & Suding, K.N. (2008) Management of novel ecosystems: are novel approaches required? *Frontiers in Ecology and Environment*, **6**, 547–553.

Sgrò, C.M., Lowe, A.J. & Hoffmann, A.A. (2011) Building evolutionary resilience for conserving biodiversity under climate change. *Evolutionary Applications*, **4**, 326–337.

Shaffer, M.L. (1987) Minimum viable populations: coping with uncertainty. In: *Viable Populations for Conservation* (ed. M.E. Soulé). Cambridge University Press, New York.

Soulé, M.E. (1987) Introduction. In: *Viable Populations for Conservation* (ed. M.E. Soulé). Cambridge University Press, New York.

Spalding, M., Wood, L., Fitzgerald, C. & Gjerde, K. (2010) The 10% target: where do we stand? In: *Global Ocean Protection: Present Status and Future Possibilities* (eds C. Toropova, I. Meliane, D. Laffoley, E. Matthews & M. Spalding). Agence des Aires Marines Protégées, Brest, France.

Speidel, J.J., Weiss, D.C., Ethelston, S.A. & Gilbert, S.M. (2007) Family planning and reproductive health: the link to environmental preservation. *Population and Environment*, **28**, 247–258.

Stokstad, E. (2010) Despite progress, biodiversity declines. *Science*, **329**, 1272–1273.

Taberlet, P., Coissac, E., Pansu, J. & Pompanon, F. (2011) Conservation genetics of cattle, sheep, and goats. *Comptes Rendus Biologies*, **334**, 247–254.

Thompson, R. & Starzomski, B.M. (2007) What does biodiversity actually do? A review for managers and policy makers. *Biodiversity and Conservation*, **16**, 1359–1378.

Turner, R.. & Daily, G.C. (2008) The ecosystem services framework and natural capital conservation. *Environmental and Resource Economics*, **39**, 25–35.

Turner, R., Morse-Jones, S. & Fisher, B. (2010) Ecosystem valuation. *Annals of the New York Academy of Sciences*, **1185**, 79–101.

Turner, W.R., Brandon, K., Brooks, T.M., Costanza, R., da Fonseca, G. & Portela, R. (2007) Global conservation of biodiversity and ecosystem services. *Bioscience*, **57**, 868–873.

Valenzuela-Galván, D., Arita, H. & Macdonald, D. (2008) Conservation priorities for carnivores considering protected natural areas and human population density. *Biodiversity and Conservation*, **17**, 539–558.

Vitousek, P.M., Ehrlich, P.R., Ehrlich, A.H. & Matson, P.A. (1986) Human appropriation of the products of photosynthesis. *Bioscience*, **36**, 368–373.

Walston, J., Robinson, J.G., Bennett, E.L., et al. (2010) Bringing the tiger back from the brink – the six percent solution. *Plos Biology*, **8**, e1000485.

Wasser, S., Poole, J., Lee, P., et al. (2010) Elephants, ivory, and trade. *Science*, **327**, 1331–1332.

Whitman, K., Starfield, A.M., Quadling, H.S. & Packer, C. (2004) Sustainable trophy hunting of African lions. *Nature*, **428**, 175–178.

Wilhere, G.F. (2008) The how-much-is-enough myth. *Conservation Biology*, **22**, 514–517.

Williams, J.W. & Jackson, S.T. (2007) Novel climates, no-analog communities, and ecological surprises. *Frontiers in Ecology and Environment*, **5**, 475–482.

Willis, K.J. & Bhagwat, S.A. (2009) Biodiversity and climate change. *Science*, **326**, 806–807.

Willis, K.J. & Birks, H.J.B. (2006) What is natural? The need for a long-term perspective in biodiversity conservation. *Science*, **314**, 1261–1265.

Willis, K.J. & Macdonald, G.M. (2011) Long-term ecological records and their relevance to climate change predictions for a warmer world. *Annual Review of Ecology, Evolution, and Systematics*, **42**, 267–287.

Willis, K.J., Bailey, R.M., Bhagwat, S.A. & Birks, H.J.B. (2010) Biodiversity baselines, thresholds and resilience: testing predictions and assumptions using palaeoecological data. *Trends in Ecology and Evolution*, **25**, 583–591.

Woodroffe, R., Donnelly, C.A., Jenkins, H.E., *et al.* (2006) Culling and cattle controls influence tuberculosis risk for badgers. *Proceedings of the National Academy of Sciences USA*, **103**, 14713–14717.

World Commission on Environment and Development. (1987) *Our Common Future*. Oxford University Press, Oxford.

WWF. 2008. Living Planet Report 2008. WWF International, Switzerland. http://awsassets.panda.org/downloads/living_planet_report_2008.pdf

Yamaguchi, N., Cooper, A., Werdelin, L. & Macdonald, D.W. (2004) Evolution of the mane and group-living in the lion (Panthera leo): a review. *Journal of Zoology*, **263**, 329–342.

Zamin, T.J., Baillie, J.E.M., Miller, R.M., Rodriguez, J.P., Ardid, A. & Collen, B. (2010) National Red Listing beyond the 2010 target. *Conservation Biology*, **24**, 1012–1020.

A synthesis

Index

Key Topics in Conservation Biology 2, First Edition. Edited by David W. Macdonald and Katherine J. Willis.

Printed and bound by CPI Group (UK) Ltd, Croydon, CR0 4YY
22/06/2021

03074524-0001